Encyclopedia of
Drugs and Alcohol

Encyclopedia of Drugs and Alcohol

VOLUME 1

Jerome H. Jaffe, M.D.

Editor in Chief
University of Maryland, Baltimore

MACMILLAN LIBRARY REFERENCE USA
SIMON & SCHUSTER MACMILLAN
NEW YORK

SIMON & SCHUSTER AND PRENTICE HALL INTERNATIONAL
LONDON   MEXICO CITY   NEW DELHI   SINGAPORE   SYDNEY   TORONTO

Macmillan Library Reference
Simon & Schuster Macmillan
1633 Broadway
New York, NY 10019-6785

Library of Congress Catalog Card Number: 94-21458

Printed in the United States of America

printing number
 2 3 4 5 6 7 8 9 10

Library of Congress Cataloging-in-Publication Data
Encyclopedia of drugs and alcohol / Jerome H. Jaffe, editor-in-chief.
    p.   cm.
    Includes bibliographical references and index.
    ISBN 0-02-897185-X (set)
    1. Drug abuse—Encyclopedias.   2. Substance abuse—Encyclopedias.
3. Alcoholism—Encyclopedias.   4. Drinking of alcoholic beverages—
Encyclopedias.   I. Jaffe, Jerome.
Ref HV5804.E53   1995 v. 1
362.29′03—dc20   95-2321
                                                 CIP

This paper meets the requirements of ANSI-NISO Z39.48-1992 (Permanence of Paper). ∞™

Contents

Preface

The Macmillan *Encyclopedia of Drugs and Alcohol* has been written as a comprehensive source of information for nonspecialists who have an interest in any of the diverse topics that are included under the broad general heading of substance use and abuse. While many of the entries are devoted to the actions of drugs on the body, the work as a whole is intended to serve the wider interests of social science and includes articles on social policy, history, politics, economics, international trafficking, law enforcement, scientific and medical research, treatment and prevention of drug abuse, and epidemiology. Contributors were encouraged to make the more technical articles understandable to readers without training beyond high school biology.

The idea of this encyclopedia evolved over several years. It began with a proposal in 1989 by David Eckroth and Elly Dickason of Macmillan for a compendium that would cover the various drugs used by people for both therapeutic and nontherapeutic purposes. However, because of my own interests this initial notion was quickly transformed and expanded beyond the topic of drugs and their actions on the body to the more general one of drugs and society. My interest and involvement in this field are long standing. During my early years of medical training I spent some time at the U.S. Public Health Service Hospital at Lexington, Kentucky, where I took care of hundreds of people addicted to a variety of drugs and had contact with the great researchers at the Addiction Research Center there, including Abraham Wikler and Harris Isbell. This experience led me into pharmacological research trying to understand the basic mechanisms of physical dependence and withdrawal. However, I soon found myself caught up in trying to find better ways to treat people who had become addicted. From that time on I was "hooked" by the field. I have now spent more than three decades conducting research, teaching at medical schools, writing in scientific journals and books, treating patients, setting up treatment systems, consulting with the World Health Organization, working in government organizations in both policymaking and research roles, all having to do with the use and misuse of drugs, alcohol, and tobacco—topics now commonly subsumed under the more general rubric of substance use and abuse.

Over the years, I have contributed chapters to respected textbooks of pharmacology and psychiatry, as well as to books that tried to bring together history, the pharmacology of drugs, treatment of addiction, and social policy. In doing so, I became acutely aware of how vast and rich is the field, how abbreviated the best textbook chapters, and how limited any single book dealing with policy, pharmacology, treatment, or prevention. The idea of a multivolume work for the general reader to which experts across all the various disciplines that comprise the broad topic of substance use and society would contribute was too appealing a notion to dismiss. An opportunity to encompass not just the drugs that people use but the ways in which society responds to that use, and not just in the United States, but elsewhere in the world. This Encyclopedia represents an effort to

convey something of the richness of the subject matter and the complexity of the problem to the general reader and to bring it all together in one place.

Such an undertaking immediately encounters the limits of knowledge of any one or even several experts. The first task was to assemble a group of editors whose collective expertise could encompass the fields of pharmacology, epidemiology, treatment, prevention, government policy, and law enforcement. I was indeed fortunate that my colleagues who agreed to take on the task of serving on the board of editors are truly internationally recognized experts in their areas: Michael Kuhar in pharmacology and neuroscience; Chris-Ellyn Johanson in behavioral pharmacology, treatment, and etiology; Edward Sellers in pharmacology, treatment, and medical complications; James Anthony in epidemiology and prevention; and Mark Moore in government policy and law enforcement.

The editors in turn made an effort to have each article written by a leading expert on that particular topic. In most instances we were successful beyond expectations. For example, most of the articles on the actions of drugs and their effects on the body have been written by leading pharmacological researchers in the United States and Canada. Similarly, leading experts on the treatment of alcohol and drug dependence were generous in their willingness to contribute. We were also successful in enlisting the participation of a number of outstanding experts from the fields of law enforcement, government, and drug policy.

The original concept called for two types of articles: broad overviews of a topic, such as treatment of drug dependence; and short articles on specific topics within the broader framework, such as pharmacological treatment of opioid dependence. There were, however, numerous pleasant surprises of receiving "minor" articles that, while longer than anticipated, were gems too interesting to cut. In this encyclopedia, length is not an invariant clue to the importance of a topic within the broader field.

Finally, our original assumption that we could include even brief descriptions of the important organizations in the United States working in the fields of treatment, prevention, and education and drug policy proved untenable when we confronted the current reality of more than 8,000 programs in the field of treatment alone. Macmillan generously agreed to develop an appendix volume, *Directory of Substance Abuse Services*, which includes many of these organizations, primarily from materials in the public domain compiled by U.S. government agencies. Despite any shortcomings, this Encyclopedia is probably the most comprehensive work of its kind in this field, and we trust that it will be a useful source of information for some time to come.

No undertaking of this scope is accomplished by the editors alone. This Encyclopedia could not have been brought to completion without the encouragement, unending dedication, and remarkable editorial skills of Senior Project Editor Martha Goldstein of Macmillan Library Reference, the forbearance of Associate Publisher Elly Dickason and Publisher Phil Friedman, and the assistance, support and tolerance of my wife, Faith Jaffe.

February 1995

JEROME H. JAFFE
Editor in Chief

List of Articles

List of Authors

Darrell R. Abernethy
Division of Clinical Pharmacology
Roger Williams General Hospital, Providence, RI
WITHDRAWAL: NONABUSED DRUGS

Alfonso Acampora
Walden House, Inc., San Francisco, CA
WALDEN HOUSE

Caroline Jean Acker
Department of History
Carnegie Mellon University, Pittsburgh, PA
OPIOIDS AND OPIOID CONTROL: HISTORY

Manuella Adrian
Addiction Research Foundation, Toronto, Canada
CANADA, DRUG AND ALCOHOL USE IN
CANCER, DRUGS, AND ALCOHOL

Marlene Aldo-Benson
Indiana University School of Medicine, Bloomington, IN
ALLERGIES TO ALCOHOL AND DRUGS

John P. Allen
National Institute on Alcoholism and Alcohol Abuse
Rockville, MD
TREATMENT: ALCOHOL

Arthur I. Alterman
Department of Psychiatry, University of Pennsylvania
Veterans Administration Medical Center
Philadelphia, PA
TREATMENT TYPES: OUTPATIENT VERSUS
INPATIENT

Robert T. Angarola
Hyman, Phelps, and McNamara, Washington, DC
EXCLUSIONARY RULE
HARRISON NARCOTICS ACT OF 1914
PSYCHOTROPIC SUBSTANCES CONVENTION
OF 1971
SHANGHAI OPIUM CONFERENCE OF 1911
SINGLE CONVENTION ON NARCOTIC DRUGS

Christopher B. Anthony
Sparks, MD
DRAMSHOP LIABILITY LAWS

James C. Anthony
School of Hygiene and Mental Health
Johns Hopkins University, Baltimore, MD
EPIDEMICS OF DRUG ABUSE
EPIDEMIOLOGY OF DRUG ABUSE
COMORBIDITY AND VULNERABILITY

H. David Archibald
Addiction Research Foundation, Toronto, Canada
JELLINEK MEMORIAL FUND

Amelia Arria
Johns Hopkins University, Baltimore, MD
ETHNICITY AND DRUGS

Ustik Avico
Istituto Superiore di Sanità, Rome, Italy
ITALY, DRUG USE IN

Thomas F. Babor
Alcohol Research Center
Department of Psychiatry
University of Connecticut Health Center
Farmington, CT
DIAGNOSIS OF DRUG ABUSE: DIAGNOSTIC
CRITERIA
DIAGNOSTIC AND STATISTICAL MANUAL (DSM)
DIAGNOSTIC INTERVIEW SCHEDULE (DIS)
INTERNATIONAL CLASSIFICATION OF
DISEASES (ICD)
MINNESOTA MULTIPHASIC PERSONALITY
INVENTORY (MMPI)
STRUCTURED CLINICAL INTERVIEW FOR DSM-
III-R (SCID)

Jerald G. Bachman
Institute for Social Research
University of Michigan, Ann Arbor, MI
RELIGION AND DRUG USE

Robert L. Balster
Department of Pharmacology and Toxicology
Medical College of Virginia, Virginia Commonwealth
University, Richmond, VA
PHENCYCLIDINE (PCP): ADVERSE EFFECTS

James E. Barrett
American Cyanamid Co., Lederle Laboratory,
Pearl River, NY
REINFORCEMENT
RESEARCH: DRUGS AS DISCRIMINATIVE
  STIMULI
RESEARCH, ANIMAL MODEL:
  ENVIRONMENTAL INFLUENCES ON DRUG
  EFFECTS
RESEARCH, ANIMAL MODEL: LEARNING,
  CONDITIONING, AND DRUG EFFECTS—AN
  OVERVIEW
RESEARCH, ANIMAL MODEL: LEARNING
  MODIFIES DRUG EFFECTS

Andrew Baum
Department of Psychiatry
Uniformed Services University of Health Science
Bethesda, MD
STRESS
VULNERABILITY AS CAUSE OF SUBSTANCE
  ABUSE: STRESS

Jim Baumohl
Graduate School of Social Work and Social Research
Bryn Mawr College, Bryn Mawr, PA
ALCOHOL- AND DRUG-FREE HOUSING
HALFWAY HOUSES
HOMELESSNESS AND DRUGS, HISTORY OF
TREATMENT: HISTORY OF, IN THE UNITED
  STATES

Jan Bays
Emmanuel Hospital, Portland, OR
CHILD ABUSE AND DRUGS

Gary Bennett
Center for Prevention Research
University of Kentucky, Lexington, KY
WOMEN'S CHRISTIAN TEMPERANCE UNION
  (WCTU)

Neal L. Benowitz
Division of Clinical Pharmacology
San Francisco General Hospital, San Francisco, CA
NICOTINE
TOBACCO: DEPENDENCE
TOBACCO: HISTORY OF

Virginia Berridge
AIDS Social History Program, Health Policy Unit
London School of Medicine, London, England
BRITAIN, DRUG USE IN

Floyd Bloom
Research Institute Scripps Clinic, La Jolla, CA
ACETYLCHOLINE
CATECHOLAMINES
DOPAMINE
ENKEPHALIN
ENDORPHINS
GAMMA-AMINOBUTYRIC ACID (GABA)
GLUTAMATE
MONOAMINE
NEURON
NEUROTRANSMISSION
NEUROTRANSMITTERS
NOREPINEPHRINE
SEROTONIN
SYNAPSE, BRAIN

Sheila B. Blume
South Oaks Hospital, Amityville, NY
GAMBLING AS AN ADDICTION
JEWS, DRUG AND ALCOHOL USE AMONG
NATIONAL COUNCIL ON ALCOHOLISM AND
  DRUG DEPENDENCE (NCADD)

Michael J. Bohn
Department of Psychiatry
University of Wisconsin Hospitals, Madison, WI
SUICIDE AND SUBSTANCE ABUSE

Richard J. Bonnie
School of Law
University of Virginia, Charlottesville, VA
CONTROLLED SUBSTANCES ACT OF 1970
LEGAL REGULATION OF DRUGS AND
  ALCOHOL
MARIHUANA COMMISSION:
  RECOMMENDATIONS ON
  DECRIMINALIZATION
NATIONAL COMMISSION ON MARIHUANA
  AND DRUG ABUSE
PARAPHERNALIA, LAWS AGAINST

G. Borges
Instituto de Psichiatra, Xochimilco, Mexico
ADVERTISING AND ALCOHOL USE

Gilbert J. Botvin
Cornell Medical Center, New York, NY
PREVENTION PROGRAMS: LIFE SKILLS
  TRAINING

Joseph V. Brady
Johns Hopkins University School of Medicine
Baltimore, MD
ABUSE LIABILITY OF DRUGS: TESTING IN
HUMANS

Marc N. Branch
Department of Psychology
University of Florida, Gainesville, FL
BEHAVIORAL TOLERANCE

Robert M. Bray
Research Triangle Institute, Triange Park, NC
MILITARY, DRUG AND ALCOHOL ABUSE IN
THE U.S.

Karen E. Bremner
Psychopharmacology Research Program
Sunnybrook Health Science Centre, Toronto, Canada
SEROTONIN-UPTAKE INHIBITORS IN
TREATMENT OF SUBSTANCE ABUSE

Gregory W. Brock
Department of Family Studies
University of Kentucky, Lexington, KY
TOUGHLOVE

Judith S. Brook
Psychiatry and Behavioral Sciences
New York Medical College, Valhalla, NY
CHILDHOOD BEHAVIOR AND LATER DRUG
USE

Kirk J. Brower
Department of Psychiatry
University of Michigan Alcohol Research Center
Ann Arbor, MI
ANABOLIC STEROIDS

Lawrence S. Brown, Jr.
Division of Medical Services Evaluation and Research
Addiction Research and Treatment Corporation
Brooklyn, NY
COMPLICATIONS: ENDOCRINE AND
REPRODUCTIVE SYSTEMS

Kathleen K. Bucholz
Department of Psychiatry
Washington University School of Medicine
St. Louis, MO
ANTISOCIAL PERSONALITY

Alan J. Budney
Department of Psychiatry
University of Vermont, Burlington, VT
CONTINGENCY CONTRACTS
TREATMENT: COCAINE, PSYCHOLOGICAL
APPROACHES

TREATMENT TYPES: BEHAVIOR
MODIFICATION

Ellen Burke
Department of Family Studies
University of Kentucky, Lexington, KY
TOUGHLOVE

Usoa E. Busto
Department of Pharmacy
Addiction Research Foundation, Toronto, Canada
PHARMACODYNAMICS
PHARMACOKINETICS: GENERAL

Robert Byck
Department of Pharmacology
Yale University School of Medicine, New Haven, CT
FREUD AND COCAINE

Howard D. Cappell
Addiction Research Foundation, Toronto, Canada
TOLERANCE AND PHYSICAL DEPENDENCE

Peter L. Carlen
Playfair Neuroscience Unit, Division of Neurology
The Toronto Hospital, University of Toronto, Canada
COMPLICATIONS: NEUROLOGICAL

Molly Carney
Department of Psychology
University of Washington, Seattle, WA
ABSTINENCE VIOLATION EFFECT
EXPECTANCIES
RELAPSE PREVENTION
TREATMENT TYPES: ASSERTIVENESS
TRAINING
TREATMENT TYPES: COGNITIVE THERAPY

Jerome F. X. Carroll
Project Return Foundation, Inc., New York, NY
PROJECT RETURN FOUNDATION, INC.

Marilyn E. Carroll
Department of Psychiatry
University of Minnesota, Twin Cities Campus
Minneapolis, MN
PHENCYCLIDINE (PCP)

Richard F. Catalano
Social Development Research Group, Seattle, WA
CHINESE AMERICANS, ALCOHOL AND DRUG
USE AMONG

Arthur I. Cederbaum
Department of Biochemistry
Mt. Sinai School of Medicine, New York, NY
DRUG INTERACTIONS AND ALCOHOL

Timmen L. Cermak
San Francisco, CA
CODEPENDENCE

John N. Chappel
Department of Psychiatry
University of Nevada School of Medicine
Reno, NV
SOBRIETY

Cheryl J. Cherpitel
Alcohol Research Group, Berkeley, CA
ACCIDENTS AND INJURIES FROM ALCOHOL

Domenic A. Ciraulo
Psychiatry Service
VA Outpatient Clinic, Boston, MA
BENZODIAZEPINES

Lorenzo Cohen
Department of Medical Psychology
Uniformed Services University of Health Sciences
Bethesda, MD
STRESS
VULNERABILITY AS CAUSE OF SUBSTANCE
ABUSE: STRESS

Patricia Cohen
New York State Psychiatric Institute
Columbia University, New York, NY
CHILDHOOD BEHAVIOR AND LATER DRUG
USE

Shirley D. Coletti
Operation PAR, St. Petersburg, FL
OPERATION PAR

James J. Collins
Research Triangle Institute, Triangle Park, NC
CRIME AND ALCOHOL

Sandra D. Comer
Department of Psychiatry
University of Minnesota, Minneapolis, MN
PHENCYCLIDINE (PCP)

Edward J. Cone
Addiction Research Center, National Institute on Drug
Abuse, Baltimore, MD
HAIR ANALYSIS AS A TEST FOR DRUG USE

Brian L. Cook
Psychiatry Service
Veterans Medical Center, Iowa City, IA
COMPLICATIONS: MENTAL DISORDERS

Philip J. Cook
Institute of Public Policy
Duke University, Durham, NC
SOCIAL COSTS OF ALCOHOL AND DRUG
ABUSE
TAX LAWS AND ALCOHOL

Richard B. Craig
Department of Political Science
Kent State University, Kent, OH
OPERATION INTERCEPT

Valerie Curran
Institute of Psychiatry, London, England
SEDATIVES: ADVERSE CONSEQUENCES OF
CHRONIC USE

Michael Darcy
Gateway Foundation, Chicago, IL
GATEWAY FOUNDATION

Robert C. Davis
Victim Services, New York, NY
PREVENTION: COMMUNITY DRUG RESISTANCE

Valina Dawson
Molecular Neurobiology Laboratory, Addiction
Research Center
National Institute on Drug Abuse, Baltimore, MD
ANTIDEPRESSANT
ANTIPSYCHOTIC
NEUROLEPTIC
RECEPTOR: NMDA

George De Leon
Center for Therapeutic Community Research
New York, NY
TREATMENT TYPES: THERAPEUTIC
COMMUNITIES

David A. Deitch
Department of Psychiatry, School of Medicine
University of California, San Diego, CA
DAYTOP VILLAGE
LE PATRIARCHE
SYNANON

Paolo DePetrillo
Division of Clinical Pharmacology
Roger Williams General Hospital, Providence, RI
WITHDRAWAL: NONABUSED DRUGS

Don C. Des Jarlais
Chemical Dependency Unit
Beth Israel Medical Center, New York, NY
NEEDLE AND SYRINGE EXCHANGES AND HIV/
AIDS

David P. Desmond
Department of Psychiatry
University of Texas Health Center, San Antonio, TX
OPIOID DEPENDENCE: COURSE OF THE
DISORDER OVER TIME

Paul Devenyi
Don Mills, Ontario, Canada
COMPLICATIONS: LIVER DAMAGE

John J. DiIulio, Jr.
Woodrow Wilson School
Princeton University, Princeton, NJ
TREATMENT IN THE FEDERAL PRISON
SYSTEM

Salvatore Di Menza
ADAMHA
National Institute on Drug Abuse, Rockville, MD
TREATMENT FUNDING AND SERVICE
DELIVERY

Lewis Donohew
Department of Communication
University of Kentucky, Lexington, KY
PREVENTION: SHAPING MASS-MEDIA
MESSAGES TO VULNERABLE GROUPS

D. Peter Drotman
Division of Chemical Research
National Institute on Drug Abuse, Rockville, MD
SUBSTANCE ABUSE AND AIDS

D. Colin Drummond
Department of Psychiatry
The Bethlem Royal Hospital and Maudsley Hospital
University of London, England
BRITISH SYSTEM OF DRUG-ADDICTION
TREATMENT
HEROIN: THE BRITISH SYSTEM

Steven I. Dworkin
Department of Physiology and Pharmacology
Bowman Gray Medical School
Wake Forest University, Winston-Salem, NC
BRAIN STRUCTURES AND DRUGS
LIMBIC SYSTEM

Linda A. Dykstra
Department of Psychology
University of North Carolina, Chapel Hill, NC
PAIN: BEHAVIORAL METHODS FOR
MEASURING ANALGESIC EFFECTS OF
DRUGS
RESEARCH: MEASURING EFFECTS OF DRUGS
ON BEHAVIOR
RESEARCH, ANIMAL MODEL: OPERANT
LEARNING IS AFFECTED BY DRUGS
SENSATION AND PERCEPTION AND EFFECTS
OF DRUGS

Griffith Edwards
Addiction Research Unit (U.K.)
Institute of Psychiatry, London, England
ADDICTION
ADDICTION RESEARCH UNIT (ARU) (U.K.)

Everett H. Ellinwood
Department of Psychiatry
Duke University Medical Center, Durham, NC
CAUSES OF SUBSTANCE ABUSE: DRUG
EFFECTS AND BIOLOGICAL RESPONSES

Margaret E. Ensminger
Department of Health Policy and Management
Johns Hopkins School of Public Health, Baltimore, MD
POVERTY AND DRUG USE
VULNERABILITY AS CAUSE OF SUBSTANCE
ABUSE: GENDER
VULNERABILITY AS CAUSE OF SUBSTANCE
ABUSE: RACE
VULNERABILITY AS CAUSE OF SUBSTANCE
ABUSE: SEXUAL AND PHYSICAL ABUSE

Jennean Everett
Johns Hopkins University School of Medicine
Baltimore, MD
VULNERABILITY AS CAUSE OF SUBSTANCE
ABUSE: GENDER
VULNERABILITY AS CAUSE OF SUBSTANCE
ABUSE: RACE

Jeffrey Fagan
School of Criminal Justice
Rutgers University, Newark, NJ
GANGS AND DRUGS

John L. Falk
Department of Psychology
Rutgers University, New Brunswick, NJ
ADJUNCTIVE DRUG TAKING

Michael P. Finnegan
Alcohol, Drug Abuse, and Mental Health
Administration
Division for State Assistance
Office for Treatment Improvement, Rockville, MD
FETUS: EFFECTS OF DRUGS ON THE
PREGNANCY AND DRUG DEPENDENCE:
OPIOIDS AND COCAINE

Loretta P. Finnegan
Women's Health Initiative
National Institute of Health, Bethesda, MD
FETUS: EFFECTS OF DRUGS ON THE
PREGNANCY AND DRUG DEPENDENCE:
OPIOIDS AND COCAINE

Marian W. Fischman
College of Physicians and Surgeons
Columbia University, New York, NY
    AMPHETAMINE
    BENZOYLECOGNINE
    COCAINE
    COCA PASTE
    COCA PLANT
    CRACK
    DEXTROAMPHETAMINE
    FREEBASING
    METHAMPHETAMINE
    METHEDRINE
    METHYLPHENIDATE
    PEMOLINE
    PSYCHOMOTOR STIMULANT

Brian R. Flay
School of Public Health, Prevention Research Center
University of Illinois, Chicago, IL
    PREVENTION PROGRAMS: WATERLOO
        SMOKING PREVENTION PROJECT

Alice B. Fredericks
Division of Clinical Pharmocology
San Francisco General Hospital,
    San Francisco, CA
    NICOTINE
    TOBACCO: DEPENDENCE
    TOBACCO: HISTORY OF

Daniel X. Freedman
Department of Pharmacology
Louisiana State University Medical Center
    New Orleans, LA
    DIMETHYLTRYPTAMINE (DMT)
    DOM
    HALLUCINOGENS
    HALLUCINOGENIC PLANTS
    LYSERGIC ACID DIETHYLAMIDE (LSD) AND
        PSYCHEDELICS
    MDMA
    MESCALINE
    MORNING GLORY SEEDS
    NUTMEG
    PEYOTE
    PSILOCYBIN

William A. Frosch
Department of Psychiatry
Cornell University Medical School, New York, NY
    ADDICTIVE PERSONALITY
    ADDICTIVE PERSONALITY AND
        PSYCHOLOGICAL TESTS
    PERSONALITY AS A RISK FACTOR FOR DRUG
        ABUSE
    PSYCHOANALYSIS

    TREATMENT TYPES: TRADITIONAL DYNAMIC
        PSYCHOTHERAPY
    VULNERABILITY AS CAUSE OF SUBSTANCE
        ABUSE: PSYCHOANALYTIC PERSPECTIVE

Richard K. Fuller
National Institute on Alcoholism and Alcohol Abuse
    Rockville, MD
    DISULFIRAM
    TREATMENT: ALCOHOL, AN OVERVIEW
    TREATMENT: ALCOHOL, PHARMACOTHERAPY

Marc S. Galanter
Department of Psychiatry
New York University School of Medicine, New York, NY
    AMERICAN SOCIETY OF ADDICTION MEDICINE
        (ASAM)
    ASSOCIATION FOR MEDICAL EDUCATION AND
        RESEARCH IN SUBSTANCE ABUSE (AMERSA)

Paul F. Gavaghan
Chevy Chase, MD
    DISTILLED SPIRITS COUNCIL

Jeffrey A. Gere
Harmans, MD
    IMMUNOASSAY

Dean R. Gerstein
National Opinion Research Center
    Washington, DC
    PRODUCTIVITY: EFFECTS OF
        ALCOHOL ON
    PRODUCTIVITY: EFFECTS OF DRUGS ON

Joseph C. Gfroerer
Office of Applied Studies
Substance Abuse and Mental Health Services
    Administration, Rockville, MD
    NATIONAL HOUSEHOLD SURVEY ON DRUG
        ABUSE (NHSDA)

Frederick B. Glaser
University of Michigan Medical Center
    Ann Arbor, MI
    ALCOHOLISM: ORIGIN OF THE TERM

Elbert Glover
Department of Psychiatry
University of West Virginia Medical School
    Morgantown, WV
    TOBACCO: SMOKELESS

Penny N. Glover
Department of Psychiatry
University of West Virginia Medical School
    Morgantown, WV
    TOBACCO: SMOKELESS

Nick E. Goeders
Department of Pharmacology
Louisiana State University Medical Center
  Shreveport, LA
  AGONIST-ANTAGONIST (MIXED)
  AGONIST
  ALKALOIDS
  APHRODISIAC
  ANTAGONIST
  DESIGNER DRUGS
  DOSE-RESPONSE RELATIONSHIP
  DRUG
  DRUG TYPES
  ETHNOPHARMACOLOGY
  ED50
  DRUG INTERACTION AND THE BRAIN
  LD50
  PHARMACOLOGY
  PSYCHOACTIVE
  RECEPTOR: DRUG
  PSYCHOACTIVE DRUG
  PSYCHOPHARMACOLOGY
  PLANTS, DRUGS FROM

Mark S. Gold
Department of Neuro-Science and Psychiatry
Brain Institute, University of Florida, Gainesville, FL
  CLONIDINE

Steven Goldberg
Addiction Research Center
National Institute on Drug Abuse, Baltimore, MD
  RESEARCH, ANIMAL MODEL: CONDITIONED
    WITHDRAWAL

David Goldbloom
Department of Psychiatry
Toronto Hospital, Toronto, Canada
  BULIMIA NERVOSA

Stephen Goldsmith
Indianapolis, IN
  DRUG LAWS, PROSECUTION OF

Donald W. Goodwin
Department of Psychology
University of Kansas Medical Center
School of Medicine, Kansas City, KS
  TREATMENT TYPES: AVERSION THERAPY

Eric Goplerud
Pennsylvania State University, State College, PA
  EDUCATION AND PREVENTION

Enoch Gordis
National Institute on Alcoholism and Alcohol Abuse
  Rockville, MD
  U.S. GOVERNMENT AGENCIES: NATIONAL
    INSTITUTE ON ALCOHOLISM
    AND ALCOHOL ABUSE (NIAAA)

David A. Gorelick
Addiction Research Center, National Institute on Drug
  Abuse, Baltimore, MD
  NALTREXONE IN TREATMENT OF DRUG
    DEPENDENCE

Louis D. Gottlieb
St. Mary's Hospital, Waterbury, CT
  WITHDRAWAL: ALCOHOL, BETA BLOCKERS

David J. Greenblatt
Department of Pharmacology and Experimental
  Therapeutics
Tufts University School of Medicine, Boston, MA
  PHARMACOKINETICS: IMPLICATIONS IN
    ABUSEABLE SYMPTOMS

Roland R. Griffiths
Department of Psychiatry and Behavioral Science
Johns Hopkins School of Medicine
  Baltimore, MD
  CAFFEINE
  COFFEE
  TEA

Steven W. Gust
National Institute on Drug Abuse, Rockville, MD
  EMPLOYEE ASSISTANCE PROGRAMS (EAPs)

Sharon Hall
Psychiatry Service
Veterans Affairs Medical Center
  San Francisco, CA
  TREATMENT TYPES: PSYCHOLOGICAL
    APPROACHES

David Halperin
Department of Psychiatry
Mt. Sinai School of Medicine, New York, NY
  CULTS AND DRUG USE

Thomas E. Hanlon
Human Behavior Associates, Inc., Annapolis, MD
  CRIME AND DRUGS

William B. Hansen
Bowman Gray School of Medicine
Wake Forest University, Winston-Salem, NC
  PREVENTION PROGRAMS: PROJECT SMART

Tracy W. Harachi
Social Development Research Group, Seattle, WA
  CHINESE AMERICANS, ALCOHOL AND DRUG
    USE AMONG

Nancy Grant Harrington
Department of Communication
University of Kentucky, Lexington, KY
PREVENTION: SHAPING MASS-MEDIA
MESSAGES TO VULNERABLE GROUPS

Louis Harris
Department of Pharmacology and Toxicology
Medical College of Virginia, Richmond, VA
WORLD HEALTH ORGANIZATION EXPERT
COMMITTEE ON DRUG DEPENDENCE

Christine R. Hartel
American Psychological Association, Washington, DC
RESEARCH: AIMS, DESCRIPTION, AND GOALS
U.S. GOVERNMENT: AGENCIES SUPPORTING
SUBSTANCE ABUSE RESEARCH

Jonas Hartelius
Department of Alcohol Toxicology
National Laboratory of Forensic Chemistry
University Hospital, Linkoping, Sweden
SWEDEN, DRUG USE IN

Dorothy Hatsukami
Department of Psychiatry
University of Minnesota, Minneapolis, MN
TREATMENT: TOBACCO, PSYCHOLOGICAL
APPROACHES

Harry W. Haverkos
Division of Chemical Research
National Institute on Drug Abuse, Rockville, MD
SUBSTANCE ABUSE AND AIDS

Dwight B. Heath
Department of Anthropology
Brown University, Providence, RI
ALCOHOL: HISTORY OF DRINKING

Thomas A. Hedrick, Jr.
Partnership for a Drug-Free America,
New York, NY
PARTNERSHIP FOR A DRUG-FREE AMERICA

Daniel S. Heit
Abraxas Group, Philadelphia, PA
ABRAXAS

Jack E. Henningfield
Addiction Research Center, Clinical Pharmacology
Branch
National Institute on Drug Abuse, Baltimore, MD
TREATMENT: TOBACCO, PHARMACOTHERAPY
WITHDRAWAL: NICOTINE (TOBACCO)

Stephen T. Higgins
Department of Psychiatry
University of Vermont, Burlington, VT
CONTINGENCY CONTRACTS
TREATMENT: COCAINE, PSYCHOLOGICAL
APPROACHES
TREATMENT TYPES: BEHAVIOR
MODIFICATION

Riley Hinson
Department of Psychology
University of Western Ontario, London, Canada
CONDITIONED TOLERANCE

Leo E. Hollister
Harris County Psychiatric Center, Houston, TX
BHANG
CANNABIS SATIVA
HASHISH
HEMP
GANJA
MARIJUANA
TETRAHYDROCANNABINOL (THC)

Ralph I. Horwitz
Yale University School of Medicine, New Haven, CT
WITHDRAWAL: ALCOHOL, BETA BLOCKERS

Robert Hubbard
Research Triangle Institute, Triangle Park, NC
DRUG ABUSE TREATMENT OUTCOME STUDY
(DATOS)
TREATMENT OUTCOME PERSPECTIVE STUDY
(TOPS)

John R. Hughes
Department of Psychiatry
University of Vermont, Burlington, VT
TOBACCO: WITHDRAWAL (ABSTINENCE)
SYNDROME

Patrick H. Hughes
Department of Psychiatry
University of South Florida College of Medicine,
Tampa, FL
YIPPIES

Peter Barton Hutt
Covington and Burling, Washington, DC
PUBLIC INTOXICATION

Ted Inaba
Department of Pharmacology
University of Toronto, Canada
DRUG METABOLISM

James A. Inciardi
Department of Sociology
University of Delaware, Newark, DE
  BOLIVIA, DRUG USE IN

M. S. Irwanto
Department of Child Development and Family Studies
Purdue University, West Lafayette, IN
  FAMILIES AND DRUG USE

Ivan Izquierdo
Center for Neurobiology of Learning
University of California, Irvine, CA
  MEMORY, EFFECTS OF DRUGS ON

James B. Jacobs
New York University Law School, New York, NY
  DRUNK DRIVING

Arthur E. Jacobson
National Institute of Health, Bethesda, MD
  COLLEGE ON PROBLEMS OF DRUG
    DEPENDENCE (CPDD), INC.

Nora Jacobson
School of Public Health
Johns Hopkins University, Baltimore, MD
  POVERTY AND DRUG USE

Faith K. Jaffe
Towson, MD
  AMERICAN ACADEMY OF PSYCHIATRISTS IN
    ALCOHOLISM AND ADDICTIONS (AAPAA)
  INSTITUTE ON BLACK CHEMICAL ABUSE
    (IBCA)
  JEWISH ALCOHOLICS, CHEMICALLY
    DEPENDENT PERSONS AND SIGNIFICANT
    OTHERS FOUNDATION, INC. (JACS)
  REMOVE INTOXICATED DRIVERS
    (RID-USA, INC.)
  SOCIETY OF AMERICANS FOR RECOVERY
    (SOAR)
  TREATMENT ALTERNATIVES TO STREET
    CRIME (TASC)
  U.S. GOVERNMENT AGENCIES: SPECIAL
    ACTION OFFICE FOR DRUG ABUSE
    PREVENTION (SAODAP)
  WOMEN FOR SOBRIETY

Jerome H. Jaffe
School of Medicine
University of Maryland, Baltimore, MD
  AMERICAN SOCIETY OF ADDICTION MEDICINE
    (ASAM)
  AYAHUASCA
  BETEL NUT
  CENTER ON ADDICTION AND SUBSTANCE
    ABUSE (CASA)
  DISEASE CONCEPT OF ADDICTION
  DRUG POLICY FOUNDATION (DPF)
  HALFWAY HOUSES
  IBOGAINE
  NARCOTIC ADDICT REHABILITATION ACT
    (NARA)
  OVEREATING AND OTHER EXCESSIVE
    BEHAVIORS
  OXFORD HOUSE
  PROHIBITION OF ALCOHOL
  PROPOXYPHENE
  RATIONAL RECOVERY (RR)
  SECULAR ORGANIZATIONS FOR SOBRIETY
    (SOS)
  TOBACCO: MEDICAL COMPLICATIONS
  TREATMENT ALTERNATIVES TO STREET
    CRIME (TASC)
  TREATMENT, HISTORY OF, IN THE UNITED
    STATES
  TREATMENT TYPES: ART OF ACUPUNCTURE
  TREATMENT TYPES: THERAPEUTIC
    COMMUNITIES
  UNITED NATIONS CONVENTION AGAINST
    ILLICIT TRAFFIC IN NARCOTIC DRUGS AND
    PSYCHOTROPIC SUBSTANCES, 1988
  U.S. GOVERNMENT: AGENCIES IN DRUG LAW
    ENFORCEMENT AND SUPPLY CONTROL
  U.S. GOVERNMENT AGENCIES: SPECIAL
    ACTION OFFICE FOR DRUG ABUSE
    PREVENTION(SAODAP)

Joni Jensen
Department of Psychiatry
University of Minnesota, Minneapolis, MN
  TREATMENT: TOBACCO, PSYCHOLOGICAL
    APPROACHES

Chris-Ellyn Johanson
Department of Psychiatry
Wayne State University School of Medicine, Detroit, MI
  AMOTIVATIONAL SYNDROME
  POLYDRUG ABUSE

Bruce Johnson
Narcotic and Drug Research, Inc., New York, NY
  HEROIN: THE BRITISH SYSTEM
  ROLLESTON REPORT OF 1926 (U.K.)

Elaine Johnson
Substance Abuse and Mental Health Services
Administration, Rockville, MD
  U.S. GOVERNMENT AGENCIES: CENTER FOR
    SUBSTANCE ABUSE PREVENTION (CSAP)
  U.S. GOVERNMENT AGENCIES: SUBSTANCE
    ABUSE AND MENTAL HEALTH SERVICES
    ADMINISTRATION (SAMHSA)

Eric O. Johnson
Drug Dependence Epidemiology
National Institute on Drug Abuse, Rockville, MD
Johns Hopkins University, Baltimore, MD
  DROPOUTS AND SUBSTANCE ABUSE
  TERRY AND PELLENS STUDY

Rolley E. Johnson
Behavioral Pharmacology Research Unit
Francis Scott Key Medical Center, Baltimore, MD
  CONTROLS: SCHEDULED DRUGS/DRUG
    SCHEDULES, U.S.

Bryan M. Johnstone
Department of Behavioral Science
University of Kentucky, Lexington, KY
  PREVENTION: PREVENTION OF
    ALCOHOLISM,THE LEDERMANN MODEL

A. W. Jones
Department of Alcohol Toxicology
National Laboratory of Forensic Chemistry
University Hospital, Linkoping, Sweden
  BLOOD ALCOHOL CONCENTRATION,
    MEASURES OF
  PHARMACOKINETICS OF ALCOHOL
  SWEDEN, DRUG USE IN

Harold Kalant
Department of Pharmacology
University of Toronto, Canada
  ADDICTION: CONCEPTS AND DEFINITIONS
  PHYSICAL DEPENDENCE
  ADDICTION RESEARCH FOUNDATION OF
    ONTARIO, CANADA

Peter Kalix
Department of Pharmacology
University of Geneva, Switzerland
  KHAT

George A. Kanuck
Malvern Preparatory School, Malvern, PA
  FETUS: EFFECTS OF DRUGS ON THE
  PREGNANCY AND DRUG DEPENDENCE:
    OPIOIDS AND COCAINE

Charles Kaplan
International Institute for Psychosocial and
  Socioecological Research
Maastricht, The Netherlands
  NETHERLANDS, DRUG USE IN THE

Bhushan K. Kapur
Department of Clinical Biochemistry
University of Toronto, Canada
CRTI Addiction Research Foundation, Toronto, Canada
  DRUG TESTING AND ANALYSIS

Clifford L. Karchmer
Police Executive Research Forum
  Washington, DC
  DRUG LAWS: FINANCIAL ANALYSIS IN
    ENFORCEMENT
  MONEY LAUNDERING

Edward Kaufman
Capistrano by the Sea Hospital, Dana Point, CA
  TREATMENT TYPES: GROUP AND FAMILY
    THERAPY

Jat M. Khanna
Department of Pharmacology
University of Toronto, Canada
  BARBITURATES: COMPLICATIONS

Ed Khantzian
Department of Psychiatry
Cambridge Hospital, Cambridge, MA
  CAUSES OF SUBSTANCE ABUSE:
    PSYCHOLOGICAL (PSYCHOANALYTIC)
    PERSPECTIVE

Michael Kidorf
Behavioral Pharmacology Research Unit
Francis Scott Key Medical Center, Baltimore, MD
  TREATMENT: HEROIN, PSYCHOLOGICAL
    APPROACHES

M. Marlyne Kilbey
Department of Psychology
Wayne State University, Detroit, MI
  PROFESSIONAL CREDENTIALING

Sewhan Kim
Drug Education Center, Charlotte, NC
  PREVENTION PROGRAMS: OMBUDSMAN
    PROGRAM

Sion Kim
Johns Hopkins University, Baltimore, MD
  VULNERABILITY AS CAUSE OF SUBSTANCE
    ABUSE: RACE

George R. King
Department of Psychiatry
Duke University Medical Center, Durham, NC
  CAUSES OF SUBSTANCE ABUSE: DRUG
    EFFECTS AND BIOLOGICAL RESPONSES

Rufus King
Berliner, Corcoran, and Rowe, Washington, DC
  ANSLINGER, HARRY J., AND U.S. DRUG POLICY

Timothy W. Kinlock
Human Behavior Associates, Inc., Annapolis, MD
  CRIME AND DRUGS

Herbert D. Kleber
Department of Psychiatry
College of Physicians and Surgeons
Columbia University, New York, NY
TREATMENT TYPES: AN OVERVIEW

Michael Klitzner
Klitzner & Associates, Vienna, VA
SADD (STUDENTS AGAINST DRIVING DRUNK)

Clifford Knapp
Department of Pharmacology and Experimental
Therapeutics
Tufts University School of Medicine, Boston, MA
BENZODIAZEPINES

Conan Kornetsky
Department of Psychiatry and Pharmacology
Boston University School of Medicine, Boston, MA
REWARD PATHWAYS AND DRUGS

Madhu R. Korrapati
Research Service
Veterans Administration Medical Center
Boise, ID
AGING, DRUGS, AND ALCOHOL

Therese A. Kosten
Substance Abuse Treatment Unit
Yale University School of Medicine
New Haven, CT
RESEARCH: DEVELOPING MEDICATIONS TO
TREAT SUBSTANCE ABUSE AND
DEPENDENCE

Thomas R. Kosten
Substance Abuse Treatment Unit
Yale University School of Medicine, New Haven, CT
AMANTADINE
COCAINE, TREATMENT STRATEGIES
TREATMENT: COCAINE, PHARMACOTHERAPY
TREATMENT TYPES: PHARMACOTHERAPY, AN
OVERVIEW
WITHDRAWAL: COCAINE

Nicholas Kozel
National Vaccine Program Office, Rockville, MD
AMPHETAMINE EPIDEMICS

Lynn T. Kozlowski
Department of Health and Human Development
Pennsylvania State University, University
Park, PA
TOBACCO: TREATMENT, AN OVERVIEW

Mark L. Kraus
Internal Medicine
Westside Medical Group, Waterbury, CT
WITHDRAWAL: ALCOHOL, BETA BLOCKERS

Michael J. Kuhar
Neuroscience Branch, Addiction Research Center
National Institute on Drug Abuse, Baltimore, MD
ANTIDOTE
CHOCOLATE
COLA/COLA DRINKS
GINSENG
POISON
THEOBROMINE
VITAMINS

Karol L. Kumpfer
Department of Health Education
University of Utah, Salt Lake City, UT
CHILD ABUSE AND DRUGS

Malcolm H. Lader
Institute of Psychiatry, London, England
BENZODIAZEPINES: COMPLICATIONS
PRESCRIPTION DRUG ABUSE

Robin A. LaDue
Fetal Alcohol and Drug Unit
Department of Psychiatry and Behavioral Sciences
University of Washington Medical School
Seattle, WA
FETAL ALCOHOL SYNDROME (FAS)

Phyllis A. Langton
Department of Sociology
George Washington University, Washington, DC
TEMPERANCE MOVEMENT

J. Clark Laundergan
Center for Addiction Studies, Duluth, MN
HAZELDEN CLINICS
TREATMENT TYPES: MINNESOTA MODEL

Scott J. Leischow
Division of Community and Environmental Health
School of Health Related Professions, Tucson, AZ
TOBACCO: SMOKING CESSATION AND WEIGHT
GAIN

Carl G. Leukefeld
Research Center on Drug and Alcohol Abuse
University of Kentucky, Lexington, KY
INJECTING DRUG USERS AND HIV

Frances R. Levin
Department of Psychiatry
College of Physicians and Surgeons
Columbia University, New York, NY
TREATMENT TYPES: AN OVERVIEW

R. A. Lewis
Department of Family Studies
Purdue University, West Lafayette, IN
FAMILIES AND DRUG USE

Barbara Lex
Harvard Medical School
McLean Hospital, Belmont, MA
    FAMILY VIOLENCE AND SUBSTANCE ABUSE

Charles S. Lieber
Alcoholism Research and Treatment Center
Bronx Veterans Administration Medical Center
    Bronx, NY
    COMPLICATIONS: LIVER (ALCOHOL)

Marsha Lillie-Blanton
School of Hygiene and Public Health
Johns Hopkins University, Baltimore, MD
    ETHNICITY AND DRUGS

Jennifer K. Lin
Drug Dependence Epidemiology
National Institute on Drug Abuse, Rockville, MD
    COMMISSIONS ON DRUGS

Walter Ling
Beverly Hills, CA
    IMPAIRED PHYSICIANS AND MEDICAL
        WORKERS

Richard Lingeman
New York, NY
    SLANG AND JARGON

Arlene R. Lissner
Abraxas Group, Philadelphia, PA
    ABRAXAS

Raye Z. Litten
National Institute on Alcoholism and Alcohol Abuse
    Rockville, MD
    DISULFIRAM
    TREATMENT: ALCOHOL, AN OVERVIEW
    TREATMENT: ALCOHOL, PHARMACOTHERAPY

Edythe D. London
Intramural Research Program
National Institute on Drug Abuse, Rockville, MD
    IMAGING TECHNIQUES: VISUALIZING THE
        LIVING BRAIN

Joyce H. Lowinson
Division of Substance Abuse
Albert Einstein College of Medicine, Bronx, NY
    TREATMENT TYPES: ART OF ACUPUNCTURE

Arnold Ludwig
Department of Psychiatry
University of Kentucky College of Medicine
    Lexington, KY
    CREATIVITY AND DRUGS

Scott E. Lukas
Harvard Medical School
McLean Hospital, Belmont, MA
    ALCOHOL: CHEMISTRY AND PHARMACOLOGY
    AMOBARBITAL
    BARBITURATES
    BEERS AND BREWS
    CHLORAL HYDRATE
    CHLORDIAZEPOXIDE
    DISTILLATION
    DISTILLED SPIRITS
    ETHCHLORVYNOL
    ETHINAMATE
    FERMENTATION
    GLUTETHIMIDE
    MEPROBAMATE
    METHANOL
    METHAQUALONE
    MOONSHINE
    PHENOBARBITAL
    RUBBING ALCOHOL
    SECOBARBITAL
    SEDATIVE
    SEDATIVE-HYPNOTIC
    SLEEPING PILLS
    STILL

Arthur J. Lurigio
Victim Services, New York, NY
    PREVENTION: COMMUNITY DRUG RESISTANCE

Doris Layton MacKenzie
Department of Criminal Justice
University of Maryland, College Park, MD
    SHOCK INCARCERATION AND BOOT CAMP
        PRISONS

David J. Mactas
Center for Substance Abuse Treatment,
    Rockville, MD
    MARATHON HOUSE

James F. Maddux
Department of Psychiatry
University of Texas Health Center, San Antonio, TX
    OPIOID DEPENDENCE: COURSE OF THE
        DISORDER OVER TIME
    NARCOTIC ADDICT REHABILITATION ACT
        (NARA)
    U.S. GOVERNMENT AGENCIES: U.S. PUBLIC
        HEALTH SERVICE HOSPITALS

Alan Marlatt
Department of Psychology
University of Washington, Seattle, WA
    ABSTINENCE VIOLATION EFFECT
    EXPECTANCIES

RELAPSE PREVENTION
TREATMENT TYPES: ASSERTIVENESS
    TRAINING
TREATMENT TYPES: COGNITIVE THERAPY

Garth Martin
Addiction Research Foundation
Clinical Research and Treatment Institute, Toronto,
Canada
TREATMENT: POLYDRUG ABUSE, AN
    OVERVIEW

Peter Martin
Department of Psychiatry and Pharmacology
Vanderbilt University School of Medicine, Nashville, TN
COMPLICATIONS: COGNITION

William R. Martin
Department of Pharmacology
University of Kentucky, Lexington, KY
OPIOID COMPLICATIONS AND WITHDRAWAL

George Mathews
Department of Psychiatry and Pharmacology
Vanderbilt University School of Medicine, Nashville, TN
COMPLICATIONS: COGNITION

Mauri J. Mattila
Department of Pharmacology and Toxicology
University of Helsinki, Finland
PSYCHOMOTOR EFFECTS OF ALCOHOL AND
    DRUGS

Armand L. Mauss
Department of Sociology
Washington State University, Pullman, WA
PREVENTION PROGRAMS: "HERE'S LOOKING
    AT YOU" SERIES

Mary Pat McAndrews
Department of Psychology
Playfair Neuroscience Unit
The Toronto Hospital, University of Toronto, Canada
COMPLICATIONS: NEUROLOGICAL

Duane C. McBride
Behavioral Sciences Department
Andrews University, Berrien Springs, MI
POLICY ALTERNATIVES: PROHIBITION OF
    DRUGS: PRO AND CON

Kevin McEneaney
Phoenix House, New York, NY
PHOENIX HOUSE

James L. McGaugh
Center for Neurobiology of Learning
University of California, Irvine, CA
MEMORY, EFFECTS OF DRUGS ON

A. Thomas McLellan
Veterans Administration Medical Center
Philadelphia, PA
ADDICTION SEVERITY INDEX (ASI)
MICHIGAN ALCOHOLISM SCREENING TEST
    (MAST)

Jonnie McLeod
University of North Carolina, Charlotte, NC
PREVENTION PROGRAMS: OMBUDSMAN
    PROGRAM

Roger E. Meyer
George Washington University, Washington, DC
DISEASE CONCEPT OF ALCOHOLISM AND
    DRUG ABUSE

Ada C. Mezzich
Western Psychiatric Institute
University of Pittsburgh School of Medicine
ATTENTION DEFICIT DISORDER
CONDUCT DISORDER AND DRUG USE

Klaus A. Miczek
Department of Psychology
Tufts University, Medford, MA
AGGRESSION AND DRUGS, RESEARCH ISSUES

Marissa Miller
National Vaccine Program Office, Rockville, MD
AMPHETAMINE EPIDEMICS

William R. Miller
Department of Psychology
University of New Mexico, Albuquerque, NM
TREATMENT: ALCOHOL, PSYCHOLOGICAL
    APPROACHES

Richard A. Millstein
National Institute on Drug Abuse, Rockville, MD
U.S. GOVERNMENT: AGENCIES SUPPORTING
    SUBSTANCE ABUSE PREVENTION AND
    TREATMENT
U.S. GOVERNMENT AGENCIES: NATIONAL
    INSTITUTE ON DRUG ABUSE (NIDA)

Alan Minsk
Hyman, Phelps, and McNamara, Washington, DC
EXCLUSIONARY RULE
HARRISON NARCOTICS ACT OF 1914
PSYCHOTROPIC DRUG CONVENTION OF 1971
SHANGHAI OPIUM CONFERENCE OF 1911
SINGLE DRUG CONVENTION OF 1961

Joan Moore
Department of Sociology
University of Wisconsin, Milwaukee, WI
HISPANICS AND DRUG USE, IN THE
    UNITED STATES

Timothy H. Moran
Department of Psychiatry
Johns Hopkins University, Baltimore, MD
ANORECTIC
ANOREXIA
OBESITY

Herbert Moskowitz
Encino, CA
DRIVING, ALCOHOL, AND DRUGS

Otto Moulton
Committees of Correspondence, Danvers, MA
COMMITTEES OF CORRESPONDENCE

Rod Mullen
Amity, Inc., Tucson, AZ
AMITY, INC.

Clare Mundell
Center for Substance Abuse Research, College Park, MD
DRUG ABUSE WARNING NETWORK (DAWN)

Margaret Murray
National Institute on Alcoholism and Alcohol Abuse
Rockville, MD
HOMELESSNESS, ALCOHOL, AND OTHER
DRUGS

Claudio A. Naranjo
Psychopharmacology Research Program
Sunnybrook Health Science Centre, Toronto, Canada
SEROTONIN-UPTAKE INHIBITORS IN
TREATMENT OF SUBSTANCE ABUSE

Anastasia E. Nasis
Behavioral Pharmacology Research Unit
Francis Scott Key Medical Center, Baltimore, MD
CONTROLS: SCHEDULED DRUGS/DRUG
SCHEDULES, U.S.

Peter E. Nathan
University of Iowa, Iowa City, IA
RUTGERS CENTER OF ALCOHOL STUDIES

Ethan Nebelkopf
Walden House, Inc., San Francisco, CA
WALDEN HOUSE

Carl A. Newcombe
U.S. Customs Service
Canine Enforcement Training Center
Front Royal, VA
DOGS IN DRUG DETECTION

John Newmeyer
Haight-Ashbury Free Clinics Inc.
San Francisco, CA
HAIGHT-ASHBURY FREE CLINIC

Esko Nuotto
National Medicines Control Lab, Helsinki, Finland
PSYCHOMOTOR EFFECTS OF ALCOHOL AND
DRUGS

David Nurco
Human Behavior Associates, Inc., Annapolis, MD
CRIME AND DRUGS

Charles P. O'Brien
Treatment Research Center
University of Pennsylvania, Philadelphia, PA
TREATMENT: COCAINE, AN OVERVIEW

Patrick M. O'Malley
Institute for Social Research
University of Michigan, Ann Arbor, MI
HIGH SCHOOL SENIOR SURVEY

Marion Olmsted
Toronto Hospital Eating Disorder Day Centre, Toronto,
Canada
BULIMIA NERVOSA

S. Victoria Otton
Addiction Research Foundation, Toronto, Canada
PHARMACOGENETICS (GENETIC FACTORS
AND DRUG METABOLISM)

Donald A. Overton
Department of Psychology
Temple University, Philadelphia, PA
MEMORY AND DRUGS: STATE DEPENDENT
LEARNING

Denise Paone
Chemical Dependency Unit
Beth Israel Medical Center, New York, NY
NEEDLE AND SYRINGE EXCHANGES AND HIV/
AIDS

Karen Parker
Addiction Research Foundation, Toronto, Canada
ANXIETY
BLOOD ALCOHOL CONTENT
BREATHALYZER
BULIMIA
DELIRIUM
DELIRIUM TREMENS (DTS)
CONDUCT DISORDER IN CHILDREN
DEPRESSION
DRIVING UNDER THE INFLUENCE (DUI)
HALLUCINATION
NICOTINE GUM
MULTIDOCTORING
OVERDOSE, DRUG (OD)
OVER-THE-COUNTER (OTC) MEDICATION
PERSONALITY DISORDER

RECIDIVISM
RELAPSE
SCHIZOPHRENIA
SUICIDE ATTEMPT
SUICIDE GESTURE
TRIPLICATE PRESCRIPTION
WOOD ALCOHOL (METHANOL)

Gavril W. Pasternak
Department of Neurology
Memorial Sloan Kettering Cancer Center
New York, NY
ANALGESIC
BUPRENORPHINE
CODEINE
DIHYDROMORPHINE
HEROIN
HYDROMORPHONE
L-ALPHA-ACETYLMETHADOL (LAAM)
LAUDANUM
MEPERIDINE
METHADONE
MORPHINE
MPTP
NALOXONE
NALTREXONE
NARCOTIC
OPIATES/OPIOIDS
OPIUM
OXYCODONE
OXYMORPHONE
PAIN: DRUGS USED IN TREATMENT OF
PAPAVER SOMNIFERUM
PAREGORIC

J. Thomas Payte
School of Medicine
University of California at San Francisco
San Francisco, CA
METHADONE MAINTENANCE PROGRAMS

John E. Peachey
Addiction Services
Royal Ottawa Hospital, Ontario, Canada
CALCIUM CARBIMIDE

Robert N. Pechnick
Department of Pharmacology
Louisiana State University Medical Center
New Orleans, LA
DIMETHYLTRYPTAMINE (DMT)
DOM
HALLUCINOGENS
HALLUCINOGENIC PLANTS
LYSERGIC ACID DIETHYLAMIDE (LSD) AND
PSYCHEDELICS

MDMA
MESCALINE
MORNING GLORY SEEDS
NUTMEG
PEYOTE
PSILOCYBIN

Stanton Peele
Lindesmith Center, New York, NY
ALCOHOLISM: ABSTINENCE VERSUS
CONTROLLED DRINKING
ALCOHOLISM: CONTROLLED DRINKING
VERSUS ABSTINENCE
VALUES AND BELIEFS: EXISTENTIAL MODELS
OF ADDICTION

Anthony G. Phillips
Department of Psychology
University of British Columbia, Vancouver, Canada
ANHEDONIA
RESEARCH: MOTIVATION

Roy W. Pickens
National Institute on Drug Abuse, Rockville, MD
VULNERABILITY AS CAUSE OF SUBSTANCE
ABUSE: AN OVERVIEW
VULNERABILITY AS CAUSE OF SUBSTANCE
ABUSE: GENETICS

Russell K. Portenoy
Memorial Sloan Kettering Cancer Center
New York, NY
IATROGENIC ADDICTION

Kenzie L. Preston
Francis Scott Key Medical Center, Baltimore, MD
RESEARCH: MEASURING EFFECTS OF DRUGS
ON MOOD

Beny J. Primm
Addiction Research and Treatment Corporation
Brooklyn, NY
U.S. GOVERNMENT AGENCIES: CENTER FOR
SUBSTANCE ABUSE TREATMENT (CSAT)

Marie Ragghianti
Center for Substance Abuse Research
The University of Maryland, College Park, MD
COERCED TREATMENT FOR SUBSTANCE
OFFENDERS

Timothy J. Regan
Division of Cardiovascular Disease
New Jersey Medical School, Newark, NJ
COMPLICATIONS: CARDIOVASCULAR SYSTEM
(ALCOHOL AND COCAINE)

Faye E. Reilly
Department of Behavioral Science
University of Kentucky, Lexington, KY
  INJECTING DRUG USERS AND HIV

Peter Reuter
Rand Corporation, Washington, DC
  DRUG INTERDICTION
  SEIZURES OF DRUGS
  STREET VALUE

Dorothy P. Rice
School of Nursing
University of California, San Francisco, CA
  ECONOMIC COSTS OF ALCOHOL ABUSE AND
    ALCOHOL DEPENDENCE

James E. Rivers
Comprehensive Drugs Research Center
University of Miami, Miami, FL
  ACCIDENTS AND INJURIES FROM DRUGS

Cynthia A. Robbins
Department of Sociology
University of Delaware, Newark, DE
  ELDERLY AND DRUG USE
  WOMEN AND SUBSTANCE ABUSE

Steven J. Robbins
Department of Psychology
Haverford College, Haverford, PA
  CAUSES OF SUBSTANCE ABUSE: LEARNING
  WIKLER'S PHARMACOLOGIC THEORY OF
    DRUG ADDICTION

Lee N. Robins
Department of Psychiatry
Washington University School of Medicine
  St. Louis, MO
  VIETNAM: DRUG USE IN
  VIETNAM: FOLLOW-UP STUDY

Daphne Roe
Department of Nutritional Sciences
Cornell University, Ithaca, NY
  COMPLICATIONS: NUTRITIONAL

Timothy A. Roehrs
Sleep Disorders Center, Detroit, MI
  SLEEP, DREAMING, AND DRUGS

Myroslava Romach
Addiction Research Foundation, Toronto, Canada
  ANXIETY
  BLOOD ALCOHOL CONTENT
  BREATHALYZER
  BULIMIA

CONDUCT DISORDER IN CHILDREN
DELIRIUM
DELIRIUM TREMENS (DTS)
DEPRESSION
DRIVING UNDER THE INFLUENCE (DUI)
HALLUCINATION
MULTIDOCTORING
NICOTINE GUM
OVERDOSE, DRUG (OD)
OVER-THE-COUNTER (OTC) MEDICATION
PERSONALITY DISORDER
RECIDIVISM
RELAPSE
SCHIZOPHRENIA
SUICIDE ATTEMPT
SUICIDE GESTURE
TRIPLICATE PRESCRIPTION
WOOD ALCOHOL (METHANOL)

Charles M. Rongey
Orlando, FL
  ADVERTISING AND THE ALCOHOL INDUSTRY
  ADVERTISING AND THE PHARMACEUTICAL
    INDUSTRY
  ADVERTISING AND TOBACCO USE

Jed E. Rose
Nicotine Research Laboratory
Veterans Administration Medical Center
  Durham, NC
  NICOTINE DELIVERY SYSTEMS FOR SMOKING
    CESSATION

Marc Rosen
Department of Psychiatry
Yale University School of Medicine
  New Haven, CT
  TREATMENT: HEROIN, PHARMACOTHERAPY

Thomas Roth
Sleep Disorders Center, Detroit, MI
  SLEEP, DREAMING, AND DRUGS

Sue Rusche
National Families in Action, Atlanta, GA
  PARENTS MOVEMENT
  PREVENTION: NATIONAL FAMILIES IN ACTION
    (NFIA)
  PREVENTION: NATIONAL FEDERATION OF
    PARENTS FOR DRUG-FREE YOUTH/
    NATIONAL FAMILY PARTNERSHIP (NFP)
  PREVENTION MOVEMENT
  PREVENTION PROGRAMS: PRIDE (NATIONAL
    PARENTS RESOURCE INSTITUTE FOR DRUG
    EDUCATION)

Carl Salzman
Massachusetts Mental Health Center
Harvard Medical School, Boston, MA
WITHDRAWAL: BENZODIAZEPINES

Herman H. Samson
Alcohol and Drug Abuse Institute
University of Washington, Seattle, WA
ALCOHOL: PSYCHOLOGICAL CONSEQUENCES
OF CHRONIC ABUSE

Robert Saporito
Division of Cardiovascular Disease
UMDNJ, New Jersey Medical School, Newark, NJ
COMPLICATIONS: CARDIOVASCULAR SYSTEM
(ALCOHOL AND COCAINE)

Sally L. Satel
Substance Abuse Treatment Unit
Yale University School of Medicine, New Haven, CT
WITHDRAWAL: COCAINE

Eric Schaps
Developmental Studies Center, Oakland, CA
PREVENTION PROGRAMS: NAPA PROJECT,
REVISITED

Charles Schindler
Addiction Research Center
National Institute on Drug Abuse, Baltimore, MD
RESEARCH, ANIMAL MODEL: CONDITIONED
WITHDRAWAL

Joyce F. Schneiderman
Addiction Research Foundation, Toronto, Canada
ADDICTED BABIES
GENDER AND COMPLICATIONS OF SUBSTANCE
ABUSE

Marc A. Schuckit
Department of Psychiatry
Veterans Administration Medical Center
San Diego, CA
ADULT CHILDREN OF ALCOHOLICS (ACOA)
BETEL NUT

Leslie M. Schuh
National Institute on Drug Abuse
Addiction Research Center, Baltimore, MD
TREATMENT: TOBACCO, PHARMACOTHERAPY
WITHDRAWAL: NICOTINE (TOBACCO)

Christian G. Schutz
Department of Psychiatry
Ludwig Maximilians University, Munich, Germany
ASIA, DRUG USE IN

Richard B. Seymour
Haight-Ashbury Free Clinics Inc.
San Francisco, CA
COMPLICATIONS: DERMATOLOGICAL
COMPLICATIONS: ROUTE OF
ADMINISTRATION
ETHNIC ISSUES AND CULTURAL RELEVANCE
TREATMENT
HAIGHT-ASHBURY FREE CLINIC
LIFESTYLE AND DRUG-USE COMPLICATIONS

Sidney Shankman
Second Genesis, Bethesda, MD
SECOND GENESIS, INC.

Carl Shantzis
Drug Education Center, Charlotte, NC
PREVENTION PROGRAMS: OMBUDSMAN
PROGRAM

Dianne Shuntich
Department for Mental Health
Cabinet for Human Resources, Frankfort, KY
MOTHERS AGAINST DRUNK DRIVING (MADD)
PREVENTION PROGRAMS: TALKING WITH
YOUR STUDENT ABOUT ALCOHOL (TWYSAA)

Kenneth Silverman
Department of Psychiatry and Behavioral Science
Johns Hopkins School of Medicine
Baltimore, MD
CAFFEINE
COFFEE
TEA

Dwayne Simpson
Institute of Behavioral Research
Texas Christian University, Fort Worth, TX
DRUG ABUSE REPORTING PROGRAM (DARP)

John Slade
University of Medicine and Dentistry of
New Jersey, New Brunswick, NJ
TOBACCO: INDUSTRY

David E. Smith
Haight-Ashbury Free Clinics Inc., San Francisco, CA
COMPLICATIONS: DERMATOLOGICAL
COMPLICATIONS: ROUTE OF
ADMINISTRATION
ETHNIC ISSUES AND CULTURAL RELEVANCE
IN TREATMENT
HAIGHT-ASHBURY FREE CLINIC
LIFESTYLE AND DRUG-USE COMPLICATIONS

James E. Smith
Department of Physiology and Pharmacology
Bowman Gray Medical School
Wake Forest University, Winston-Salem, NC
    BRAIN STRUCTURES AND DRUGS
    LIMBIC SYSTEM
    NUCLEUS ACCUMBENS

Linda C. Sobell
Addiction Research Foundation, Toronto, Canada
    TREATMENT TYPES: LONG-TERM APPROACHES
    TREATMENT TYPES: TREATMENT STRATEGIES

Mark B. Sobell
Addiction Research Foundation, Toronto, Canada
    TREATMENT TYPES: LONG-TERM APPROACHES
    TREATMENT TYPES: TREATMENT STRATEGIES

Robin Solit
Department of Psychology
University of California, San Diego, CA
    DAYTOP VILLAGE
    SYNANON

David Spiegel
Department of Psychiatry and Behavioral Science
Stanford University School of Medicine, Stanford, CA
    TREATMENT TYPES: HYPNOSIS

John Stang
National Addiction Centre, London, England
    BRITISH SYSTEM OF DRUG-ADDICTION
    TREATMENT

June M. Stapleton
Addiction Research Center
National Institute on Drug Abuse, Baltimore, MD
    IMAGING TECHNIQUES: VISUALIZING THE
    LIVING BRAIN

Mark S. Steinitz
Department of State
Bureau of Intelligence and Research
Washington, DC
    TERRORISM AND DRUGS

Verner Stillner
Department of Psychiatry
University of Kentucky, Lexington, KY
    DOVER'S POWDER

Amy L. Stirling
Department of Psychology
Wayne State University, Detroit, MI
    PROFESSIONAL CREDENTIALING

Maxine L. Stitzer
Behavioral Pharmacology Research Unit
Francis Scott Key Medical Center, Baltimore, MD
    TREATMENT: HEROIN, PSYCHOLOGICAL
    APPROACHES

Tim Stockwell
National Centre for Research into the Prevention of
Drug Abuse
Curtin University, Perth, Australia
    TREATMENT TYPES: NONMEDICAL
    DETOXIFICATION

Carla L. Storr
Center for Substance Abuse Research
The University of Maryland, College Park, MD
    ADOLESCENTS AND DRUG USE
    EPIDEMICS OF DRUG ABUSE

John T. Sullivan
Center for Chemical Dependence
Francis Scott Key Medical Center, Baltimore, MD
    DETOXIFICATION: AS ASPECT OF TREATMENT
    COMPLICATIONS: MEDICAL AND BEHAVIORAL
    TOXICITY OVERVIEW
    WITHDRAWAL: ALCOHOL

Nancy L. Sutherland
Alcohol and Drug Abuse Institute
University of Washington, Seattle, WA
    ALCOHOL: PSYCHOLOGICAL CONSEQUENCES
    OF CHRONIC ABUSE

Dace S. Svikis
National Institute on Drug Abuse, Rockville, MD
    VULNERABILITY AS CAUSE OF SUBSTANCE
    ABUSE: AN OVERVIEW
    VULNERABILITY AS CAUSE OF SUBSTANCE
    ABUSE: GENETICS

Neil Swan
ROW Sciences, Rockville, MD
    INHALANTS: EXTENT OF USE AND
    COMPLICATIONS

Robert M. Swift
Department of Psychiatry
Roger Williams Hospital, Providence, RI
    TREATMENT: POLYDRUG ABUSE,
    PHARMACOTHERAPY

John D. Swisher
Pennsylvania State University, State College, PA
    EDUCATION AND PREVENTION

Ralph E. Tarter
Western Psychiatric Institute
University of Pittsburgh School of Medicine
  Pittsburgh, PA
  ATTENTION DEFICIT DISORDER
  CONDUCT DISORDER AND DRUG USE
  COPING AND DRUG USE

W. Kenneth Thompson
Bureau of International Narcotics Matters
  Chevy Chase, MD
  FOREIGN POLICY AND DRUGS

Steven T. Tiffany
Department of Psychological Sciences
Purdue University, West Lafayette, IN
  CRAVING

Michael Tonry
University of Minnesota Law School
  Minneapolis, MN
  MANDATORY SENTENCING
  ROCKEFELLER DRUG LAW

Harrison M. Trice
New York School of Industry and Labor Relations
Cornell University, Ithaca, NY
  AL-ANON
  ALATEEN
  ALCOHOLICS ANONYMOUS (AA)
  COCAINE ANONYMOUS (CA)
  NARCOTICS ANONYMOUS (NA)
  TREATMENT TYPES: SELF-HELP AND
    ANONYMOUS GROUPS
  TREATMENT TYPES: TWELVE STEPS, THE

Alison M. Trinkoff
School of Nursing
University of Maryland, Baltimore, MD
  ADOLESCENTS AND DRUG USE

George R. Uhl
Molecular Neurobiology Laboratory, Addiction
  Research Center
National Institute on Drug Abuse, Baltimore, MD
  ANTIDEPRESSANT
  ANTIPSYCHOTIC
  NEUROLEPTIC
  RECEPTOR: NMDA
  SUBSTANCE ABUSE, GENETIC FACTORS AND
    VULNERABILITY

James Van Wert
Small Business Administration, Silver Spring, MD
  COLUMBIA AS DRUG SOURCE
  CROP-CONTROL POLICIES (DRUGS)
  GOLDEN TRIANGLE AS DRUG SOURCE

  INTERNATIONAL DRUG SUPPLY SYSTEMS
  MEXICO AS DRUG SOURCE
  SOURCE COUNTRIES FOR ILLICIT DRUGS
  TRANSIT COUNTRIES FOR ILLICIT DRUGS

Jane Velez
Project Return Foundation, Inc., New York, NY
  PROJECT RETURN FOUNDATION, INC.

Robert E. Vestal
Clinical Pharmacology and Gerontology Research Unit
Veterans Administration Medical Center
  Boise, ID
  AGING, DRUGS, AND ALCOHOL

David Vlahov
School of Hygiene and Mental Health
Johns Hopkins University, Baltimore, MD
  ARGOT

Alexander C. Wagenaar
School of Public Health, Division of Epidemiology
University of Minnesota, Minneapolis, MN
  ALCOHOL: MINIMUM DRINKING AGE LAWS

Elizabeth Wallace
Substance Abuse Treatment Unit
Yale University School of Medicine
  New Haven, CT
  TREATMENT TYPES: PHARMACOTHERAPY, AN
    OVERVIEW

John M. Wallace, Jr.
Institute for Social Research
University of Michigan, Ann Arbor, MI
  RELIGION AND DRUG USE

Michael Walsh
Drug Advisory Council
Executive Office of the President, Washington, DC
  INDUSTRY AND WORKPLACE, DRUG USE IN

Sharon L. Walsh
Francis Scott Key Medical Center, Baltimore, MD
  RESEARCH: MEASURING EFFECTS OF DRUGS
    ON MOOD

Ronald R. Watson
Alcohol Research Center
University of Arizona, Tucson, AZ
  ALCOHOL AND AIDS
  ALCOHOL: COMPLICATIONS
  BETTY FORD CENTER
  COCAETHYLENE
  COMPLICATIONS: IMMUNOLOGIC
  PARASITE DISEASES AND ALCOHOL

Andrew T. Weil
Tucson, AZ
POLICY ALTERNATIVES: SAFE USE OF DRUGS

Meryle Weinstein
Psychiatry Service
Veterans Affairs Medical Center
San Francisco, CA
TREATMENT TYPES: PSYCHOLOGICAL
APPROACHES

Donald R. Wesson
Beverly Hills, CA
IMPAIRED PHYSICIANS AND MEDICAL
WORKERS

Harry K. Wexler
The Psychology Center, Laguna Beach, CA
CALIFORNIA CIVIL COMMITMENT PROGRAM
CIVIL COMMITMENT
NEW YORK STATE CIVIL COMMITMENT
PROGRAM
PRISONS AND JAILS
PRISONS AND JAILS: DRUG TREATMENT IN
PRISONS AND JAILS: DRUG USE AND AIDS IN
RATIONAL AUTHORITY

Richard L. Williams
Alexandria, VA
BORDER MANAGEMENT
U.S. GOVERNMENT: DRUG POLICY OFFICES IN
THE EXECUTIVE OFFICE OF THE
PRESIDENT
U.S. GOVERNMENT: THE ORGANIZATION OF
U.S. DRUG POLICY
U.S. GOVERNMENT AGENCIES: BUREAU OF
NARCOTICS AND DANGEROUS DRUGS
(BNDD)
U.S. GOVERNMENT AGENCIES: OFFICE OF
DRUG ABUSE LAW ENFORCEMENT (ODALE)
U.S. GOVERNMENT AGENCIES: OFFICE OF
DRUG ABUSE POLICY (ODAP)
U.S. GOVERNMENT AGENCIES: OFFICE OF
NATIONAL DRUG CONTROL POLICY (ONDCP)
U.S. GOVERNMENT AGENCIES: U.S. CUSTOMS
SERVICE

Amy Windham
Department of Mental Hygiene
Johns Hopkins School of Hygiene and Mental Health
Baltimore, MD
ZERO TOLERANCE

George Winokur
Department of Psychiatry
University of Iowa College of Medicine, Iowa City, IA
COMPLICATIONS: MENTAL DISORDERS

Eric D. Wish
Center for Substance Abuse Research
University of Maryland, College Park, MD
DRUG USE FORECASTING PROGRAM (DUF)

Friedner D. Wittman
KLEW Associates, Berkeley, CA
ALCOHOL- AND DRUG-FREE HOUSING

Ronald W. Wood
New York University Medical Center
Tuxedo Park, NY
INHALANTS

William Woolverton
Department of Pharmacological Sciences
Prizker School of Medicine, Chicago, IL
ABUSE LIABILITY OF DRUGS: TESTING IN
ANIMALS
RESEARCH, ANIMAL MODEL: AN OVERVIEW
OF DRUG ABUSE
RESEARCH, ANIMAL MODEL: CONDITIONED
PLACE PREFERENCE
RESEARCH, ANIMAL MODEL: DRUG
DISCRIMINATION STUDIES
RESEARCH, ANIMAL MODEL: DRUG SELF-
ADMINISTRATION

Colleen J. Yoo
Department of Health Policy and Management
Johns Hopkins School of Public Health, Baltimore, MD
VULNERABILITY AS CAUSE OF SUBSTANCE
ABUSE: SEXUAL AND PHYSICAL ABUSE

Robert Zaczek
Du Pont Merck Pharmaceuticals
Wilmington, DE
FLY AGARIC
JIMSONWEED
KAVA
SCOPOLAMINE AND ATROPINE

Marvin Zuckerman
Department of Psychology
University of Delaware, Newark, DE
VULNERABILITY AS CAUSE OF SUBSTANCE
ABUSE: SENSATION SEEKING

Joan Ellen Zweben
School of Medicine
University of California at San Francisco
San Francisco, CA
Clinic and Medical Group and East Bay Community
Recovery Project, Oakland, CA
METHADONE MAINTENANCE PROGRAMS

Encyclopedia of
Drugs and Alcohol

AA *See* Alcoholics Anonymous.

ABRAXAS The Abraxas Foundation was started in Pennsylvania in 1973, in response to Requests for Proposals (RFP) from the Governor's Council on Drug and Alcohol Abuse. Abraxas's founder, Arlene Lissner, had been the deputy clinical director for the State of Illinois drug-abuse treatment system. There were two mandates to the RFP: (1) that a drug-treatment program be devised to directly serve the juvenile and adult justice system, and (2) that the program would utilize a then-abandoned U.S. forest-service camp, Camp Blue Jay, in a remote northwestern section of Pennsylvania, within the Allegheny National Forest. The original proposal stressed the development of a comprehensive program incorporating intensive treatment, education, and, of particular importance, a continuum of care to assist residents to reenter through regional reentry facilities. After an initial attempt to use only a behavioral approach, a THERA-PEUTIC COMMUNITY (TC) model was implemented. The TC model proved to be effective in using positive peer pressure to motivate and redirect this population. Abraxas's willingness to evaluate its program model and operation, and to work in partnership with its referring courts, were cornerstones of the program, facilitating dramatic growth and development.

Since Abraxas was a pioneer project, from inception it has maintained an evaluation and research department. Studies over the years have validated the program design, which has brought about dramatic reductions in drug abuse and recidivism and increases in prosocial behavior. The process of self-evaluation and the focus on key quality factors—such as retention, treatment goal achievement, meeting educational objectives, and monitoring staff turnover—have facilitated and guided program evolution over the years.

The Abraxas program has served as a resource for all of Pennsylvania's sixty-seven counties and has been used by a dozen other states. In 1984, Abraxas was cited as a model program by ADAMHA and the Office of Juvenile Justice and Delinquency Prevention. The program has received citations of merit and appreciation by the Pennsylvania State Senate and House and the cities of Pittsburgh and Philadelphia, has been the recipient of the Benjamin Rush Award through the Allegheny County Medical Society, and has been featured at national and international conferences.

In 1983, a supporting philanthropy, the Foundation for Abraxas, was created to assist in program development and fund raising. This resulted in the complete rebuilding of the original facility, Abraxas I, that now provides treatment capacity for more than 170 male adolescents.

In 1987, the Abraxas Foundation of West Virginia, Inc., initiated service to delinquent youth in that state. A sixty-five-bed facility was developed at White Oak Village, a fifty-acre complex within a very large county park outside Parkersburg. In 1990,

Abraxas of West Virginia was certified as a behavioral health center and became a participant in Medicaid funding with a continued focus on treating delinquent males. In 1991, a specialized and discrete treatment unit was developed for delinquents with sexual offense histories. Recently, nonresidential services were developed for adolescents and their families in six locations throughout the state.

By 1988, all Abraxas facilities had focused their target populations solely on adolescents and had become gender specific. For example, Abraxas V in Pittsburgh was developed as an all-female residential facility; recently, a forty-four-bed facility was purchased to house the female program, now known as the Abraxas Center for Adolescent Females. In 1990, an intensive project known as Non-Residential Care was developed to provide community-based transitional services to youngsters returning to Philadelphia after placement in state institutions. The success of this project led to its expansion to Pittsburgh. Inspired by the Non-Residential Care model, Supervised Home Services was developed later that year as a nonresidential reentry service for youngsters returning to Philadelphia from Abraxas's residential programs.

During 1991, a specialized assessment unit at the Abraxas I facility began utilizing standardized adolescent assessment tools and a computer network to develop more precise and comprehensive treatment planning; also, a discrete specialized treatment unit was started at Abraxas I, for delinquents with extensive drug-selling histories.

Shorter-term community-based programs operating in Erie and Pittsburgh offer a four-month program in group settings of twenty to twenty-five residents. Pittsburgh's Abraxas Center for Adolescent Females offers intensive residential services for up to forty-four females in a rigorous community-based setting. In Erie, a halfway house offers residential service to delinquents returning from state institutions. A specialized program has been developed with Northampton County coupling four months of residential stay in the Pittsburgh community-based residential program with six months of nonresidential transitional service via the Supervised Home Services program in the client's home community. In Philadelphia, a recently opened residential facility, Michael's House, offers an alternative home environment for fifteen to twenty delinquent and homeless males who participate in the Non-Residential Care or Supervised Home Services programs in Philadelphia.

Abraxas has dramatically expanded its nonresidential services, now grouped under the rubric Abraxas Community Treatment. Supervised Home Services are now available in the Wyoming and Butler County areas of Pennsylvania; Family Preservation and Sex Offender programs are also available in Wyoming County. Abraxas developed a Supervised Home Services program in Washington, as part of an effort to deinstitutionalize its burgeoning juvenile offender population.

Education has been an integral part of the philosophy of treatment since Abraxas's inception in 1973. The Abraxas School, a private high school on the Abraxas I treatment campus, offers a full curriculum of courses and special educational services for the resident population. Alternative schools have been developed in Erie and Pittsburgh in recognition of the tremendous difficulty troubled adolescents have returning to public high schools. Abraxas has also aggressively extended its programming to include families of origin: The Abraxas Family Association meets in chapters throughout Pennsylvania and West Virgina to offer education, group counseling, intervention, and referral work to the families of clients.

Most recently, Abraxas inaugurated a large treatment campus in the state of Ohio for delinquent drug-abusing youth; the program is jointly funded by Ohio's Department of Youth Services and Rehabilitation Services Commission. Abraxas now offers a continuum of programs and activities at twenty-one locations in sixteen cities and towns in Pennsylvania, West Virginia, Ohio, and Washington, D.C.

Abraxas has offered care to over 7,000 individuals to date; by spring 1993, Abraxas was providing service to over 300 residential clients and more than 100 nonresidential clients; it was operating with a combined budget in excess of $26 million and a total staff of more than 550.

(SEE ALSO: *Civil Commitment*)

BIBLIOGRAPHY

FALCO, M. (1992). *The making of a drug-free America: Programs that work.* New York: Times Books.

ARLENE R. LISSNER
DANIEL S. HEIT

ABSTINENCE *See* Addiction: Concepts and Definitions; Alcoholism; Sobriety; Withdrawal.

ABSTINENCE VIOLATION EFFECT (AVE)

The abstinence violation effect occurs when an individual, having made a personal commitment to abstain from using a substance or to cease engaging in some other unwanted behavior, has an initial lapse—whereby the substance or behavior is engaged in at least once. Some individuals may then proceed to uncontrolled use. The AVE occurs when the person attributes the cause of the initial lapse (the first violation of abstinence) to internal, stable, and global factors within (e.g., lack of willpower or the underlying addiction or disease). In RELAPSE PREVENTION, the aim is to teach people how to minimize the size of the relapse (i.e., to counter the AVE) by directing attention to the more controllable external or situational factors that triggered the lapse (e.g., stress in the environment), so that the person can quickly return to the goal of abstinence and not "lose control" of the behavior.

(SEE ALSO: *Treatment*)

BIBLIOGRAPHY.

CURRY, S. J., MARLATT, G. A., & GORDON, J. R. (1987). Abstinence violation effect: Validation of an attributional construct with smoking cessation. *Journal of Consulting and Clinical Psychology, 55*, 145–149.

MARLATT, G. A., & GORDON, J. R. (1985). *Relapse prevention: Maintenance strategies in the treatment of addictive behaviors.* New York: Guilford Press.

ALAN MARLATT
MOLLY CARNEY

ABUSE *See* Addiction: Concepts and Definitions

ABUSE LIABILITY OF DRUGS: TESTING IN ANIMALS

One approach to decreasing the prevalence of drug abuse is to restrict the availability of drugs that are liable to be abused. To do this, before drugs become widely available, we must determine if they are likely to be abused. The task is not simply detecting abuse liability and making the drug unavailable, however. Abuse liability must be considered in the context of any potential therapeutic use of the drug. For example, MORPHINE has high abuse liability, but is often the only appropriate ANALGESIC for intense PAIN. Making it completely unavailable would cause undue suffering. In short, a cost-benefit analysis that weighs liability for abuse against therapeutic benefits should be made. Therefore, it is important not only to be able to predict whether a drug is liable to be abused but also to be able to estimate the abuse liability of a new drug relative to the abuse liability of drugs we already know something about. The key to this is determining whether a drug has reinforcing effects (i.e., the effect that maintains drug self-administration and estimating the magnitude or strength of those effects.

Although prediction of abuse liability used to be based upon results of experiments in human subjects (former drug-addict volunteers in prison were used), most of the research today is done on animal subjects. Research conducted since the early 1960s has demonstrated that animals (e.g., rats, monkeys) will, with very few exceptions, *self-administer the same drugs that humans are likely to abuse. Moreover, they do not self-administer drugs that humans do not abuse.* As a result, drug self-administration studies in animals play a critical role in the prediction of the abuse liability of new drugs.

Most often, abuse liability is evaluated in what has been termed a *substitution procedure.* Each day an animal is allowed to self-administer a baseline drug of known abuse potential, for example COCAINE or CODEINE, during sessions that last several hours. When session-to-session intake of the baseline drug is stable, the vehicle in which the unknown drug will be dissolved is substituted for the baseline drug for several consecutive sessions. Since the vehicle is usually inert, responding declines to low levels over several sessions. The subjects are then briefly returned to baseline conditions, followed by a substitution period during which a dose of the test drug is made available—in at least as many sessions as were required for responding—for the drug vehicle to decline. This process is repeated with several doses of the unknown drug until the experimenter is assured of having tested an active range of doses.

Rates of responding for the test drug are compared to rates of responding for its vehicle and for the baseline drug. A drug that maintains self-admin-

istration above vehicle levels is considered to be a positive reinforcer and would be predicted to have abuse liability.

Substitution procedures allow us to predict whether a drug is liable to be abused, but do not allow an estimate of abuse liability relative to other drugs. The reason for this is that they measure the rate of responding for injections, a measure that is determined not only by the reinforcing effect but also by direct effects of the drug itself that generally decrease drug self-administration. To do this, we need some way to measure the reinforcing effect of a drug that is not confounded with its other effects. To compare drugs it is useful to know the magnitude of the maximum reinforcing effect—termed its *reinforcing efficacy*. Several procedures have been developed to measure reinforcing efficacy. Most either allow an organism to choose between the drug and another drug or nondrug reinforcer (choice procedures), or measure the amount of behavior that an organism will emit to obtain an injection (progressive-ratio procedures). In choice procedures, the measure of the efficacy of a reinforcer is how often it is chosen relative to the other reinforcer. In progressive-ratio procedures, the number of lever presses required to receive a drug injection is increased until it reaches a point at which the animal no longer responds to receive an injection. This is the *break point*—a measure of the reinforcing efficacy of that dose of drug.

That these procedures provide a valid measure of reinforcing efficacy is demonstrated by the fact that animals given a choice between different doses of a drug choose the higher dose most often. In addition, break points in progressive ratio experiments are higher for higher doses than they are for lower doses. When comparing across drugs, results are consistent with what is know about abuse liability of drugs in humans—that is, COCAINE is a highly preferred drug in choice studies, maintains higher break points in progressive-ratio studies than other drugs, and is frequently abused by humans.

Since discriminative-stimulus effects of drugs in animals predict their subjective effects in humans, and subjective effects play a major role in drug abuse, drug-discrimination experiments are also used in the evaluation of abuse liability of new drugs. A drug that has subjective effects that are similar to those of a drug of abuse is likely, although not certain, to have abuse liability itself. In addition, drug discrimination experiments can be used to clas-

sify drugs of abuse according to predicted similarities in subjective effects, something that drug self-administration experiments cannot do. Thus, drug discrimination provides additional information relevant to the comparison between the new drug and drugs that we already know something about. For example, a self-administered drug with cocaine-like discriminative stimulus effects is likely to have abuse liability and cocaine-like subjective effects.

CONCLUSION

Restricting access to new drugs that are liable to be abused is one approach to decreasing drug abuse. To do this, it is necessary to predict the abuse liability of a new drug. Current research techniques allow the evaluation of a new compound; we may reliably predict whether a drug is liable to be abused and to which known drugs it is likely to be similar, both in terms of relative abuse liability and subjective effects. Such information is clearly valuable in deciding how much to restrict a new drug.

(SEE ALSO: *Abuse Liability of Drugs: Testing in Humans*; *Controlled Substances Act of 1970*; *Reinforcement*; *Research: Animal Model*)

BIBLIOGRAPHY

BRADY, J. V., & LUKAS, S. E. (1984). Testing drugs for physical dependence potential and abuse liability. NIDA Research Monograph no. 52. Washington, DC: U.S. Government Printing Office.

THOMPSON, T., & UNRA, K. R. (EDS.) (1977). *Predicting dependence liability of stimulant and depressant drugs.* Baltimore: University Park Press.

WILLIAM WOOLVERTON

ABUSE LIABILITY OF DRUGS: TESTING IN HUMANS There is probably no drug used to treat illness that does not also pose certain risks. One such risk, generally limited to drugs that have actions on the central nervous system, is that the drug will be misused or abused because of these effects. Drugs such as these are said to have abuse potential or abuse liability. If the drugs have important therapeutic use, they may still be made available, but they will be subject to certain legal controls

under various federal and state laws (see CON-TROLLED SUBSTANCES ACT). Over the past 50 years, a number of methods have been developed to test new drugs to determine their abuse liability, so that both the public and the medical profession can be warned about the need for appropriate caution when using certain drugs. These methods involve both testing in animals (preclinical) and testing in humans (clinical).

Several important reasons exist for why testing with humans is useful and necessary in the development of safer and more effective pharmacological agents. The research on laboratory animals demonstrating greater or lesser degrees of the abuse liability of drugs must be validated with humans; this reduces the likelihood of error in assessing potential risks. Moreover, certain self-reported changes associated with the subjective effects of medicinal drugs can be more readily evaluated in the humans for whom they were developed. Human clinical studies are also important in determining appropriate dose levels and dosage forms to ensure safety and efficacy while minimizing unwanted side effects. Finally, comprehensive and effective testing with humans helps to reduce the availability of abusable drugs to those who are likely to misuse them and to provide for the legitimate medical and scientific needs for such pharmacological agents.

HUMAN VOLUNTEER SELECTION

One of the most important factors in drug-abuse-liability testing with humans is the way the volunteer subjects are chosen to participate in the assessment procedures. In most studies, the human volunteer subjects are experienced drug users, but wide variations exist in the nature and extent of their drug use and abuse. Some studies, for example, use students and other volunteers whose misuse and abuse of drugs has been mostly "recreational"; other studies involve people with histories of more intensive drug use and abuse over extended periods. Also, the settings in which the tests are conducted vary widely, from residential laboratory environments, where the subjects live for several days or weeks at a time, to laboratories, where the subjects do not remain in residence but continue their daily routine after drug ingestion. Variations also occur in the age of the subjects tested and the time of day that the drug is administered. Often subjects have been selected for

certain human drug-abuse-liability tests on the basis of some special features (e.g., anxiety levels, level of alcohol consumption) to determine the extent to which such factors influence the outcome of the tests.

Convincing evidence now exists that many of these factors—particularly the prior experience of the subject with respect to drug and alcohol use and misuse—play an important role in the assessment of abuse liability. The obvious value in using such subjects lies in the fact that these individuals are similar to those most likely to abuse drugs with abuse liability—for example, drug abusers who help determine whether a new drug has a greater or lesser chance for abuse than the one they already know. It is also important to carry out abuse-liability testing with people who, for example, do not usually abuse drugs but are light social drinkers—to assess the likelihood of abuse of certain generally available medications, such as sleeping pills or appetite suppressants.

DRUG COMBINATIONS

The prediction of a drug's abuse liability, based on a wide variety of testing procedures with humans, is further complicated by the fact that drugs of abuse are often used in combination and simultaneously with other pharmacological agents. This creates some very difficult problems for the testing of abuse liability, because of the large number of possible drug combinations that need to be tested and their unknown, potentially toxic, effects. While it has long been known that drugs such as COCAINE and HEROIN or MARIJUANA and ALCOHOL are used simultaneously by drug abusers, few testing procedures have been developed for assessing their interactions. Even more puzzling is that some drugs with opposite effects (e.g., stimulants like the AMPHETAMINES and depressants like the BARBITURATES) are known to be used simultaneously by POLYDRUG ABUSERS, suggesting that unique subjective-effect changes may be important factors in such abuse patterns.

PRINCIPLES OF ABUSE-LIABILITY TESTING

Based on extensive research undertaken over the past several decades, some important general principles governing abuse-liability testing with humans

have been established. In the first instance, for example, a meaningful assessment requires that the test drug be compared with a drug of known abuse liability to provide a standard for evaluation. Second, the assessment procedure must involve the indicated comparison over a range of doses of both the test drug and the standard drug of abuse. This permits both a quantitative and a qualitative comparison of the drugs, while guarding against the possibility of overlooking some unique high- or low-dose effects. Third, the testing procedures should include measures of drug effects in addition to those like drug taking in the lab, which directly predict the likelihood of abuse. With these additional measures, it is often possible to obtain reliable estimates of abuse liability by comparing test drugs with a drug of abuse across a range of effects as a standard for evaluation. Fourth, confidence in conclusions regarding the abuse liability of a drug compound can be enhanced by utilizing a multiplicity of measures and experimental procedures. This is the case because our present level of knowledge in this area does not permit a firm determination of the best or most valid predictor of the likelihood of abuse. Finally, a population of test subjects with histories of drug use appears to be the most appropriate selection for predicting the likelihood of abuse of a new test drug, since this is the population who might use such a drug in that way.

DEVELOPMENT OF ABUSE-LIABILITY TESTING PROCEDURES

The origins of assessing drug-abuse liability with humans can be found in some of the earliest writings of civilization, describing the subjective effects of naturally occurring substances, such as wine. Since the mid-nineteenth century, literary accounts of the use and misuse of opium, marijuana, and cocaine, among other substances, have emphasized their mood-altering effects and their potential for abuse. Only in recent years, however, have systematic methods for measuring such subjective effects been refined through the use of standardized questionnaires. Volunteers who are experienced drug users complete the questionnaires after they have taken a drug; their answers to the subjective-effects questions—how they feel, their likes and dislikes—readily distinguish between the various drugs and doses, as well as between drug presence or absence (i.e., placebo).

This basic subjective-effects methodology has been further refined in recent years by using a training procedure to ensure that the human volunteer can differentiate a given drug (e.g., morphine) from a placebo (i.e., nondrug). Then new drugs are tested to evaluate their similarity to the trained reference drug of abuse. This behavioral drug discrimination method permits the volunteer to compare a wide range of subjective and objective effects of abused drugs with those of new drugs. These highly reliable procedures have proven very useful in identifying drugs that may have abuse liability.

Among the most important factors in assessing abuse liability is the determination of whether humans will take the drug when it is offered to them and whether such drug taking is injurious to the individual or society. These cardinal signs of drug abuse have provided an important focus for laboratory animal self-injection experiments, but systematic studies in which humans self-administer drugs of abuse have been less common. Methods have been developed with humans, however, for comparing the behavioral and physiological changes produced by self-administration of a known drug of abuse with the changes produced by other self-administered drugs.

The measure that has proven most useful in this approach to human drug-abuse-liability assessment is the ability of a drug to reinforce and maintain self-administration behaviors much like the behaviors used to obtain food and water. Such reinforcing effects of drugs are an important determinant influencing the likelihood that a particular drug will be abused. Laboratory studies with volunteers who are experienced drug users, for example, have shown that they will perform bicycle-riding exercises to obtain doses of abused drugs (e.g., pentobarbital). There is a systematic relationship between the amount of exercise performed and the amount of drug available (i.e., higher doses and shorter intervals between doses produce more exercise behavior than lower doses and longer interdose intervals). When a placebo or a drug that is not abused (e.g., Thorazine) is made available for bicycle riding, on the other hand, the rate of self-administration declines to near zero.

Differences between drugs in abuse liability can also be assessed by determining whether humans prefer one drug of abuse to another. During a training period, for example, experienced drug users sample coded capsules containing different drugs or

different doses of a drug. Then, during subsequent test sessions, they are presented with the coded capsules and allowed to choose the one containing the drug or drug dose they prefer. This "blind" procedure (i.e., the volunteers are not told what drugs the capsules contain) prevents biases that might be introduced by using the drug names. When neither the volunteer subject nor the person conducting the test knows what drug the capsules contain, the procedure is referred to as "double blind."

Not surprisingly, it has also been shown that the preference for one drug over another or one drug dose over another agrees well with ratings of "drug liking" made independently of the choice tests just described. In self-administration studies in which volunteers show a preference for one drug of abuse over another, subjective ratings of "liking" and positive mood changes were clearly more frequent for the preferred drug than for the drug chosen less often. Such self-reports have inherent limitations, however, because of variations in individual verbal skills, which make it necessary to confirm such findings with other measures.

In addition to the self-administration and subjective-effects measures of obvious value in testing the abuse liability of drugs in humans, other quantitative drug-effect measurements have proven useful. When, for example, observer ratings (e.g., nurses watching the patients) and performance tests (e.g., speed of movement) are measured after different drugs are administered to volunteers, the results can be compared to determine whether the behavioral changes produced by a test drug are the same as or different from those of a known drug of abuse. When a number of different performance tests (e.g., arithmetic calculations, memory for numbers and letters, speed of reaction) are given following such drug administration, it is possible to construct a behavioral profile showing the performance effects of different drugs. Comparisons between different drugs and test drugs with regard to the similarity of such profiles across their respective dose ranges increase confidence in assessments made of the abuse liability of unknown drugs.

EFFECTIVENESS OF ABUSE-LIABILITY TESTING

Because of the availability of procedures for abuse-liability testing in humans, it seems reasonable to ask how well they work. That is, has it been possible to predict from the results of these tests whether a new drug will be abused when it becomes generally available? The two major sources available for checking the effectiveness of human abuse-liability testing procedures are case reports by clinicians of patient drug abuse, and EPIDEMIOLOGICAL surveys of large numbers of individuals as well as of specific target sites (e.g., hospital emergency rooms). Both of these approaches have many shortcomings, since they lack the precision and focus that human laboratory testing can provide. But despite the drawbacks, they can detect abuse-liability problems in both specific groups of individuals and the population at large, in a manner that has generally validated the results of human laboratory-testing procedures.

ETHICAL CONSIDERATIONS

A number of codes and regulations agreed on by scientists and the lay public provide norms for the conduct of research and testing with human volunteers. In general, they require a clear statement, understandable to the volunteer, of the risks and benefits of the testing procedure, as well as an explicit consent document in written form. After it is clear that the participant thoroughly appreciates all that is involved and the potential consequences of participation, the volunteer signs the consent form in the presence of a witness who is not associated with the research. These required procedures ensure both the autonomy and the protection of volunteers for drug-abuse-liability testing.

(SEE ALSO: *Abuse Liability of Drugs: Testing in Animals; Research; Research, Animal Model*)

BIBLIOGRAPHY

BRADY, J. V., & LUKAS, S. E. (1984). Testing drugs for physical dependence potential and abuse liability. NIDA Research Monograph 52. Washington, DC: U.S. Government Printing Office.

FISCHMAN, M. W., & MELLO, N. K. (1989). Testing for abuse liability of drugs in humans. *NIDA Research Monograph, 92*, No. 89-1613. Washington, DC: U. S. Government Printing Office.

JOSEPH V. BRADY

ACCIDENTS AND INJURIES FROM ALCOHOL

Trauma (bodily injury) is a major social and medical problem in both developed and developing countries. In developed countries it is the leading cause of death between the ages of one and forty, and in the U.S. population it is the fourth leading cause of death (exceeded only by heart disease, stroke, and cancer). Of all deaths from injury in the United States, about 65 percent are classified as unintentional (which excludes deaths from suicide, homicide, and other criminal offenses); of these, about half are from motor vehicle accidents. Trauma also accounts for high rates of morbidity (number of sick to well). In the United States, the rate of serious injury is estimated to be at least three hundred times the death rate.

The first documentation of alcohol's involvement in injury occurrence dates to 1500 B.C., where an Egyptian papyrus warns that excessive drinking leads to falls and broken bones. The scientific study of alcohol and injuries has been the subject of investigation throughout the twentieth century. Data from both coroner and emergency-room studies indicate that a large proportion of victims of both fatal and nonfatal injuries are positive for blood alcohol—this proportion is larger than one would expect to find in the general population on any given day. The consumption of alcohol (ethanol) has been found to be highly associated with fatalities and serious injuries, but this may be the result of other high-risk behaviors on the part of the drinking accident victim, such as not using seat belts or motorcycle helmets. While alcohol cannot be said to actually *cause* the accident to occur in most cases, alcohol consumption is thought to contribute to both fatal and nonfatal injury occurrence, primarily because it is known to diminish motor coordination and balance and to impair attention, perception, and judgment with regard to behavior, placing the drinker at a higher risk of accidental injury than the nondrinker.

ESTIMATES OF ALCOHOL'S INVOLVEMENT

In emergency-room (ER) studies, such as those conducted by Cherpitel (1988), patients testing positive for alcohol had levels that ranged from 6 percent to 32 percent—established either directly or from a breath sample taken at the time of admission to the ER. In a review of ER studies, Cherpital (1993) found this variation in blood alcohol level (BAL) or BLOOD ALCOHOL CONCENTRATION (BAC) is due to differences in the time that passed between the injury and arrival in the ER; to individual characteristics of the particular ER populations studied (such as age, sex, and socioeconomic status—all known to be associated with alcohol consumption in the general population); and to the mix of various types of injury in the ER caseload. In studies that have been restricted to weekend evenings, when one would expect a large proportion of the population to be consuming alcohol, the proportion of those positive for alcohol at the time of ER admission has been found to be close to 50 percent. In coroner studies, such as those conducted by Haberman and Baden (1978), alcohol-related fatalities were estimated to be about 43 percent of all unintentional injuries. (However, the distinction between intentional and unintentional injury is not always readily apparent among victims of fatal injuries.) It is well known that many who drink also consume psychoactive drugs, so the independent effect of alcohol on both fatal and nonfatal accidents is not possible to ascertain.

TRAUMA RELATED TO MOTOR VEHICLES

Motor vehicle crashes are the leading cause of death from injury—and the greatest single cause of all deaths for those between the ages of fifteen and thirty-four. Almost 50 percent of these fatalities are believed to be alcohol related, and alcohol's involvement is even greater for drivers involved in single-vehicle nighttime fatal crashes. The risk of a fatal crash is estimated to be from three to fifteen times higher for a drunk driver (one with a BAL of at least .10—100 milligrams of alcohol for each 100 milliliters of blood—the legal limit in most U.S. states) than for a nondrinking driver, according to Roizen (1982). Alcohol is more frequently present in fatal than in nonfatal crashes. It is estimated that about 25 to 35 percent of those drivers requiring ER care for injuries resulting from such crashes have a BAL of .10 or greater.

Motorcyclists are at a greater risk of death than automobile occupants, and it has been estimated that up to 50 percent of fatally injured motorcyclists have a BAL of at least .10. Pedestrians killed or injured by motor vehicles have also been found more likely to have been drinking than those not involved

in such accidents. Estimates of 31 to 44 percent of fatally injured pedestrians were drinking at the time of the accident. According to Giesbrecht et al. (1989), 14 percent of fatal pedestrian accidents involved an intoxicated driver, but 24 percent involved an intoxicated pedestrian.

HOME ACCIDENTS

Among all nonfatal injuries occurring in the home, an estimated 22 to 30 percent involve alcohol, with 10 percent of those injured having a BAL at the legally intoxicated level at the time of the accident. Coroner data suggest that alcohol consumption immediately before a fatal accident occurs more often in deaths from falls and fires than in motor vehicle deaths.

Falls. Falls are the most common cause of nonfatal injuries in the United States (accounting for over 60%) and the second-leading cause of fatal accidents, according to Baker, O'Neill, and Karpf (1984). Alcohol's involvement in fatal falls has been found to range from 21 to 48 percent (with an average of 33%), according to Roizen; for nonfatal falls, alcohol's involvement has been estimated from 17 to 53 percent (with an average of 30%). Alcohol may increase the likelihood of a fall as much as sixtyfold in those well over the legal limit for intoxication, compared with those having no alcohol exposure.

Fires and Burns. Fires and burns are the fourth-leading cause of accidental death in the United States, according to Baker et al. Alcohol involvement has been estimated in 12 to 83 percent of these fatalities (with a median value of 46%), and between 0 and 50 percent among nonfatal burn injuries (with a median value of 17%). In a review of studies of burn victims, Hingson and Howland (1993) estimated that about 50 percent of burn fatalities were intoxicated and that alcohol exposure is most frequent among victims of fires caused by cigarettes.

RECREATIONAL ACCIDENTS

Drownings. Drownings rank as the third-leading cause of accidental death in the United States. Haberman and Baden (1978) reported that 68 percent of drowning victims had been drinking, but other estimates have ranged from 30 to 54 percent (with an average of 38%). Alcohol is consumed in relatively large quantities by many of those involved in water-recreation (especially boating) activities, and studies suggest that those involved in aquatic accidents are more likely to be intoxicated than those not involved in such accidents. In a review of the literature on those who came close to drowning, Roizen (1982) found that about 35 percent had been drinking at the time.

WORK-RELATED ACCIDENTS

Alcohol's involvement in work-related accidents varies greatly by type of industry, but the proportion of those positive for blood alcohol following a work-related accident is considerably lower than for other kinds of injuries, particularly in the United States, since drinking on the job is not a widespread or regular activity. Among work-related fatalities, an estimated 15 percent have been found positive for blood alcohol, and a range of 1 to 16 percent has been estimated for nonfatal injuries, according to Giesbrecht et al. (1989).

ALCOHOLISM VERSUS UNWISE DRINKING

The available literature on the role of alcoholism as opposed to unwise drinking in injury occurrence suggests that problem drinkers and those diagnosed as alcoholics are at a greater risk of both fatal and nonfatal injuries than those in the general population who may drink prior to an accident. Alcoholics and problem drinkers are significantly more likely to be drinking and to be drinking heavily prior to an accident than others. Haberman and Baden (1978) found among fatalities from all causes that alcoholics and heavy drinkers were more than twice as likely as nonproblem drinkers to have a BAL at the legal limit. Alcoholics have also been found to experience higher rates of both fatal and nonfatal accidents, even when sober.

Chronic alcohol abuse has long-term physiologic and neurologic effects that may increase the risk of accidents. Chronic drinking also impairs liver function, which plays an important part in injury recovery. A damaged liver compromises the immune system, predisposing the alcoholic to bacterial infections following injury. The risk of accidental death has been estimated to be from three to sixteen times greater for alcoholics than for nonalcoholics, with

the highest risk being for death from fires and burns. Haberman and Baden (1978) found that among all fatalities from fires, 34 percent were alcoholics.

(SEE ALSO: *Alcohol; Driving, Alcohol, and Drugs; Driving under the Influence; Industry and Workplace, Drug Use in; Social Costs of Alcohol and Drug Abuse*)

BIBLIOGRAPHY

BAKER, S. P., O'NEILL, B., & KARPF, R. (1984). *The injury fact book.* Lexington, MA: Lexington Books.

CHERPITEL, C. J. (1993). Alcohol and injuries: A review of international emergency room studies. *Addiction, 88,* 923–937.

CHERPITEL, C. J. (1988). Alcohol consumption and casualties: A comparison of emergency room populations. *British Journal of Addiction, 83,* 1299–1307.

GIESBRECHT, N., ET AL. (EDS.) (1989). *Drinking and casualties: Accidents, poisonings and violence in an international perspective.* London: Croom Helm.

HABERMAN, P. W., & BADEN, M. M. (1978). *Alcohol, other drugs, and violent death.* New York: Oxford University Press.

HINGSON, R., & HOWLAND, J. (1993). Alcohol and non-traffic unintended injuries. *Addiction, 88,* 877–883.

ROIZEN, J. (1982). Estimating alcohol involvement in serious events. In *Alcohol consumption and related problems.* DHHS Publication no. ADM 8201190. Washington, DC: U.S. Government Printing Office.

CHERYL J. CHERPITEL

ACCIDENTS AND INJURIES FROM DRUGS

Throughout history, humans have taken substances other than food into their bodies in ways that usually were socially accepted. The most common form has been as medicine, in attempts to change some feeling of ill-being or dis-ease, such as PAIN, fatigue, or tension. Some cultures distinguish between socially approved and disapproved uses of substances by labeling those approved as "medicine" and those disapproved as "drugs." Although the word *medicine* derives from a Latin word meaning "of a physician," throughout much of recorded history and even today, "folk" medicine and "home remedies" are widely practiced. Early medicines were taken exclusively from nature, and PLANTS are still an important source of medicinal products (e.g., foxglove for heart problems, bread mold for penicillin).

Organic HALLUCINOGENIC substances (plants with perception-altering properties, such as peyote and mescaline, and fungi, such as psilocybin mushrooms) have been used in various cultures in the context of religious rituals, with the dramatic visual and aural hallucinations induced being interpreted in spiritual terms. Intoxicating beverages (with alcohol and/or other drugs) also have a long history of use, usually recreationally (i.e., for their relaxing and disinhibiting effects in social situations), but sometimes also for supposed medicinal purposes, as in elixirs and tonics that were marketed as patent medicines in the United States in the late 1800s and early 1900s.

With the development of modern chemistry and scientific growing methods, there has been an increase in the number, types, and strength of both organically derived (principally MARIJUANA) and chemically synthesized (laboratory-created) substances, which has prompted the implementation of measures to regulate their processing or manufacture, distribution, and dispensing (by pharmacists and physicians). In the United States, this regulatory scheme is called the Federal CONTROLLED SUBSTANCES ACT (Public Law 91-513, H.R. 18583, October 27, 1970). This act, which is regularly updated, classifies substances into five categories according to the potential for abuse, accepted medical effectiveness and use, and potential for creating physiological or psychological dependence. The Controlled Substances Act is the basis for federal and state drug laws that specify the conditions making specific substances illicit (illegal) drugs and define differential criminal penalties for their manufacture, distribution, sale, and possession. Substance abuse, for the purposes of this discussion, is defined as the use of illicit drugs, the misuse of medicines (particularly those which must be prescribed by physicians, but also those which are available OVER THE COUNTER), and the excessive or prohibited (e.g., underage) use of other legal drugs, such as alcoholic beverages.

There have been increases since the mid-1960s in both substance abuse and public awareness of it. Here the emphasis will be on an aspect of substance abuse—the unintended, negative consequences— that is relatively well known to researchers but less familiar to the general public and their representatives in government.

UNINTENDED AND NEGATIVE CONSEQUENCES

Data do not exist to document in a comprehensive or detailed manner the extent to which the negative consequences of substance abuse exist worldwide, for specific nations (including the United States), or for given states or cities. Consequently, I have chosen to present mainly summary information that is based on the best evidence available rather than partial statistical data of questionable accuracy, which would soon be out of date.

The following discussion is divided into four categories of drug problems; acute, chronic, drug-caused, and drug-related. Acute problems are defined as those which usually occur suddenly and often can be remedied in a relatively short time. Chronic problems typically have a relatively gradual onset and tend to persist, sometimes indefinitely. Drug-caused problems, for the purposes of this discussion, are defined as those which have an obvious and/or demonstrated direct causal connection between the use of a substance and a negative effect. In drug-related problems, negative outcomes result from drug-diminished capacities and their effects on user behaviors.

Acute Drug-Caused Problems. The evidence seems to support alcohol-related medical diagnoses as the most-common acute drug-caused problem. There is no single definitive measure with which to document this, but the fact that in 1985, 1.1 million discharge reports from short-term hospitalizations in the United States involved alcohol-related diagnoses is illustrative. The National Institute on Drug Abuse's DRUG ABUSE WARNING NETWORK (DAWN) estimates that alcohol-in-combination (alcohol used with another drug, the criterion for reporting this substance to DAWN) represented the most frequently reported category in 1990 drug-related hospital emergency department visits (over 31% of the 371,208 reported visits).

"Other drugs" (illicit drugs, accidentally misused, and intentionally abused legal medications) collectively represent an acute drug-caused problem that may equal or surpass ALCOHOL in this category. The DAWN system is the best source of documentation of these problems, capturing data on substance abusers who come or are brought to hospital emergency departments, particularly for negative and/or unexpected reactions. These cases are usually thought of as overdoses, attributable to (a) tolerance effects (the

need to use increasingly larger doses to achieve the same PSYCHOACTIVE effects), (b) inexperienced users with panic reactions, or (c) the use of a substance of greater strength than intended or expected. There also is increasing evidence that users of some drugs, such as COCAINE, can experience medical emergencies and deaths from seizure disorders and allergic reactions. This is true not only for first-time users but also experienced users at their regular (and sometimes low) dosage levels. Other frequently noted hospital emergency department (and medical examiner) cases involve SUICIDE ATTEMPTS (and successes) where medications and/or drugs are the means chosen. In 1990, DAWN reports documented over 256,000 hospital emergency department visits for drugs other than alcohol, with over 520,000 drug "mentions" (i.e., the typical case involves two or more drugs).

Chronic Drug-Caused Problems. The most common chronic drug-caused problems are probably alcohol-related diseases (particularly cirrhosis and chronic liver disease). There were more than 26,000 deaths attributed to these causes in 1986. Intravenous drug users can suffer from chronic cardiovascular problems that may be primarily attributable to infections and damage from "fillers"—talcum powder, cornstarch, or baking soda added to drugs to increase volume, unit sales, and profits—that have been injected. Some drugs—especially the DESIGNER DRUGS, where an easily added molecule can produce a deadly variant of the intended substance—are neurotropic/neuropathic (have an affinity for and do damage to nerve endings and tissue). Researchers and clinicians are studying and treating individuals who are afflicted with Parkinson's disease caused by such "party" drugs.

The final chronic drug-caused problem discussed is particularly tragic because the individuals most damaged are totally innocent and defenseless against the substances that can cause them permanent disabilities or even death. I refer here to teratogenic drug use (use by pregnant women that causes abnormal fetal development). A phenomenon known as FETAL ALCOHOL SYNDROME (FAS), characterized by malformed facial features and various other developmental problems, has long been recognized as a problem among infants born to mothers who consume alcohol during pregnancy, and educational campaigns have been used extensively as prevention. Given the unavailability of current or even reliable historical data, the magnitude of the FAS

problem is unknown. As of the mid-1990s the most current estimate found for FAS rates is approximately 1.9 cases per 1,000 live births in 1987, but the rates are highly variable among population subgroups, and precise estimates of overall numbers and rates are difficult.

When the use of CRACK-cocaine exploded in the mid-1980s, fetal developmental damage resembling FAS began to be noted among infants born to women who had used this drug during the pregnancy. Although accurate counts are difficult to obtain, estimates cited in a 1990 government report—which are not based on representative national samples—range from 100,000 infants annually exposed to cocaine alone to as many as 375,000 annually exposed to drugs in general.

Acute Drug-Related Problems. Acute drug-related problems are typified by physical trauma; because of inebriation- or intoxication-impaired judgment or motor control/coordination, substance-abusing individuals can and do accidentally injure themselves. The evidence seems to support alcohol-related vehicle accidents as the most-common acute drug-related problem. According to National Highway Transportation Safety Administration statistics, alcohol was implicated in the deaths of approximately half of the 46,386 individuals who died in traffic crashes in 1985.

Other common examples in this category are accidental weapon discharges and motor vehicle and boating accidents; drownings, falls, and electrocutions are less common but not rare. It also should be noted that although intoxication from drugs other than alcohol is highly probable, there are few statistical data (apart from DAWN) to document it.

Even more unfortunate are instances where others—clean and sober—are injured or killed by the impaired substance abuser. (These cases often go unnoted; for example, DAWN records only cases where the injured or deceased had drugs "on board.") In addition to the types of accidental injuries noted immediately above, there are anecdotal (verbally reported but not documented) accounts which suggest that spouses, children, other relatives, and friends often are the victims of drug-impaired individuals. With diminished emotional control, inhibitions, and judgment, substance abusers often inflict physical trauma (e.g., gunshot, blunt force, or penetrating injuries) in chance encounters with strangers (other drivers, businesspeople), in disputes with coworkers or friends, in domestic disputes, or in physical abuse of their children. Moreover, analysis by Brookoff and his colleagues conclude that DAWN reporting procedures seriously underreport drug-involved emergency department cases, especially those with serious trauma.

Chronic Drug-Related Problems. Some of the most common chronic drug-related problems are similar to those resulting from in utero exposure of fetuses to cocaine, alcohol, and other development-impairing substances, in that severe illness and death are frequently involved and individuals other than the users themselves are victims. This class of drug-related problems is most closely associated with injecting drug use because the category is defined by infectious diseases that can be transmitted via sharing of unsanitized hypodermic needles. The most deadly infectious agent spread in this manner is the HUMAN IMMUNODEFICIENCY VIRUS (HIV), which causes ACQUIRED IMMUNODEFICIENCY SYNDROME (AIDS). Another potentially deadly infection spread in this manner is the liver disease hepatitis B. Various sexually transmitted diseases (including AIDS, syphilis, and gonorrhea), as well as tuberculosis, are among the other debilitating and often fatal diseases that are chronic problems related to substance abuse.

DRUG-SPECIFIC NEGATIVE CONSEQUENCES

Undeniably, the substance producing the largest number of individuals who suffer negative consequences from its use is alcohol. The aggregate social costs from alcohol abuse and alcoholism—health care, social functioning and productivity, family stability, death/years of productive life lost—are horrific, and very well could surpass those from illicit drugs. But these alcohol-caused and -related problems are well known; alcohol is a legally available, socially acceptable (in moderation) drug, which makes it easier to publicly acknowledge and address associated problems than is the case with illicit drugs (a condition that is beginning to change).

Therefore, discussion in this section is limited to the specific negative consequences of a few of the most prevalent illicit drugs: marijuana, cocaine, and heroin. Other illicit substances that could have been discussed here include LYSERGIC ACID DIETHYLAMIDE

(LSD), PHENCYCLIDINE (PCP), and other "alphabet" or "designer" drugs ("ecstacy," etc.), AMPHETAMINE, and METHAMPHETAMINE (and its smokable form, "ice"). This discussion also could well include legal substances that are abused by inhalation, such as gasoline, airplane glue, and various solvents, which are mundane but widely used—especially in economically depressed areas of the United States and in developing nations. These are very harmful to lungs and brain cells, and are often deadly.

Marijuana. The smoking of marijuana, probably the most widely used illicit drug, may well have more serious acute and chronic consequences than once thought. Recent and continuing research is casting new light on the chronic health risks posed by marijuana smoking, contradicting the conventional wisdom that it is less harmful than either drinking alcoholic beverages or smoking tobacco. For example, it is reported that three times the tar is delivered (and four times more is deposited) to the mouth and lungs per puff from a marijuana joint than from a filter-tipped cigarette. Smoking marijuana also produces up to five times more carbon monoxide in the user's blood than does tobacco. Knowledge is also being accumulated regarding the specific health-damaging mechanisms from the 426 known chemicals contained in *Cannabis sativa* (which are transformed into over 2,000 when ignited).

Among the pertinent facts already established is that 70 of these chemicals are fat-soluble and accumulate in fatty body tissue, notably the brain, lungs, liver, and reproductive organs. This represents a persistence-of-residue effect in which portions of THC (delta-9-TETRAHYDROCANNABINOL), the most potent psychoactive chemical in marijuana, not only remain in the body (and are thus detectable) for several weeks following use, but also accumulate with repeated use. This THC buildup is particularly noteworthy when one considers that the potency (THC content) of marijuana has increased dramatically. During the 1960s, the average potency was about 0.2 percent, a stark contrast with the 1980s average of around 5 percent—even higher for certain varieties, such as 14 percent for California sensimilla.

Regular marijuana smoking can contribute to emotional and other behaviorally defined mental-health problems through degraded interpersonal relationships and arrested development. The mechanism for this seems to be a drug-induced perception of well-being and problem abatement that may not reflect reality and contributes to avoidance rather than coping with life situations.

Research findings from the 1980s and 1990s highlight a marijuana health risk that is largely unmeasured but may be much greater than generally considered. Researchers examined blood samples from over 1,000 individuals brought to a hospital trauma unit with severe injuries. Two-thirds of these individuals had accidental injuries associated with the operation of motor vehicles (drivers, passengers, and pedestrians injured by cars, trucks, or motorcycles). Using a blood test that normally ceases to detect THC around 4 hours after use, the researchers found about 34 percent of these accident victims had psychoactive levels of THC in their blood when they arrived at the hospital. (Similar figures were reported for alcohol.) A more recent study found that 45 percent stopped for reckless driving tested positive for marijuana. Given that there currently are no simple, legally recognized tests to detect marijuana use (analogous to breath-analysis tests used to detect alcohol consumption), the extent of marijuana-related vehicular injuries and deaths is an unknown but potentially sizable statistic.

Cocaine. Many readers undoubtedly have heard that Sigmund FREUD, the famous Viennese psychoanalyst, was an avid user and proponent of cocaine. The initial account of his observations of the drug's effects for himself and some of his patients indeed was glowing. It was translated from German and reprinted in an American medical journal in the mid-1880s, thus popularizing the drug in the United States and prompting its incorporation into products from patent medicines to soft drinks. Less well reported is the fact that Freud and his colleagues had discovered the significant negative effects of cocaine by the end of that same decade and had withdrawn their support for its applications in medical therapy.

Cocaine has had several periods of popularity in the United States as a drug of abuse, with the most recent beginning in the early 1980s. Touted as a safe, nonaddicting, recreational drug, cocaine hydrochloride (in powder form) was "snorted" (inhaled) by millions who liked the absence of hypodermic needles, the lack of lung-cancer risk, and the rapid high, with its feelings of alertness, wittiness, and sexual prowess.

Unfortunately, cocaine users often progress from casual to compulsive patterns of use. The grandiose

perceptions of heightened mental and physical abilities inevitably wane (typically within 20 minutes following use), and the resulting dysphoria (opposite of euphoria) is so marked in contrast that it resembles depression. Trying to relieve the depression and regain the euphoria, cocaine abusers repeat this cycle over long periods (called binges) until their supplies, resources, and/or stamina are exhausted. In the study of reckless drivers noted earlier, 25 percent tested positive for cocaine.

The risk of infection with HIV and other sexually transmitted diseases is high among compulsive cocaine users, particularly female crack users. Often forsaking socially acceptable means of earning income, they live an existence that revolves around crack use. Many maintain their crack supply by repeatedly selling or trading their sexual services, with each unprotected sexual contact increasing the chance that they will have an HIV-infected partner.

Other manifestations of negative effects of cocaine abuse include hyperstimulation; digestive disorders, nausea, loss of appetite and weight; tooth erosion; nasal mucous membrane erosion, including perforations of the nasal septum (holes in the membrane separating the nostrils); cardiac irregularities; stroke (from vascular constriction); convulsions (especially among individuals prone to seizure disorders); and paranoid psychoses and delusions of persecution. Cocaine is a notoriously fickle drug—some experts say it behaves as though it belongs in other pharmacological categories besides stimulant. A highly publicized case is the 1986 cocaine-induced death of the athlete Len Bias, who reportedly was a first-time user of a small amount. In addition, cutting-edge research by Dr. Deborah Mash and others indicates that the concurrent use of cocaine and alcohol (a common practice) produces a new, liver- and brain-accumulating and -damaging drug (COCA-ETHYLENE) within the user's body. It is implicated in puzzling low-dosage "excited delirium" fatalities and increased mortality risks for individuals with existing heart problems.

Some of the fetal damage from maternal cocaine use occurs because cocaine is a vasoconstrictor, a useful characteristic for topical application in delicate medical procedures, such as eye surgery, but a decided negative as it concerns the placenta of a pregnant woman. The restriction of blood flow through the placenta limits nutrients and oxygen to the fetus, leading to retarded growth and develop-

ment of vital organs. Heavy cocaine use during pregnancy also can cause spontaneous abortion, and anecdotal reports of cocaine being used intentionally for this purpose are not uncommon. Premature separation of the placenta from the uterus, another common medical complication among cocaine-using pregnant women, results in either a premature birth or a stillbirth. Surviving infants usually have low (sometimes very low) birthweights, and low birthweight itself increases risk for a variety of problems. Cocaine-exposed underweight newborns have been documented to be at greater risk for stroke and respiratory ailments, and at much greater risk for sudden infant death syndrome (SIDS or crib death). Research studies are being conducted to confirm anecdotal and preliminary studies that indicate higher rates of retarded emotional, motor, and cognitive development, including ATTENTION DEFICIT DISORDERS, among such children entering school.

Heroin. Paradoxically, the negative direct physiological consequences to the user that are attributable to heroin itself are less than from the use of tobacco, alcohol, cocaine, or many prescription drugs. This does not mean that heroin is a drug whose use is without negative consequences, however. Heroin is highly addictive, and once its central nervous system depressive effects wear off, (typically in 4 to 6 hours), users tend compulsively to seek sources and means for another "hit." This often leads to socially unproductive, self-neglectful lifestyles, not uncommonly involved with income-producing crime committed to maintain the addiction. Fetuses exposed to heroin from their mothers' use during pregnancy suffer many of the same negative effects as those exposed to cocaine.

In addition, heroin users frequently experience negative reactions and overdoses because of TOLERANCE effects, since the drug purity and type of filler may vary widely among dealers. Often, they are unknown with certainty by the user. However, the greatest threat to current heroin users' health and lives undoubtedly stems from the risks of hepatitis and HIV infection from sharing contaminated hypodermic syringes and needles. Their risks of infection with HIV and other diseases through sexual activity are elevated in ways similar to those described for cocaine users. The risks of fetal HIV infection among pregnant heroin-using women is also increased because of their own needle use and the likelihood that they have had at least one intravenous-drug-using sexual partner.

ECONOMIC COSTS
OF SUBSTANCE ABUSE

Numerous studies since 1980 have been designed to estimate the cost or burden of drug abuse. This presentation will focus on two of the most recent and authoritative reports.

Gerstein and Harwood (1990) produced a set of estimated societal ECONOMIC COSTS of the illicit drug problem in the United States totaling 71.9 billion dollars. This total is broken down into categories; almost half ($33.3 billion) of the total estimated costs was attributed to productivity losses resulting from substance-abuse-impaired workers. More than half of the total estimated cost was attributed to the criminal aspects of drug abuse ($5.5 billion to tangible losses to crime victims; $12.8 billion to law enforcement; $17.6 billion to lost economic productivity because of time spent in crime or in prison). A minor portion of the estimated costs was assigned to drug prevention and treatment ($1.7 billion) and drug-related AIDS ($1.0).

The Gerstein and Harwood estimate is an update incorporating "a number of statistical updates and revisions" of a similar estimate for 1980 (Harwood et al., 1984), which totaled $46.9 billion dollars. (The earlier analysis provided an additional estimate of $90 billion as the societal costs for alcohol problems; there was no such 1990 estimated cost provided by Gerstein and Harwood.) The authors note that the fairly standard "human capital" approach used in these estimates is "conservative," focusing on potentially productive property and labor, and unable to assign dollar values to pain, suffering, fear, demoralization, and so on.

Rice et al. (1990) produced what is probably the most authoritative work currently available on this topic: *The Economic Costs of Alcohol and Drug Abuse and Mental Illness: 1985.* Among the multiple objectives of the study were the following: to estimate as precisely as possible the economic costs to society of alcohol abuse, drug abuse, and mental illness (ADM) for 1985, the most recent year for which reliable data were available; to update previous cost estimates, using new data sources and a revised methodology; and to develop an improved approach to deal with the issues of COMORBIDITY (the tendency for cases to overlap, that is, for individuals to have problems in more than one ADM category). Briefly, Rice et al. produced ADM cost estimates for 1980, 1985, and 1988. Estimated costs for illicit drug problems (in billions of dollars) were 1980, 46.9; 1985, 44.1; 1988, 58.3. The comparable figures for alcohol drug problems were 1980, 89.5; 1985, 70.3; 1988, 85.8.

Their cost estimates are considered to be the best available, but even the analysts who produced them will readily agree that they are probably not very precise. It is probable, in my opinion, that the health and social care costs for AIDS-infected drug abusers is underestimated even for the years addressed, and are undoubtedly significantly higher in the mid-1990s. Similarly, health and social service costs for infants and children who have suffered prenatal exposure to drugs undoubtedly have mushroomed since 1985, the approximate beginning of the crack-cocaine epidemic (these costs could become astronomical if the worst fears regarding permanent learning disabilities are confirmed). And if there were significant increases in public dollars spent for substance-abuse treatment, as there would have to be to serve the approximately 5.5 million drug abusers (1987/1988) estimated to need it, the estimated costs in this category would also be higher.

(SEE ALSO: *Accidents and Injuries from Alcohol; Alcohol: History of Drinking; Dover's Powder; Driving, Alcohol, and Drugs; Fetus: Effects of Drugs on; Social Costs of Alcohol and Drug Abuse*)

BIBLIOGRAPHY

BARINAGE, MARCIA. (1990). Miami Vice metabolite. *Science, 250,* 758.

BROOKOFF, D., CAMPBELL, E. A., & SHAW, L. M. (1993). The underreporting of cocaine-related trauma: Drug Abuse Warning Network reports vs. hospital toxicology tests. *American Journal of Public Health, 83* (March), 369.

BROOKOFF, D., COOK, C. S., WILLIAMS, C., & MANN, C. S. (1994). Testing reckless drivers for cocaine and marijuana. *New England Journal of Medicine, 331* (August), 518.

CENTERS FOR DISEASE CONTROL. (1991). The HIV/AIDS epidemic: The first 10 years. *Morbidity and Mortality Weekly, 40*(22), 357–365.

CHIANG, C. N., & HAWKS, R. L. (EDS.). (1990). *Research findings on smoking of abused substances.* Rockville, MD: National Institute on Drug Abuse.

CHITWOOD, D. D., ET AL. (1991). *A community approach to AIDS intervention.* Westport, CT: Greenwood Press.

GERSTEIN, D. R., & HARWOOD, H. J. (EDS.). (1990). *Treating drug problems*, vol. 1. Washington, DC: National Academy Press.

HARWOOD, H. J., NAPOLITANO, D. M., KRISTIANSEN, P., & COLLINS, J. J. (1984). *Economic costs to society of alcohol and drug abuse and mental illness: 1980.* Research Triangle Park, NC: Research Triangle Institute.

INCIARDI, J. A. (1992). *The war on drugs II.* Mountain View, CA: Mayfield.

MACDONALD, D. I. (1988). Marijuana smoking worse for lungs. *Journal of the American Medical Association, 259* (June), 3384.

NATIONAL HIGHWAY TRAFFIC SAFETY ADMINISTRATION. (1987). *Fatal accident reporting system, 1985.* Washington, DC: U.S. Department of Transportation.

NATIONAL INSTITUTE ON DRUG ABUSE. (1991). *Annual emergency room data, 1990: Data from the Drug Abuse Warning Network (DAWN), series I, no. 10-A.* DHHS Publication no. (ADM) 90-1839. Rockville, MD: U.S. Department of Health and Human Services.

NATIONAL INSTITUTE ON DRUG ABUSE. (1991). *Epidemiologic trends in drug abuse: Proceedings of the Community Epidemiology Work Group.* DHHS Publication no. (ADM)91-1849. Rockville, MD: U.S. Department of Health and Human Services.

RICE, D. P., KELMAN, S., MILLER, L. S., & DUNMEYER, S. (EDS.). (1990). *The economic costs of alcohol and drug abuse and mental illness: 1985.* San Francisco: Institute for Health and Aging.

SODERSTROM, C. A., ET AL. (1988). Marijuana and alcohol use among 1023 trauma patients: A prospective study. *Archives of Surgery, 123,* 733–737.

TURNER, C. F., & MILLER, H. G. (EDS.). (1989). *AIDS: Sexual behavior and intravenous drug use.* Washington, DC: National Academy Press.

U.S. DEPARTMENT OF HEALTH AND HUMAN SERVICES. (1990). *Alcohol and health: Seventh special report to the U.S. Congress.* Alexandria, VA: Editorial Experts.

U.S. DEPARTMENT OF JUSTICE, DRUG ENFORCEMENT ADMINISTRATION. (Annual). *Drugs of abuse.* P. E. Fitzgerald (Ed.). Washington, DC: U.S. Government Printing Office.

U.S. GENERAL ACCOUNTING OFFICE. (1990). *Drug-exposed infants: A generation at risk.* HRD-90-138. Washington, DC: General Accounting Office.

JAMES E. RIVERS

ACETALDEHYDE   *See* Complications: Liver (Alcohol); Disulfiram.

ACETAMINOPHEN   *See* Analgesic; Drug Metabolism; Pain, Drugs Used in Treatment of.

ACETYLCHOLINE   Acetylcholine (ACh) is a NEUROTRANSMITTER released at autonomic nerve endings, active in the transmission of the nerve impulse. It is the ester of acetate and choline, formed by the enzyme choline acetyltransferase, from choline and acetyl-CoA. This was the first substance (1906) to meet the criteria of identification for being a neurotransmitter. Later, ACh was shown to be the general neurotransmitter for the neuromuscular junctions in all vertebrate species, the major neurotransmitter for all autonomic ganglia, and the neurotransmitter between parasympathetic ganglia and their target cells. Acetylcholine neurotransmission occurs widely within the central nervous system. Collections of NEURONS arising within the brain—the medulla, the pons, or the anterior diencephalon—innervate a wide set of targets; some of these circuits are destroyed in Alzheimer's disease.

(SEE ALSO: *Scopolomine and Atropine*)

BIBLIOGRAPHY

COOPER, J. R., BLOOM, F. E., & ROTH, R. H. (1991). *The biochemical basis of neuropharmacology*, 6th ed. New York: Oxford University Press.

FLOYD BLOOM

ACETYLMETHADOL   *See* L-Alpha-Acetylmethadol (LAAM).

ACID   *See* Lysergic Acid Diethylamide (LSD) and Psychedelics; Slang and Jargon.

ACQUIRED IMMUNODEFICIENCY SYNDROME   *See* Alcohol and AIDS; Complications: Route of Administration; Injecting Drug Users and HIV; Substance Abuse and AIDS; Needle and Syringe Exchanges and HIV/AIDS.

ACUPUNCTURE   *See* Treatment Types: Art of Acupuncture.

ADAMHA   *See* Treatment, History of; U.S. Government.

ADDICT   *See* Addiction: Concepts and Definitions.

ADDICTED BABIES   Technically, the term *addicted babies* should refer to infants who are born passively physically dependent on drugs. In practice, it is used to refer to all babies extensively exposed to drugs before birth. A 1988 survey estimated that in the United States each year some 375,000 babies are born who were exposed to illicit drugs while in the uterus; a far larger number have been exposed, in utero, to alcohol, sedatives, and nicotine. The increased recognition of such drug-exposed babies parallels the dramatic increase in drug use, both licit and illicit, by women since the beginning of the 1970s.

Drug-addicted women often use multiple substances—including ALCOHOL, NICOTINE, MARIJUANA, TRANQUILIZERS, COCAINE, and OPIOIDS (e.g., HEROIN and METHADONE). The drugs are carried across the placenta from mother to fetus. The clinical presentation of the newborn (neonate) depends on the substance, the amount and frequency used during pregnancy, and the time since last use. Infants of regular heavy users usually have a low birthweight, because of intrauterine growth retardation and frequent premature births.

If the mother has used a large dose of depressant drug (alcohol, any number of sedative-hypnotics, or heroin) immediately before delivery, the neonate may have respiratory depression and may require resuscitation. If the mother has used one of these drugs regularly during pregnancy, there may be a neonatal abstinence or withdrawal syndrome, which has as key features irritability, tremor, and increased muscle tone. Other symptoms include poor feeding, vomiting, and diarrhea; high-pitched cry; difficulty in sleeping; and sneezing, sweating, yawning, nasal stuffiness, rapid breathing, and seizures. Withdrawal from heroin generally occurs within forty-eight hours of birth, but it can be somewhat delayed with the longer-acting methadone. Alcohol-exposed infants may develop a very similar withdrawal syndrome, except with the more frequent occurrence of seizures.

Cocaine, a stimulant, constricts blood vessels, thereby decreasing oxygen delivery to the fetus. Consequently, neonatal stoke has occurred. Although cocaine withdrawal symptoms have been reported in neonates, they probably reflect acute cocaine intoxication when the mother's last use was close to the time of birth. Such infants often appear less alert and less responsive to external stimuli than noncocaine-affected newborns, which may represent true withdrawal—comparable to the so-called crash seen in adults.

MANAGEMENT

A thorough alcohol and drug history should be obtained from the expectant mother—and this should be corroborated by testing the urine of both mother and newborn for alcohol and other drugs. Newborns should be closely monitored for signs of withdrawal for a minimum of forty-eight to seventy-two hours, and longer when the mother has been on METHADONE-maintenance treatment. Since symptoms of withdrawal are nonspecific and may be confused with a variety of infections or metabolic disturbances, a search for concurrent illness to explain any symptoms is mandatory.

Most nurseries use a standardized neonatal abstinence-syndrome scoring system. The earliest withdrawal symptoms are treated by swaddling, holding, rocking, and a low-stimulation environment. If symptoms continue or increase, medication may be initiated. Common medications include PAREGORIC (camphorated tincture of opium) or phenobarbital for opioid withdrawal; PHENOBARBITAL or DIAZEPAM for alcohol withdrawal. Diazepam is also used to help with cocaine hyperexcitability.

Interviewing the mother is essential in reviewing the anticipated home environment. Unfortunately, addicted babies are often at high risk for either abuse or neglect or both. Normal maternal-infant bonding is difficult in the case of an irritable poorly responding neonate and a mother dealing with the guilt, low self-esteem, poverty, inadequate housing, and an

abusive or absent partner or parent—which often accompany her own drug addiction. A referral to child protection services may therefore be indicated.

OUTCOME

Some studies have indicated that addicted babies have an increased risk for breathing abnormalities and sudden infant death syndrome (SIDS). Many studies of opioid-exposed children up to school age show few differences from nonexposed children of a similarly disadvantaged environment. Although no long-term adverse effects of prenatal cocaine exposure have been shown to date, this topic remains controversial, especially with an expanding range of cocaine-based products being used illicitly (e.g., crack and freebase). The outcome for babies with prenatal alcohol exposure depends on the extent of the damage to the fetus born with FETAL ALCOHOL SYNDROME (FAS).

The drug-addicted mother's lifestyle is often characterized by inadequate or no prenatal care, poor nutrition, and prostitution, any or all of which may result in a high risk for medical and obstetrical complications. Needle use may result in infection with hepatitis B and HIV. Methadone-maintenance programs for heroin-addicted mothers generally offer medical and social services to help mitigate these negative influences and contribute to the improved outcome seen in their babies, despite a continuation of opioid drug addiction on their part.

Drug WITHDRAWAL, in the absence of other problems, is now readily managed in hospitals. The outcome for addicted babies depends on any permanent medical sequelae as well as the quality of the postnatal environment. These babies often require ongoing medical, school, and social services to ensure that they reach their maximal potential.

(SEE ALSO: *Fetus: Effects of Drugs on; Pregnancy and Drug Dependence*)

BIBLIOGRAPHY

CHASNOFF, I. J. (Ed.). (1988). *Drugs, alcohol, pregnancy and parenting.* Boston: Kluwer Academic Publishers.

COOK, P. SHANNON, PETERSEN, R. C., & MOORE, D. T. (1990). *Alcohol, tobacco, and other drugs may harm the unborn.* DHHS Publication no. (ADM)90-1711. Rock-ville, MD: U.S. Department of Health and Human Services.

JOYCE F. SCHNEIDERMAN

ADDICTION (formerly the *British Journal of Addictions*) is the oldest specialist journal in its field, originating in 1884 as the *Proceedings for the Society for the Study and Cure of Inebriety.* The bound volumes provide a unique perspective on the historical development of clinical practice, policy debates, and the emergence of a scientific tradition. *Addiction* is today among the most international of journals focusing on addiction. In addition to publishing refereed research reports, editorial policy has been directed at establishing it as a leading forum for informed debate—specially commissioned "commentary" series contribute to this purpose. The prestigious *Addiction* Book Prize is awarded annually. In furtherance of its role as an international medium of scientific exchange, the journal, which has its head office in Britain, in 1993 established regional offices in the United States and Australia.

GRIFFITH EDWARDS

ADDICTION: CONCEPTS AND DEFINITIONS This article deals with a number of concepts related to the basic nature of addiction, that are widely used but often misused, and that have undergone significant changes since the term *addiction* first came into the common vocabulary. In the following discussion, the terms are grouped according to themes, rather than being arranged in alphabetic order.

Abuse and Misuse. In everyday English, *abuse* carries the connotations of improper, perverse, or corrupt use or practice, as in child abuse, or abuse of power. As applied to drugs, however, the term is difficult to define and carries different meanings in different contexts. In relation to therapeutic agents such as BENZODIAZEPINES or MORPHINE, the term *drug abuse* is applied to their use for other than medical purposes, or in unnecessarily large quantities. With reference to licit but non-therapeutic substances such as ALCOHOL, it is understood to mean a level of use that is hazardous or damaging, either to

the user or to others. When applied to illicit substances that have no recognized medical applications, such as PHENCYCLIDINE (PCP) or MESCALINE, any use is generally regarded as abuse. The term *misuse* refers more narrowly to the use of a therapeutic drug in any way other than what is regarded as good medical practice.

Substance abuse means essentially the same as drug abuse, except that the term "substance" (shortened form of psychoactive substance) avoids any misunderstanding about the meaning of "drug." Many people regard drugs as exclusively compounds that are, or could be, used for the treatment of disease, whereas "substances" would also include materials such as organic solvents, MORNING-GLORY SEEDS, or toad poisons, that have no medical applications but are "abused" in one or more of the senses defined above.

The best general definition of drug abuse is the use of any drug in a manner that deviates from the approved medical or social patterns within a given culture at a given time. This is probably the concept underlying the official acceptance of the term *abuse* in such instances as the names of the National Institute on Drug Abuse (USA) and the Canadian Centre on Substance Abuse. Such official acceptance, however, does not prevent the occurrence of ambiguities such as those mentioned in the next section.

Recreational or Casual Drug Use. These two terms are generally understood to refer to drug use that is small in amount, infrequent, and without adverse consequences, but these characteristics are not in fact necessary parts of the definitions. In the terminology recommended by the World Health Organization (WHO), the two terms are synonymous. However, *recreational use* really refers only to the motive for use, which is to obtain effects that the user regards as pleasurable or rewarding in some way, even if that use also carries some potential risks. *Casual use* refers to occasional as opposed to regular use, and therefore implies that the user is not addicted or dependent (see below), but it carries no necessary implications with respect to the amount used on any occasion. Thus, a casual user might become intoxicated (see below) or suffer an acute adverse effect on occasions, even if these are infrequent.

Occasional use may also be *circumstantial* or *utilitarian*, if employed to achieve some specific short-term benefit under special circumstances. The use

of AMPHETAMINES to increase endurance and postpone fatigue by students studying for examinations, truck drivers on long hauls, athletes competing in endurance events, or military personnel on long missions, are all instances of such utilitarian use. Most observers also consider the first three of these to be abuse or misuse, but many would not regard the fourth example as abuse because it is or was prescribed by military authorities under unusual circumstances, for necessary combat goals. Nevertheless, in all four instances the same drug effect is sought for the same purpose. This illustrates the complexities and ambiguities of definitions in the field of drug use.

Intoxication. This is the state of functional impairment resulting from the actions of a drug. It may be acute, i.e., caused by consumption of a high dose of drug on one occasion; it may be chronic, i.e., caused by repeated use of large enough doses to maintain an excessive drug concentration in the body over a long period of time. The characteristic pattern of intoxication varies from one drug to another, depending upon the mechanisms of action of the different substances. For example, intoxication by alcohol or barbiturates typically includes disturbances of neuromuscular coordination, speech, sensory functions, memory, reaction time, reflexes, judgment of speeds and distances, and appropriate control of emotional expression and behavior. In contrast, intoxication by amphetamine or cocaine usually includes raised blood pressure and heart rate, elevation of body temperature, intense hyperactivity, mental disturbances such as hallucinations and paranoid delusions, and sometimes convulsions. The term may be considered equivalent to *overdosage*, in that the signs of intoxication usually arise at higher doses than the pleasurable subjective effects for which the drug may be taken.

Habit and Habituation. In everyday English, a *habit* is a customary behavior, especially one that has become largely automatic or unconscious as a result of frequent repetition of the same act. In itself, the word is simply descriptive, carrying no fixed connotation of good or bad. As applied to drug use, however, it is somewhat more judgmental. It refers to regular persistent use of a drug, in amounts that may create some risk for the user, and over which the user does not have complete voluntary control. Indeed, an *alcohol habit* has been defined in terms very similar to those used to define dependence (see

below). In older writings, *habit strength* was used to characterize the degree of an individual's habitual drug use, in terms of the average amount of the drug taken daily. Reference to a *drug habit* implies that the drug use is the object of some concern on the part of the user or of the observer, but that it may not yet be sufficiently strongly established to make treatment clearly necessary.

Habituation refers either to the process of acquiring a drug habit, or to the state of the habitual user. Since habitual users frequently show increased tolerance (see below)—decreased sensitivity to the effects of the drug—habituation is also used in the earlier literature to mean an acquired increase in tolerance. In its early reports, the WORLD HEALTH ORGANIZATION EXPERT COMMITTEE ON DRUG DEPENDENCE (as it is now known, after several changes of name) used the term *habituation* to refer to a state arising from repeated drug use that was less serious than addiction, in the sense that it included only psychological and not physical dependence, and that harm, if it occurred, was only to the user and not to others. Drugs were classified according to whether they caused habituation or addiction. These distinctions were later recognized to be based on misconception, because (1) psychological (or psychic) dependence is even more important than physical dependence with respect to the genesis of addiction; (2) any drug that can damage the user is also capable of causing harm to others and to society at large; and (3) the same drug could cause effects that might be classed as "habituation" in one user and "addiction" in another. The WHO Expert Committee later recommended that both terms be dropped from use and that *dependence* be used instead.

Problem Drinking. In an effort to avoid semantic arguments and value judgments about abuse or addiction, clinical and epidemiological researchers have increasingly made use of objective operational definitions and measures. *Problem drinking* is alcohol consumption at an average daily level that causes problems, regardless of whether these are of medical, legal, interpersonal, economic, or other nature, to the drinker or to others. The actual level, in milliliters of absolute alcohol per day, will obviously vary with the individual, the type of problem, and the circumstances. The advantage of this term is that a drinker who may not meet the criteria of dependence or who is reluctant to accept a diagnostic label of alcoholism or addiction can, nevertheless, often

acknowledge that a problem has arisen requiring intervention.

Addiction and Dependence. The term *addiction* was used in everyday and legal English long before its application to drug problems. In the sixteenth century, it was used to designate the state of being legally bound or given over (e.g., bondage of a servant to a master) or, figuratively, of being habitually given over to some practice or habit; in both senses, it implied a loss of liberty of action. At the beginning of this century it came to be used more specifically for the state of being given over to the habitual excessive use of a drug, and the person who was "given over" to such drug use was described as an *addict*. By extension from the original meanings of addiction, drug addiction meant a practice of drug use which the user could not voluntarily cease, and loss of control over drinking was considered an essential feature of alcohol addiction. The emphasis was placed upon the degree to which the drug use dominated the person's life, in such forms as constant preoccupation with obtaining and using the drug, and inability to discontinue its use even when harmful effects made it necessary or strongly advisable to do so.

During the first half of the twentieth century, however, the pharmacological and social consequences of such use came increasingly to be the defining criteria. In 1957, the WHO Expert Committee defined addiction as "a state of periodic or chronic intoxication produced by the repeated consumption of a drug (natural or synthetic). Its characteristics include (1) an overpowering need (compulsion) to continue taking the drug and to obtain it by any means; (2) a tendency to increase the dose [later said to reflect tolerance]; (3) a psychic (psychological) and generally a physical dependence on the effects of the drug; and (4) detrimental effect on the individual and on society". *Physical dependence* is an altered physiological state arising from the regular heavy use of a drug, such that the body cannot function normally unless the drug is present. This state is recognizable only by the physical and mental disturbances that occur when drug use is abruptly discontinued or "withdrawn," and the constellation of these disturbances is known as a *withdrawal syndrome*. The specific pattern of the withdrawal syndrome varies according to the type of drug that has been used, and usually consists of changes opposite in direction to those originally produced by the ac-

tion of the drug. For example, if opiate drugs cause constipation, a withdrawal symptom is typically diarrhea; if cocaine causes prolonged wakefulness and euphoria, the withdrawal syndrome will include profound sleepiness and depression; if alcohol decreases the reactivity of nerve cells, the withdrawal syndrome will include signs of over-reactivity, such as exaggerated reflexes or convulsions. In all cases, however, the withdrawal syndrome is quickly abolished by resumption of administration of the drug.

It is now well recognized that a person can become physically dependent on a drug given in high doses for medical reasons (e.g., morphine given repeatedly for relief of chronic pain) and yet not show any subsequent tendency to seek and use the drug for nonmedical purposes. The WHO Expert Committee therefore revised its definitions and concepts in 1973, substituting the single term *dependence* for the two terms addiction and habituation.

In essence, dependence is a state in which the individual cannot function normally—physically, mentally, or socially—in the absence of the drug. Unfortunately, this change has not led to uniform terminology or concepts.

A simple definition given in the *Diagnostic and Statistical Manual of the American Psychiatric Association* (DSM-III-R) includes only two fundamental elements: (1) impaired control of use of the drug, and (2) persistence in its use despite the occurrence of adverse consequences. In contrast, a more detailed description of the *dependence syndrome* includes—increased tolerance (see below) to the drug; repeated experience of withdrawal symptoms; increasing prominence of drug-seeking behavior, even at the cost of disruption of other important parts of the user's daily life; use of the drug to prevent or relieve withdrawal symptoms; awareness by the user of frequent craving (see below) or compulsive need to take the drug; increasing frequency of use, and growing tendency to use the drug in a fixed or stereotyped pattern; and a high tendency to relapse if the user does succeed in stopping drug use. *Psychic dependence* or *psychological dependence* refers to those components of the dependence syndrome other than tolerance and withdrawal symptoms, in particular the urgency of drug-seeking behavior, craving, inability to function in daily life without repeated use of the drug, and the inability to maintain prolonged abstinence. It has been attributed to a distress or tension, especially during periods of absti-

nence from the drug, that the user seeks to relieve by taking the drug again. This is, however, really a description, rather than an explanation.

Because of these differences in definition of dependence by different authorities, the term has proven to be less clear than intended, and has not displaced the term addiction from common use. The latter carries a clearer emphasis on the behavior of the individual, rather than the consequences of that behavior, as in the concept of nicotine addiction. A committee report of the Academy of Sciences of the Royal Society of Canada concluded that the only elements common to all definitions of addiction are a strongly established pattern of repeated self-administration of a drug in doses that reliably produce reinforcing psychoactive effects, and great difficulty in achieving voluntary long-term cessation of such use, even when the user is strongly motivated to stop.

Craving and Related Concepts. *Craving* refers to an intense desire for the drug, expressed as constant, obsessive thinking about the drug and its desired effects, a sense of acute deprivation that can be relieved only by taking the drug, and an urgent need to obtain it. This state is probably induced by exposure to bodily sensations and external stimuli that have in the past been linked to circumstances and situations in which drug has been necessary, such as self-treatment of early withdrawal symptoms by taking more drug. *Drug hunger* is essentially synonymous with craving, and *urge* represents the same phenomenon but of lesser intensity.

The behavioral consequence of an urge or craving is usually a redirecting of the person's thoughts and activities toward obtaining and using a new supply of drug. All the behaviors directed toward this end, such as searching drawers and cupboards for possible remnants of drug, getting money (whether by legal or illegal means), contacting the sources of supply, purchasing the drug, and preparing it for use, are included under the term *drug-seeking behavior*. The more intense the craving, the more urgent, desperate or irrational this behavior tends to become.

Tolerance and Sensitization. The term *tolerance*, which has long held a prominent place in the literature on drug dependence, has a number of different meanings. All of them relate to the degree of sensitivity or susceptibility of an individual to the effects of a drug. *Initial tolerance* refers to the degree of sensitivity or resistance displayed on the first exposure to the drug; it is expressed in terms of the

degree of effect (as measured on some specified test) produced by a given dose of the drug, or by the concentration of drug in the body tissues or fluids resulting from that dose: the smaller the effect produced by that dose or concentration, the greater is the tolerance. Initial tolerance can vary markedly from one individual to another, or from one species to another, as a result of genetic differences, constitutional factors, or environmental circumstances.

The more frequent meaning of tolerance, however, is *acquired tolerance* (or *acquired increase in tolerance*)—the increased resistance or decreased sensitivity to the drug as a result of adaptive changes produced in the body by previous exposure to that drug. This is expressed in terms of the degree of reduction in the magnitude of effect produced by the same dose or concentration, or (preferably) the increase in dose or concentration required to produce the same magnitude of effect. Acquired tolerance can be due to two quite different processes. *Metabolic tolerance* (also known as *pharmacokinetic* or *dispositional tolerance*) is produced by an adaptive increase in the rate at which the drug is inactivated by metabolism in the liver and other tissues. This results in lower concentrations of drug in the body after the same dose, so that the effect is less intense and of shorter duration. *Functional tolerance* (also known as *pharmacodynamic* or *tissue tolerance*) is produced by a decrease in the sensitivity of the tissues on which the drug acts, primarily the central nervous system, so that the same concentration of drug produces less effect than it did originally.

Acquired functional tolerance can occur in three different time frames. *Acute tolerance* is that which is displayed during the course of a single drug exposure, even the first time it is taken. As soon as the brain is exposed to the drug, compensatory changes begin to develop and become more marked as time passes. As a result, the degree of effect produced by the same concentration of drug is greater at the beginning of the exposure than it is in the later part; this phenomenon is sometimes called the *Mellanby effect*. A second pattern of tolerance development is known in the experimental literature as *rapid tolerance*. This refers to an increased tolerance seen on the second exposure to the drug, if this occurs not more than one to two days after the first exposure. *Chronic tolerance* is that form of acquired tolerance that develops progressively over an extended period of time in which repeated exposure to the drug takes place. Although acquired tolerance involves impor-

tant physiological changes in the nervous system, it is also markedly influenced by learning. Tolerance develops much more rapidly if the individual is required to perform tasks under the influence of the drug, than if the same dose of the same drug is experienced without any performance requirement. Similarly, environmental stimuli that regularly accompany drug administration can come to serve as Pavlovian conditional stimuli that elicit tolerance as a conditional response, so that tolerance is demonstrated much more rapidly in the presence of these stimuli than in their absence.

Sensitization refers to a change opposite to tolerance, that occurs with respect to certain effects of a few drugs (most notably, central stimulant drugs such as cocaine and amphetamine) when these are given repeatedly. The degree of effect produced by the same dose or concentration grows larger rather than smaller. For example, after repeated administration of amphetamine, a dose that initially produced only a slight increase in physical activity can come to elicit very marked hyperactivity, and a convulsion can be produced by a dose that did not initially do so. This does not apply to all effects of the drug, however; tolerance can occur toward some effects at the same time that sensitization develops to others. The reason for this difference is not yet known.

Cross-tolerance and Cross-dependence. The term acquired tolerance is applied to tolerance developing to the actions of the same drug that has been administered repeatedly. If, however, a second drug has actions similar to those of the first, an individual who becomes tolerant to the first drug is usually also tolerant to the second drug, even on the first occasion when the latter is used. This phenomenon is called *cross-tolerance*, and it may be partial or complete—it may extend to all the effects of the second drug or only to some of them. The adaptive changes in the nervous system that give rise to acquired tolerance are believed by most researchers to be responsible also for the development of physical dependence. Thus, an adaptive change in cell function, opposite in direction to the effect of the drug, will offset the latter when the drug is present (tolerance), but will give rise to a withdrawal sign or symptom when the drug is removed. The term *neuroadaptive state* has been proposed to designate all the physiological changes underlying the development of tolerance and physical dependence. If the second drug, to which cross-tolerance is present, is

given during withdrawal from the first, it can prevent or suppress the withdrawal effect; this is known as *cross-dependence*. A related concept is that of *transfer of dependence*, from a first drug on which a person has become dependent to a second drug with similar effects that has been given therapeutically to relieve the withdrawal signs produced by the first.

(SEE ALSO: *Diagnosis of Drug Abuse; Disease Concept of Alcoholism and Drug Abuse*)

BIBLIOGRAPHY

AMERICAN PSYCHIATRIC ASSOCIATION. (1987). *Diagnostic and statistical manual of mental disorders-3rd edition-revised* (DSM-III-R). Washington, DC: Author.

EDWARDS, G., & GROSS, M. M. (1976). Alcohol dependence: Provisional description of a clinical syndrome. *British Medical Journal, 1,* 1058–1061.

GLASS, I. B. (ED.). (1991). *The international handbook of addiction behaviour.* London: Routledge.

GOUDIE, A. J., & EMMETT-OGLESBY, M. (1989). *Psychoactive drugs: Tolerance and sensitization.* Clifton, NJ: Humana Press.

KELLER, M., & McCORMICK, M. (1968). *A dictionary of words about alcohol.* New Brunswick, NJ: Rutgers Center of Alcohol Studies.

ROYAL SOCIETY OF CANADA. (1989). *Tobacco, nicotine, and addiction.* Ottawa: Royal Society of Canada.

WORLD HEALTH ORGANIZATION. *Reports of the Expert Committee on problems of drug dependence* (various). Published in the Technical Report Series of the World Health Organization. Geneva: WHO, 1952, 1957, 1973, and others.

HAROLD KALANT

ADDICTION RESEARCH FOUNDATION OF ONTARIO (CANADA)
The Alcoholism Research Foundation was created by an act of the legislature of the Province of Ontario, Canada, in 1949, with the mandate to conduct research into the causes, prevention, and treatment of alcoholism—and to help the relevant professional and community groups apply the knowledge gained from that research. The act clearly set out an "arm's-length" relationship to the government, similar to that of the universities and other major health research foundations. In 1961, the mandate was broadened by another act of the provincial legislature, to include research on "addiction to substances other than alcohol". The name was changed to the Alcoholism and Drug Addiction Research Foundation, in practice shortened to Addiction Research Foundation (ARF). ARF was the first major North American institution to study and treat addiction to alcohol and other drugs within the same facilities.

In 1951, H. David Archibald was appointed the executive director, a position he held until 1976, when he became director of international relations of ARF. His successors, designated president of the foundation, have been John B. Macdonald (1976–1981), Joan A. Marshman (1981–1989) and Mark R. F. Taylor (1989–1994). Archibald presided over a period of steady and at times rapid growth, during which the foundation staff increased from fewer than 10 people to a peak of more than 700, including scientists, clinicians, community-development workers, educators, and administrators. E. M. Jellinek and J. R. Seeley played important roles in shaping the research programs in the early years. In the 1960s, in response to growing public alarm about the spread of drug use among young people, ARF established a network of treatment and education centers in all the major cities and towns of Ontario. Responsibility for treatment was later turned over to the general health-care system of the province, and the role of ARF was limited to research and transmission of knowledge. In 1971, the Clinical Institute was established as the center for ARF clinical research, and it became a teaching hospital affiliated with the Faculty of Medicine of the University of Toronto.

ARF research has made major contributions in various areas. Epidemiological work has included the first household surveys of alcohol and other drug use in Canada, biennial surveys since 1977 of drug use by Ontario students in grades 7 to 13, detailed analysis of the effects of price and availability on alcohol consumption, and intensive analysis of the changing patterns of cigarette use as affected by age, sex, and educational level. The best-known ARF medical advances include the hypoxic theory of alcoholic liver damage and its treatment by propylthiouracil, the recognition of reversible cortical atrophy in alcoholics, and the development of a clinical-trials unit for the evaluation of medications used in reducing alcohol consumption. Much of its basic science research has dealt with mechanisms of reinforcement, tolerance, and dependence, in relation to alcohol, opioids, nicotine, benzodiazepines, and other drugs. ARF scientists were among the first to

set up and evaluate nonmedical detoxication units, employee assistance programs, and minimal intervention programs using self-help manuals; to explore the use and limitations of controlled-drinking programs; to conduct systematic evaluation of the components of resident and outpatient treatment programs; to analyze the roles of alcohol and other drugs in driving accidents and develop community programs for their prevention; and to conduct an early detection-and-intervention study involving the primary-care physicians of a whole city. In addition, ARF publications include many monographs, annotated bibliographies, and critical reviews that have contributed much, not only to science and education, but also to ARF's role as policy advisor on alcohol, marijuana, stimulants, and other drugs to governments, commissions of enquiry, and regulatory bodies.

In 1977, ARF became the first officially designated Collaborating Center of the World Health Organization in the addictions field. It has close ties with other federal and provincial addiction agencies in Canada; with international bodies such as the United Nations (UN), ICAA, ISBRA and the Kettil Bruun Society; and with national bodies such as the Finnish Foundation for Alcohol Studies; the NATIONAL INSTITUTE ON ALCOHOLISM AND ALCOHOL ABUSE (NIAAA), the NATIONAL INSTITUTE ON DRUG ABUSE (NIDA), and CPDD in the United States; the National Addiction Centre in the United Kingdom; and numerous institutions in Mexico, the Caribbean area, Southeast Asia, Australia, and New Zealand. Through these links, ARF has contributed greatly to professional training for the staffs of addiction programs in various developing countries, and ARF continues to collaborate in international research activities.

(SEE ALSO: *Addiction Research Unit; College on Problems of Drug Dependence*)

BIBLIOGRAPHY

ARCHIBALD, H. D. (1990). *The Addiction Research Foundation: A voyage of discovery.* Toronto: ARF Books.

CAPPELL, H. D. (1987). Reports from the research centres—4. The Addiction Research Foundation of Ontario. *British Journal of Addiction, 82,* 1081–1089.

HAROLD KALANT

ADDICTION RESEARCH UNIT (ARU) (U.K.)

The Addiction Research Unit of the Institute of Psychiatry, University of London, was set up in 1967 on the Joint Maudsley Hospital/Institute of Psychiatry campus in Camberwell, South East London, England. Its funding has come from many different sources (principally the Medical Research Council), but its fundamental identity has always been that of a university center, which has close ties and valued links with a postgraduate psychiatric medical school (the Institute) and a teaching hospital (the Maudsley). Its present scientific staff number thirty, with the mix of psychiatrists, psychologists, statisticians, and social scientists reflecting the ARU's interdisciplinary commitment. The ARU's field of study embraces TOBACCO as well as ALCOHOL and other drugs.

The ready access to hospital facilities has, over the years, greatly aided the ARU's ability to conduct clinical research. One line of investigation continuously developed from this base has, for example, concentrated on definition, description, measurement, and validation of the dependence-syndrome concept—in relation to both alcohol and opiates. The Smoking Section has done much to demonstrate that the cigarette habit is indeed nicotine dependence. A sustained effort has also been directed at the development of methods for assessing treatment efficacy through controlled trials. As regards both alcohol dependence and smoking, results have generally tended to support research in fairly simple, minimalist interventions, often delivered in the primary-care setting. Another line of research has focused on long-term follow-up studies and the determinants of "natural history and career."

If the above paragraph identifies some of the ARU's core activities, much else has also gone on. For example, the ARU, for a number of years, employed a professional historian, who did much to open up an understanding of the history of opiate use in Britain. Epidemiological research has at times been undertaken. As of the 1990s, the ARU has been developing a computerized method for handling interview texts. In the psychophysiological laboratories, studies of cue responsiveness are being conducted with patients and normal subjects. A new line of research is focusing on the relationship between the two axes of problems and dependence. The Smoking Section has contributed to studies on the impact of passive smoking.

Besides the research, a great deal of clinical and research training is being undertaken, and the ARU runs a full-time one-year course leading to a Master of Science (MSc) in clinical and public health aspects of addiction. The ARU enjoys the benefits of an extensive national and international network of friendships and professional contacts through its many former staff and students who today hold influential positions. There are particularly strong links with the developing world, and support has often been given by the ARU to the World Health Organization.

In April 1991, the ARU moved to a purpose-built office and laboratory accommodation on the same campus, and it became associated with the newly established National Addiction Centre. The ARU's Smoking Section has also been strengthened by involvement with the recently established Health Behaviour Research Unit, funded by the Imperial Cancer Research Fund.

(SEE ALSO: *Addiction Research Foundation of Ontario; U.S. Government Agencies*)

GRIFFITH EDWARDS

ADDICTION SEVERITY INDEX (ASI)

This is a semistructured interview designed to provide important information about aspects of the life of patients that may contribute to their substance-abuse problems. Developed by McLellan and co-workers in 1981, the ASI has been translated into seventeen languages—Japanese, French, Spanish, German, Dutch, and Russian among them—and was designed to be administered by a technician or counselor. Consistent guidelines for each question on the ASI have been compiled in training materials including two videos and three instructional manuals. Self-training can be accomplished by using the video along with the administration manual, although a one-day formal training seminar is recommended. (Since the instrument is in the public domain, there is no charge for it; only a minor fee is charged for copies of the administration materials and the computer scoring disk.)

The interview is based on the idea that addiction to drugs or alcohol is best considered in terms of life events that preceded, occurred at the same time

as, or resulted from the substance-abuse problem. The ASI focuses on seven functional areas that have been widely shown to be affected by the substance abuse: medical status, employment and support, drug use, alcohol use, legal status, family and social status, and psychiatric status. Each of these areas is examined individually by collecting information regarding the frequency, duration, and severity of symptoms of problems both historically over the course of the patient's lifetime and more recently during the thirty days prior to the interview. Within each of the problem areas, the ASI provides both a 10-point, interviewer-determined severity rating of lifetime problems as well as a multi-item composite score that indicates the severity of the problems in the past thirty days.

The ASI is widely used clinically for assessing substance-abuse patients at the time of their admission for treatment. It takes about an hour to gather the basic information that forms the first step in the development of a patient profile for subsequent use by the staff in planning treatment. Researchers have found the ASI useful because the composite scores and the individual variables can be compared within groups over time as a measure of improvement, or between groups of patients at a posttreatment follow-up point as a measure of outcome of treatment. The ASI has shown excellent reliability and validity across a range of types of patients and treatment settings in this country and abroad.

BIBLIOGRAPHY

BALL, J. C., ET AL. (1986). Medical services provided to 2,394 patients at methadone programs in three states. *Journal of Substance Abuse Treatment, 3,* 203–209.

HENDRICKS, V. M. (1989). The Addiction Severity Index: Reliability and validity in a Dutch addict population. *Journal of Substance Abuse Treatment,*

KOSTEN, T. R., ROUNSAVILLE, B., & KLEBER, H. D. (1983). Concurrent validity of the Addiction Severity Index. *Journal of Nervous & Mental Disease, 171,* 606–610.

LAPORTE, D., ET AL. (1981). Treatment response in psychiatrically impaired drug abusers. *Comprehensive Psychiatry, 22*(4), 411–419.

McLELLAN, A. T., ET AL. (1992). The fifth edition of the Addiction Severity Index: Cautions, additions and normative data. *Journal of Substance Abuse Treatment, 9*(5), 461–480.

McLELLAN, A. T., ET AL. (1991). Using the ASI to compare cocaine, alcohol, opiate and mixed substance abusers. In L. Harris (Ed.), *Problems of drug dependence 1990* (NIDA Research Monograph). Washington, DC: U.S. Government Printing Office.

McLELLAN, A. T., ET AL. (1985). New data from the Addiction Severity Index: Reliability and validity in three centers. *Journal of Nervous and Mental Disease, 172,* 84–91.

McLELLAN, A. T., ET AL. (1983). Increased effectiveness of substance abuse treatment: A prospective study of patient-treatment "matching." *Journal of Nervous and Mental Disease, 171*(10), 597–605.

McLELLAN, A. T., ET AL. (1983). Predicting response to alcohol and drug abuse treatments: Role of psychiatric severity. *Archives of General Psychiatry, 40,* 620–625.

McLELLAN, A. T., ET AL. (1981). Are the addiction-related problems of substance abusers really related? *Journal of Nervous and Mental Disease, 169*(4), 232–239.

McLELLAN, A. T., ET AL. (1981). Psychological severity and response to alcoholism rehabilitation. *Drug and Alcohol Dependence, 8*(1), 23–35.

McLELLAN, A. T., ET AL. (1980). An improved diagnostic instrument for substance abuse patients: The Addiction Severity Index. *Journal of Nervous and Mental Disease, 168,* 26–33.

NIAAA. (1991). Alcohol alert: Assessing alcoholism. ADAMHA Press, no. 12, PH 294, April. Washington, DC: U.S. Government Printing Office.

ROGALSKI, C. J. (1987). Factor structure of the Addiction Severity Index in an inpatient detoxification sample. *International Journal of Addiction, 22*(10), 981–992.

ROUNSAVILLE, B. J., ET AL. (1987). Psychopathology as a predictor of treatment outcome in alcoholism. *Archives of General Psychiatry.*

SAXE, L., ET AL. (1983). *The effectiveness and costs of alcoholism treatment* (Health Technology Case Study 22). Washington, DC: Office of Technology Assessment.

STOFFELMAYR, B., HARWELL, M., & MAVIS, B. (1991). Interrater reliability and validity of the Addiction Severity Index. *Journal of Educational Measurement,*

A. THOMAS McLELLAN

ADDICTIVE PERSONALITY

The term *addictive personality* has been used in various ways, most commonly to refer to a recurrent pattern observed in many alcoholics and other substance abusers: impulsivity, immaturity (dependency and neediness), poor frustration tolerance, anxiety, and depression. Many of these features disappear during extended periods of abstinence, however, suggesting that they may be either related directly to the drug use, to the life it imposes, or to social response, rather than to personality. Addictive personality—more accurately preaddictive personality—has also been used to refer to personality characteristics presumed to predate drug use and as such are predictive of such use. These aspects of personality are likely to include early difficulties in impulse control and submission to authority—and sensitivity to anxiety and depresssion.

(SEE ALSO: *Addictive Personality and Psychological Tests; Causes of Substance Abuse; Childhood Behavior and Later Drug Use; Coping and Drug Use*)

BIBLIOGRAPHY

KANDEL, D. B. (1978). Convergences in prospective longitudinal surveys of drug use in normal populations. In D. B. Kandel (Ed.), *Longitudinal research on drug abuse: Empirical findings and methodological issues,* (pp. 3–38). Washington, DC: Hemisphere Publishing.

SMITH, G. M., & FOGG, C. P. (1978). Psychological predictors of early youth, late youth, and nonuse of marihuana among teenage students. In D. B. Kandel (Ed.), *Longitudinal research on drug abuse: Empirical findings and methodological issues* (pp. 101–113). Washington, DC: Hemisphere Publishing.

VAILLANT, G. E. (1983). *The natural history of alcoholism.* Cambridge; Harvard University Press.

WILLIAM A. FROSCH

ADDICTIVE PERSONALITY AND PSYCHOLOGICAL TESTS

Psychological tests and measurements (psychometrics) are structured ways of evaluating an individual's inner mental life and external behaviors. They present subjects with more or less standard stimuli to which the subjects respond. Depending on the test, these responses tell us something about their intelligence, abilities and skills, educational and vocational interests and achievements, and personality. Often, the tests are especially helpful in diagnosing organic brain disease—its presence, presumed location, and the particular resulting functional deficits. The tests

themselves range from structured questionnaires or interviews, to pen-and-pencil tasks, to obtaining responses to purposely ill-defined stimuli such as ink blots (Rorschach test). They have been used (1) to evaluate the probability of the presence of a substance-abuse problem, (2) to examine the impact of substance use on behavior and brain function both acutely and chronically, and (3) to assess personality features—profiling which ones are predisposed to use and abuse or which are the result of such use.

Historically, we have moved from the search for a single trait as *cause* to looking for a cluster of traits (the *addictive personality*), to the recognition that a number of different pathways lead to addictive behavior(s); we now understand that different types of people may use drugs for different reasons. Some underlying trait or combination of traits, however, may predispose individuals to problems with drugs. In addition, the "addictive life" may so structure behavior that it imposes similarities in attitudes, responses, and the like; it may present us with a state-related personality—that is, a pattern of typical response and behavior that is present while living in the drugged state but that disappears, in part or in whole, after a period of abstinence. For example, Vaillant's long-term population studies (1983) have shown that people diagnosed with personality disorders or other psychopathologies while drinking often appear to lose these "illnesses" after they have been alcohol-free for some time.

A review of studies of personality in alcoholics concluded that there appeared to be six constellations: (1) those who drink to escape the pain of frustration; (2) those for whom drink gratifies childish dependency; (3) those who drink to reduce guilt and anxiety; (4) those who escape disappointment into fantasy; (5) social isolates for whom alcohol supplies a pseudo-life; (6) social context–driven alcoholics. Other studies have defined additional groups, such as unsocial aggressive, psychopathic, and inhibited-conflicted. Similar findings and classifications have been described in users of other drugs. The overlapping, but not identical, or at times contradictory descriptions can be accounted for at least partly. Each study has differed from the others in obvious ways: different measures of personality (e.g., the MINNESOTA MULTIPHASIC PERSONALITY INVENTORY or the Rorschach Inkblot test); different subject populations (e.g., ALCOHOLICS ANONYMOUS members or hospitalized patients); different comparison groups; or

different statistical analyses of the data. Future research in the understanding of addictive personality(ies) needs to define in advance each of these parameters and build in true replications.

While psychometric studies of personality have so far led to contradictory or confusing findings, they have proven useful in other ways: A variety of structured questionnaires are good screening devices and help the clinician toward further inquiry concerning alcohol or other drug use; tests of organic brain function have identified both acute and lasting effects of a wide variety of drugs; and the tests have helped us identify some of the common accompanying psychopathologies, such as antisocial personality disorder and depressive illness.

(SEE ALSO: *Causes of Substance Abuse; Conduct Disorder and Drug Use*)

BIBLIOGRAPHY

BARR, H. L. (1977). In *Summary report of technical review on the psychiatric aspects of opiate dependence*. Arlington: National Institute on Drug Abuse.

NEURINGER, C. (1982). Alcoholic addiction: Psychological tests and measurements. In E. M. Pattison & E. Kaufman (Eds.), *Encyclopedia handbook of alcoholism* (pp. 517–528). New York: Gardner Press.

VAILLANT, G. E. (1983). *The natural history of alcoholism*. Cambridge: Harvard University Press.

WILLIAM A. FROSCH

ADJUNCTIVE DRUG TAKING Drug abuse is usually viewed as a behavior that is motivated directly by the attractive effects produced by the drugs—their physiological and subjective effects. There is another, less direct, way in which drugs can become overwhelmingly important to people and dominate their lives. Excessive drug taking can develop as a side effect of other strongly motivated behavior that, for various reasons, cannot be engaged in or completed. For example, for some reason people may not be able to do something they are motivated to do: They may not be permitted to do it, the goals of the behavior may not be available to them, or they may not have the skills or knowledge required to perform the behavior successfully. When blocked in any of these ways from performing, a per-

son may turn aside to some sort of easy, satisfying alternative—an adjunctive behavior.

The first kind of adjunctive behavior that was produced in an experimental laboratory did not involve drug taking, but it had many of the exaggerated and compulsive characteristics usually thought to be a result of the attractiveness of a drug.

Normal, adult laboratory rats were reduced in body weight. It was arranged for them to receive much of their daily food ration in individual chambers. The food was given as small (45 milligram [mg]) pellets, each of which was available on the average of one per minute over a 3-hour period. Although water was always freely available back in each animal's home cage, this schedule of intermittent food availability, in which the eating of small portions was spread out over the 3-hour period, produced a curious result. When an animal received a food pellet, it quickly ate it and then took a rather large drink of water. Since an animal received a total of about 180 pellets during each daily, 3-hour session, and drinking occurred after almost every pellet, by the end of a session an animal had drunk an amount of water equal to about one half its body weight. That is an excessive amount, and the overdrinking occurred day after day during each daily session.

It was easy to prove that the excessive drinking was not a physiological effect of reducing the daily intake of food: If, instead of being given pellets spread across the session, animals were given the same 180 pellets all at once as a single ration, then they drank about 10 milliliters (ml) of water over the next 3 hours, rather than almost 100 ml. So animals do not really need the excessive amount of water they drink under such schedules of intermittent pellet delivery. There is something about doling out bits of food to them over time that drives the exaggerated drinking behavior. They get what they need (in this case, food), but the individual portions are small and there is a delay between portions.

The details of how the size of the portions and the time between them, as well as the state of food deprivation, affect the excessive drinking have been worked out in many experiments. As described, the phenomenon is referred to as schedule-induced polydipsia, meaning that the food schedule induces excessive drinking of many sorts of fluids. More generally, it is known as adjunctive behavior, and it is not limited to rats, or to food schedules, or to drinking. Excessive behavior can be induced in many animal species—for example, mice, monkeys, and pigeons, as well as in humans—and it can be induced by a generator schedule that is not based on doling out food; other kinds of incentives are effective in inducing excessive behavior. For example, in studies, humans reinforced intermittently with money, by the opportunity to gamble, or even by maze solving, showed adjunctive increases in fluid intake, general activity, eating, or smoking. In animal experiments, excessive aggression, overactivity, and eating were a few of the behaviors that were demonstrated to occur adjunctively owing to the intermittent availability of some important commodity or activity.

The adjunctive behavior of greatest interest with respect to the problem of drug abuse is, of course, the excessive seeking and taking of drugs. As indicated in the description of the laboratory rats, it is important to understand that adjunctive drug taking is just one kind of excessive behavior that can be driven by a generator schedule. It is one sort of adjunctive behavior rather than a special problem driven exclusively by the pharmacological actions of drugs that are abused.

A number of studies here explored adjunctive drug taking. In them, once the conditions inducing adjunctive drinking had been established, it was of interest to determine whether fluids other than water would be drunk to excess, especially drug solutions. Researchers found that if the drug concentration of a solution was not too high, a much greater unforced (voluntary) oral drug intake could be induced by a generator schedule than was possible by other means. Several classes of drugs have been investigated, and excessive intakes of ALCOHOL, opioids (e.g., morphine), SEDATIVES (e.g., barbiturates), anxiolytics (e.g., BENZODIAZEPINES), stimulants (e.g., COCAINE), and other agents (e.g., NICOTINE) have been sustained for many months under these inducing conditions. For some of these agents, PHYSICAL DEPENDENCE resulted from the excessive intakes. However, if the generator schedule was discontinued, drug intakes decreased immediately to control levels. Thus, without the generator schedule, the high drug intakes did not continue. (Similar generator schedules, or intermittent food delivery, also could induce excessive self-administration of IV drugs.)

These and related studies clarify what sustains the excessive intake of drugs (i.e., drug abuse). It is important to note that, by this analysis, drug abuse

stems from a background of excessive behavior induced by a generator schedule. These schedules are similar to conditions in natural and social environments that provide what we need, but only in bits, and with delay intervals. The adjunctive behavior generated may be largely noninjurious, like water drinking, or it may be creative, like an intense hobby. It also can result in aggression or drug taking, depending upon one's personal history, skills, and currently available opportunities. The pharmacological consequences of drug taking constitute only one factor that sustains drug abuse. The environmental generating conditions and the context of alternative opportunities are of crucial importance. Drug abuse occurs under conditions that are already generating behavioral excesses. The problem of drug abuse is often described as if it were a direct consequence of exposure to a drug with liability for abuse, but the great majority of people who experiment with these drugs do not become abusers—they simply lose interest and turn to other pursuits. (This is not to suggest that such hazardous experimentation is acceptable.) There are also people who self-administer these drugs for medical reasons (e.g., opioids are prescribed on a long-term basis for controlling chronic pain) and almost never become dependent on them or motivated to abuse them.

Adjunctive behavior studies demonstrate that drug abuse stems more from environmental generating conditions, together with a lack of, or poor utilization of, other opportunities, than from any overwhelming intrinsic attractiveness of agents with abuse liability. An example frequently given is that of a poorly educated youth living in an urban ghetto, with minimal job prospects, who deals in drugs for economic reasons and starts to use them because of the sparse schedule of conventional opportunities and satisfactions available in that environment. Less extreme conditions can also lead to drug abuse. Consider overprivileged young persons sent to superior boarding schools by parents with great expectations (i.e., demands) but little time in which to give their offspring direct social reinforcement. The lean schedule of social reinforcement, the often competitive nature of social interactions in these schools, and the persons' migrant status, together with a high disposable income, can lead to engaging in drug abuse, particularly if this activity bestows local power and social status while the threat of enduring legal consequences remains negligible. Although perhaps expected to take over the family business, together

with its social obligations, these young persons are not yet empowered to do so, or to have their opinions taken seriously, so that their current social status is weak even if they appear privileged.

Consider also a sales representative who travels for a corporation, making intermittent sales agreements (which may be subject to cancellation), with little effective influence on company policies or politics and with little to do in brief stopovers in strange towns. Given the demand characteristics of the job, the uncertainty and intermittence of reinforcement, and the sparse opportunities for creative efforts when on the road, a person so exposed may be vulnerable to drinking too much in the easy ambience of hotel bars.

Studies on the schedule-induced production and chronic maintenance of excessive intake for a variety of drugs indicate that the drug abuse problem has its roots in the behavioral effects induced by environmental conditions as well as in the pharmacological consequences of these elevated intakes. The exaggerated and problematic behavior designated as drug abuse is not behavior that is specific to, or the outcome of, a person's interaction with drugs. It is one possible adjunctive outcome of a set of conditions comprising economically or socially restricted schedules of reinforcement that generate and sustain a host of possible exaggerated and persistent behaviors.

(SEE ALSO: *Causes of Substance Abuse; Research, Animal Model*)

BIBLIOGRAPHY

FALK, J. L. (1993). Schedule-induced drug self-administration. In F. van Haaren (Ed.), *Methods in behavioral pharmacology*. Amsterdam: Elsevier.

FALK, J. L. (1981). The environmental generation of excessive behavior. In S. J. Mule (Ed.), *Behavior in excess: An examination of the volitional disorders*. New York: Free Press.

FALK, J. L. (1977). The origin and functions of adjunctive behavior. *Animal Learning and Behavior, 5,* 325–535.

FALK, J. L. (1971). The nature and determinants of adjunctive behavior. *Physiology and Behavior, 6,* 577–588.

FALK, J. L., & LAU, C. E. (1995). Stimulus control of addictive behavior: Persistence in the presence and ab-

sence of a drug. *Pharmacology, Biochemistry and Behavior, 50,* 71–75.

FALK, J. L., & TANG, M. (1988). What schedule-induced polydipsia can tell us about alcoholism. *Alcoholism: Clinical and Experimental Research, 12,* 576–585.

JOHN L. FALK

ADOLESCENTS AND DRUG USE

As individuals pass through adolescence, they undergo many physical, cognitive, social, and emotional changes. Most learn to adapt to these changes in healthy ways. For others, turmoil, conflict, and deviant behavior lead to upheaval and disorganization as they attempt to cope. Drug use as a behavior may serve many functions in this attempt to cope, and it can have many consequences. A single episode of drug use does not necessarily lead to further use—but several episodes may lead to ever increasing use, with abuse and dependence the result.

Use of a drug, age of first use, and reasons for use are all factors related to continued drug use. Early adolescents who try one type of drug may venture on to sample a diverse number of substances. This can lead to regular use of certain drugs (e.g., daily cigarette or MARIJUANA smoking); it may become part of a pattern of multiple drug use (e.g., weekend drinking and smoking or daily uppers and downers) that by late adolescence becomes dependence or abuse. Factors related to initiation and progression into other drug phases—regular drug use, abuse, and dependency—or into the use of multiple drugs are important to understand if we are to develop appropriate PREVENTION programs aimed at reducing all drug use—whether legal or illicit.

REASONS FOR FURTHER DRUG USE

People who try a particular type of drug are more likely to use that substance again if they like the results—the drug's effects. These effects can be both pharmacological and psychological. If one enjoys the taste or the feeling created by the drug, one is more likely to continue to use it. However, if unpleasant experiences are associated with the use, trying it again is less likely.

Because the body gets used to the effect of a drug, often the amount will need to be increased in order to get some effect. This phenomenon is known as *tolerance,* and once tolerance to a drug develops, the level of drug use may escalate into the taking of larger and larger doses. Continued use of drugs may also occur because of unpleasant symptoms—withdrawal—that may appear as the drug (e.g., heroin, nicotine, caffeine) begins to wear off. To avoid these withdrawal symptoms, a user may feel compelled to establish a regular pattern of use.

The way the drug is used or administered is also a factor in developing tolerance and an increased need for continuing use of the drug. For example, an adolescent who sniffs COCAINE may find that the amount one has to inhale to get the desired effects becomes enormous. Because of this, the user may switch to injecting the cocaine instead of inhaling it. This new route of administration exposes the user to a more potent form of the drug as well as to increased medical complications.

Other reasons that adolescents continue using a particular drug may be socially and environmentally driven. Teenagers looking for peer acceptance or wanting to appear "cool" or mature might decide to use drugs. For example, although the use of TOBACCO and ALCOHOL is illegal for adolescents, it is both legal and socially acceptable for adults in certain situations. ADVERTISING, the media, and role models portray drinking and smoking as desirable. Associating with peers who are using drugs and enjoying the company of this group provide an opportunity for access to drugs that can encourage experimentation and ongoing use.

DRUG-USE SEQUENCE

The use of one drug is often related to the subsequent use of another. Certain drugs are usually tried prior to others. Typically, drug use begins with alcohol and cigarettes, which are followed by marijuana and other illicit drugs. This typical sequence of drug use was established in the 1970s (Kandel & Faust, 1975) and has been found to continue into the 1990s, in different populations and in different ethnic and cultural groups. Problem drinking typically fits into the pattern between ongoing marijuana use and the use of other illicit drugs (Donovan & Jessor 1983).

Cocaine use tends to follow marijuana use, with crack-cocaine use occurring after cocaine use (Kandel & Yamaguchi, 1993). For example, it is likely that someone who smokes CRACK has already tried tobacco, alcohol, marijuana, and cocaine. Many adolescents who use drugs in one category, however, do

not necessarily progress to drug use in a "higher" category; many stop before becoming involved in further use or habitual use.

An important factor in the progression through the sequence of drug use is age of onset or initiation. The use of alcohol and cigarettes usually but not always begins at an earlier age than does the use of illicit drugs. Adolescents who progress to using illicit drugs such as crack generally begin smoking and drinking earlier than those who do not. Early drug use (before age fifteen) is highly correlated with the development of drug and alcohol abuse in adulthood (Robins & Przybeck 1985).

Studies of adult populations provide additional support for a connection between regular adolescent drug use and later, further drug use. For example, illicit drug use during adolescence and early adulthood has been found to occur more often in adults who have used psychotherapeutic medicines (e.g., tranquilizers, sedatives) (Trinkoff, Anthony, & Muñoz, 1990). Studies of people going to drug-treatment centers often demonstrate that these persons are not only entering treatment for use of substances like cocaine or heroin but that they are also addicted to caffeine, tobacco, and/or alcohol, the very substances they first started using.

MULTIPLE-DRUG USE

The use of one type of drug may lead to experimenting with other drugs as users add to or find substitutes for their original drug of choice. Some of this progression may be an effect of maturation—that is, of drug-using adolescents moving on to other drugs as they grow older—or it may be attributable to the cost and availability of different types of drugs, introduction to a new substance by drug-using peers, or drug-seeking behavior in which individuals continue trying different drugs until they achieve the desired effect. Polydrug (multiple-drug) use can also occur when people try to counteract the effect of one drug by the use of another—for example, by taking tranquilizers (which relieve symptoms of anxiety) in order to counteract the anxiety-producing effects of cocaine.

RESEARCH METHODS

Data for the studies referred to here were collected either by self-administered questionnaires or during interviews. Often the questionnaire was given out in classrooms for students to complete anonymously. This type of study, which obtains data from all respondents at the same time, is known as a cross-sectional survey, and it provides important clues about how the use of different substances relates to such factors as age, gender, and ethnicity. In the drug-sequencing studies, researchers collected information as to whether a substance was used and the age of the person at the time of first use. Then, using a statistical technique called Guttman scaling, they combined each drug category and the ages of first use for the entire sample to establish a predominant sequence of use, by age, for the different substances.

Longitudinal cohort study is the term used for the study design whereby researchers test the progression of drug users to stronger substances. In these studies, people are interviewed or given a questionnaire to fill out repeatedly over time. For example, the same youths may be contacted annually to provide information on their drug use during the year. Although this method may help establish the correct timing and order of drug-use initiation for any given individual, it is a very difficult approach because of the cost and time involved in tracking people for many years.

Since illegal drug use has an antisocial connotation, people may underreport their use, although some teenagers may exaggerate reports of their drug use to create an impression. The biggest hurdle a researcher studying drug use must face is obtaining accurate information. Reports are assumed to be honest and correct, based on the respondents' memory. Researchers try to promote honesty and accuracy by providing memory aids (e.g., pictures of drugs) as well as by assurances of anonymity (i.e., privacy) and confidentiality.

Another concern of researchers is that reports of drug use will be affected by the way the population is sampled or by those participating. For example, a survey conducted in an inner-city public school may not reflect all adolescents. High school dropouts who are not in class when the data are collected and students enrolled in private schools may have levels of drug use that are different from those of students attending public schools.

PREVENTION IMPLICATIONS

Adolescent drug use must be considered in relation to the normal developmental challenges of adolescence. Because individuals use drugs in different

ways for many reasons, no single prevention program will be effective with all groups at all ages. Understanding the factors that determine the link between the usage of one drug to the usage of another has important policy implications for developing prevention and educational programs. The sequential nature of drug use, as it is now understood, would indicate that prevention efforts targeted toward reducing or delaying adolescents' initiation into use of alcohol and cigarettes would reduce these adolescents' use of marijuana and other drugs. Similarly, efforts targeted toward reducing adolescents' marijuana use might reduce the rates of these adolescents' progression to "higher" stages of drug involvement. Prior drug use is a risk factor for progression; that is, the use of one drug may increase the likelihood of use of another drug, but it is not in itself a cause of further progression.

(SEE ALSO: *Conduct Disorder and Drug Use; Coping and Drug Use; High School Senior Survey*)

BIBLIOGRAPHY

DONOVAN, J. E., & JESSOR, R. (1983). Problem drinking and the dimension of involvement with drugs. *American Journal of Public Health, 73,* 543–552.

KANDEL, D. B., & FAUST, R. (1975). Sequences and stages in patterns of adolescent drug use. *Archives of General Psychiatry, 32,* 923–932.

KANDEL, D. B., & YAMAGUCHI, K. (1993). From beer to crack: Developmental patterns of drug involvement. *American Journal of Public Health, 83,* 851–855.

ROBINS, L. N., & PRZYBECK, T. R. (1985). Age of onset of drug use as a factor in drug and other disorders. In N. Kozel & E. Adams (Eds.), *Cocaine technical review.* Rockville, MD: National Institute on Drug Abuse.

TRINKOFF, A. M., ANTHONY, J. C., & MUÑOZ, A. (1990). Predictors of the initiation of psychotherapeutic medicine use. *American Journal of Public Health, 80,* 61–65.

ALISON M. TRINKOFF
CARLA L. STORR

ADULT CHILDREN OF ALCOHOLICS (ACDA)

Taken literally, this term indicates people over the age of eighteen, who have at least one biological parent with severe and repetitive life problems with alcohol. Because of their genetic and familial relationship to an alcoholic, these people carry an increased risk of severe alcohol problems themselves (a probability of two to four times that of children of nonalcoholics). Probabilities also indicate that they are not more vulnerable to severe psychiatric disorders (such as schizophrenia or manic depressive disease) and that they do not carry a heightened risk for severe problems with some drugs of abuse (such as heroin). Nevertheless, it is possible that when children of alcoholics reach adolescence or adulthood, they might be slightly more likely to have problems with marijuana-type drugs or with stimulants (such as cocaine or amphetamines). It has also been observed that if their childhood home has been disrupted by alcohol-related problems in either or both parents, the children may have greater difficulties with a variety of areas of life adjustment as they mature or go off on their own.

The label Adult Children of Alcoholics has been given to a self-help group, often abbreviated as ACOA. Within the group, people with at least one alcoholic parent can meet with others for discussion, the sharing of old and current experiences, and the chance to find interpersonal support—which helps place their own individual experiences into perspective. People who join this voluntary organization are likely to be those who both feel impaired and seek help toward coping with their past and/or present problems.

The term ACOA has also been thought to describe a group of characteristics of individuals who grew up with alcoholic parents in the home. There are research projects that do indicate that such men and women have a greater than chance likelihood of having problems expressing their feelings, feeling comfortable with intimacy, or in maintaining long-term relationships including marriages. There are less data to support conclusions regarding a possible association between an ACOA status and problems with developing trust, impairment in feelings of well-being and achievement, enhanced feelings of a need to rescue other people when they are in emotional or psychological distress, excessive feelings of a need to control a situation, and a more global dissatisfaction with life and current situations than would be expected from chance alone.

While common sense would dictate that being raised in the home of an alcoholic individual might contribute to problems with intimacy and cause levels of psychological discomfort, most of the existing studies do not control for important factors in attempting to draw conclusions. For example, it is well established that children of alcoholics are much

more likely to develop alcoholism themselves, with the possibility that many of the characteristics being described relate to the consequences of their own alcohol problems as they developed. A second problem is the variety of backgrounds that can contribute to an alcoholism risk, with the possibility that severe impulse-control disorders or the presence of an ANTISOCIAL PERSONALITY disorder in the parents could have been associated with passing on several biological characteristics to *some*, but certainly not all, children of alcoholics. In this instance, the characteristics described would relate to the associated disorder, such as the antisocial personality disorder, rather than the childhood experiences. In considering these factors, it is also important to remember that a substantial proportion, perhaps a majority, of children of alcoholics never develop alcoholism, are not likely to join ACOA groups, and appear to demonstrate many personality and behavioral characteristics that resemble those of individuals who do not have alcoholic parents.

In summary, the term adult children of alcoholics has a variety of meanings. First, biological children of alcoholic parents carry a two-to-four-fold increased risk for alcoholism themselves. Thus, this designation is an important risk factor for the future development of alcoholism. Second, the abbreviation ACOA relates to a self-help group where a minority of children of alcoholics, especially those who expressed levels of discomfort, have joined together to share experiences and offer support. The third and least meaningful definition of adult children of alcoholics relates to a variety of inadequately studied personality characteristics that might relate to the childhood environment in which an individual was raised, might be a result of additional psychiatric conditions among the parents, might reflect general factors associated with a disordered childhood home but have nothing specifically to do with alcoholism, or might relate to specific alcohol-related experiences in the childhood home.

(SEE ALSO: *Al-Anon; Codependence; Conduct Disorder and Drug Use*)

BIBLIOGRAPHY

BURK, J. P. & SHER, K. J. (1990). Labeling the child of an alcoholic: Negative stereotyping by mental health professionals and peers. *Journal of the Study of Alcohol, 51,* 156–163.

DAWSON, D. A. (1992). The effect of parental alcohol dependence on perceived children's behavior. *Journal of Substance Abuse, 4,* 329–340.

FISHER, G. L., ET AL. (1992). Characteristics of adult children of alcoholics. *Journal of Substance Abuse, 4,* 27–34.

FULTON, A. I. & YATES, W. R. Adult children of alcoholics. *Journal of Nervous and Mental Disorders, 178,* 505–509.

GREENFIELD, S. F., ET AL. (1993). Long-term psychological effects of childhood exposure to parental problem drinking. *American Journal of Psychiatry, 150,* 608–613.

SCHUCKIT, M. (1994). Low level of response to alcohol as a predictor of future alcoholism. *American Journal of Psychiatry, 151,* 184–189.

WERNER, E. E. (1986). Resilient offspring of alcoholics: A longitudinal study from birth to age 18. *Journal of the Study of Alcohol, 47,* 34–40.

MARC A. SCHUCKIT

ADVERTISING AND ALCOHOL USE

Factors that influence the development of substance use and abuse range from individual characteristics (some of them presumably based on genetics) to social factors (educational level, cultural background, and/or political considerations such as war). All these factors can influence one another and, at times, create the principal pressures that affect a complex behavior like alcohol consumption. For example, during wars and strikes and other times of trouble, whole populations may refrain from using alcohol or may not have access to it. Advertising, in such circumstances, may not affect either sale or use patterns.

Under more normal conditions, however, the advertising of alcoholic beverages is part of a complex situation that may initiate alcohol use and its possible progression to a detrimental stage of abuse. Nevertheless, research in this area does not support the idea that the human mind is so malleable that liquor advertising can radically affect it. According to a Canadian researcher: "Econometric exposure, experimental studies and advertising bans ... indicate that advertising bans do not reduce alcohol sales, total advertising expenditures have no reliable correlation with sales of alcoholic beverages, and that experimental studies typically show no effect of advertising on actual consumption. However, one set

of studies does show that drinkers are exposed to more television alcohol advertisements, without making the causal connection clear. In general, the evidence indicates little impact of alcohol advertising on alcohol sales or drinking." Other researchers have come to similar conclusions.

Since the mass media are quite new phenomena but alcohol use and alcoholism are quite ancient phenomena in most societies, it may safely be concluded that the media role in the initiation and persistence of alcohol use is, at the least, limited. Nevertheless, the mass media also have a great impact on social conduct, so the advertisement of such products as tobacco and alcohol, which can be health-compromising or even lethal, should be a matter of great social concern. Both the products and the advertisements are often controlled for the public good by law.

Since tobacco and alcohol companies are reluctant to accept or admit to any health-related consequences from the use of their products, their advertisements should be carefully screened for overt or hidden messages that could lead to detrimental effects. This issue is especially relevant to the problems that can affect specific populations—such as pregnant women and their unborn children or young people, still minors. Since advertisements do not usually target only one audience, the public should be cautioned to view them with care. In addition, certain audiences should be warned against certain products or behaviors—perhaps by clever counter-advertisements such as the anti-smoking campaign, the designated-driver campaign, or the safe-sex campaign.

In addition to advertisements that constitute warnings, warning labels have also been added to substances such as alcohol and tobacco, which are legal but controlled by government regulations and state and local laws. Since 1989, for example, U.S. law has required that alcoholic beverage containers carry a warning label (Public Law 100-690, 1988), but early evidence indicates that a substantial proportion of the population has not noticed their existence or taken them seriously. The warning labels must also be shown in any corporate advertising for the product.

(SEE ALSO: *Advertising and Tobacco Use; Legal Regulation of Drugs and Alcohol; Prevention: Shaping Mass Media Messages to Vulnerable Groups*)

BIBLIOGRAPHY

GRUB, J. W. (1993). Alcohol portrayals and alcohol advertising on television: Content and effects on children and adolescents. *Alcohol Health & Research World*, 17(1), 45–60.

SMART, R. G. (1988). Does alcohol advertising affect overall consumption? A review of empirical studies. *Journal of Studies on Alcohol*, 49(4), 314–323.

SOBALL, L. C., ET AL. (1986). Effect of TV programming and advertising on alcohol consumption in normal drinkers. *Journal of Studies on Alcohol*, 47(4), 333–340.

G. BORGES

ADVERTISING AND THE ALCOHOL INDUSTRY

The beverage alcohol industry includes companies that market beers and brews (malt liquors), wines and sparkling wines (fermented), and distilled spirits—whisky, vodka, scotch, gin, rum, and flavored liquors. Sales of these products, usually through distributors, are limited to those businesses that have obtained special licenses to sell one or more of the above categories of products. For example, if a restaurant has a license to serve beer and wine, they cannot serve other types of alcoholic beverages.

In the United States, alcoholic beverages and tobacco products are the only consumer goods that are legally restricted for sale *only* to those who are not minors—at least 21 years in the case of alcohol or 18 in the case of tobacco. Sales to anyone under those ages, respectively, are illegal, yet every day thousands of minors buy beer, wine coolers, cigarettes, and snuff with no questions asked by store clerks or owners. Even if a store refuses to sell to them, they can usually find a vending machine or get an older friend to buy for them.

In a survey of the Class of 1990, 57.1 percent of senior high school students reported using alcohol in the previous month and 32.2 percent of seniors reported having five or more drinks in a row within the last two weeks. Also, in 1990, 24.5 percent of youths aged 12 to 17 reported using alcohol in the previous month. Such illegal use of alcoholic beverages by such a high percentage of teenagers has generated a high level of concern on the part of health-care professionals, police, parents, activist groups such as MOTHERS AGAINST DRUNK DRIVING

(MADD), and groups such as the Center for Science in the Public Interest (CSPI).

This creates an advertising problem for the companies that market alcoholic beverages. How do you advertise to the over-21 group and not also appear to be appealing to the under-21 group? Since teenagers have a very strong desire to grow up fast, or at least participate in activities they view as *adult*, they are very vulnerable to anything they think would help them achieve their dreams.

Critics accuse the alcoholic-beverage companies of making their advertising and promotional programs inviting to teenagers, who are already very receptive to the ideas of "engaging in adult activities," "being successful," "being more confident," and "being more attractive to the opposite sex." The alcoholic-beverage companies respond that they follow the industry voluntary advertising guidelines and do *not* target teenagers. They also point with pride to the Public Service Initiatives Sponsored by America's Beer Industry, which encourage drinkers to "know when to say when," "drink smart or don't start," "think when you drink," or "drink safely." Coors has started a "Now, Not Now" campaign.

WHO MINDS THE STORE?

The U.S. Bureau of Alcohol, Tobacco, and Firearms (BATF) in the Department of the Treasury is the federal agency with responsibility for overseeing the alcohol industry. Its rules discourage advertising claims that are obscene or misleading, as well as those that associate athletic ability with drinking. Until the 1990s, the alcohol content of beer could not be included on the labeling of the container or in any associated advertising. As a result of a suit by Coors Brewing, a federal court decision overturned this restriction on labeling, and so companies are now permitted to label their beers and malt liquors with the alcohol content—although a few states still have laws that prohibit listing the alcohol content.

The Federal Trade Commission (FTC) also reviews advertising, with emphasis on instances of false or misleading ads. Neither the BATF nor the FTC has been aggressive in challenging ads that seem to be targeted at young drinkers or ads that seem to encourage heavy drinking. The BATF did order a halt to advertising for Cisco, an alcohol fortified wine (with 20% alcohol) that was packaged like the low-alcohol wine coolers (4% alcohol), and

ordered the company to change their packaging so it would not look like a wine cooler, to prevent serious injury to unsuspecting drinkers.

The Food and Drug Administration (FDA) in the Department of Health and Human Services has no jurisdiction over alcohol advertising. Unlike pharmaceuticals, there is no mandate that labels or advertising materials for alcohol should provide fair balance of the risks/consequences, as well as the benefits, to be expected from using one of their products. Americans see ads and company logos that encourage people to drink, but the ads and billboards fail to provide information about the down side of drinking, especially excessive drinking.

As of November 1989, all alcoholic beverages sold in the United States must carry a warning on the container that states: "GOVERNMENT WARNING: (1) According to the Surgeon General, women should not drink alcoholic beverages during pregnancy because of the risk of birth defects. (2) Consumption of alcoholic beverages impairs your ability to drive a car or operate machinery, and may cause health problems."

PROPOSED LAWS TO PROVIDE HEALTH WARNINGS

The Coalition for the Prevention of Alcohol Problems lobbied in 1993 for the passage of the Sensible Advertising and Family Education Act, which would have required health messages in alcoholic-beverage ads. Coverage would include any ad broadcast on radio and television (including cable) or printed in magazines, newspapers, and promotional displays. A series of seven rotating health messages about alcohol would have included the following:

HEALTH AND SAFETY WARNING MESSAGES (broadcast and cable)

1. SURGEON GENERAL'S WARNING: If you are pregnant, don't drink alcohol. Alcohol may cause mental retardation and other birth defects.
2. SURGEON GENERAL'S WARNING: If you are under the age of 21, it's illegal to buy alcoholic beverages.
3. SURGEON GENERAL'S WARNING: Alcohol is a drug and may be addictive.
4. SURGEON GENERAL'S WARNING: Drive sober. If you don't, you could lose your driver's license.

5. SURGEON GENERAL'S WARNING: Don't mix alcohol with over-the-counter, prescription, or illicit drugs.
6. SURGEON GENERAL'S WARNING: If you drink too much alcohol too fast, you can die of alcohol poisoning.
7. SURGEON GENERAL'S WARNING: Drinking increases your risk of high blood pressure, liver disease, and cancer.

If passed as proposed, the health-and-safety warning message would be required to be read as part of the alcoholic beverage advertisement. For television, the words to each message would be flashed on the screen as the warning is read. Print advertising and promotional materials would require more comprehensive messages.

WHAT IS ADVERTISING?

Merriam-Webster's Collegiate Dictionary, tenth edition (1993), defines the verb *advertise*: "to call public attention to especially by emphasizing desirable qualities so as to arouse a desire to buy or patronize." The noun *advertising* includes "by paid announcements." The broad umbrella of advertising—in addition to the usual television, radio, and print—uses billboards, point-of-purchase signs and displays, and, increasingly, sponsorship of special events such as music festivals, auto, bicycle, and boat racing, and other sports.

ROLE OF ADVERTISING

Advertising is used as a major tool in marketing. When a company first introduces a new product, the goals generally are

1. To inform potential purchasers that a particular product is available and why they might like to try this new product
2. To persuade people that they should go out and buy the product
3. To let people know where the product can be purchased
4. To reassure people who buy the product that they have made a wise choice in doing so.

Where more than one company has products in a given category, the goals generally become

1. To increase market share by taking business away from a competitive product. This can be done by offering a better product or a better value and/or by increasing the level of advertising and promotion to out-shout the competition.
2. To increase the size of the market by inducing more people to start using the product. In the case of alcoholic beverages, this can be done by aggressively promoting features which will appeal to the potential purchaser, that is, makes you more confident, more outgoing, more appealing to the opposite sex, and, in the case of minors, participating in adult-type activities.
3. To increase the size of the market by inducing people to increase their usage of your product(s). This can be done by tying the product to occasions such as Spring Break and by promoting the product heavily to the target audiences.
4. To keep reassuring heavy drinkers that they are in good company by drinking the particular brand of beer or liquor being advertised. Since the 10 percent of those who drink most heavily account for fully 50 percent of all alcohol consumed in the United States, this is a very important reason to advertise.

ADVERTISING EXPENDITURES

The alcoholic-beverage companies spend more than 1 billion dollars each year in the print and broadcast media to advertise their products. In 1991, beer marketers spent over 740 million dollars, liquor marketers over 280 million dollars, and wine marketers over 92 million dollars on print, radio, and television. Anheuser Busch spent 270 million dollars just on television ads in 1991.

An estimated 1 billion dollars more was spent on other advertising and promotional programs. Brewers and beer distributors spent many millions of dollars sponsoring sporting events, rock concerts, Spring Break promotions, and other activities heavily oriented to college campuses. They also were heavy advertisers and supporters of baseball, football, racing events, and music or other cultural events.

Examples of the costs of advertising campaigns for 1993 include 110 million dollars for the new "Proud to Be Your Bud" campaign for Budweiser and 83 million dollars on the "Can Your Beer Do This?" campaign for Miller Beer.

Event spending by Anheuser Busch (A-B) in 1992 was between 85 and 90 million dollars—with 70 mil-

lion of this spent on baseball; most of this in local advertising. For 1993, A-B agreed to be the major-league sponsor for the preseason period as well as for the world series in the fall. In addition to owning the St. Louis Cardinals baseball team, A-B has advertising agreements with all but four of the major-league teams. The exceptions are the Chicago Cubs, the Florida Marlins, the Milwaukee Brewers, and the Oakland A's.

CORRELATION OF ADVERTISING TO CONSUMPTION

In August 1993, the NATIONAL INSTITUTE ON ALCOHOL ABUSE AND ALCOHOLISM (NIAAA) had this to say about alcohol, media, and advertising:

> The effects of mass communications in either promoting or preventing alcohol consumption and the problems associated with it are equivocal. Alcohol advertisements and broadcast media programming have been found to encourage a favorable view of alcohol use. Yet studies provide only modest support for the hypothesis that favorable presentations lead to positive attitudes and distorted perceptions, and consequently to increased consumption, particularly among youthful viewers. The effects of advertising bans and linkages between advertising expenditures and per capita consumption also appear to be weak and inconsistent.

There is a general feeling that advertising, including all of the other promotions associated with it, plays a significant part in creating an image of desirability toward the use of alcohol, especially for beer. It is almost impossible, however, to conduct a meaningful study to separate this effect of advertising from the many other factors involved in alcohol consumption. For example, L. R. Lieberman and M. A. Orlandi (1987) said, "It is clear that, due to methodological difficulties, a direct causal relationship between exposure to alcohol advertising and alcohol consumption is difficult to establish." They continued: "This, however, should not keep alcohol educators from continuing to pursue research that focuses on the contribution of alcohol advertising to adolescent alcohol consumption." R. G. Smart (1988) said: "In general, the evidence indicates little impact of alcohol advertising on alcohol sales or drinking. However, some results are suggestive and there is a need for more sophisticated econometric, exposure

and experimental studies that take into account a wider range of variables." G. Frank and G. Wilcox (1988) reported: "Analysis of the results reveals no significant relationship between total advertising expenditures and consumption of beer. Significant relationships were found, however, between consumption of wine and distilled spirits and their advertising. It is emphasized that the relationships are correlational, not necessarily causal." C. K. Atkin (1990) said: "It is concluded that alcohol commercials on television contribute to a slight increase in overall consumption by adolescents and may therefore have an effect on misuse of alcohol and drinking and driving."

BEVERAGE ALCOHOL PER CAPITA CONSUMPTION

In the United States, per capita consumption of all alcoholic beverages combined reached its peak in 1980–81; that of wine did not reach its peak until 1986. During the 1980s (from 1980 to 1989), there was a 12 percent decrease in per capita ethanol (drinking alcohol) consumption—the only sustained decrease since Prohibition—down to 2.43 gallons per person. The greatest decrease was seen in the consumption of distilled spirits.

As of 1992, beer ranks fourth (behind soft drinks, milk, and coffee) in per capita consumption of any kind of beverage products, a position it has held for many years. The U.S. Department of Health and Human Services has an objective for the year 2000 to decrease per capital alcohol consumption to no more than 2 gallons of ethanol per person.

BEVERAGE ALCOHOL SALES

Sales at the retail level of the major categories of alcoholic beverages were as follows in 1992:

Beer	$51.3 billion
Spirits	$30.9 billion
Wine	$12.9 billion
Total	$95.1 billion

In the beer category, Anheuser Busch captured 44.9 percent of the market, Miller Brewing (part of Phillip Morris) had 16.9 percent, Coors had 13.4 percent, John Labatt had 5.4 percent, and G. Heileman had 2.6 percent.

BEER INSTITUTE ADVERTISING AND MARKETING CODE

The Beer Institute published, in March 1992, the latest version of their Advertising and Marketing Code, which they cite when trying to deflect some of their critics' charges. Selected portions are listed below. A copy of the entire Advertising and Marketing Code may be obtained from the Beer Institute.

Guidelines

3. Beer advertising and marketing materials are intended for adults of legal purchase age who choose to drink.

   a. Beer advertising and marketing materials should not employ any symbol, language, music, gesture, or cartoon character that is intended to appeal primarily to persons below the legal purchase age. Advertising or marketing material has a "primary appeal" to persons under the legal purchase age if it has special attractiveness to such persons above and beyond the general attractiveness it has for persons above the legal purchase age, including young adults above the legal purchase age.

   b. Beer advertising and marketing materials should not employ any entertainment figure or group that is intended to appeal primarily to persons below the legal purchase age.

   c. Beer advertising and marketing materials should not depict Santa Claus.

   d. Beer advertising and marketing materials should not be placed in magazines, newspapers, television programs, radio programs, or other media where most of the audience is reasonably expected to be below the legal purchase age.

   e. To help insure that the people shown in beer advertising are and appear to be above the legal purchase age, models and actors employed should be a minimum of 25 years old, substantiated by proper identification, and should reasonably appear to be over 21 years of age.

   f. Beer should not be advertised or marketed at any event where most of the audience is reasonably expected to be below the legal purchase age. This guideline does not prevent brewers from erecting advertising marketing materials at or near facilities that are used primarily for adult-oriented events, but which oc-

casionally may be used for an event where most attendees are under age 21.

   g. No beer identification, including logos, trademarks, or names should be used or licensed for use on clothing, toys, games or game equipment, or other materials intended for use primarily by persons below the legal purchase age.

LIMITATIONS OF VOLUNTARY BEER INDUSTRY ADVERTISING CODES

The beer industry's voluntary advertising codes are vaguely or narrowly written so as to restrict advertising practices as little as possible. In addition, they are not legally enforceable and only cover media advertising, not promotions. By qualifying most statements: "that advertising should not be used where *most* of the audience is reasonably expected to be below the legal purchase age" or "should not employ any entertainment figure or group that is intended to appeal *primarily* to persons below the legal purchase age," the code permits advertising even if almost half of the audience is below the legal purchase age.

THE COSTS OF ALCOHOLISM

The National Institute on Alcohol Abuse and Alcoholism estimates that $85 billion goes to health care, lost employment, and reduced productivity as a result of alcoholism in the United States.

(SEE ALSO: *Advertising and Alcohol Use; Advertising and the Pharmaceutical Industry; Advertising and Tobacco Use; Social Costs of Alcohol and Drug Use*)

BIBLIOGRAPHY

Advertising Age. May 11, 1992; March 8, 1993; March 15, 1993; May 10, 1993; June 14, 1993; June 21, 1993.

ATKIN, C. K. (1990). *Journal of Adolescent Health Care, 11*(1), 10–24.

BEER INSTITUTE. (1992). *Advertising and Marketing Code.*

Beverage World. May 1992; May 1993.

CENTER FOR SCIENCE IN THE PUBLIC INTEREST. Washington, DC.

COALITION FOR THE PREVENTION OF ALCOHOL PROBLEMS. Washington, DC.

FRANK, G., & WILCOX, G. (1988). *Journal of Advertising, 16*(3), 22–30.

LIEBERMAN, L. R., & ORLANDI, M. A. (1987). *Alcohol Health and Research World, 12*(1), 30–33.

NATIONAL INSTITUTE OF DRUG ABUSE. (1991). *NIDA national high school senior survey, class of 1990.* Rockville, MD: U.S. Department of Health and Human Services.

NATIONAL INSTITUTE OF DRUG ABUSE. (1991). *NIDA national household survey on drug abuse, 1990.* Rockville, MD: U.S. Department of Health and Human Services.

SMART, R. G. (1988). *Journal of Studies on Alcohol, 49*(4), 214–323.

WILLIAMS, G. D., ET AL. (1992). *Surveillance report no. 23.* Washington, DC: National Institute on Alcohol Abuse and Alcoholism.

CHARLES M. RONGEY

ADVERTISING AND THE PHARMACEUTICAL INDUSTRY

The pharmaceutical industry, which researches, develops, produces, and markets prescription drugs in the United States, is the most heavily regulated of all industries when it comes to the advertising and promotion of its products. Through its Drug Marketing, Advertising, and Communications Division, the Food and Drug Administration (FDA) regulates all advertising and promotional activities for prescription drugs, including statements made to physicians and pharmacists by pharmaceutical sales representatives. Advertising of over-the-counter (OTC) drugs, which is not regulated by the FDA, is under the jurisdiction of the Federal Trade Commission (FTC).

Before a new prescription drug is approved for marketing, the FDA and the pharmaceutical company must agree on the "Full Prescribing Information" that will accompany the product and that must be included in all ads, brochures, promotional pieces, and samples. This Full Prescribing Information must include, in the correct order, the following information about the drug: its trade name, its assigned name, the strength of its dosage form, a caution statement (stating that a prescription is required), a description of its active ingredient, the clinical pharmacology of the drug, indications for its usage, contraindications for usage, precautions, adverse reactions, instructions on what to do in case of overdosage, dosage and administration, and how the drug is supplied.

Typically, this information is very detailed, and even when it is given in six-point type, it can run to two printed pages. The majority of pharmaceutical companies pay to have this information published in the *Physician's Desk Reference,* which is sent to physicians free of charge. The book is also sold in bookstores or on library reference shelves for use by consumers who want to know more about specific drugs.

All promotional pieces and ads to be used when a new drug is marketed must first be approved by the FDA before marketing begins—to ensure that the statements being made are consistent with those in the official labeling. After the introduction of a new drug has been completed, copies of all subsequent ads and promotional pieces must be sent to the FDA at the time of their first use, too, but they do not have to be preapproved. The FDA reviews ads, brochures, direct-mail pieces, and sales aids to ensure that a "fair balance" has been maintained in presenting both the benefits and the risks of a medication. In the 1990s, the FDA has directed its attention to "scientific symposia" and other medical meetings at which information about new drugs, or new indications for drugs, are presented. This ensures that they are not just promotional programs for a single drug. In no other industry are advertising and promotion required to meet such strict standards.

THE CHANGING ROLE OF PHARMACEUTICAL ADVERTISING

Traditionally, advertising and promotion of pharmaceutical products were directed primarily to physicians, with some limited advertising and promotion being directed to pharmacists. With the expiration of patents on some of the major drugs in the 1960s and 1970s, generic versions of the drugs became available from competing manufacturers. The generic drugs were priced lower than the brand-name products, so pharmacists worked to get laws passed allowing them to substitute a generic product for the brand-name product. Since this gave pharmacists more discretion in selecting which product to purchase and dispense, advertising and promotion to pharmacists increased. By the 1980s, when formulary committees selected which drugs could be prescribed or reimbursed under third-party payment

programs (medical insurance), advertising and promotion were also directed to these decision makers.

DIRECT-TO-CONSUMER ADVERTISING

In the mid-1980s, two pharmaceutical companies began direct-to-consumer advertising (DTC). Pfizer led the way with its health-care series of ads to the general public. Merrell Dow, now Marion Merrell Dow, used DTC ads to inform the public that physicians had a new treatment to help smokers who wanted to stop smoking. When the company's new, nonsedating antihistamine became available, it used DTC ads to tell allergy sufferers that physicians now had a new treatment for allergies. The ads did not mention the name of the products; rather, they asked patients with specific problems or symptoms to see their physician. By the late 1980s, other prescription products, such as Rogaine, Actigall, and Estraderm were promoted in this manner, as more and more ethical drug manufacturers came to the conclusion that DTC advertising could help their sales and profits (Medical Marketing and Media, 1991).

The next phase of DTC advertising led to ads in magazines and newspapers that mentioned the name of the product *and* its indication for use. To meet FDA requirements, such ads must always include a "brief summary of prescribing information." To date, the advertising of prescription drugs on television or radio has remained greatly restricted, since it is not possible to include the necessary brief summary of prescribing information on the air. Because of this limitation, the ads on television or radio must focus on either the name of the product *or* the indication for the product.

Marion Merrell Dow promoted its nonsedating antihistamine on television by portraying allergy sufferers and suggesting that if viewers suffered from allergies, they should see their doctor because their doctor had new treatments to relieve their symptoms. They could not mention the name of the product, Seldane®, since they had mentioned the indication for the product. When a company has the only—or the major—product in the market, this approach can be very effective in increasing sales because it increases awareness among patients that a new treatment is available and gets them to see their doctors. Marion Merrell Dow has also advertised Seldane in consumer magazines, in which, with the inclusion of the brief summary of information about the product, it could tell the full story about the drug's nonsedat-

ing properties. In 1990, more than 11 million dollars was spent to advertise Seldane.

To promote Nicorette, a NICOTINE-containing gum designed to help smokers stop smoking, Marion Merrell Dow showed a picture on television of the package with the message that this was Nicorette and that it was available only by prescription. According to FDA rules, the company could not say that the gum was useful in helping smokers who wanted to stop smoking, since it had named the Marion Merrell Dow product. Dow also started using consumer ads in magazines for Nicorette so that, with the inclusion of the brief summary of information, it could tell the full story of how the gum could help a smoker become a nonsmoker. In 1990, the company spent more than 9 million dollars to advertise Nicorette to the general public.

The pharmaceutical companies that use DTC advertising are attempting to assume the role of communicator of truthful health-care information. They provide an important service to the public while avoiding the tactics used in OTC advertising, tactics to which we are exposed daily on television and radio and in the print media (Medical Marketing and Medicine, 1991). The FDA is very carefully following the activities of the pharmaceutical companies, and it will almost certainly keep a tight control over the advertising of prescription drugs.

PROMOTIONAL EXPENDITURES

The research-based pharmaceutical companies, which are members of the Pharmaceutical Manufacturers Association (PMA), estimated that the prescription drug companies spent about 10 billion dollars in 1991 on advertising, marketing, and selling expenses. This figure represents about 25 percent of sales (Forbes, 1991). The major part of this sum was to be spent on the more than 10,000 men and women who call on physicians, pharmacists, and other health-care professionals to educate them about their products—that is, how to use them, the side effects to anticipate, and the different dosage forms available for each product.

Some members of Congress feel that the pharmaceutical companies are spending too much on advertising and promotion, and some would even like to limit these expenditures (US Senate, 1991). Such restrictions are already in effect in Great Britain. Proponents of spending limits feel that the limits would result in lower prices for prescription drugs

and refuse to believe that the dissemination of information about new drugs and new treatment procedures would suffer. However, many in the health industry think that if physicians had to depend on their medical journals for information about new drugs, they might not know about them for several years. Meanwhile, their patients who need the new drugs would be deprived of the latest advances in medical care. In 1991, in the United States, only 7 percent of health-care dollars was spent on drugs. By comparison, in Japan, 30 percent of the health-care bill is attributable to drugs (Forbes, 1991). The advertising and promotion of prescription drugs, including the cost for more than 10,000 pharmaceutical representatives who call on the nation's physicians and other health-care professionals, make up less than 2 percent of the health-care expenditures in the United States.

CODE OF PHARMACEUTICAL MARKETING PRACTICES

The member companies of the PMA have worked together to create guidelines for the ethical promotion of prescription pharmaceutical products. The following guidelines were revised in December 1990 (Pharmaceutical Manufacturers Association).

The Pharmaceutical Manufacturers Association and its member companies recognize their responsibilities and obligations to establish and implement good pharmaceutical marketing practices that ensure appropriate communication with their audiences and reflect the industry's special role in the health-care delivery system. The member companies of the Pharmaceutical Manufacturers Association hereby declare their intention to conform voluntarily to the following code of pharmaceutical marketing practices. The obligations of the industry are identified below.

The pharmaceutical industry, conscious of its special position arising from its involvement in public health, and justifiably eager to fulfill its obligations in a free and fully responsible manner, undertakes:

to ensure that all products it makes available for prescription purposes to the public are backed by the fullest technological service and have full regard for the needs of public health; to produce pharmaceutical products under adequate procedures and strict quality assurance;

to base the claims for substances and formulations on valid scientific evidence, thus determining the therapeutic indications and conditions of use;

to provide scientific information with objectivity and good taste, with scrupulous regard for truth, and with clear statements with respect to indications, contraindications, tolerance, and toxicity;

to use complete candor in dealing with public health officials, health-care professionals, and the public, and to comply with the regulations and policies issued by the Food and Drug Administration and other government agencies in implementing United States laws regarding pharmaceuticals.

I. General Principles

1. The term "pharmaceutical product" means any prescription pharmaceutical or biological product intended for use in the diagnosis, cure, mitigation, treatment, or prevention of disease in humans, or to affect the structure or any function of the human body, which is promoted and advertised to the medical profession rather than directly to the lay public.

2. Information on pharmaceutical products should be accurate, fair, and objective, and presented in such a way as to conform not only to legal requirements but also to ethical standards and to standards of good taste.

3. Information should be based on an up-to-date evaluation of all the available scientific evidence and should reflect this evidence clearly.

4. No public communication should be made with the intent of promoting a pharmaceutical product as safe and effective for any use before the required approval of the pharmaceutical product for marketing for such product is obtained. However, this provision is not intended to abridge the right of the scientific community and the public to be fully informed concerning scientific and medical progress. It is not intended to restrict a full and proper exchange of scientific information concerning a pharmaceutical product, including appropriate dissemination of investigational findings in scientific or lay

communications media, nor to restrict public disclosure to stockholders and others concerning any pharmaceutical product as may be required or desirable under law, rule, or regulation.

5. Statements in promotional communications should be based upon substantial scientific evidence or other responsible medical opinion. Claims should not be stronger than such evidence warrants. Every effort should be made to avoid ambiguity.

6. Particular care should be taken that essential information as to pharmaceutical products' safety, contraindications, and side effects or toxic hazards is appropriately and consistently communicated subject to the legal, regulatory, and medical practices of the United States. The word "safe" should not be used without qualification. Compliance with FDA regulations with respect to full disclosure in pharmaceutical product labeling and to fair balance in promotional communications satisfies the need for qualification.

7. Promotional communications should have medical clearance before their release.

II. Medical Representatives

Medical representatives should be adequately trained and possess sufficient medical and technical knowledge to present information on their company's products in an accurate and responsible manner.

III. Symposia, Congresses, and Other Means of Verbal Communication

Symposia, congresses, and the like are indispensable for the dissemination of knowledge and experience. Scientific objectives should be the principal focus in arranging such meetings, and entertainment and other hospitality should not be inconsistent with such objectives.

IV. Printed Promotional Material

Scientific and technical information should fully disclose properties of the pharmaceutical product as approved in the United States based on current scientific knowledge and FDA regulations including:

-   the active ingredients, using the approved names where such names exist;

-   at least one approved indication for use together with the dosage and method of use;

-   a succinct statement of the side effects, precautions, and contraindications.

Promotional material, such as mailings and medical journal advertisements, should not be designed to disguise their real nature, and the frequency and volume of such mailings should not be offensive to the health-care professionals.

V. Samples

Samples may be supplied to the medical and allied professions to familiarize them with the products or to enable them to gain experience with the product in their practice. The requirements of the Prescription Drug Marketing Act of 1987 should be observed.

VI. Inquiries and Code Administration

Inquiries concerning this code and its administration may be referred to the Pharmaceutical Manufacturers Association.

VII. Endorsement of the Position Statements of the American College of Physicians

The Pharmaceutical Manufacturers Association Board of Directors, in May 1990, endorsed the four position statements on relationships between physicians and the pharmaceutical industry that had been adopted by the Board of Regents of the American College of Physicians on September 13, 1989.

The PMA Board also determined to include these four position statements as an adjunct to the PMA *Code of Pharmaceutical Marketing Practices*. The four position statements are:

1. Gifts, hospitality, or subsidies offered to physicians by the pharmaceutical industry ought not to be accepted if acceptance might influence or appear to others to influence the objectivity of clinical judgment. A useful criterion in determining accept-

able activities and relationships is: Would you be willing to have these arrangements generally known?

2. Independent institutional and organizational continuing medical education providers that accept industry-supported programs should develop and enforce explicit policies to maintain complete control of program content.

3. Professional societies should develop and promulgate guidelines that discourage excessive industry-sponsored gifts, amenities, and hospitality to physicians at meetings.

4. Physicians who participate in practice-based trials of pharmaceuticals should conduct their activities in accord with basic precepts of accepted scientific methodology.

[These four positions have also been endorsed by the American Diabetes Association, the American Society of Hematology, and the American Thoracic Society.]

VIII. **Adoption of the American Medical Association Guidelines on Gifts to Physicians from Industry**

The Pharmaceutical Manufacturers Association Board of Directors on December 6, 1990, adopted as part of the PMA *Code of Pharmaceutical Marketing Practices* the guidelines set for them in the Opinion of the Council on Ethical and Judicial Affairs filed by the House of Delegates of the American Medical Association on December 4, 1990. The guidelines are:

1. Any gifts accepted by physicians individually should primarily entail a benefit to patients and should not be of substantial value. Accordingly, textbooks, modest meals, and other gifts are appropriate if they serve a genuine educational function. Cash payments should not be accepted.

2. Individual gifts of minimal value are permissible as long as the gifts are related to the physician's work (e.g., pens and notepads).

3. Subsidies to underwrite the costs of continuing medical education conferences or

professional meetings can contribute to the improvement of patient care and therefore are permissible. Since the giving of a subsidy directly to a physician by a company's sales representative may create a relationship which could influence the use of the company's products, any subsidy should be accepted by the conference's sponsor, who in turn can use the money to reduce the conference's registration fee. Payments to defray the costs of a conference should not be accepted directly from the company by the physicians attending the conference.

4. Subsidies from industry should not be accepted to pay for the costs of travel, lodging, or other personal expenses of physicians attending conferences or meetings, nor should subsidies be accepted to compensate for the physicians' time. Subsidies for hospitality should not be accepted outside of modest meals or social events held as a part of a conference or meeting. It is appropriate for faculty at conferences or meetings to accept reasonable honoraria and to accept reimbursement for reasonable travel, lodging, and meal expenses. It is also appropriate for consultants who provide genuine services to receive reasonable compensation and to accept reimbursement for reasonable travel, lodging, and meal expenses. Token consulting or advisory arrangements cannot be used to justify compensating physicians for their time or their travel, lodging, and other out-of-pocket expenses.

5. Scholarship or other special funds to permit medical students, residents, and fellows to attend carefully selected educational conferences may be permissible as long as the selection of students, residents, or fellows who will receive the funds is made by the academic or training institution.

6. No gifts should be accepted if there are strings attached. For example, physicians should not accept gifts if they are given in relation to the physician's prescribing practices. In addition, when companies underwrite medical conferences or lectures other than their own, responsibility for and con-

trol over the selection of content, faculty, educational methods, and materials should belong to the organizers of the conferences or lectures.

(SEE ALSO: *Advertising and Alcohol Use; Advertising and the Alcohol Industry; Advertising and Tobacco Use*)

BIBLIOGRAPHY

Forbes. April 15, 1991, p. 49.

Medical Marketing and Media. May 1991, p. 38.

PHARMACEUTICAL MANUFACTURERS ASSOCIATION. (1990). *Code of pharmaceutical marketing practices.* Washington, DC: Author.

U.S. SENATE. (1991). *The drug manufacturing industry: A prescription for profits.* Special Committee on Aging Report. Washington, DC: U.S. Government Printing Office.

CHARLES M. RONGEY

ADVERTISING AND TOBACCO USE

Tobacco companies spend more than 3.2 billion dollars annually to advertise and promote cigarettes and other tobacco products. They have sales of tobacco products of some 40 billion dollars a year. Tobacco companies claim that the purpose and desired effect of marketing are merely to provide information and to influence brand selection among current smokers, although only about 10 percent of smokers switch brands in any one year. Since more than 1 million adult smokers stop smoking every year and almost half a million other adult smokers die from smoking-related diseases, the tobacco companies must recruit more than 3,000 new young smokers every day to replace those who die or otherwise stop smoking. Tobacco companies contend that smoking is an "adult habit" and that adult smokers "choose" to smoke. However, many medical researchers assert that cigarette smoking is primarily a *childhood addiction or disease* and that most of the adults who smoke started as children and could not quit.

Unlike the pharmaceutical companies, which are tightly regulated as to their advertising and promotion, the tobacco industry is essentially unregulated. The only restrictions are that companies cannot use paid advertising on television or radio, they cannot claim what they cannot prove (e.g., that low-tar cig-

arettes are less hazardous to health), and they must include one of four warnings on cigarette packages and ads. The fact that warning labels are printed on a pack of cigarettes has been used by the tobacco companies as a defense against tobacco victims' lawsuits.

UNDERSTANDING THE SMOKING HABIT

Almost all smokers started before the age of 21—most before the age of 18. Young people who learn to inhale the cigarette smoke and experience the mood-altering effects from the inhaled nicotine quickly become dependent on cigarettes to help them cope with the complexities of everyday life. Having developed a nicotine dependence, they find they must continue smoking to avoid the downside of nicotine withdrawal. The earlier they start to smoke, the more dependent they seem to become—and the sooner they start to experience smoking-related health problems.

A survey conducted by the U.S. Department of Health and Human Services among high school students who smoked half a pack of cigarettes a day, found that 53 percent said they had tried to quit but could not. When asked whether they would be smoking 5 years later, only 5 percent said they would be—but 8 years later, 75 percent were still smoking.

CIGARETTE ADVERTISING

The tobacco companies are very experienced in using advertising and different kinds of promotional programs to help them accomplish several major objectives:

1. To Reassure Current Smokers. To offset the effect of thousands of studies showing the adverse health effects of smoking and of the warning labels on the cigarette packages, the tobacco industry continues to claim that no one has yet "proven" that smoking "causes" health problems—that there are just "statistical associations." Therefore, they assert, since the purported harmful effects have not been conclusively established, it is okay for smokers to continue smoking.

2. To Associate Smoking with Pleasurable Activities. In their ads, tobacco companies show healthy young people enjoying parties, dancing, attending sporting events, having a picnic at the beach, sailing, and so on. The implication is that if

you smoke, you too will have the kind of good times enjoyed by the smokers shown in the ads.

3. To Associate Smoking with Other Risk-Taking Activities. Since as indicated by the warning labels on every package of cigarettes, smoking involves risk to one's health, the tobacco companies attempt to counter this by showing in their ads such risk-taking activities as ballooning, mountain climbing, sky diving, skiing, and motorcycle riding. This is the industry's not-so-subtle way of saying: Go ahead and take a risk by smoking; you are capable of deciding the level of risk you want to assume. The tobacco companies are betting on the fact that most young people consider themselves to be immortal and do not think any bad effects of smoking would ever happen to them.

4. To Associate Cigarette Smoking with Becoming an Adult. Realizing that teen-agers desire to be considered adults, to be free to make their own decisions, and to be free from restrictions on what they can or cannot do, the tobacco companies go to great lengths to stress that smoking is an "adult habit"—that only adults have the right to choose whether or not to smoke. Since teenagers are in a hurry to grow up and be free, the simple act of smoking cigarettes can become their perfect way to show to the world that they are adult.

5. To Associate Cigarette Smoking with Attractiveness to the Opposite Sex. Many ads for cigarettes imply that if you smoke, you will also be attractive to members of the opposite sex. In fact, surveys of young people and adults show that most people prefer to date nonsmokers.

6. To Associate Smoking with Women's Liberation. "You've come a long way baby" is the theme of the ads for Virginia Slims cigarettes. What the ads don't say is that women who smoke like men will die like men who smoke. The slogan "Torches of Freedom" coupled with women smoking cigarettes while marching down Fifth Avenue in the Easter Parade, was a cigarette company's public relations ploy years ago. In the 1990s, lung cancer has become the number-one cancer found in women, exceeding the incidence of breast cancer.

7. To Show That Smoking Is an Integral Part of Our Society. The sheer number of cigarette ads—on billboards, on articles of clothing, on signs at sporting events—leave the impression that smoking is socially acceptable by the majority of people. Many events sponsored by tobacco companies include the name of a major brand of cigarettes or smokeless tobacco, such as the Kool Jazz Festival, the Benson & Hedges Blues Festival, the Magna Custom Auto Show, the Winston Cup (stock car racing), the Marlboro Cup (soccer), the Marlboro Stakes (horse racing), and the Virginia Slims Tennis Tournament, just to name a few.

Although tobacco advertising is legally prohibited on television, the ban is ignored by the strategic placement of tobacco-product ads in baseball and football stadiums, basketball arenas, and hockey rinks, around auto racetracks, and at tractor pulls and other sporting events. In *USA Today*, it was reported that tobacco company sponsorship of sports and other cultural events amounted to 150 million dollars in 1989, with about 70 percent spent on sports. Dr. Alan Blum, of Doctors Ought to Care, an organization critical of the tobacco industry and its advertising and promotional strategies, stated, "You watch an auto race and you see cars going around tracks with cigarette logos. You watch tennis and you see the logos in the background."

8. To Discourage Articles in Magazines about Health Risks of Smoking. So important to magazines and newspapers are the ads they carry for cigarettes, beer, food, and other products, which are marketed by the major tobacco companies or their parent companies, that many publishers are very reluctant to antagonize them by running articles on the health risks of smoking.

9. To Gain Legitimacy. This tobacco companies do by supporting groups and programs. Many groups receive significant amounts of funding from the tobacco companies to support their programs. One especially large grant, from RJR Nabisco, was a contribution of 30 million dollars for "innovative education programs" to schools across the country. In 1989, Philip Morris made arrangements to sponsor the Philip Morris Bill of Rights Exhibit to tour the United States in celebration of the 200th anniversary of the Bill of Rights. In this way, Philip Morris tried to associate its companies—including its tobacco company—with the Bill of Rights, to reap positive press coverage as the exhibit went on display in each city.

HISTORY OF TOBACCO ADVERTISING AND PROMOTION

Cigarettes constitute the most heavily advertised category of products in outdoor advertising (e.g., on billboards), the second most heavily advertised cat-

egory in magazines, and the third most heavily advertised subcategory in newspapers. Expenditures for tobacco ads increased consistently from 1970 to 1990. The expenditures totaled 361 million dollars in 1970, 1,242 million dollars in 1980, and 3,200 million dollars in 1990.

The tobacco companies' diverse promotional activities include the distribution of free product samples, coupons for price reductions, and offerings of discounted products that bear the name of cigarette brands. Promotional activities also encompass the industry sponsorship of cultural, sporting and entertainment events as well as sponsorship of community or political organizations. Incentives paid to retailers or distributors are another form of promotion. Since 1975, promotional expenditures have steadily increased from 25 percent of the tobacco marketing dollars to more than 60 percent in 1990. According to *Advertising Age*, the R.J. Reynolds Tobacco Company's plans for 1992 called for reducing their magazine and outdoor ads in favor of more free-gifts-with-purchase and buy-two-get-one-free deals, because of increased competition from low-price generic brands of cigarettes.

There were no restrictions on cigarette advertising in the United States until the first *Report of the Surgeon General* was released on January 11, 1964. Because of the health hazards described therein, the report led to the Federal Cigarette Labeling and Advertising Act of 1965, and beginning in 1966, a congressional mandated health warning appeared on all cigarette packages, but not on advertisements. On June 2, 1967, the Federal Communications Commission (FCC) ruled that the Fairness Doctrine in advertising applied to cigarette ads on television and radio and required broadcasters who aired cigarette commercials to provide "a significant amount of time" to citizens who wished to point out that smoking "may be hazardous to the smoker's health." The rule went into effect on July 1, 1967. The FCC required that there be one free public-service announcement (PSA) for every three paid cigarette commercials. During the three-year period of 1968 to 1970, in which PSA's were mandated by the Fairness Doctrine, per capita cigarette sales decreased by 6.9 percent.

In January 1970, the cigarette industry offered voluntarily to end all cigarette advertising on television and radio by September 1970—a move that would also eliminate any PSA's, which were hurting

sales. Ultimately, Congress approved the Public Health Cigarette Smoking Act of 1969, which prohibited cigarette advertising in the broadcast media as of January 1, 1971.

In September 1973, the Little Cigar Act of 1973 banned broadcast advertising of little cigars (cigarette-sized cigars). During the three-year period of 1971 to 1973, following the end of the PSAs required by the Fairness Doctrine and the beginning of the broadcast advertising ban, cigarette sales increased by 4.1 percent.

Over a decade later, smokeless tobacco advertising in the broadcast media was banned by the Comprehensive Smokeless Tobacco Health Education Act of 1986. This ban took effect on August 27, 1986. The Federal Trade Commission (FTC) Bureau of Consumer Protection ruled in 1991 that the Pinkerton Tobacco Company violated the 1986 statute banning the advertising of smokeless tobacco and prohibited it from "displaying its brand name, logo, color or design during televised (sports) events" of its Red Man Chewing Tobacco and Snuff. This was the first action of its kind by the FTC.

EXPOSING UNETHICAL TOBACCO ADVERTISING

STAT (Stop Teenage Addiction to Tobacco), at their 1991 STAT-91 Conference, addressed the problem of the tobacco companies' efforts to encourage tobacco addiction in young people. It was learned that the RJR Nabisco cartoon camel may have become the center of the most viciously effective advertising campaign ever created to influence the values and behavior of young people. Camel's share of the teenage market has been boosted from almost nothing to almost 35 percent in just three years by "this sleazy dromedary."

It might be instructive to compare the cigarette companies' current U.S. advertising programs to the Tobacco Institute's December 1990 *Cigarette Advertising and Promotion Code*, which follows.

TOBACCO INSTITUTE'S CIGARETTE ADVERTISING AND PROMOTION CODE

Cigarette smoking is an adult custom. Children should not smoke. Laws prohibiting the sale of cigarettes to minors should be strictly enforced. The cigarette man-

ufacturers advertise and promote their products only to adult smokers. They support the enactment and enforcement of state laws prohibiting the sales of cigarettes to persons under 18 years of age.

The cigarette manufacturers have adopted the following Code to emphasize their policy that smoking is solely for adults.

Advertising

1. Cigarette advertising shall not appear—

   (a) in publications directed primarily to those under 21 years of age, including school, college or university media (such as athletic, theatrical or other programs), comic books or comic supplements; or

   (b) on billboards located within 500 feet of any elementary school, junior high school, or high school or any children's playground.

2. No payment shall be made by any cigarette manufacturer or any agent thereof for the placement of any cigarette, cigarette package, or cigarette advertisement as a prop in any movie produced for viewing by the general public.

3. No one depicted in cigarette advertising shall be or appear to be under 25 years of age.

4. Cigarette advertising shall not suggest that smoking is essential to social prominence, distinction, success or sexual attraction, nor shall it picture a person smoking in an exaggerated manner.

5. Cigarette advertising may picture attractive, healthy looking persons provided there is no suggestion that their attractiveness and good health is due to cigarette smoking.

6. Cigarette advertising shall not depict as a smoker anyone who is or has been well known as an athlete, nor shall it show any smoker participating in, or obviously just having participated in, a physical activity requiring stamina or athletic conditioning beyond that of normal recreation.

7. No sports or celebrity testimonials shall be used or those of others who would have special appeal to persons under 21 years of age.

Sampling

1. Persons who engage in sampling shall refuse to give a sample to any person whom they know to be under 21 years of age or who, without reasonable identification to the contrary, appears to be less than 21 years of age.

2. Sampling shall not be conducted in or on public streets, sidewalks or parks, except in places that are open only to persons to whom cigarettes lawfully may be sold.

3. Cigarette product samples shall not otherwise be distributed in any public place within two blocks of any centers of youth activities, such as playgrounds, schools, college campuses, or fraternity or sorority houses.

4. The mails shall not be used to distribute unsolicited cigarette samples.

5. Cigarette samples shall not be distributed by mail without written, signed certification that the addressee is 21 years of age or older, a smoker, and wishes to receive a sample.

6. Cigarette samples shall not be distributed in direct response to requests by telephone.

7. Persons who engage in sampling shall not urge any adult 21 years of age or older to accept a sample if the adult declines or refuses to accept such sample.

8. Persons who engage in sampling shall indicate by oral or written means that samples are intended only for smokers.

9. No cigarette samples shall be distributed by a sampler in a public place to any person in a vehicle.

10. Persons distributing cigarette samples shall secure their stocks of samples in safe locations to avoid inadvertent distribution of samples contrary to these provisions.

11. Persons distributing cigarette samples shall avoid blocking or otherwise significantly impairing the flow of pedestrian traffic.

12. In the event that circumstances arise at a particular location that make it unlikely that sampling can be conducted in a manner consistent with the provisions of this Code, sampling shall be stopped at that location until circumstances abate.

13. Persons distributing samples shall promptly dispose of empty sample boxes and shall take rea-

sonable steps to ensure that no litter remains in the immediate area of sampling as a result of sampling activities.

Other Promotional Activities

1. There shall be no mail distribution of nontobacco premium items bearing cigarette brand names, logos, etc., without written, signed certification that the addressee is 21 years of age or older, a smoker, and wishes to receive the premium.

2. There shall be no other distribution of nontobacco premium items bearing cigarette brand names, logos, etc., except with the purchase of a package or carton of cigarettes or to persons 21 years of age or older.

3. Clothing bearing cigarette brand names or logos shall be in adult sizes only.

Definitions

1. "Advertising" means all forms of advertising including vehicle decals, posters, pamphlets, matchbook covers, and point-of-purchase materials in the United States, Puerto Rico, and U.S. territorial possessions.

2. "Sampling" means giving or distributing without charge packages of cigarettes in a public place for commercial advertising purposes ("cigarette samples"), but does not include isolated offerings of complimentary packages or the distribution of such packages to wholesale or retail customers or to company shareholders or employees in the normal course of business.

3. "Public Place" includes any street, sidewalk, park, plaza, public mall, and the public areas of shopping centers and office buildings.

After reviewing the Tobacco Institute's *Cigarette Advertising and Promotion Code* and comparing it to the actual practices of the tobacco companies, many would conclude that the tobacco companies make a less than serious effort to observe the code. This is unfortunate, since in 1994, the Food and Drug Administration testified before the U.S. Congress about the appropriateness of regulating tobacco as a drug. The logic rested on the evidence that tobacco is smoked for the nicotine it delivers and that the nicotine in cigarettes is a potent psychoactive drug. Moreover, it is a drug that causes addiction.

In his memoirs, former Surgeon General C. Everett Koop said this about the tobacco industry: "After studying in depth the health hazards of smoking, I was dumbfounded—and furious. How could the tobacco industry trivialize extraordinarily important public-health information: the connection between smoking and heart disease, lung and other cancers, and a dozen or more debilitating and expensive diseases? The answer was—it just did. The tobacco industry is accountable to no one."

(SEE ALSO: *Advertising and the Alcohol Industry; Tobacco: Dependence; Tobacco: Industry*)

CHARLES M. RONGEY

AFRICAN AMERICANS   *See* Ethnic Issues and Cultural Relevance in Treatment; Ethnicity and Drugs; Vulnerability As Cause of Substance Abuse: Race.

AGGRESSION AND DRUGS, RESEARCH ISSUES   ALCOHOL, narcotics, HALLUCINOGENS, and PSYCHOMOTOR STIMULANTS differ markedly one from the other in terms of pharmacology and neurobiological mechanisms, dependence liability, legal and social restraints, expectations and cultural traditions. No general and unifying principle applies to all these substances, and it would be misleading to extrapolate from the conditions that promote violence in individuals under the influence of alcohol to those with other drugs. Different types of drugs interact with aggressive and violent behavior in several direct and indirect ways from (1) direct activation of brain mechanisms that control aggression, mainly in individuals who have already been aggressive in the past; (2) drug states, such as alcohol or hallucinogen intoxication, serving as license for violent and aggressive behavior; (3) drugs such as heroin or cocaine serving as commodities in an illegal distribution system, drug trafficking, that relies on violent enforcement tactics; to (4) violent behavior representing one of the means by which an expensive cocaine or heroin habit is financed. Systematic experimental studies in animals represent the primary means to investigate the proximal and distal *causes* of aggressive behavior, whereas studies in humans most often attempt to infer causative relation-

ships mainly from *correlating* the incidence of violent and aggressive behavior with past alcohol intake or abuse of other drugs. The ethical dilemma of research on aggression in animals and humans is the demand for reducing harm and risk to the research subject, on the one hand, and on the other, to validly capture the essential features of human violence that is by definition injurious and harmful.

Methodologically, aggression research stems from several scientific roots, the experimental-psychological, ethological, and neurological traditions being the most important. The use of aversive environmental manipulations in order to produce defensive and aggressive behavior has been the focus of the experimental-psychological approach. During the 1960s, "models" of aggression were developed that rely upon prolonged isolated housing or crowding; exposure to noxious, painful electrical shock pulses; omission of scheduled rewards; or restricted access to limited food supplies as the major aversive environmental manipulations. The behavioral endpoints in these models are defensive postures and bites in otherwise placid, domesticated laboratory animals. The validity of such experimental preparations in terms of the ethology of the animal, i.e. how animals normally react outside of the laboratory, and in their relation to human aggressive and violent behavior remains to be determined. Aggression research using human subjects studied under controlled laboratory conditions has employed aversive environmental manipulations, such as the administration of electric shocks, noxious noise, or loss of prize money to a fictitious opponent. This type of experimental aggression research highlights the dilemma of attempting to model essential features of valid violence under controlled laboratory conditions without risking the harm and injury that are characteristic of such phenomena. While it is unethical to demand experimental studies that involve "realistic" violent behavior, the relation between competitive behavior in laboratory situations to violence outside the laboratory remains to be validated.

In addition to environmental manipulations, histopathological findings of brain tumors in violent patients prompted the development of experimental procedures that ablate and destroy tissue in areas of the brain such as the septal forebrain, medial hypothalamus, or certain midbrain regions of laboratory rats and other animals. Such experimental manipulations most often result in ragelike defensive pos-

tures and biting, often called rage, hyperreactivity, hyperdefensiveness. Alternatively, electrical stimulation of specific brain regions can evoke predatory attack, aggressive and defensive responses in certain animal species. When animals are given very high, near-toxic amphetamine doses and similar drugs, bizarre, rage-like responses may emerge. Similarly, aggressive and defensive behavioral elements are induced by exposure to very high doses of hallucinogens and during withdrawal from opiates. The inappropriate context, the unusually fragmented behavioral response patterns, and the limitation to domesticated laboratory rodents make aggressive and defensive reactions that are induced by lesions, electrical brain stimulation, drugs and toxins difficult to interpret or generalize to the human situation.

In contrast to the emphasis on aversive environmental determinants or on neuropathologies, the ethological approach to the study of animal aggression has focused on adaptive forms of aggressive behavior. Defense of a territory, rival fighting among mature males during the formation and maintenance of a group, defense of the young by a female, and anti-predator defense are examples of these types of aggressive, defensive, and submissive behavior patterns, often referred to as agonistic behavior. Sociobiological analysis portrays these behavior patterns as having evolved as part of reproductive strategies ultimately serving the transmission of genetic information to the next generation. The focus on aggressive behavior as it serves an adaptive function in the reproductive strategies, however, complicates the extrapolation to violent behavior as it is defined at the human level. How the range of human violent acts relate to the various types of animal aggression and how they may share common biological roots remains to be specified.

How have these ethological, neurological and experimental-psychological research traditions contributed to our understanding of the link between drugs of abuse and alcohol to human aggression and violence? Epidemiological and criminal statistics link alcohol to aggressive and violent behavior in a human pattern large in magnitude, consistent over the years, widespread in types of aggressive and violent acts, massive in cost to individual, family, and society, and serious in suffering and harm. Systematic experimental studies have identified the early phase after a low acute (short-term) alcohol dose as a con-

dition that increases the probability of many types of social interactions, including aggressive and competitive behaviors, and high-dose alcohol intoxication as the condition most likely to be linked to many different kinds of violent activities. Yet, most alcohol drinking is associated with acceptable social behavior. This is because individuals differ markedly in their propensity to become intoxicated with alcoholic beverages and to subsequently engage in violent and aggressive behavior, rendering population averages poor representations of how alcohol causes individuals to behave violently. The sources for the individual differences may be genetic, developmental, social, and environmental. Genetic association between antisocial personality, possibly diagnosed with the aid of certain electrophysiological measures, and alcoholism remains to be firmly established. In the early 1990s, the neurobiological mechanisms of alcohol action for a range of physiological and behavioral functions began to be identified; it appears that the actions of alcohol on brain serotonin and the benzodiazepine/$GABA_A$ receptor complex are particularly relevant to alcohol's effects on aggressive and violent behavior. For example, studies in rodents and primates indicate that benzodiazepine-receptor antagonists prevent the aggression-heightening effects of alcohol. Similarly, the actions of alcohol on neuroendocrine events that control testosterone and adrenal hormones appear important in the mechanisms of alcohol's aggression-heightening effects. Among the environmental determinants of alcohol's effects on violence that are of paramount significance are social expectations and cultural habits as well as the early history of the individual in situations of social conflict. Impaired appraisal of the consequences, inappropriate sending and receiving of socially significant signals, disrupted patterns of social interactions are characteristic of alcohol intoxication that contribute to the violence-promoting effects. A particularly consistent observation is the high prevalence of alcohol in victims and targets of aggression and violence. In contrast to heroin and cocaine, since alcohol is not an illicit drug, its link to violence is not a characteristic of the economic distribution network for this substance.

Violence in the context of drug addiction is due largely to securing the resources to maintain the drug habit as well as to establishing and conducting the business of drug dealing. Neither animal nor human data suggest a direct, pharmacological asso-

ciation between violence and acute or chronic administration of opiates. Although measures of hostility and anger are increased in addicts seeking methadone treatment, these feelings usually do not lead to aggressive or violent acts. Rather, the tendency to commit violent crimes correlates with pre-addiction rates of criminal activity. However, experimental studies in animals point to the phase of withdrawal from chronic opiates as the most vulnerable period to be provoked to heightened levels of aggressive behavior. Nevertheless, although humans undergoing opiate withdrawal may experience increased feelings of anger, there is no evidence suggesting that they are more likely to become violent as a result.

The most serious link of amphetamine to violence is in individuals who, after taking intravenous amphetamine—most often chronically—develop a paranoid psychotic state during which they commit violent acts. Most psychiatric reports and police records do not support a psychiatric opinion of the early 1970s that "amphetamines, more than any other group of drugs, may be related specifically to aggressive behavior." The prevalence of violence by individuals who experience amphetamine paranoid psychosis may be less than 10 percent in general population samples and as high as 67 percent among individuals who showed evidence of psychopathology prior to amphetamine use. Low acute amphetamine doses can increase various positive and negative social behaviors; higher doses often lead to disorganizing effects on social interactions and to severe social withdrawal. At present, the neurobiological mechanisms for the range of amphetamine effects on aggressive and social behavior remain unknown.

There are surprisingly few pharmacological and psychiatric studies on cocaine's effects on aggression and violence; the available evidence points to psychopathological individuals who may develop the propensity to engage in violent acts. However, the far more significant problem is the violence associated with the supplying, dealing, and securing of crack-cocaine, as documented in epidemiological studies.

Most experimental studies with animals and humans, as well as most data from chronic users, emphasize that *Cannabis* preparations (e.g., marijuana, hashish) or the active agent tetrahydrocannabinol (THC) decrease aggressive and violent behavior. Owing to the relatively widespread access, lower cost,

and characteristic pattern of use, socioeconomic causes of violence in *Cannabis* dealing and procuring are less significant than they are with cocaine or heroin.

LSD is not of significance in the early 1990s, but older data suggest that certain psychopathological individuals who begin using LSD may engage in violent acts; however, this phenomenon is rare.

Phencyclidine (PCP) cannot be causally linked to violent or assaultive behavior in the population as a whole. Generally, personality traits and a history of violent behavior appear to determine whether or not PCP intoxication leads to violence. PCP violence is a relatively rare phenomenon, although when it occurs, it stands out by its highly unusual form and intensity. It depends on the individual's social and personal background.

The impact of genetic predispositions to be susceptible to becoming involved with dependence-producing drugs—such as alcohol, heroin, or cocaine—and to act violently has, as of yet, not been delineated in terms of specific neural mechanisms. Similarly, the modulating influences of learning, social modeling, or parental physical abuse on the neural substrate for drug action and for aggressive behavior have not been specified. Since these critical connections remain poorly understood, it is not possible at present to support specific modes of intervention on the basis of neurobiological data.

(SEE ALSO: *Alcohol: Psychological Consequences of Chronic Abuse; Crime and Alcohol; Crime and Drugs*)

BIBLIOGRAPHY

EVANS, C. M. (1986). Alcohol and violence: Problems relating to methodology, statistics and causation. In P. F. Brain (Ed.), *Alcohol and aggression*. London: Croom Helm.

MICZEK, K. A. (1987). The psychopharmacology of aggression. In L. L. Iversen, S. D. Iversen, & S. H. Snyder (Eds.), *New directions in behavioral pharmacology (Handbook of psychopharmacology*, vol. 19). New York: Plenum.

MICZEK, K. A., ET AL. (1994). Alcohol, drugs of abuse, aggression and violence. In A. Reis & J. Roth (Eds.), *Understanding and preventing violence*, vol. 3. Washington, DC: National Academy of Sciences Press.

KLAUS A. MICZEK

AGING, DRUGS, AND ALCOHOL One of the most overwhelming demographic developments of the twentieth century has been the enormous rise in worldwide population, including the survival of an estimated 600 million people aged 60 or older (Ikels, 1991). The increase in the proportion of elderly results from medical, economic, and social factors plus a decline in the birthrate. According to 1989 U.S. Bureau of the Census figures, those over 65 represented 12 percent of the U.S. population, and it is projected that this proportion will almost double by the year 2030—since the baby-boom generation, born after 1945, will reach 65 between 2010 and 2030.

This fastest growing segment—the elderly—uses pharmacological and health services more often than any other segment of the population (Brock, Guralnik, & Brody, 1990). Aging people are more susceptible to infectious disease, and they suffer from multiple chronic diseases and conditions that have progressed throughout their lifetimes. Some are the result of accidents and some are degenerative diseases. These include cancer; immune-function disease such as lupus; cardiovascular disease such as coronary-artery disease and stroke; endrocrine-function disease such as diabetes; bone and joint disease such as arthritis and osteoporosis; and respiratory disease such as emphysema. Like the rest of the population, they also suffer from psychiatric disorders, some of which may respond to medication. Hence physicians (sometimes multiple physicians) often prescribe multiple drugs for treatment. Although they comprised only 12 percent of the U.S. population in 1988, the aging accounted for 35 percent of prescription-drug expenditures (Health Care Financing Administration, 1990). Furthermore, the elderly—like the rest of the population—also take over-the-counter drugs such as aspirin or allergy tablets, may smoke tobacco, drink caffeine-laden and alcoholic beverages, and may even use illicit drugs. Because of certain age-related physiological changes affecting drug disposition, their responses to all drugs and drug interactions may differ from those in younger people (Montamat, Cusack, & Vestal, 1989).

The use and abuse of ALCOHOL (ethanol) is a public health problem. Among people 65 to 74, 42.5 percent report using some amount of alcohol; only 30 percent do after 75. About 6 percent of the elderly are considered heavy drinkers (more than two drinks

per day) but about 5 to 12 percent of men and 1 to 2 percent of women in their sixties are problem drinkers (Atkinson, 1984). Alcoholism and prescription-drug abuse do take their toll in physical, psychological, and social illness and eventual death among the elderly from either severe withdrawal symptoms, medical complications, or suicide. Psychotropic drugs (antipsychotics, antidepressants, anxiolytics, and sedative-hypnotics) are commonly prescribed for the elderly, and studies suggest that the BENZODIAZEPINES or other SEDATIVE-HYPNOTICS are the most commonly prescribed classes of these drugs. The effects of these drugs are additive to and interactive with those of alcohol. All these factors taken together—alcohol, old age, multiple diseases, and multiple medications—can lead to the toxic interactions of two or more drugs. The complexities of alcohol, age, and drug interactions are discussed in the sections that follow.

NORMAL AGING AND BODILY CHANGES

Today, there is great interest in gerontology, the study of aging, because there are now more older individuals in society than ever before, and the number is expected to rise dramatically. The present goal of gerontology is not necessarily to increase the life span but rather to increase the health span—that is, the number of years that an individual will enjoy the function of all body parts and processes. Aging comprises multiple ongoing processes; disease and disability are disruptions. Several factors once thought part of normal human aging have now been shown to be diseases. With aging, the immune system no longer performs as it once did. For example, the thymus gland, one of the central pacemakers of the immune system, diminishes in weight over a person's life span. It gradually decreases in size, and eventually most of it is replaced by fat and connective tissue. The incidence of cancer and autoimmune diseases increases among the elderly. An older person exhibits a weaker response to bacterial antigens and also produces autoantibodies (antibodies against his or her own tissues). Instead of defending against foreign parasites and aggressors, the body begins to produce antibodies against its own tissues (Weksler, 1990). This may signify that the immune system is no longer functioning normally. Perhaps these immune changes are responsible for the increased mor-

bidity and mortality of the elderly to infectious diseases. A decline in the hormonal system may affect many different organs of the body. For example, glucose intolerance and overt diabetes mellitus is a common development in older individuals. The pancreas makes insulin, but cells in the body are resistant to its effects. In addition to age-related changes in glucose tolerance, both thyroid-stimulating hormone produced in the pituitary and thyroid hormone secreted by the thyroid gland itself show a decline with advancing age. This process is functionally reflected in a steady decrease in the basal metabolic rate by as much as 20 percent from age 30 to age 70. Thus aging may result in part from the loss of hormonal activities and a decline in the functions they control.

With advancing age, homeostatic regulation tends to be less effective, resulting in slower reactions to stimuli, wider variations in function, and slower return to resting states. This decline in homeostasis is exhibited in a number of body systems. For example, the sensitivity of the baroreceptors, which help maintain a normal blood pressure by changing vascular tone and heart rate, declines with age. Likewise, the elderly are prone to hyperthermia and hypothermia because of an impaired ability to regulate body temperature. A large proportion of age-related gastrointestinal problems, such as constipation, are induced and aggravated by chronic laxative abuse, poor eating habits, inadequate fluid intake, and lack of exercise. Some elderly are not aware of the importance to general health of diet and exercise. For example, atherosclerotic diseases are less prevalent in populations that eat no meat and little fat.

For large populations, increased age is associated with increased variability in most dimensions of health. Thus it is difficult to discriminate between normal and abnormal states. Moreover, even those aging changes considered usual or normal within a defined population do not necessarily represent the general outcome for a particular aging individual.

AGING AND ALTERED DRUG RESPONSE

Some drugs act differently in older people than they do in young or middle-age people. The difference stems from age-related changes in PHARMACOKINETICS, a process that includes absorption, distribution, metabolism, and excretion of drugs (Vestal & Cusack, 1990).

All these factors can affect the levels of drugs in blood and target tissues. For example, with aging, the percent of water and lean tissue (mainly muscle) in the body decreases, while the percent of fat tissue increases. These changes can affect the distribution and the length of time that a drug stays in the body, as well as the amount that is absorbed by body tissues. One reason that alcohol has a greater dose-related effect in the elderly is that there is a smaller volume of distribution (total body water), resulting in higher blood alcohol levels than in the young (Vestal et al., 1977). Most drugs are eliminated from the body by metabolism in the liver and excretion by the kidney. To a limited extent metabolism occurs in other organs as well, including the gastrointestinal tract, kidneys, and lungs. Although the intrinsic activity of drug-metabolizing enzymes in general does not decline with age, liver mass as a percentage of body weight and blood flow through the liver both decrease with aging (Loi & Vestal, 1988). As a result, the overall capacity of the liver to convert some drugs to their inactive metabolites declines with age. For example, some studies show that drugs such as diazepam (Valium), alprazolam (Xanex), chlordiazepoxide (Librium), propranolol, valproic acid, lidocaine, and theophylline are metabolized at a slower rate in older people than in young people. This is highly variable, however, and not all drugs metabolized by the liver show an age-related decline in the rate of metabolism. In fact, the metabolism of alcohol by the liver does not decline with age (Vestal et al., 1977). The most consistent physiological change with aging is a decline in renal function. Both glomerular filtration rate and the renal blood flow decline with age. As a result, drugs that are substantially excreted by the kidney regularly exhibit decreased plasma clearance or accumulation in the elderly. This is particularly important for drugs with narrow therapeutic indices (a ratio of effective to toxic drug levels) such as digoxin, aminoglycoside antibiotics, lithium, and chlorpropamide (Greenblatt, Sellars, & Shader, 1982).

Another mechanism of age-related changes in the response to some drugs is an apparent change in receptor sensitivity. In general, drugs acting on the central nervous system produce an enhanced response. Any drug that affects alertness, coordination, and balance will likely cause more falls and accidents in elderly people than in younger people. Thus hangover effects of sedative-hypnotic drugs and other psychotropics (e.g., ANTIPSYCHOTICS, ANTIDEPRESSANTS, anxiolytics) are common and often more prominent in the elderly. The serious consequences of the hangover effects, such as falls and hip fractures, suggest, in part, that neurotransmitter receptors are more sensitive or supersensitive in the elderly. In contrast to psychotropic medications, the response of the heart to stimulation by catecholamines is diminished in the elderly. A larger dose of isoproterenol is needed to achieve the same increase in heart rate in the elderly as in the young (Vestal, Wood, & Shand, 1979).

AGING AND ADVERSE DRUG REACTIONS

In general, because of multiple and chronic diseases, geriatric patients often take multiple prescription and over-the-counter drugs. Persons over the age of 65 may take as many as seven or more prescription drugs in addition to some over-the-counter drugs. Multiple drug therapy predisposes the elderly to an increased risk of adverse drug reactions (ADRs). The overall incidence of ADRs in this age group is two to three times that of young adults. Although the results of studies vary, about 20 percent of all adverse drug reactions occur in the elderly (Korrapati, Loi, & Vestal, 1992); these may result from drug overuse or misuse, slowed drug metabolism or elimination because of age-related chronic diseases, intake of alcohol, and/or food-drug incompatibilities. Furthermore, ADRs are more severe than among young adults. Risk factors include female gender, living alone, multiple diseases, multiple drugs, poor nutritional status, and impaired sensorium. Some of the age-related physiological causes for increased plasma levels and examples of altered receptor sensitivity to drugs have already been discussed. The elderly who drink regularly, even if they are not alcoholic, place themselves at risk for drug-alcohol interactions. Thus, since the elimination capacity of both the kidneys and the liver can be reduced in old age, drugs should generally be taken at lower initial doses by geriatric patients.

AGING AND ALCOHOLISM

Alcohol is a drug that has addictive properties in susceptible individuals. Although its victims often do not recognize that alcoholism is a disease, it does meet the criteria for a disease. It has definite symptoms, is chronic, and is often progressive to fatal—

but it is treatable. It destroys its victims not only physically but mentally, emotionally, and spiritually. Many people with this disease die from accidents, suicide, or physical complications. In Western society, excessive consumption of alcohol and the smoking of cigarettes are two of the most insidious forms of drug abuse. Yet they are often considered socially acceptable. In the United States, two-thirds of all adults use alcohol occasionally. It is estimated that between 2 and 10 percent of individuals over the age of 60 suffer from heavy drinking that interferes with health and well-being. These persons by definition suffer from alcoholism (Jinks & Raschko, 1990). If cigarette smoking is excluded, alcoholism is by far the most serious drug problem in the United States and most other countries. These two forms of drug abuse cause thousands of premature deaths, and the cost of complications contributes billions of dollars to any nation's health expenditure.

Men in their sixties continue to drink at a rate that is almost equal to that of the prior decades of life, but problem drinking decreases in the mid-seventies. The prevalence of alcoholism and problem drinking is lower in women than in men. A large majority of male alcoholics have strong family histories of alcoholism, an early onset, and a fulminant course. These are the early-onset or problem drinkers or type II alcoholics (Atkinson, 1984). It is postulated that the disorder in this group has a strong genetic component. The other main group is comprised of later-onset alcoholics who may drink from grief, loneliness, or poor health. The many losses and stresses of later life make the elderly especially vulnerable to alcoholism and suicide (Schonfeld & Dupree, 1991). The principal risk factor for suicide in the elderly is depression, which is heightened by alcohol and drug abuse.

ALCOHOL AND ITS COMPLICATIONS IN THE ELDERLY

Health-care costs for a family with an alcoholic member are twice those for other families, and up to half of all emergency-room admissions are alcohol-related. Alcohol (ethanol) abuse contributes to the high health-care costs of elderly beneficiaries of government-supported health programs. In general, the medical complications of alcohol abuse observed in older individuals are the same as in younger alcoholics. These include alcoholic liver disease, acute and chronic pancreatitis, gastrointestinal (GI) bleeding and other GI tract diseases, and susceptibility to infections and other metabolic disturbances. The elderly tolerate gastrointestinal bleeding and infection less well than do younger individuals. They are particularly prone to vitamin deficiencies, protein/calorie malnutrition, anemia, osteopenia, disease of the central and peripheral nervous system, cardiomyopathy, and cancer. Finally, alcohol-induced peripheral neuropathy and cerebral degeneration will be superimposed on the normal loss of neurons that occurs with age.

A number of studies have shown that alcohol in moderate amounts is a useful therapeutic agent for the elderly and that it improves social interaction, alertness, and a variety of physical indices. Alcohol is primarily a depressant drug in the central nervous system (CNS), although in moderate doses it has mood-elevating effects that account for its popularity. The depression and disinhibition it causes contribute to feelings of relaxation, confidence, and euphoria. However, alcohol abuse can result in serious neurologic damage, including brain atrophy with dementia, cerebellar ataxia, and peripheral neuropathy. Large doses of alcohol cause inflammation of the stomach, pancreas, and intestine that can impair the digestion of food and the absorption of nutrients into the blood stream. The adult population appears less knowledgeable about the many adverse effects of alcohol on health than about the effects of smoking. For example, although many people recognize that heavy alcohol consumption often leads to cirrhosis of the liver, only about one-third are aware of the association between alcohol use and cancers of the mouth and throat. Alcohol use can impair the effectiveness of routine drug therapy or can create new medical problems requiring additional therapy. Excessive alcohol use in association with medications in the elderly can severely compromise and complicate a well-planned therapeutic program. Thus even casual use of alcohol may be a problem for the elderly, particularly if they are taking medications that interact with ethanol. Difficulties can also arise from the interaction of alcohol and over-the-counter (OTC) medications. The combination of alcohol and prescribed or OTC sleeping pills, for example, could decrease intellectual function by producing an organic brain syndrome; frequent results include confusion, falls, emotional lability, and adverse drug interactions.

PRECAUTIONS WHILE USING ALCOHOL

Patients with hepatic disease and GI ulcer disease should not use alcohol. Alcohol should be avoided by patients with alcoholic skeletal or cardiac myopathy. Clearly, it should be taken only in strict moderation or not at all. For older individuals who have no medical contraindications and take no drugs (prescription or over-the-counter) that interact with alcohol, one drink a day is a prudent level of alcohol consumption. In general, the use of alcohol in the presence of any particular disease or medication is a matter that the physician and patient must decide.

ALCOHOL AND DRUG ABUSE IN THE ELDERLY

Alcohol, itself a drug, mixes unfavorably with many other drugs, including those purchased over the counter. In addition, use of certain prescription drugs may intensify the older person's reaction to alcohol, leading to more rapid intoxication. Alcohol, when combined acutely with certain groups of drugs, can dangerously slow down performance skills such as driving and walking. It impairs judgment, and reduces alertness when taken with drugs such as antipsychotics, sedative-hypnotics, anxiolytics, opioid analgesics, antihistamines, and certain antihypertensive drugs (Table 1). Acutely ingested, alcohol impairs the clearance of some drugs by the liver. In contrast, chronically ingested alcohol induces the synthesis of enzymes, leading to accelerated metabolism and increased clearance of some drugs, including anticonvulsants, anticoagulants, and oral hypoglycemic agents. Thus these therapeutic drugs can become less effective, which increases the need for monitoring. Alcohol-drug interactions do not generally result in death. However, there is evidence for a contributory role of alcohol in drug-related fatalities. Anyone who drinks even moderately should ask a physician or pharmacist about possible drug–drug interactions.

It is very difficult to determine the actual incidence of combined drug and alcohol use by the elderly, but it is likely to be reasonably high for the following reasons: (1) The average adult over 65 takes two to seven prescription medicines daily in addition to over-the-counter medications; (2) Most

elderly persons do not view alcohol as a drug and assume that modest amounts of alcoholic beverages can do little harm to an already aged body; (3) Few elderly persons hold to the traditional notion that mixing alcohol and medications will have deleterious consequences. Certainly not every medication reacts with alcohol; however, a variety of drugs interact consistently. The most dangerous of these reactions occurs when alcohol is combined with another CNS depressant. Since alcohol itself is a potent CNS depressant, its use with antihistamines, barbiturates, sedative-hypnotics, or other psychotropic drugs produces additive or synergistic CNS depressant effects that may inhibit the sensorium as well as central and motor functions (Gerbino, 1982). In one study, diazepam, codeine, meprobamate (Equanil), and fluorazepam (Dalmane) were the top four agents responsible for drug-alcohol interactions (Jinks & Raschko, 1990). Antihistamines, including diphenhydramine (Benadryl), dimenhydrinate (Dramamine), and most cold medications and anticholinergics such as scopolamine, which are found in over-the-counter medications, can also cause confusion in the elderly. An important consideration in the elderly is the confused and altered behavior that so regularly follows excessive consumption of alcohol. Many times, elderly alcoholics show symptoms of falls, confusion, and self-neglect. Such changes may impair the elderly patient's ability to adhere to a prescribed treatment regimen, and the risks of mistakes or mishaps in dosage become far greater (Gerbino, 1982). Some of the well-described interactions are discussed in the following sections.

ALCOHOL AND ANALGESICS

Aspirin is the active ingredient in many over-the-counter arthritis pain formulas and in numerous nonprescription combination headache and minor pain products. The ability of aspirin to cause gastric inflammation, GI erosion, and frank GI bleeding is well recognized. Alcohol not only produces gastritis but also increases the risk of GI bleeding caused by aspirin and other nonsteroidal antiinflammatory drugs (Bush, Sholtzhauer, and Imai, 1991). Elderly people at high risk for bleeding should avoid regular use of either alcohol or aspirin. Chronic alcohol abuse can cause hepatotoxicity in a patient taking acetaminophen (Tylenol), probably because of en-

TABLE 1
Drug–Alcohol Interactions and Adverse Effects

Drug	Adverse effects with alcohol
Acetaminophen	Severe hepatotoxicity with therapeutic doses of acetaminophen in chronic alcoholics
Anticoagulants, oral	Decreased anticoagulant effect with chronic alcohol abuse
Antidepressants, tricyclic	Combined CNS depression decreases the psychomotor performance, especially in the first week of treatment
Aspirin and other nonsteroidal antiinflammatory drugs	Increased the possibility of gastritis and GI hemorrhage
Barbiturates	Increased CNS depression (additive effects)
Benzodiazepines	Increased CNS depression (additive effects)
Beta-adrenergic blockers	Masked signs of delirium tremens
Bromocriptine	Combined use increases the GI side effects
Caffeine	Possible further decreased reaction time
Cephalosporins	Antabuse-like reaction with some cephalosporins
Chloral hydrate	Prolonged hypnotic effect and adverse cardiovascular effects
Cimetidine	Increased CNS depressant effect of alcohol
Cycloserine	Increased alcohol effect or convulsions
Digoxin	Decreased digitalis effect
Disulfiram (Antabuse)	Abdominal cramps, flushing, vomiting, hypotension, confusion, blurred vision, and psychosis
Glutethimide	Additive CNS depressant effect
Guanadrel	Increased sedative effect and orthostatic hypotension
Heparin	Increased bleeding
Hypoglycemics, sulfonylurea	Acute ingestion-increased hypoglycemic effect of sulfonylurea drugs
	Chronic ingestion-decreased hypoglycemic effect of these drugs
	Antabuse-like reaction
Isoniazid	Increased liver toxicity
Ketoconazole	Antabuse-like reaction
Lithium	Increased lithium toxicity
Meprobamate	Synergistic CNS depression
Methotrexate	Increased hepatic damage in chronic alcoholics
Metronidazole	Antabuse-like reaction
Nitroglycerin	Possible hypotension
Phenformin	Lactic acidosis (synergism)
Phenothiazines	Additive CNS depressant activity
Phenytoin	Acutely ingested, alcohol can increase the toxicity of phenytoin
	Chronically ingested, alcohol can decrease the anticonvulsant effect of phenytoin
Quinacrine	Antabuse-like reaction
Tetracyclines	Decreased effect of doxycycline

Adapted from M.A. Rizack & C.D.M. Hillman (1987), Adverse interactions of drugs. In *The Medical Letter handbook of adverse drug interactions.* New York: Medical Letter.

zyme induction leading to the formation of toxic intermediary metabolites of acetaminophen.

ALCOHOL AND CENTRAL NERVOUS SYSTEM DEPRESSANTS

Much controversy exists as to whether the observed enhancement of CNS depression is simply additive or synergistic (greater than merely additive). When combined with CNS depressants, alcohol—even in small quantities—produces undesirable and sometimes dangerous additive effects. The interaction of alcohol with benzodiazepine drugs, however, may be much greater in the elderly than in the other age groups. This is especially true for diazepam (Valium) and chlordiazepoxide (Librium). Commonly observed side effects include hypotension, sedation, confusion, and CNS depression that may progress to respiratory depression. Two or more drinks can be enough to induce a drug-alcohol interaction with a CNS depressant medication (Hartford & Samorajski, 1982). Therefore, as a general rule, elderly patients should be instructed to refrain from alcohol while taking CNS depressant medications, including benzodiazepines, barbiturates, muscle relaxants, and antihistamines (both by prescription and in the form of over-the-counter cold remedies or sleeping aids). Alcohol increases the clinical effects of these drugs, which already are hazardous in a segment of the population with decreased agility and greater danger of serious complications from falls and accidents.

ALCOHOL AND PSYCHOTROPIC DRUGS

When alcohol is combined with psychotropic drugs such as antipsychotics and antidepressants, the effects are less predictable than with other drugs. Antipsychotic drugs inhibit the metabolism of alcohol and may markedly enhance its effects on the CNS in the elderly. Antidepressants exaggerate the response to alcohol and impair the motor skills. This could be a significant hazard in the elderly. Depression of the CNS may range from drowsiness to coma, because acute alcohol consumption may increase the CNS effects of antidepressants. Alcohol may also increase the risk of hypothermic reactions in the elderly taking tricyclic antidepressants. Hence the avoidance of alcohol in elderly patients taking any of these drugs is a prudent recommendation (Scott & Mitchell, 1988).

ALCOHOL AND OTHER DRUGS

Many elderly patients with adult-onset diabetes take orally effective antidiabetic agents (sulfonylureas). When alcohol is taken along with sulfonylureas, it may cause additive hypoglycemia, especially in patients with restricted carbohydrate intake. Another problem associated with this combination is the infrequent potential for an Antabuse-like reaction, which is usually mild, causing nausea, vomiting, headache, blurred vision, and flushing. However, symptoms of severe Antabuse-like reactions include tachycardia, abdominal distress, sweating, hypotensive episodes, myocardial infarction, and laceration of the esophagus induced by vomiting; psychosis may also occur, and fatal reactions have been reported. Concomitant use of a variety of other drugs (Table 2) can also lead to an Antabuse-like reaction. Cough medicines may contain a narcotic analgesic such as codeine in combination with antihistamines. When taken together with alcohol, these drugs are hazardous and can cause altered sensorium and respiratory depression. Despite the fact that cardiac disorders are very common in older individuals, few of those who suffer from these problems modify their

TABLE 2
Drugs Producing Antabuse-like Reactions with Alcohol

Disulfiram (Antabuse)
Hypoglycemic agents
  chlorpropamide (Diabinese)
  tolbutamide (Orinase)
Other drugs
  cefamandole (Mandole)
  cefmetazole (Zefazone)
  cefoperazone (Cefobid)
  cefotetan (Cefotan)
  chloramphenicol (Chloromycetin)
  furazolidone (Furoxone)
  griseofulvin (Fulvicin)
  ketoconazole (Nizoral)
  metronidazole (Flagyl)
  monoamine oxidase inhibitors (e.g., phenelzine
    and tranylcypromine)
  moxalactam (Moxam)
  procarbazine (Matulane)
  quinacrine (Atabrine)
Alcohol sensitizing mushrooms
  Coprinus atramentarius (inky cap mushroom)
  Clitocybe clavipes

drinking patterns. This may be dangerous, since as little as one cocktail can adversely affect cardiac efficiency in the presence of heart disease. For example, alcohol consumption in a person suffering from angina (cardiac pain on exertion) can mask the pain that might otherwise serve as a warning signal (Horowitz, 1975).

ALCOHOL AND ILLICIT DRUGS

Abuse of hallucinogens, illicit psychomotor stimulants and sedatives, and marijuana is uncommon in old age; use of these drugs is almost exclusively by longstanding opioid users and aging criminals. The low incidence of this type of substance abuse in old age may result from early mortality and underreporting. However, problem drinkers may abuse drugs such as sedatives, opioids, marijuana, and amphetamines. Sometimes these drugs are used in combination with alcohol; at other times, such drugs are taken in preference to alcohol, and alcohol is used only when the drug of choice is not available.

SUMMARY

Elderly people are the fastest-growing segment of world population and consume about 25 percent of all the drugs prescribed. Their capacity to handle drugs differs from the young because of age-related physiological changes in various systems of the body. Alcohol abuse among older people (or any other) can lead to falls, fractures, and other similar medical complications. The addition of medications (pre-

scription and over-the-counter) to alcohol drinking can lead to disastrous complications and even death. However, in the absence of any contraindications or concomitant medications, a small quantity of alcohol consumption may be beneficial in some elderly persons. In case of doubt, the elderly and their families or caregivers are encouraged to seek the advice of the pharmacist or family physician and to follow the guidelines given in Table 3.

(SEE ALSO: *Social Costs of Alcohol and Drug Abuse*)

BIBLIOGRAPHY

ATKINSON, R. M. (1984). Substance use and abuse in late life. *Alcohol and drug abuse in the old age.* Washington, D.C.: American Psychiatric Press.

BROCK, D. B., GURALNIK, J. M., & BRODY, J. A. (1990). Demography and epidemiology of aging in the United States. In E. L. Schneider & J. N. Rowe (Eds.), *Handbook of the biology of aging*, 3rd ed. San Diego: Academic Press.

BUSH, T. M., SHLOTZHAUER, T. L., & IMAI, K. (1991). Nonsteroidal anti-inflammatory drugs: Proposed guidelines for monitoring toxicity. *Western Journal of Medicine, 155*(1), 39–42.

DUFOUR, M. C., ARCHER, L., & GORDIS, E. (1992). Alcohol and the elderly. *Clinical Geriatric Medicine, 8*(1), 127–141.

GERBINO, P. P. (1982). Complications of alcohol use combined with drug therapy in the elderly. *Journal of the American Geriatric Society, 30*(11s), 88–93.

GREENBLATT, D. J., SELLARS, E. M., & SHADER, R. I. (1982). Drug disposition in old age. *New England Journal of Medicine, 306*(18), 1081–1088.

TABLE 3
Guidelines for Use of Alcohol by the Elderly

- Elderly patients are advised to avoid alcohol consumption just before going to bed in order to avoid sleep disturbances.
- Because of the potential for alcohol–drug interaction, alcohol ingestion should be avoided before driving.
- Abstinence from alcohol by elderly patients receiving CNS depressants, analgesics, anticoagulants, antidiabetic drugs, and some cardiovascular drugs is recommended.
- A doctor or pharmacist should be consulted about alcohol–drug interactions.
- Any side effect or loss of energy should be immediately reported to the physician.
- Older individuals who want to drink, have no medical contraindications, and take no medications that interact with alcohol may consider one drink a day to be a prudent level of alcohol consumption. Alcohol when taken in moderation may be useful.

Adapted from M. C. Dufour, L. Archer, & E. Gordis (1992), Alcohol and the elderly. *Clinical Geriatric Medicine, 8*(1), 127–141.

HARTFORD, J. T., & SAMORAJSKI, T. (1982). Alcoholism in the geriatric population. *Journal of the American Geriatric Society, 30*(1), 18–24.

HEALTH CARE FINANCING ADMINISTRATION, OFFICE OF NATIONAL COST ESTIMATES. (1990). National health expenditures, 1988. *Health Care Financing Review, 11*, 1–41.

HOROWITZ, L. D. (1975). Alcohol and heart disease. *Journal of the American Medical Association, 232*(9), 959–960.

IKELS, C. (1991). Aging and disability in China: Cultural issues in measurement and interpretation. *Social Science Medicine, 32*(6), 649–665.

JINKS, M. J., & RASCHKO, R. R. (1990). A profile of alcohol and prescription drug abuse in a high-risk community-based elderly population. *Drug Intelligence Clinical Pharmacology, 24*(10), 971–975.

KORRAPATI, M. R., LOI, C. M., & VESTAL, R. E. (1992). Adverse drug reactions in the elderly. *Drug Therapy, 22*(7), 21–30.

LOI, C. M., & VESTAL, R. E. (1988). Drug metabolism in the elderly. *Pharmacological Therapy, 36*, 131–149.

MONTAMAT, S. C., CUSACK, B. J., & VESTAL, R. E. (1989). Management of drug therapy in the elderly. *New England Journal of Medicine, 321*(5), 303–309.

RIZACK, M. A., & HILLMAN, C. D. M. (1987). Adverse interactions of drugs. In *The Medical Letter handbook of adverse drug interactions*. New York: Medical Letter.

SCHONFELD, L., & DUPREE, L. W. (1991). Antecedents of drinking for early and late-onset elderly alcohol abusers. *Journal of the Study of Alcoholism, 52*(6), 587–592.

SCOTT, R. B., & MITCHELL, M. C. (1988). Aging, alcohol, and the liver. *Journal of the American Geriatric Society, 36*(3), 255–265.

SPENCER, G. (1989). Projections of the population of the United States, by age, sex, and race: 1988 to 2080. *Current Population Reports*, series P-25, no. 1018, U.S. Bureau of the Census. Washington, D.C.: U.S. Government Printing Office.

VESTAL, R. E., & CUSACK, B. J. (1990). *Pharmacology and aging*. In E. L. Schneider & J. W. Rowe (Eds.), *Handbook of the biology of aging*, 3rd ed. San Diego: Academic Press.

VESTAL, R. E., WOOD, A. J. J., & SHAND, D. J. (1979). Reduced β-adrenoreceptor sensitivity in the elderly. *Clinical Pharmocology Therapy, 26*(2), 181–186.

VESTAL, R. E., ET AL. (1977). Aging and ethanol metabolism. *Clinical Pharmacologic Therapy, 21*(3), 343–354.

WEKSLER, M. E. (1990). Protecting the aging immune system to prolong quality of life. *Geriatrics, 45*(7), 72–76.

MADHU R. KORRAPATI
ROBERT E. VESTAL

AGONIST　An agonist is a drug or an endogenous substance that binds to a RECEPTOR (it has affinity for the receptor binding site) and produces a biological response (it possesses intrinsic activity). The binding of a drug agonist to the receptor produces an effect that mimics the physiological response observed when an endogenous substance (e.g., hormone, NEUROTRANSMITTER) binds to the same receptor. In many cases, the biological response is directly related to the concentration of the agonist available to bind to the receptor. As more agonist is added, the number of receptors occupied increases, as does the magnitude of the response. The potency (strength) of the agonist for producing the physiological response (how much drug is needed to produce the effect) is related to the strength of binding (the affinity) for the receptor and to its intrinsic activity. Most drugs bind to more than one receptor; they have multiple receptor interactions.

(SEE ALSO: *Agonist/Antagonist (Mixed); Antagonist*)

BIBLIOGRAPHY

ROSS, E. M. (1990). Pharmacodynamics: Mechanisms of drug action and the relationship between drug concentration and effect. In A. G. Gilman et al (Eds.), *Goodman and Gilman's the pharmacological basis of therapeutics*, 8th ed. New York: Pergamon

NICK E. GOEDERS

AGONIST-ANTAGONIST (MIXED)　A mixed agonist-antagonist is a drug or receptor ligand that possesses pharmacological properties similar to both AGONISTS and ANTAGONISTS for certain RECEPTOR sites. Well-known mixed agonist-antagonists are drugs that interact with OPIOID (morphine-like) receptors. Pentazocine, nalbuphine, butorphanol, and BUPRENORPHINE are all mixed agonist-antagonists for opioid receptors. These drugs bind to the μ (mu) opioid receptor to compete with other substances (e.g., MORPHINE) for this binding site; they either block the binding of other drugs to the μ receptor (i.e., competitive antagonists) or produce a much smaller effect than that of "full" agonists (i.e., they are only partial agonists). Therefore, these drugs block the effects of high doses of morphine-like drugs at μ opioid receptors, while producing partial agonist effects at κ (kappa) and/or δ (delta) opioid

receptors. Some of the mixed opioid agonist-antagonists likely produce analgesia (pain reduction) and other morphine-like effects in the CENTRAL NERVOUS SYSTEM by binding as agonists to κ opioid receptors.

In many cases, however, there is an upper limit (ceiling) to some of the central nervous system effects of these drugs (e.g., respiratory depression). Furthermore, in people physically addicted to morphine-like drugs, the administration of a mixed opioid agonist-antagonist can produce an abstinence (WITHDRAWAL) syndrome by blocking the μ opioid receptor and preventing the effects of any μ agonists (i.e., morphine) that may be in the body. Pretreatment with these drugs can also reduce or prevent the euphoria (high) associated with subsequent morphine use, since the μ opioid receptors are competitively antagonized. Therefore, the mixed opioid agonist-antagonists are believed to have less ABUSE LIABILITY than full or partial opioid receptor agonists.

As more and more subtypes of receptors are discovered in other NEUROTRANSMITTER systems (there are now more than five serotonin receptor subtypes and five dopamine receptor subtypes), it is quite likely that mixed agonist-antagonist drugs will be identified that act on these receptors as well.

BIBLIOGRAPHY

GILMAN, A. G., ET AL. (EDS.). (1990). *Goodman and Gilman's the pharmacological basis of therapeutics*, 8th ed. New York: Pergamon.

NICK E. GOEDERS

AIDS *See* Alcohol and AIDS; Complications: Route of Administration; Injecting Drug Users and HIV; Substance Abuse and AIDS; Needle and Syringe Exchanges and HIV/AIDS.

AL-ANON Al-Anon is a fellowship very similar to ALCOHOLICS ANONYMOUS (AA), but it is for family members and friends of alcoholics. Although formally totally separate from the fellowship of AA, it has incorporated into its groups the AA Twelve Steps and Twelve Traditions and AA's beliefs and organizational philosophy, but it has directed them toward helping families of alcoholics cope with the baffling and disturbing experiences of living in close interaction with an active alcoholic. In this sense, it is a satellite organization of AA (Rudy, 1986). Proselytizing organizations, such as AA, of necessity attempt to reduce, even eliminate, the ties of newcomers with other significant persons and groups who are not members. Rather than attempt to sever those bonds for prospective AA members, AA evolved Al-Anon as a way to include families into a parallel organization, and thus also initiate them into the beliefs and practices of AA. In addition, as AA expanded and more alcoholics became "recovering" ones, close relatives became aware that their own personal problems could be reduced by applying AA principles to themselves and working the Twelve Step program, even though they were not alcoholic. In 1980, there were 16,500 Al-Anon groups worldwide, including 2,300 ALATEEN groups of children of alcoholics (Maxwell, 1980).

BRIEF HISTORY

Early in the 1940s, wives started attending AA meetings and soon began to informally meet together. By the late 1940s, there were so many family members at AA affairs that the AA Board of Trustees had to decide how to manage this valuable but perplexing influx. Since relatives of AA members had already begun to hold their own meetings, the board recommended that AA meetings be only for alcoholics but that whenever family members asked to participate they should be listed at the AA General Service office as a resource. Several AA wives began their own clearinghouse committee to coordinate the approximately ninety groups already in existence. Soon there was a separate network distinct and apart from AA itself. Because they were closely related to AA, however, they decided to shorten the first two letters of "alcoholic," and the first four letters of "anonymous" into Al-Anon. This has been their name ever since.

In 1950, the anonymous Bill W., a founder of AA, persuaded his wife Lois to get involved with the fledgling Al-Anon. The rapidly accumulating lists in the General Service office were turned over to her and to an associate, Anne B., who contacted those on the list, "and soon they had more work than they could handle" (Wing, 1992:136). For two years, the two conducted their activities at Stepping Stones, the suburban home of Bill W. and Lois. In 1952 they moved to New York City, where volunteer workers

could be more easily recruited. By 1989, there were over 28,000 weekly Al-Anon Family Groups, which included Alateen, worldwide. "With each meeting averaging 12–15 members, an estimated total of 336,000–420,000 visits to Al-Anon meetings occur each week" (Cermak, 1988:92). Using data from a 1984 random sample of groups conducted by the Al-Anon fellowship, it was estimated that a quarter of a million people visit an Al-Anon meeting each week.

AL-ANON'S STRATEGY

Al-Anon strives to direct its members' attention away from the active alcoholic with whom they attempt to interrelate, and toward their own behavior and emotions. In many ways, their personalities resemble those of alcoholics: they repeatedly attempt to control the feelings and behaviors of the alcoholics in their midst by simple force of personal will— much as alcoholics attempt to control their drinking by the sheer force of their individual will. In both instances, a denial syndrome emerges in their emotional makeup that protects their compulsive drive toward continued control. In sum, family members often become codependent—as obsessed with the alcoholics' behavior as he or she is with the bottle (Huppert, 1976).

For example, the alcoholic's spouse or partner has often vainly attempted to control the drinking. Except for brief periods, most pleas have been rejected and most promises have not been kept. Often, while the alcoholic has continued to drink and enjoy the brief emotional payoffs of intoxication, the spouse or other caretaker must try to cope for both of them by running the household, rearing the children, and working steadily to earn a living. If the alcoholic does show signs of improvement in a treatment center, the spouse or life partner may resent it deeply, since strangers have done more in a short period than all the partner's efforts over the years. To alcoholics' partners and relatives it appears as if they have not been wise enough, or determined enough, or superhuman enough, to get the alcoholics in their lives to stop drinking.

Al-Anon attempts to introduce the Twelve Steps of AA into the lives of family members as a way of minimizing the resentments and obsessive-control behavior they typically display. Al-Anon emphasizes an adaptation of AA's first step: "We admit we are powerless to control an alcoholic relative, that we are

not self-sufficient." Such a step is an admission that it is a waste of time to try to control what is beyond their capacities. According to this strategy, it is no longer necessary for them to deny that their control efforts are powerless, and this relieves them from the enormous sense of accumulated burden and guilt. In addition, it allows acceptance of outsider treatment and its possible success.

(SEE ALSO: *Adult Children of Alcoholics; Codependence; Families and Drug Use; Treatment Types: Twelve Steps*)

BIBLIOGRAPHY

AL-ANON FAMILY GROUP HEADQUARTERS. (1984). *Al-Anon faces alcoholism* (2nd ed.) New York: Author.

CERMAK, T. (1989). Al-Anon and recovery. In M. Galanter (Ed.), *Recent developments in alcoholism*, Vol. 7 (pp. 91–104). New York: Plenum Press.

HUPPERT, S. (1976). The role of Al-Anon groups in the treatment program of a V.A. alcoholism unit. *Hospital and Community Psychiatry, 27*, 693–697.

MAXWELL, M. (1980). Alcoholics Anonymous. In E. Gomberg, H. R. White, & J. Carpenter (Eds.), *Alcohol, science, and society revisited* (pp. 295–305). Ann Arbor: University of Michigan Press.

RUDY, D. R. (1986). *Becoming alcoholic: Alcoholics Anonymous and the reality of alcoholism*. Carbondale: Southern Illinois University Press.

WING, N. (1992). *Grateful to have been there: My 42 years with Bill and Lois and the evolution of Alcoholics Anonymous*. Park Ridge, IL: Parkside Publishing.

HARRISON M. TRICE

ALATEEN Alateen is a division of the Al-Anon Family Group. Its members typically are teenagers whose lives have been impacted by someone else's problem drinking. Roughly, 59 percent are age 14 or younger, while 26 percent are ages 15 to 16 and 15 percent are age 17 or more. The problem drinkers in their lives are predominantly one or both parents, but brothers and sisters are not uncommon.

The prevailing story about the origin of Alateen is quite straightforward. Legend has it that in 1957 a 17 year old in California was attending ALCOHOLICS ANONYMOUS (AA) and AL-ANON meetings with his parents. His father had just gotten sober in AA and his mother was an active member of Al-Anon. Al-

though the teenager decided that the Twelve Steps of AA were helping him, his mother suggested that instead of attending AA meetings he start a teenaged group and pattern it after Al-Anon. The young man found five other teenaged children of alcoholic parents and, while the adult groups met upstairs, he goth them together downstairs.

As other teenagers came forward from Al-Anon groups, the idea spread and it is estimated that today about 3,500 Alateen groups meet worldwide. In formal terms, however, these groups are an important and an integral part of Al-Anon Family Groups. They are coordinated from the Al-Anon Family Group Headquarters in New York City and tied closely to their public-information programs. Thus, Alateen uses AA's Twelve Steps, but alters step twelve to simply read "carry the message to others," rather than "to other alcoholics." Alateen groups meet in churches and schoolrooms, often in the same building as Al-Anon, but in a different room.

Although there are a few exceptions, an active, adult member of Al-Anon usually serves as a sponsor. Also, members of Alateen can choose a personal sponsor from other Alateen members or from Al-Anon members.

Alateen enables its members to openly share their experiences and to devise ways of coping with the problem of living closely with a relative who has a drinking problem. The strategy is to change their own thinking about the problem-drinking relative. Alateen teaches that alcoholism is like diabetes—it cannot be cured, but it can be arrested. Members learn that they did not cause it, and they cannot control it or cure it. Scolding, tears, or persuasion, for example, are useless. Rather, "they learn to take care of themselves whether the alcoholic stops or not" (Al-Anon Family Groups, 1991:5). They apply the Twelve Steps to themselves—to combat their often obsessive thinking about controlling alcoholic relatives and to help them stop denying those relatives' alcoholism. In addition, they adapt and apply AA's Twelve Traditions to the conduct of their groups. For example, they practice anonymity, defining it not as secrecy, but as privacy and the lowering of competitiveness among members. A 1990 survey of Alateen members indicated an increase in the number of black, Hispanic, and other minority members.

In essence, Alateen uses the strategy of AA itself to learn how to deal with obsession, anger, feelings of guilt, and denials. Newcomers, like newcomers in

AA, gain hope when they bond with other teenagers to help one another cope with alcoholic parents and other relatives with drinking problems (Al-Anon Group Headquarters, Inc., 1973).

(SEE ALSO: *Adult Children of Alcoholics; Codependence; Families and Drug Use; Treatment Types: Twelve Steps*)

BIBLIOGRAPHY

AL-ANON FAMILY GROUPS. (1991). *Youth and the alcoholic parent.* New York: Author.

AL-ANON FAMILY GROUPS. (1973). *Alateen—hope for children of alcoholics.* New York: Author.

HARRISON M. TRICE

ALCOHOL   This section contains articles on some aspects of alcohol, and the following topics are covered: *Chemistry and Pharmacology; Complications; History of Drinking;* and *Psychological Consequences of Chronic Abuse.* For discussions of alcoholism, its treatment, and withdrawal symptoms, see the sections entitled *Alcoholism; Treatment;* and *Withdrawal.* See also the articles *Alcoholics Anonymous (AA)* and *Treatment Types: Twelve Steps.* Other articles on related topics are listed throughout the Encyclopedia.

Chemistry and Pharmacology   Alcohol is a general central nervous system depressant. In addition to drinking alcohol (ethanol, also called ethyl alcohol), there are many other alcohols—such as METHANOL, propanol, isopropyl, and so on—but ethanol was probably among the first substances to be misused by humans.

Alcoholic beverages were first consumed by prehistoric peoples as a result of natural fermentation—of honey, fruit, or grain. For thousands of years, fermented juices were commonly used for rituals, celebrations, and dining; in the civilizations of the Mediterranean (Near East, Egypt, Ancient Greece, and Rome) and in China, both wines and beers were produced in quantity for special occasions. Once the process of distillation was developed—about 800 A.D. by Jabir (or Geber) ibn Hayyan—alcoholic bev-

erages with a higher ethanol content were made, distilled off the fermented wines and beers that already existed (the original was the Dutch *brantwijn* of the 1600s). Eventually, special fermented mashes were made from local produce that, with multiple distilling and careful filtration, yielded high ethanol (high proof) concentrations. By the 1700s, whisky (today whiskey) from Scotland and Ireland were sold by the expanding British Empire; and American bourbon was being made in Kentucky. Today's liquor industry is based on carefully controlled and often blended products, which are sold to consumers with the cachet of "tradition" attached to each variation. Of the many types of alcohols, ethanol is the most intoxicating, the least toxic, and therefore the most commercially exploited. It is also the world's most widely used and abused psychoactive drug.

Social attitudes differ greatly among the world's societies as to the acceptable use of ethanol. Because it is kept in a great many households, the worldwide concept of a "drunk" has become far less threatening than that of an "addict"—an abuser of HEROIN, COCAINE, or AMPHETAMINES. In fact, most alcoholics are not typical "skid-row bums"; they come from all walks of life, and some are well-respected working citizens. For this reason, economists have estimated that the cost of alcoholism exceeds 25 billion dollars per year in the United States alone—with costs measured not only in loss of life but in loss of productivity and in medical fees as well.

CHEMISTRY

Ethanol has a very simple molecular structure, C_2H_6O. It is composed of only two carbon atoms, six hydrogen atoms, and one oxygen atom, yet its precise mechanism of action is not fully understood. Although it is commonly believed that ethanol is useful in a number of physical ailments (as medicinal alcohol, the medieval elixir of life), in reality its uses are not therapeutic—and its chronic use is toxic.

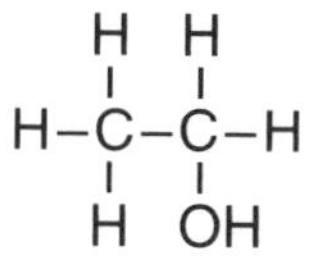

Figure 1
Ethyl alcohol

EFFECTS ON THE BODY AND THERAPEUTIC USES

Ethanol is a general central nervous system depressant, producing sedation and even sleep at higher doses. The degree of this depression is proportional to its concentration in the blood; however, this relationship is more predictable when ethanol levels are rising than three or four hours later, when blood levels are the same but ethanol levels are falling. This variance occurs because during the first fifteen or twenty minutes after an ethanol dose, the peripheral venous blood is losing ethanol to the tissues while the brain has equilibrated with arterial blood supply. Thus, brain levels are initially higher than the venous blood levels, and since all blood samples for ethanol determinations are taken from a peripheral vein, the ethanol concentrations are appreciably lower than a few hours later, when the entire system has achieved equilibrium.

The reticular activating system of the brain stem is the most sensitive area to ethanol's effects; this accounts for the loss of integrative control of the brain's higher functions. Anecdotal reports of a stimulating effect, especially at low doses, are likely due to the depression of the mechanisms that normally control speech and other behaviors that evolved from training or prior experiences. However, there may be a genetic basis for this initial stimulating effect, since rodents differing genetically show differences in the degree of initial stimulation or excitement. Upon drinking a moderate amount of ethanol, humans may quickly pass through the "stimulating" phase. Memory, the ability to concentrate, and insight are affected next whereas confidence often increases as moods swing from one extreme to another. If the dose is increased, then neuromuscular coordination becomes impaired. It is at this point that drinkers may be most dangerous, since they are still able to move about but reaction times and judgment are impaired—and sleepiness must be fought. The ability to drive an automobile or operate machinery is compromised. With higher doses, general (sleep) or surgical (unconsciousness) anesthesia may develop, but respiration is dangerously depressed.

Ethanol is believed by many to have a number of medicinal (therapeutic) uses; these are mostly based on anecdotal reports and have few substantiated claims. One example of a well-known but misguided

use is to treat hypothermia—exposure to freezing conditions. Although the initial effects of an alcoholic beverage appear to "warm" the patient, ethanol actually dilates blood vessels, causing further loss of body heat. Another example is its effects on sleep—it is believed that a nightcap relaxes one and puts one to sleep. Acute administration of ethanol may decrease sleep latency, but this effect dissipates after a few nights. In addition, waking time during the latter part of the night is increased, and there is a pronounced rebound insomnia that occurs once the ethanol use is discontinued. Except as an emergency treatment to reduce uterine contractions and delay birth, the therapeutic use of oral ethanol is confined to treating poisoning from methanol and ethylene glycol. Most of ethanol's therapeutic benefits are derived from applying it to the skin, since it is an excellent skin disinfectant. Ethanol can lessen the severity of dermatitis, reduces sweating, cools the skin during a fever and, when added to ointments, helps other drugs penetrate the skin. These therapeutic uses for ethanol are for acute problems only.

Until recently, it had been felt that the chronic drinking of ethanol led only to organ damage. Recent evidence suggests that low or moderate intake of ethanol (1–2 drinks per day) can indirectly reduce the risk of heart attacks. The doses must be low enough to avoid liver damage. This beneficial effect is thought to be due to the elevation of high-density lipoprotein cholesterol (HDL-C) in the blood which, in turn, slows the development of arteriosclerosis and, presumably, heart attacks. This relationship has not been proven, but has been culled from the results of several epidemiological studies.

Several mechanisms have been proposed to explain how oral ethanol exerts its effects. One is thought to be its ability to alter the fluidity of cell membranes—particularly neurons. This disturbance alters ion channels in the membrane resulting in a reduction in the propagation of neuronal transmission. The anesthetic gases share this property with ethanol. Furthermore, it has been shown that the degree of membrane disordering is directly proportional to the drug's lipid solubility. It has also been argued that such membrane effects occur only at very high doses. More recently, scientists have reported that ethanol may augment the activity of the neurotransmitter GABA by its actions on a receptor site close to the GABA receptor. The effect of this action is to increase the movement of chloride across

biological membranes. Again, this effect would alter the degree to which neuronal transmission is maintained.

PHARMACOKINETICS AND DISTRIBUTION

Ethanol is quickly and rapidly absorbed from the stomach (about 20%) and from the first section of the small intestines (called the duodenum). Thus the onset of action is related in part to how fast it passes through the stomach. Having food in the stomach can slow absorption because the stomach does not empty its contents into the small intestines when it is full. However, drinking on an empty stomach leads to almost instant intoxication because the ethanol not absorbed in the stomach passes directly to the small intestines. Maximal blood levels are achieved about thirty to ninety minutes after ingestion. Ethanol mixes with water quite well, and so once it enters the body it travels to all fluids and tissues, including the placenta in a pregnant woman. After about twenty to thirty minutes for equilibration, blood levels are a good estimate of brain levels. Ethanol freely enters all blood vessels, including those in the small air sacs of the lungs. Once in the lungs, ethanol exchanges freely with the air one breathes, making a breath sample a good estimate of the amount of ethanol in one's body. A breathalyzer device is often used by police officers to detect the presence of ethanol in an individual.

Between 90 and 98 percent of the ethanol dose is metabolized. The amount of ethanol that can be metabolized per unit of time is roughly proportional to the individual's body weight (and probably the weight of the liver). Adults can metabolize about 120 mg/kg/hr which translates to about 30 ml (1 oz) of pure ethanol in about 3 hours. Women generally achieve higher alcohol blood concentrations than do men, even after the same unit dose of ethanol, because women have a lower percentage of total body water but also because they may have less activity of alcohol-metabolizing enzymes in the wall of the stomach. The enzymes responsible for ethanol and acetaldehyde metabolism—alcohol dehydrogenase and aldehyde dehydrogenase, respectively—are under genetic control. Genetic differences in the activity of these enzymes account for the fact that different racial groups metabolize ethanol and acetaldehyde at different rates. The best-known example

is that of certain Asian groups who have a less active variant of the aldehyde dehydrogenase enzyme. When they consume alcohol, they accumulate higher levels of acetaldehyde than do Caucasian males, for example; this causes a characteristic response called "flushing," actually a type of hot flash with reddening of the face and neck. Some experts believe that the relatively low levels of alcoholism in such Asian groups may be linked to this genetically based aversive effect.

TOXIC EFFECTS

Chronic consumption of excessive amounts of ethanol can lead to a number of neurological disorders, including altered brain size, permanent memory loss, sleep disturbances, seizures, and psychoses. Some of these neuropsychiatric syndromes, such as Wernicke's encephalopathy, Korsakoff's psychosis, and polyneuritis can be debilitating. Other, less obvious problems also occur during chronic ethanol consumption. The chronic drinker usually fails to meet basic nutritional needs and is often deficient in a number of essential vitamins, which can also lead to brain and nerve damage.

Chronic drinking also causes damage to a number of major organs. Permanent alterations in brain function have already been discussed. By far, one of the most important causes of death in alcoholics (other than by accidents) is liver damage. The liver is the organ that metabolizes ingested and body toxins; it is essential for natural detoxification. Alcohol damage to the liver ranges from acute fatty liver to hepatitis, necrosis, and cirrhosis. Single doses of ethanol can deposit droplets of lipids, or fat, in the liver cells (called hepatocytes). With an accumulation of such lipid, the liver's ability to metabolize other body toxins is reduced. Even a weekend drinking binge can produce measurable increases in liver fat. It was found that liver fats doubled after only two days of drinking; blood ethanol levels ranged between 20 and 80 mg/dl, suggesting that one need not be drunk in order to experience liver damage.

Alcohol-induced hepatitis is an inflammatory condition of the liver. The symptoms are anorexia, fever, and jaundice. The size of the liver increases, and its ability to cleanse the blood of other toxins is reduced. Cirrhosis is the terminal and most dangerous type of liver damage. Cirrhosis results after many years of intermittent bouts with hepatitis or other liver damage, resulting in the death of liver cells and the formation of scar tissue in their place. Fibrosis of the blood vessels leading to the liver can result in elevated blood pressure in the veins around the esophagus, which may rupture and cause massive bleeding. Ultimately, the cirrhotic liver fails to function and is a major cause of death among alcoholics. Although only a small percentage of drinkers develop cirrhosis, it appears that a continuous drinking pattern results in greater risk than does intermittent drinking, and an immunological factor may be involved.

The role of poor nutrition in the development of some of these disorders is well recognized but not very well understood. Ethanol provides 7.1 kilocalories of energy per gram. Thus, a pint of whiskey provides around 1,300 kilocalories, which is a substantial amount of raw energy, although devoid of any essential nutrients. These nutritional disturbances can exist even when food intake is high, because ethanol can impair the absorption of vitamins B_1 and B_{12} and folic acid. Ethanol-related nutritional problems are also associated with magnesium, zinc, and copper deficiencies. A chronic state of malnutrition can produce symptoms that are indistinguishable from chronic ethanol abuse.

Fetal alcohol syndrome (FAS) was recognized and described in the 1980s. Children of chronic drinkers are born deformed; the abnormality is characterized by reduced brain function as evidenced by a low IQ and smaller than usual brain size, slower than normal growth rates, characteristic facial abnormalities (widely spaced eyes and flattened nasal area), other minor malformations, and developmental and behavioral problems. Fetal malnutrition caused by ethanol-induced damage to the placenta can also occur, and fetal immune function appears to be weakened, resulting in the child's greater susceptibility to infectious disease. Depending on the population studied, the rate of FAS ranges from 1 in 300 to 1 in 2,000 live births; however, the incidence is 1 in 3 infants of alcoholic mothers. As of the mid-1990s, it is not known if there is a safe lower limit of ethanol that can be consumed by pregnant women without risk of having a child with FAS. The lowest reported level of ethanol that resulted in FAS was about 75 ml (2.5 oz) per day during pregnancy. Among alcoholic mothers, if drinking during pregnancy is reduced, then the severity of the resulting syndrome is reduced.

TOLERANCE, DEPENDENCE, AND ABUSE

Tolerance, a feature of many different drugs, develops rather quickly to many of ethanol's effects after frequent exposure. When tolerance develops, the dose must be increased to achieve the original effect. Ethanol is subject to two types of tolerance: tissue (or functional) tolerance and metabolic (or dispositional) tolerance. Metabolic tolerance is due to alterations in the body's capacity to metabolize ethanol, which is achieved primarily by a greater activity of enzymes in the liver. Metabolic tolerance only accounts for 30 to 50 percent of the total response to alcohol in experimental conditions. Tissue tolerance, however, decreases the brain's sensitivity to ethanol and may be quite extensive. The development of tolerance can take just a few weeks or may take years to develop, depending on the amount and pattern of ethanol intake. As with other central nervous system depressants, when the dose of ethanol is increased to achieve the desired effects (e.g., sleep), the margin of safety actually decreases, as the dose comes closer to producing toxicity and the brain's control of breathing becomes depressed.

Like tolerance, dependence on ethanol can develop after only a few weeks of consistent intake. The degree of dependence can be assessed only by measuring the severity of the withdrawal signs and symptoms observed when ethanol intake is terminated. Victor and Adams (1953) provided perhaps one of the best descriptions of the clinical aspects of ethanol dependence. Patients typically arrive at the hospital with the "shakes," sometimes so severe that they cannot perform simple tasks by themselves. During the next twenty-four hours of their stay in the hospital, an alcoholic might experience hallucinations, which typically are not too distressing. Convulsions, however, which resemble those in people with epilepsy, may occur in susceptible individuals about a day after the last drink. Convulsions usually occur only in those who have been drinking extremely large amounts of ethanol. If the convulsions are severe, the individual may die. Many somatic effects, such as nausea, vomiting, diarrhea, fever, and profuse sweating are also part of alcohol withdrawal. Some sixty to eighty-four hours after the last dose, there may be confusion and disorientation; more vivid hallucinations may begin to appear. This phase of withdrawal is often called the delirium tremens,

or DTs. Before the days of effective treatment, a mortality rate of 5 to 15 percent was common among alcoholics whose withdrawal was severe enough to cause DTs.

TREATMENT FOR ALCOHOL DEPENDENCE

The first step in treating alcoholics is to remove the ethanol from the system, a process called detoxification. Since rapid termination of ethanol (or any other central nervous system depressant) can be life threatening, people who have been using high doses should be slowly weaned from the ethanol by giving a less toxic substitute depressant. Ethanol itself cannot be used because it is eliminated from the body too rapidly, making it difficult to control the treatment. Although barbiturates were once employed in this capacity, the safer benzodiazepines have become the drugs of choice. Not only do they prevent the development of the potentially fatal convulsions, but they reduce anxiety and help promote sleep during the withdrawal phase. New medications are constantly being tested for their abilities to aid in the treatment of alcohol withdrawal.

Once a person has become abstinent, various methods can be used to maintain abstinence and encourage sobriety—some are pharmacologic and others are through social-support networks or formal psychological therapies. One type of treatment involves making drinking an adverse toxic event for the individual, by giving a drug such as DISULFIRAM (Antabuse) or citrated CALCIUM CARBIMIDE, which inhibits the metabolism of acetaldehyde and causes facial flushing, nausea, and rapid heartbeat. When ethanol is ingested by someone on disulfiram, the acetaldehyde levels rise very high, very quickly. Disulfiram has not been successful in maintaining abstinence in all patients, however.

Many support groups are available to help people remain abstinent. ALCOHOLICS ANONYMOUS (AA) is one of the most widely known and available; it is structured around a self-help philosophy. The AA program emphasizes total avoidance of alcohol and any medication. Instead it relies on a "buddy" or "sponsor" system, providing support partners who are personally experienced with alcoholism and alcoholism recovery. A number of other types of psychological and behavioral approaches to treatment also exist.

(SEE ALSO: *Accidents and Injuries from Alcohol; Alcoholism; Fetus, Effects of Drugs on; Complications; Social Costs of Alcohol and Drug Abuse*)

BIBLIOGRAPHY

GILMAN, A. G., ET AL. (EDS.). (1990). *Goodman and Gilman's the pharmacological basis of therapeutics*, 8th ed. New York: Pergamon.

GOLDSTEIN, A., ARONOW, L., & KALMAN, S. M. (1974). *Principles of drug action: The basis of pharmacology*. New York: Wiley.

GOLDSTEIN, D. B. (1983). *Pharmacology of alcohol*. New York: Oxford University Press.

HOFFMAN, F. G. (1975). *A handbook on drugs and alcohol abuse: The biomedical aspects*. New York: Oxford University Press.

WEST, L. J. (ED.). (1984). *Alcoholism and related problems: Issues for the American public*. Englewood Cliffs, NJ: Prentice-Hall.

SCOTT E. LUKAS

Complications Through their ethanol (alcohol) content, alcoholic beverages significantly affect the body's cellular function as well as its cognitive actions. Many of these effects are the consequence of a complex set of biochemical reactions, long-term exposure to ethanol with an accumulation of damage that is manifested in diverse ways, or the result of increased incidence or severity of major disease states, including AIDS, CANCER, or heart disease. However, some effects of ethanol are immediate and do not require prolonged exposure, nor are they induced as the end product of many physiological changes. For example, ethanol induces changes in cell membranes' fluidity by mixing with the lipids there. The membrane changes inhibit neurological functions and thus can cause car ACCIDENTS. All of these can occur with a single exposure and thus could be considered a direct effect of the ethanol in alcoholic beverages.

ALCOHOL METABOLISM

Ethanol Absorption and Metabolism. Because the ethanol molecule has a hydroxyl group, its metabolism involves dehydrogenase enzymes. After some metabolism in the stomach and intestine, it is transported to the liver for further metabolism. Alcohol dehydrogenase produces acetaldehyde, which causes many of the indirect effects attributed to ethanol. Because females metabolize alcohol less efficiently in the stomach wall than males, their exposure can be higher, with more direct consequences, from the same amount of alcohol consumption. Ethanol is also metabolized by the liver cells' MEOS system. Ethanol also affects the transportation of proteins across membranes in the cell. Thus aldehyde dehydrogenase's transportation into the mitochondria from the cell's cytoplasm is retarded. This reduces the oxidation of acetaldehyde to acetic acid, and increases ethanol's indirect effects by altering its metabolism and that of its metabolites. Acetaldehyde is very reactive with proteins. Thus increased levels result in damage to proteins with which it reacts. As many are vital for cell function, cell death or dysfunction occurs. This damage persists for the life of the protein or cell.

Alcohol and Nutrition. Alcohol has major effects when consumed frequently or in high amounts by affecting the frequency and quality of foods consumed. This directly affects the amounts of vitamins and minerals that are consumed and available for absorption. The long-term consequences involve undernutrition, nutritional deficiencies, and ultimately malnutrition. Ethanol also directly affects the absorption of vitamin A, betacarotene (a vitamin A precursor), vitamin B_1 (thiamine), folate, vitamin E, vitamin D, and folate. Vitamins are critical for many enzymatic reactions, so ethanol causes indirect effects by altering vitamin levels. Acute alcohol ingestion changes many vitamin metabolic pathways. Folate and vitamin A metabolism can cause increased urinary excretion. Thiamine deficiency is responsible for a severe neurological consequence of excessive alcohol use—WERNICKE'S SYNDROME.

ACTIONS OF ALCOHOL ON THE BRAIN

The molecular site of alcohol's action on neurons is unknown. Alcohol may work by perturbing lipids in the cell membrane of the NEURON, interacting directly with the hydrophobic region of neuronal membrane proteins, or interacting directly with a lipid-free enzyme protein in the membrane. Ethanol alters the function of neuron-specific proteins. For example, evidence suggests that the activity of the chloride ion channel linked to the A-type receptor

67

of the GABA NEUROTRANSMITTER increases during exposure to intoxicating amounts of alcohol. Acute exposure to alcohol effects the actions of GLUTA-MATE, the major excitatory transmitter in the mammalian central nervous system. Chronic exposure to alcohol can result in TOLERANCE for and PHYSICAL DEPENDENCE on the drug. Tolerance is recognized as a chronic drinker's ability to consume increasing amounts of alcohol without displaying gross signs of intoxication. Alcohol's effects on stress may be regulated by the combination of its effect on information processing. Thus it can decrease internal conflicts and block inhibitions, thereby making social behaviors more extreme.

Free Radical Generation by Alcohol. Free radicals are a highly reactive oxygen species. They are important components of the body's host defense, yet in high levels can cause tissue damage. Cytochrome P-450 is an oxidizing system that generates free radicals from ethanol. The reactive oxygen species include superoxide and hydrogen peroxide. They react with DNA, protein, and lipids. Products of the free radical reactions include lipid peroxides; thus alcohol's production of free radicals indirectly initiates cancer, heart disease, and other major health problems. Free radicals are produced in higher levels when ethanol and acetaldehyde begin to accumulate in cells and saturate dehydrogenases. Then other products, such as free radicals and cocaethylene (when cocaine is present), are produced.

Cholesterol and Fatty-Acid Production from Alcoholic Beverages. Excessive ethanol intake leads to formation of ethanol- and fatty-acid-containing ethyl esters, produced by synthases. Thus tissues containing large amounts of synthases, such as the heart, would be more likely to be damaged. These products can adversely affect protein synthesis, alter cell membranes that contain large amounts of normal lipids, and suppress energy production by the cells' mitochondria. Cholesterol esterase connects cholesterol to fatty acids, thus producing fatty-acid cholesterol esters. When ethanol is present, the esterase produces fatty-acid ethyl esters with a reduction of cholesterol. Ethanol consumption modifies components of cell membranes, phospholipids, through the phospholipase D. The importance of these changes is poorly defined and understood.

Cocaethylene and Drug Metabolism. When alcohol and cocaine are ingested together, the "high"

is accentuated. Ethanol can react with COCAINE via the enzyme cocaine esterase, producing a potentially toxic product, COCAETHYLENE. This enzyme inactivates cocaine in the absence of ethanol. Metabolism of cocaine and other drugs occurs in large part via cytochrome P-450 IIEl. It is increased by chronic alcohol consumption. This cytochrome oxidizes ethanol in the liver as well as many other compounds, including cocaine and the pain killer acetaminophen. Oxidative products of cytochrome P-450 are more toxic than the parent compounds, and thus can accentuate liver damage.

Metabolism of Protein. Consumption of alcoholic beverages affects the metabolism of ethanol and other alcohols, and alters the NADH/NAD ratio—the ratio of reduced nicotinamide adenine dinucleotide to oxidized nicotinamide adenine dinucleotide—which influences lipid, vitamin, and protein metabolism, membrane composition and function, and energy production. Such changes lead to indirect effects including cell damage, undernutrition, and weight loss. Chronic alcohol beverage use reduces type II muscle fibers, reducing the capacity for prolonged muscle activity and thus the ability to exercise, run, or do physical work. Loss of this fiber produces muscle pain, weakness, and damage. Reduced type II fibers may be due to lower RNA, which would indicate less protein synthesis.

Metabolism of Lipids and Fats. Fat and lipid functions and metabolism are altered by alcohol consumption. High alcohol intakes result in changes in the ratio of NADH/NAD+, which rduces breakdown of fats and lipids. The accumulated lipids are stored in the liver, producing a fatty liver. The NADH/NAD+ ratio also inhibits synthesis of cholesterol and related steroid hormones. Thus production of progesterone and androstenedione are reduced by alcohol use. Such changes may be the cause of hypogonadism in males who consume alcohol chronically. Lipoprotein lipase is inhibited by ethanol, thus reducing removal of long acyl chains from lipids. In heart muscle this reduces available energy and could be a component of heart disease. Lipoproteins are transport molecules for fats, including cholesterol, in plasma fluids. Alcohol increases both low- and high-density lipoproteins, which could be beneficial and damaging, respectively, to the heart.

Lipids in the Function and Composition of Cell Membranes. Membranes have lipids and pro-

teins as major components. Ethanol clearly affects lipids and membranes directly and indirectly. Alcohol affects cell membranes directly by its entry into them. Its physical characteristics modify arrangement of lipids in the cell membrane, and hence should affect cell function directly. For example, electrolyte balance within all cells is produced by sodium and potassium ion transportation. High alcohol intake reduces the ion transporters, which causes cells to take up water and thus to swell, affecting function. In addition, cells respond to hormones and other chemicals in the plasma outside the cell membrane by signal transduction. These signals regulate the functions of the various cell types, affecting overall physiology of the body. Important enzymes in this process include phospholipases. Ethanol acts like hormones and signal molecules, changing membrane phospholipases, which should modify cell function.

ALCOHOL TRAUMA, ACCIDENTS, AND BEHAVIORAL EFFECTS

Alcohol is directly involved in injuries by altering neurological function in ways that lead to motor vehicle ACCIDENTS, plane crashes, drownings, SUICIDE, and homicide. It appears to play a role in both unintentional and intentional injuries. Nearly one-fourth of suicide victims, one-third of homicide victims, and one-third of unintentional injury victims have high BLOOD ALCOHOL CONCENTRATIONS. Alcohol was a factor in half of fatal traffic crashes and 5 percent of all deaths. It causes premature mortality (not including deaths from indirect, biochemical changes induced by long-term exposure).

Alcohol and Auto Accidents. Alcohol consumption directly and promptly impairs many perceptual, cognitive, and motor skills needed to operate motor vehicles safely. Although in 1989 traffic fatalities involving at least one intoxicated driver or nonoccupant (pedestrian or other) decreased by half, 22,413 people were killed in alcohol-related motor vehicle crashes, representing approximately half of all traffic fatalities. The decrease in alcohol's involvement may be partially attributed to changes in MINIMUM DRINKING AGE LAWS. WOMEN drivers are involved in half as many alcohol-related car accidents as men. Impaired drivers arrested are significantly more hostile; they have greater psychopathic deviance, nontraffic arrests, and frequency of impaired driving, and they drink more than drunk drivers caught in roadblocks. Thus, impaired driving and alcohol-related accidents are part of problematic behaviors that can be directly modified by ethanol.

Alcohol and Airplane Accidents. Alcohol has not been shown to have caused a U.S. commercial airline accident. However, it plays a direct and prominent role in general aviation accidents. Pilot function is impaired by cognitive, perceptual, and psychomotor changes due to ethanol use. Positional alcohol nystagmus may contribute to many aviation crashes involving spatial disorientation.

Alcohol and Water Accidents. Alcohol is associated with between half and two-thirds of adult drownings. Alcohol is also important in water-related spinal cord injuries.

Alcohol and Sexual Behavior. Via neurological changes, alcohol impairs rational thought, thus decreasing behavioral inhibitions. Alcohol is an excuse for behavior that violates social norms. Problem drinking behavior is associated with sexually transmitted disease.

Alcohol and Violence. High alcohol consumption reduces inhibition, impairs moral judgment, and increases aggression; thus there is greater likelihood of homicide or assault resulting from fights. Frequently, alcohol use has occurred in situations that emerge spontaneously from personal disputes. Alcohol is linked to a high proportion of violence, with perpetrators more often under the influence of alcohol than victims. Very high rates of problem drinking are reported among both property and violent offenders.

(SEE ALSO: *Accidents and Injuries from Alcohol; Complications*)

BIBLIOGRAPHY

SECRETARY OF HEALTH AND HUMAN SERVICES. (1993). *Eighth special report to the U.S. Congress on alcohol and health.* Washington, DC: U.S. Government Printing Office.

WATSON, R. R. (1995). *Alcohol and accidents.* Totowa, N.J.: Humana Press.

WATSON, R. R. (1995). *Alcohol and hormones.* Totowa, N.J.: Humana Press.

WATSON, R. R. (1992). *Alcohol and neurobiology. I, Receptors, membranes and channels.* Boca Raton, FL: CRC.

RONALD R. WATSON

History of Drinking The key to the importance of alcohol in history is that this simple substance, presumably present since bacteria first consumed some plant cells nearly 1.5 billion years ago, has become so deeply embedded in human societies that it affects their religion, economics, age, sex, politics, and many other aspects of human life. Furthermore, the roles that alcohol plays differ, not only from one culture to the next but even within a culture over time. A single chemical compound, used (or sometimes emphatically avoided) by a single species, has resulted in a complex array of customs, attitudes, beliefs, values, and effects. A brief review of the history of this relationship illustrates both unity and diversity in the ways people have thought about and treated alcohol. Special attention is paid to the United States as a case study of particular interest to many readers.

THE QUESTION OF ORIGINS

Although chemists distinguish five kinds of alcohols on the basis of small but important molecular differences, only ethyl alcohol, also called ethanol (C_2H_6OH), is deliberately included in beverages. Ethanol is a compound that occurs naturally, as a by-product, when yeasts that are free-floating in the air consume the sugars present in almost all fruits, vegetables, and grains. This process can occur without human intervention, and we often see birds or insects getting drunk from eating fruits or berries past their prime.

In one sense, people do not really drink ethanol—since most beers, meads, and other home brews rarely exceed an 8 percent concentration, and few wines go over 15 percent. Even distilled liquors are below 50 percent (100 proof in U.S. terms) by weight or by volume, and such drinks may be further diluted by the addition of water or other mixers. In recent scientific terminology, however, any beverage with as little as 2 percent tends to be called "alcoholic," and it is noteworthy that in the United States in the 1990s, a normal measure of beer contains about the same amount of ethanol as a normal measure of wine (in a smaller container) or a normal measure of liquor (in a still smaller container). If we assume that it is ethanol that produces a host of presumed favorable effects, as well as alcohol-related problems, then the logic of labeling some drinks "alcoholic" can be justified. It is important to remember, however, that labels are merely a social

convention. No matter how great its alcohol content may be, wine is thought of as "food" in much of France and Italy—as is beer in Scandinavia and Germany. Similarly, in the United States, many people who regularly drink beer in considerable quantities do not think of themselves as using alcohol. Some fruit juices, candies, and desserts come close to having enough alcohol to be so labeled, but they are not. Thus many of the concerns that people have about alcohol relate more to their expectations than to the actual pharmacological or biochemical impact that the substance would have on the human body.

According to the Bible, one of the first things Noah did after the great Flood was to plant a vineyard (Genesis 9:21). According to the predynastic Egyptians, the great god Osiris taught people to make beer, a substance that had great religious as well as nutritional value for them. Similarly, early Greeks credited the god Dionysus with bringing them wine, which they drank largely as a form of worship. In Roman times, the god Bacchus was thought to be both the originator of wine and always present within it. It was a goddess, Mayahuel, with 400 breasts, who supposedly taught the Aztecs how to make pulque from the sap of the century plant; that mild beer is still important in the diet of many Indians in Mexico, where it is often referred to as "the milk of our Mother." In each of these instances, whether the giver was male or female, alcohol was viewed as supernatural, reflecting deep appreciation of its important roles in nourishing and comforting people.

Anthropologists often treat myths as if they were each people's own view of history, but clearly it would be difficult to take all myths at face value. We cannot know when or where someone first sampled alcohol, but we can imagine that it might well have been just an attempt to make the most of an overripe fruit or a soured bowl of gruel. The taste, or the feeling that resulted, or both, may have been pleasant enough to prompt repetition and then experimentation. Probably it happened not just once but various times, independently, at a number of different places, just as did the beginnings of agriculture.

PREHISTORY AND ARCHEOLOGY

Although it is impossible to say where or when *Homo sapiens* first sampled alcohol, there is firm evidence, from chemical analysis of the residues found in pots dating from 3500 B.C., that wine was already

being made from grapes in Mesopotamia (now Iran). This discovery makes alcohol almost as old as farming, and, in fact, beer and bread were first produced at the same place at about the same time from the same ingredients. We know little about the gradual process by which people learned to control fermentation, to blend drinks, or to store and ship them in ways that kept them from souring, but the distribution of local styles of wine vessels serves as a guide to the flow of commerce in antiquity.

It would be misleading to think of early wines and beers as similar to the drinks we know today. In a rough sense, the distinction between them is that a wine is generally derived from fruits or berries, whereas a beer or ale comes from grain or a grain-based bread. Until as recently as A.D. 1700, both were often relatively dark, dense with sediments, and extremely uneven in quality. Usually handcrafted in small batches, home-brewed beers tend to be highly nutritious but to last only a few days before going sour (i.e., before all the fermenting sugars and alcohol are depleted and become vinegar). By contrast, homemade wines have relatively little in the way of vitamins or minerals but can last a long time if adequately sealed.

In Egypt between 2700 and 1200 B.C., beer was not only an important part of the daily diet; it was also buried in royal tombs and offered to the deities. Many of the paintings and carvings in Egyptian tombs depict brewing and drinking; early papyri include commercial accounts of beer, a father's warning to his student son about the danger of drinking too much, praises to the god who brought beer to earth, and other indications of its importance and effects.

The earliest written code of laws we know, from Hammurabi's reign in Babylon around 2000 B.C., devoted considerable attention to the production and sale of beer and wine, including regulations about standard measures, consumer protection, and the responsibilities of servers.

In ancient Greece and Rome (roughly 800 B.C.– A.D. 400), there was wider diffusion of grape-growing north and westward in Europe, and wine was important for medicinal and religious purposes, although it was not yet a commonplace item in the diet of poor people. The much-touted sobriety of the Greeks is presumably based on their custom of diluting wine with water and drinking only after meals, in contrast to neighboring populations who often sought drunkenness through beer as a transcendental state of altered consciousness. Certainly heavy drinking was an integral part of the religious orgies that, commemorating their deities, we now call "Dionysiac" (or, in the case of Rome, "Bacchic"). The temperate stereotype also overlooks the infamous chronic drunkenness of Alexander the Great. Born in Macedon, in 356 B.C., he managed to conquer most of the known world in his time, by 325 B.C., bringing what are now Egypt and most of the Middle East under the rule of Greece before he died in 323 B.C.

Romans were quick to point out how their relative temperance contrasted with the boisterous heavy drinking of their tribal neighbors in all directions, whom they devalued as the bearded ones, "barbarians." To a remarkable degree, the geographic spread of Latin-based languages and grape cultivation coincided with the spread of the Roman Empire through Europe and the accompanying diffusion of the Mediterranean diet—rich in carbohydrates and low in fats and protein—with wine as the usual beverage. In striking contrast were non-Latin speakers, who were less reliant on bread and pasta and without olive oil; they drank beers and meads, with drunkenness more common. Plato considered wine an important adjunct to philosophical discussion, and St. Paul recommended it as an aid to digestion.

The Hebrews established a new pattern around the time of their return from the Babylonian exile, and the construction of the Second Temple (c. 500 B.C.). Related to a new systematizing of religious practices was a strong shift toward family rituals, in which the periodic sacred drinking of wine was accompanied by a pervasive ethic of temperance, a pattern that persists today and often marks drinking by religious Jews as different from that of their neighbors. Early Christians (many of whom had been Jews), praised the healthful and social benefits of wine while condemning drunkenness. A majority of the many biblical references to drinking are clearly favorable, and Jesus' choice of wine to symbolize his blood is perpetuated in the solemn rite of the *Eucharist*, which has become central to practice in many Christian churches as *Holy Communion.*

In the Iron Age in France (c. 600 B.C.), distinctive drinking vessels found in tombs strongly suggest that political leadership involved the redistribution of goods to one's followers, with wine an important symbol of wealth. Archeologists have learned so much about the style and composition of pots made in any given area that they can often trace routes and times of trade, military expansion, or migrations by

noting where fragments of drink containers are found. Although we know little about Africa at that time, we assume that mild fermented home brews (such as banana beer) were commonplace, as they were in Latin America. In Asia, we know most about China, where as early as 2000 B.C. grain-based beer and wine were used in ceremony, offered to the gods, and included in royal burials. Most of North America and Oceania, curiously, appear not to have had any alcoholic beverages until contact with Europeans.

Alcohol in classical times served as a disinfectant and was thought to strengthen the blood, stimulate nursing mothers, and relieve various ills, as well as to be an ideal offering to both gods and ancestral spirits. Obviously, drink and drinking had highly positive meanings for early peoples, as they do now for many non-Western societies.

FROM A.D. 1000 TO 1500

The Middle Ages was marked by a rapid spread of both Christianity and Islam. Large-scale political and economic integration spread with them to many areas that had previously seen only local warring factions, and sharp social stratification between nobles and commoners was in evidence at courts and manors, where food and drink were becoming more elaborate. National groups began to appear, with cultural differences (including preferred drinks and ways of drinking) increasingly noted by travelers, of whom there were growing numbers. Excessive drinking by poor people was often criticized but may well have been limited to festive occasions. With population increases, towns and villages proliferated, and taverns became important social centers, often condemned by the wealthy as subverting religion, political stability, and family organization. But for peasants and craftspeople, the household was still often the primary economic unit, with home-brewed beer being a major part of the diet.

During this period, hops, which enhanced both the flavor and durability of beer, were introduced. In Italy and France, wine became even more popular, both in the diet and for expanding commerce. Distillation had been known to the Arabs since about 800, but among Europeans, a small group of clergy, physicians, and alchemists monopolized that technology until about 1200, producing spirits as beverages for a limited luxury market and for broader use as a medicine. Gradual overpopulation was halted by

the Black Death (a pandemic of bubonic plague), and schisms in the Catholic church resulted in unrest and political struggles later in this period.

Across northern Africa and much of Asia, populations, among whom drinking and drunkenness had been lavishly and poetically praised as valuable ways of altering consciousness, became temperate and sometimes abstinent, in keeping with the tenets of Islam and the teachings of Buddha and of Confucius. China and India both had episodes of prohibition, but neither country was consistent. In the Hindu religion, some castes drank liquor as a sacrament, whereas others scorned it—vivid proof that a culture, in the anthropological sense (as a set of beliefs and practices that guide one through living), is often much smaller than a religion or a nation, although we sometimes tend to think of those larger entities as more homogenous than they really are.

As the Middle Ages gave way to the Renaissance, both the population and the economy expanded throughout most of Europe. Because the Arabs (who had ruled from 711 to 1492) had been expelled from Spain and Portugal, they cut off overland trade routes to Asia; European maritime exploration therefore resulted in increasing commerce all around the coasts of Africa. The so-called Age of Exploration led to the startling encounter with high civilizations and other tribal peoples who had long occupied North America, Central America, and South America. Ironically, alcoholic beverages appear to have been totally unknown north of Mexico, although a vast variety of beers, chichas, pulques, and other fermented brews were important in Mexico as foods, as offerings to the gods and to ancestral spirits, and as shortcuts to religious ecstasy—if we

assume that Indians then lived much as those who were soon to be described by the European conquerors and missionaries.

Throughout sub-Saharan Africa, we assume, home-brewed beers were plentiful, nutritious, and symbolically important, as they came to be described in later years.

During the Middle Ages, drinking was treated as a commonplace experience, little different from eating, and drunkenness appears to have been infrequent, tolerated in association with occasional religious festivals and of little concern in terms of health or social welfare. Alcoholic beverages themselves were becoming more diverse but still were thought to be invigorating to humans, appreciated by spirits, and important to sociability.

FROM A.D. 1500 TO 1800

Wealth and extravagance were manifest in the rapidly growing cities of Europe, but so were poverty and misery, as class differences became even more exaggerated. The Protestant Reformation, which set out to separate sacred from secular realms of life, seemed to justify an austere morality that included injunctions against celebratory drunkenness. If the body was the vessel of the spirit, which itself was divine, one should not desecrate it with long-term heavy drinking. Puritans viewed intoxication as a moral offense—although they drank beer as a regular beverage and appreciated liquor for its supposed warming, social, and curative properties. Public drinking establishments evolved, sometimes as important town meeting places and sometimes as the workers' equivalent of social clubs, with better heat and lighting than at home, with news and gossip, games and companionship. COFFEE, TEA, and CHOCOLATE were also introduced to Europe at this time, and each became popular enough to be the focus of specialized shops. But each was also suspect for a time, while physicians debated whether they were dangerous to the health; clergy debated their effects on morality; and political and business leaders feared that retail outlets would become breeding places of crime, labor unrest, and civil disobedience. Brandies (*brantwijns*, liquor distilled from wines to be shipped as concentrates) spread among the aristocracy, and champagne was introduced as a luxury beverage (wine), as were various cordials and liqueurs. Brewing and wine-making grew from cottage industries to

major commercial ventures, incorporating many technical innovations, quality controls, and other changes.

The "gin epidemic" in mid-eighteenth-century London is sometimes cited as showing how urban crowding, cheap liquor, severe unemployment, and dismal working conditions combined to produce widespread drinking and dissolution, but the vivid engravings by William Hogarth may exaggerate the problem. At the same time, the artist extolled beer as healthful, soothing, and economically sound. In France, even peasants began to drink wine regularly. In 1760, Catherine the Great set up a state monopoly to profit from Russia's prodigious thirst, and Sweden followed soon after.

Throughout Latin America and parts of North America, the Spanish and Portuguese conquistadors found that indigenous peoples already had home brews that were important to them for sacred, medicinal, and dietary purposes. The Aztecs of Mexico derived a significant portion of their nutritional intake from pulque but reserved drunkenness as the prerogative of priests and old men. Cultures throughout the rest of the area similarly used chicha or beer made from maize, manioc, or other materials. The Yaqui (in what is now Arizona) made a wine from cactus as part of their rain ceremony, and specially made chicha was used as a royal gift by the Inca of Peru. Religious and political leaders from the colonial powers were ambivalent about what they perceived as the risks of public drunkenness and the profits to be gained from producing and taxing alcoholic beverages. A series of inconsistent laws and regulations, including sometime prohibition for Indians, were probably short-lived experiments, affected by such factors as local revolts and different opinions among religious orders.

As merchants from various countries competed to gain commercial advantage in trading with the various Native American groups of North America, liquor quickly became an important item. It has become popular to assume that Native Americans are genetically vulnerable to alcohol, but some tribes (such as Hopi and Zuni) never accepted it, and others drank with moderation. The Seneca are an interesting case study, because they went from having no contact with alcohol through a series of stages culminating in a religious ban. When brandy first arrived, friends would save it for an unmarried young man, who would drink it ceremoniously to help in

his required ritual quest for a vision of the animal that would become his guardian spirit. In later years, drinking became secular, with anyone drinking and boisterous brawling a frequent outcome. In 1799, when a tribal leader, who was already alcoholic, had a very different kind of vision, he promptly preached abstention from alcohol, an end to warfare, and devotion to farming—all of which remain important today in the religion that is named after him, Handsome Lake.

Throughout the islands of the Pacific, local populations reacted differently to the introduction of alcohol, sometimes embracing it enthusiastically and sometimes rejecting it. Eskimos were generally quick to adopt it, as were Australian Aborigines, to the extent that some interpret their heavy drinking as an attempt to escape the stresses of losing valued parts of their traditional ways of life. Detailed information about the patterns of belief and behavior associated with drinking among the diverse populations of Asia and Africa vividly illustrates that alcohol results in many kinds of comportment—depending more on sociocultural expectations than any qualities inherent in the substance.

In what is now the United States, colonial drinking patterns reflected those of the countries from which immigrants had come. Rum (distilled from West Indies sugar production) became an important item in international trade, following routes dictated by the economic rules of the British Empire. In the infamous Triangle Trade, captive black Africans were shipped to the West Indies for sale as slaves. Many worked on plantations there, producing not only refined sugar, a sweet and valuable new faddish food, but also molasses, much of which was shipped to New England. Distillers there turned it into rum, which was in turn shipped to West Africa, where it could be traded for more slaves. During the American Revolution (1775/6–1783), however, that trade was interrupted and North Americans shifted to whiskey. Farmers along what was then the frontier, still east of the Mississippi, were glad to have a profitable way of using surplus corn that was too bulky to bring to distant markets. After the war, when the first federal excise tax was imposed (on whiskey) in 1790, to help cut the debt of the new United States, producers' anger about a tax increase was expressed in the Whiskey Rebellion of 1794. To quell the uprising, federal troops (militia) were used for the first time. At about the same time, Benjamin Rush, a noted physician and signer of the Declaration of In-

dependence, started a campaign against long-term heavy drinking as injurious to health.

Evidently, alcohol plays many roles in the history of any people, and changes in attitudes can be abrupt, illustrating again the importance that social constructions of reality have in relation to drinking.

THE 1800s

The large-scale commercialization of beer, wine, and distilled liquor spread rapidly in Europe as many businesses and industries became international in scope. Large portions of the European proletariat were no longer tied to the land for subsistence, and new means of transportation facilitated vast migrations. The industrial revolution was not an event but a long process, in which, for many people, work became separated from home. The arbitrary pace imposed by wage work contrasted markedly with the seasonal pace of traditional agrarianism.

In some contexts neighbors still drank while helping each other—as, for example, in barn-raising or reciprocal labor exchange during the harvest. But for the urban masses, leisure and a middle class emerged as new phenomena. Drinking, which became increasingly forbidden in the workplace as dangerous or inefficient, gradually became a leisure activity, often timed to mark the transition between the workday and home life. As markets grew, foods became diverse, so that beers and ciders (usually hard) lost their special value as nourishing and energizing.

In Europe, political boundaries were approximately those of the twentieth century; trains and steamships changed the face of trade; and old ideas about social inequality were increasingly challenged. Alcohol lost much of its religious importance as ascetic Protestant groups, and even fervent Catholic priests in Ireland, associated crime, family disruption, unemployment, and a host of other social ills with it, and taxation and other restrictions were broadly imposed. In Russia, the czar ordered prohibition, but only briefly as popular opposition mounted and government revenues plummeted. Those who paid special attention to physical and mental illnesses were quick to link disease with long-term heavy drinking, although liquor remained an important part of medicine for various curative purposes. A few institutions sprang up late in the nineteenth century to accommodate so-called inebriates,

although there was little consensus about how or why drinking created problems for some people but not for others, nor was there any systematic research.

A wave of mounting religious concern that has been called the "great awakening" swept over the United States early in the 1800s, and, by 1850, a dozen states had enacted prohibition. Antialcohol sentiment was often associated with opposition to slavery. The local prohibition laws were repealed as the Civil War and religious fervor abated, and hard drinking became emblematic of cowboys, miners, lumberjacks, and other colorful characters associated with the expanding frontier. Distinctions of wealth became more important than those of hereditary social status, and a wide variety of beverages, of apparatus associated with drinking, and even of public drinking establishments accentuated such class differences.

Near the end of the century, another wave of sentiment against alcohol grew, as large numbers of immigrants (many of them Catholic and anything but ascetic) were seen by Protestant Yankees as trouble—competing for jobs, changing the political climate, and challenging old values. Coupled with this attitude was enthusiasm for "clean living," with an emphasis on natural foods, exercise, fresh air and water, loose-fitting clothing, and a number of other fads that have recently reappeared on the scene.

Native American populations, in the meantime, suffered various degrees of displacement, exploitation, and annihilation, sometimes as a result of deliberate national policy and sometimes as a result of local tensions. The stereotype of the drunken Indian became embedded in novels, news accounts, and the public mind, although the image applied to only a small segment of life among the several hundred native populations. Some Indians remained abstinent and some returned to abstinence as part of a deliberate espousal of indigenous values—for example, in the Native American Church, using PEYOTE as a sacrament, or in the sun dance or the sweat lodge, using asceticism as a combined religious and intellectually cleansing precept.

From Asia, Africa, and Oceania, explorers, traders, missionaries, and others brought back increasingly detailed descriptions of non-Western drinking practices and their outcomes. It is from such ethnographic reports—often sensationalized—that we can guess about the earlier distribution of native drinks and can recognize new alcoholic beverages as major commodities in the commercial exploitation of pop-

ulations. Although some of the sacramental associations of traditional beverages were transferred to new ones, the increasing separation of brewing from the home, the expansion of a money-based economy, and the apparent prestige value of Western drinks all tended to diminish the significance of home brews. In African mines, Latin American plantations, and even some U.S. factories, liquor became an integral part of the wage system, with workers required to accept alcohol in lieu of some of their cash earnings. In some societies where drinking had been unknown before Western colonization, the rapid spread of alcohol appears to have been an integral part of a complex process that eroded traditional values and authority.

THE TWENTIETH CENTURY

It has been said that the average person's life in 1900 was more like that of ancestors thousands of years earlier than like that of most people today. The assertion certainly applies to the consumption of liquor. Pasteurization, mass production, commercial canning and bottling, and rapid transport all transformed the public's view of beer and wine in the twentieth century. The spread of ideas about individualism and secular humanism loosened the hold of traditional religions on the moral precepts of large segments of the population. New assumptions about the role of the state in support of public health and social welfare now color our expectations about drinking and its outcomes. Mass media and international conglomerates are actively engaged in the expansion of markets, especially into developing countries.

World War I prompted national austerity programs in many countries that curtailed the diversion of foodstuffs to alcoholic beverages but didn't quite reach the full prohibition for which the United States became famous. Absinthe was thought to be medically so dangerous that it was banned in several European countries, and Iceland banned beer but not wine or liquor. Sweden experimented with rationing, and the czar again tried prohibition in Russia. The worldwide economic depression of the 1930s appears to have slowed the growth of alcohol consumption, which grew rapidly during the economic boom that followed World War II. The Scandinavian countries, beset by a pattern of binge drinking, often accompanied by violence, tried a variety of systems of regulation, including state monopolies, high taxa-

tion, and severely restricted places and times of sale, before turning to large-scale social research.

While several Western countries were expanding their spheres of influence in sub-Saharan Africa, they agreed briefly on a multinational treaty that outlawed the sale of alcoholic beverages there, although they did nothing to curtail production of domestic drinks by various tribal populations. A flurry of scientific analyses of indigenous drinks surprised many by demonstrating their significant nutritional value, and more detailed ethnographic studies showed how important they were in terms of ideology, for vows, communicating with supernatural beings, honoring ancestors, and otherwise building social and symbolic credit—among native societies not only in Africa but also in Latin America and Asia. Closer attention to the social dynamics of drinking and other aspects of culture showed that the impact of contact with Western cultures is not always negative and that for many peoples the role of alcohol remained diverse and vital.

In the United States, a combination of religious, jingoistic, and unsubstantiated medical claims resulted in the enactment of nationwide prohibition in 1919. Often called "the noble experiment," the Eighteenth Amendment to the Constitution was the first amendment to deal with workaday behavior of people who have no important public roles. It forbade commercial transaction but said nothing about drinking or possession. Most authorities agree that, during the early years, there was relatively little production of alcoholic beverages and not much smuggling or home production. It was not long, however, before illegal sources sprang up. Moonshiners distilled liquor illegally, and bootleggers smuggled it within the U.S. or from abroad. Speakeasies sprang up as clandestine bars or cocktail lounges, and a popular counterculture developed in which drinking was even more fashionable than before prohibition. Some entrepreneurs became immensely wealthy and brashly confident and seemed beyond the reach of the law, whether because of superior firepower or corruption or both. The government had been suffering from the loss of excise taxes on alcohol, which accounted for a large part of the annual budget. The stock-market crash, massive unemployment, the crisis in agriculture, and worldwide economic depression aggravated an already difficult situation, and civil disturbances spread throughout the country. Some of the same influential people who had

pressed strongest for prohibition reversed their stands, and the Twenty-first Amendment, the first and only repeal to affect the U.S. Constitution, did away with federal prohibition in 1933. Although the national government retained close control over manufacturing and distribution to maximize tax collection, specific regulations about retail sales were left up to the states. An odd patchwork of laws emerged, with many states remaining officially dry, others allowing local option by counties or towns, some imposing a state monopoly, some requiring that drinks be served with food and others expressly prohibiting it, some insisting that bars be visible from the street and others the opposite, and so forth. The last state to vote itself wet was Mississippi, in 1966, and many communities remain officially dry today. The older federal law prohibiting sales to Indians was not repealed until 1953, and many Native American reservations and Alaska native communities remain dry under local option.

The experience of failed prohibition in the U.S. is famous, but a similar combination of problems with lawlessness, corruption, and related issues led to repeal, after shorter experiments, in Iceland, Finland, India, Russia, and parts of Canada, demonstrating again that such drastic measures seem not to work except where supported by consensus and religious conviction (e.g., Saudi Arabia, Iran, and Ethiopia). It is ironic that some Indian reservations with prohibition have more alcohol-related deaths than those without. A more salutary recent factor is the growth of culturally sensitive programs of prevention and treatment that are being developed, often by the communities themselves, for Indian and other minority populations.

In the middle decades of the twentieth century, a number of alcoholics formed a mutual-help group, modeled on the earlier Washingtonians, and ALCOHOLICS ANONYMOUS has grown to be an international fellowship of individuals whose primary purpose is to keep from drinking. At about the same time, scientists from a variety of disciplines started studying various aspects of alcohol, and our knowledge has grown rapidly. Because of the large constituency of recovering alcoholics, the subject has become politically acceptable, and the disease concept has overcome much of the moral stigma that used to attach to alcoholism. Establishment of a National Institute on Alcohol Abuse and Alcoholism in 1971 signaled a major government commitment to the field, and its

incorporation among the National Institutes of Health in 1992 indicates that concerns about wellness have largely displaced theological preoccupations.

Consumption of all alcoholic beverages increased gradually in the U.S. from repeal until the early 1980s, with marked increase following World War II, although it never reached more than one-third of what is estimated for the corresponding period a century earlier. Around 1980, sales of spirits started dropping and have continued to do so. A few years later, wine sales leveled off and have gradually fallen since; beer sales also appear to have passed their peak even more recently. These reductions occurred, despite increasing advertising, along with a return of the "clean living" movement and another shift toward physical exercise, less-processed foods, and concern for health. Linked with the reduction in drinking, what some observers call a "new temperance movement" has emerged, in which individuals not only drink less but call for others to do the same; the decline would be enforced by laws and regulations that would increase taxes, index liquor prices to inflation, diminish numbers and hours of sales outlets, require warning labels, ban or restrict advertising, and otherwise reduce the availability of alcohol. Such a "public health approach" is by no means limited to the U.S.; its popularity is growing throughout Europe and among some groups elsewhere, even as alcohol consumption continues to rise in Asia and many developing countries.

CURRENT IMPLICATIONS

A quick review of the history of alcohol lends a fresh perspective to the subject. The vast literature on ethnographic variation among populations demonstrates the different ways in which peoples, widely separated geographically and historically, have used and thought about alcohol. The idea of alcohol as being implicated in a set of problems is peculiar to the recent past and is not yet generally accepted in many areas.

What some observers call the "new temperance movement" and others call "neoprohibitionism" is a recent phenomenon that grew out of Scandinavian social research. The conclusion, on the basis of transnational comparisons, was that there appeared to be some relationship between the amount of alcohol people drink and a broad range of what the researchers called "alcohol-related problems" (including spouse abuse, child neglect, social violence, psychiatric illness, a variety of organic damages, traffic fatalities). The vague and general findings gradually came, through a process of misquotation and paraphrasing, to be treated as a pseudoscientific iron-clad law, to the effect that problems are invariably proportionate to consumption, so that the most effective way to diminish problems would be to cut drinking. This approach is sometimes called the "control of consumption model," or the "single distribution model" (referring to the fact that heavy drinkers are on the same distribution-of-consumption curve as low and moderate drinkers, with no clear points that would objectively divide the groups).

This movement is not restricted to the U.S. and Scandinavia, however. The World Health Organization of the United Nations called for a worldwide reduction, by 25 percent, of alcohol consumption during the last decade and a half of the twentieth century, recommending that member countries adopt similar policies. Throughout most of central and western Europe and North America, sales have fallen markedly, although the opposite trend can be seen in much of the third world. An ironic development has been recent loosening of controls in Scandinavian countries, traditionally the exemplars of that approach, while controls are being introduced and progressively tightened in southern Europe, where drinking has traditionally been an integral part of the culture.

The European Community standardization of tariffs may result in further changes soon. A more realistic way of lessening whatever problems may be related to alcohol consumption would appear to be the "sociocultural model" of prevention, emphasizing, on the basis of cross-cultural experience, that people can learn to drink differently, to expect different outcomes from drinking, and actually to find their expectancies fulfilled. This program would not be quick or easy, requiring intensive public education, but it seems more feasible than simply curtailing availability—in which case those who enjoy moderate drinking would be inconvenienced but those who insist on drinking heavily would continue to do so. Concern over policy is not only directed at helping individuals who may have become dependent; it also has the aim of making life safer and more pleasant for all. The history of alcohol indi-

cates that problems are by no means inherent in the substance but, rather, are mediated by the individual user and by social norms.

(SEE ALSO: *Beers and Brews; Temperance Movement; Treatment, History of*)

BIBLIOGRAPHY

AUSTIN, G. A. (1985). *Alcohol in Western society from Antiquity to 1800: A chronological history.* Santa Barbara, CA: ABC-Clio Information Services.

BADRI, M. B. (1976). *Islam and alcoholism.* Indianapolis: American Trust Publications.

BARROWS, S., & ROOM, R. (EDS.). (1991). *Drinking: Behavior and belief in modern history.* Berkeley: University of California Press.

BARROWS, S., ET AL. (EDS.). (1987). *The social history of alcohol: drinking and culture in modern society.* Berkeley, CA: Alcohol Research Group, Medical Research Institute of San Francisco.

BLOCKER, J. S. (1980). *American temperance movements: Cycles of reform,* 2nd ed. Boston: Twayne.

BLUM, R. H. (1974). *Society and drugs I. Social and cultural observations.* San Francisco: Jossey-Bass.

CHANG, K. C. (ED.). (1977). *Food in Chinese cultures: Anthropological and historical perspectives.* New Haven: Yale University Press.

EAMES, A. (1993). *Blood, sweat, and beer.* Berkeley, CA: Milk and Honey Press.

GOMBERG, E. L., ET AL. (EDS.). (1982). *Alcohol, science, and society revisited.* Ann Arbor: University of Michigan Press.

GUSFIELD, J. (1986). *Symbolic crusade: Status politics and the American temperance movement,* 2nd ed. Urbana: University of Illinois Press.

HAMER, J., & STEINBRING, J. (EDS.). (1980). *Alcohol and native peoples of the north.* Lanham, MD: University Press of America.

HATTOX, R. S. (1985). *Coffee and coffeehouses: The origins of a social beverage in the medieval Near East.* University of Washington Near Eastern Studies 3, Seattle.

HEATH, D. B. (1994). *International handbook on alcohol and culture.* Westport, CT: Greenwood.

HEATH, D. B., & COOPER, A. M. (1981). *Alcohol use in world cultures: A comprehensive bibliography of anthropological sources.* Toronto: Addiction Research Foundation.

LENDER, M., & MARTIN, J. (1987). *Drinking in America: A social-historical explanation,* rev. ed. New York: Free Press.

MACANDREW, C., & EDGERTON, R. B. (1969). *Drunken comportment: A social explanation.* Chicago: Aldine.

MAIL, P. D., & McDONALD, D. R. (1980). *Tulapai to Tokay: A bibliography of alcohol use and abuse among Native Americans of North America.* New Haven: Human Relations Area Files Press.

MARSHALL, M. (1979). *Beliefs, behaviors, and alcoholic beverages: A cross-cultural survey.* Ann Arbor: University of Michigan Press.

McGOVERN, P., ET AL. (EDS.). (1993). *The origins and ancient history of wine.* New York: Gordon & Breach.

MUSTO, D. (1989). *The American disease: Origins of narcotic control,* rev. ed. New York: Oxford University Press.

O'BRIEN, J. M., & ALEXANDER, T. W. (1992). *Alexander the Great: The invisible enemy.* New York: Routledge.

PAN, L. (1975). *Alcohol in colonial Africa.* Helsinki: Finnish Foundation for Alcohol Studies.

PARTANEN, J. (1991). *Sociability and intoxication: Alcohol and drinking in Kenya, Africa, and the modern world.* Helsinki: Finnish Foundation for Alcohol Studies.

RORABAUGH, W. (1987). *The alcoholic republic: An American tradition.* New York: Oxford University Press.

ROUCHÉ, B. (1960). *Alcohol: The neutral spirit.* Boston: Little, Brown.

ROYCE, J. E. (1989). *Alcohol problems and alcoholism: A comprehensive survey,* 2nd ed. New York: Free Press.

SEGAL, B. M. (1990). *The Drunken society: Alcohol abuse and alcoholism in the Soviet Union.* New York: Hippocrene.

SEGAL, B. M. (1987). *Russian drinking: Use and abuse of alcohol in pre-revolutionary Russia.* New Brunswick, NJ: Rutgers Center of Alcohol Studies.

SIEGEL, R. K. (1989). *Intoxication: Life in pursuit of artificial paradise.* New York: Dutton.

SNYDER, C. R. (1958). *Alcohol and the Jews: A cultural study of drinking and sobriety.* Glencoe, IL: Free Press.

WADDELL, J. O., & EVERETT, M. W. (EDS.). (1980). *Drinking behavior among southwestern Indians: An anthropological perspective.* Tucson: University of Arizona Press.

WALLACE, F. C. (1970). *The death and rebirth of the Seneca.* New York: Knopf.

WEIL, A. (1986). *The natural mind: A new way of looking at drugs and the higher consciousness,* rev. ed. Boston: Houghton Mifflin.

DWIGHT B. HEATH

Psychological Consequences of Chronic Abuse
Chronic alcohol abuse (heavy drinking over a long period) can lead to numerous adverse

effects—to direct effects such as impaired attention, increased ANXIETY, depression, and increased risk-taking behaviors—and to indirect affects such as impaired cognitive abilities, which may be linked to nutritional deficiencies from long-term heavy drinking.

A major difficulty in describing the effects of chronic alcohol abuse is that many factors interact with such consumption, resulting in marked individual variability in the psychological consequences. In addition, defining both what constitutes chronicity and abusive drinking in relation to resulting behavioral problems is not simply a function of frequency and quantity of alcohol consumption. For some individuals, drinking three to four drinks per day for a few months can result in severe consequences, while for others, six drinks per day for years may not have any observable effects. One reason for this variability is related to genetic differences in the effects of alcohol upon an individual. While not all of the variability can be linked to genetic predispositions, it has been demonstrated that the interactions between individual genetic characteristics and environmental factors are important in determining the effects of chronic alcohol consumption.

Other factors to consider when assessing the effects of chronic drinking relate to the age and sex of the drinker. In the United States, heavy chronic drinking occurs with the greatest frequency in white men, ages nineteen to twenty-five. For the majority of individuals in this group, heavy drinking declines after age twenty-five to more moderate levels and then decreases to even lower levels after age fifty. As might be expected, the type and extent of psychological consequences depend on the age of the chronic drinker. Research has indicated that younger problem drinkers are more likely to perform poorly in school, have more arrests, and be more emotionally disturbed than older alcoholics. Also, younger drinkers have more traffic accidents, which may result from a combination of their heavy drinking and increased risk-taking behavior. Many of the more serious consequences of chronic alcohol use occur more frequently in older drinkers—individuals in their thirties and forties; these include increased cognitive and mental impairments, divorce, absenteeism from work, and suicide. Chronic drinking in women tends to occur more frequently during their late twenties and continuing into their forties—but the onset of alcohol-related problems appears to develop more rapidly in women than in men. In a study

of ALCOHOLICS ANONYMOUS members, women experienced serious problems only seven years after beginning heavy drinking, as compared to an average of more than eleven years for men.

Black and Hispanic men in the United States tend to show prolonged chronic drinking beyond the white male's reduction period during his late twenties. Thus, for many of the effects of chronic drinking discussed below, age, sex, and duration of drinking are important factors that mediate psychological consequences.

NEGATIVE CONSEQUENCES

In the early 1990s, it is currently estimated that between 7 and 10 percent of all individuals drinking alcoholic beverages will experience some degree of negative consequences as a result of their drinking pattern. Most people believe that chronic excessive drinking results in a variety of behavioral consequences, including poor work/school performance and inappropriate social behavior. These two behavioral criteria are used in most diagnostic protocols when determining if a drinking problem exists. Several surveys have found that heavy chronic drinking does produce a variety of school- and job-related problems. A survey of personnel in the U.S. armed services found that for individuals considered heavy drinkers, 22 percent showed job-performance problems. Health professionals also show high rates of alcohol problems, with a late 1980s British survey indicating that physicians experience such problems at a rate of 3.8 times that of the general population. A variety of surveys have consistently shown that chronic excessive drinking leads to loss of support by moderate-drinking family and friends. The dissolution of marriage in couples in which only one member drinks is estimated to be over 50 percent. Often the interpersonal problems that surround a problem drinker can lead to family violence; a 1980s study found that more than 44 percent of men with alcohol problems admitted to physically abusing their wives, children, or significant-other living partners. Survey data also indicate that people who use alcohol frequently are more likely to become involved with others who share their drinking patterns—particularly those who do not express concern about the individual's excessive and altered behavior that results from drinking. This increased association with fellow heavy drinkers as one's main

social-support network can itself result in increased alcohol use.

The interaction between the social setting and the individual, the current level of alcohol intoxication, and past drinking history all play a role in the psychological consequences of chronic heavy drinking. It is impossible to determine which changes in behavior result only from the use of alcohol.

Depression. One major psychological consequence resulting from heavy chronic drinking for a subpopulation of alcohol abusers (predominantly women) is the feeling of loss of control over one's life, commonly manifested as depression. (While not conclusive, some studies suggest that the menstrual cycle may be an additional factor for this population.) In many cases, increased drinking occurs as the depression becomes more intense. It has been postulated that the increased drinking is an attempt to alleviate the depression. Unfortunately, since this "cure" usually has little success, a vicious drinking cycle ensues. While no specific causality can be assumed, research on suicide has indicated that chronic alcohol abuse is involved in 20 to 36 percent of reported cases. The level of suicide in depressed individuals with no alcohol abuse is somewhat lower—about 10 percent. At this time, it is not clear if the chronic drinking results in depression or if the depression is a pre-existing psychopathology, which becomes exacerbated by the drinking behavior. The rapid improvement of depressive symptoms seen in the majority of alcoholics within a few weeks of detoxification (withdrawal) suggests that, for many, depressive symptoms are reflective of toxic effects of alcohol. Regardless of the mechanism, it appears that the combination of depression and drinking can be a potent determinant for increasing the potential to commit suicide.

Aggression. For another subpopulation of chronic alcohol abusers (mainly young men), an increase in overall aggressive behaviors has been reported. Again, there is an indication that these individuals represent a group that has an underlying antisocial personality disorder, which is exacerbated by chronic alcoholic drinking.

Sex Drive. Although it is often assumed that alcohol increases sexual behavior, chronic excessive use has been found to decrease the level of sexual motivation in men. In some gay male populations, where high alcohol consumption is also associated with increased high-risk sexual activity, this decrease

in sex drive does not appear to result; however, for many chronic male drinkers, a long-term consequence of heavy drinking is reduced sexual arousal and drive. This may be the combined result of the decreased hormonal levels produced by the heavy drinking and the decline of social situations where sexual opportunities exist.

Nutritional Deficiency. One common result of prolonged chronic alcohol use is an alteration in daily eating habits (the skipping of meals and/or the exclusion of certain food groups), which can result in nutritional deficiencies. The most common problems are B-complex deficiencies of folic acid, thiamin (B_1), and niacin (B_5). Prolonged nutritional imbalance of these vitamins can lead to a variety of physical and mental consequences, some of which may be exacerbated by the actions of alcohol.

Cognitive Changes. Perhaps the best-documented changes in psychological function resulting from chronic excessive alcohol use are those related to cognitive functioning. While no evidence exists for any overall changes in basic intelligence, specific cognitive abilities become impaired by chronic alcohol consumption. These most often include visuospatial deficits, language (verbal) impairments, and in more severe cases, memory impairments (alcoholic amnestic syndrome). A specific form of dementia, alcoholic dementia, has been described as occurring in a small fraction of chronic alcohol abusers. The pattern and nature of the cognitive effects, as measured on neuropsychiatric-assessment batteries in chronic alcohol abusers, exhibit a wide variety of individual patterns. Also, up to 25 percent of chronic alcoholics tested show no detectable cognitive deficits. Although excessive alcohol use has been clearly implicated in such deficits, a variety of coexisting lifestyle behaviors might be responsible for the cognitive impairments observed. For example, poor eating habits leading to vitamin deficiencies result in cognitive deficits similar to those observed in some alcohol abusers. Head trauma from accidents, falls, and fights (behaviors frequent in heavy drinkers) may also produce similar cognitive deficits. Therefore, it is extremely difficult to determine the extent to which alcohol abuse is directly responsible for the impairments—or if they are a result of the many alterations in behaviors that become part of the heavy-drinker lifestyle.

The specific psychological consequences of chronic drinking are complex and variable, but there

is clear evidence that chronic abuse of alcohol results in frequent and often disastrous problems for the drinker and for those close to him or her.

(SEE ALSO: *Aggression and Drugs, Research Issues; Complications*)

BIBLIOGRAPHY

AKERS, R. (1985). *Deviant behavior: A social learning approach.* Belmont, CA: Wadsworth.

CAHALAN, D. (1970). *Problem drinkers: A national survey.* San Francisco: Jossey-Bass.

FISHBURNE, P., ABELSON, H. I., & CISIN, I. (1980). *National survey on drug abuse: Main findings.* Washington, DC: U.S. Government Printing Office.

ROYCE, J. E. (1989). *Alcohol problems and alcoholism*, rev. ed. New York: Free Press.

VAILLANT, G. (1983). *The natural history of alcoholism.* Cambridge: Harvard University Press.

HERMAN H. SAMPSON
NANCY L. SUTHERLAND

ALCOHOL AND AIDS   Some of the most interesting questions about infection by the HUMAN IMMUNODEFICIENCY VIRUS (HIV) include the following: Why does progression to acquired immunodeficiency syndrome (AIDS) after HIV infection vary in length of time from under 1 year to 15 years or more? Does inhibition of mental functions by alcohol use increase risky sexual behaviors—and thus the chance of becoming infected with HIV? Does alcohol use or abuse affect the production of the virus or the invasion of cells by the virus?

Conventional wisdom based on long experience is that alcohol's relaxation of sexual inhibitions will increase risky sexual behaviors in those who use alcohol even in moderation (2–3 drinks at a party, for example). Risky behaviors include unprotected vaginal or anal intercourse with one or more potentially infected partners; unprotected vaginal or anal intercourse with one or more potentially infected intravenous (IV) drug users or abusers; unprotected oral sex with one or more potentially infected partners. Potentially infected partners are any who do not know for certain that they are not infected with HIV—in other words, any who have not been tested, who have tested positive but say nothing or lie about it, who have had unprotected sex since their test, or who have developed HIV since the test (because they tested too soon and have not been retested in the prescribed fashion). *Anyone*, therefore, *might be* a potentially infected partner.

COCAINE and other drugs of abuse are commonly used along with alcoholic beverages (which contain ethanol, sometimes called ethyl alcohol). They appear to synergize, in terms of affecting behavior, thereby enhancing the risk of HIV infection should one be exposed. In addition, in vitro studies in the laboratory have shown that alcohol use increases the susceptibility of lymphocytes (white blood cells) to HIV infection; this increases the likelihood that an invading HIV virus will successfully cause disease.

While epidemiological studies in humans have been inconclusive, animal models and test-tube studies with human cells show clearly that alcohol increases the immune damage resulting from retrovirus infection—and HIV is a retrovirus. Alcohol, then, would decrease disease and tumor resistance further in AIDS patients, may accelerate development of AIDS by those who are HIV infected, may increase HIV production by cells, and may increase cellular susceptibility to HIV infection. Alcohol use by AIDS patients should therefore by discouraged.

AIDS patients suffer from various opportunistic pathogens, which normally do not reproduce sufficiently to infect immunologically normal people or even HIV-positive people who have not progressed to the AIDS stage, with its severe immune suppression. Alcohol and cocaine exacerbate the immune dysfunction in studies with mice; they also increase the extent of colonization in mice with AIDS by two opportunistic parasites—*Giardia* and *Cryptosporidium*. Normal mice and humans quickly clear such parasites, being essentially resistant. Alcohol is a cofactor that accentuates immune damage, therefore, resulting from the HIV retrovirus.

BIBLIOGRAPHY

WATSON, R. R. (1993). Resistance to intestinal parasites during murine AIDS: Role of alcohol and nutrition in immune dysfunction. *Parasitology, 5,* 69–74.

WATSON, R. R. & GOTTESFELD, Z. (1993). Neuroimmune effects of alcohol and its role in AIDS. *Advances in Neuroimmunology, 3,* 151–162.

WATSON, R. R., ET AL. (1994). Alcohol, immunomodulation, and disease. *Alcohol and Alcoholism, 29,* 131–139.

RONALD R. WATSON

ALCOHOL- AND DRUG-FREE HOUSING

Alcohol- and drug-free (ADF) housing, also called *sober housing,* or *sober living environments,* or *alcohol-free living centers,* provides domestic accommodation for people who choose to live in an environment that is free of alcohol and/or drugs. ADF housing is ordinary housing, located in residentially zoned areas, distinguished only by the residents' shared commitment not to use alcohol or other drugs.

By definition, ADF housing excludes formal treatment or recovery services on the site. The philosophic premise of ADF housing is that the sober living environment is itself the "service" for residents. ADF housing provides a setting for daily living that supports residents' efforts to maintain sobriety among themselves.

As a practical matter, the presence of on-site human services would subject ADF residences to state or local licensing of staff and to certification of their facilities. Under such circumstances, residences would be treatment facilities and no longer ADF housing. As such, they would lose protections afforded to ADF housing by the Fair Housing Amendments Act of 1988 and become subject to zoning laws that prohibit treatment programs in residential neighborhoods. The result would be the systematic exclusion of such residences from the safe and economically stable areas most conducive to recovery. Under provisions of the Fair Housing Amendments Act, *no* regulation of any ADF residence is legal unless such requirements are imposed on *all* private residences in the surrounding community with the same zoning. The Department of Justice is vigorously enforcing the act and has a number of suits pending across the country that bear directly on sober housing.

Such protections aside, ADF housing is a creature of the marketplace. For it to be affordable, public sponsorship of some sort must close the gap between market rents or mortgage costs and what residents can reasonably pay. A bewildering variety of affordability strategies for rent, property, and construction-cost subsidies has been worked out for individual ADF housing projects, but very few cities or other local jurisdictions have established formal policies to create and to sustain ADF housing.

Even so, for three principal reasons, interest in affordable ADF housing has increased remarkably in the last few years. First, local units of government, special boards, and districts, are under pressure to make provisions for low-income housing. Second, recent studies indicate that affordable ADF housing helps homeless and very low-income people maintain sobriety following initial successes in treatment/recovery programs. Third, ADF housing now figures prominently in discussions of using "social-model" recovery programs as vehicles for the cost-efficient deployment of treatment and recovery services. The search for economical ways to provide such services has led health-care system planners to reduce or eliminate the use of expensive residential-treatment programs. In conjunction with outpatient treatment and adjunct health and social services, affordable ADF housing increasingly is viewed as an alternative.

ADF housing follows three simple tenets: (1) residents must remain alcohol and drug free; (2) rent must be paid on time; and (3) residents must abide by provisions of the landlord-tenant agreement. This agreement may stipulate only that tenants must refrain from disruptive behavior that provides grounds for eviction for cause in local ordinances (typically these are violence or threats of violence, illegal activity, destruction of property, and perpetuation of undue nuisances). However, it may also impose house rules that dictate curfews, limit overnight guests, restrict automobile ownership, delegate house chores, and so forth. As long as the tenant's participation in the regulated household is voluntary, and as long as the rules do not violate civil rights, ADF housing may be highly structured and closely governed.

An ADF residence can be program-affiliated or free-standing. Program-affiliated sober houses are tied to the treatment and recovery orientations of particular organizations. Residents are likely to come from the sponsoring program, and so will have been exposed to the sponsor's procedures and values. Accordingly, the residence will reflect the philosophy and practices of the parent organization. Free-standing houses operate more along the lines of conventional residences and rely exclusively on self-government.

Sober housing can be run by staff or residents. Staff-run houses operate under the direct management of owners, program operators, or housing management firms. Site managers are compensated and often are recovering people who have several more years of sobriety than the residents of the house. (It should be emphasized that for obvious legal reasons, they are not "treatment personnel" and their activities do not comprise formal service interventions.) In resident-run houses, residents take full responsibility for all aspects of house operation related to maintaining sobriety: admissions, maintenance of the house's social environment, disciplinary action, community relations, and physical maintenance.

Resident-run ADF houses may be democratic or oligarchic. Democratic houses may be highly egalitarian—the residents have equal votes and share equally in houses duties, as in OXFORD HOUSES—or they may be stratified, formally or informally, by residents' seniority in sobriety or by other measures of status. Some therapeutic communities, such as Delancey Street in San Francisco, California, are oligarchic. In general, larger resident-run programs tend to be more oligarchic in nature, though some provide many opportunities for resident participation in management and operation of the house, as does Beacon House in San Pedro (Los Angeles), California.

The interests of landlords, owners, and program operators notwithstanding, the rules of ADF households seek to protect a sober environment. For their own peace of mind, residents often prefer places that impose restrictions in the service of restraint, predictability, and good order. Entrance and eviction policies are critical in this respect.

ADF housing is not magically exempt from our meaner or more self-serving impulses toward exclusiveness. However, in well-run residences, entrance decisions focus only on the capacity of an individual to benefit from the house milieu. Such decisions, in which residents usually play an important part (if only to exercise rights of refusal), consider the structure and character of the house "program" in relation to the needs of a potential resident. Thus, a prospective resident with an expressed need or desire for a highly structured environment would be discouraged from entering a house with little structured activity. All recovering people can benefit from ADF housing, regardless of their treatment histories or other circumstances. But as ADF housing repre-

sents a spectrum of possibilities for group living, particularly concerning the extent to which the environment is regulated, its potential consumers must find a good fit. Variety in ADF housing is therefore essential.

Residents are free to live in sober housing as long as they follow house rules. Any fixed time limit on length of tenancy is contrary both to the spirit and (in most communities) to the laws of residency. However, those few who violate basic house rules must leave. Although management or a residents' council makes the final determination that basic house rules have been broken, violations usually are obvious in a well-run house. If formal proceedings are necessary, the only question to be settled is whether the violation actually occurred; thereafter, commencement of the eviction process is automatic. Residents usually understand well the penalties for violations; those who know they will be evicted often choose to leave immediately.

Violation of sobriety policy is the most common reason for a tenant's eviction or voluntary separation from ADF housing. ADF houses vary somewhat in their toleration of drinking and drug use. Nearly all take a "no-slip" approach, in which a single episode of drinking or using means that the person must leave. Some permit individuals to slip once or twice before being evicted. Residents generally find that firm no-drinking, no-using policies promote a sense of equity and maintain tranquility in the house.

Some houses prefer a "client-centered" policy that permits the drinker/user to remain in contact with counselors and otherwise receive help without having to leave the residence. The risk in such a policy is that repeated drinking or using episodes among residents will disturb the social environment of the house. Some multicomponent programs with very large facilities handle this issue by asking the drinker/user to move from the sober housing component of the program to the detoxification or primary-recovery unit. Some ADF residences that permit off-site but not on-site drinking or using have found this policy to work well to reduce chronic intoxication among those for whom continuous sobriety is not a realistic expectation.

Eviction proceedings are time consuming and filled with legal procedures that have been designed to protect and extend the rights of the resident. A signed landlord-tenant agreement that specifically proscribes drinking and/or using offers the strongest

starting point for eviction; but an eviction process can sometimes drag on for weeks or months even in the most clear-cut cases. Successful ADF houses rely on resident participation in management. (A residents' manual provides a model for peer-based response to a resident's drinking or drug use.)

ADF housing works best when residents themselves actively maintain the collective sobriety of their home. Management's task is to create a living environment in which sobriety is respected and maintained by the residents. Frequent contact between management and residents, both informally and through regular house meetings, provides a medium for interaction that quickly identifies a resident who has been drinking or using. Secretiveness and the absence of communication are important signs that something isn't right.

Architectural design plays a central role in creating an ADF housing environment that promotes both open social interaction and mutual accountability. The basic floor plan of the residence makes a critical contribution. "Open" circulation systems in buildings bring people into contact with one another. Examples are open-plan houses in which space flows from one room into another; areas that have nooks and side areas where people can sit or stop to chat; and centrally located corridors with wide openings directly into the rooms they serve.

Spaces that invite social contact are called "sociopetal." They subtly but powerfully encourage people to socialize, to greet each other, to notice one another during the day. Sociopetal spaces are lively, engaging places, in stark contrast with "sociofugal" spaces whose circulation systems emphasize separation and isolation. Sociofugal circulation systems keep people apart by using long corridors such as those found in hotels and rooms isolated by stairs and such. Sociofugal spaces are dull, depressing, and sometimes disorienting or frightening to anxious people who cannot easily see what is going on in the building. Developers of successful ADF residences understand the influences of architecture. Specialized design will play an important role in the future development of ADF housing, particularly affordable ADF housing for single parents or couples with children, who require functionally different and far more space than do single people.

It is not clear how quickly or in which specific directions ADF housing development will proceed. The National Institute on Alcohol Abuse and Alcoholism (NIAAA) promoted sober residences in its

homelessness demonstration grant program, but this is winding down (as of the mid-1990s) and no new grants will be made. The SUBSTANCE ABUSE AND MENTAL HEALTH SERVICES ADMINISTRATION (SAMHSA) has no program in the offing to support development of ADF housing, nor does the Department of Housing and Urban Development (HUD)—though pending homelessness legislation may change the situation at HUD. The Anti-Drug Abuse Act of 1988 provided that every state receiving federal block grant funds for alcohol and other drug programs establish a revolving fund of at least 100,000 dollars to make start-up loans for sober housing. This was a foot in the door for ADF projects, although the relatively paltry funds involved do not provide much leverage by themselves.

In addition, government at all levels—federal, state, or local—has a strong tendency to regulate and standardize the activities it supports and to demand accountability to its agencies rather than to other relevant constituencies, such as consumers of sober housing. Any governmental attempt to regulate the environment of sober housing raises far-reaching questions about the invasion of privacy, for ADF housing is by definition *ordinary* and protected from oversight not extended to other domestic households. The philosophy of ADF housing, and more practically, its necessary and salutary diversity, is not compatible with intrusive regulation.

Still, considerable potential exists for an explosion of interest and activity. As noted, ADF residences may play an important role in the reform of the system of services for people with alcohol and drug problems. Although interest in sober housing originated in the search for ways to support homeless and very low-income people completing residential treatment and recovery programs, the useful scope of sober housing may be much broader. Combined with various services in the surrounding community, perhaps it is a good alternative to expensive, traditional forms of residential treatment. Private insurance companies and housing entrepreneurs already are developing sober living arrangements that are attractive to the middle and upper classes, as well as to low-income people.

It is also possible that sober residences appeal to many more people than those actively engaged in treatment or recovery programs. Just as some university dormitories have become sober housing for self-selected students, perhaps sober residences will become part of a larger trend to reconfigure domestic

living arrangements to fit our changed family demography and our changing styles of life. In the heyday of the temperance movement, the United States was littered with dry hotels and boarding houses that catered to a preference of style, not merely or only to prohibitionist sentiment. It is not hard to imagine a future in which likeminded citizens cause "dry" households to reappear.

(SEE ALSO: *Treatment, History of*)

BIBLIOGRAPHY

McCARTY, D., ET AL. (1993). Development of alcohol and drug-free housing. *Contemporary Drug Problems, 20,* 521–539.

MOLLOY, J. P. (1990). *Self-run, self-supported houses for more effective recovery from alcohol and drug addiction.* DHHS Publication no. ADM 90-1678. Rockville, MD: Office of Treatment Improvement.

OXFORD HOUSE, INC. (1988). *The Oxford House manual.* Great Falls, VA: Author.

WITTMAN, F. D. (1993). Affordable housing for people with alcohol and other drug problems. *Contemporary Drug Problems, 20,* 541–609.

FRIEDNER D. WITTMAN
JIM BAUMOHL

ALCOHOL BEVERAGE CONTROL   *See* Distilled Spirits Council

ALCOHOL DEPENDENCE SYNDROME *See* Diagnosis of Drug Abuse; Disease Concept of Alcoholism and Drug Abuse.

ALCOHOLICS ANONYMOUS (AA)   This is a fellowship of problem drinkers, both men and women, who voluntarily join in a mutual effort to remain sober. It was started in the United States in the 1930s and has been maintained by alcohol-troubled people who had themselves "hit bottom"—they had discovered that the troubles associated with their drinking far outweighed any pleasures it might provide. AA serves, without professional guidance, a significant minority of the population of alcoholics in the United States. Various professionally oriented treatments serve other significant minorities of alcoholics.

AA is not the only hope for alcoholics; nor is it everything they need. Nevertheless, its program and meetings have restored thousands of alcoholics to abstinence, both in the United States and in many other countries. In 1992, the General Service Office of AA, located in New York City, reported a worldwide total of 87,403 AA groups, 48,747 of them in the United States, with an additional 1,783 in U.S. correctional facilities, and 5,173 in Canada, leaving 31,700 in other countries. The report estimated there were almost 2 million individual members in these groups worldwide; over half (1,079,719) lived in the United States.

THE TWELVE STEPS OF ALCOHOLICS ANONYMOUS

AA's program for remaining sober is called the Twelve Steps. They are:

1. We admitted we were powerless over alcohol—that our lives had become unmanageable.
2. Came to believe that a Power greater than ourselves could restore us to sanity.
3. Made a decision to turn our will and our lives over to the care of God *as we understood Him.*
4. Made a searching and fearless moral inventory of ourselves.
5. Admitted to God, to ourselves, and to another human being the exact nature of our wrongs.
6. Were entirely ready to have God remove all these defects of character.
7. Humbly asked Him to remove our shortcomings.
8. Made a list of all persons we had harmed, and became willing to make amends to them all.
9. Made direct amends to such people wherever possible, except when to do so would injure them or others.
10. Continued to take personal inventory and when we were wrong promptly admit it.
11. Sought through prayer and meditation to improve our conscious contact with God *as we understood Him*, praying only for knowledge of His will for us and power to carry that out.
12. Having had a spiritual awakening as the result of these steps, we tried to carry this message to

alcoholics and to practice these principles in all our affairs.

The steps are based on suggestions gleaned from the collective experiences of members about how they achieved sobriety—and then maintained it. In this sense, AA is a collectivity of mutual help groups more than it is discrete individuals engaging in self-help. At meetings both open to the public and "closed" (for members only), the Twelve Steps are closely examined, and members frankly tell their own versions of their drinking histories—their AA "stories"—and describe how the AA program helped them to achieve sobriety.

Membership in AA depends on an individual's declaration of intention to stop drinking. An AA group comes into being when two or more "drunks" join together to practice the AA program. "Loners" are relatively few, but some exist. There are no dues or fees for membership; AA is self-supporting and is not associated with any sect, denomination, political group, or other organization. It neither endorses nor opposes any causes. These points, and other basic descriptions of AA, appear on the first page of AA's monthly magazine, *The Grapevine*. Although AA is not set up as a centralized organization, a commonly shared set of traditions guides their meetings and treatment strategies. For example, one of the Twelve Traditions sets forth AA's singleness of purpose—to help alcoholics achieve and sustain sobriety; another tradition underscores the necessity for the anonymity of members, as a way to avoid personality inflation and to promote humility. Over time, the Twelve Traditions have come to be as vital a part of AA as the Twelve Steps. They are:

1. Our common welfare should come first; personal recovery depends upon A.A. unity.
2. For our group purpose there is but one ultimate authority . . . a loving God as He may express Himself in our group conscience. Our leaders are but trusted servants . . . They do not govern.
3. The only requirement for A.A. membership is a desire to stop drinking.
4. Each group should be autonomous except in matters affecting other groups or A.A. as a whole.
5. Each group has but one primary purpose . . . to carry its message to the alcoholic who still suffers.
6. An A.A. group ought never endorse, finance, or lend the A.A. name to any related facility or out-

side enterprise, lest problems of money, property and prestige divert us from our primary purpose.
7. Every A.A. group ought to be fully self-supporting, declining outside contributions.
8. Alcoholics Anonymous should remain forever nonprofessional, but our service centers may employ special workers.
9. A.A., as such, ought never be organized; but we may create service boards or committees directly responsible to those they serve.
10. Alcoholics Anonymous has no opinion on outside issues; hence the A.A. name ought never be drawn into public controversy.
11. Our public relations policy is based on attraction rather than promotion; we need always maintain personal anonymity at the level of press, radio, and films.
12. Anonymity is the spiritual foundation of all traditions ever reminding us to place principles before personalities.

Written in 1939, *Alcoholics Anonymous* (Alcoholics Anonymous World Services, 1939, 1955, 1976), is the basic text that outlines the experiences of the first 100 members in staying sober. It is fondly referred to as "The Big Book."

ORIGINS

The unlikely coming together of the two cofounders of AA, "Bill W." and "Dr. Bob" (William Griffith Wilson, a stockbroker, and Robert Holbrook Smith, a surgeon)—both pronounced alcoholics—is probably the most concrete event in AA's origins. Anonymity was basic, but there were other factors, among them Bill W.'s experiences prior to contact with Dr. Bob. Had it not been for the readiness these experiences generated in Bill W. to interact in a unique way with Dr. Bob, their initial meeting might have turned out to be the fifteen-minute encounter the doctor had initially planned.

First, Bill W. had reached a point of profound hopelessness. Second, he had had a chance encounter with an old drinking friend—Ebby T.—who despite strong and similar feelings of hopelessness had achieved considerable sobriety. Third, Ebby T. attributed this accomplishment to the Oxford Group, a nondenominational movement with no membership lists, rules, or hierarchy. It embraced specific ideas that would soon find their way into AA practice: For example, members alone were powerless to solve

their own problems; they must carefully examine their behavior and try to make restitution to others they had damaged; and they practiced helping others, resisting personal prestige in the process.

Next, a severe relapse had forced Bill W. into a hospital, where he had visits from Ebby T. and where Bill W. longed for the sobriety Ebby seemed to have. Following a cry of lonely desperation and agony, he reports that "the result was instant, electric, beyond description. The place lit up, blinding white ... came the tremendous thought, 'You are a free man'" (W.W., 1949, p. 372). As a result of these accumulated experiences, Bill W. decided on two strategies. One was to take his story to other alcoholics and the other was to become an evangelist, because of his spiritual experience. Even though he brought many alcoholics home with him and preached at them, he utterly failed and almost returned to drinking himself. But his account underscores how he came to realize that "to talk with another alcoholic, even though I failed with him, was better than to do nothing" (W.W., 1949, p. 374).

Two other experiences accumulated. He discovered that some medical authorities considered alcoholism a disease. Almost instantly, he replaced evangelism with science. Finally, in an effort to recoup some financial losses, he pursued a slim business opportunity in Akron, Ohio, on May 11 and 12, 1935. An Episcopalian minister there put him in contact with an Oxford Group member who, in turn, arranged for a meeting the next day with Dr. Bob. Abandoning his evangelical approach, he used his newly found scientific/disease approach with the doctor. An immediate rapport developed between the two men, and they talked until late into the night. Dr. Bob was, in effect, Bill W.'s first follower (Trice & Staudenmeier, 1989). Dr. Bob had only one "slip" during the next month, but soon thereafter the two began working together on other alcoholics, using the sickness/scientific approach. By August 1936, AA meetings, within an Oxford Group context, were being held both in Akron and New York City. Soon, however, both Bill W. and Dr. Bob decided to sever their relationship with the Oxford Groups, and the small AA groups were on their own. By 1940, their newly formed board of trustees listed twenty-two cities in which groups were well established and holding weekly meetings. Soon thereafter, *The Saturday Evening Post*, a popular magazine with a wide circulation, published an article simply entitled "Alcoholics Anonymous." It proved to be a compelling media event for the AA program, and a flood of positive responses and new groups resulted. Ever since then, AA growth has steadily expanded.

Two coordinating groups have acted to link together the thousands of AA local groups in the United States and abroad. In AA's first year the founders, along with members of the first New York City group, formed a tax-free charitable trust with a board of trustees composed of both alcoholic and nonalcoholic members. It acted as a mechanism for the collection and management of voluntary contributions and as a general repository of the collective experience of all AA groups.

Today the board of trustees consists of fourteen alcoholic and seven nonalcoholic members who meet quarterly. At an annual conference, specific regions elect the alcoholic board members for four-year terms. The board appoints the nonalcoholic members for a maximum of three terms of three years each. An annual conference was established in 1955 at AA's Twentieth Anniversary Convention. It expresses to the trustees the opinions and experiences of AA groups throughout the movement. A General Service office (GSO) in New York City interprets and implements the deliberations of these two groups on a daily basis.

THE PROCESS OF AFFILIATION

Affiliation is a process, not a single, unitary happening within AA. Its elements and phases act to select and make ready certain alcoholics and problem drinkers for affiliation, leaving behind others with less readiness. The process begins before the problem drinker ever goes to a meeting (Trice, 1957). If the person has heard favorable hearsay about AA; if long-time drinking friendships have faded; if no will-power models of self-quitting have existed in the immediate background; and if the drinker has formed a habit of often sharing troubles with others—the stage is set for affiliation. It is further enhanced if, upon first attending meetings, the person has had experiences leading to the decision that the troubles associated with drinking far outweigh the pleasures of drinking (i.e., "hitting bottom"). Typically, this means that affiliates, contrasted with nonaffiliates, had a longer and more severe history of alcoholism—and those with more severe alcohol problems are more likely to make consistent efforts to affiliate than are those with less severe problems (Trice & Wahl, 1958; Emerick, 1989b).

Five other specific phases follow from those forces that make for commitment to the AA program: (1) first-stepping, (2) making a commitment, (3) accepting one's problem, (4) telling one's story, and (5) doing twelfth-step work (Rudy, 1986). First-stepping involves the initial contact with AA; it often entails orientation meetings that dwell on the group's notions of alcoholism as a disease and on step one in the twelve-step program: "We admitted we were powerless over alcohol . . . that our lives had become unmanageable." The newcomers also will probably become associated with an AA guide, who may soon become the newcomer's sponsor. Quick action by the AA group—closeness of initial contact—to include the newcomer increases the likelihood of affiliation (Sonnenstuhl & Trice, 1987). This expresses itself in pressing any obviously interested newcomer into a challenge of "ninety meetings in ninety days." In effect, the receiving group seeks to keep a close watch over the newcomer, gently forcing the person to forego other commitments and increase commitments to the AA program.

Decisive third and fourth phases are the acceptance of telling one's drinking story, with the beginning phrase, "I'm Chris X and I'm an alcoholic." Throughout the initial weeks and months, newcomers are gently and sometimes bluntly pressed to realize that they *are* alcoholics. They are encouraged to "go public" and tell their stories before the entire group at an open meeting. In numerous instances, newcomers may already have decided that they are alcoholic. In other cases, it may be a lengthy process of self-examination before this identity transformation occurs. In still others, it may never transpire, making them suspect as real AAs. In any event, the public telling of one's story is an act of commitment that symbolizes a conversion of self into a genuine AA member. Members counsel newcomers on the appropriate first time to tell their stories, and their narrations are cause for many congratulations. Much applause typically attends this open act of commitment.

A final phase involves the literal execution of the program's twelfth step: "Having had a spiritual awakening as a result of these Steps, we tried to carry this message to alcoholics, and to practice these principles in all our affairs." In essence, doing twelfth-step work exemplifies one of AA's basic philosophies, namely, one is never recovered from the disease; one is only "recovering." As a consequence, a member

can maintain sobriety only by remaining active in AA and by steadily engaging in carrying the program to those who are still active alcoholics. In short, by doing twelfth-step work, members reinforce their membership and the new definition of self.

Throughout this affiliation process, another dynamic is also at work—"slipping,"—a relapse into drinking by a recovering AA member. After reviewing six relevant studies, plus a summary of his own fieldwork with AA, Rudy (1980, p. 728) reported that among both newly committed and longer-term members "slipping is a common occurrence, but it is possible that it serves a function in A.A. . . . [It is] a deviant behavior and the function of this deviance is boundary maintenance." The response of most AA members to another's slipping is sympathy and understanding, sentiments that in turn enhance group solidarity. In essence, "their abstinence is dependent on interaction with those who slip" (Rudy, 1980, p. 731).

WHO AFFILIATES?

What are the characteristics of those who do undergo the affiliation processes, contrasted with those who do not, even though exposed to the possibility? Demographic variables such as age, social class, race, employment status, and parental socioeconomic status have been consistently found to be *unrelated* to membership (Trice & Roman, 1970; Emerick, 1989). These findings provide considerable certainty about the existence of often-alleged demographic barriers to AA affiliation—they, in effect, fail to deter affiliation.

Less certainty can be attributed to significant psychological characteristics that have been found to encourage affiliation. As in evaluations of psychotherapies in general (Eysenck, 1952; Rachman & Wilson, 1980), researchers cannot predict with any certainty who will affiliate. Despite this, certain personality features have been systematically found to distinguish between affiliates and nonaffiliates (Ogborne & Glaser, 1981; Ogborne, 1989).

Several studies have suggested that, among other things, A.A. members can be distinguished from other heavy drinkers with respect to personality and perceptual characteristics. . . . The authors suggest that A.A. affiliation is associated with authoritarianism and conformist tendencies, high affiliation needs, proneness to

guilt, religiosity, external control and field dependency [Ogborne, 1989, p. 59].

Ogborne (1989) also reports on two additional studies that support the belief that AA attracts individuals with certain emotional makeups. Ogborne's overall findings were that alcohol-troubled persons who expressed group adherence, extroversion, submissiveness, and conservatism were attracted to AA and its program. Overall, these findings appear to be consistent with the role demands made on members. For example, the sociability and affiliativeness themes found among those who do affiliate, as compared to nonaffiliates, seem to match the heavy group interactions expected of members.

All things considered, affiliation with AA is a distinctively selective process that fits only a distinct minority of those in the alcohol-abusing population. Although the exact proportion of the population helped by AA is unknown, even AA's critics recognize that it is substantial. Other specific types of therapies may do proportionally somewhat better or worse, but a reasonable estimate would be that AA is associated with fairly typical improvement rates.

Current studies strongly suggest that AA appeals to a highly specific and select segment and, by doing so, further suggest that other therapies are also selective as to their appeal. These points underscore the need for service providers to be aware of the diverse makeup of the problem-drinking population. Assessment and services need to be far more individualized than they have been, so that assignments may be made to the most appropriate organizations, institutions, or therapies.

THERAPEUTIC MECHANISMS

As with psychotherapies in general, the effectiveness of AA has not been convincingly established. For example, some problem drinkers drop out after the first two or three AA meetings. Nevertheless, for those who remain, AA has unique and distinctive features that contribute to its therapeutic effectiveness.

By definition, as problem drinkers move into addiction, alcohol comes to be central to their lives. How can this centrality be reduced and a new conception of alcohol be put in its place? AA experiences provide a new orientation, not only to alcohol, but to self and to others. "One of A. A.'s great

strengths lies in the quality of its social environment: the empathetic understanding, the acceptance and concern which alcoholics experience there which, along with other qualities, make it easier to internalize new ways of feeling, thinking, and doing" (Maxwell, 1982). Even brief exposure to AA introduces the alcoholic to the idea that self-regulation seems to be rarely achieved alone by self-reliance and willpower. Its basic premise describes the compelling sense of ego powerlessness—but immediately offers the potent substitute of a viable community that provides individual attention, an explanation of alcoholism, and simple prescriptions for sobriety. "In a community that shares the same distresses and losses, accepts its members' vulnerabilities and applauds and rewards successes, A.A. provides a stabilizing, sustaining, and ultimately, transforming group experience" (Khantzian & Mack, 1989, 76).

Within the AA community, there are group-based specific therapeutic strategies unlikely to exist in professionally directed psychotherapy. Examples include (1) empirically based hope; (2) direct attack on denials; (3) practical guidelines for achieving sobriety; and (4) one-to-one sponsorship. When problem drinkers first attend meetings they are immediately aware of others who have confronted their very problem, and they hear these people speak about dramatic improvements in their lives via the AA program. Moreover, the AA program consistently reminds them that denial of the realities surrounding their drinking is a major barrier to any change. Telling one's story, either publicly or in closed sessions, helps to dissolve the entrenched denial systems that ward off therapeutic changes.

The Twelve Steps are structured phases that provide an organized approach to the confusions and frustrations of an individual's attempt to cope with alcohol problems. This is especially so when members reach the developmental stage where twelfth-step work is indicated. They dramatically see themselves as they once had been, and this reinforces their need to "work the program." In addition, simple, practical guidelines are repeated as group-tested ways to avoid using alcohol as an adjustment technique: "First things first," "one day at a time," "easy does it," and practical advice about how to work the program.

AA typically arranges for the informal sponsorship of newcomers—who often identify with and closely relate to their sponsors. Sponsors are recover-

ing alcoholics typically available at all hours of the day and night for phone or in-person discussions and crises. These valuable treatment strategies are voluntary and free of any monetary cost or financial obligation. Drinkers who drop out or who reject active membership in AA may nevertheless be substantially helped by primary or secondary exposure to AA and its unique but widely publicized or modified therapeutic mechanisms. (Several organizations in the alcohol or drug recovery field have been working along similar but modified lines.) It is impossible to estimate the numbers of those helped by such exposure, but they are surely numerous and should not be discounted.

CRITICAL EVALUATIONS

Probably the most widespread and long-standing critical assessment of AA centers around the question of the selective nature of membership. Thus, critic Stanton Peele (1989, 57) bluntly insists: "In fact, research has not found A.A. to be an effective treatment for general populations of alcoholics." Again, however, neither has any given professional psychotherapeutic method been found to be effective for general populations of the psychoneurotic. In 1981, Ogborne and Glaser predicted that evidence will soon be found for the effectiveness of AA, but "it will be limited to a particular, identifiable subgroup of persons with alcohol problems." This concern had been expressed since the 1950s (Trice & Staudenmeier 1989), and more evidence of AA's selective nature came from Walsh et al. (1991).

Walsh and her colleagues randomized 227 employed problem drinkers into compulsory inpatient treatment, compulsory attendance at AA, and a choice of options. During a two-year follow up, the researchers used measures of performance with drinking and drug use to gauge effectiveness. They concluded that the hospital group fared best and that the group assigned to AA fared least well. Many of those randomly assigned to AA probably lacked the readiness and the emotional makeup that appear to be required for affiliation. Matching patients to specific treatments has been advocated for many years (Ogborne & Glaser 1981; Pattison 1982), and the Walsh study certainly indicates the importance of matching in the case of AA.

Emerick (1989) has broadened the array of criticisms of AA to include (1) that AA has denounced in the media scientific discoveries that contradict its "formal ideology" and dogma; (2) AA has brought pressures to bear in an effort to suppress various psychological findings. Emerick quickly acknowledges, however, that "it is these very characteristics . . . that provide for AA's strength and effectively preserve its boundaries and identity" (p. 5). This criticism boils down to charges that AA is anti-intellectual and antiprofessional. Second, harmful effects may come to those who "do not fit comfortably in the organization" (p. 9). These harmful effects include beliefs that "slipping" inevitably leads to loss of control, making for more problems than otherwise. Some members may despair and lose hope when they discover they do not mesh with AA's norms and beliefs. Third, there is a risk of becoming "AA addicts" who spend so much time and energy on the AA program that they neglect other areas of life such as family and job. Fourth, AA groups may contain alcohol-troubled persons who also suffer from other psychiatric disorders—i.e., schizophrenia and anxiety disorders—that should be directly treated, but are covered up by AA ideologies. Finally, AA members may develop dual overlapping relationships inside AA that are ultimately harmful, i.e., a newcomer becomes the lover of an established member, or a sponsor enters into an unfortunate business partnership with a sponsoree.

Other negative judgments of AA that have been voiced at one time or another are that it is tilted toward being a religion by too much emphasis on a "higher power"; local groups are not nearly as accepting of drunks as advertised; it suffers from too much adulation and consequently often becomes a "dumping ground" at which companies and courts require compulsory attendance; and it insists that members come to accept the label of "alcoholic"—a label that continues to be highly stigmatic outside AA and tends to repel many of those who are inclined to affiliate.

Understandably, these criticisms have fueled alternative groups that claim to help members cope with alcohol problems but without many of the beliefs and rituals of AA. Apparently, many of the members of these groups are AA dropouts. Beyond this observation, no systematic research efforts have been mounted to determine how affiliation is achieved, and with what kinds of problem drinkers these alternatives to AA are effective. For example, RATIONAL RECOVERY (RR), SECULAR ORGANIZATION

FOR SOBRIETY (SOS), and WOMEN FOR SOBRIETY (WFS) have all claimed to be alternative self-help groups for alcoholics. They tend to reject the notion that alcoholism is a disease and advocate instead personal responsibility. Also, they underscore individual willpower rather than AA's belief in a higher power (Gelman, Leonard, & Fisher, 1991). Regarding the charge that AA is overly religious, numerous close observers, including the present writer, have concluded that religion plays a minor role in the practical day-to-day effort of AAs to "work the program."

ADAPTATION OF AA TO OTHER DISORDERS

Despite the criticisms that have been directed against AA, its format and beliefs have nevertheless been applied to a wide variety of other addictions and behavior disorders. For example, NARCOTICS ANONYMOUS (NA) (estimated in 1979 to have about 700 groups in practically every U.S. state and in several other countries) first applied the AA pattern to drug addicts at the U.S. PUBLIC HEALTH SERVICE HOSPITAL at Lexington, Kentucky, in 1947. In 1948, and in 1953, groups of AA members who were also drug addicts formed an independent NA group in New York City and in Sun Valley, California. The resemblance to AA was made even more remarkable by the fact that the magazine that had made AA well-known (*The Saturday Evening Post*) also gave NA a national audience through a lengthy piece that played up similarities with AA (Ellison, 1954).

Similarly, AA's beliefs and strategies have been adapted to help people with a broad spectrum of other problems, including excessive buying, sexual excesses or deviations, gambling, child abuse, overdependence on others, eating disorders, and excessive shame and guilt. In addition, AL-ANON family groups and ALATEEN groups have adapted AA's philosophy to family, children, and friends of problem drinkers. Many others could be cited. Veteran AA members point to this great proliferation as evidence that AA's influence goes well beyond its impact on AA members. They argue that this widespread adaptation to other disorders demonstrates the essential value and appeal of the AA program.

(SEE ALSO: *Alcoholism; Gambling As an Addiction; Treatment*)

BIBLIOGRAPHY

ALCOHOLICS ANONYMOUS. (1939, 1955, 1976). *Alcoholics Anonymous: The story of how many thousands of men and women have recovered from alcoholism.* New York: Alcoholics Anonymous World Service.

ALCOHOLICS ANONYMOUS WORLD SERVICE. (1993). *A.A. directory, 1992–1993.* New York: Author.

ELLISON, J. (1954). These drug addicts cure one another. *Saturday Evening Post, 227* (#6, Aug. 7), 22–23; 48–52.

EMERICK, C. (1989a). Overview: Alcoholics Anonymous. In M. Galanter (Ed.). *Recent developments in alcoholism: Treatment research.* Vol. 7. New York: Plenum.

EMERICK, C. (1989b). Alcoholics Anonymous: Membership characteristics and effectiveness as treatment. In M. Galanter (Ed.), *Recent developments in alcoholism: Treatment and research.* Vol. 7. New York: Plenum.

EYSENCK, J. (1952). The effects of psychotherapy: An evaluation. *Journal of Consulting Psychology, 16,* 319–324.

GELMAN, D., LEONARD, E., & FISHER, B. (1991). Clean and sober and agnostic. *Newsweek,* July 8, 62–63.

KHANTZIAN, E. J., & MACK, J. E. (1989). Alcoholics Anonymous and contemporary psychodynamic theory. In M. Galanter (Ed.) *Recent developments in alcoholism: Treatment research.* New York: Plenum.

MAXWELL, M. (1982). Alcoholics Anonymous. In E. Gomberg, H. R. White, & A. Carpenter (Eds.), *Alcohol, science and society revisited.* Ann Arbor: University of Michigan Press.

OGBORNE, A. C. (1989). Some limitations of Alcoholics Anonymous. In M. Galanter (Ed.), *Recent developments in alcoholism: Treatment research.* New York: Plenum.

OGBORNE, A. C., & GLASER, F. B. (1981). Characteristics of affiliates of Alcoholics Anonymous: A review of the literature. *Journal of Studies on Alcohol, 42* (7), 661–679.

PATTISON, E. M. (1982). *Selection of treatment for alcoholics.* New Brunswick, NJ: Rutgers Center of Alcohol Studies.

PEELE, S. (1989). *Diseasing of America: Addiction treatment out of control.* Lexington, MA: Lexington Books.

RACHMAN, S. J., & WILSON, G. T. (1980). *The effects of psychological therapy,* 2nd ed. New York: Pergamon.

RUDY, D. R. (1986). *Becoming alcoholic: Alcoholics Anonymous and the reality of alcoholism.* Carbondale, IL: Southern Illinois University Press.

RUDY, D. R. (1980). Slipping and sobriety: The functions of drinking in A.A. *Journal of Studies on Alcohol, 41,* 727–732.

SONNENSTUHL, W. J., & TRICE, H. M. (1987). The social

construction of alcohol problems in a union's peer counseling program. *Journal of Drug Issues, 17* (3), 223–254.

TRICE, H. M. (1957). A study of the process of affiliation with A.A. *Quarterly Journal of Studies on Alcohol, 18,* 38–54.

TRICE, H. M., & STAUDENMEIER, W. J., JR. (1989). A sociocultural history of Alcoholics Anonymous. In M. Galanter (Ed.), *Recent developments in alcoholism: Treatment research.* Vol. 7. New York: Plenum.

TRICE, H. M., & WAHL, R. (1958). A rank order analysis of the symptoms of alcoholism. *Quarterly Journal of Studies on Alcohol, 19,* 636–648.

WALSH, D. C., & ASSOCIATES (1991). A randomized trial of treatment options for alcohol-abusing workers. *New England Journal of Medicine, 325,* 775–782.

W., W. (1949). The society of Alcoholics Anonymous. *American Journal of Psychiatry, 106,* 370–376.

HARRISON M. TRICE

ALCOHOLISM This section contains articles on some aspects of chronic drinking: *Abstinence versus Controlled Drinking; Controlled Drinking versus Abstinence;* and *Origin of the Term.* For further information on this subject, see *Disease Concept of Alcoholism and Drug Abuse* and the sections on *Complications,* on *Treatment,* and on *Withdrawal.*

Abstinence versus Controlled Drinking
Abstinence is the total avoidance of an activity. It is the dominant approach in the United States to resolving alcoholism and drug abuse (e.g., "Just Say No"). Abstinence was at the base of Prohibition (legalized in 1919 with the Eighteenth Amendment) and is closely related to prohibitionism—the legal proscription of substances and their use.

Although temperance originally meant moderation, the nineteenth-century TEMPERANCE MOVEMENT's emphasis on complete abstinence from alcohol and the mid-twentieth century's experience of the ALCOHOLICS ANONYMOUS movement have strongly influenced alcohol- and drug-abuse treatment goals in the United States. Moral and clinical issues have been irrevocably mixed.

The disease model of alcoholism and drug addiction, which insists on abstinence, has incorporated new areas of compulsive behavior—such as overeat-

ing and sexual involvements. In these cases, redefinition of *abstinence* to mean "the avoidance of excess" (what we would otherwise term moderation) is required.

Abstinence can also be used as a treatment-outcome measure, as an indicator of its effectiveness. In this case, abstinence is defined as the number of drug-free days or weeks during the treatment regimen—and measures of drug in urine are often used as objective indicators.

(SEE ALSO: *Disease Concept of Alcoholism and Drug Addiction*)

BIBLIOGRAPHY

HEATH, D. B. (1992). Prohibition or liberalization of alcohol and drugs? In M. Galanter (Ed.), *Recent developments in alcoholism: Alcohol and cocaine.* New York: Plenum.

LENDER, M. E., & MARTIN, J. K. (1982). *Drinking in America.* New York: Free Press.

PEELE, S., BRODSKY, A., & ARNOLD, M. (1991). *The truth about addiction and recovery.* New York: Simon & Schuster.

STANTON PEELE

Controlled Drinking versus Abstinence
The position of ALCOHOLICS ANONYMOUS (AA) and the dominant view among therapists who treat alcoholism in the United States is that the goal of treatment for those who have been dependent on alcohol is total, complete, and permanent abstinence from alcohol (and, often, other intoxicating substances). By extension, for all those treated for alcohol abuse, including those with no dependence symptoms, moderation of drinking (termed *controlled drinking,* or *CD*) as a goal of treatment is rejected (Peele, 1992). Instead, providers claim, holding out such a goal to an alcoholic is detrimental, fostering a continuation of denial and delaying the alcoholic's need to accept the reality that he or she can never drink in moderation.

In Britain and other European and Commonwealth countries, controlled-drinking therapy is widely available (Rosenberg et al., 1992). The following six questions explore the value, prevalence, and clinical impact of controlled drinking versus abstinence outcomes in alcoholism treatment; they are

intended to argue the case for controlled drinking as a reasonable and realistic goal.

1. *What proportion of treated alcoholics abstain completely following treatment?*

At one extreme, Vaillant (1983) found a 95 percent relapse rate among a group of alcoholics followed for 8 years after treatment at a public hospital; and over a 4-year follow-up period, the Rand Corporation found that only 7 percent of a treated alcoholic population abstained completely (Polich, Armor, & Braiker, 1981). At the other extreme, Wallace et al. (1988) reported a 57 percent continuous abstinence rate for private clinic patients who were stably married and had successfully completed detoxification and treatment—but results in this study covered only a 6-month period.

In other studies of private treatment, Walsh et al. (1991) found that only 23 percent of alcohol-abusing workers reported abstaining throughout a 2-year follow-up, although the figure was 37 percent for those assigned to a hospital program. According to Finney and Moos (1991), 37 percent of patients reported they were abstinent at all follow-up years 4 through 10 after treatment. Clearly, most research agrees that most alcoholism patients drink at some point following treatment.

2. *What proportion of alcoholics eventually achieve abstinence following alcoholism treatment?*

Many patients ultimately achieve abstinence only over time. Finney and Moos (1991) found that 49 percent of patients reported they were abstinent at 4 years and 54 percent at 10 years after treatment. Vaillant (1983) found that 39 percent of his surviving patients were abstaining at 8 years. In the Rand study, 28 percent of assessed patients were abstaining after 4 years. Helzer et al. (1985), however, reported that only 15 percent of all surviving alcoholics seen in hospitals were abstinent at 5 to 7 years. (Only a portion of these patients were specifically treated in an alcoholism unit. Abstinence rates were not reported separately for this group, but only 7 percent survived and were in remission at follow-up.)

3. *What is the relationship of abstinence to controlled-drinking outcomes over time?*

Edwards et al. (1983) reported that controlled drinking is more unstable than abstinence for alcoholics over time, but recent studies have found that controlled drinking increases over longer follow-up periods. Finney and Moos (1991) reported a 17 percent "social or moderate drinking" rate at 6 years and a 24 percent rate at 10 years. In studies by McCabe (1986) and Nordström and Berglund (1987), CD outcomes exceeded abstinence during follow-up of patients 15 and more years after treatment (see Table 1). Hyman (1976) earlier found a similar emergence of controlled drinking over 15 years.

4. *What are legitimate nonabstinent outcomes for alcoholism?*

The range of nonabstinence outcomes between unabated alcoholism and total abstinence includes (1) "improved drinking" despite continuing alcohol abuse, (2) "largely controlled drinking" with occasional relapses, and (3) "completely controlled drinking." Yet some studies count both groups (1) and (2) as continuing alcoholics and those in group (3) who engage in only occasional drinking as abstinent. Vaillant (1983) labeled abstinence as drinking less than once a month and including a binge lasting less than a week each year.

The importance of definitional criteria is evident in a highly publicized study (Helzer et al., 1985) that identified only 1.6 percent of treated alcoholism patients as "moderate drinkers." Not included in this category were an additional 4.6 percent of patients who drank without problems but who drank in fewer than 30 of the previous 36 months. In addition, Helzer et al. identified a sizable group (12%) of former alcoholics who drank a threshold of 7 drinks 4 times in a single month over the previous 3 years but who reported no adverse consequences or symptoms of alcohol dependence and for whom no such problems were uncovered from collateral records. Nonetheless, Helzer et al. rejected the value of CD outcomes in alcoholism treatment.

While the Helzer et al. study was welcomed by the American treatment industry, the Rand results (Polich, Armor, & Braiker, 1981) were publicly denounced by alcoholism treatment advocates. Yet the studies differed primarily in that Rand reported a higher abstinence rate, using a 6-month window at assessment (compared with 3 years for Helzer et al.). The studies found remarkably similar nonabstinence outcomes, but Polich, Armor, and Braiker (1981) classified both occasional and continuous moderate drinkers (8%) and sometimes heavy drinkers (10%) who had no negative drinking consequences or de-

TABLE 1
Selected Alcoholism Outcome Studies

Study	Years of Follow-up	No. of Assessed Subjects	Percent Abstinent	Percent Controlled Drinking	Percent Remission Survivors[a]
Untreated					
Goodwin, Crane, & Guze (1971)	8	93	8	33[b]	41
1989 Canadian National Survey	≥1	497	49	51	100[c]
Treated					
Rand (Polich, Armor, & Braiker 1981)	4	548	28	18[d]	46
Vaillant (1983)	8	100	39	6	45
Helzer et al. (1985)[e]	5–7	387	15[f]	18[g]	33
McCabe (1986)	16	31	26	35	61
Nordström & Berglund (1987)	18–24	55	20	38	58
Rychtarik et al. (1987)	5–6	43	23	21	44
Wallace et al. (1988)	0.5	169	68[h]	—[h]	68[h]
Finney and Moos (1991)	10	83	54	24	78
Walsh et al. (1991)	2	200	23 (37)[i]	10 (7)[i]	33 (44)[i]

[a]Since only survivors are included, successful remission rates are overstated.

[b]Nonabstinence remission included 18 percent "moderate drinkers," 9 percent getting "drunk about once a week," 6 percent "switched from whiskey to beer, . . . drank almost daily and sometimes excessively, [but] had experienced no problems from drinking since making the change."

[c]The Canadian National Survey data concern only recovered alcohol abusers.

[d]Defined as nonproblem drinking, with either low quantities of consumption (8%) or some heavy drinking (10%).

[e]Although all subjects in this study were hospital patients, only one group was treated in an alcohol unit. This group had the worst outcomes of any group, but these outcomes were not reported separately.

[f]Reported data were weighted by Helzer et al.

[g]Controlled drinking outcomes include occasional drinkers (4.6%), moderate drinkers (1.6%), and heavy, nonproblem drinkers (12%).

[h]Wallace et al. reported 57 percent continuous abstinence over 6 months and an additional 11 percent currently abstinent. Although Wallace et al. reported no controlled drinking, a small group (4%) had "one brief, contained return to drinking or drug use" in the 180-day period.

[i]Figures are for all treated groups, with assigned hospital patients in parentheses. No controlled-drinking category was included, but this column comprises those in the study who drank without ever becoming intoxicated during the 2-year follow-up (the latter data are not fully reported in the published article).

pendence symptoms in a nonabstinent remission category. (Rand subjects had been highly alcoholic and at intake were consuming a median of 17 drinks daily.)

The harm-reduction approach seeks to minimize the damage from continued drinking and recognizes a wide range of improved categories (Heather, 1992). Minimizing nonabstinent remission or improvement categories by labeling reduced but occasionally excessive drinking as "alcoholism" fails to address the

morbidity associated with continued untrammeled drinking.

5. *How do untreated and treated alcoholics compare in their controlled-drinking and abstinent-remission ratios?*

Alcoholic remission many years after treatment may depend less on treatment than on posttreatment experiences, and in some long-term studies, CD outcomes become more prominent the longer subjects

are out of the treatment milieu, because patients unlearn the abstinence prescription that prevails there (Peele, 1987). By the same token, controlled drinking may be the more common outcome for untreated remission, since many alcohol abusers may reject treatment because they are unwilling to abstain.

Goodwin, Crane, & Guze (1971) found that controlled-drinking remission was four times as frequent as abstinence after eight years for untreated alcoholic felons who had "unequivocal histories of alcoholism" (see Table 1). Results from the 1989 Canadian National Alcohol and Drug Survey confirmed that those who resolve a drinking problem without treatment are more likely to become controlled drinkers. Only 18 percent of 500 recovered alcohol abusers in the survey achieved remission through treatment. About half (49%) of those in remission still drank. Of those in remission through treatment, 92 percent were abstinent. But 61 percent of those who achieved remission without treatment continued drinking (see Table 2).

6. For which alcohol abusers is controlled-drinking therapy or abstinence therapy superior?

Severity of alcoholism is the most generally accepted clinical indicator of the appropriateness of CD therapy (Rosenberg, 1993). Untreated alcohol abusers probably have less severe drinking problems than clinical populations of alcoholics, which may explain their higher levels of controlled drinking. But the less severe problem drinkers uncovered in nonclinical studies are more typical, outnumbering those who "show major symptoms of alcohol dependence" by about four to one (Skinner, 1990).

Despite the reported relationship between severity and CD outcomes, many diagnosed alcoholics do control their drinking, as Table 1 reveals. The Rand study quantified the relationship between severity of alcohol dependence and controlled-drinking outcomes, although, overall, the Rand population was a severely alcoholic one in which "virtually all subjects reported symptoms of alcohol dependence" (Polich, Armor, and Braiker, 1981).

Polich, Armor, and Braiker found that the most severely dependent alcoholics (11 or more dependence symptoms on admission) were the least likely to achieve nonproblem drinking at 4 years. However, a quarter of this group who achieved remission did so through nonproblem drinking. Furthermore, younger (under 40), single alcoholics were far more likely to relapse if they were abstinent at 18 months than if they were drinking without problems, even if they were highly alcohol-dependent (Table 3). Thus the Rand study found a strong link between severity and outcome, but a far from ironclad one.

Some studies have failed to confirm the link between controlled-drinking versus abstinence outcomes and alcoholic severity. In a clinical trial that included CD and abstinence training for a highly dependent alcoholic population, Rychtarik et al. (1987) reported 18 percent controlled drinkers and 20 percent abstinent (from 59 initial patients) at 5 to 6 year follow-up. Outcome type was not related to severity of dependence. Nor was it for Nordström and Berglund (1987), perhaps because they excluded "subjects who were never alcohol dependent."

Nordström and Berglund, like Wallace et al. (1988), selected high-prognosis patients who were socially stable. The Wallace et al. patients had a high level of abstinence; patients in Nordström and Berglund had a high level of controlled drinking. Social stability at intake was negatively related in Rychtarik et al. to consumption as a result either of abstinence or of limited intake. Apparently, social stability predicts that alcoholics will succeed better whether they choose abstinence or reduced drinking. But other research indicates that the pool of those who achieve remission can be expanded by having broader treatment goals.

TABLE 2

Controlled Drinking and Abstinence in Relation to Treated and Untreated Remission: The 1989 Canadian National Alcohol and Drug Survey (Data Weighted to Represent National Population)

	Treated Remission (n = 89) %	Untreated Remission (n = 408) %
Abstinent (51%)	92	39
Nonabstinent (49%)	8	61

Data presented by L. C. Sobell & M. B. Sobell (November 1991). Cognitive mediators of natural recoveries from alcohol problems: Implications for treatment. Paper presented as part of a symposium, "Therapies for Substance Abuse: A View towards the Future," 25th Annual Meeting of the Association for the Advancement of Behavior Therapy, New York.

TABLE 3

Relapse Rates at 4-Year Follow-up According to Remission Category at Eighteen Months, by Alcohol Dependence Category, Marital Status, and Age

	Age < 40 at Admission		Age > 40 at Admission	
	Abstaining 18 Months	Nonproblem Drinking 18 Months	Abstaining 18 Months	Nonproblem Drinking 18 Months
High Dependence Symptoms				
Married	12	17	14	50
Single	21	7	24	28
Low Dependence Symptoms				
Married	16	7	19	28
Single	29	3	32	13

J. M. Polich, D. J. Armor, & H. B. Braiker (1981). *The course of alcoholism: Four years after treatment.* New York: Wiley.

Rychtarik et al. found that treatment aimed at abstinence or controlled drinking was not related to patients' ultimate remission type. Booth, Dale, and Ansari (1984), on the other hand, found that patients did achieve their *selected* goal of abstinence or controlled drinking more often. Three British groups (Elal-Lawrence, Slade, & Dewey, 1986; Heather, Rollnick, & Winton, 1983; Orford & Keddie, 1986) have found that treated alcoholics' beliefs about whether they could control their drinking and their commitment to a CD or an abstinence-treatment goal were more important in determining CD versus abstinence outcomes than were subjects' levels of alcohol dependence. Miller et al. (in press) found that more dependent drinkers were less likely to achieve CD outcomes but that desired treatment goal and whether one labeled oneself an alcoholic or not independently predicted outcome type.

SUMMARY

Controlled drinking has an important role to play in alcoholism treatment. Controlled drinking as well as abstinence is an appropriate goal for the majority of problem drinkers who are not alcohol-dependent. In addition, while controlled drinking becomes less likely the more severe the degree of alcoholism, other factors—such as age, values, and beliefs about

oneself, one's drinking, and the possibility of controlled drinking—also play a role, sometimes the dominant role, in determining successful outcome type. Finally, reduced drinking is often the focus of a harm-reduction approach, where the likely alternative is not abstinence but continued alcoholism.

(SEE ALSO: *Alcohol; Disease Concept of Alcoholism and Drug Abuse; Relapse Prevention; Treatment*)

BIBLIOGRAPHY

BOOTH, P. G., DALE, B., & ANSARI, J. (1984). Problem drinkers' goal choice and treatment outcomes: A preliminary study. *Addictive Behaviors, 9,* 357–364.

EDWARDS, G., ET AL. (1983). What happens to alcoholics? *Lancet, 2,* 269–271.

ELAL-LAWRENCE, G., SLADE, P. D., & DEWEY, M. E. (1986). Predictors of outcome type in treated problem drinkers. *Journal of Studies on Alcohol, 47,* 41–47.

FINNEY, J. W., & MOOS, R. H. (1991). The long-term course of treated alcoholism: 1. Mortality, relapse and remission rates and comparisons with community controls. *Journal of Studies on Alcohol, 52,* 44–54.

GOODWIN, D. W., CRANE, J. B., & GUZE, S. B. (1971). Felons who drink: An 8-year follow-up. *Quarterly Journal of Studies on Alcohol, 32,* 136–147.

HEATHER, N. (1992). The application of harm-reduction principles to the treatment of alcohol problems. Paper

presented at the Third International Conference on the Reduction of Drug-Related Harm, Melbourne, Australia, March.

HEATHER, N., ROLLNICK, S., & WINTON, M. (1983). A comparison of objective and subjective measures of alcohol dependence as predictors of relapse following treatment. *Journal of Clinical Psychology, 22,* 11–17.

HELZER, J. E. ET AL. (1985). The extent of long-term moderate drinking among alcoholics discharged from medical and psychiatric treatment facilities. *New England Journal of Medicine, 312,* 1678–1682.

HYMAN, H. H. (1976). Alcoholics 15 years later. *Annals of the New York Academy of Science, 273,* 613–622.

MCCABE, R. J. R. (1986). Alcohol-dependent individuals 16 years on. *Alcohol & Alcoholism, 21,* 85–91.

MILLER, W. R. ET AL. (1992). Long-term follow-up of behavioral self-control training. *Journal of Studies on Alcohol, 53,* 249–261.

NORDSTRÖM, G., & BERGLUND, M. (1987). A prospective study of successful long-term adjustment in alcohol dependence. *Journal of Studies on Alcohol, 48,* 95–103.

ORFORD, J., & KEDDIE, A. (1986). Abstinence or controlled drinking: A test of the dependence and persuasion hypotheses. *British Journal of Addiction, 81,* 495–504.

PEELE, S. (1992). Alcoholism, politics, and bureaucracy: The consensus against controlled-drinking therapy in America. *Addictive Behaviors, 17,* 49–61.

PEELE, S. (1987). Why do controlled-drinking outcomes vary by country, era, and investigator?: Cultural conceptions of relapse and remission in alcoholism. *Drug and Alcohol Dependence, 20,* 173–201.

POLICH, J. M., ARMOR, D. J., & BRAIKER, H. B. (1981). *The course of alcoholism: Four years after treatment.* New York: Wiley.

ROSENBERG, H. (1993). Prediction of controlled drinking by alcoholics and problem drinkers. *Psychological Bulletin, 113,* 129–139.

ROSENBERG, H., MELVILLE, J., LEVELL, D., & HODGE, J. E. (1992). A ten-year follow-up survey of acceptability of controlled drinking in Britain. *Journal of Studies on Alcohol, 53,* 441–446.

RYCHTARIK, R. G., ET AL. (1987). Five-six-year follow-up of broad spectrum behavioral treatment for alcoholism: Effects of training controlled drinking skills. *Journal of Consulting and Clinical Psychology, 55,* 106–108.

SKINNER, H. A. (1990). Spectrum of drinkers and intervention opportunities. *Journal of the Canadian Medical Association, 143,* 1054–1059.

VAILLANT, G. E. (1983). *The natural history of alcoholism.* Cambridge: Harvard University Press.

WALLACE, J., ET AL. (1988). 1. Six-month treatment outcomes in socially stable alcoholics: Abstinence rates. *Journal of Substance Abuse Treatment, 5,* 247–252.

WALSH, D. C., ET AL. (1991). A randomized trial of treatment options for alcohol-abusing workers. *New Enland Journal of Medicine, 325,* 775–782.

STANTON PEELE

Origin of the Term The term *alcoholism* is of relatively recent date; knowledge of the adverse effects of heavy alcohol (ethanol) consumption is not. A proverb describes alcohol as "both mankind's oldest friend and oldest enemy." Alcohol occurs in nature, and humans have long known how to ferment plants to create it; both its moderate and excessive use have therefore occurred since prehistory. The Bible cautions: "Do not look at wine when it is red, when it sparkles in the cup and goes down smoothly. At the end it bites like a serpent and stings like an adder" (Proverbs 23:31–32). A drunken Noah (Genesis 9:20–28) is one of a long line of such literary descriptions. In the classical era of the Greeks and the Romans we have drunks in the *Character Sketches* of Theophrastus, in the *Satyricon* of Petronius Arbiter, and in the *Epistles* of Seneca. In the 1600s, we have Shakespeare's porter in *Macbeth* (Act II, Scene 3) and others.

Viewing the long-term adverse effects of alcohol as a disease is a concept that also predates the term *alcoholism.* Benjamin Rush (1745–1813) and Thomas Trotter (1760–1832), both physicians, wrote extensively in this vein, using words such as *drunkenness;* their elder contemporary Benjamin Franklin (1706–1790) produced a glossary of 228 synonyms in use in 1737 for "being under the influence of alcohol." It was not until 1849 that the Swedish physician and temperance advocate Magnus Huss (1807–1890) first used the word *alcoholism* in his book *Alcoholismus Chronicus (The Chronic Alcohol-disease).* Huss's term, used originally in a descriptive sense to denote the consequences *of* the prolonged consumption of large quantities of alcohol, has come to connote a disease, believed by some to result *in* such consumption.

Huss meant by the term *chronic alcoholism* "those pathologic symptoms which develop in such persons who over a long period of time continually use wine

or other alcoholic beverages in large quantities" and stated that it "corresponds with chronic poisoning." His book is filled with detailed case histories illustrating the various symptoms that might occur. Sweden was at that time highest in the list of countries that consumed liquors, and Huss, as attending physician to the Serafim Clinic In Stockholm, had ample opportunity to observe cases. The London *Daily News* of December 8, 1869, carried a story on "the deaths of two persons from alcoholism," which according to the *Oxford English Dictionary* was the first popular use of the word in English. From that time on, its use in both the professional and the popular literature greatly expanded. This is partly because of the natural process that popularizes usage of certain words and partly because of deliberate activities on behalf of the term *alcoholism*.

The period of national prohibition in the United States (1919–1933) was accompanied by a lack of attention to the consequences of alcohol consumption, for understandable reasons. Such consumption was illegal—permanently, it was assumed—and as a result, it was thought that there would be little in the way of consequences. Indeed, such consequences as cirrhosis of the liver did decline abruptly during this period. But as enthusiasm for prohibition waned, and especially after it was repealed, a need to promote treatment became increasingly evident. One group involved in this promotion used *alcoholism* as the key word in their efforts, and accordingly were called *the alcoholism movement* by sociologists who subsequently studied their work. In an early statement of this movement, Anderson (1942) predicted that "When the dissemination of these ideas is begun through the existing media of public information, press, radio, and platform, which will consider them as news, a new public attitude can be shaped." It was also felt that the term, together with the disease connotations attached to it, would encourage the involvement of physicians in its study and treatment. The medical profession was viewed as critical to the success of the effort to increase the nation's concern about the consequences of alcohol consumption. The formation of the NATIONAL COUNCIL ON ALCO-HOLISM, the largest public interest group in this area, was a project of the same movement. Their successful efforts may be the reason that the term *alcoholism* developed and sustained a popularity in the United States beyond anything it achieved in Europe and even in Scandinavia, where it was first used.

MEANING BROADENS

As the term *alcoholism* became widely used, its meaning broadened. In a 1941 review of treatment, ten definitions of chronic alcoholism and sixteen definitions of alcohol addiction were collected from the international literature. Originally used by Huss to refer to a disease that consisted of the *consequences* of alcohol consumption, *alcoholism* came in time to represent a disease that *caused* high levels of alcohol consumption (Jellinek, 1960). A variant theory attempts to preserve the original meaning: High levels of alcohol consumption resulted in consequences of various kinds, particularly in terms of damage to the central nervous system, which damage in turn caused the high levels of consumption to continue (Vaillant, 1983). That is, the term *alcoholism* evolved over time from a primarily descriptive term to a largely explanatory concept. An example of a definition of *alcoholism* with clear explanatory intent is one that R. C. Rinaldi and colleagues produced in 1988 through an elaborate consensus exercise (a Delphi process) among eighty American experts, who defined the term as "a chronic, progressive, and potentially fatal biogenetic and psychosocial disease characterized by tolerance and physical dependence manifested by a loss of control, as well as diverse personality changes and social consequences." As a counterpoint to this line of development, a growing and increasingly influential literature holds that problems developing in the context of alcohol consumption do not constitute a disease at all (Fingarette, 1988).

The greater interest taken in alcohol consumption and its consequences as a result of the popularization of the term *alcoholism* has been gratifying as well as useful. But the broadening of meaning of the term, with much attendant controversy among the advocates of various definitions, has become problematic. For example, in a review of alternative definitions, Babor & Kadden (1985) concluded: "Clearly, the past and present lack of consensus concerning the definition of alcoholism and the criteria for its diagnosis does not provide a solid conceptual basis to design screening procedures for early detection or casefinding." Because of its imprecise meaning, the term *alcoholism* has for some time now been dropped from the two major official systems of diagnosis of diseases, the INTERNATIONAL CLASSIFICATION OF DISEASES of the World Health Organization and the DIAGNOSTIC AND STATISTICAL MANUAL *of*

Mental Disorders of the American Psychiatric Association. A recent comprehensive study of treatment deliberately avoided the use of the word *alcoholism* as too narrow in its focus, while suggesting that the word was not incompatible with the phrase that it chose to use—*alcohol problems*—to refer to any problem occurring in the context of alcohol consumption (Institute of Medicine, 1990, pp. 30–31).

COMPLEX PROBLEMS

These recent attempts to be precise in the use of words represent a return to the more straightforward, descriptive use of *alcoholism* by its originator, Huss. Two major realities contributing to this change of direction have been widely recognized since Huss first used the term in the 1840s. One is that the problems people experience are complex, including those that may arise in the context of alcohol use. Although alcohol may be a factor in some such problems—even an important factor—it is not often the full explanation for them. Multiple factors, including heredity, early environment, cultural factors, personality factors, situational factors, and others, contribute to the development of human problems and must be considered in their resolution. This formulation should not be taken to minimize the important role of alcohol in such problems or to say that the reduction or elimination of alcohol consumption may not be a critical factor in the resolution of problems in particular individuals. The other reality has to do with the extremely broad spectrum of problems that arise in the context of alcohol consumption. Although a substantial proportion of these problems arise from those who drink too much over a long period of time and who usually have multiple problems (those to whom the term *alcoholism* is usually applied), an even greater burden of problems arises from those who drink too much over *short* periods of time, and who have only a *few* problems. The simple reason is that there are more of the latter than of the former (Institute of Medicine, 1990, chapter 9). To reduce the burden upon society effectively, both kinds of populations must be dealt with. An exclusive concentration on *alcoholism* may cause this reality to be overlooked.

Costly Consequences. The term *alcoholism* retains, and probably will always retain, its place in general, nontechnical speech as an indicator of serious problems that are the consequences of prolonged heavy alcohol consumption. Its continued popularity has some advantages, for the public-health consequences of such alcohol consumption are horrendous. The presence of a convenient shorthand term for this fact in the public consciousness—*alcoholism*—serves as a continuing reminder of this major unfinished item on the public-health agenda. Certainly, there is a legitimate place in Western society for the use of alcohol. But with equal certainty, too many individuals fail to use alcohol wisely or well.

The ravages that prolonged exposure to alcohol produces in the human body are manifold, as Huss well understood; they include neurological problems (damage to the central and peripheral nervous systems), cirrhosis (fibrosis and shrinking) of the liver, hypertension (high blood pressure), and many forms of cancer, particularly of the digestive tract, to name but a few. If to these are added the consequences of short-term but intense exposure to alcohol and the intoxication it produces, one can include a high proportion of all accidents, burns, all types of violence including suicide, and especially automobile crashes, as well as the common behavioral effects of intoxication with which we are all too familiar. Small wonder that almost 30 percent of all admissions to hospitals in the United States are of persons with severe alcohol problems; yet most of these problems go unrecognized, and the individuals go untreated. About 50 percent of American women have or have had a parent, blood relative, or spouse to whom they would apply the term *alcoholism*; the figure is closer to 40 percent for men. The difficulties that this creates are legion—and its remediation would be a remarkable step forward.

(SEE ALSO: *Addiction: Concepts and Definitions; Disease Concept of Alcoholism; Treatment, History of*)

BIBLIOGRAPHY

ANDERSON, D. (1942). Alcohol and public opinion. (1942). *Quarterly Journal of Studies on Alcohol, 3*, 376. An early manifesto of the alcoholism movement.

BABOR, T. F., & KADDEN, R. (1985). Screening for alcohol problems: Conceptual issues and practical considertiions. In N. C. Chong & H. M. Cho (Eds.), *Early identification of alcohol abuse.* Washington, DC: U.S. Government Printing Office.

BYNUM, W. F. (1968). Chronic alcoholism in the first half of the 19th century. *Bulletin of the History of Medicine,*

42, 160–185. An excellent review of medical thought about alcohol problems at the time of Magnus Huss.

FINGARETTE, H. (1988). *Heavy drinking: The myth of alcoholism as a disease.* Berkeley: University of California Press. A comprehensive summary of the evidence against the disease concept of alcoholism.

HUSS, M. (1852). *Chronische Alkoholskrankheit oder Alcoholismus Chronicus* (Alcoholismus chronicus or the chronic alcohol disease). Translated by G. van dem Busch into German from the Swedish, with revisions by the author. Stockholm and Leipzig: C. E. Fritze. Original published 1849.

INSTITUTE OF MEDICINE. (1990). *Broadening the base of treatment for alcohol problems.* Washington, DC: National Academy Press. A detailed contemporary review of all aspects of treatment.

INSTITUTE OF MEDICINE. (1987). *Causes and consequences of alcohol problems: An agenda for research.* Washington, DC: National Academy Press. A comprehensive look at some basic issues in the field that includes extensive chapters on the medical, social, and psychological consequences of alcohol consumption.

JELLINEK, E. M. (1960). *The disease concept of alcoholism.* Highland Park, NJ: Hillhouse Press. The classic book on the disease concept. If read carefully, it is more skeptical than credulous.

RINALDI, R. C., STEINDLER, E. M., WILFORD, B. B., & GOODWIN, D. (1988). Clarification and standardization of substance abuse terminology. *Journal of the American Medical Association, 259,* 555–557.

TURNER, T. B., BORKENSTEIN, R. F., JONES, R. K., & SANTORA, P. B. (EDS.) (1985). *Alcohol and highway safety.* Supplement no. 10 to the *Journal of Studies on Alcohol.* New Brunswick, NJ: Rutgers University Center of Alcohol Studies. Covers multiple aspects of this complex and highly important area.

VAILLANT, G. E. (1983). *The natural history of alcoholism: Causes, patterns, and paths to recovery.* Cambridge: Harvard University Press. An ingenious attempt to discover what happens to people with severe alcohol problems by tracing their histories over long periods of time.

FREDERICK B. GLASER

ALKALOIDS This is the general term for any number of complex organic bases that are found in nature in seed-bearing plants. These substances are usually colorless but bitter to the taste. Alkaloids often contain nitrogen and oxygen and possess important physiological properties.

Examples of alkaloids include not only quinine, atropine, and strychnine but also CAFFEINE, NICOTINE, MORPHINE, CODEINE, and COCAINE. Therefore, many drugs that are used by humans for both medical and nonmedical purposes are produced in nature in the form of alkaloids. Naturally occurring receptors for many alkaloids have also been identified in humans and other animals, suggesting an evolutionary role for the alkaloids in physiological processes.

NICK E. GOEDERS

ALLERGIES TO ALCOHOL AND DRUGS

In addition to ALCOHOL, OPIATES, and BARBITURATES, some street drugs have been reported to induce allergic reactions. These allergic phenomena are most frequently mediated by reactions of the immune system known as immediate hypersensitivity and delayed hypersensitivity. *Immediate hypersensitivity* is mediated by the serum protein immunoglobulin E(IgE), whereas *delayed hypersensitivity* is mediated by thymus-derived lymphocytes (the white blood cells called T cells).

Immediate Hypersensitivity. The symptoms and signs associated with IgE-mediated immune reactions are urticaria (hives); bronchospasm that produces wheezing; angioedema (swelling) of face and lips or full-blown *anaphylaxis* (a combination of all the above symptoms and lowering of blood pressure). Abdominal pain and cardiac arrhythmias (irregular heartbeat) may also occur with anaphylaxis. Any or all of these symptoms occur when IgE, which has previously been synthesized by a sensitized lymphocyte, fixes to mast cells or basophils in the skin, bronchial mucosa, and intestinal mucosa. This cell-fixed IgE then binds the antigen that triggers the release of the following—the histamine, the slow-reacting substance of anaphylaxis (SRSA), the bradykinin, and the other mediators that induce these symptoms. Examples of this type of allergic reaction are the allergic responses to either bee stings or to penicillin.

Similar symptoms may also occur when mediators are released by mast cells in response to chemical or physical stimuli. This is called an *anaphylactoid reaction.* In this instance, the mast cell or basophil is directly activated by the chemical to release mediators without having to bind to IgE. Examples of this type of reaction are responses to intravenous contrast

material, such as IVP dye, or the hives induced by exposure to cold.

Delayed Hypersensitivity. Reactions occur when antigenic chemicals stimulate T lymphocytes and induce their proliferation. T effector cells are then recruited into the tissue site. These effector cells bind the antigen and subsequently release effector molecules, such as the interleukins, the chemotactic factors, and the enzymes. These effector molecules induce an inflammatory response in the area and may also induce formation of granuloma (a mass of inflamed tissue) by macrophages and inflammatory cells. Symptoms of delayed hypersensitivity reactions are skin rashes, which may be red, pruritic (itchy), or bullous (blistered) in nature. Granulomas can cause lymph node enlargement and nodules in the skin or in organs. Examples of this response are poison ivy, cosmetic allergies, Erythema Nodosum or Sarcoidosis.

ALLERGIC RESPONSE TO ALCOHOL

True anaphylactic or anaphylactoid reactions to ALCOHOL (ethanol) are rare. Most reactions to ingested alcoholic beverages are secondary to other chemicals in the beverage such as yeasts, metabisulfite, papain, or dyes. However, there are reports of true allergic reactions in which the offending agent was shown to be the ethanol itself.

Symptoms of anaphylaxis have been reported to occur in several subjects following ingestion of beer and/or wine, and these symptoms were reproduced in one patient by administration of 95 percent ethanol. Hives have been reported with ethanol ingestion, and hives on contact with ethanol have been reported for some Asian patients. Bronchospasm was precipitated in some asthmatic patients by administration of ethanol, and contact hypersensitivity to 50 percent ethanol solution was produced in 6 percent of subjects tested. These allergic responses differ from the "flush" reaction exhibited in individuals (especially Asians) with acetaldehyde dehydrogenase abnormalities.

ALLERGIES TO OPIATES, BARBITURATES, AND STREET DRUGS

There have been reports of MORPHINE-induced hives in some people, and studies show that morphine can cause histamine release directly from cells without binding to specific receptors on cells. Anaphylaxis may also occur with either morphine or CODEINE, and IgE antibodies against morphine and codeine have been found in patients experiencing anaphylaxis. Thus, the OPIATES can mediate allergic reaction by either mechanism, and the antagonist drug NALOXONE will not reverse these reactions. There are also reports of HEROIN causing bronchospasm.

Some instances of anaphylaxis associated with the medical administration of opiates or local anesthesia during surgery are due to the often included preservative methylparaben, rather than to the opiate itself. Anaphylaxis may occur with more than one local anesthetic and/or analgesic compound in the same patient because of the methylparabens.

Numerous reports exist for anaphylactoid reactions following the use of BARBITURATES for the induction of anesthesia. The drugs themselves may induce histamine release. This may also be mediated through a true allergic IgE mediated response in some patients. Skin rashes also occur frequently following barbiturate usage. This may be a hypersensitivity reaction, or it may be a pseudo-allergic reaction.

Street drugs have been reported to induce asthma and or anaphylaxis. Bronchospasm may occur in patients smoking COCAINE or in those injecting heroin. This may occur more often in patients who have a previous history of asthma. The asthma may persist after the subjects have stopped smoking cocaine. Pulmonary edema (fluid in the lungs) may also occur with FREEBASING cocaine. These side effects are not likely to be mediated by the immune system. However, a hypersensitivity pneumonitis to cocaine has been described and is associated with elevated levels of IgE. MARIJUANA does not appear to increase the incidence of either asthma or anaphylaxis.

(SEE ALSO: *Complications*)

BIBLIOGRAPHY

ETTINGER, N. A., & ALBIN, R. J. (1989). A review of the respiratory effects of smoking cocaine. *American Journal of Medicine, 87,* 664.

FAKUDA, T., & DOHI, S. (1986). Anaphylactic reaction to fentanyl or preservative. *Canadian Anesthetists' Society Journal, 33,* 826.

HICKS, R. (1968). Ethanol, a possible allergen. *Annals of Allergy, 26,* 641.

KARVOREN, J., & HANNUKSELA, M. (1976). Urticaria from alcoholic beverages. *Acta Allergan, 31,* 167.

KELSO, J., ET AL. (1990). Anaphylactoid reaction to ethanol. *Annals of Allergy, 64,* 452.

McLELLAND, J. (1986). The mechanism of morphine induced urticaria. *Archives of Dermatology, 122,* 138.

PRZYBILLA, B., & RING, J. (1983). Anaphylaxis to ethanol and sensitization to acetic acid. *Lancet, 1,* 483.

REBHUN, J. (1988). Association of asthma and freebase smoking. *Annals of Allergy, 60,* 339.

RILBET, A., HUNZIKER, N., & BRAUN, R. (1980) Alcohol contact urticaria syndrome. *Dermatologica, 161,* 361.

RUBIN, R., & NEUGARTEN, J. (1990). Codeine associated asthma. *American Journal of Medicine, 88,* 438.

SETO, A., ET AL. (1978). Biochemical correlates of ethanol-induced flushing in Orientals. *Journal in the Studies of Alcohol, 39,* 1.

SHAIKH, W. A. (1970). Allergy to heroin. *Allergy, 45,* 555.

SHANABAN, E. C., MARSHALL, A. G., & GARRETT, C. P. U. Adverse reactions to intravenous codeine phosphate. *Anesthesia, 38,* 40.

TING, S., ET AL. (1988). Ethanol induced urticaria: A case report. *Ann. Allergy, 60,* 527.

WILKINS, J. K., & FORTNER, G. (1985). Ethnic contact urticaria to alcohol. *Contact Dermatology, 12,* 118.

MARLENE ALDO-BENSON

AMANTADINE Amantadine is a medication (Symmetrel) that is believed to be an indirect DOPAMINE agonist; this means that it releases the neurotransmitter dopamine from nerve terminals in the brain. Since some of the symptoms of COCAINE withdrawal and cocaine dependence are thought to be related to abnormalities in the dopamine systems of the brain, and these are thought to contribute to relapse, amantadine has been examined as a treatment possibility.

After chronic cocaine use, many patients' dopamine systems either fail to release sufficient dopamine or are insensitive to the dopamine that is released. This relative dopamine deficit is believed to be responsible for the dysphoria of cocaine withdrawal. It was hoped that amantadine would relieve their dysphoria and reduce relapse back to cocaine abuse by increasing the release of dopamine in the brains of cocaine-dependent patients. Amantadine has been effective in reducing depressive symptoms in patients with neurological disorders such as Parkinson's disease, which is due to the death of dopamine-producing cells in the brain; however, no solid evidence exists that it is helpful in preventing continued cocaine use or relapse to cocaine use after detoxification.

(SEE ALSO: *Treatment: Cocaine; Withdrawal: Cocaine*)

THOMAS R. KOSTEN

AMERICAN ACADEMY OF PSYCHIATRISTS IN ALCOHOLISM AND ADDICTIONS (AAPAA)

This is a not-for-profit scientific professional organization of psychiatrists who specialize in treating addictive disorders. The organization's purpose is to stimulate and encourage scientific and medical research into the problem of addiction; to develop standards of professional practice; to disseminate and publish information for treatment professionals and the general public; and to educate and inform the public about the role of psychiatry and psychiatrists in caring for patients with addictive disorders.

Psychiatrists who work with alcoholism and addiction in their practices and who are members of the American Psychiatric Association and or the Canadian Psychiatric Association are eligible for membership in the AAPAA. There are additional categories of membership for medical students, members in training, and honorary members. Psychiatrists who have made significant contributions to the field through clinical work, teaching, research, or administration may, upon recommendation of the executive committee, be elected fellows. Physicians already certified as specialists in psychiatry by the American Board of Psychiatry and Neurology can earn further credentials in the subspecialty of addiction psychiatry by passing a certification examination written and administered by the AAPAA. A peer reviewed publication, the *American Journal on Addictions,* is the official journal of the AAPAA.

(SEE ALSO: *American Society of Addiction Medicine*)

FAITH K. JAFFE

AMERICAN INDIANS AND DRUG USE
See Ethnicity and Drugs

AMERICAN SOCIETY FOR THE PRO-MOTION OF TEMPERANCE   *See* Temperance Movement; Women's Christian Temperance Union (WCTU).

AMERICAN SOCIETY OF ADDICTION MEDICINE (ASAM)   The American Society of Addiction Medicine (5225 Wisconsin Avenue, Suite 409, Washington, DC, 20015; 202-244-8948) is a society of physicians in all medical specialties and subspecialties who devote a significant part of their practice to treating patients addicted to, or having problems with, alcohol and other drugs. Many of its members are also involved in the medical education and research having to do with the treatment of addiction.

ASAM's roots can be traced to the early 1950s, when Dr. Ruth Fox organized regular meetings at the New York Academy of Medicine with other physicians interested in alcoholism and its treatment. These meetings led to the establishment, in 1954, of the New York City Medical Society on Alcoholism, which eventually became the American Medical Society on Alcoholism (AMSA). Another state medical society devoted to addiction as a subspecialty, the California Society for the Treatment of Alcoholism and Other Drug Dependencies, was established in the 1970s. By 1982, the American Academy of Addictionology was incorporated, and all these groups united within AMSA the following year. Because the organization was concerned with all drugs of addiction, not only alcohol, and was interested in establishing addiction medicine as part of mainstream medical practice, the organization was renamed the American Society on Alcoholism and Other Drug Dependencies (AMSAODD), which was soon changed to the American Society of Addiction Medicine (ASAM) in 1989. By the early 1990s, membership in the society exceeded 3,000, with chapters in all 50 states, as well as overseas.

The stated mission and goals of ASAM are to increase access to and improve the quality of addictions treatment; educate physicians, medical students, and the public; and establish addiction medicine as a medical subspecialty—and a specialty recognized by the American Board of Medical Specialties. Educational activities are carried out through publications, courses, and clinical and scientific conferences. Publications of the Society include, among others, the *Journal of Addictive Diseases*, published quarterly; and the *ASAM Patient Placement Criteria for the Treatment of Psychoactive Substance Use Disorders*, a clinical guide for matching patients to appropriate levels of care. Courses include overviews and updates of basic and clinical material, in-depth study of the principles of addiction medicine, and the clinical application of research in addiction. Scientific conferences explore new concepts, review general principles, or focus on specific problems, such as the interrelationship between ACQUIRED IMMUNODEFICIENCY SYNDROME (AIDS) and drug use.

In its continued effort to establish the legitimacy of the subspecialty within medicine, ASAM administers a six-hour certification examination, is a primary sponsor of medical postgraduate fellowships in alcoholism and drug abuse, and has developed guidelines for the training of physicians in this area of medical practice.

MARC GALANTER
JEROME H. JAFFE

AMITY, INC.   Founded in 1969, Amity is a nonprofit organization that provides a wide variety of services for substances abusers. Amity provides services for pregnant addicts, addicted mothers, and those mothers' infants and children; prevention programs for high-risk school-aged youngsters; short- and long-term intensive teaching and therapeutic community programs for addicted adolescents and adults; services for incarcerated adolescent and adult substance abusers; HIV prevention programs for drug users and their sexual partners; and services for homeless substance abusers.

Amity (from the Latin, *amicus*, for friend) began in Tucson, Arizona, in 1969 when a local high school teacher, Ann Knowles, spoke with a community leader, Debbie Jacquin, about her worries concerning students' drug use. Jacquin rallied community support, formed a nonprofit organization with a local board of directors, looked for outside expertise, and found a California organization called Awareness House (a spin-off of the Mendocino Family—itself modeled after the first therapeutic community, SYNANON). Awareness House sent trained counselors to begin Tucson Awareness House, the first drug program in Tucson. Tucson Awareness House pro-

mulgated several small programs over the next years—nonresidential programs for school-aged youth and both outpatient and residential programs for adult substance abusers.

In 1981, Naya Arbiter, Bette Fleishman, and Rod Mullen, all trained at Synanon, took over the leadership of the then economically troubled Tucson Awareness House. They began transforming it into an innovative teaching and therapeutic community that emphasized outreach and enhanced services to women, adolescents, racial and ethnic minorities, drug users at high risk for HIV infection and transmission, and drug users whose abuse has brought them into frequent contact with the criminal justice system. In 1986, the name of the organization was changed to Amity.

Amity defines its mission as conducting research into the root causes of substance abuse, criminality, and other dysfunctional behaviors; developing model programs; and disseminating the results of its work through training, presentations, and publications. Amity currently provides eighteen separate substance abuse–related services. Most are located in Tucson, Arizona, but programs also exist in Phoenix, Payson (Arizona), and in San Diego, California, where Amity has a 200-bed prison teaching community in a designated unit within a medium-security correctional facility. Amity has received many federal grants to develop demonstration models of national significance and was recognized in 1988 by the White House Conference for a Drug Free America as one of three exemplary therapeutic communities in the United States.

SIGNIFICANT INNOVATIONS AND FINDINGS

Emphasizing women's services, particularly increasing the number of female staff and allowing addicted mothers to bring their children with them into the residential setting, improved retention for both women and men. Amity's length of stay increased to roughly twice the national average for therapeutic community programs. Since length of stay appears to be the most important predictor of posttreatment success across all treatment modalities, this suggests an important avenue of exploration at other substance abuse programs nationally.

In addition to a traditional staff training regimen, all staff members who work directly with program participants are expected to participate in a week-long intensive retreat each year to improve understanding of family and cultural dynamics, the changing street culture, and continued personal growth. They are also expected to participate in regular encounter groups not only with residents but with each other to improve communication and cohesiveness. Encounter groups and retreats are seen not only as therapy but as intensive educational processes; thus participation by the staff provides role model expectations for participants.

In developing several model programs working with the criminal justice system to habilitate drug users, Amity has used cross-training as the basis for successful programs. Criminal justice (CJ) personnel train Amity staff before project initiation and continue regularly throughout the life of the project; Amity trains CJ personnel in effective interventions in a similar manner. Regular realignment of the treatment and CJ organization improves communication, effectiveness, participant and staff satisfaction, and reduces recidivism to drug use and criminal behavior.

(SEE ALSO: *Prisons and Jails; Treatment Types: Therapeutic Communities*)

BIBLIOGRAPHY

DELEON G., & ROSENTHAL, M. (1989). Treatment in residential therapeutic communities. *Treatments of Psychiatric Disorders, Vol. 1.* Washington, DC: American Psychiatric Press.

MULLEN, R. (1992). Therapeutic communities in prisons: Dealing with toxic waste. XIV World Conference of Therapeutic Communities, Montreal, Canada.

MULLEN, R., & ARBITER, N. (1992). Against the odds: Therapeutic community approaches to underclass drug abuse. In *Drug policy in the Americas.* Boulder: Westview Press.

MULLEN, R., ARBITER N., & GILDER, P. (1991). A comprehensive therapeutic community approach for chronic substance-abusing juvenile offenders: The Amity model. In *Intensive interventions with high-risk youths: Promising approaches in juvenile probation and parole.* Monsey, NY: Willow Tree Press.

STEVENS, S., ARBITER, N. & GLIDER, P. (1989). Women residents: Expanding their role to increase treatment effectiveness in substance abuse programs. *The International Journal of the Addictions, 24*(5):425–434.

WEXLER, H., FALKIN, G. P., & LIPTON, D. S. (1988). *A model prison rehabilitation program: An evaluation of the Stay'n Out Therapeutic Community.* New York: Narcotic and Drug Research.

ROD MULLEN

AMOBARBITAL Amobarbital (Amytal) is one of the many different members of the BARBITURATE family of central nervous system depressants used to produce relaxation, sleep, anesthesia, and anticonvulsant effects. In terms of the duration of its effects, it is considered an intermediate-acting barbiturate. When taken by mouth, its sedating effects take about one hour to develop and last about six to eight hours, although it takes considerably longer for all the drug to leave the body.

The adult dosage for sedation is 15–50 milligrams but 65–200 milligrams for sleep. Like other barbiturates, amobarbital was used to treat sleep disorders and anxiety. Unfortunately, most barbiturates have a high potential for abuse, result in physical dependence with continued use, and are very dangerous—even lethal—after high doses. Since the introduction of the BENZODIAZEPINES, amobarbital is rarely used.

BIBLIOGRAPHY

HARVEY, STEWART C. (1980). Hypnotics and sedatives. In A. G. Gilman, L. S. Goodman, & A. Gilman (Eds.), *Goodman and Gilman's the pharmacological basis of therapeutics,* 6th ed. New York: Macmillan.

S. E. LUKAS

AMOTIVATIONAL SYNDROME This term refers to a hypothetical effect produced by drugs, especially MARIJUANA, whereby individuals lose interest or the ability to engage in activities motivated by normal psychological processes. It is associated with lethargy, a severe reduction in activities, unwillingness to work, failure to meet responsibilities, and neglect of personal needs including hygiene and nutrition (despite efforts by others to help and despite statements by the individuals that they wish they could get started in these activities).

The experimental evidence of the existence of this syndrome has been mixed, but it is generally be-

lieved that amotivation is rarely caused by a drug alone; it is instead the result of a complex interaction among the effects of the drug, the personality and experience of the individual, and the context in which the drug is repeatedly administered. In addition, there is some confusion as to whether this syndrome is seen only during drug intoxication and is therefore transient or whether it is a more permanent consequence that persists for a long period of time following cessation of drug use.

(SEE ALSO: *Cannabis sativa; Complications*)

CHRIS-ELLYN JOHANSON

AMPHETAMINE Amphetamine was first synthesized in 1887, but its central nervous system (CNS) stimulant effects were not noted at that time. After rediscovery, in the early 1930s, its use as a respiratory stimulant was established and its properties as a central nervous system stimulant were described. Reports of abuse soon followed. As had occurred with cocaine products when they were first introduced in the 1880s, amphetamine was promoted as being an effective cure for a wide range of ills without any risk of addiction. The medical profession enthusiastically explored the potentials of amphetamine, recommending it as a cure for everything from alcohol hangover and depression to the vomiting of pregnancy and weight reduction. These claims that it was a miracle drug contributed to public interest in the amphetamines, and they rapidly became the stimulant of choice—since they were inexpensive, readily available, and had a long duration of action.

Derivatives of amphetamine, such as METHAMPHETAMINE, were soon developed and both oral and intravenous preparations became available for therapeutic uses. Despite early reports of an occasional adverse reaction, enormous quantities were consumed in the 1940s and 1950s, and their liability for abuse was not recognized. During World War II, the amphetamines, including methamphetamine, were widely used as stimulants by the military in the United States, Great Britain, Germany, and Japan, to counteract fatigue, to increase alertness during battle and night watches, to increase endurance, and to elevate mood. It has been estimated that approxi-

mately 200 million Benzedrine (amphetamine) tablets were dispensed to the U.S. armed forces during World War II. In fact, much of the research on performance effects of the amphetamines was carried out on enlisted personnel during this period, as the various countries sought ways of maintaining an alert and productive armed force. Although amphetamine was found to increase alertness, little data were collected supporting its ability to enhance performance.

Since 1945, use of the amphetamines and CO-CAINE appears to have alternated in popularity, with several stimulant epidemics occurring in the United States. There was a major epidemic of amphetamine and methamphetamine abuse (both oral and intravenous) in Japan right after the war. The epidemic was reported to have involved, at its peak, some half-million users and was related to the release with minimal regulatory controls of huge quantities of surplus amphetamines that had been made for use by the Japanese military. Despite this experience, there were special regulations governing their manufacture, sale, or prescription in the United States until 1964 (Kalant, 1973).

The first major amphetamine epidemic in the United States peaked in the mid-1960s, with approximately 13.5 percent of the university population estimated, in 1969, to have used amphetamines at least once. By 1978, use of the amphetamines had declined substantially, contrasting with the increase of cocaine use by that time. The major amphetamine of concern in the United States in the 1990s is methamphetamine, with pockets of "ice" (smoked methamphetamine) abuse.

Amphetamines are now controlled under Schedule II of the CONTROLLED SUBSTANCES ACT. Substances classified within this schedule are found to have a high potential for abuse as well as currently accepted medical use within the United States. Amphetamine, methamphetamine, cocaine, METHYLPHENIDATE, and phenmetrazine are all stimulants included in this schedule.

MEDICAL UTILITY

Amphetamines are frequently prescribed for the treatment of narcolepsy, obesity, and for childhood ATTENTION DEFICIT DISORDER. They are clearly efficacious in the treatment of narcolepsy, one of the first conditions to be successfully treated with these drugs. Although patients with this disorder can require large doses of amphetamine for prolonged periods of time, attacks of sleep can generally be prevented. Interestingly, tolerance does not seem to develop to the therapeutic effects of these drugs, and most patients can be maintained on the same dose for years.

Although the amphetamines have been used extensively in the treatment of obesity, considerable evidence exists for a rapid development of tolerance to the anorectic (appetite loss) effects of this drug, with continued use having little therapeutic effect. These drugs are extremely effective appetite suppressants, but after several weeks of use the dose must be increased to achieve the same appetite-suppressant effect. People remaining on the amphetamines for prolonged periods of time to decrease food intake can reach substantial doses, resulting in toxic side effects (e.g., insomnia, irritability, increased heart rate and blood pressure, and tremulousness). Therefore, these drugs should only be taken for relatively short periods of time (4–6 weeks). In addition, long-term follow-up studies of patients who were prescribed amphetamines for weight loss have not found any advantage in using this medication to maintain weight loss. Data indicate that weight lost under amphetamine maintenance is rapidly gained when amphetamine use is discontinued. In addition to the lack of long-term efficacy, the dependence-producing effects of amphetamines make them a poor choice of maintenance medication for this problem.

The use of amphetamines in the treatment of attention deficit disorders in children, remains extremely controversial. It has been found that the amphetamines have a dramatic effect in reducing restlessness and distractibility as well as lengthening attention span, but there are side effects. These include reports of growth impairment in children, insomnia, and increases in heart rate. Those promoting their use point to their potential benefits and they advocate care in limiting treatment dose and duration. Opponents of their use, while agreeing that they provide some short-term benefits, conclude that these do not outweigh their disadvantages. Amphetamine therapy has also been attempted, but with little success, in the treatment of Parkinson's disease, and both amphetamine and cocaine have been suggested for the treatment of depression, although the evidence to support their efficacy does not meet cur-

rent standards demanded by the U.S. Food and Drug Administration.

PHARMACOLOGY

The amphetamines act by increasing concentrations of the neurotransmitters DOPAMINE and NOREPINEPHRINE at the neuronal synapse, thereby augmenting release and blocking uptake. It is the augmentation of release that differentiates amphetamines from cocaine, which also blocks uptake of these transmitters. Humans given a single moderate dose of amphetamine generally show an increase in activity and talkativeness, and they report euphoria, a general sense of well-being, and a decrease in both food intake and fatigue. At higher doses repetitive motor activity (i.e., stereotyped behavior) is often seen, and further increases in dose can lead to convulsions, coma, and death. This class of drugs increases heart rate, respiration, diastolic and systolic blood pressure, and high doses can cause cardiac arrhythmias. In addition, the amphetamines have a suppressant effect on both rapid eye movement sleep (REM)—the stage of sleep associated with dreaming—and total sleep. The half-life of amphetamine is about ten hours, quite long when compared to a stimulant like cocaine, which has a half-life of approximately one hour, or even methamphetamine which has a half-life of about five hours.

The amphetamine molecule has two isomers: the d-$(+)$ and l-$(-)$ isomers. There is marked stereoselectivity in their biological actions, with the d-isomer (dextroamphetamine) considerably more potent. For example, it is more potent as a locomotor stimulant, in inducing stereotyped behavior patterns, and in eliciting central nervous system excitatory effects. The isomers appear to be equipotent as cardiovascular stimulants. The basic amphetamine molecule has been modified in a number of ways to accentuate various of its actions. For example, in an effort to obtain appetite suppressants with reduced cardiovascular and central nervous system effects, structural modifications yielded such medications as diethylproprion and fenfluramine, while other structural modifications have enhanced the central nervous system stimulant effects and reduced the cardiovascular and anorectic actions, yielding medications such as methylphenidate and phenmetrazine. These substances share, to a greater or lesser degree, the properties of amphetamine.

TOXICITY

A major toxic effect of amphetamine in humans is the development of a schizophrenia-like psychosis after repeated long-term use. The first report of an amphetamine psychosis occurred in 1938, but the condition was considered rare. Administration of amphetamine to normal volunteers with no histories of psychosis (Griffith et al., 1968) resulted in a clear-cut paranoid psychosis in five of the six subjects who received d-amphetamine for one to five days (120–220 mg/day), which cleared when the drug was discontinued. Unless the user continues to take the drug, the psychosis usually clears within a week, although the possibility exists for prolonged symptomology. This amphetamine psychosis has been thought to represent a reasonably accurate model of schizophrenia, including symptoms of persecution, hyperactivity and excitation, visual and auditory hallucinations, and changes in body image. In addition, it has been suggested that there is sensitization to the development of a stimulant psychosis—once an individual has experienced this toxic effect, it is readily reinitiated, sometimes at lower doses and even following long drug-free periods.

Amphetamine abusers taking repeated doses of the drug can develop repetitive behavior patterns which persist for hours at a time. These can take the form of cleaning, the repeated dismantling of small appliances, or the endless picking at wounds on the extremities. Such repetitive stereotyped patterns of behavior are also seen in nonhumans administered repeated doses of amphetamines and other stimulant drugs, and they appear to be related to dopaminergic facilitation. Cessation of amphetamine use after high-dose chronic intake is generally accompanied by lethargy, depression, and abnormal sleep patterns. This pattern of behavior, opposite to the direct effects of amphetamine, does not appear to be a classical abstinence syndrome. The symptoms may be related to the long-term lack of sleep and food intake that accompany chronic stimulant use as well as to the catecholamine depletion that occurs as a result of chronic use.

Animals given unlimited access to amphetamine will self-administer it reliably, alternating days of high intake with days of low intake. They become restless, tremulous, and ataxic, eating and sleeping little. If allowed to continue self-administering the drug, most will take it until they die. Animals main-

tained on high doses of amphetamines develop tolerance to many of the physically and behaviorally debilitating effects, but they also develop irreversible damage in some parts of the brain, including long-lasting depletion of dopamine. It has been suggested that the prolonged anhedonia seen after long-term human amphetamine use may be related to this, although the evidence for this is not very strong.

BEHAVIORAL EFFECTS

Nonhumans. As with all PSYCHOMOTOR STIMULANT drugs, at low doses animals are active and alert, showing increases in responding maintained by other reinforcers, but often decreasing food intake. Higher doses produce species-specific repetitive behavior patterns (stereotyped behavior), and further increases in dose are followed, as in humans, by convulsions, hyperthermia, and death. Tolerance (loss of response to a certain dose) develops to many of amphetamine's central effects, and cross-tolerance among the stimulants has been demonstrated in rats. Thus, for example, animals tolerant to the anorectic effects of amphetamine also show tolerance to cocaine's anorectic effects. Although there is tolerance development to many of amphetamine's effects, sensitization develops to amphetamine's effects on locomotor activity. Thus, with repeated administration, doses of amphetamine that initially did not result in hyperactivity or stereotypy can, with repeated use, begin to induce those behaviors when injected daily for several weeks. In addition, there is cross-sensitization to this effect, such that administration of one stimulant can induce sensitization to another one. In contrast to cocaine, however (in which an increased sensitivity to its convulsant effects develops with repeated use), amphetamines have an anticonvulsive effect.

Learned behaviors, typically generated by operant schedules of reinforcement, are generally affected by the amphetamines in a rate-dependent fashion. Thus, behaviors that occur at relatively low rates in the absence of the drug tend to be increased at low-to-moderate doses of amphetamine, while behaviors occurring at relatively high frequencies tend to be suppressed by those doses of amphetamine. In addition, with high doses most behaviors tend to be suppressed. As is seen with other stimulants, such as cocaine, environmental variables and behavioral context can play a role in modulating these effects.

For example, behavior under strong stimulus control shows tolerance to repeated amphetamine administration much more rapidly than does behavior under weak stimulus control. In addition, if the amphetamine-induced behavioral disruption has the effect of interfering with reinforcement delivery, tolerance to that effect develops rapidly. Tolerance does not develop to the amphetamine-induced disruptions when reinforcement density is increased or remains the same.

Amphetamines can serve as reinforcers in nonhumans and, as described above, can produce severely toxic consequences when available in an unlimited fashion. However, when available for a few hours a day, animals will take them in a regular fashion, showing little or no tolerance to their reinforcing effects.

Humans. A substantial number of studies have been carried out evaluating the effects of amphetamines on learning, cognition, and other aspects of performance. The data indicate that under most conditions the amphetamines are not general performance enhancers. When there is improvement in performance associated with amphetamine administration, it can usually be attributed to a reduction in the deterioration of performance due to fatigue or boredom. Attention lapses that impair performance after sleep deprivation appear to be reduced by amphetamine administration; however, as sleep deprivation is prolonged, this effect is reduced. A careful review of the literature in this area (Laties & Weiss, 1981) concluded that improvement is more obvious with complex, as compared with simple, tasks.

In addition, in trained athletes, whose behavior shows little variability, only very small improvements can be seen. Laties and Weiss have argued persuasively, however, that the small changes in performance induced by amphetamines can result in the 1 to 2 percent improvement that may make the difference in a close athletic competition. Although the facilitation in performance after amphetamine does not appear to be substantial, it is sufficient to "spell the difference between a gold medal" and any other. Unfortunately, such data have led athletes to take stimulants prior to athletic events, particularly those in which strenuous activity is required over prolonged periods (e.g., bicycle racing), leading to hyperthermia, collapse, and even death in some cases.

The mood-elevating effects of the amphetamines are generally believed to be related to their abuse.

Their use is accompanied by reports of increased self-confidence, elation, frequently euphoria, friendliness, and positive mood. When amphetamine is administered repeatedly, tolerance develops rapidly to many of its subjective effects (such that the same dose no longer exerts much of an effect). This means that the user must take increasingly larger amounts of amphetamine to achieve the same effect. As with nonhuman research subjects, there is however, little or no evidence for the development of tolerance to amphetamine's reinforcing effects.

Experienced stimulant users, given a variety of stimulant drugs, often cannot differentiate among cocaine, amphetamine, methamphetamine, and methylphenidate—all of which appear to have similar profiles of action. Since these drugs have different durations of action, however, it becomes easier to make this differentiation over time.

ABUSE

In the United States in the 1950s, nonmedical amphetamine use was prevalent among college students, athletes, truck drivers, and housewives. The drug was widely publicized by the media when very little evidence of amphetamine toxicity was available. Pills were the first form to be widely abused. Use of the drug expanded as production of amphetamine and methamphetamine increased significantly, and abusers began to inject it. An extensive black market in amphetamines developed, and it has been estimated that 50 to 90 percent of the quantity commercially produced was diverted into illicit channels. In the 1970s, manufacture of amphetamines was substantially curtailed, amphetamines were placed in Schedule II of the Controlled Substances Act, and abuse of these substances was substantially reduced. Perhaps only by coincidence, as amphetamine use declined, cocaine use increased.

The amphetamines, as with other stimulants, are generally abused in multiple-dose cycles (i.e., binges), in which people take the drug repeatedly for some period of time, followed by a period in which they take no drug. Amphetamines are often taken every three or four hours for periods as long as three or four days, and dosage can escalate dramatically as tolerance develops. Like cocaine binges, these amphetamine-taking occasions are followed by a "crash" period in which the user sleeps, eats, and does not use the drug. Abrupt cessation from amphetamine use is usually accompanied by depression. Mood generally returns to normal within a week, although craving for the drug can last for months.

There is little evidence for the development of physical dependence to the amphetamines. Although some experts view the "crash" (with lethargy, depression, exhaustion, and increased appetite) that can follow a few days of moderate-to-high dose use as meeting the criteria for a withdrawal syndrome, others believe that the symptoms can also be related to the effects of chronic stimulant use. When using stimulants people do not eat or sleep very much and, as well, catecholamine depletion may well be contributing to these behavioral changes.

TREATMENT

As of the mid-1990s, little information is available about the treatment of amphetamine abusers, and no reports of successful pharmacological interventions exist in the treatment literature. As with cocaine abuse, the most promising nonpharmacological approaches include behavioral therapy, RELAPSE PREVENTION, rehabilitation (e.g., vocational, educational, and social-skills training), and supportive psychotherapy. Unlike cocaine, however, minimal clinical trials with potential treatment medications for amphetamine abuse have been carried out. The few that have been attempted report no success in reducing a return to amphetamine use.

(SEE ALSO: *Amphetamine Epidemics; Pharmacokinetics; Treatment*)

BIBLIOGRAPHY

ANGRIST, B., & SUDILOVSKY, A. (1978). Central nervous system stimulants: Historical aspects and clinical effects. In L. L. Iversen, S. D. Iverson, & S. H. Snyder (Eds.), *Handbook of psychopharmacology.* New York: Plenum.

GRIFFITH, J. D., ET AL. (1970). E. Costa and S. Garattini (Eds.), *Amphetamines and related compounds.* New York: Raven Press.

GRILLY, D. M. (1989). *Drugs and human behavior.* Needham, MA: Allyn & Bacon.

KALANT, O. J. (1973). *The amphetamines: Toxicity and addiction.* Springfield, IL: Charles C. Thomas.

LATIES, V. G., & WEISS, B. (1981). *Federation proceedings, 40,* 2689–2692.

MARIAN W. FISCHMAN

AMPHETAMINE EPIDEMICS The amphetamine class of drugs refers to a group of sympathomimetic amines or compounds that mimic the action of existing chemicals within the sympathetic nervous system. Every drug in the amphetamine group acts as a PSYCHOMOTOR STIMULANT, increasing the activity of the brain. These drugs can be taken orally, intranasally (snorted), intravenously (injected), or by inhalation (when they have been vaporized by heat). Unlike COCAINE, these drugs do not occur in nature but can only be synthesized in a laboratory. For this reason, amphetamine, METHAMPHETAMINE, and related compounds have relatively brief abuse histories, dating from the 1930s and 1940s, whereas cocaine and other PLANT-derived drugs have histories of use dating back to at least A.D. 500. Similar to cocaine, the amphetamines are addictive, and a number of cycles of epidemic use have occurred in the United States and in other countries.

Cocaine and amphetamine are chemically distinct stimulants that can produce similar subjective, behavioral, physiologic, and toxic effects. Users of either drug may experience euphoria, anxiety, enhanced energy, increased locomotor activity—that is, restlessness—hypertension, tachycardia (rapid heartbeat), ANOREXIA, pupillary dilation, insomnia, and such stereotypical behavior as hair pulling and lip picking. Users' behavior is characterized as hyperactive, talkative, irritable, and restless; distortion in thinking and perception frequently occurs. Large single doses or chronic use of either drug can result in amphetamine psychosis, clinically similar to paranoid schizophrenia, and neurotoxicity.

The major differences between cocaine and amphetamine relate to their duration of effect and pharmacological consequences. Cocaine remains active in the bloodstream for fewer than sixty minutes—the elimination half-life in plasma has been measured, with subjective effects like euphoria occurring for fewer than forty-five minutes. Cook et al. (1991) demonstrated that methamphetamine and its effects, by comparison, last eight to twelve times longer. There is evidence from animal studies that methamphetamine causes damage to the neurotransmitter system in the brain and that this damage is long lasting and, probably, irreversible. Cocaine exhibits similar acute effects but, to date, has not been demonstrated to cause chronic neurotoxicological consequences. Cocaine has local anesthetic effects and is more likely than the amphetamines to cause cardiovascular toxicity.

EARLY USE IN THE UNITED STATES

Amphetamines were initially synthesized in 1887. The related substance methamphetamine was developed approximately thirty years later by modifying the amphetamine structural formula but retaining similar effects on the central nervous system (CNS) and similar pharmaceutical uses. The first pharmacological studies showed hypertensive and bronchodilator properties, as well as the ability to reverse barbiturate anesthesia. The utility of the drug in treating lung congestion was recognized in 1933. Amphetamines such as Benzedrine were first marketed by the pharmaceutical firm Smith, Kline and French as a nasal inhaler for relief of nasal and bronchial congestion associated with colds and hay fever.

In 1933, CNS stimulant actions of amphetamines were reported, about the same time that reports of their effectiveness in treating narcolepsy and Parkinson's disease were released. When the use of amphetamines for these disorders was approved by the American Medical Association (AMA), a mild warning was added that "continuous doses higher than recommended" might cause "restlessness and sleeplessness," but physicians were assured that "no serious reactions had been observed." It was not until several decades later that the addictive properties and psychiatric complications were fully recognized by the medical community.

U.S. PATTERNS AND TRENDS

The rise in the popularity of amphetamines parallels that experienced during the introduction of cocaine. Exaggerated publicity and fallacious claims about amphetamines, combined with medical optimism concerning potential uses and a lack of understanding of abuse, contributed to a dramatic increase in public interest in amphetamines. Between 1932 and 1946 the pharmaceutical industry developed more than three dozen generally accepted clinical uses for amphetamines, among them the treatment of schizophrenia, morphine and codeine addiction, tobacco smoking, heart block, head injuries, infantile cerebral palsy, radiation sickness, low blood pressure, seasickness, and persistent hiccups. Amphetamines were then promoted as being effective and nonaddictive. During this period, amphetamine and its derivatives, such as methamphetamine, became available in both oral and intravenous preparations.

In the 1940s and 1950s amphetamines were prescribed liberally and soon replaced cocaine as an illicit stimulant widely available on the street. The increase in the popularity of amphetamines was influenced by easy availability, low cost, and long duration of effect. Amphetamine-type drugs were in widespread use before their addictive qualities were fully recognized and adverse consequences characterized. Between the 1930s and the 1970s the public could obtain amphetamine-like drugs in a variety of over-the-counter (OTC) nasal inhaler preparations. Abuse involved breaking open the inhalers and ingesting directly or soaking the fillers in alcohol or coffee. Although inhaler use may have introduced hundreds of thousands of Americans to amphetamine abuse, this type of abuse was most prevalent in prison populations and among deviant groups. The ability to cause euphoria, dysphoria, and psychic stimulation resulted in removal of amphetamine-like drugs from OTC inhaler preparations in 1971. However, amphetamine products remained available in pill, capsule, or injectable form.

During World War II, methamphetamine and amphetamines were widely used by the U.S., British, German, and Japanese military as an insomniac and as a stimulant to increase alertness during battle and night watches; they were used as well by war-related industries. Perhaps as many as 200 million tablets and pills were supplied to American troops during the war. The U.S. armed services authorized the issue of amphetamines on a regular basis beginning with the Korean conflict, escalating to well over 225 million standard-dose tablets dispensed between 1966 and 1969.

Monroe and Drell (1947) reported that at the end of both World War II and the Korean conflict, some soldiers who had used amphetamines returned home with drug habits. In addition, during the 1940s and 1950s enormous quantities of these drugs were prescribed without concern for any addictive effects. College students, athletes, truck drivers, and housewives began using amphetamines for nonmedical purposes. Amphetamine use expanded during the 1950s. These drugs were being marketed to treat obesity, narcolepsy, hyperkinesis, and depression. People were taking them illicitly primarily to increase energy, decrease the need for sleep, and elevate mood. Pharmaceutical production reached 3.5 billion tablets (about 20 standard dosage units per U.S. citizen) in 1958 and 10 billion tablets by 1970.

One consequence of excessive production and widespread popularity of amphetamines was the diversion of pharmaceutical-grade drugs to illegal traffic and use. Drugs sold on the black market came from or would otherwise have gone to pharmaceutical companies, wholesalers, druggists, and physicians. Probably over half (and potentially 90%) of the total commercial product was diverted into the black market. In 1966, the Food and Drug Administration (FDA) estimated that more than 25 tons of amphetamine were illegally distributed (Fischman, 1990). One market for the product was composed of long-distance truck drivers who found that amphetamines allowed them to work for extended periods without resting. The all-night restaurants and truck stops served as a distribution network that spanned the entire country.

By the mid 1960s the need for intervention and legislative controls over amphetamine production and distribution was clear. The Drug Abuse Control Amendments of 1965, passed by Congress, required increased record keeping throughout the system of manufacture, distribution, prescription, and sale. However, diversion of pharmaceutical amphetamine to illicit use continued. In 1971 the Justice Department began imposing quotas on legal amphetamine production.

A significant shift from abuse of oral preparations to abuse of the intravenous form occurred during the 1960s. Intravenous methamphetamine abuse described by Pittel and Hofer (1970), was particularly prevalent in the Haight-Ashbury district of San Francisco, where "speed," the street name for amphetamine and methamphetamine, began to replace hallucinogenic drugs, such as LSD, in popularity. Escalating doses of methamphetamine were taken, often as a series of injections over several days or weeks—what came to be known as a "speed run." Exhaustion, then depression, accompanied the end of a run, followed by readministration of the drug to mitigate the unpleasant side effects and regain the previous euphoria and high—thus the cycle of high to low to high. Initially the drugs were diverted from pharmaceutical supplies. Later, some unscrupulous physicians who were already prescribing intravenous methamphetamine to treat heroin addiction became involved in illegal prescriptions. In 1963, injectable ampules of methamphetamine were voluntarily removed by manufacturers from sale to retail pharmacies in California.

Speed use escalated during the 1960s, with the Haight-Ashbury district serving as a focal point.

With this escalation came an increase in violence and the diffusion of manufacturing and distribution of speed from Haight-Ashbury to other areas along the West Coast (Smith, 1970). Outlaw motorcycle gangs were reported to be heavily involved in methamphetamine manufacture and distribution. Within the subculture, serial speed users became known as "speed freaks." A public campaign was initiated to inform users of the hazards associated with speed use. Partly as a result of the "speed kills" campaign, amphetamine and methamphetamine use dropped sharply after 1972. From 1972 through 1977, the characteristics of the drug-taking population changed from heavy users to predominantly light-to-moderate users, and a growing proportion were women.

Because of increasing controls on the prescribing and marketing of amphetamines, the clandestine manufacture of methamphetamine became more widespread. One such control was the passage of the CONTROLLED SUBSTANCES ACT (CSA) of 1970, the legal foundation of the government's fight against drug abuse, which placed amphetamine and some related stimulant drugs in Schedule II—reflecting high potential for abuse, development of psychological or physical dependency, and restricted medical use. The availability of illicitly synthesized methamphetamine varied greatly during the 1970s and 1980s. Analyses of street samples of drugs purported to be methamphetamine revealed that until 1974, specimens were on average less than 30 percent methamphetamine. From 1975 through 1983 the composition of methamphetamine in samples increased from 60 percent to over 95 percent. For the street samples submitted as stimulants, including those submitted as amphetamine, methamphetamine, or speed, methamphetamine made up a relatively small percentage between 1972 and 1979 but increased to approximately 60 percent in 1983. These data demonstrate the increasing predominance of methamphetamine in the speed market during this time period.

Prior to the increase in quality of street speed, the products sold as methamphetamine or speed were usually a combination of phenylpropanolamine hydrochloride, ephedrine, and caffeine and referred to as "look-alike" speed. The term referred to the similarity of appearance of these drugs and of central nervous system effects. Other constituents also found in products purported to be speed included pseudoephedrine and cocaine.

Since the mid-1980s, virtually all substances marketed illicitly as amphetamine or by street terms, such as *speed, crystal, crank, go, go-fast, zip,* or *cristy*, contain methamphetamine. By analyzing contaminants found in street methamphetamine samples, researchers have determined that clandestine manufacture of methamphetamine, rather than diversion of pharmaceutical products, now supplies the illicit marketplace. According to U.S. Drug Enforcement Administration (DEA), methamphetamine is the most prevalent clandestinely manufactured controlled substance in the United States. In 1991, 84 percent of all clandestine laboratories seized were producing methamphetamine. In descending frequency were clandestine laboratories producing HALLUCINOGENS (PCP, MDA, MDMA, and LSD), methaqualone, and OPIOIDS (fentanyl, heroin) with the latter group making up only 1 percent of the laboratories seized. Figure 1 shows that the number of laboratories seized rose dramatically during the 1980s, from 88 in 1981 to 652 in 1989. Most of the laboratories were in California, Texas, or Oregon. Along with the increase in methamphetamine laboratory seizures was a localized resurgence of methamphetamine abuse—since the clandestine manufacture of methamphetamine in a community facilitates the development of a market for the drug. Clandestine labs also create other hazards for the

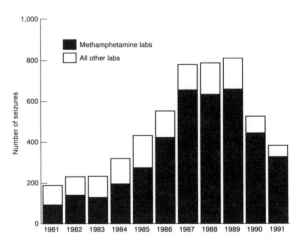

Figure 1

Number of Clandestine Methamphetamine Laboratories Seized in the United States, 1981–1991

SOURCE: U.S. Department of Justice (1992).

community since the materials used (precursors and solvents) are hazardous in the hands of inexperienced chemists, who may cause explosions and fires. Also, the operators (who rarely own the property) commonly discard the wastes on or near the site, creating long-lasting chemical contamination of the area. The number of laboratories seized declined in the early 1990s, largely because of the passage and enforcement of the Chemical Diversion and Trafficking Act of 1988, which placed under federal control the distribution of twelve precursor and eight essential chemicals used in the production of illicit drugs.

The DRUG ABUSE WARNING NETWORK (DAWN), a nationally based surveillance system that monitors emergency medical consequences and deaths related to drug use, reflected a stable trend across the U.S. from the mid-1970s until the mid-1980s. Over 700 hospitals in 21 metropolitan areas and a panel of hospitals outside of these areas report to DAWN. During the mid-1980s sharp increases in nonfatal emergency-room episodes began to appear, largely in metropolitan areas on the West Coast. Increases in drug-use indicators were also reported for methamphetamine through the Community Epidemiology Work Group (CEWG), a network of state and local drug-abuse experts representing 20 cities and metropolitan areas across the United States.

The DAWN emergency-room data from 1986 to 1988 show statistically significant increases in mentions of methamphetamine/speed in Atlanta, Dallas, Los Angeles, Phoenix, San Diego, and Seattle. Total methamphetamine mentions in DAWN peaked during 1988, plateaued at this level in 1989, then decreased during 1990 and 1991 (figure 2). Among DAWN emergency-room cases, the most common route of administration of methamphetamine was in-

travenous. The methamphetamine emergency-room mentions per 100,000 population in 1989 show San Francisco and San Diego to have the highest rates, 77.3 and 47.2 respectively, with the DAWN average rate at 4.0. Methamphetamine accounted for slightly more than 1 percent of the total DAWN drug mentions in 1989 and just under 1 percent in 1990, compared with 15 percent and 13 percent for cocaine during 1989 and 1990, respectively.

A 1988 field study of drug-treatment clients in San Diego, Portland, Oregon, and Dallas who had abused methamphetamine demonstrated the increase in abuse of the substance during the middle to late 1980s, particularly on the West Coast. The demographic profile of communities tracked in this study showed an abusing population that was predominantly white, low to middle income, high-school educated, young adults generally ranging in age from 20 to 35. Most of the clients in this study reported intravenous methamphetamine administration.

At the same time that increases were being noted in methamphetamine use on the mainland of the United States, a new phenomenon was developing in Hawaii. A sharp rise in law-enforcement activity and

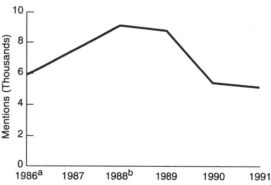

Figure 2

Trends in Methamphetamine-Related Emergencies, 1986–1991

SOURCE: The Drug Abuse Warning Network (DAWN).
[a]Estimates for 1986–1987 are provisional. The estimates are based on a nonrandom sample of hospital emergency rooms in the coterminous U.S.
[b]Estimates for 1988–1991 are based on a representative sample of nonfederal short-stay hospitals with 24-hour emergency rooms in the coterminous U.S.

TABLE 1
Drugs and Their Precursors

Controlled Substance	Precursor
Methamphetamine	Phenyl-2-propanone
Amphetamine	Phenylacetic acid
Methamphetamine	Ephedrine
Phencyclidine (PCP)	Piperidine
Methaqualone	Anthranilic acid
Lysergide (LSD)	Ergotamine tartrate

clients entering treatment because they smoked a new dosage form of methamphetamine were recorded between 1986 and 1989. The street names for this drug were *ice, crystal, shabu* (Japanese), and *batu* (Filipino for *rock*), and it looked like a large, usually clear, crystal resembling broken fragments of glass or rock candy. Ice is of high purity (90–100%) and the d-isomer (a more active pharmacological form) of methamphetamine hydrochloride salt. In Hawaii it is almost always smoked in a glass pipe. The hydrochloride salt is sufficiently volatile to vaporize in a pipe so that it can be inhaled. This route of administration allows rapid absorption into the bloodstream, with onset of effects similar to those experienced with intravenous administration.

The use of ice was first detected by Hawaiian treatment programs during the summer of 1986, with more widespread use occurring in 1988 and 1989. This epidemic, which was described in an outbreak investigation and follow-up field study conducted by the NATIONAL INSTITUTE ON DRUG ABUSE, involved a population widely varying in age and ethnic background and included both sexes. The ice-using treatment population was studied and reflected a younger population, with a higher representation of women and a larger proportion of Hawaiian/part Hawaiian than other drug users in treatment in the state. Ice was typically smoked in runs, or periods of continuous use, averaging three to eight days, with one or two days between runs, during which the user would "crash" into deep, prolonged sleep. Users reporting this pattern became rapidly addicted and experienced numerous adverse medical, social, and physiologic consequences.

Methamphetamine is produced in the United States for domestic consumption, but it is also produced in Asia, principally in Korea. Until the late 1980s the ice form of methamphetamine came only from Asia, specifically Hong Kong, Korea, Japan, Taiwan, Thailand, and the Philippines. Attempts to smuggle ice from Taiwan and Korea into Hawaii can be documented back to the mid-1980s by the DEA. The importation and distribution of ice in Hawaii has been linked to Asian and Hawaiian criminal organizations and gangs. By 1989, limited distribution of ice had occurred on the West Coast of the United States. In the following year, increased amounts of ice were found in California and subsequently in other limited locations. The increase in availability of ice was believed to stem from clandestine labora-

tories operating in California. During 1990, seven clandestine ice laboratories were seized nationwide, six of them in California, and domestically manufactured ice began to be supplied to distributors in Hawaii. This domestic production was compensating for a disruption of the major Asian trafficking organizations smuggling ice from South Korea.

INTERNATIONAL PATTERNS AND TRENDS

The use and abuse of amphetamines has also occurred in countries outside the United States, although the absence of significant epidemiologic information in many countries and the lack of standardization in data collection and analysis make multinational comparisons difficult. Based on available data, amphetamine use and abuse appear to be an endemic problem worldwide and, as in the United States, other countries reflect patterns that are episodic, localized, population-specific, and rooted in multiple etiologies. Senay (1991) cites a number of studies conducted principally in the 1970s and first half of the 1980s that document this phenomenon. Amphetamine epidemics have been evident in several countries during the past fifty years and continue to be the primary concern in some. The experience of two of these, Japan and Sweden, will be described briefly, but are described at greater length by Kalant (1973).

Japan. Methamphetamine has been the single most prevalent drug of abuse in Japan since the 1940s. Based on indicators from law-enforcement data, epidemic patterns appeared at several periods during that time. Figure 3 shows this trend.

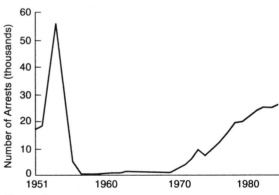

Figure 3

Methamphetamine Abuse in Japan, 1951–1985

The increase of methamphetamine abuse in Japan during the 1940s and early 1950s has been attributed to the wholesale appearance of the drug in the black market following World War II. Both amphetamine and methamphetamine were available to Japanese forces during the war and became widely used by the beleaguered civilian population following military defeat. In response to the escalation in abuse, Japan's Stimulants Drug Law of 1951 was enacted and the eventual decline in abuse was attributed to the effectiveness of the penal provisions of the law and subsequent control of the raw materials used to produce the drug. After a hiatus of fifteen years, Japan's methamphetamine abuse began to increase again and continued at relatively elevated levels through the mid-1980s. That epidemic was associated with illicit production of the drug and trafficking by criminal organizations—Yakuza and Boryokudan. The downturn in recent years is being attributed to the implementation of a stimulant-abuse prevention campaign that was begun in 1979. The stimulant epidemics in Japan have been described by Kato (1990) and Fukui et al. (1991).

Sweden. Amphetamines also have been at the forefront of the drug-abuse problem in Sweden since the 1930s. According to a 1988 report from the Swedish Council for Information on Alcohol and Other Drugs, abuse of amphetamines increased during the three decades following their introduction to the market as an over-the-counter (OTC) medication and their widespread promotion as a treatment for a variety of health conditions. While the prevalence of heavy, dysfunctional use during that era has been disputed, amphetamine abuse continues as a predominant problem in Sweden, especially among injection drug users.

CONCLUSION

In the United States, the overall magnitude of use and abuse of amphetamines, including methamphetamine, is relatively minor compared with the prevalence of other illicit drugs, such as marijuana and cocaine. The National Household Survey on Drug Abuse (NHSDA), a nationally representative survey of the household population age 12 and older, estimates that during 1991, 7 percent had ever used stimulants—including amphetamines, methamphetamine, and other prescription stimulants (National

Institute on Drug Abuse, 1991). The NHSDA also estimated that 1.3 percent of the household population had used stimulants during 1991 and 0.3 percent had used them during the month prior to the interview. These numbers are in contrast to 37.1 percent estimated to have ever used any illicit drug, 33.4 percent who reported any use of marijuana, and 11.7 percent who had ever used cocaine. However, data from the national HIGH SCHOOL SENIOR SURVEY indicate that among twelfth graders surveyed each year since 1975, annual and past-30-day use of stimulants such as methamphetamines have always been substantially higher than use of cocaine. These differences in magnitude in no way diminish the impact of the health consequences and social problems, both at the individual and at the community level, resulting from amphetamine epidemics in the United States and in other countries.

In the United States, abuse of amphetamine and methamphetamine dates back to the early part of the twentieth century. During the ensuing years, abuse of amphetamine and methamphetamine has become endemic throughout this country, with focal problematic areas. Survey data and ethnographic information indicate a concentration of abuse in cities along the West Coast and in Hawaii. Historically, the typical composite methamphetamine user was white, male, young adult, and with a low to middle income. As was experienced in the Hawaii "ice" outbreak, and seen recently in California, methamphetamine users can include diverse ethnic and socioeconomic groups. Methamphetamine is reported by the Drug Enforcement Administration to be the most common product of illicit drug laboratories in the United States. With extensive production and distribution systems in place—and potentially serious medical, psychological, and social consequence to abuse—these drugs continue to pose a significant public health threat.

The literature suggests that abuse of amphetamines has been and remains an endemic problem among diverse populations in countries throughout the world, at times reaching epidemic proportion. Measurement of the scope of drug abuse, trend analysis, and valid cross-cultural comparison are fraught with difficulties. However, based on history, it is clear that the prevention of future epidemics requires the implementation of an effective program of international drug-abuse surveillance, communication, and early intervention.

(SEE ALSO: *Epidemics of Drug Abuse; Ethnicity and Drugs*)

BIBLIOGRAPHY

AMA COUNCIL ON DRUGS. (1963). New drugs and developments in therapeutics. *Journal of the American Medical Association, 183,* 362–363.

BRECHER, E. M. (1972). *Licit and illicit drugs.* Boston: Little, Brown.

CALDWELL, J. (ED.) (1980). *Amphetamines and related stimulants: Chemical, biological, clinical, and sociological aspects.* Boca Raton: CRC Press.

CHO, A. K. (1990). Ice: A new dosage form of an old drug. *Science, August 10,* 631–634.

COOK, C. E., ET AL. (1991). Plasma levels of methamphetamine after smoking of methamphetamine hydrochloride. In *Problems of drug dependence, 1990.* Proceedings of the 52nd annual meeting of The Committee on Problems of Drug Dependence. NIDA Research Monograph No. 105. Washington, DC: U.S. Government Printing Office.

DRUG ENFORCEMENT ADMINISTRATION. (1989). A Special report on "Ice." In *Epidemiologic trends in drug abuse.* DHHS Publication no. 721-757:20058. Washington, DC: U.S. Government Printing Office.

ELLINWOOD, E. H. (1970). Amphetamine psychosis: Individuals, settings, and sequences. In E. H. Ellinwood & S. Cohen (Eds.), *Current concepts on amphetamine abuse.* National Institute of Mental Health. DHEW Publication no. (HSM)72-9085. Washington, DC: U.S. Government Printing Office.

FISCHMAN, M. W. (1990). History and current use of methamphetamine in the United States. In "Cocaine and Methamphetamine: Behavioral toxicology, clinical psychiatry and epidemiology." Proceedings from Japan–U.S. Scientific Symposium '90 on Drug Dependence and Abuse, Tokyo.

FUKUI, S., WADA, K., & IYO, M. (1991). History and current use of methamphetamine in Japan. In S. Fukui et al. (Eds.), *Cocaine and methamphetamine: Behavioral toxicology, clinical pharmacology and epidemiology.* Tokyo: Drug Abuse Prevention Center.

GRINSPOON, L., & HEDBLOM, P. (1975). *The speed culture, amphetamine use and abuse in America.* Cambridge, MA: Harvard University Press.

JAFFE, J. H. (1985). Drug addiction and drug abuse: CNS sympathomimetics—Amphetamine, cocaine and related drugs. In A. G. Gilman et al. (Eds.), *The pharmacological basis of therapeutics,* 7th ed. New York: Macmillan.

KALANT, D. J. (1973). *The amphetamines: Toxicity and addiction.* Springfield, IL: Charles C. Thomas.

KATO, N. (1991). General trends and pattern of drug abuse in Japan. In S. Fukui et al. (Eds.), *Cocaine and methamphetamine: Behavioral toxicology, clinical psychiatry and epidemiology.* Tokyo: Drug Abuse Prevention Center.

KEUP, W. (1986). Use, indications and distribution in different countries of the stimulant and hallucinogenic amphetamine derivatives under consideration by WHO. *Drug and Alcohol Dependence, 17,* 169–192.

LUKAS, S. E. (1985). *The encyclopedia of psychoactive drugs: Amphetamines—Danger in the fast lane.* New York: Chelsea House.

MILLER, M. A. (1991). Trends and patterns of methamphetamine smoking in Hawaii. In M. A. Miller & N. J. Kozel (Eds.), *Methamphetamine abuse: Epidemiologic issues and implications.* NIDA Monograph Series No. 115, DHHS Publication no. (ADM)91-1836. Washington, DC: U.S. Government Printing Office.

MONROE, R. R., & DRELL, H. H. (1947). Oral use of stimulants obtained from inhalers. *Journal of the American Medical Association, 135,* 909–914.

NATIONAL INSTITUTE ON DRUG ABUSE. (1991). *National Household Survey on Drug Abuse: Population estimates 1991.* DHHS Publication no. (ADM)92-1887. Washington, DC: U.S. Government Printing Office.

NATIONAL INSTITUTE ON DRUG ABUSE. (1990). *Statistical Series. Data from the Drug Abuse Warning Network (DAWN) Semiannual Report, Series 6, No 24.* Washington, DC: U.S. Government Printing Office.

NATIONAL INSTITUTE ON DRUG ABUSE. (1989). *Methamphetamine abuse in the United States.* DHHS Publication no. (ADM)89-1608. Washington, DC: U.S. Government Printing Office.

NATIONAL NARCOTICS INTELLIGENCE CONSUMERS COMMITTEE (NNICC). (1992). *The NNICC Report 1991.* U.S. Department of Justice, Drug Enforcement Administration.

OSCARSSON, L. (1987). Drug abuse trends in Sweden. In National Institute on Drug Abuse: *Patterns and Trends of Drug Abuse in the United States and Europe.* DHHS Publication no. 181-322:60368. Washington, DC: U.S. Government Printing Office.

PITTEL, S. M., & HOFER, R. (1970). The transition to amphetamine abuse. In E. H. Ellinwood & S. Cohen (Eds.), *Current concepts on amphetamine abuse.* Washington, DC: U.S. Government Printing Office.

PUDER K. S., KAGAN D. V., & MORGAN, J. P. (1988). Illicit methamphetamine: Analysis, synthesis, and availability. *American Journal of Drug and Alcohol Abuse, 14,* 463–473.

SEIDEN, L. S. (1991). Neurotoxicity of methamphetamine: Mechanisms of action and issues related to aging. In M. A. Miller & N. J. Kozel (Eds.), *Methamphetamine abuse: Epidemiologic issues and implications.* DHHS Publication no. (ADM)91-1836. Washington, DC: U.S. Government Printing Office.

SEIDEN, L. S., & KLEVEN, M. S. (1989). Methamphetamine and related drugs: Toxicity and resulting behavioral changes in response to pharmacological probes. In K. Asghar & E. D. Souza (Eds.), *Pharmacology and toxicology of amphetamine and related designer drugs.* DHHS Publication no. (ADM)89-1640. Washington, DC: U.S. Government Printing Office.

SENAY, E. C. (1991). Drug abuse and health: A global perspective. *Drug Safety, 6*(1), 1–65.

SMITH, R. C. (1970). Compulsive methamphetamine abuse and violence in the Haight-Ashbury district. In E. H. Ellinwood & S. Cohen (Eds.), *Current concepts on amphetamine abuse.* Washington, DC: U.S. Government Printing Office.

U.S. DEPARTMENT OF JUSTICE. (1991). *Clandestine laboratory seizures in the United States 1990.* Washington, DC: U.S. Department of Justice, Drug Enforcement Administration, Office of Intelligence.

MARISSA MILLER
NICHOLAS KOZEL

AMSTERDAM, DRUG USE IN *See* Netherlands, Drug Use in the.

AMYL NITRITE *See* Inhalants.

AMYTAL *See* Amobarbital

ANABOLIC STEROIDS Anabolic steroids are synthetic versions of the naturally occuring male sex hormone testosterone. They are more properly called anabolic-androgenic steroids (AASs), because they have both bodybuilding (anabolic) effects and masculinizing (androgenic) effects. The masculinizing effects of testosterone cause male characteristics to appear during puberty in boys, such as enlargement of the penis, hair growth on the face and pubic area, muscular development, and deepened voice. Females also produce natural testosterone, but ordinarily in much smaller amounts than males.

AASs are sometimes referred to simply as steroids. Steroid means only that a substance either resembles cholesterol in its chemical structure or is made from cholesterol in the body. Thus, AASs are one kind of steroid. (They are not to be confused with an entirely different group of steroids called corticosteroids—of which prednisone and cortisone are examples— which are commonly used to treat illnesses such as arthritis, colitis, and asthma. In contrast to anabolic steroids, corticosteroids can cause muscle tissue to be wasted.) AASs are also referred to as ergogenic drugs—which means performance-enhancing. Street or slang terms for AASs include "roids" and "juice."

Soon after testosterone was first isolated and synthesized in the laboratory in 1935, a number of synthetics were created to be used as medicines. The synthetic forms were developed because natural testosterone did not work for very long when given as a pill or injection (it is subject to rapid breakdown in the body). Bodybuilders may have begun using AASs to build muscle size and strength as early as the 1940s. Olympic athletes started to use these drugs in the 1950s. Most of this use went undetected, however, because the technology of drug testing did not allow reliable detection of AASs in the urine until the 1976 Olympic Games. Even so, anabolic steroids did not become a household word until 1988, when Ben Johnson, a track champion, tested positive for AASs during that year's Olympic Games. In the same year, a study reported that 6.6 percent of U.S. male high school seniors had tried AASs. This study made it clear that elite athletes were not the only ones taking these drugs. In 1991, AASs were added by federal law to the list of Schedule III of the CONTROLLED SUBSTANCES ACT (Brower, 1992). Schedule III controlled substances are recognized to have value as medicines, but also to have a potential for abuse that may lead to either low to moderate physical dependence or high psychological dependence. Thus, parents, teachers, coaches, and students need to be knowledgeable about AASs. Table 1 lists the names of some AASs that bodybuilders have used. Hundreds of these drugs have been synthesized, and

TABLE I
Anabolic Steroids Used by Bodybuilders

Generic Name	*Representative Brand Names*
Injectable testosterone esters[a]	
Testosterone cypionate	Depo-Testosterone (Slang name: Depo-T), Virlon IM
Testosterone enanthate	Delatestryl
Testosterone propionate	Testex, Oreton propionate
Other injectables	
Nandrolone decanoate	Deca-Durabolin (Slang names: Deca, Deca-D)
Nandrolone phenpropionate	Durabolin
Methenolone enanthate	Primobolan Depot
Veterinary injectables used by humans	
Trenbolone acetate	Finaject (Finajet) 30, Parabolan
Boldenone undecylenate	Equipoise
Stanozolol	Winstrol V
Pills (17-alkylated AASs)[b]	
Ethylestrenol	Maxibolan
Fluoxymesterone	Halotestin
Methandrostenolene	Dianabol (Slang names: D-bol, D-ball)
Methenolone	Primobolan
Methyltestosterone	Android (10 & 25), Metandren, Oreton Methyl, Testred, Virilon
Oxandrolone	Anavar
Oxymetholone	Adroyd, Anadrol-50
Oxymesterone	Oranabol
Stanozolol	Winstrol

[a]Least toxic to liver and cholesterol levels; cause estrogen levels to increase.
[b]Most toxic to liver and cholesterol levels.

more comprehensive lists exist (Goldman, Bush, & Klatz, 1987; Wright & Cowart, 1990).

GENERAL CHEMICAL STRUCTURE

Testosterone has a four-ring structure composed of nineteen carbon atoms. Accordingly, the carbon atoms are labeled by number from 1 to 19 (see Figure 1). Many synthetic forms of testosterone are made by adding either an alkyl group or an ester to the 17-carbon atom (a chain of carbon and hydrogen atoms). An ester is formed by reacting an acidic chain of carbon and hydrogen atoms to the −OH group on the 17-carbon atom. In general, when an

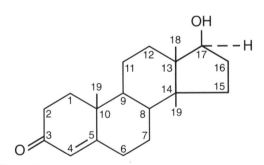

Figure 1
Testosterone Molecule. The numbers refer to carbon atoms, and the hydrogen and hydroxyl groups are at carbon 17.

group is added to the 17-carbon atom, the resulting drug can be taken as a pill; however, these so-called 17-alkylated AASs are relatively toxic to the liver and are more likely to cause negative effects on cholesterol levels. By contrast, when an ester is formed at the 17-carbon atom, an injectable form of testosterone is created that is less toxic to the liver and cholesterol levels. Other AASs are created by making modifications at other carbon atoms.

MEDICAL AND NONMEDICAL USES

AASs are prescribed by physicians to treat a variety of medical conditions (Brower, 1992). The most accepted use is for treating men unable to produce normal levels of their own testosterone, a condition known as testosterone deficiency or hypogonadism. AASs are also used to treat a rare skin condition called hereditary angioedema, certain forms of anemia (deficiency of red blood cells), advanced breast cancer, and endometriosis (a painful condition in females in which uterus (lining) tissue develops in other body parts). AASs are also combined with female hormones to treat distressing symptoms that can accompany menopause. Experimentally, AASs have been used to treat a condition in which bone loss occurs (osteoporosis), to treat impotency and low sexual desire, and as a male birth control pill. All these medical uses are uncommon, however, either because the conditions are rare or because other treatments are preferred. Nevertheless, AASs are important medicines to have available.

Nonmedically, AASs are used to enhance athletic performance, physical appearance, and fighting ability. Since society endows people who look physically fit and attractive with many benefits and much recognition, some individuals see AASs as a means to those benefits. Three groups of AAS users have been described:

1. The athlete group, aims to win at any cost; the athlete also believes, sometimes correctly, that the competition is using AASs. The anticipated rewards to the athlete are the glory of victory, social recognition and popularity, and financial incentives (college scholarships, major league contracts).

2. The aesthete group aims to create a beautiful body, as if to make the body into a work of art. Aesthetes may be competitive bodybuilders, or aspiring models, actors, or dancers. They put their bodies on display to obtain admiration and financial rewards.

3. This group of AAS users seek to enhance their ability to fight or intimidate; they include bodyguards, security guards, prison guards, police, soldiers, bouncers, and gang members. These people depend on fighting for their very survival.

Whether AASs actually work to improve performance and appearance has been debated. Invariably, users believe AASs do work, but scientific studies are divided between those that show an effect and those that do not. Limitations exist for how these studies can be designed and what they can show. In general, most researchers agree that AASs can work in some individuals to enhance muscle size and strength when combined with a proper exercise program and diet. Alternatively, AASs probably do not improve performance of aerobic or endurance activities. In any case, most researchers are shifting attention from whether AASs work to whether they are harmful.

PATTERNS AND CONSEQUENCES OF USE

AASs have been associated with a variety of undesirable effects. The most severe consequence attributed to their use is death, although fortunately the number of reported fatalities among steroid-using athletes has been low. The distinction between fatalities that occur among relatively healthy athletes who use AASs and among patients with illnesses (such as anemia) who are prescribed AASs is important—because ill patients are already at a higher risk for an early death. Nevertheless, deaths in nonmedical steroid users (such as athletes and aesthetes) have occurred from liver disease, cancer, and suicide; near-deaths have occurred from heart attacks and strokes (Brower, 1992).

Psychiatric Effects. Another serious, life-threatening consequence has been violent aggression toward other people. Both the medical literature and newspapers contain reports of previously mild-mannered individuals who committed murder and lesser assaults while taking AASs (Pope & Katz, 1992). Although reports of severe violence generate both alarm and widespread attention, the total number of such reports is small. Moreover, the effects of

AASs on violent behavior vary widely depending on the social circumstances and the characteristics of the individual. Nevertheless, animal studies clearly show that AASs increase aggressive behavior, and most but not all studies in humans have found that high doses of AASs increase feelings and thoughts of aggressiveness. Although an increase in feelings and thoughts of violence does not always lead to violent behavior, it can be very distressing to the individual and to those around him or her. "Roid rage" is a slang expression used to describe the aggressive feelings, thoughts, and behaviors of AAS users.

Other psychiatric effects of AASs include mood swings and psychosis (Brower, 1992; Pope & Katz, 1992). AAS users commonly report that they feel energetic, confident, and even euphoric during a cycle of use. They may have a decreased need for sleep or find it difficult to sleep because of their high energy level. Such feelings may give way to feeling down, depressed, irritable, and tired between cycles of use. With continued use of high AAS doses, moods may shift suddenly, so that the user feels on top of the world one moment, irritable and aggressive the next, and then depressed or nervous. The appetite may also swing widely with cycles of use (Wright & Cowart, 1990). During a cycle on AASs, huge quantities of food may be consumed to support the body's requirements for muscle growth and energy. During the off cycles, appetite may diminish.

The term *psychosis* means that a person cannot distinguish between what is real and what is not. For example, a person may believe that other people intend harm when no real threat exists; or a person may believe that an impossible life-threatening stunt can be performed with no problem. Such false beliefs are called delusions. The psychotic person may also experience hallucinations, such as hearing a voice that is not there. Psychotic effects were reported in 12 percent of AAS users in one study, although other studies have reported lower rates (Brower, 1992; Pope & Katz, 1992). Fortunately, the psychiatric effects of AASs tend to disappear soon after AASs are stopped, although a depressed mood may last for several months. Obviously, when suicides, homicides, or legal consequences from assault have occurred, they cannot be reversed simply by stopping one's use of AASs.

Effects on the Liver. AASs can affect the liver in various ways, but the 17-alkylated AASs are more toxic to the liver than other AASs (Friedl, 1993).

Most commonly, AASs cause the liver to release extra amounts of enzymes into the bloodstream that can be easily measured by a blood test. The liver enzymes usually return to normal levels when AASs are stopped. The liver also releases a substance called bilirubin, which in high amounts can cause the skin and eyes to turn yellow (a condition called jaundice). As many as 17 percent of patients treated with the 17-alkylated AASs develop jaundice (Friedl, 1993). Nonmedical AAS users can also develop jaundice. Although untreated jaundice can be dangerous and even fatal, jaundice usually disappears within several weeks of stopping AASs. Jaundice can also signal other dangerous conditions of the liver, such as hepatitis, so it should always be treated by a physician. Another condition that occurs among patients treated with AASs is peliosis hepatis, in which little sacs of blood form in the liver. Death can occur from bleeding if one of the sacs ruptures. Finally, liver tumors may occur in 1 to 3 percent of individuals (including athletes) using high doses of the 17-alkylated AASs for more than two years (Friedl, 1993). Rare cases of liver tumors have been reported with other types of AAs as well. Some of the tumors are cancerous, and although half or more of the tumors disappeared when AASs were stopped, others resulted in death.

Potential to Affect the Heart. AASs can cause changes in cholesterol levels (Friedl, 1993). Low amounts of a certain kind of cholesterol (high-density lipoprotein cholesterol) in the blood are known to increase the risk of heart attacks. AASs, especially the 17-alkylated ones, cause a lowering of this so-called good form of cholesterol. When AASs are stopped, however, cholesterol levels return to normal. Another risk factor for heart attacks and strokes is high blood pressure. Studies have shown that AASs can cause small increases in blood pressure (Friedl, 1993); this also returns to normal when AASs are stopped. As a result of strenuous exercise, many athletes develop an enlarged heart that is not harmful. Some, but not all studies, suggest that AAS users can develop a harmful enlargement of the heart (Brower, 1993). As noted previously, heart attacks and strokes have been reported in AAS users, but as of the early 1990s studies are needed to determine if AAS users have a higher risk of heart attacks and strokes than nonusers (Friedl, 1993).

Sexual Side Effects. Not surprisingly, AASs can alter the levels of several sex-related hormones

in the body, resulting in the following effects (Brower, 1992; Wright & Cowart, 1990). In males, the prostate gland can enlarge, making it difficult to urinate; the testes shrink; and sterility can occur. The effects on the prostate, the testes, and sterility reverse when AASs are stopped; however, at least one case of prostate cancer has been reported, an exception to reversibility. Males can also develop enlarged breast tissue from taking AASs, an effect medically termed gynecomastia (it is referred to by male users as "bitch tits"). Gynecomastia occurs because testosterone is chemically changed in the body to the female hormone estrogen. Thus, the male user experiences higher amounts of estrogen than normal. Painful lumps in the male breast may persist after stopping AASs, and they sometimes require surgical removal. Females, however, may undergo shrinkage of their breasts, as a response to higher amounts of male hormone than normal. Menstrual periods become irregular, and sterility can occur in females as well. Deepened voice and an enlarged clitoris are effects in females that do not always reverse after stopping AASs. Women may also develop excessive hair growth in typically masculine patterns, such as on the chest and face. Finally, both males and females may experience increases or decreases in their desire for sex.

Other Effects. In children of both sexes before the onset of puberty, AASs can initiate the characteristics of male puberty and cause the bones to stop growing prematurely. The latter effect can result in shorter adult heights than would otherwise occur. AASs can cause premature baldness in some individuals, and it can cause acne. The acne is reversible with cessation of AASs. Other possible effects include small increases in the number of red blood cells, worsening of a condition called sleep apnea (in which afflicted persons stop breathing for short intervals during sleep), and worsening of muscle twitches (known as tics) in those who are predisposed (Brower, 1992).

PATTERNS OF ILLICIT USE

AASs are commonly smuggled from Mexico and other countries, where they are easily obtained over the counter without a prescription. Between 70 and 90 percent of nonmedical steroid users purchase their drugs from this flourishing black market, with sales totals at some 300 to 500 million U.S. dollars

yearly (Brower, 1992). Dealers and users typically connect in weight-lifting gyms. Without a doubt, these drugs are easy to obtain for anyone who wants them.

Steroids are taken both as pills and by injection. Injection occurs into large muscle groups (buttocks, thigh, or shoulder) or under the skin, but not into veins. Cases of acquired immune deficiency syndrome (AIDS) have been reported in steroid users because of needle sharing (Brower, 1992). Steroids are often taken in cycles of six to twelve weeks on the drugs, followed by six to twelve weeks off. At the beginning of a cycle, small doses are taken with the intent to build to larger doses, which are then tapered off at the end of a cycle. Illicit users typically consume 10 to 100 times the amounts ordinarily prescribed for medical purposes, requiring them to combine or "stack" multiple steroid drugs. The dose cannot always be determined, however, because illicit steroids may contain both falsely labelled and veterinary preparations. (Drugs purchased on the illicit market do not always contain what the labels indicate; vials contaminated with bacteria have been confiscated by law-enforcement officials.)

Steroid users commonly take other drugs, each with their own risks, to manage the unpleasant side effects of steroids, to increase body-building effects, and/or to avoid detection in drug testing (Wright & Cowart, 1990). For example, estrogen blockers, such as tamoxifen or clomiphene, are taken to prevent breast enlargement. Water pills (diuretics) are taken both to dilute the urine prior to drug testing and to eliminate fluid retention so that muscles will look more defined. Human chorionic gonadotropin (HCG) is an injectable, nonsteroidal hormone that stimulates the testicles to produce more testosterone and to prevent them from shrinking. Human growth hormone is another nonsteroidal hormone that is taken to increase muscle and body size.

ADDICTIVE POTENTIAL

As with other drugs of abuse, dependence on AASs occurs when a user reports several of the following symptoms—inability to stop or cut down use, taking more drugs than intended, continued use despite having negative effects, tolerance, and withdrawal. *Tolerance* refers to the needing of more drug to get the same effect as was previously obtained with smaller doses—or of having diminished effects with

the same dose. In terms of the anabolic effects, tolerance was demonstrated in animals in the 1950s. In a 1990s study, 18 percent of nonmedical AAS users reported tolerance (Brower, 1992). Whether tolerance develops to the mood-altering effects of AASs is unknown. *Withdrawal* refers to the uncomfortable effects users experience when they stop taking AASs. As noted previously, many of the undesirable effects reverse when AASs are stopped; however, others can begin—such as depressed mood, fatigue, loss of appetite, difficulty sleeping, restlessness, decreased sex drive, headaches, muscle aches, and a desire for more AASs (Brower, 1992; 1993). The depression can become so severe that suicidal thoughts occur. The risk of suicide described previously is thought to be highest during the withdrawal period.

Studies indicate that between 14 and 57 percent of nonmedical AAS users develop dependence on AASs (Brower, 1992). These studies support the addition of AASs to the list of Schedule III controlled substances. Nevertheless, AASs may differ from other drugs of abuse in several ways. First, physical dependence on AASs has not been reported to occur when AASs are prescribed for treating medical conditions. This differentiates the AASs from the opioid painkillers and the sedative-hypnotics. Second, dependence may be primarily on the muscle-altering effects of AASs, rather than the mood-altering effects. Some researchers have questioned whether AASs produce dependence in the true sense of the word (psychological dependence), because most definitions require that drugs be taken primarily for their mood-altering effects (Brower, 1992). Third, AAS users appear more preoccupied with their bodies and the way they look than do users of other drugs of dependence.

SUMMARY

The anabolic-androgenic steroids are synthesized forms of the male sex hormone testosterone. They have both masculinizing and bodybuilding effects. AASs are useful to treat a variety of somewhat uncommon medical conditions. Increasingly, they are used for the nonmedical purposes of enhancing athletic performance and physical appearance. Most researchers agree with users that AASs can increase muscle size and strength in some individuals when combined with a proper exercise program and diet. Many are also concerned about the potential for harmful effects with AASs, especially when the pat-

terns of illicit use are considered. Drugs obtained on the illicit market may be contaminated, may be falsely labeled, or may contain substances not approved for human use. Multiple steroid and nonsteroidal drugs are combined, and AAS doses may exceed therapeutic doses by 10 to 100 times. Although the 17-alkylated AASs are commonly used, because pills are more convenient than injections, they are more toxic to the liver and to cholesterol levels than are the injectable testosterone esters. Nevertheless, injections carry their own risks from improper injection techniques to dirty and shared needles.

The most serious side effects of AASs seem relatively uncommon, such as deaths or near-deaths from liver disease, heart attacks, strokes, cancer, suicide, and homicidal aggression. Most other side effects appear to be reversible when AASs are stopped, such as altered cholesterol levels, some liver effects, most psychiatric effects, testicular shrinkage, sterility, high blood pressure, and acne. Exceptions to reversibility include lumps in the male breast, deepened voice and enlarged clitoris in females, and cessation of bone growth in children. Moreover, some individuals may develop dependence on AASs, making it difficult for them to stop using. Stopping use can also produce distressing withdrawal symptoms, the worst of which is suicidal depression. Finally, studies of the long-term effects of using AASs are lacking, so safety cannot be guaranteed with the high-dose use of these drugs.

(SEE ALSO: *Lifestyle and Drug-Use Complications*)

BIBLIOGRAPHY

BROWER, K. J. (1992). Anabolic steroids: Addictive, psychiatric, and medical consequences. *American Journal on Addictions, 1,* 100–114.

GOLDMAN, B., BUSH, P., & KLATZ, R. (1987). *Death in the locker room.* Tuscon, AZ: Body Press.

POPE, H. G., & KATZ, D. L. (1992). Psychiatric effects of anabolic steroids. *Psychiatric Annals, 22*(1), 24–29. This entire issue is devoted to anabolic steroids.

WRIGHT, J. E., & COWART, V. S. (1990). *Anabolic steroids: Altered states.* Carmel, IN: Benchmark Press.

YESALIS, C. E. (ED.). (1993). *Anabolic steroid use in exercise and sport.* Champaign, IL: Human Kinetics Publishers. Contains chapters by Brower and by Friedl, as cited in text.

KIRK J. BROWER

ANALGESIC Analgesics are drugs used to control pain without producing anesthesia or loss of consciousness. Analgesics vary in terms of their class, chemical composition, and strength. Mild analgesics, such as aspirin (e.g., Bayer, Bufferin), acetominophen (e.g., Tylenol), and ibuprofen (e.g., Advil), work throughout the body. More potent agents, including the OPIATES codeine and morphine, work within the central nervous system (the brain and spinal cord). The availability of the more potent analgesics is more carefully regulated than that of aspirin and other similar analgesic/anti-inflammatory agents that are sold in drugstores OVER-THE-COUNTER. The more potent opiate agents typically require prescriptions to be filled by pharmacists.

An important aspect of analgesics is that they work selectively on pain, but not on other types of sensation, such as touch. In this regard, they are easily distinguished from anesthetics, which block all sensation. Local anesthetics, such as those used in dental work, make an area completely numb for several hours. General anesthetics typically are used to render patients unconscious for surgery.

(SEE ALSO: *Pain: Drugs Used in Treatment of*)

BIBLIOGRAPHY

GILMAN, ALFRED G., ET AL. (EDS.). (1990). *Goodman and Gilman's the pharmacological basis of therapeutics*, 8th ed. New York: Pergamon.

GAVRIL W. PASTERNAK

ANESTHETIC *See* Inhalants.

ANGEL DUST *See* Phencyclidine (PCE): Slang and Jargon.

ANHEDONIA This term refers to a clinical condition in which a human or an experimental animal cannot experience positive emotional states derived from obtaining a desired or biologically significant stimulus. Generally, certain stimuli serve as positive reinforcers in normal individuals (e.g., food, water, the company of friends). *Positive reinforcement* is a descriptive term used by behavioral scientists to denote an increase in the probability of a behavior that is contingent on the presentation of biologically significant stimuli, such as food or water.

Anhedonia may be idiopathic (of unknown cause), may occur as a side effect of certain drugs (for example, the NEUROLEPTICS), which act as dopamine-receptor antagonists, or may be an aspect of certain psychiatric disorders, such as depression. It is conjectured that a state of anhedonia may occur during the "crash" that follows a prolonged bout of drug self-administration, particularly COCAINE or amphetamine-like stimulants.

BIBLIOGRAPHY

DACKIS, C. A., & GOLD, M. S. (1985). New concepts in cocaine addiction: The dopamine depletion hypothesis. *Neuroscience Biobehavioral Review, 9*, 469–477.

WISE, R. A. (1985). The anhedonia hypothesis: Mark III. *Behavior and Brain Science, 8*, 178–186.

WISE, R. A. (1982). Neuroleptics and operant behavior: The anhedonia hypothesis. *Behavior and Brain Science, 5*, 39–87.

ANTHONY PHILLIPS

ANIMAL RESEARCH. *See* Research, Animal Model.

ANORECTIC This term derives from the Greek (*a + oregein*, meaning "not to reach for"; later, *anorektos*), and it refers to a substance that reduces food intake. It came into use in English about 1900. Anorectic agents (also referred to as anorexics, anorexegenics, or appetite suppressants) fall into a number of categories according to the brain neurotransmitter system through which they work.

Central nervous system (CNS) stimulants that act through the noradrenergic and dopaminergic systems include COCAINE, amphetamine-like compounds, mazindol, and phenylpropanalamine. Serotonergic compounds include fenfluramine, fluoxetine, and sertraline. Several endogenous peptides (within the body) also have anorectic actions, in that they inhibit food intake—these include cholecystokinin, glucagon, and the bombesin-like peptides.

Not *all* agents that can suppress appetite are medically approved for such use. For example, cocaine is approved *only* as a local anesthetic.

(SEE ALSO: *Amphetamine*)

TIMOTHY H. MORAN

ANOREXIA This term means a "loss of appetite," especially when prolonged, and comes into English in the 1620s from Latin usage, based on Greek stems (*a* [no] + *orexis* [appetite]). Anorexia generally leads to loss of weight because of loss of appetite; anorexia nervosa is an appetite disorder associated with severe weight loss. Eating disorders of this type and those associated with compulsive eating are, in some ways, behavioral equivalents of drug abuse.

(SEE ALSO: *Overeating and Other Excessive Behaviors*)

TIMOTHY H. MORAN

ANSLINGER, HARRY J., AND U.S. DRUG POLICY From the mid-1920s until the early 1970s, one man, Harry J. Anslinger (1892–1975), had the dominant role in shaping U.S. policy about the use of drugs—other than alcohol and tobacco. Understanding his life and work is, therefore, a necessity for understanding the evolution of current drug and alcohol policies. Anslinger was commissioner of narcotics, U.S. Treasury Department Bureau of Narcotics, from 1930 to 1962; chief U.S. delegate to international drug agencies until 1970; and a leading proponent of repressive antidrug measures in the United States—and worldwide.

He was born in Altoona, Pennsylvania, May 20, 1892, the eighth child in a Swiss immigrant family. After a broad education supported by many jobs—railroad policeman to silent-movie pianist and noncombat military service in World War I—Anslinger entered the U.S. diplomatic service. His first post was Holland, where he was assigned as liaison to deposed Kaiser Willhem's entourage; then he was sent to Germany and on to Venezuela for a frustrating three-year stint as U.S. consul.

THE FEDERAL BUREAU OF NARCOTICS

In 1926, Anslinger became U.S. consul in the Bahamas, which were then a principal gateway for bootlegging operations. Consul Anslinger quickly came to notice for his good work in persuading the British to cooperate in curbing the flow of contraband. The Volstead Act and the Harrison Act, aimed respectively at enforcing Prohibition and controlling distribution of so-called narcotic drugs, were both tax measures and hence within the jurisdiction of the U.S. Treasury Department. Treasury soon borrowed Anslinger from the Department of State to serve in its Prohibition Bureau, which then enforced both acts. In 1930, the drug-regulation functions were shifted to a new Bureau of Narcotics; Anslinger was named acting commissioner; and when other candidates were disqualified by scandal, President Herbert C. Hoover made Anslinger's appointment permanent.

REPRESSIVE POLICIES

Drug addiction, particularly genuine addiction to the OPIATES, had already been demonized when Anslinger came on the scene: Addicts had been labeled dope fiends in the public mind, and the illicit traffic had been attributed first to sinister German agents during World War I, then to terrifying Chinese tongs (secret societies). Clinics, set up to relieve the plight of addicts cut off from their supplies by providing maintenance doses of HEROIN, were curbed by a series of U.S. Supreme Court rulings between 1918 and 1922, interpreted by the Treasury Department to prohibit heroin maintenance as a form of medical treatment. The clinics were closed, and by 1925 doctors had stopped prescribing to addicts. A black market had begun to flourish.

Anslinger, a skillful bureaucratic empire builder, espoused an extreme punitive approach from the outset. He cultivated members of Congress and other politicians, providing all who served his interests with material to portray themselves as fierce drug fighters. He relentlessly opposed "education" about the realities of drug use—on the ground that it would encourage "youthful experimenters." His bureau sounded one alarm after another: For example, drugs caused users to commit violent crimes (it was solemnly claimed, for a while, that dangerous crim-

inals took drugs to sharpen their vicious courage in committing crimes) and that addicts induced others to become addicted (each addict made seven others in his career), so they were "infectious" in the community.

Sometimes the Bureau of Narcotics warned that an entire generation of American youth, including young children, was imperiled; then, suddenly, drugs would be revealed as responsible for a wave of juvenile delinquency and the menace of "young hoodlums." The bureau announced its enforcement success in terms of the total number of years' sentences imposed on drug offenders annually ("3,248 years, 10 months, 18 days" for 1933); SEIZURES OF DRUGS were announced as STREET VALUE, which grossly inflated their price. Because formal systems to measure levels of drug abuse were lacking, drug statistics could be and were manipulated—skyrocketing when Anslinger wanted support or appropriations, plummeting when he wanted credit and praise. These enforcement activities resulted in an increase in federal prisoners and led to the establishment of two U.S. Public Health Service Hospitals to treat addicted inmates, but the major thrust of U.S. policy was control of supply, and punishment of users.

MARIJUANA TAX ACT OF 1937

During his tenure as commissioner, Anslinger dominated the enactment of U.S. narcotics laws. In the mid-1930s, to puff the menace his bureau was combatting, he turned his attention to MARIJUANA (*Cannabis sativa* [hemp]), used at the time by a few Spanish Americans, Caribbean blacks, and in special circles, such as jazz musicians. At the time, a number of responsible studies of the effects of marijuana (such as by the Indian Hemp Commission in 1895) and its more potent form, hashish, had pronounced it relatively harmless—but that gave Anslinger and a few other sensationalists of the day no pause. Shocking accounts of heinous crimes induced by marijuana began emanating from the bureau; the myth that pot smoking was a dangerous "gateway" to other addictions gained credence; and a bureau-sponsored film, *Reefer Madness*, was produced to popularize Anslinger's visions of the hazards of drug use. Viewed from today, this film convulses audiences as classic kitsch.

Anslinger orchestrated the passage of a bill in Congress to place marijuana in the same highly re-

stricted categories as HEROIN and COCAINE. President Franklin D. Roosevelt signed the bill on August 2, 1937. The drug soon came to account for more enforcement activity than any other. Honest research on its toxic properties was stifled because the bureau would not license its use by researchers outside of government. Although its therapeutic value in alleviating nausea due to chemotherapy for cancer patients or for treating glaucoma are now generally recognized, it remains in the most strictly prohibited "dangerous drug" classification today.

BOGGS ACT OF 1951

In the late 1940s, Anslinger launched an attack on judges, claiming that the drug problem was caused by too-lenient sentences imposed on drug offenders. This was picked up by Anslinger disciples in Congress—and resulted in legislation—the Boggs Act, signed November 2, 1951, and amended by the Narcotics Control Act of 1956, signed July 18, 1956. These acts introduced severe mandatory minimum punishments following conviction, *at least* 2, 5, and 10 years for repeated convictions, without probation or parole, and a mandatory life sentence, or death, at a jury's discretion, for sale of heroin by an adult to a minor.

UNIFORM STATE LAWS

The narcotics bureau similarly pressed state legislatures to enact extreme drug laws, promulgating a Uniform Narcotic Drug Act through the National Conference of Commissioners on Uniform State Laws in 1932, inducing the passage of tough marijuana measures after 1937, and promoting "Little Boggs Acts" in the late 1950s. Some of the latter contained penalties even more severe than the federal law, and it became common practice for Anslinger's men and local prosecutors to shuffle drug offenders into either state or federal courts, depending on where they would receive the harsher sentences.

Federal lawmakers were somewhat inhibited, in Anslinger's day, by the fact that federal drug laws were based solely on Congress's power to tax; the states could enact penalties based on their general powers to punish crime, and some even prescribed punishment for the mere status of being an addict—until the Supreme Court ruled that out in *Robinson v. California* (1963). Little attention was paid to

"treatment" or "rehabilitation"; addicts should either give up their wicked habits, and their spreading of the vice to others, or they should be isolated from society by "quarantine" somewhere for life. Toward the end of Anslinger's tenure, attention turned to "civil commitment," which sometimes accomplished the quarantine result when lawmakers provided scanty resources for rehabilitating those thus committed.

INTERNATIONAL DRUG POLICIES

By 1930, when Anslinger became commissioner, the patterns of international controls had also been largely set, with the United States urging stringent repression and most of the rest of the world remaining indifferent or resistant (the basic Hague Convention of 1912 would never have been ratified by more than a few nations had not the United States insisted upon its inclusion in the Paris peace treaties, which created the League of Nations in 1921). Although the United States never joined the League of Nations, U.S. representatives were always given a voice in drug matters—and Anslinger dominated international deliberations, leading the U.S. delegations to first League of Nations and then United Nations (U.N.) drug-control agencies, even after his resignation as U.S. commissioner.

Anslinger's annual bureau reports to the U.S. Treasury were also submitted as his official annual reports to the League Opium Advisory Committee and its successor, the U.N. Commission on Narcotic Drugs; he could thus push his views in the United States as recommendations endorsed by the international bodies and, simultaneously, present them to the latter as official statements of U.S. positions.

Anslinger participated in the drafting of the 1931 Narcotics Limitation Convention, which imposed controls on the production of drugs for legitimate medical uses; he pressed for the 1936 Convention for Suppression of Illicit Traffic, which sought to induce other nations to impose criminal sanctions on domestic distribution and consumption. When World War II isolated Geneva and ended most of the functions of the League of Nations based there, he arranged for moving the international drug agencies to New York City, where they continued to operate. After the war, he was the leading proponent of a Single Convention, finally approved in 1961, after ten years of drafting. It incorporated much of the U.S. law-

enforcement orientation, including obligations upon members to control crops and production, to standardize identification and packaging, and to impose severe criminal penalties on drug offenders.

Although the Single Convention was widely ratified, many adherents ignored its requirements. Lacking enforcement sanctions, it had small effect. Some of Anslinger's more radical proposals, such as including the promotion of opium addiction in the definition of *genocide*, which he charged to enemies like the People's Republic of China and the Soviet Union, won no international support, but they played well at home.

THE ANSLINGER LEGACY

Anslinger resigned as commissioner in 1962, when, for the first time, the U.S. war on drugs was preempted as a White House operation. Over Anslinger's protests, President John F. Kennedy and his brother Attorney General Robert F. Kennedy called for redefining *addiction* as "drug abuse" and *narcotics* as "dangerous drugs"; they pressed for the addition of new categories to the repressive list—amphetamines, barbiturates, and hallucinogens. After the Kennedy assassination, President Lyndon B. Johnson moved much of the federal drug-control apparatus from Treasury to the Department of Justice.

Yet Anslinger's perspective seems to live on. Presidents Richard M. Nixon, Ronald W. Reagan, and George H. Bush intensified the drug war, justifying such efforts with arguments initially developed by Anslinger. Congress, too, continues to be influenced by Anslinger's views. Congress's rhetoric and penal statutes are more extreme than those of the Boggs era. Marijuana is still lumped with heroin and cocaine, and new alarms—the discovery of a CRACK-baby epidemic and the menace of "ice"—are periodically trumpeted. Anslinger created and set a pattern of aggressive drug suppression—contrary to the initial purposes of the Harrison Act—and kept drug prohibition alive when the alcohol ban, Prohibition, which had truly been intended, was repealed in 1933. Anslinger may indeed have been, as critical Congressman John M. Coffee labeled him in 1938, "far and away the costliest man in the world."

(SEE ALSO: *Methadone Maintenance Programs; Treatment, History of, in the United States; U.S. Government Agencies*)

BIBLIOGRAPHY

ANSLINGER, H. J. (1953). *The traffic in narcotics*. New York: Funk & Wagnalls.

ANSLINGER, H. J., WITH OURSLER, W. (1961). *The murderers: The story of the narcotics gangs*. New York: Farrar, Straus & Cudahy.

BRECHER, E. M., ET AL. (1972). *Licit and illicit drugs: The consumers union report*. Boston: Little, Brown.

KING, R. (1974). *The drug hang-up: America's fifty-year folly*, 2nd ed. Springfield, IL: Charles C. Thomas.

LINDESMITH, A. R. (1965). *The addict and the law*. Bloomington: Indiana University Press.

MCWILLIAMS, J. C. (1990). *The protectors: Harry J. Anslinger and the Federal Bureau of Narcotics*. Newark, DE: University of Delaware Press.

MUSTO, D. F. (1973). *The American disease: Origins of narcotics control*. New Haven: Yale University Press.

RUFUS KING

ANTABUSE   *See* Disulfiram.

ANTAGONIST   An antagonist is a drug that binds to a RECEPTOR (i.e., it has affinity for the receptor binding site) but does not activate the receptor to produce a biological response (i.e., it possesses no intrinsic activity). Antagonists are also called receptor "blockers" because they block the effect of AGONISTS. The pharmacological effects of an antagonist therefore result from preventing agonists (e.g., drugs, hormones, neurotransmitters) from binding to and activating the receptor. A competitive antagonist competes with an agonist for binding to the receptor. As the concentration of antagonist is increased, the binding of the agonist is progressively inhibited, resulting in a decrease in the physiological response. High antagonist concentrations can completely inhibit the response. This inhibition can be reversed, however, by increasing the concentration of the agonist, since the agonist and antagonist compete for binding to the receptor. A competitive antagonist, therefore, shifts the dose-response relationship for the agonist to the right, so that an increased concentration of the agonist in the presence of a competitive antagonist is required to produce the same biological response observed in the absence of the antagonist.

A second type of receptor antagonist is an irreversible antagonist. In this case, the binding of the antagonist to the receptor (its affinity) may be so strong that the receptor is unavailable for binding by the agonist. Other irreversible antagonists actually form chemical bonds (e.g., covalent bonds) with the receptor. In either case, if the concentration of the irreversible antagonist is high enough, the number of receptors remaining that are available for agonist binding may be so low that a maximum biological response cannot be achieved even in the presence of high concentrations of the agonist.

(SEE ALSO: *Naloxone; Naltrexone*)

BIBLIOGRAPHY

ROSS, E. M. (1990). Pharmacodynamics: Mechanisms of drug action and the relationship between drug concentration and effect. In A. G. Gilman, T. W. Rall, A. S. Nies, & P. Taylor (Eds.), *Goodman and Gilman's the pharmacological basis of therapeutics*, 8th ed. New York: Pergamon.

NICK E. GOEDERS

ANTIDEPRESSANT   Antidepressants are a diverse group of drugs that demonstrate a capacity to produce improvement in the symptoms of clinical depression, and they are used to treat the abnormal mood states that characterize depressive illnesses. The word *depression* is used commonly to describe a state of sadness; but health professionals use the term in a more restricted or defined manner to describe several psychiatric disorders characterized by abnormal moods. One of these is bipolar disorder, in which periods of depression (marked by dejection, lack of energy, inactivity, and sadness), alternate with periods of manic behavior (marked by abnormally high energy levels and increased activity). Another is major depression, which is often a recurring problem characterized by severe and prolonged periods of depression without the manic swing. A third is dysthymia, a chronic mood state characterized by depression and irritability, which was once referred to as depressive neurosis. The signs and symptoms of depressive mood disorders may occur as part of other medical and psychiatric disorders (i.e., following stroke) as a result of endocrine disorders, or as a

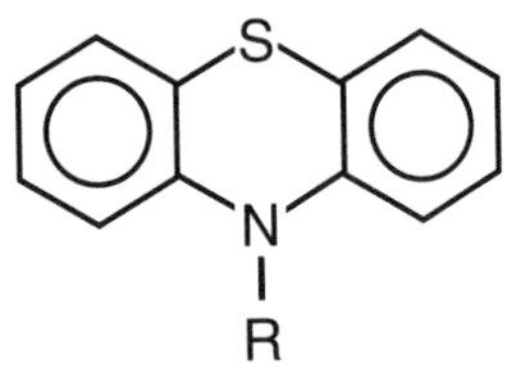

Figure 1
Tricyclic Antidepressant (TCA)

consequence of excessive use of drugs. Often these abnormal mood states may not meet established criteria for one of the major psychiatric mood disorders, but they may nevertheless respond to one of the antidepressant drugs.

Antidepressants can also be useful in a number of medical and psychiatric disorders where depression is not the major feature. For example, some categories of antidepressants can be used to treat anxiety and panic disorders, and they are often useful as adjunctive medications in certain types of chronic pain. Antidepressant drugs are not generally helpful for many kinds of short-term depressed moods that are part of everyday life or for the normal period of grief that follows loss of a loved one.

New categories of antidepressants are being continuously developed and tested. There are now at least five categories in use. These include tricyclic antidepressants, monoamine oxidase (MAO) inhibitors, lithium, nontricyclic antidepressants, and serotonin-uptake inhibitors. The chemical structures of some of these are shown below. Although lithium is

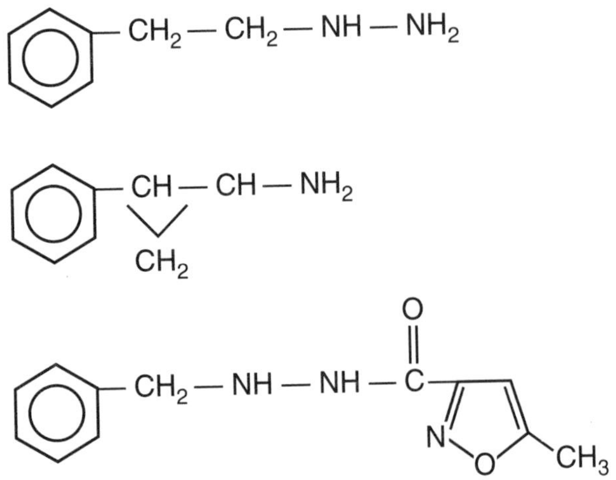

Figure 2
Monoamine Oxidase (MAO) Inhibitors

useful in treating manic states and in preventing depression in bipolar disorders, it is not generally used for other types of depression. It also seems to have no significant value in treating cocaine dependence or alcoholism. Fluoxetine (Prozac) is the bestselling antidepressant of the mid-1990s.

BIBLIOGRAPHY

AMERICAN PSYCHIATRIC ASSOCIATION. (1987). *Diagnostic and Statistical Manual of Mental Disorders-3rd ed., rev.* (DSM-III-R). Washington, DC: Author.

BALDESSARINI, R. J. (1991). Drugs and the treatment of psychiatric disorders. In A. G. Gilman et al. (Eds.), *Goodman and Gilman's the pharmacological basis of therapeutics,* 8th ed. New York: Pergamon.

GEORGE R. UHL
VALINA DAWSON

ANTIDOTE   A medication or treatment that counteracts a poison or its effects. An antidote may work by reducing or blocking the absorption of a poison from the stomach. It might counteract its effects directly, as in taking something to neutralize an acid. Or an antidote might work by blocking a poison at its receptor site. For example, a medication called naloxone will block opiates such as heroin at its receptors and prevent deaths that occur because of heroin overdose. In a sense, drug ANTAGONISTS can all be antidotes under some circumstances, but not all antidotes are drug antagonists.

Many cities have a telephone "poison hot line," where information on antidotes is given. In case of drug overdose or poisoning, it is advisable to call for expert medical help immediately.

BIBLIOGRAPHY

KLAASSEN, C. D. (1990). Principles of toxicology. In A. G. Gilman, T. W. Rall, A. S. Nies, & P. Taylor (Eds.), *Goodman and Gilman's the pharmacological basis of therapeutics,* 8th ed. (p. 58). New York: Pergamon Press.

MICHAEL J. KUHAR

ANTIPSYCHOTIC   Any of a group of drugs, also termed *neuroleptics,* used medicinally in the therapy of schizophrenia, organic psychoses, the manic

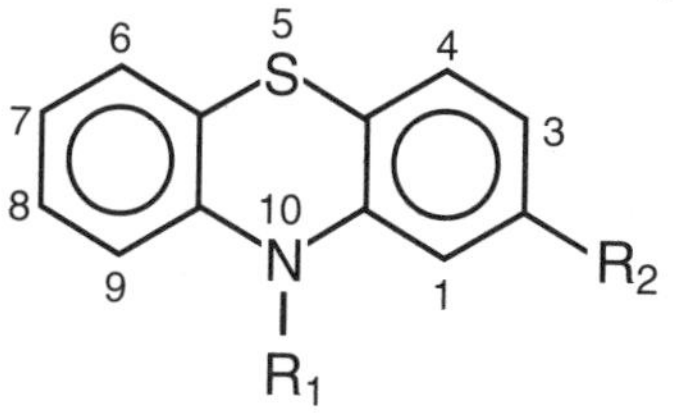

Figure 1
Antipsychotics

phase of manic-depressive illness, and other acute psychotic illnesses. The prototype antipsychotics are the phenothiazines, such as chlorpromazine (Thorazine), and the butyrophenones such as haloperidol (Haldol). The antipsychotics are tricyclic compounds, with chemical substitution at R_1 and R_2, which determine the selectivity and potency of the neuroleptic.

The "positive" symptoms of psychotic disorders, such as hallucinations, can often be effectively treated with antipsychotics; the "negative" symptoms, such as withdrawal, are less effectively managed by these drugs. Most of these drugs also have effects on movement, and a good response to the drugs' antipsychotic effects must often be balanced against motor side effects.

BIBLIOGRAPHY

BALDESSARINI, R. J. (1990). Drugs and the treatment of psychiatric disorders. In A. G. Gilman et al. (Eds.), *Goodman and Gilman's the pharmacological basis of therapeutics*, 8th ed. New York: Peragamon.

GEORGE R. UHL
VALINA DAWSON

ANTI-SALOON LEAGUE *See* Temperance Movement; Women's Christian Temperance Union.

ANTISOCIAL PERSONALITY Antisocial personality disorder (ASP) is particularly germane to alcohol and drug abuse because it co-occurs in a large proportion of those who abuse alcohol or drugs, and it confounds the diagnosis of, influences the course of, and is an independent risk factor for the development of alcohol or drug abuse disorder. Additionally, scientific evidence suggests that alcohol and drug abuse complicated by ASP is more heritable than is substance abuse without ASP.

In the latest diagnostic classification system DIAGNOSTIC AND STATISTICAL MANUAL-4th edition (DSM-IV), ASP is defined as a disorder that begins in childhood or early adolescence and continues into adulthood; it is characterized by a general disregard for and violation of the rights of others. At least three of the following behaviors must have occurred in any twelve-month period of time *before the age of 15, with two before age 13:* running away from home overnight twice, staying out late at night despite parental rules to the contrary (before age 13), truancy (beginning before age 13), initiating physical fights, using weapons in fights, cruelty to animals and to people, vandalism, forcing someone into sexual activity, arson, frequent lying to obtain favors or goods, frequenty bullying, breaking into someone's house or car, and stealing from others (either passively like shoplifting, or aggressively, like mugging). In addition, three of the following behaviors must have occurred *since the age of 15:* consistent irresponsibility (e.g., inability to sustain consistent work behavior or to honor financial obligations), failure to conform to social norms by repeatedly engaging in arrestable behaviors, irritability and aggressiveness, deceitfulness (e.g., frequent lying or conning of others), reckless behaviors indicating disregard for safety of oneself or of others, impulsivity or failure to plan ahead, and lack of remorse for hurtful or manipulative behaviors.

A large percentage of alcohol and drug abusers meet criteria for ASP. For example, in one multisite study of 20,000 community respondents, 15 percent of the alcoholic participants compared to 2.6 percent of the population as a whole, met criteria for ASP. Comparable data from clinical samples indicate that from 16 percent to 49 percent of treated alcoholics met criteria for ASP.

A substantial proportion of those with ASP also abuse alcohol or drugs—three times as many men with a diagnosis of ASP as without abuse alcohol and five times as many abuse drugs. For females, the association is even stronger—twelve times as many women with ASP as without abuse alcohol and thirteen times as many abuse drugs. The strong association between ASP and alcohol or drug abuse may

actually be caused by antisocial behaviors that occur when under the influence of alcohol or drugs judgment is impaired. It is possible, however, to distinguish *primary* abusers, those for whom the antisocial behaviors are a result of their substance use, from *secondary* abusers, for whom substance abuse is just one manifestation of a wide spectrum of antisocial behaviors. This differentiation is particularly important in understanding the genetic transmission of disease, where genetic factors responsible for the comorbid state might be transmitted separately from those that cause substance abuse.

The course of alcohol or drug abuse is affected by ASP. Alcoholics with ASP have a more chronic and more severe course, an earlier onset of alcohol symptoms (for example, average age of onset of 20 compared to nearly 30 for those without ASP), as well as a significantly longer history of problem drinking. Furthermore, evidence is mounting that ASP alcoholics have poorer response to treatment—relapsing much earlier than alcoholics without ASP—and they may respond only to certain therapies.

Antisocial behavior problems in childhood have been identified as an independent risk factor in the development of alcoholism. One of the first studies to document this was carried out by Robins (1962) in a follow-up study of child-guidance clinic attendees; a marked excess of alcoholism was observed among those with antisocial behavior in childhood. This finding has subsequently been replicated in numerous other studies.

Data from several studies indicate that alcoholism complicated by ASP or ASP-like behaviors may be more heritable than non-ASP alcoholism. Evidence from a Swedish adoption study indicated the adopted-out sons of fathers with ASP-like alcoholism had a risk of alcoholism nine times that of adopted-out sons of other fathers.

The causes of ASP are not known, but data are accumulating from neurochemical studies that provide some clues. Neuropharmacological studies have established associations of aggressive, impulsive, hostile, and low socialization behaviors with low SEROTONIN levels. Another neurotransmitter, DOPAMINE, is linked to novelty-seeking behavior. The interactions among neurotransmitters have led researchers to postulate an association of multiple neurotransmitter dysfunctions, with a loss of impulse control and an increased appetite for novel experiences.

SUMMARY

ASP is a common concomitant of alcohol and drug abuse that affects their course and treatment; it may represent a highly heritable subtype of such abuse. Although its etiology is unknown, evidence from neuropharmacology studies has provided some leads. As with many personality disorders, there is no known treatment for ASP. It is an important disorder to consider in alcohol and drug abuse.

(SEE ALSO: *Addictive Personality; Childhood Behavior and Later Drug Use; Conduct Disorder and Drug Use*)

BIBLIOGRAPHY

HESSELBROCK, V. M., MEYER, R., & HESSELBROCK, M. (1992). Psychopathology and addictive disorders: The specific case of antisocial personality disorder. In C. P. O'Brien & J. H. Jaffe (Eds.), *Addictive states*. New York: Raven Press.

ROBINS, L. N., TIPP, J., & PRZYBECK, T. (1991). Antisocial personality. In L. N. Robins & D. A. Regier (Eds.), *Psychiatric disorders in America*. New York: Free Press.

ROBINS, L. N., BATES, W. M., & O'NEAL, P. (1962). Adult drinking patterns of former problem children. In D. J. Pitman & C. R. Snyder (Eds.), *Society, culture and drinking patterns*. New York: Wiley.

KATHLEEN K. BUCHOLZ

ANXIETY Anxiety refers to an unpleasant emotional state, a response to anticipated threat or to specific psychiatric disorders. In anxiety, the anticipated threat is often imagined. Anxiety consists of physiological and psychological features. The physiological symptoms can include breathing difficulties (hyperventilation, shortness of breath), palpitations, sweating, light-headedness, diarrhea, trembling, frequent urination, and numbness and tingling sensations. The anxious person is usually hypervigilant and startles easily. The subjective psychological experience of anxiety is characterized by feelings of apprehension or fear of losing control, depersonalization and derealization, and difficulties in concentration. Strains around the performance of social roles (e.g., spouse, parent, wage earner) and certain life situations (e.g., separating from parents when starting school or leaving home, illness) can

generate anxiety symptoms. Other factors can contribute to the etiology of anxiety, such as use of alcohol, caffeine and other stimulant drugs (e.g., amphetamine), a family history of anxiety symptoms, or a biological predisposition. In certain cases, recurrent anxiety symptoms will lead an individual to avoid certain situations, places, or things (phobias). In many cases, an anxious emotional state can motivate positive coping behaviors (e.g., as in anxiety that leads to studying for an exam). When the anxiety becomes excessive and impairs functioning, it can lead to the development of psychiatric illness. Individuals differ in their predisposition to anxiety.

Different constellations of anxious mood, physical symptoms, thoughts, and behaviors, when maladaptive, constitute various anxiety disorders. Panic disorder is characterized by brief, recurrent, anxiety attacks during which individuals fear death or losing their mind and experience intense physical symptoms. People with obsessive compulsive disorder experience persistent thoughts that they perceive as being senseless and distressing (obsessions) and that they attempt to neutralize by performing repetitive, stereotyped behaviors (compulsions). The essential feature of phobic disorders (e.g., agoraphobia, social phobia, simple phobia), is a persistent fear of one or more situations or objects that leads the individual to either avoid the situations or objects or endure exposure to them with great anxiety. Generalized anxiety disorder is diagnosed in individuals who persistently and excessively worry about several of their life circumstances and experience motor tension and physiologic arousal. Anxiety disorders are the psychiatric illness most frequently found in the general population.

Anxiety states can result from underlying medical conditions, and therefore these conditions should always be looked for when evaluating problematic anxiety. When anxiety develops into a psychiatric illness, various forms of treatment are available to reduce it. The choice of treatment often depends on the specific disorder. Medications may be used, including anxiolytics (e.g., BENZODIAZEPINES, buspirone) and ANTIDEPRESSANTS (e.g., imipramine, fluoxetine). Psychotherapies offered generally consist of cognitive-behavioral interventions (e.g., exposure therapy), but they can include psychotherapy of a supportive nature or more psychodynamically oriented approaches. Some people with severe anxiety may turn to alcohol or nonprescribed sedative-

hypnotics for symptom relief, and this in turn may exacerbate the underlying condition.

(SEE ALSO: *Causes of Substance Abuse: Psychological* (*Psychoanalytic*) *Perspective*; *Prescription Drug Abuse*)

BIBLIOGRAPHY

AMERICAN PSYCHIATRIC ASSOCIATION. (1987). *Diagnostic and statistical manual of mental disorders-3rd edition-revised*. Washington, DC: Author.

UHDE, T. W., & NEMIAH, J. C. (1989) Panic and generalized anxiety disorders. In H. I. Kaplan & B. J. Sadock (Eds.), *Comprehensive textbook of psychiatry*, 5th ed. Baltimore: Williams & Wilkins.

MYROSLAVA ROMACH
KAREN PARKER

APHRODISIAC An aphrodisiac is a substance that can be administered topically, internally, by injection, or by inhalation to stimulate sexual arousal or to enhance sexual performance. The term is based on Aphrodite, the ancient Greek goddess of love and beauty, and it came into the English language during the early 1800s. Although no solid scientific evidence exists for any substances that have selective effect on sexual function, many foods and food combinations have a long-standing reputation as aphrodisiacs—such as oysters, caviar, champagne, and truffles (a subterranean fungus uprooted by pigs in the oak forests of France).

Alcoholic drinks have also been considered to be an aphrodisiac, since sexual behavior often occurs after "cocktails," during or after parties, or during periods of alcohol intoxication—but only if not too much alcohol has been consumed. Objective measurements have demonstrated that ALCOHOL (a depressant) actually decreases sexual responsiveness in both men and women. This paradoxical effect was best expressed about 1605 by William Shakespeare in *Macbeth*, Act 2, scene 3:

MACDUFF: What three things does drink especially provoke?

PORTER: Marry, sir, nose-painting, sleep, and urine.
Lechery, sir, it provokes, and

unprovokes; it
provokes the desire, but it takes
away the performance...

Since the 1980s, COCAINE has gained popularity as a potential aphrodisiac, since its use purportedly enhances the sexual experience; MARIJUANA and AMYL NITRITE have had this reputation, in general, since the 1960s. Nevertheless, chronic cocaine users, like chronic heroin users, often report a loss of sexual interest and capability; therefore no rationale exists for the use of alcohol or other drugs as sexual stimulants. Quite the contrary, the use of these substances can lead to a loss of sexual desire and excitement and the development of a physical and/or psychological dependence.

A prescription drug—yohimbine, derived from the African yohimbe tree—seems to help cure some men of impotence. The data suggest that it may work as a placebo (psychologically), but urologists prescribe it nonetheless in the hope that the patient can avoid more invasive treatments. The treatment takes three to six months before there is an effect, and the natural form (available in health-food stores), is not the form used therapeutically.

BIBLIOGRAPHY

CHITWOOD, D. D. (1985). Patterns and consequences of cocaine use. In *Cocaine use in America: Epidemiologic and clinical perspectives*. NIDA Research Monograph No. 61. DHHS Publication no. (ADM)85-1414. Washington, DC: U.S. Government Printing Office.

ROOT, W. (1980). *Food: An authoritative and visual history and dictionary of the foods of the world*. New York: Simon & Schuster.

WILSON, G. T. (1977). Alcohol and human sexual behavior. *Behavioral Research and Therapy, 15*, 239–252.

NICK E. GOEDERS

ARGOT   When we talk about the argot of drug users, we mean their vocabulary or the collection of slang words and phrases used by one drug user to communicate with another—often to the exclusion of non-users. In some instances, argot extends to the intonation or pitch used to speak words and phrases.

Argot fascinates sociologists, anthropologists, and others who study human behavior because the use of argot is an example of learned behavior that helps identify members of one social group as opposed to another. Argot also marks off the boundaries of group membership: those of the in-group use argot with ease, while outsiders use argot ineptly or not at all.

In one study, we showed more than 2,000 drug users in Baltimore, Maryland, a photograph of someone injecting a drug into a vein, and we asked "What do you call this?" The vast majority, well over 50 percent, said that it was a picture of someone "firing up" and most of the others called it "shooting." Most other people in Baltimore speak of "firing up" their furnaces or "shooting" baskets on a basketball court, but they do not think about drug use when these terms are used.

The argot of drug users also varies from place to place and from time to time. A very small minority of the drug users in our study spoke of "mainlining" the drug, "spiking," or "oiling" when they saw the picture of a drug injection into a vein. These are older terms for the same injecting behavior now called "firing up" and "shooting" by younger drug users.

In some ways, argot reflects the social structure of groups: in-group members use the argot, while others do not. However, if argot was used to serve as a badge of membership in an in-group, then we might expect to hear argot in general conversations, no matter who is present. Nonetheless, sociologists studying the use of argot often have been surprised to find that argot is spoken mainly between group members but not as often when nonmembers are present.

Arguing from evidence of this type, some observers claim that argot serves more to convey and reinforce identities within groups than to distinguish one group from another. That is, the process of learning and using drug-related argot reinforces the process of joining in with others who use drugs. In some ways, this process might serve to supplement the reinforcing functions of drug use, making continued drug use more likely rather than less likely.

Many of the terms that originally were part of the argot of drug users have entered into more common usage. For example, "dope" has become a general purpose term, widely used in relation to many types of drug; many people know that "weed" or "reefer" refers to marijuana while "acid" is LSD (lysergic acid diethylamide). This encyclopedia includes a glossary

of terms in the article SLANG AND JARGON, which lists many examples of argot that have become part of American slang usage.

(SEE ALSO: *Slang and Jargon*)

BIBLIOGRAPHY

SMITH, A. M. ET AL. (1992). Terminology for drug injection practices among intravenous drug users in Baltimore. *International Journal of the Addictions, 27*, 435–453.

VOGEL, V. H. (in press). The argot of narcotic addicts. In *Narcotics and Narcotic Addiction.*

DAVID VLAHOV

ARTS, DRUG USE AND THE   *See* Creativity and Drugs; Lifestyle and Drug Use Complications.

ASAM   *See* American Society of Addiction Medicine.

ASIA, DRUG USE IN   Asia is the world's largest continent; India and China are its most populous countries. More than half the world's population is living in Asia. Thus we find considerable variation in drug use and drug problems there, not only among the various countries but also within them. Unfortunately, the available information about drug use in Asia is sketchy and fragmentary. Few good studies on drug use in Asia have been published. Epidemiologic data are almost completely absent. The rapid social, economic, cultural, and political transformations are adding to the complexity of drug-use patterns and associated drug-related problems in Asia and worldwide.

TEA

Most people know the tea plant *Camellia sinensis* in the brewed form of TEA. Tea has been part of Asian culture for thousands of years. Its use seems to have originated in southeast China. It is mentioned in the very early Chinese medical literature. To a large extent, the medical benefits of tea can be ascribed to the chemical theophylline, which depending on its use can have either mildly calming or stimulating effects. The use of tea as a popular beverage and its production in large quantities has only been documented since the sixth century. The history of tea is also a history of international trade. Japan was one of the first countries to import tea from China, and tea became part of the Japanese culture. *Chanoyu* (the way of the tea) is a meditation ritual introduced in Japan by Zen Buddhist monks several hundred years ago, and elaborate tea ceremonies developed there. This tea ceremony is still taught and practiced in modern Japan.

Tea became the primary stimulant beverage not only in China and Japan but also in India, Malaysia, the Russian empire, and other Asian countries.

In the 1700s, tea was imported directly to Great Britain and to the British colonies by the East India Company. Even today, there are tea-preferring countries like Britain and coffee-preferring countries like Spain. The difference in preference goes back to the time of colonial trading: Those countries with tea-producing colonies drank tea, because it was cheaper than coffee; countries with coffee-producing colonies drank coffee, because for them it was cheaper than tea.

OPIUM

After tea, the drug most often associated with Asia is OPIUM. Opium is prepared from the opium poppy (*Papaver somniferum*), which grows well in the alkaline limestone soil of the countries from Turkey and Iran, east through Afghanistan and Pakistan to the northern mountainous areas of Myanmar (formerly Burma), Thailand, and Laos. The area forms a crescent, thus the name GOLDEN CRESCENT: The mountainous areas of Myanmar, Thailand, and Laos are known as the Golden Triangle.

Medical historians have been able to document that Arabian physicians of Asia Minor extracted raw opium from the seed pods of the poppy and used it to treat pain and diarrhea before A.D. 1000. Arabian traders began exporting opium to India and China about that time, and it also appeared in trade shipments to Europe. Although accurate documentation is scarce, some observers claim that opium use spread faster in precolonial and colonial India, than in China. A British royal commission investigated

Indian opium use in 1895 and claimed that the people of India had not suffered detrimental effects from the taking of opium. The situation was different in China. The British traded Indian-grown opium for Chinese tea and porcelain. This led to an increasing supply of opium in China, associated with an increasing use of opium for recreational purposes. During the nineteenth century a raging epidemic of opium smoking in China led to a situation of great concern to the Chinese government. In an attempt to cut the supply of opium, the Chinese government tried to close its ports to British trade. This resulted in the Opium wars (1839–1842), but Britain won the war and the right to continue trading opium to China.

The different responses of India and China to the availability of opium might be explained, to some degree, by the way this drug was introduced to the population. In India, opium was introduced as a medical plant, to be taken by mouth and swallowed. In contrast, in China during the 1500s, Portuguese sailors had just introduced New World tobacco smoking as a form of a recreational drug use. Many Chinese, who had just picked up tobacco smoking substituted opium for tobacco. Thus opium was not only introduced as a nonmedical recreational drug, but it was also introduced in a different route of administration. Drugs inhaled through the lungs seem to produce faster and more severe dependence than those ingested through the gastrointestinal tract.

Effective government control of opium smoking in China did not become possible until in the late nineteenth and early twentieth centuries Britain, the United States, and other world powers signed international agreements to help curb worldwide supply and distribution networks. They cooperated because opium abuse spread and started to affect these countries directly. In 1930, the League of Nations Commission of Inquiry into the Control of Opium Smoking in the Far East reported that opium use has not been prohibited in any Asian country except the Philippines. By 1950, this situation had changed dramatically. Many Asian countries placed high priority on narcotic-control policies. Harsh penalties, including the death penalty had been reinstated for drug trafficking and possession of opium and derivatives, like MORPHINE and HEROIN.

Despite these government actions, opium and its derivatives are still used widely in regions where they

are grown. In 1985, an analysis of available information about drug use in 152 countries classified 14 as having high abuse problems. Eight of these were Asian countries: Afghanistan, Myanmar, Malaysia, Pakistan, the Philippines, Singapore, Thailand, and Vietnam.

CANNABIS

Known in the United States mainly as the MARIJUANA plant, *Cannabis sativa* may first have been cultivated in Asia in a region just north of Afghanistan. From there it seems to have spread to China and India. It is mentioned in the early medical literature of China (e.g., in the *Shenmong bencao*) as well as in India (e.g., in the *Sushruta samhita*). Early nonmedical use has also been documented.

Cannabis use seems to have become popular especially in India and the Islamic countries. The many social rules associated with its use are evidence of the long-standing integration into Indian culture. Traditional Indian society was divided into hereditary classes, the so-called castes. The highest caste was to use white-flowered cannabis; the Kshatriya, the warriors, used the red-flowered plants; the farmers and traders, the Vaishya caste, were to use the yellow-flowered plant; and the Shudra, servant caste, used plants with dark flowers.

The earliest Indian medical text, *Sushruta samhita*, apparently dating from pre-Christian times, differentiated three major ways of preparing and administering *Cannabis*—BHANG, GANJA, and *charas*. *Bhang* was a sweet drink prepared from the leaves and flower shoots, which also might be brewed as a tea. *Ganja* was the dried flowers, which was smoked. *Charas* was a cake compound from the most resinous parts of the plant; this seems to have been the upper-class favorite. While *bhang*, *ganja*, and *charas* are still used in India today, the form of preparation may not be quite the same as the recipes in the *Sushruta samhita*.

BETEL NUT

In southern parts of Asia, mainly in India, Indonesia, Malaysia, southern China, and also in East Africa, many people chew BETEL NUT (*Areca catechu*). The nut is prepared by wrapping it in a betel pepper leaf (*Piper betle*) with a compound of lime (calcium

hydroxide or calcium carbonate) and spices. Chewing this preparation produces mild stimulating effects. At the same time, the saliva becomes red and the mouth and teeth are stained red. Mouth cancer may result.

The ancient Greek traveler and historian Herodotus wrote about betel-nut chewing in 340 B.C. Although its use seems to be declining, an estimated 400 million persons are still dependent on this substance.

OTHER NATIVE DRUGS

Students interested in ETHNOPHARMACOLOGY and cultural practices associated with drug use will find many fascinating accounts in Asian history. One modern example involves the consumption of a drink called KAVA, which is prepared from the roots of *Piper methysticum*. In Polynesia, Micronesia, and Melanesia this drink is taken for recreational purposes, to calm and sedate the user.

There are ancient drug-taking practices connected to FLY AGARIC, a sometimes deadly mushroom (*Amanita mascaria*) found in several countries. One way to reduce the toxicity of this mushroom is to feed it to a reindeer and drink the reindeer urine, which contains intoxicating metabolites of the chemicals found in the mushroom.

STIMULANTS

Some Asian countries have suffered epidemics of drug taking in connection with legally produced drug products. An especially widespread epidemic of AMPHETAMINE use started in Japan during World War II and continued into the 1950s. A second wave of amphetamine use was reported in the late 1970s. Recently an epidemic of METHAMPHETAMINE "(ice)" smoking spread across the Pacific into Hawaii and other American states after earlier micro-epidemics in Asia.

An intriguing question for future historians will involve COCAINE. Will Asia become an important market for cocaine now that American demand has fallen? To date, Asia has not been an important market for international cocaine trafficking even though there existed a very early cultivation of coca plants on the island of Java, introduced by Dutch colonialists. This situation may change with increasing commerce between Asia and South America, where most of the world's cocaine supply is now produced.

ALCOHOL

The account of drug use in Asia would be incomplete without mention of alcoholic beverages. At present, Asia is the continent with the lowest overall per-capita consumption of ALCOHOL. In many Asian countries, alcohol consumption is prohibited on religious grounds—because of the prohibitions of Islam: the Koran forbids its use. Nonetheless, even in the most conservative Islamic countries, there is some alcohol dependence. Saudi Arabia for example, has an ALCOHOLICS ANONYMOUS (AA) organization and a modern hospital for drug and alcohol treatment.

In addition to religious and social restrictions on alcohol consumption, there are some important biological factors known to be related to genetic variation within the Asian population. For example, many Asian people have the "flushing syndrome" in response to alcohol that is associated with their particular configuration of aldehyde dehydrogenase, an alcohol-metabolizing enzyme. One prominent sign is that their facial skin becomes flushed. Although this response might work to discourage alcohol use, and thus protect against alcohol dependence, many Asian people—especially men—are known to "drink through" the flushing response to become intoxicated. In fact, South Korean males suffer from the highest recorded prevalence rates of alcohol abuse and dependence: An estimated 44 percent of adult men have a history of currently active or former alcohol abuse and/or dependence. The reasons for this very high rate are a matter of speculation and should be a topic of intense study. As evidence of the considerable variation in alcohol problems in Asia, Taiwan has one of the lowest rates of alcohol abuse and dependence in the world for both adult men and women. This variation cannot be explained by differences in research methods, because the same methods have been used in surveys of Taiwan and South Korea. The difference must involve fundamental social and cultural differences, or fundamental biological differences in vulnerability to alcohol-related problems, or a combination.

Alcohol use is not a new phenomenon in Asia. The drinking of fermented beverages has been part of Asian cultures since antiquity, as documented in

the early classical literature of China (in the *Shujing* and the *Liji*), India (in the *Susruta samhita*), and other countries. The *Susruta samhita* describes various stages of intoxication. In China, the fall of the Shang Dynasty in the eleventh century B.C. was attributed to excessive use of alcohol by the emperor and his followers. The same explanation was given for the fall of later dynasties. In China, different forms of alcohol have been fermented from various kinds of grain. In other parts of Asia, alcoholic beverages were based on a large variety of different substances, including rice in the case of Japanese sake; horse milk in the case of Kumys, an alcoholic beverage prepared by northern and central Asian nomads; and toddy-palm sap in the case of arrack prepared in southern India and Indonesia.

An early epidemic of drug use combining alcohol with a drug called *hanshi* can be traced in the ancient writings of the time of the fall and overthrow of the Chinese Han Dynasty—a time of rapid changes in society (second and third century A.D.). The use of *hanshi* was associated with an unconventional "bohemian" lifestyle, disregard of social norms, "disheveled hair," and "incorrect clothing." The *hanshi* users were reported to claim that the drug helped open their minds and clarify their thinking. Although reports of this early epidemic are sketchy, *hanshi* is mentioned in several later medieval texts, mainly in relation to remedies that can be used to help treat its detrimental side effects. At present it is not clear which chemical compound was present in *hanshi*.

TOBACCO

Probably the most widespread twentieth-century epidemic in Asia is TOBACCO smoking. Today, in most Asian countries, local, international, and especially American tobacco manufacturers are marketing their products aggressively—in part because of declining demand in North America and in part because of the increasing economic strength of the Asian countries. One result has been an increase in the consumption of tobacco products since the 1960s, especially the smoking of cigarettes.

Tobacco became a part of Asian culture from the time it was imported by Europeans from their colonies in the Americas during the 1600s. The "hubbly-bubbly," or hookahs, of the Middle East and India were used for smoking tobacco. This was centuries

before modern advertisement techniques were applied by the tobacco industry. But recently tobacco-related diseases and deaths are becoming more prominent in the health statistics of Asia. This toll is connected directly to an increasing consumption of tobacco products. Part of the tobacco is imported from the United States and other international suppliers. Some observers noticed similarities to the situation in the nineteenth-century, when British traders aggressively fought to keep the lucrative opium trade from being interrupted. Some thus call for international agreements concerning tobacco trade, similar to those which helped curb the opium trade from being interrupted. Some thus call for international agreements concerning tobacco trade, similar to those which helped curb the opium problem at the beginning of the twentieth century. International support seems to be needed to help these countries reduce tobacco-related problems.

THE FUTURE

As commerce between countries has increased, so has the traffic in drugs. For centuries, Asia has had trading partners for its tea, opium, and *Cannabis*. In return it has received shipments of other goods, including pharmaceuticals. Sometimes these exchanges have been within Asia, as in the early introduction of opium into China by Arabian traders, and the later commerce in opium between colonial India and China. Now trading is done on a worldwide scale, whether it is the legal trade with tea or the illegal traffic of opium. Recently some countries in Asia have reported an increase in POLYDRUG use among their younger population. If current and future changes will be associated with an increase of drug use may yet be seen.

Since the 1950s, a number of Asian countries have also experienced a growth of what might best be called "drug tourism." Travelers, mainly from the Western Hemisphere, have come to Asia to purchase and consume such drugs as opium, *Cannabis*, heroin, and magic mushrooms. For many, it has come as a surprise that Asian countries respond with harsh penalties, as did Singapore in 1994, when a man from the Netherlands was hanged for possessing a large amount of heroin. It must be kept in mind, that a long history of harsh penalties and social sanctions against those who violate social conventions, including local drug regulations, are part of Asian heri-

tage—as well as the seemingly exotic custom of drug taking.

(SEE ALSO: *Source Countries for Illicit Drugs*)

BIBLIOGRAPHY

HELZER, J. E., & CANINO, G. J. (1992). *Alcoholism in North America, Europe and Asia.* New York: Oxford University Press.

HOBHOUSE, H. (1987). *Seeds of change: Five plants that transformed mankind.* New York: Harper & Row.

SMART, R. G., & MURRAY, G. F. (1985). Narcotic drug abuse in 152 countries: Social and economic conditions as predictors. *The International Journal of the Addictions, 20*(5), 737–749.

SPENCER, C. P., & NAVARATNAM, V. (1981). *Drug abuse in East Asia.* Kuala Lumpur: Oxford University Press.

WESTENMEYER, J. (1982). *Poppies, pipes and people: Opium and its use in Laos.* Berkeley: University of California Press.

CHRISTIAN G. SCHUTZ

ASPIRIN *See* Pain: Drugs Used in Treatment of.

ASSERTIVENESS TRAINING *See* Treatment.

ASSOCIATION FOR MEDICAL EDUCATION AND RESEARCH IN SUBSTANCE ABUSE (AMERSA)

This a national organization of some 500 medical and allied faculty, founded in 1977 for the promotion of education and research in the addiction field. The organization derived from an informal coalition of U.S. Federal Career Teachers in alcoholism and drug abuse. These career teachers, one on the faculty of each of fifty-five medical schools, were funded by the National Institute on Drug Abuse (NIDA) and the National Institute on Alcohol Abuse and Alcoholism (NIAAA) to promote enhanced teaching at their respective medical campuses. The career teachers program, established in 1972, regarded as a highly successful vehicle for highlighting an issue of considerable importance in the medical curriculum. As the program wound down (it came to an end in 1981), the participants felt it important to secure the continuation of their mission and established AMERSA as a national membership organization open to all medical faculty and faculty in allied health programs.

In the year of its establishment, the AMERSA held its first national meeting, which was followed by meetings of increasing attendance in each succeeding year. The national meetings have been the focus of federal participation in teaching programs and have focused on curriculum techniques and new research findings.

AMERSA established a quarterly publication, *Substance Abuse*, in 1979, presenting educational and research findings; it served as a vehicle for broadening the base of teaching in their field. In addition, a variety of curricula were established by members, with coordination through the AMERSA national headquarters (currently located at Brown University) augmented by the Center for Medical Fellowships in Alcoholism and Drug Abuse, located at New York University.

The organization's members work in a variety of ways to effect their educational ends. Much effort is invested in developing curriculum and curriculum outlines for courses directed at a variety of disciplines and various educational levels. In addition, most members work actively within their respective departments to develop subspecialty expertise—as in psychiatry and internal medicine. Efforts are also directed at schoolwide initiatives, as with programs organized through the deans of medical schools.

MARC GALANTER

ATHLETES AND DRUGS *See* Anabolic Steroids; Lifestyle and Drug Use Complications.

ATROPINE *See* Jimsonweed; Scopolamine and Atropine.

ATTENTION DEFICIT DISORDER

Children and adolescents who "act out" their feelings, frustrations, and emotional conflicts are said to exhibit externalizing behavior. Within the framework of the American Psychiatric Association's DIAGNOSTIC AND STATISTICAL MANUAL, 3rd edition,

revised (DSM-III-R), once a certain level of severity is demonstrated, these youth qualify for the umbrella diagnosis of *disruptive behavior disorder*. Three disorders are encompassed within this general diagnostic category: (1) attention deficit/hyperactivity disorder (AD-HD); (2) CONDUCT DISORDER (CD); and (3) oppositional defiant disorder.

Children who qualify for a diagnosis of AD-HD often, but not invariably, meet criteria also for a diagnosis of conduct disorder or oppositional defiant disorder. In addition, attention deficit disorder does not always occur in conjunction with behavioral hyperactivity. These latter individuals are assigned the diagnosis of *undifferential attention deficit disorder*.

Characteristically, youth with AD-HD demonstrate an excessively high level of behavioral activity across a wide variety of situations. The behavioral disturbance is most apparent, however, where the appropriate level of behavioral activity should be low, such as in a classroom where focused task-oriented cognitive demands are placed on the youngster. In naturalistic settings, AD-HD children, although not overtly hyperactive, typically are impulsive, emotionally labile, restless, and distractable. They often act in a socially intrusive and inappropriate manner such that they are considered to be immature. Consequent to poor foresight and poor impulse control, combined with a high behavior level, AD-HD children frequently get into social difficulties with adults and peers, which subsequently lead to rejection. Physical injury is common in AD-HD because of poor self-control (e.g., dashing across the street without looking).

Because AD-HD symptomatology is often present in children with poor coordination, reading and learning problems, and other neurodevelopmental disturbances, in the past labels such as "minimal brain damage" and "minimal brain dysfunction" were assigned, although no neurologic pathology or injury was detectable. The terms "hyperactivity" and "hyperkinesis" were subsequently introduced; however, these labels emphasize the motor aspects of the disorder. The current diagnostic label—attention deficit disorder—circumvents these problems by focusing on the core etiological determinant for the multifaceted expression of cognitive, behavioral, and emotional disturbances. For this diagnosis to be assigned, the disorder must first be expressed before age 7, have at least a 6-month duration, and not be the consequence of a pervasive developmental disorder (American Psychiatric Association, 1987).

EPIDEMIOLOGY

Approximately 10 percent of boys and 3 percent of girls in the general population qualify for a diagnosis of AD-HD. The symptom presentation is different between the genders, with girls being somewhat older than boys at the time of first diagnosis. Girls manifest more mood changes, fears, and social withdrawal than boys but less aggressivity and impulsivity. Among children receiving psychiatric treatment, AD-HD is estimated to be present in 40 to 70 percent of inpatient cases and 30 to 50 percent of outpatient cases.

GENETIC ETIOLOGY

Behavior-activity level is a heritable trait of temperament characterizing the human species. Individuals who are at the high end of this trait, compared to the average, are behaviorally highly active or hyperactive. Thus, although not necessarily implying pathology or a disorder, but rather normal variation in behavioral disposition, high-end active individuals demonstrate a rapid behavioral tempo and greater vigor and forcefulness than the average. For clinicians, it is important to distinguish between individuals with behavior-activity level as a temperament trait and extreme cases that comprise psychological and psychiatric disorder.

For many individuals, extremely high manifestations of behavior-activity level has a genetic basis. AD-HD aggregates in families and twin studies show high concordance rates for this disorder. The neurobiological mechanisms underlying AD-HD is an active topic of research. As of the mid-1990s, we assume that the neurochemistry in such people is likely to be disturbed; however, a candidate neurotransmitter system has yet to be identified. Most likely, multiple neurochemical systems are involved. Neuroimaging studies have revealed lowered brain metabolism, particularly in the frontal regions, in AD-HD children (Zametkin et al., 1990).

NONGENETIC ETIOLOGY

Injury to the brain, particularly in the anterior region, can produce the symptoms of AD-HD. For the developing fetus, malnutrition, exposure to toxins (especially alcohol), and medical illness during pregnancy augment the risk for AD-HD symptoms to appear. Circumstances around birth, especially the

occurrence of toxemia (toxins in the blood) or hypoxemia (not enough oxygen in the blood), increase the risk for neurological injury in the newborn, which ultimately could result in symptoms of AD-HD. During childhood development, many factors, particularly head trauma (by accident or maltreatment), infection, toxic poisoning, and malnutrition can produce AD-HD symptoms. Neurologic conditions (e.g., epilepsy) and neurodevelopmental disorders (e.g., dyslexia, autism) are also commonly associated with some symptoms of AD-HD. Although all these latter conditions produce AD-HD–types of symptoms, according to DSM-III-R they must be *excluded* as etiologic factors to make the diagnosis of the AD-HD syndrome. In other words, the diagnosis of AD-HD is assigned only where there is *no* neurodevelopmental disability or injury.

NATURAL HISTORY

Recognizing that there is substantial variability in the etiology of AD-HD, it is obvious that the lifetime course and outcome is also highly variable. Symptoms persist into adulthood in about 50 percent of cases. Under these circumstances, the person is assigned a diagnosis of *attention deficit disorder–residual type.*

RELATION OF AD-HD TO DRUG ABUSE

Serious psychiatric disorder is common among adults with a history of AD-HD. ANTISOCIAL PERSONALITY disorder, alcohol and substance abuse, depression, and anxiety are the most common associated disorders. These associated disorders should not be viewed as invariant outcomes of AD-HD but rather as disturbances for which AD-HD youth are at increased risk. Whether any of these psychiatric outcomes are manifested depends on a variety of factors besides AD-HD, including the child's self-esteem, opportunity for normal socialization with peers, success in school, and level of social and family support (Tarter, 1988).

With respect to alcohol and other drug abuse, augmented risk appears to be circumscribed to youth who have both AD-HD and a conduct disorder. The association, however, between AD-HD and substance abuse is complex. Alcohol and other drugs may be more subjectively rewarding for AD-HD youth and adults than in the general population.

Drug use is commonly tied to a general pattern of social deviancy and nonnormative peer affiliation. Where the AD-HD person has been ostracized by the normative peer group, the use of alcohol/drugs may be just another manifestation of generalized maladjustment. Furthermore, alcohol/drug use may be mediated by a coexisting psychiatric disorder, such as anxiety and depression, and thus reflect an attempt at self-medication. Therefore, although there is substantial evidence demonstrating an increased risk for alcohol/drug abuse in AD-HD youth, this association is complex and is contingent on many factors.

TREATMENT

Because the psychological manifestations of AD-HD are multifaceted, it is necessary that comprehensive treatment interventions that encompass multiple components be implemented (Danforth, Barkley, & Stokes, 1991).

Pharmacotherapy.   PSYCHOSTIMULANTS are therapeutically beneficial for approximately 75 percent of AD-HD children and adults. The most commonly used medications are METHYLPHENIDATE (Ritalin), *d*-AMPHETAMINE, and PEMOLINE (Cylert). Tricyclic antidepressants are also effective in many cases. These medications have been shown to be useful for reducing problem behavior of AD-HD in well over 100 research studies; however, they do not improve school performance or eliminate a conduct disorder where this disturbance is present.

Lifestyle.   Coordinated effort should be made to promote a healthy lifestyle. This includes scheduled regulation of bedtime, meals, homework, and recreation. Nutrition is important; however, contrary to popular belief and anecdotal reports, there is no substantive evidence linking diet or food allergies to the cause of AD-HD. There is no scientific evidence indicating that a special diet or nutritional supplements can ameliorate AD-HD.

Education.   Informing parents and school personnel about the causes of AD-HD and the nature of the behavior disorder can constructively assist the child by evoking empathy rather than anger. Family counseling and teacher education are thus integral components of treatment to help maximize the child's adjustment in the home and at school.

Environmental Engineering.   Structuring the environment so that the child is not easily distracted

is an important intervention. In the home, this entails minimizing distracting stimulation from radio or television, especially while the youngster is doing homework. In the classroom, consideration should be given to the child's seat location, to enable the teacher to ensure that the child persists at tasks, is not distracted by other students, or has no opportunity to be disruptive.

Behavior Modification. Behavior-modification strategies are effective for training the youngster to control impulses and to monitor behavior cognitively. Behavior-modification methods, which help the child, are also useful in teaching effective parenting skills.

(SEE ALSO: *Psychomotor Stimulant; Vulnerability As Cause of Substance Abuse*)

BIBLIOGRAPHY

AMERICAN PSYCHIATRIC ASSOCIATION. (1987). *Diagnostic and statistical manual of mental disorders*, 3rd ed. Washington, DC: Author.

DANFORTH, R., BARKLEY, R., & STOKES, T. (1991). Observations of parent-child interactions with hyperactive children: Research and clinical implications. *Clinical Psychology Review, 11*, 703–727.

TARTER, R. (1988). Are there inherited behavioral traits which predispose to substance abuse? *Journal of Consulting and Clinical Psychology, 56*, 189–196.

ZAMETKIN, A., ET AL. (1990). Cerebral glucose metabolism in adults with hyperactivity of childhood onset. *New England Journal of Medicine, 323*, 1361–1366.

RALPH E. TARTER
ADA C. MEZZICH

AVERSION THERAPY *See* Treatment.

AYAHUASCA
In 1851, the botanist Richard Spruce observed natives along the Rio Negro in Brazil preparing a beverage from the roots of a vine, which he called *Banisteria caapi*, of the family Malpighiaceae (it was recently designated *Banisteriopsis caapi*.) He later observed the use of a similar drink in the Ecuadorian Amazon basin, where it was called *ayahuasca* (from the Quechua language, spoken in the Andes). He noted that the brew was often a mix-

ture of *Banisteria caapi* with the roots of another indigenous plant. There were apparently several variations in the recipe for *caapi* and most of those who have studied it believe that each recipe produces somewhat different psychic effects. In 1929, the great pioneer of psychopharmacology, Louis Lewin, published a monograph describing the pharmacological actions and possible therapeutic uses of *Banisteria caapi*, whose actions he believed to be due to an active alkaloid, harmine. In early studies in patients with Parkinsonism, harmine produced improvements in chewing, swallowing, and movement that lasted from two to six hours. Curiously, it was reported to have little or no psychic effects. It was later shown that harmine acts to inhibit the enzyme monoamine oxidase, thereby raising levels of the neurotransmitters DOPAMINE and NOREPINEPHRINE.

Mixtures containing *Banisteriopsis caapi* are still in use among the indigenous peoples of the Amazon. A tea brewed from it and the leaves of *Psychotria viridis* has been used in shamanistic rituals for hundreds of years in Colombia, Brazil, and Peru. In recent years, a number of people seeking alternatives to Western medicine, and for other reasons, have participated in Santo Daime rituals, in which drinking ayahuasca is a central feature. The tea is said to induce ecstatic states, during which the participants claim to experience great insight. In southern Brazil, some psychotherapists and homeopaths have been known to bring clients or patients to participate in such rituals.

(SEE ALSO: *Hallucinogenic Plants; Hallucinogens*)

BIBLIOGRAPHY

DEULOFEU, V. (1967). Chemical compounds isolated from *Banisteriopsis* and related species. In D. H. Efron, B. Holmstedt, & N. S. Kline (Eds.), *Ethnopharmacologic search for psychoactive drugs*. Washington, DC: Public Health Service Publication no. 1645.

LEWIN, L. (1964). *Phantastica: Narcotic and stimulating drugs*. New York: Dutton.

SCHULTES, R. E. (1967). The place of ethnobotany in the ethnopharmacologic search for pychotomimetic drugs. In D. H. Efron, B. Holmstedt, & N. S. Kline, (Eds.), *Ethnopharmacologic search for psychoactive drugs*. Washington, DC: Public Health Service Publication no. 1645.

JEROME H. JAFFE

BAC *See* Blood Alcohol Concentration, Measures of.

BARBITURATES Barbiturates refer to a class of general central nervous system depressants that are derived from barbituric acid, a chemical discovered in 1863 by the Nobel Prize winner in chemistry (1905) Adolf von Baeyer (1835–1917). Barbituric acid itself is devoid of central depressant activity; however, German scientists Emil Hermann Fischer and Joseph von Mering made some modifications to its structure and synthesized barbital, which was found to possess depressant properties. Scientists had been looking for a drug to treat anxiety and nervousness but without the dependence-producing effects of OPIATE drugs such as OPIUM, CODEINE, and MORPHINE. Other drugs such as bromide salts, CHLORAL HYDRATE, and paraldehyde were useful sedatives, but they all had problems such as toxicity or they left such a bad taste in patients' mouths that they preferred not to take them. Fischer and von Mering noted that barbital produced sleep in both humans and animals. It was introduced into chemical medicine in 1903 and was soon in widespread use.

By 1913, the second barbiturate, PHENOBARBITAL, was introduced into medical practice. Since that time, more than 2,000 similar chemicals have been synthesized but only about 50 of these have been marketed. Although the barbiturates were quickly used to treat a number of disorders effectively, their side effects were becoming apparent. The chief problem, an overdose, can result in respiratory depression, which can be fatal. By the mid-1950s, more than 70 percent of admissions to a poison-control center in Copenhagen, Denmark, involved barbiturates. Additionally, it became apparent that the barbiturates were subject to abuse, which could lead to dependence, and that a serious withdrawal syndrome could ensue when the drugs were abruptly discontinued. In the 1960s, the introduction of a safer class of hypnotic drugs, the BENZODIAZEPINES reduced the need for barbiturates.

Barbiturates are dispensed in distinctly colored capsules making them very easy to identify by the lay public. In fact, users within the drug culture often refer to the various barbiturates by names associated with their physical appearance. Examples of these names include blue birds, blue clouds, yellow jackets, red devils, sleepers, pink ladies, and Christmas trees. The term *goofball* is often used to describe barbiturates in general. All barbiturates are chemically similar to barbital, the structure of which is shown in Figure 1.

All barbiturates are general central nervous system depressants. This means that sedation, sleep, and even anesthesia will develop as the dose is increased. Some barbiturates also are useful in reducing seizure activity and so have been used to treat some forms of epilepsy. The various barbiturates differ primarily in their onset and duration of action, ability to enter the brain, and the rate at which they

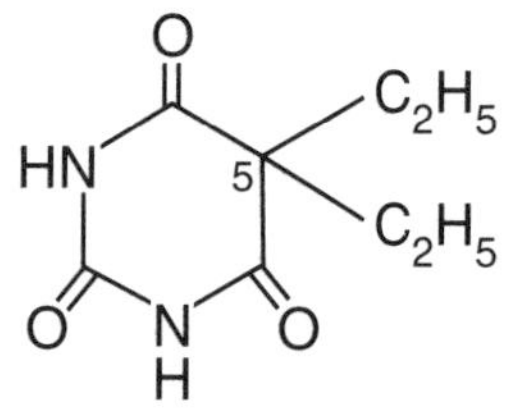

Figure 1
Barbital

twelve hours. Table 1 lists the common barbiturates, their trade names, typical route of administration, and plasma half-life. The plasma half-life is a measure of how long the drug remains in the blood, but not how long the effects last, although it does provide a general indication of when to expect the effects to wane (a half-life of five hours means that one-half of the drug will be removed from the system in five hours; one-half of the remaining drug will be removed during the next five hours, etc.).

are metabolized. These differences are achieved principally by adding or subtracting atoms to the two branches on position #5 in Figure 1. The barbiturates are classified on the basis of their duration of action, which ranges from ultrashort-acting to long-acting. The onset of action of the ultrashort-acting barbiturates occurs in seconds and lasts a few minutes. The short-acting compounds take effect within a few minutes and can last four to eight hours, while the intermediate- and long-acting barbiturates can take almost an hour to take effect but last six to

EFFECTS ON THE BODY AND THERAPEUTIC USES

Barbiturates affect all excitable tissues in the body. However, NEURONS are more sensitive to their effects than other tissues. The depth of central nervous system depression ranges from mild sedation to coma and depends on many factors including which drug is used, its dose, the route of administration, and the level of excitability present just before the barbiturate was taken. The most common uses for

TABLE 1
Classification of Barbiturates on the Basis of Duration of Action

Drug Class and Generic Names	Trade Names	Routes of Administration*	Half-Life (in hours)
Ultrashort-Acting:			
methohexital sodium	Brevital	IV	3.5–6**
thiamylal sodium	Surital	IV	†
thiopental sodium	Pentothal	IV	3–8**
Short-Acting:			
butalbital	‡	PO	35
hexobarbital	Sombulex	PO; IV	3–7
pentobarbital	Nembutal	PO; IM	15–48
secobarbital	Seconal	PO; IM	15–40
Intermediate-Acting:			
amobarbital	Amytal	PO; IM	8–42
aprobarbital	Alurate	PO	14–34
butabarbital	Butisol	PO	34–42
talbutal	Lotusate	PO	†
Long-Acting:			
phenobarbital	Luminal	PO; IV	24–96
mephobarbital	Mebaral	PO	11–67

*IV = intravenous; IM = intramuscular; PO = oral.
**Values are for whole body, half-life in the brain is less than 30 minutes.
†Half-life data not available for human subjects.
‡Various preparations in combination with acetaminophen.
SOURCE: Rall, 1990; Csáky, 1979.

the barbiturates are still to promote sleep and to induce anesthesia. Barbiturate-induced sleep resembles normal sleep in many ways, but there are a few important differences. Barbiturates reduce the amount of time spent in rapid eye movement or REM sleep—a very important phase of sleep. Prolonged use of barbiturates causes restlessness during the late stages of sleep. Since the barbiturates remain in our bodies for some time after we awaken, there can be residual drowsiness that can impair judgment and distort moods for some time after the obvious sedative effects have disappeared. Curiously, some people are actually excited by barbiturates, and the individual may even appear inebriated. This paradoxical reaction often occurs in the elderly and is more common after taking phenobarbital.

The general use of barbiturates as hypnotics (SLEEPING PILLS) has decreased significantly, since they have been replaced by the safer benzodiazepines. Phenobarbital and butabarbital are still available, however, as sedatives in a number of combination medications used to treat a variety of inflammatory disorders. These two drugs also are used occasionally to antagonize the unwanted overstimulation produced by ephedrine, AMPHETAMINE, and theophylline.

Since epilepsy is a condition of abnormally increased neuronal excitation, any of the barbiturates can be used to treat convulsions when given in anesthetic doses; however, phenobarbital has a selective anticonvulsant effect that makes it particularly useful in treating grand mal seizures. This selective effect is shared with mephobarbital and metharbital. Thus, phenobarbital is often used in hospital emergency rooms to treat convulsions such as those that develop during tetanus, eclampsia, status epilepticus, cerebral hemorrhage, and poisoning by convulsant drugs. The benzodiazepines are, however, gradually replacing the barbiturates in this setting as well.

It is not completely understood how barbiturates work but, in general, they act to enhance the activity of GABA on GABA-sensitive neurons by acting at the same receptor on which GABA exerts its effects (see Figure 2). GABA is a NEUROTRANSMITTER that normally acts to reduce the electrical activity of the brain; its action is like a brake. Thus, barbiturates enhance the braking effects of GABA to promote sedation. There is an area in the brain called the reticular activating system, which is responsible for

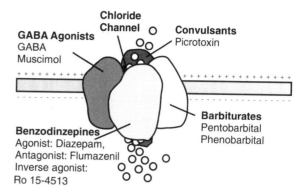

Figure 2
Barbiturates

maintaining wakefulness. Since this area has many interconnecting or polysynaptic neurons, it is the first to succumb to the barbiturates, and that is why an individual becomes tired and falls asleep after taking a barbiturate.

PHARMACOKINETICS AND DISTRIBUTION

The ultrashort-acting barbiturates differ from the other members of this class mainly by the means by which they are inactivated. Methohexital and its relatives are very soluble in lipids (i.e., fatty tissue). The brain is composed of a great deal of lipid; when the ultrashort-acting barbiturates are given intravenously, they proceed directly to the brain to produce anesthesia and unconsciousness. After only a few minutes, however, these drugs are redistributed to the fats in the rest of the body so their concentration is reduced in the brain. Thus, recovery from IV barbiturate anesthesia can be very fast. For this reason, drugs such as methohexital and thiopental are used primarily as intravenous anesthetic agents and not as sedatives.

The other longer-acting barbiturates must be metabolized by the liver into inactive compounds before the effects wane. Since these metabolites are more soluble in water, they are excreted through the kidneys and into the urine. As is the case with most drugs, metabolism and excretion is much quicker in young adults than in the elderly and infants. Plasma half-lives are also increased in pregnant women because the blood volume is expanded due to the development of the placenta and fetus.

TOLERANCE, DEPENDENCE, AND ABUSE

Repeated administration of any number of drugs results in eventual compensatory changes in the body. These changes are usually in the opposite direction of those initially produced by the drug such that more and more drug is needed to achieve the initial desired effect. This process is called TOLERANCE. There are two basic mechanisms for tolerance development: tissue tolerance and metabolic or pharmacokinetic tolerance. Tissue tolerance refers to the changes that occur on the tissue or cell that is affected by the drug. Metabolic tolerance refers to the increase in the processes that metabolize or break down the drug. This process generally occurs in the liver. Barbiturates are subject to both types of tolerance development.

Tolerance does not develop equally in all effects produced by barbiturates. Barbiturate-induced respiratory depression is one example. Barbiturates reduce the drive to breathe and the processes necessary for maintaining a normal breathing rhythm. Thus, while tolerance is quickly developing to the desired sedative effects, the toxic doses change to a lesser extent. As a result, when the dose is increased to achieve the desired effects (e.g., sleep), the margin of safety actually decreases as the dose comes closer to producing toxicity. A complete cessation of breathing is often the cause of death in barbiturate poisoning (Rall, 1990).

If tolerance develops and the amount of drug taken continues to increase, then PHYSICAL DEPENDENCE can develop. This means that if the drug is suddenly stopped, the tissues' compensatory effects become unbalanced and withdrawal signs appear. In the case of barbiturates, mild signs of withdrawal include apprehension, insomnia, excitability, mild tremors, and loss of appetite. If the dose was very high, more severe signs of withdrawal can occur, such as weakness, vomiting, decrease in blood pressure regulatory mechanisms (so that pressure drops when a person rises from a lying position, called orthostatic hypotension), increased pulse and respiratory rates, and grand mal (epileptic) seizures or convulsions. DELIRIUM with fever, disorientation, and HALLUCINATIONS may also occur. Unlike withdrawal from the opioids, withdrawal from central nervous system depressants such as barbiturates can be life threatening. The proper treatment of a barbiturate-dependent individual always includes a slow reduction in the dose to avoid the dangers of rapid detoxification.

Few, if any, illegal laboratories manufacture barbiturates. Diversion of licit production from pharmaceutical companies is the primary source for the illicit market. Almost all barbiturate users take it by mouth. Some try to dissolve the capsules and inject the liquid under their skin (called skin-popping) but the toxic effects of the alcohols used to dissolve the drug and the strong alkaline nature of the solutions can cause lesions of the skin. Intravenous administration is a rare practice among barbiturate abusers.

Many barbiturate users become dependent to some degree during the course of treatment for insomnia. This type of problem is called iatrogenic, because it is initiated by a physician. In some instances the problem will be limited to continued use at gradually increasing doses at night, to prevent insomnia that is in turn due to withdrawal. However, some individuals who are susceptible to the euphoric effects of barbiturates may develop a pattern of taking increasingly larger doses to become intoxicated, rather than for the intended therapeutic effects (for example, to promote sleepiness). To achieve these aims, the person may obtain prescriptions from a number of physicians and take them to a number of pharmacists—or secure their needs from illicit distributors (dealers). If the supply is sufficient, the barbiturate abuser can rapidly increase the dose within a matter of weeks. The upper daily limit is about 1,500 to 3,000 milligrams; however, many can titrate their daily dose to the 800 to 1,000 milligram range such that the degree of impairment is not obvious to others. The pattern of abuse resembles that of ethyl (drinking) ALCOHOL, in that it can be daily or during binges that last from a day to many weeks at a time. This pattern of using barbiturates for intoxification is more typically seen in those who, from the beginning, obtain barbiturates from illicit sources rather than those who began by seeking help for insomnia.

Barbiturates are sometimes used along with other drugs. Often, the barbiturate is used to potentiate, or boost, the effects of another drug upon which a person is physically dependent. Alcohol and HEROIN are commonly taken together in this way. Since barbiturates are "downers," they also are used to counteract the unwanted overstimulation associated with stimulant-induced intoxication. It is not uncommon for stimulant abusers (on COCAINE or amphetamines) to use barbiturates to combat the continued "high"

and the associated motor disturbances associated with heavy and continued cocaine use. Also, barbiturates are used to ward off the early signs of withdrawal from alcohol.

Treatment for barbiturate dependence is often conducted under carefully controlled conditions, because of the potential for severe developments, such as seizures. Under all conditions, a program of supervised withdrawal is needed. Many years ago, pentobarbital was used for this purpose and the dose was gradually decreased until no drug was given. More recently, phenobarbital or the benzodiazepines—CHLORDIAZEPOXIDE and diazepam—have been used for their greater margin of safety. The reason that the benzodiazepines sometimes work is because the general central nervous system depressants—barbiturates, alcohol, and benzodiazepines—develop cross-dependence to one another. Thus a patient's barbiturate or alcohol withdrawal signs are reduced or even eliminated by diazepam.

(SEE ALSO: *Addiction: Concepts and Definitions; Withdrawal*)

BIBLIOGRAPHY

CSÁKY, T. Z. (1979). *Cutting's handbook of pharmacology: The actions and uses of drugs*, 6th ed. New York: Appleton-Century Crofts.

HENNINGFIELD, J. E., & ATOR, N. A. (1986). Barbiturates: Sleeping potion or intoxicant? In *The encyclopedia of psychoactive drugs*. New York: Chelsea House.

MENDELSON, J. H., & MELLO, N. K. (1992). *Medical diagnosis and treatment of alcoholism*. New York: McGraw-Hill.

RALL, T. W. (1990). Hypnotics and sedatives: Ethanol. In A. G. Gilman et al. (Eds.), *Goodman and Gilman's the pharmacological basis of therapeutics*, 8th ed. New York: Pergamon.

WINGER, G., HOFFMAN, F. G., & WOODS, J. H. (1992). *A handbook of drugs and alcohol abuse: the biomedical aspects*, 3rd ed. New York: Oxford University Press.

S. E. LUKAS

BARBITURATES: COMPLICATIONS

Barturates are central nervous system (CNS) DEPRESSANTS ("downers"). These drugs produce sedative, hypnotic, and anesthetic effects. Depending on the dose used, any single drug in this class may produce sedation (decreased responsiveness), hypnosis (sleep), and anesthesia (loss of sensation). A small dose will produce sedation and relieve ANXIETY and tension; a somewhat larger dose taken in a quiet setting will usually produce sleep; an even larger dose will produce unconsciousness. The sleep produced by barbiturates, however, is not identical with normal sleep. Normal sleep consists of alternating phases of slow-wave sleep (SWS)—when the electroencephalogram (EEG) shows a high-voltage and low-frequency pattern—and rapid-eye-movement (REM) sleep. In the REM sleep phase, the EEG shows an arousal pattern and skeletal muscles relax, eyes move rapidly and frequently, and dreaming is thought to take place. Barbiturates decrease REM (or dreaming) sleep and thereby disturb the balance between SWS and REM sleep.

As is true for most drugs that act on the CNS, the effects of these drugs are also influenced markedly by the user's previous drug experience, the circumstances in which the drug is taken, and the route of administration of the drug. For example, a dose taken at bedtime may produce sleep, whereas the same dose taken during the daytime may produce a feeling of euphoria, incoordination, and emotional response. This, in many ways, is what happens with alcohol intoxication. In fact, the behavioral effects of this class of drugs is very similar to those observed after drinking ALCOHOL, and the user may experience impairment of skills and judgment not unlike that experienced with alcohol. It is therefore not surprising that the effects of barbiturates are enhanced when taken in combination with alcohol, antianxiety drugs (BENZODIAZEPINES), and other CNS depressants such as opioids, antihistamines, and OVER-THE-COUNTER cough and cold medications containing these drugs. Barbiturates, however, differ from some other SEDATIVE-HYPNOTIC drugs in that they do not elevate the PAIN threshold. In fact, patients experiencing severe pain may become agitated and delirious if they are given barbiturates without also receiving ANALGESICS.

Barbiturates are generally classified as being long, intermediate, short, and ultra-short acting on the basis of their duration of effect. The long-acting barbiturates, such as phenobarbital, which were at one time mainly employed as daytime sedatives for the treatment of anxiety, produce sedation that lasts from twelve to twenty-four hours. Phenobarbital is still one of the drugs used for the treatment of grand mal epilepsy. The short- and intermediate-acting drugs such as pentobarbital and secobarbital, which

were once mainly employed as hypnotics, produce CNS depression that ranges from three to twelve hours, depending on the compound used. The ultra-short-acting barbiturates (e.g., thiopental) are used for the induction of anesthesia, because of the ease and rapidity with which they induce sleep when given intravenously. The effects of barbiturates on judgment and other mental as well as motor skills, however, may persist much longer than the duration of the hypnotic effect. For this and other reasons, for the treatment of anxiety or insomnia, barbiturates have largely been replaced by the generally safer group of drugs called benzodiazepines.

The respiratory system is significantly depressed by the administration of barbiturate doses that are larger than those usually prescribed. Furthermore, there is a synergistic effect when barbiturates are combined with alcohol and other central nervous system depressants—often with a fatal outcome. Barbiturates are frequently used for suicides. For this reason, too, the barbiturates have been displaced by the less toxic benzodiazepines. The symptoms of acute barbiturate toxicity resemble the effects observed after excessive alcohol ingestion. Although repeated administration of barbiturates results in CNS tolerance, thus producing less intoxication, tolerance does not appear to develop to the same extent in regard to the respiratory depressant and lethal effects of the barbiturates; the person addicted to barbiturates may therefore be at a greater risk of respiratory toxicity because of less pronounced CNS euphoric effects with higher doses. Tolerance to barbiturates also affects metabolism; the administration of these drugs speeds up not only their own metabolism (i.e., shortens their effectiveness) but also the metabolism of a large number of other drugs. This property has been of use in some special cases (as in jaundice of the newborn), but it can be hazardous in others when it decreases the effectiveness of another drug (e.g., an anticoagulant used to treat thrombosis).

Long-term users experience withdrawal symptoms when the barbiturate is stopped abruptly. Abrupt cessation also leads to an increase in the amount and intensity of REM sleep (REM rebound). The intensity of the withdrawal symptoms varies with the degree of abuse and may range from sleeplessness and tremor in mild cases to delirium and convulsions in severe cases. Fatalities have occurred as a result of barbiturate WITHDRAWAL, usually from withdrawal of short-acting barbiturates.

In some individuals, barbiturates may produce CNS excitement rather than CNS depression. This type of idiosyncratic reaction occurs most frequently in elderly people. Among the side effects sometimes seen, there may be rashes and muscle and body aches.

(SEE ALSO: *Expectancies; Sleep, Dreaming, and Drugs*)

BIBLIOGRAPHY

GOODMAN, L. S., & GILMAN, A. G. (1975). *The pharmacological basis of therapeutics*, 5th ed. New York: Macmillan.

KALANT, H., & ROSCHLAU, W. H. E. (1989). *Principles of medical pharmacology*, 5th ed. Toronto: B. C. Decker.

JAT M. KHANNA

BEERS AND BREWS Beers and brews are beverages produced by yeast-induced fermentation of malted cereal grains, usually barley malt, to which hops and water have been added. They generally contain 2 to 9 percent ethyl ALCOHOL, although some may contain as much as 15 percent. Various types and flavors are created by adding different combinations of malts and cereals and allowing the process to continue for varying lengths of time.

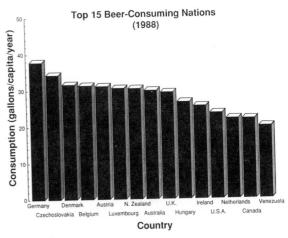

Figure 1

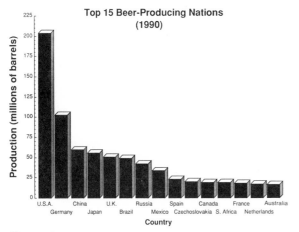

Figure 2

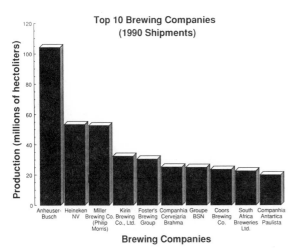

Figure 3

SOURCE: *Impact Databank, M. Shanken Communications, Inc., New York, N.Y.*

BREWING HISTORY

The origin of beer is unknown, but it was an important food to the people of the Near East, probably from Neolithic times, some 10,000 years ago. The making of beer and of bread developed at the same time. In Mesopotamia (the ancient *land between the rivers* as the Greeks called it), an early record from about 5,000 years ago describing the recipe of the "wine of the grain" was found written in Sumerian cuneiform on a clay tablet. In ancient Egypt, at about the same time, barley beer was brewed and consumed as a regular part of the diet. It was known as hek and tasted like a sweet ale, since there were no hops in Egypt. Egyptians continued to drink it for centuries, although the name was changed to hemki. More than 3,200 years ago, the Chinese made a beer called kiu that was most likely made from two parts millet and one part rice. With water added, the concoction was heated in clay pots; flour and various plants were added to provide the yeast and flavors, respectively.

The ancient Greeks, however, preferred wine and considered beer to be a drink of barbarians. Beer was drunk on special occasions in ancient Rome; Plutarch wrote of a feast in which Julius Caesar served his officers beer as a special reward after they had crossed the Rubicon river. Once the art of brewing reached England, beers and ales became the preferred drink of the rich and poor alike. King Henry VIII of England was said to consume large quantities during breakfast. It was soon discovered that sailors who drank beer avoided scurvy, a disease caused by a lack of adequate amounts of Vitamin C. Thus, beer was added to each ship's provisions and was even carried on the Mayflower during the crossing of the Atlantic Ocean in 1620. American colonists quickly learned to make their beer with Indian corn (maize), and much U.S. beer is still made with corn, although rice and wheat are also used in the mix with barley malt.

MAKING BEER

The first step in making beer is to allow barley to sprout (germinate) in water, a process that releases an important enzyme, amylase. Germinated barley seeds are called malt. Once the malt is crushed and suspended in water, the amylase breaks down the complex starch into more basic sugars. The reaction is stopped by boiling, and the concoction is filtered.

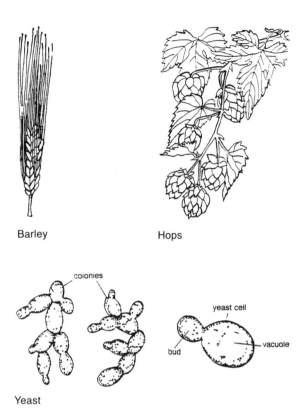

Barley

Hops

colonies

yeast cell

bud

vacuole

Yeast

This clear solution is mixed with hops (to provide the bitter flavor) and a starter culture of yeast (to begin the alcoholic fermentation process). Carbon dioxide gas (the fizz or bubbles) is produced, along with ethyl alcohol (ethanol or drinking alcohol). The malt and hops are then removed (and generally sold for cattle feed) while the yeast is skimmed off as fermentation proceeds. After the desired effect is achieved, the beer is filtered and bottled, or it is stored in kegs for aging. During the aging process (two to twenty-four weeks) proteins settle to the bottom or are digested by enzymes. The carbonation (fizz) that occurred during fermentation is then drawn off and is forced back in during the bottling process.

BEER TYPES

There are two major types of beers: top-fermenting and bottom-fermenting. Top fermentation occurs at room temperature 59° to 68°F (15° to 20°C) and is so named because the yeast rises to the top of the vessel during fermentation. This older process produces beers that have a natural fruitiness and in-

clude the wheat beers, true ales, stouts, and porters. Their flavor is most completely expressed when served at moderate (i.e., room) temperatures. The development of yeasts that sank during this process resulted in brews that were more stable between different batches. Most of the major brewers have switched to the bottom yeasts and cold storage (lagering). The significance of using yeast that sinks during fermentation is that airborne yeasts cannot mix with the special yeast and contaminate the process.

The most popular type of beer in the United States is lager, a pale, medium-hop-flavored beer. It is mellowed several months at 33°F (0.5°C) to produce its distinctive flavor. Lager beers average 3.3 to 3.4 percent ethyl alcohol by weight and are usually heavily carbonated. Pilsner is a European lager (that originated in medieval Pilsen, now the Czech city of Plzen) that is stored longer than other lagers and has a higher alcohol content and a rich taste of hops. Dark beers are popular in Europe but are not generally produced in the United States. The dark color is achieved by roasting the malt; dark beer has a heavier and richer taste than lager beer. British beers are many and varied, both pale and dark; some have a number of unique additives, including powdered eggshell, crab claws or oyster shells, tartar salts, wormwood seeds, and horehound juice. Porter, popular in England, is another dark beer—originally called porter's beer, it was a mixture of ale and beer. The porters of today are a sweet malty brew and contain 6 to 7 percent alcohol. Malt liquors are beers that are made using a higher percentage of fermentable sugars, resulting in a beverage with 5 to 9 percent alcohol content; the mild fruity flavor has a spicy taste and lacks the bitterness of hops. Low-calorie (sometimes called "light" or "lite") beers are produced by decreasing the amount of grain used in the initial brew (using more water per unit of volume) or by adding an enzyme that reduces the amount of starch in the beer. These light beers contain only about 2.5 to 2.7 percent alcohol.

Brewing is subject to national laws concerning allowable ingredients in commercial products. Although chemical additives are allowed by some countries (e.g., the United States), German and Czech purity laws consider beers and brews a natural historical resource and disallow anything that was not part of the original (medieval) brewing tradition. Individuals sensitive to U.S. or Canadian beers are often able to drink pure beers.

WORLDWIDE CONSUMPTION OF BEERS AND BREWS

Almost 800 million barrels of beer are produced in the world each year for commercial purposes—more are produced for local and private consumption, especially by traditional cultures in the third world. Germany is the country that drinks the most beer, consuming about 40 gallons (152 liters) per person per year. The United States is by far the largest brewing nation, producing more than 200 million barrels of beer in 1990. Figures 1–3 show the relative distribution of the top drinking and brewing nations, as well as the top brewing companies.

(SEE ALSO: *Alcohol: History of Drinking*)

BIBLIOGRAPHY

ABLE, B. (1976). *The book of beer.* Chicago: Henry Regnery.

JACKSON, M. (1988). *The New World guide to beer.* Philadelphia: Running Press

S. E. LUKAS

BEHAVIORAL MODIFICATION AS A TREATMENT   *See* Treatment Types: Behavior Modification.

BEHAVIORAL TOLERANCE   In everyday language, TOLERANCE implies the ability to withstand something. In pharmacology, the term *tolerance* is close to this meaning. To understand the technical meaning of the word, however, requires an understanding of the concept of the *potency* of a drug. A drug's potency is expressed in terms of the amount (the dose) of the drug needed to produce a certain effect. To illustrate, drugs may be compared with respect to potency. For example, relief from headache may be achieved with 650 milligrams of aspirin or with 325 milligrams of ibuprofen; in this case ibuprofen is said to be more potent, because less drug is needed to produce a particular effect (relief of headache). Tolerance is said to occur when a drug becomes *less potent* as a result of prior exposure to

that drug. That is, following exposure (usually repeated or continuous administrations) to a drug, it may take more of the drug to get the same effect as originally produced.

The expression *behavioral tolerance* often is used simply to refer to a drug's decreased potency in affecting a specified behavior after repeated or continuous exposure to the drug. In other contexts, however, the expression has taken on a more restricted and special meaning; it is employed only when behavioral factors have been shown experimentally to have contributed to the development of tolerance.

This special meaning is applied when either of two sets of circumstances are encountered. In the first, drug tolerance is shown to be specific to the context in which the drug is administered; in the second, drug tolerance is shown to occur only if drug administration precedes particular behavioral circumstances. Examples of each may help clarify the distinctions between them and between "simple" tolerance and behavioral tolerance.

Context-specific tolerance has been researched extensively by Siegel and his colleagues (see Siegel, 1989, for an overview). In a typical experiment two groups of subjects are compared; subjects in both groups receive the same number of repeated exposures to the drug (e.g., morphine) and then are tested for their response to the drug (e.g., alleviation of pain). For one group, the test occurs in the environment where drugging took place; for the other the test occurs in a novel environment. Typically, only those from the group tested in the familiar environment show tolerance. Siegel's theory is that subjects develop, via the principles of Pavlovian conditioning, a conditioned compensatory response that is elicited by the drug-administration context—and that this response counteracts the effect of the drug (see Baker & Tiffany, 1985, for a different view). The phenomenon of context-specific tolerance helps explain why many overdoses of abused drugs occur when the drug is taken in a novel situation—the new context does not elicit compensatory responses that counteract the effects of the drug.

The importance of the temporal relationship between drug administration and behavior is illustrated by the phenomenon of "contingent" tolerance. The basic technique for identifying contingent tolerance was pioneered by Chen (1968), and a clear example is provided by Carlton and Wolgin (1971). Three

groups of rats had the opportunity to drink milk for 30 minutes each day. For each group, injections of a drug or just a saline vehicle were made twice each day: For Group 1, each session was preceded by an injection of 2 milligrams per kilogram of AMPHETAMINE, followed by an injection of just the saline vehicle; for Group 2, the order of injections was reversed: saline before drinking, amphetamine after; Group 3 (the control group) received saline both before and after each session.

For Group 1 the drug initially decreased drinking, but during the course of several administrations, drinking recovered to control levels (i.e., tolerance developed). For Group 2, no effect on drinking was observed as a function of receiving the drug after sessions, so after several days (by which time subjects in Group 1 were tolerant) these subjects were given amphetamine before (rather than after) and saline after sessions (i.e., the conditions for Group 1 were implemented). Even though these subjects had received amphetamine just as frequently as the subjects in Group 1, when it was given before the session, drinking was suppressed just as much as it had been for Group 1 initially. Following repeated presession exposure to amphetamine the subjects in Group 2 became tolerant. These findings and many others like them show that, in many cases, for tolerance to develop to a drug's behavioral effects, mere repeated exposure to the drug is not enough. In addition, the drug must be active while the behavior of interest is occurring. (See Goudie & Demellweek, 1986, and Wolgin, 1989, for reviews.)

Contingent tolerance is sometimes called *learned* tolerance because it appears that it is a manifestation of learning to behave accurately while under the influence of a drug. An influential theory about the origin of contingent tolerance is the "reinforcement loss" theory of Schuster, Dockens and Woods (1966; for a review see Corfield-Sumner & Stolerman, 1978). Loosely stated, the theory is that contingent tolerance will emerge in situations where the initial effect of the drug is to produce a loss of reinforcement (e.g., result in a failure to meet the demands of the task). Although there are limits to the generality of the theory (Genovese, Elsmore, & Witkin, 1988), it has an excellent predictive record.

(SEE ALSO: *Addiction: Concepts and Definitions; Reinforcement; Tolerance and Physical Dependence; Wikler's Pharmacologic Theory of Drug Addiction*)

BIBLIOGRAPHY

BAKER, T. B., & TIFFANY, S. T. (1985). Morphine tolerance as habituation. *Psychology Review, 92,* 78–108.

CARLTON, P. L., & WOLGIN, D. L. (1971). Contingent tolerance to the anorexigenic effects of amphetamine. *Physiology of Behavior, 7,* 221–223.

CHEN, C. S. (1968). A study of the alcohol-tolerance effect and an introduction of a new behavioral technique. *Psychopharmacologia, 12,* 433–440.

CORFIELD-SUMNER, P. K., & STOLERMAN, I. P. (1978). Behavioral tolerance. In D. E. Blackman & D. J. Sanger, (Eds.), *Contemporary research in behavioral pharmacology.* New York: Plenum.

GENOVESE, R. F., ELSMORE, T. R., & WITKIN, J. M. (1988). Environmental influences on the development of tolerance to the effects of physostigmine on schedule-controlled behavior. *Psychopharmacology, 96,* 462–467.

GOUDIE, A. J., & DEMELLWEEK, C. (1986). Conditioning factors in drug tolerance. In S. R. Goldberg & I. P. Stolerman (Eds.), *Behavioral analysis of drug dependence.* New York: Academic.

SCHUSTER, C. R., DOCKENS, W. S., & WOODS, J. H. (1966). Behavioral variables affecting the development of amphetamine tolerance. *Psychopharmacologia, 9,* 170–182.

SIEGEL, S. (1989). Pharmacological conditioning and drug effects. In A. J. Goudie & M. W. Emmett-Oglesby (Eds.), *Psychoactive drugs: Tolerance and sensitization.* Clifton, NJ: Humana.

WOLGIN, D. L. (1989). The role of instrumental learning in behavioral tolerance to drugs. In A. J. Goudie & M. W. Emmett-Oglesby (Eds.), *Psychoactive drugs: Tolerance and sensitization.* Clifton, NJ: Humana.

MARC N. BRANCH

BENZEDRINE/BENZEDRINE INHALERS *See* Amphetamine Epidemics.

BENZENE *See* Inhalants.

BENZODIAZEPINES The benzodiazepines were introduced into clinical practice in the 1960s for the treatment of anxiety and sleep disorders. Members of this class of drug were classified initially as *minor tranquilizers* although this term has fallen

into disfavor. These agents have proven to be safe and effective alternatives to older SEDATIVE-HYPNOTIC agents such as BARBITURATES, CHLORAL HYDRATE, glutethimide, and carbamates. Benzodiazepines are widely prescribed drugs, with 8.3 percent of the U.S. population reporting medical use of these agents in 1990.

BASIC PHARMACOLOGY

All benzodiazepines produce similar pharmacologic effects, although the potency for each effect may vary with individual agents. They decrease or abolish ANXIETY, produce sedation, induce and maintain sleep, control certain types of seizures, and relax skeletal muscles. The basic chemical structure is shown in Figure 1.

Dissimilarity in the effects of different benzodiazepines tend to be more quantitative than qualitative in nature. Many of these differences are attributable to how benzodiazepines are absorbed, distributed, and metabolized in the body. A few benzodiazepines—clorazepate for example—are pro-

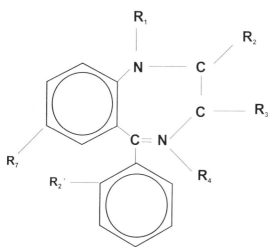

Figure 1
Outline of the basic structure of the benzodiazepines. R denotes substituent groups such −H, −O, −OH, −NO₂, and −Cl that are attached to the core benzodiazepine structure. These groups determine the precise physicochemical and pharmacologic properties of each benzodiazepine.
SOURCE: Adapted from Rall, T. W. (1990).
Hypnotics and sedatives: Ethanol. Figure created by Rebecca Bulotsky.

drugs; that is, they become active only after undergoing chemical transformation in the body. The extent to which a benzodiazepine is soluble in fatlike substances—that is, the degree to which it is lipophilic—determines the rate at which it crosses the tissue barriers that protect the brain. Drugs that are highly lipophilic such as DIAZEPAM (Valium) rapidly enter and then leave the brain. Benzodiazepines are metabolized in the body in a number of ways (see Table 1). Many benzodiazepines are transformed in the liver into compounds that possess pharmacologic activity similar to that of the originally administered drug. Diazepam, prazepam, and halazepam are all converted to the active metabolite desmethyldiazepam, which is eliminated from the plasma at a very slow rate. Oxazepam (Serax) and lorazepam (Ativan), in contrast, are conjugated with glucuronide, a substance formed in the liver, to form inactive metabolites that are readily excreted into the urine.

Most of the effects that result from the administration of benzodiazepines are a consequence of the direct action of these agents on the central nervous system. Benzodiazepines interact directly with proteins that form the benzodiazepine receptor. Benzodiazepine receptors exist as part of a larger receptor complex (Figure 2). The interaction of the NEUROTRANSMITTER gamma-amino butyric acid (GABA) with this complex leads to the enhanced flow of chloride ions into neurons (Kardos, 1993). This complex is referred to as the GABA$_A$ receptor-chloride ion channel complex. Much of the available evidence indicates that the action of benzodiazepines involves a facilitation of the effects of GABA and similarly acting substances on the GABA$_A$ receptor complex, thus leading to an increased movement of chloride ions into nerve cells. Entry of chloride ions into neurons tends to diminish their responsiveness to stimulation by other nerve cells, and consequently substances that produce an increase in chloride flow into cells depress the activity of the central nervous system. This depressant effect becomes manifested as either sedation or sleep. Agents that increase chloride ion inflow include not only the benzodiazepines but also other central nervous system depressant agents such as ETHANOL (alcohol) and the barbiturates. Benzodiazepines differ from barbiturates in that they require the release of GABA to affect the movement of chloride, whereas at higher doses barbiturates, through their own direct effects, can act to increase chloride inflow into cells.

TABLE 1
Benzodiazepines Available in the United States

Anxiolytics

Generic Name	Trade Name	Usual Dose (mg/day)	Half-life (hours)	Transformation Pathway	Metabolites (half-life, hours)
Alprazolam	Xanax	0.75–4	8–15	Oxidation	Alphahydroxyalprazolam Benzophenone
Bromazepam			20–30	Oxidation	
Chlordiazepoxide	Librium	15–100	5–30		Desmethylchlordiazepoxide Demoxepam Desmethyldiazepam (36–96)
Clonazepam	Klonopin	1.5–20	18–50	Nitroreduction	Inactive 7-amino or 7-acetyl amino derivatives
Clobazam	Frisium	20–30	18	Oxidation	Desmethylclobazam (up to 77)
Clorazepate	Tranxene	15–60	30–100	Oxidation	Desmethyldiazepam (36–96)
Diazepam	Valium	4–40	20–70	Oxidation	Desmethyldiazepam (36–96)
Halazepam	Paxipam	60–160	14	Oxidation	Desmethyldiazepam (36–96) 3-hydroxyhalazepam
Lorazepam	Ativan	2–4	10–120	Conjugation	Inactive glucuronide conjugate
Oxazolam				Oxidation	Desmethyldiazepam (36–96)
Oxazepam[a]	Serax	30–120	5–15	Conjugation	Inactive glucuronide conjugate
Prazepam	Centrax	20–60	30–100	Oxidation	Desmethyldiazepam (36–96)

Hypnotics

Generic Name	Trade Name	Usual Dose (mg/day)	Half-life (hours)	Transformation Pathway	Metabolites (half-life, hours)
Brotizolam			4–7	Oxidation	
Estazolam	ProSom	1–2	8–24	Oxidation	
Flunitrazepam			10–40	Oxidation nitro-reduction	Desmethylflunitrazepam
Flurazepam	Dalmane	15–30	.5–3.0	Oxidation	Desalkylflurazepam (36–120) Hydroxyethyl flurazepam (1–4) Flurazepam aldehyde (2–8)
Lormetazepam			8–20	Conjugation	
Nitrazepam	Mogadon	2.5–10	20–30	Nitroreduction	
Quazepam	Doral	7.5–15	20–40	Oxidation	Oxoquazepam (25–35) Desalkylflurazepam (36–120)
Temazepam	Restoril	15–30	8–20	Conjugation	
Triazolam	Halcion	.125–.5	2–6	Oxidation	

Perioperative Hypnotic

Generic Name	Trade Name	Usual Dose (mg/day)	Half-life (hours)	Transformation Pathway	Metabolites (half-life, hours)
Midazolam	Versed	1–2.5 mg/ml	1–4	Oxidation	Hydroxymethylmidazolam

NOTE: The half-life of a compound is the amount of time that must pass for the level of that agent in the plasma to be reduced by half.

[a]Oxazepam is also a metabolite of diazepam, clorazepate, prazepam, halazepam, and temazepam.

SOURCES: *Drug Facts and Comparisons.* (1994). St. Louis, MO: Facts and Comparisons. Greenblatt, D. J. (1991). Benzodiazepine hypnotics: Sorting the pharmacokinetic facts. *Journal of Clinical Psychiatry, 52* (Suppl. 9), 4–10. Greenblatt, D. J., & Shader, R. I. (1987). Pharmacokinetics of antianxiety drugs. In H. Y. Meltzer (Ed.), *Psychopharmacology: The Third Generation of Progress.* NY: Raven Press.

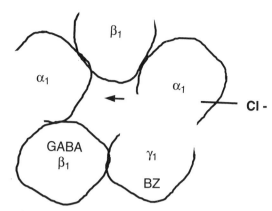

Figure 2

Schematic of one possible form of the GABA$_A$ receptor-chloride ion channel complex. Chloride ions enter through the center channel formed by alpha (α), beta (β), and gamma (γ) subunits. GABA receptors on β subunits regulate the flow of chloride ions through the channel. The activity of GABA receptors can be modulated by benzodiazepine receptors located on the γ subunit.

SOURCE: *Figure Created by Rebecca Bulotsky*

The GABA$_A$ receptor complex is composed of alpha, beta, and gamma subunits (Zorumski & Isenberg, 1991). Each subunit consists of a chain of twenty to thirty amino acids. Multiple subtypes of the alpha, beta, and gamma subunits have been shown to exist, and the types of subunits that form a single receptor complex appear to vary in different areas of the central nervous system. Some researchers have proposed that different drugs selectively interact with benzodiazepine receptors composed of a particular kind of α subunit, thereby leading to differences in drug effects. Although there is little evidence to support this hypothesis, future research should clarify the issue.

New compounds, such as the imidazopyridines, have been developed that act at the benzodiazepine receptor but are chemically distinct from the benzodiazepines. Zolpidem is an imidazopyridine used in clinical practice as a hypnotic agent. Other new drugs have been synthesized that can stimulate the benzodiazepine receptor but do not produce the maximal effects that result from the administration of higher doses of benzodiazepines. These drugs are classified as partial AGONISTS. The drug abecarnil, which belongs to the beta-carboline class of compounds, is an example of such an agent that has been used experimentally to treat anxiety.

Flumazenil is a benzodiazepine derivative that has no activity of its own but acts to antagonize the actions of benzodiazepines at the benzodiazepine receptor. It is used to reverse the effects of these drugs during anesthesia or in benzodiazepine overdoses. Other compounds, including some of the beta-carbolines such as methyl-beta-carboline-3-carboxylate, act on the benzodiazepine receptor to produce effects that are opposite to those of benzodiazepines (Kardos, 1993; Zorumski & Isenberg, 1991). Administration of these inverse agonists can lead to the appearance of anxiety and convulsions.

THERAPEUTIC USE

Benzodiazepines are used for a variety of therapeutic purposes. Anxiety is the experience of fear that occurs in a situation where no clear threat exists. Numerous studies have demonstrated that anxiety disorders, including generalized anxiety disorder and many phobias, can be treated effectively with benzodiazepines. Panic disorder is a psychiatric illness in which patients experience intense sporadic attacks of anxiety often accompanied by the avoidance of open spaces and other places or objects that are associated with panic. High-potency benzodiazepines such as alprazolam (Xanax) or clonazepam (Klonopin) can prevent the occurrence of panic attacks in patients suffering from panic disorder. Flurazepam (Dalmane), triazolam (Halcion), and the other benzodiazepines listed in the Table are used in the treatment of insomnia and other sleep disorders. All rapidly acting benzodiazepines marketed in the United States have hypnotic effects. Classification of a benzodiazepine as a hypnotic is often more a marketing strategy than it is a decision based on pharmacologic differences among the class of drugs.

Status epilepticus is a seizure or a series of seizures that occurs over an extended period of time. This condition can lead to irreversible brain damage and is often successfully managed by the intravenous infusion of diazepam. Clonazepam is used either alone or in combination with other anticonvulsant medications to treat absence seizure and other types of seizure disorders. Clorazepate is used to control some types of partial seizures—that is, seizures that occur in a limited area of the brain. The increase in central nervous system excitability, seizures, and anxiety that may appear during alcohol withdrawal

can be treated with any benzodiazepine. Midazolam (Versed) is a benzodiazepine that is rapidly metabolized in the body and is used to help induce anesthesia during surgical procedures. The skeletal-relaxant properties of benzodiazepines make them useful for the treatment of back pain due to muscle spasms.

ADVERSE EFFECTS

Benzodiazepines have proven to be exceptionally safe agents. The dose at which these agents are lethal tends to be exceedingly high. Fatalities are more apt to occur when these drugs are taken in combination with other central nervous system depressant agents such as ethanol. Sedation is a common adverse effect associated with benzodiazepine use. Light-headedness, confusion, and loss of motor coordination may all result following the administration of benzodiazepines. MEMORY impairment may be detected in individuals treated with benzodiazepines, and this effect may prove to be particularly troublesome to ELDERLY patients who are experiencing memory-related problems. PSYCHOMOTOR impairment can be hazardous to individuals when they are driving. This problem can be exacerbated in individuals who consume ethanol while they are being treated with benzodiazepines. Hypnotic agents that are converted into active metabolites that are slowly eliminated from the body, such as flurazepam, may produce residual daytime effects that can impair tasks such as driving. The adverse effects of benzodiazepines on performance tend to be more of a problem in elderly people than in younger individuals. Patients with cirrhosis, a liver degenerative disease, are also more likely to experience benzodiazepine toxicity than are those with normal liver function. The appearance of the adverse effects associated with benzodiazepine administration in both elderly people and in cirrhotic patients can be minimized by treating them with agents such as oxazepam and lorazepam, which tend not to accumulate in the blood because they are excreted rapidly into the urine as glucuronide conjugates.

A small number of patients may exhibit paradoxical reactions when they are treated with benzodiazepines (Rall, 1990). These may include low-level anxiety, restlessness, depression, paranoia, hostility, and rage. Sleep patterns may be disrupted by benzodiazepine administration, and nightmares may increase in frequency. Benzodiazepines suppress two stages of the sleep cycle—the stage of deepest sleep, stage IV, and the rapid eye movement (REM) stage in which dreaming occurs.

TOLERANCE AND PHYSICAL DEPENDENCE

TOLERANCE to a drug involves either a decrease in the effect of a given dose of a drug during the course of repeated administration of the agent or the need to increase the dose of a drug to produce a given effect when it is administered repeatedly. Chronic treatment of animals with benzodiazepines leads to a reduction in potency of these agents as enhancers of chloride ion uptake. These effects at the cellular level are paralleled by the appearance of tolerance to the sedative effects of benzodiazepines. Tolerance also develops to the impairment of motor coordination that is produced by these drugs. Limited evidence suggests that the antianxiety effects of benzodiazepines may not diminish with time, or at the very least that benzodiazepines retain their effectiveness as antianxiety agents for several months.

PHYSICAL DEPENDENCE results from adaptive changes in the nervous system that may be related to the development of tolerance. Dependence of this sort can be detected by the appearance of a characteristic abstinence or WITHDRAWAL syndrome when chronic administration of a drug is either abruptly discontinued or after the administration of an antagonist to the drug that has been taken for a prolonged period of time (Ciraulo & Greenblatt, in press). Individuals who are treated chronically with benzodiazepines may exhibit signs and symptoms of withdrawal when the administration of these drugs is discontinued. Minor symptoms of withdrawal include ANXIETY, insomnia, and nightmares. Less common and more serious symptoms include psychosis, death, and generalized seizures. Signs of withdrawal may become evident twenty-four hours after the discontinuation of a benzodiazepine that is rapidly eliminated from the blood. Peak abstinence symptoms may not appear until two weeks after discontinuation of a benzodiazepine that is removed from the body slowly. Some of the symptoms that appear after benzodiazepine treatment is discontinued may be due to the recurrence of the anxiety disorder for which the drug had been originally prescribed.

In animals, the severity of withdrawal can be directly related to the dose and length of time of administration of a benzodiazepine. This kind of relationship has been harder to demonstrate in clinical studies. Many patients who are treated with benzodiazepines for prolonged periods of time may experience at least some symptoms of withdrawal, but most of these individuals should not be viewed as benzodiazepine "addicts" because they have relied on their medications for medical reasons, have taken the medications as directed by their physicians, and will not continue to compulsively seek out benzodiazepines once their prescribed course of treatment with these medications has been discontinued. The intensity of abstinence symptoms that may be seen in patients who are physically dependent on benzodiazepines can be markedly reduced if patients are allowed to gradually taper off their medications. There may be a risk of physical withdrawal from benzodiazepines in some patients who abruptly stop the medication following as few as four weeks after treatment. Patients who discontinue taking rapidly metabolized hypnotic drugs such as triazolam may be at risk for experiencing rebound insomnia, even if they have been under treatment for a few days to one week. Serious problems associated with benzodiazepine withdrawal are more likely to be a problem for patients who have been treated with high doses of these medications for four or more months.

ABUSE AND DEPENDENCE

Although no consensus exists as to the definition of drug addiction, diagnostic criteria for drug abuse and dependence have been developed by both the American Psychiatric Association and the World Health Organization. Drug abuse can be viewed as the use of a pharmacological substance in a manner that is not consistent with existing medical, social, or legal standards and practice. Alternatively, drug abuse has been defined in the DIAGNOSTIC AND STATISTICAL MANUAL of Mental Disorders of the APA as involving a "maladaptive pattern of substance use manifested by recurrent and significant adverse consequences related to repeated use" (American Psychiatric Association, 1994). Abuse of drugs may involve the use of drugs for recreational purposes—that is, drugs are administered to experience their mood-elevating (euphoric) effects. For some individuals, self-administration of drugs for these purposes

may lead to compulsive drug-seeking behavior and other extreme forms of drug-controlled behavior. These behavior patterns may become further reinforced by the effects of withdrawal symptoms that dependent individuals attempt to reduce by the administration of the abused agent. The APA specifies that individuals can be classified as being drug dependent if they exhibit signs of drug tolerance, symptoms of withdrawal, cannot control their drug use, feel compelled to use a drug, and/or continue to use a substance even if the consequences of this use may prove harmful to them (American Psychiatric Association, 1994).

Abuse of drugs may sometimes represent self-medication. COCAINE and AMPHETAMINE users sometimes rely on benzodiazepines to relieve the jitteriness that may result from the administration of PSYCHOMOTOR STIMULANTS. Some abusers of benzodiazepines may be medicating themselves with these agents to treat preexisting conditions of anxiety and DEPRESSION.

The ABUSE LIABILITY of benzodiazepines—that is, the likelihood that they will be misused—has been assessed in studies of the tendency of either human beings or animals to administer these agents to themselves and studies of the subjective effects that result from the administration of different benzodiazepines. When provided access to cocaine and other psychomotor stimulants, animals will consistently self-administer these agents at high rates over time. Primates will intravenously self-administer benzodiazepines at moderate rates that are below those observed for the administration of BARBITURATES or COCAINE. This finding and the results of a number of additional animal studies indicate that the benzodiazepines have a lower abuse liability than do the barbiturates or the psychomotor stimulants (Ciraulo & Greenblatt, in press).

Individuals with a history of sedative-hypnotic abuse will self-administer triazolam and diazepam (Roache & Griffiths, 1989). In contrast, normal volunteers do not prefer diazepam to placebo. Subjective responses to drugs can be assessed through the use of instruments such as the Addiction Research Center Inventory–Morphine Benzedrine Group Scale and the Profile of Moods States that help to standardize the reports of subjects concerning their drug-induced experiences. Investigations in which subjective responses of normal subjects to benzodiazepine administration have been assessed indicate

that these agents tend not to produce mood elevations in normal populations. On the other hand, individuals with a history of either alcoholism or sedative-hypnotic abuse are more likely to experience euphoria after the administration of a single dose of either diazepam or other benzodiazepines. Adult children of alcoholics experience mood elevation after the ingestion of either alprazolam or diazepam, thus suggesting that these individuals may have a predisposition to benzodiazepine abuse.

Studies suggest that benzodiazepines are less likely to be abused than the barbiturates, opiates, or psychomotor stimulants, but that they carry more risk for abuse than do medications such as the antianxiety agent buspirone or drugs that have sedating effects such as the antihistamine diphenhydramine (Preston et al., 1992). There also may be differences among the benzodiazepines themselves. Some authorities believe that diazepam has greater abuse liability than halazepam, oxazepam, chlordiazepoxide, or clorazepate, although others believe that there is little difference among them. Diazepam, lorazepam, alprazolam, and triazolam all produce mood effects that are similar to those of known drugs of abuse. The rate at which these drugs reach the brain after administration may be a major determining factor in the onset of euphoria or pleasant effects associated with abuse. Inferences about abuse potential are made on the basis of subjective effects and self-administration in drug abusers and alcoholics. Many experts question the applicability of these findings to the general population.

Studies that accurately reflect the extent of benzodiazepine abuse in the United States are not available. A survey of American households produced by the National Institute on Drug Abuse suggested that the nonmedical use of tranquilizers was not a major health problem (Ciraulo & Greenblatt, in press). Only 2.4 percent of individuals between the ages of 18 and 24 and 1.3 percent of survey respondents who were older than 26 reported using tranquilizers for non-medical purposes. This type of survey does not take into account benzodiazepine usage among groups such as homeless people, prisoners, and migrant workers, and so it cannot convey a complete picture of how benzodiazepines are misused at the nationwide level (Cole & Chiarello, 1990).

Benzodiazepines are frequently used by individuals who abuse other drugs, but they are rarely used as either initial or primary drugs of abuse. Benzodiazepine abusers often take these drugs in combination with other agents. In Scotland, drug abusers have often injected temazepam in combination with the OPIOID drug BUPRENORPHINE (Ruben & Morrison, 1992). Large percentages of methadone-clinic patients have urine tests that are positive for benzodiazepines. METHADONE-MAINTENANCE patients have indicated that diazepam, lorazepam, and alprazolam can produce desirable pleasurable effects (Sellers et al., 1993). Whether methadone patients use benzodiazepines to increase the effects of methadone or as self-medication for anxiety is not clear.

The percentage of alcoholics admitted for treatment who also concurrently use benzodiazepines ranges between 12 to 23 percent. High rates of benzodiazepine abuse have been found in alcoholics who have experienced failure in treatment programs for alcohol abuse. Clinical experience suggests that benzodiazepine abuse occurs with the greatest frequency in alcoholics with severe dependence and in alcoholics who abuse multiple types of drugs.

Individuals with a history of either alcohol abuse or alcohol dependence often have anxiety disorders. The issue of treating alcoholics with benzodiazepines is complex because some of these patients can take the medications without abusing them or relapsing to alcohol use whereas others take them in higher than prescribed doses and find that their desire to drink alcohol is increased.

SUMMARY

A large number of benzodiazepines are available for clinical use. These agents all share a set of pharmacologic properties that result from enhanced chloride flux at the $GABA_A$-receptor complex, which in turn results in the inhibition of neuronal activity in many regions of the central nervous system. Differences in activity among the benzodiazepines appear to be related primarily to differences in rates of absorption and metabolism, although recent research has suggested that intrinsic activity at benzodiazepine receptor subtypes also may influence drug effects. These drugs have been used extensively to treat anxiety, insomnia, seizures, and other disorders. They are safe and effective and their use has rarely been associated with irreversible adverse effects. Both physical and psychological dependence may be problematic for some individuals who are treated on a long-term basis with these agents or who have abused alcohol or other drugs.

(SEE ALSO: *Addiction: Concepts and Definitions; Benzodiazepines: Complications; Sleep, Dreaming, and Drugs*)

BIBLIOGRAPHY

AMERICAN PSYCHIATRIC ASSOCIATION. (1994). *Diagnostic and statistical manual of mental disorders-4th ed.* Washington, DC: Author.

CIRAULO, D. A., & GREENBLATT, D. J. (in press). *Sedative-, hypnotic-, or anxiolytic-related disorders.* Baltimore: Williams & Wilkins.

COLE, J. O., & CHIARELLO, R. J. (1990). The benzodiazepines as drugs of abuse. *Journal of Psychiatric Research, 24*(Suppl. 2), 135–144.

KARDOS, J. (1993). The GABA-A receptor channel mediated chloride ion translocation through the plasma membrane: New insights from 36 Cl-ion flux measurements. *Synapse, 13,* 74–93.

PRESTON, K. L., ET AL. (1992). Subjective and behavioral effects of diphenhydramine, lorazepam and methocarbamol: Evaluation of abuse liability. *Journal of Pharmacology and Experimental Therapeutics, 262*(2), 707–720.

RALL, T. W. (1990). Hypnotics and sedatives: Ethanol. In A. G. Gilman et al. (Eds.), *Goodman and Gilman's the pharmacological basis of therapeutics,* 8th ed. New York: Pergamon.

ROACHE, J. D., & GRIFFITHS, R. R. (1989). Diazepam and triazolam self-administration in sedative abusers: Concordance of subject ratings, performance and drug self-administration. *Psychopharmacology, 99,* 309–315.

RUBEN, S. M., & MORRISON, C. L. (1992). Temazepam misuse in a group of injecting drug users. *British Journal of Addiction, 87,* 1387–1392.

SELLERS, E. M., ET AL. (1993). Alprazolam and benzodiazepine dependence. *Journal of Clinical Psychiatry, 54*(Suppl. 10), 64–75.

ZORUMSKI, C. F., & ISENBERG, K. E. (1991). Insights into the structure and function of GABA-Benzodiazepine receptors: Ion channels and psychiatry. *American Journal of Psychiatry, 148,* 162–173.

DOMENIC A. CIRAULO
CLIFFORD KNAPP

BENZODIAZEPINES: COMPLICATIONS

As medicines, BENZODIAZEPINES have been widely used—as tranquilizers, to allay anxiety. Until the 1990s, they were believed to be both effective and extremely safe; however, beginning in the early 1980s, problems with these drugs started to become evident. Currently, the medical profession in many countries is trying to inculcate a cautious attitude toward their prescription and use. Lay people and the media have also become increasingly critical of the widespread use of these medicines for apparently trivial indications. To understand these problems, some aspects of the different types and effects of these medicines will be outlined.

WHAT ARE BENZODIAZEPINES?

These medicines are used to lessen a patient's anxiety; they include such drugs as CHLORDIAZEPOXIDE (Librium), diazepam (Valium), lorazepam (Ativan) and oxazepam (Serenid; Serox). The term *benzodiazepine* describes a basic chemical structure. Some, like diazepam, are long acting and can be taken once daily; others, like lorazepam and alprazolam (Xanax), need to be taken more often. Most sleeping tablets (hypnotics) are benzodiazepines, and these include short-acting drugs such as triazolam (Halcion), medium-acting drugs such as temazepam (Restoril), and long-acting drugs such as flurazepam (Dalmane) and nitrazepam (Mogadon).

Other medicines are used in psychiatry, such as ANTIDEPRESSANTS, ANTIPSYCHOTICS, and lithium. These have effects that differ from the benzodiazepines, effects both therapeutic and unwanted.

WHAT DO BENZODIAZEPINES DO?

Tranquilizers promote calming, soothing, and pacifying—without sedating or depressant effects. They are effective in lessening ANXIETY whatever its context. Thus, they are useful in treating *generalized anxiety*, which is often quite severe and comes on without apparent cause. Tranquilizers can also be used to deaden the upset of *normal anxiety*, the anxiety felt by people under stress, feeling threatened by life's problems. In these instances, the reasons for feeling anxious are clear, the degree of anxiety seems in line with the stress experienced—but despite this, help is sought for the symptoms. Unfortunately, the borderline between the medical disorder of clinical generalized anxiety and the normal response to stress is not always clear. The professional consulted will usually try to make the distinction and avoid using tranquilizers to treat people upset by adverse cir-

cumstances—thereby "medicalizing" everyday social and personal problems.

Similar considerations apply to the use of benzodiazepines as sleeping tablets. Short-term use of these drugs—for example, for disturbed sleep with jet travel across time zones, severe stress, or shift work—is generally accepted. Long-term use in the chronically poor sleeper is not usually encouraged, however.

Benzodiazepines can also be used as sedatives before surgical operations, as light anesthetics during operations, and to lessen muscle spasms, such as occur with sports injuries. Some benzodiazepines can be used to treat some forms of epilepsy. Benzodiazepines are prescribed mainly by general family practitioners, although they vary greatly in how often they use these medicines. Some still prescribe them widely, some hardly at all. Other doctors who use benzodiazepines include psychiatrists, orthopedic specialists, and gynecologists.

HOW MUCH ARE THEY USED?

An international survey at the beginning of the 1980s showed that tranquilizers and sedatives of any type had been used at some time during the previous year by 12.9 percent of U.S. adults, 11.2 percent in the United Kingdom (U.K.), 7.4 percent in the Netherlands, and 15.9 percent in France. Persistent long-term users comprised 1.8 percent of all U.S. adults, 3.1 percent in the U.K., 1.7 percent in the Netherlands and 5.0 percent in France. The proportion of repeat prescriptions for tranquilizers has increased steadily since about 1970 in many countries, the U.K. in particular. This suggests that fewer people are being newly started on tranquilizers but that a large group of long-term users is accumulating. People starting tranquilizers have at least a 10 percent chance of going on to long-term use, that is for more than 6 months. Some of these chronic users have chronic medical or social problems, and the tranquilizer blunts the unpleasant feelings of tension, anxiety, insomnia and, to a lesser extent, depression.

UNWANTED (SIDE) EFFECTS

Side effects are reactions to drugs that are not therapeutic or helpful, and they are therefore unwanted. The most common side effects from taking benzodiazepine are drowsiness and tiredness, and they are most marked within the first few hours after large doses. Other complaints of this type include dizziness, headache, blurred vision, and feelings of unsteadiness. The elderly are particularly sensitive to tranquilizers and may become unsteady on their feet or even mentally confused.

The feelings of drowsiness are, of course, what is wanted with a sleeping tablet. With the longer-acting benzodiazepines and with higher doses of medium-duration or with short-acting drugs, drowsiness can still be present the morning after taking a sleeping tablet; the drowsiness may even persist into the afternoon. The elderly are more likely to experience such residual, or "hang-over," effects.

As well as these feelings of sedation, special testing in a psychology laboratory indicates that alertness, coordination, performance at skilled work, mental activities, and memory can all be impaired. Patients should be warned about this, and advised not to drive or operate machinery, at least initially until the effects of the benzodiazepine can be assessed and the dosage adjusted if necessary. If driving is essential—that is, to the patient's livelihood—small doses of the benzodiazepines should be taken at first and the amount built up gradually under medical supervision. Judgment and memory are often impaired early in treatment, so important decisions should be deferred.

As with many drugs affecting the brain, benzodiazepines can interact with other drugs, especially ALCOHOL. People taking tranquilizers or hypnotics should not also drink alcoholic beverages. Other drugs whose effects may be enhanced include antihistamines (such as for hay fever), painkillers, and antidepressants. Cigarette smoking may lessen the effect of some benzodiazepines.

Patients taking benzodiazepines may show so-called paradoxical responses—that is to say, the effects produced are the opposite of those intended. Feelings of anxiety may heighten rather than lessen, insomnia may intensify, or, more disturbing, patients may feel hostile and aggressive. They may engage in uncharacteristic criminal activities, sexual improprieties or offenses such as importuning or self-exposure, or show excessive emotional responses such as uncontrollable bouts of weeping or giggling. All these are signs of the release of inhibitions, and they are also characteristic of alcohol effects in some people. Although these paradoxical effects may not last long, it is better to stop the benzodiazepine.

Benzodiazepines can affect breathing in individuals who already have breathing problems, such as with bronchitis. Other side effects that may be occasionally encountered include excessive weight gain, rash, impairment of sexual functioning, and irregularities of menstruation. Benzodiazepines should be avoided during pregnancy whenever possible, as there may be a risk to the fetus. Given during childbirth, benzodiazepines pass into the unborn infant and may depress the baby's breathing after birth. They also pass into the mother's milk and may sedate the suckling baby too much. Many people have taken an overdose of a tranquilizer as a suicidal attempt or gesture. Fortunately, these drugs are usually quite safe and the person wakes up unharmed after a few hours' sleep.

SIDE EFFECTS VERSUS MAIN EFFECTS

There are more subtle side effects of benzodiazepines, effects that interfere in various ways with the treatment of the anxiety or sleep disorder. The benzodiazepine lessens the symptom but does not alter the underlying problem—say, an unhappy marriage or a precarious job. Indeed, by lessening the symptoms, the individual may lose his or her motivation to identify, confront, and tackle the basic problems. Giving benzodiazepine medicalizes the problem by making the nervous or sad person into a patient, implying that there is something physically wrong. Finally, some events like bereavement need "working through"—typically by grieving—but benzodiazepines can stop this normal process and actually prevent the bereaved individual from coming to terms with loss.

LONG-TERM EFFECTS OF BENZODIAZEPINES

It is not clear whether benzodiazepines and hypnotics continue to be effective after months or years of daily use. Undoubtedly, many patients believe that they continue to benefit in being less anxious, or in sleeping better. The effect of the drug may be more to stop the anxiety or insomnia that follows withdrawal, however, than to combat any continuing, original anxiety. Most of the side effects lessen over time, a process known as *tolerance*. Some impairments, however, such as memory disturbances, may persist indefinitely, but patients usually come to

terms with this—for example, by resorting to written reminders.

REBOUND

Rebound occurs when stopping the drug makes the underlying condition worse. Most is known about rebound in insomnia. Sleeping tablets may improve sleep by inducing it more rapidly, making it sounder, and prolonging it. When the sleeping tablet is stopped, rebound may occur on the following night or two, with the insomnia being worse than ever. Eventually, the rebound insomnia subsides, but the patient may have been so distressed as to resume medication, thereby running the risk of indefinite use. The risk of rebound is greatest with short-acting benzodiazepines, especially in higher dose.

A similar problem follows stopping a daytime tranquilizer, particularly lorazepam. Anxiety and tension rebound to levels higher than those experienced on treatment and often higher than the initial complaints. Tapering off the tranquilizer over a week or two lessens or avoids this complication. Rebound may even be seen in the daytime between doses of tranquilizer. The patient, increasingly anxious as the effect of the earlier dose wears off, watches the clock until his or her next dose is due. Rebound may also occur later in the day after taking a short-acting sleeping tablet the night before.

WITHDRAWAL

In withdrawal, symptoms occur which the patient has not previously experienced. They come on a day or two after stopping alprazolam or lorazepam, after a week or so on stopping diazepam or chlordiazepoxide. The symptoms rise to a crescendo and then usually subside over two to four weeks. In an unfortunate few, the symptoms seem to persist for months on end—sometimes called the *post-withdrawal syndrome*. The existence of this condition is disputed by some doctors, who ascribe the symptoms to return of the original anxiety for which the drug was given.

Patients commonly experience bodily symptoms of anxiety such as tremor, palpitations, dry mouth, or hot and cold feelings. Insomnia is usually marked. Some complain of unpleasant feelings of being out of touch with reality or with their own bodies. Severe headaches and muscle aches and pains can occur, sleep is greatly disturbed, appetite is lost as is several

pounds of weight. Disturbances of perception are characteristic of benzodiazepine withdrawal and include intolerance to loud noises or bright lights, numbness or pins and needles, unsteadiness, a feeling of being in motion (as on a ship at sea), and a sensing of strange smells and tastes. Some people become quite depressed; rarely, some experience epileptic fits or a paranoid psychosis (with feelings of persecution and loss of contact with reality).

HOW BIG IS THE PROBLEM?

The withdrawal symptoms are evidence of physical dependence—that is, the body has become so used to the effects of the benzodiazepine that it cannot manage without. About a third of long-term (over a year) steady users show withdrawal, even when the tranquilizer or hypnotic is tapered off. Some users have tried to stop and have encountered problems. Many others have never tried to stop and so are unaware whether they are dependent. Because these people continued to take the doses prescribed by their doctors, the medical profession was reluctant for a long time to admit the scale of the problem—perhaps 500,000 people dependent on tranquilizers in the U.K. alone. In addition, the similarity between some withdrawal symptoms and features of the original anxiety has led to confusion in the mind of both the patient and the doctor. True withdrawal symptoms, however, arise at a predictable time after stopping the benzodiazepine and are new experiences for the patient; the old anxiety and insomnia symptoms are familiar to the patient and may return at any time, depending on external stresses.

HOW TO WITHDRAW

Essentially, the patient must be prepared for withdrawal by being told what to expect; he or she should be taught other ways of combatting anxiety; and withdrawal should be by graded tapering off the dose over six to twelve weeks, occasionally longer. Many people experience little or no upset, a few undergo much distress. Sometimes substituting diazepam in the lorazepam or alprazolam user helps. Antidepressants may be needed if the patient becomes very depressed, but by and large, other drugs are unhelpful.

Family and social support is essential. Usually the family doctor can supervise the withdrawal quite safely, but occasionally specialist advice is sought. A self-help group may provide useful continued advice and support.

It is important that tranquilizers are never stopped abruptly. There is a greatly increased risk of severe complications such as seizures or convulsions.

ABUSE OF TRANQUILIZERS

Only a few patients prescribed benzodiazepines push the dose up above recommended levels. If this happens, the user may become intoxicated, with slurred speech and incoordination. Some people with alcohol problems also abuse benzodiazepines. Intravenous (IV) injection of benzodiazepines and hypnotics has become an increasing problem and has led to controls on these drugs concerning manufacture and prescription in various countries, including the United States and the U.K. Some addicts abuse benzodiazepines alone; others combine it with heroin-type drugs. Injection of benzodiazepines can result in clotting of the veins. It also carries the risk of getting infectious diseases from sharing dirty syringes, such as hepatitis and the human immunodeficiency virus (HIV or the AIDS virus).

ALTERNATIVES TO THE TRANQUILIZERS

Dissatisfaction with the benzodiazepine tranquilizers and hypnotics has led to numerous initiatives to find better alternatives. Some drugs have been developed that are better benzodiazepines, in that they are less sedative and perhaps less likely to induce dependence. Others are chemically not benzodiazepines but share many of their properties, both therapeutic and unwanted. Other compounds seem to act in a totally different way in the brain and are less sedative and probably much less likely to induce dependence. One such compound—buspirone (Buspar)—has been available for a few years, but many others are in the process of development. Finally, interest has been rekindled in the use of other types of older drugs to treat anxiety; examples include the antihistamines and the beta blockers.

Problems with the benzodiazepines has led to a reevaluation of the whole role of prescribed medicines in the management of anxiety, insomnia, and stress-related disorders. Numerous nondrug methods have been developed and improved, among them relaxation training; cognitive therapy, in which pa-

tients learn to think less anxious thoughts; behavior therapy, in which the patient learns to confront stressful situations; and sleep counseling. Alternative medicine, like ACUPUNCTURE, is enjoying a vogue and helps some anxious people.

CONCLUSIONS

Hailed as wonder drugs, prescribed widely and for long periods of time, the benzodiazepines have now been shown to be problematic medicines with undoubted benefits but definite risks. For short-term treatment in the severely anxious and sleepless, they are still useful—although other drugs are beginning to supplement and even supplant them. For the bulk of anxious people, though, nondrug treatments are increasingly popular.

(SEE ALSO: *Addiction: Concepts and Definitions; Complications; Iatrogenic Addiction; Sleep, Dreaming, and Drugs; Tolerance and Physical Dependence; Withdrawal*)

BIBLIOGRAPHY

AMERICAN PSYCHIATRIC ASSOCIATION. (1990). *Task force report: Benzodiazepine dependence, toxicity, and abuse.* Washington, DC: Author.

CURRAN, H. V., & GOLOMBOK, S. G. (1985). *Pill popping: How to get clear.* Boston: Faber & Faber.

WOODS, J. H., KATZ, J. L., & WINGER, G. (1987). Abuse liability of benzodiazepines. *Pharmacological Reviews, 39*(4), 251–419.

MALCOLM H. LADER

BENZOYLECOGNINE Cocaine is metabolized by plasma and liver enzymes (cholinesterases) to water-soluble metabolites that are excreted in the urine. The two major metabolites are benzoylecognine and ecognine methyl ester, with only benzoylecognine reported to have behavioral activity. Since COCAINE has a relatively short half-life and may only be present in the urine for twenty-four to thirty-six hours, benzoylecognine levels in urine are useful markers of cocaine use, because it is present for a longer time in urine, two to four days, depending on the quantity of cocaine ingested. Assays for this metabolite are frequently employed in treatment programs, to evaluate compliance with the program, and

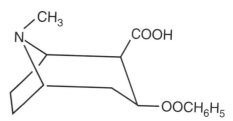

Figure 1
Benzoylecognine

in workplace drug testing to indicate cocaine use. Under these conditions, it is important to keep in mind that benzoylecognine in the urine is an indication of prior cocaine use, but reflects neither current use nor impairment.

(SEE ALSO *Cocaethylene; Drug Testing and Analysis*)

MARIAN W. FISCHMAN

BETA BLOCKERS FOR ALCOHOL WITHDRAWAL SYNDROME See Withdrawal: Alcohol, Beta Blockers.

BETEL NUT Betel nut, the seed of the betel palm (*Areca catechu*), is one of the most widely used substances in areas of the western Pacific and parts of Africa and Asia. It is prepared with other substances as a mixture for chewing and is used as a mild stimulant by more than 200 million people.

References to betel nut appear in ancient Greek, Sanskrit, and Chinese texts from more than a cen-

Figure 1
Betel Palm and Betel Nut

tury B.C. Ancient historic documents of Ceylon refer to its use, and its prevalence in Persia by 600 A.D. is documented by Persian historians. Its use in different parts of the Arab world by the eighth and ninth centuries is also well documented, and it had become an important aspect of the economy and social life in India, Malaysia, the Philippines, and New Guinea. Betel was probably brought to Europe by Marco Polo, around 1300; it soon proved to be an important commodity in the western Pacific and a source of tax revenue for the Dutch in the mid-1600s.

PATTERNS OF USE

The mottled brown-and-gray betel seeds are gathered before they ripen during the period between August and November. They are boiled in water, cut into slices and dried in the sun, becoming dark brown or reddish in color. This betel seed, or nut, becomes the primary ingredient in the betel nut chewing mixture ("quid"), which is made up of several ingredients. While the component substances differ in different parts of the world, all the preparations contain fresh chunks or dried powdered forms of the betel nut. Mixtures prepared for children frequently contain only the husk of the nut, while the full strength form for adults always has the nut itself.

The second ingredient of the quid is usually a form of peppermint or mustard, or the leaf, bean and/or bark of the shrub-like or climbing pepper plant vine (*Piper betel*). The third component is slaked lime, which is usually produced from limestone or by burning sea shells or coral stones in the presence of water. This process produces calcium hydroxide, usually used as a white powder. While all betel-nut quids contain some of the three main components, other ingredients, such as spices, dyes, and aromatics are frequently added. In India, tobacco is mixed with the quid. The combination of nut, mustard or vine, lime, and other ingredients create an alkaline, bitter-tasting mixture that is chewed, forming a red paste which stains the teeth, mouth, gums, and lips, and generating large amounts of saliva. Like tobacco chewers, betel-nut chewers spit out the excess juices.

Some habitués chew betel nut all day long. Others use it as part of social custom, not unlike the use of KAVA, or of the consumption of ALCOHOL in Western

countries. In some areas of the world, such as Papua New Guinea, betel-nut mixtures are often offered as a before-dinner "appetizer" or as an after-dinner treat. Like kava, in many places, sharing betel-nut mixture is important in courtship and marriage customs and in establishing friendships.

ACTIVE INGREDIENTS

The major active ingredient of betel nut is arecoline, present in a concentration estimated to be 0.25 percent. The mixture also contains small amounts of pilocarpine and muscarine. These three ingredients are all natural plant products that act in the body in a manner similar to the normal brain NEUROTRANSMITTER acetylcholine. In the presence of calcium hydroxide, arecoline is also converted to another psychoactive substance, aredaidine.

Chewing betel nut produces immediate effects that in some ways resemble those of NICOTINE, but which are likely to continue for hours. These include euphoria and feelings of general arousal and activation, perceived by the user as a decrease in tiredness and a blunting of feelings of irritability.

Other prominent effects are also related to the acetylcholine-like actions. These include sweating, increased production of saliva, and an increase in breathing rate and lacrimation (tearing of the eyes). Effects on the digestive tract include a decrease in appetite and, especially if the drug is taken on an empty stomach, diarrhea. All these effects can be blocked by atropine, a type of anti-acetylcholine drug.

Some of the active ingredients in betel nut are used in modern pharmacological treatment in the Western world. Betel-nut preparations have been used in Western society as a purgative and in veterinary medicine as an agent for treating worm infestation in animals. Probably the most interesting use appears around 1842, when betel nut was included in toothpastes in England. It was touted as an important way to prevent decay, a claim that may or may not be accurate, although much larger doses than those found in toothpastes would be required to really be clinically effective. It was also said that this ingredient would help strengthen the tooth enamel and remove tartar, claims of questionable value. In view of the fact that betel, as it is commonly used, stains the teeth dark red to black, is thought to cause tooth decay, and can cause serious lesions of

the mouth and throat, it is curious that it should have appeared in Western society in preparations for dental care.

SOME DANGERS

The acetylcholine-type drugs, especially muscarine, can be deadly when taken in high doses. In fact, muscarine is the active ingredient causing some forms of lethal mushroom poisoning, but it is unlikely that the mixture of any of the plant products in betel-nut preparations are potent enough to cause lethal overdose. Regular "recreational" use of betel nut is, however, responsible for a number of adverse health consequences that can contribute to the risk for early death.

The most prominent dangers associated with betel-nut chewing are probably the result of a combined effect of the active ingredients and the lime on the gums. The first and most frequently observed physical changes are white plaques appearing on the mucosal lining of the mouth or on the tongue. These are precancerous lesions (leukoplakia) that often lead to the development of very aggressive and serious tumors (squamous cell carcinoma), which can subsequently invade muscles and bone tissue. The prevalence of this cancer among regular betel-nut users is estimated to be as high as 7 percent. Potentially lethal cancers may also develop in the esophagus. Chronic use may also cause oral submucous fibrosis—a form of fiber formation (fibrosis) that usually starts just beneath the gums and may involve the back of the throat and the pharynx. The problem is estimated to be seen in at least mild form in up to 50 percent of chronic betel-nut chewers. This condition usually has a very slow onset, and if use continues it is irreversible, untreatable, and likely to become progressively more severe. The major finding involves a loss of elasticity of the tissue lining the mouth, which causes stiffness that can become so severe as to interfere with eating. Associated problems are a burning sensation in the mouth, ulcers or blisters on the lining of the mouth, decreased sense of taste, and dryness of the mouth lining.

There is little doubt that betel-nut substance can produce fairly intense psychological dependence. Individuals can develop a pattern of constant use, feeling unhappy and incomplete if they cannot get their betel nut. They are also likely to feel they cannot work properly without it, and may spend a great deal of money and time obtaining and using betel-nut mixtures. It is not clear, however, that there is a prominent and identifiable form of physical withdrawal associated with cessation of use.

Betel-nut consumption can be viewed as a public-health hazard in parts of the world where its use is prevalent, because, at least theoretically, the habit of spitting the juice on the street can increase the spread of diseases such as tuberculosis.

(SEE ALSO: *Plants, Drugs from*)

BIBLIOGRAPHY

BEECHER, D., ET AL. (1985). Betel nut chewing in the United States. *Journal of the Indiana Dental Association, 64*, 42–44.

FORD, C. S. (1967). Ethnographical aspects of kava. In D. H. Efron, B. Holmstedt, & N. S. Kline, (Eds.), *Ethnopharmacologic search for psychoactive drugs.* PHS Publ. No. 1645. Washington, DC: U.S. Department of Health, Education, and Welfare.

LEWIN, L. (1964). *Phantastica: Narcotic and stimulating drugs.* New York: Dutton.

SCHUCKIT, M. A. (1992). Betel Nut: A widespread drug of abuse. *Drug Abuse & Alcoholism Newsletter, 21*(1). San Diego: Vista Hill Foundation.

SCHULLIAN, D. M. (1984). Toothpastes containing betel nut from England of the 19th century. *Journal of History of Medicine, 39*, 65–68.

TALONU, N. T. (1989). Observations of betel nut use, habituation, addiction and carcinogenesis. *Papua New Guinea Medical Journal, 32*, 195–197.

TAUFA, T. (1988). Betel nut chewing and pregnancy. *Papua New Guinea Medical Journal, 31*, 229–233.

MARC A. SCHUCKIT
JEROME H. JAFFE

BETTY FORD CENTER This eighty-bed hospital for recovery from chemical dependency consists of six buildings in a landscaped setting. It was named in honor of President Gerald Ford's wife, who was treated successfully and who promotes such therapy. The center is located southeast of Palm Springs, California, on the campus of the Eisenhower Medical Center, an acute-care hospital.

The Betty Ford Center's administration building contains reception, a lecture hall, dining hall, bookstore, administrative offices, and a serenity room where patients can meditate or pray. Additionally, there are four twenty-bed residential and therapeutic units, each of which includes a living room, lounge, group-therapy rooms, and patient bedrooms.

Treatment is carried out within the patient's assigned residential unit. For family treatment, children's aftercare, outpatient, and alumni programs, the center uses the Cork Family Pavilion's therapy rooms, a classroom/lecture hall, therapy offices for one-to-one conferences, an alumni office/lounge, a staff library and a minigymnasium.

The staff at the center views ALCOHOLISM and other drug dependencies as chronic progressive diseases that will be fatal if they are not treated. With proper treatment, however, recovery rates are high. The program at Betty Ford is designed so that patients learn to become responsible for their own actions and recovery. Following any hospitalization for detoxification or other medical complications, patients are admitted to the center to begin a series of clinical assessments that verify the diagnosis of chemical dependency and to design individual treatment programs.

Chemical dependency affects all those who have a close relationship with the patient, and it is therefore important that all family members be included in treatment. To fill this need, the center has created the family-treatment program, a five-day intensive process that includes education and individual and group therapy.

The center's staff is addressing the fact that women have traditionally been hidden chemically dependent people, so their treatments for women differ from those for men. Prior to discharge from the center, the patient develops an aftercare plan with a counselor.

Betty Ford's observation has become a guiding force for the center: "Since I began practicing abstinence, I have become more valuable to myself and to others."

BIBLIOGRAPHY

BETTY FORD CENTER. (1994). *Brochure.* Palm Springs, CA: Author.

RONALD R. WATSON

BHANG   This is one of the many names given to the HEMP plant, *Cannabis sativa*, and its products. Bhang is of Hindi origin (from *bhāg*, which came into English about 1563) and refers to the leaves and flowering tops of uncultivated hemp plants. In 1895, the Indian Hemp Commission took the position that bhang was not a major health hazard. Bhang is taken in a beverage in India called *thandaii*, may be served in sweetmeats, or is used in making ice cream. It is often served at weddings or religious festivals and is freely available from sidewalk stands in the major cities. Generally, in India, the use of bhang and other *Cannabis* products has been considered lower class. Probably as a result of continuing British-based influence, the upper-class drugs are ALCOHOL and OPIUM.

(SEE ALSO: *Cannabis sativa; Marijuana; Plants, Drugs from*)

LEO E. HOLLISTER

BLOOD ALCOHOL CONCENTRATION, MEASURES OF   The first analytical methods for measuring ALCOHOL (ethanol) in blood and other body fluids were developed in the nineteenth century. Although by modern standards these pioneer efforts were fairly crude, they were sufficiently reliable to establish a quantitative relationship between blood-alcohol concentration (BAC) and the various signs and symptoms of inebriation. A significant advance in methodology came in 1922 when Erik M. P. Widmark published his micromethod for analyzing ethanol in specimens of capillary blood.

Blood was drawn by pricking a fingertip or earlobe. The specimen for analysis, 100–150 milligrams, was collected with specially prepared S-shaped glass capillaries that contained a thin film of potassium oxalate and sodium fluoride on the walls of the tube. In Widmark's day, the small amounts (aliquots) of blood needed for each analysis could be measured more accurately by weight than by volume, since constriction pipettes were not yet available. Widmark therefore weighed the amount of blood required to the nearest milligram (0.001 g) with the aid of a torsion balance. The results of ethanol determinations were then reported in terms of mass per mass units, actually milligram of ethanol per gram of whole blood (mg/g), sometimes referred to as per mille (meaning, parts per thousand). This

way of reporting BAC survives today in Scandinavian countries where Widmark's method became widely used for legal purposes.

Widmark's micromethod of blood-ethanol analysis involved the following four steps: (1) separation of ethanol from blood by diffusion in specially blown glassware; (2) oxidation of ethanol with a mixture of potassium dichromate and sulfuric acid; (3) addition of potassium iodide to the reaction mixture after oxidation of ethanol; and (4) back titration of liberated iodine with standard sodium thiosulfate and a starch indicator to detect an endpoint.

Many later modifications to this basic procedure appeared, such as using a different endpoint indicator for the titration (e.g., methyl orange), or another kind of oxidizing agent (e.g., ferrous salts), or separation of ethanol from the biological matrix in a different way. It became common practice to refer to these modified methods with the name of the scientist who first published the report—including Harger, Kozelka-Hine, Smith, Southgate-Carter, and Cavette, to name just a few.

Later developments in methods of ethanol analysis (such as gas chromatography) plus the availability of modern clinical laboratory equipment made it more convenient to dispense the aliquots of blood needed for analysis by volume rather than by weight. Micropipettes and more recently diluter-dispenser devices are now widely used for dilution of blood prior to the analysis. The term "concentration" has little meaning when used alone, because it can be expressed in many different ways. The choice of units for reporting BAC differs among countries: for example, milligrams per hundred milliliters (mg/100 ml) in Great Britain (unfortunately often appearing as the ambiguous mg%); gram percent weight per

volume (g% w/v) in the United States; and milligrams per milliliter (mg/ml) in many European countries. Other ways of reporting BAC in clinical medicine are milligrams per deciliter (mg/dl), grams per liter (g/liter), or micrograms per liter (μg/liter). When countries outside Scandinavia enacted legal limits of ethanol in the blood of motorists, the concentrations were defined in units of mass of ethanol per unit volume; whether it was grams, milligrams, or micrograms of ethanol in a volume of milliliters, deciliters, or liters seems chosen arbitrarily.

Because the specific gravity of whole blood is greater than water (on average, 1 ml of whole blood weighs 1.055 g), BAC expressed in terms of mass per mass (w/w) is not the same as mass per volume (w/v). In fact, a concentration of 0.10% w/v equals 0.095% w/w. This difference of about 5.5 percent could mean punishment or acquittal in borderline cases of driving while under the influence of alcohol. With the current trend toward "per se" ethanol limits in many U.S. states, great care is needed to ascertain whether w/v or w/w units were intended by the legislature when the statute was drafted. Table 1 gives examples of concentration units commonly used to report BAC for legal purposes. Note that if ethanol were determined in plasma or serum, the concentration would be about 10 to 15 percent higher than for the same volume of whole blood, because there is more water in the sample after the erythrocytes (red blood cells) are removed.

In clinical chemistry laboratories, the Système International d'Unités (SI) has gained worldwide acceptance. According to the SI system, the amount of substance implies "mole" rather than mass. The mole or a submultiple thereof replaces mass units such as grams or milligrams. Accordingly, the con-

TABLE 1
Concentrations of Alcohol (Ethanol) in Whole Blood for Legal Purposes

Concentration Unit	Country	Legal Limit
Percent weight/volume (% w/v)	United States*	0.10 g/100 ml
Milligrams per 100 milliliter (mg/dl)	Britain	80 mg/100 ml
Milligrams per milliliter (mg/ml)	Netherlands	0.50 mg/ml
Milligrams per gram (mg/g)	Sweden**	0.20 mg/g
Milligrams per gram (mg/g)	Norway**	0.50 mg/g

*The Uniform Vehicle Code of the National Committee on Uniform Traffic Laws and Ordinances recommends 0.08 grams per 100 milliliters of blood or per 210 liters of breath; at least six U.S. states have adopted this recommendation.
**1 milliliter whole blood weighs 1.055 grams.

TABLE 2
Concentrations of Alcohol (Ethanol) in Breath for Legal Purposes

Concentration Unit	Country	Legal Limit
Grams per 210 liters (g/210 l)	United States*	0.10 g/210 l
Micrograms per 100 milliliter (μg %)	Britain**	35 μg/100 ml
Micrograms per liter (μg/l)	Netherlands**	220 μg/l
Milligrams per liter (mg/l)	Sweden*	0.10 mg/l
Milligrams per liter (mg/l)	Norway*	0.25 mg/l

*Blood/breath ratio of ethanol is assumed as 2,100:1.
**Blood/breath ratio of ethanol is assumed as 2,300:1.

centration of a substance of known molecular weight might appear as mole/liter or millimole per liter (mmol/l) or micromole per liter (μmol/liter). Note that liter is the preferred unit of volume when reporting concentrations of a substance in solution in the SI system. The molecular weight of ethanol is 46.06, and therefore a concentration of 1.0 mol/l corresponds to 46.06 g of ethanol in 1 liter of solution. Likewise 1.0 mmol/l contains 46.06 mg; 1.0 μmol/l contains 46.06 μg, and so on. Publications in the field of biomedical alcohol research often report BAC in this way. It follows that 0.1 g% w/v or 100 mg/dl is the same as 21.7 mmol/l.

Statutory limits of BAC existed in several countries before methods of analyzing the breath were developed. It therefore became a standard practice to convert the concentration of ethanol measured in the breath (BrAC) into the presumed concentration in the blood. For this purpose, a conversion factor, usually 2,100:1 was used. Presumably, it was less troublesome to make this conversion than to rewrite the statute to include both BAC and BrAC as evidence of impairment. Accordingly, breath-ethanol analyzers were calibrated in such a way that the readout was obtained directly in terms of the presumed BAC. This conversion of breath to blood ethanol created the dilemma of a constant blood breath ratio existing for all subjects under all conditions of testing. In the United States and elsewhere, a blood/breath factor of 2,100:1 was approved for legal purposes with the understanding that this gives a margin of safety (about 10%) to the

TABLE 3
Effects of Blood Alcohol Levels

BAL	(BAC)	Effects
50	(0.05%)	There may be no observable effects on behavior, but thought, judgment, and restraint may be more lax and vision is affected. Significantly more errors in tasks that require divided attention; more steering errors; and increased likelihood of causing an accident.
80	(0.08%)	Reaction time for deciding and acting increases. Motor skills are impaired. The likelihood of a crash increases to three to four times the likelihood when sober.
100	(0.10%)	Six times as likely to be involved in a crash. Reaction time to sights and sounds increases. Physical and mental coordination are impaired; movement becomes noticeably clumsy.
150	(0.15%)	Twenty-five times as likely to be involved in a crash. Reaction time increases significantly, especially in tasks that require divided attention. Difficulty performing simple motor skills. Physical difficulty in driving.
200	(0.20%)	One hundred times as likely to be involved in a crash. Motor area of brain significantly depressed, and all perception and judgment distorted. Difficulty standing, walking, and talking. Driving erratic.
300	(0.30%)	Confusion and stupor; inability to track a moving object with the eyes. Passing out is likely.
400	(0.40%)	Coma is likely.
450–500	(0.45–0.50%)	Death is likely.

SOURCE: Mothers Against Drunk Driving (MADD) and the National Safety Council.

accused. Indeed, more recent research suggests that the blood/breath factor should be 2,300:1 for closer agreement between direct BAC and the result derived from BrAC. In the Netherlands and Great Britain, 2,300:1 was chosen to set the legal limit of BrAC when evidential breath-ethanol analyzers were introduced in these countries. Similarly, in some U.S. states, a legal limit of 0.1 g/210 liters in breath is considered equivalent to 0.1 g% w/v in blood for law-enforcement purposes. Table 2 gives the statutory limits of breath-ethanol concentrations in several countries.

Both the prescribed BAC or BrAC limits for motorists and the units of concentration used differ among countries and even within regions of the same country. The notion of reaching an international agreement about one common BAC or BrAC limit for motorists is an attractive one but hardly attainable.

(SEE ALSO: *Blood Alcohol Content; Breathalyzer; Driving Under the Influence; Drug Testing and Analysis*)

BIBLIOGRAPHY

JONES, A. W. (1989). Measurement of alcohol in blood and breath for legal purposes. In K. E. Crow & R. D. Batt (Eds.), *Human metabolism of alcohol*, vol. I. Boca Raton, FL: CRC Press.

WIDMARK, E. M. P. (1922). Eine Mikromethode zur Bestimmung von Äthylalkohol im Blut. *Biochemical Zeitschrift, 131,* 473.

A. W. JONES

BLOOD ALCOHOL CONTENT

The consumption of alcoholic beverages results in the absorption into the bloodstream of ALCOHOL (ethanol, also called ethyl alcohol) from the stomach and small intestine. The amount of alcohol distributed in the blood is termed blood alcohol concentration (BAC) and is proportional to the quantity of ethanol consumed. It is expressed as the weight of alcohol in a fixed volume of blood, for example, grams per liter (g/l) or milligrams per deciliter (mg/dl). The measurement of blood alcohol concentrations has both clinical and legal applications.

Consuming food with alcohol generally decreases the amount of alcohol that can be quickly absorbed

into the bloodstream. Consuming more than one drink per hour causes the BAC to increase rapidly, because it is exceeding the rate at which the body can metabolize alcohol. The percentage of body fat that contributes to a person's total weight also affects BAC. A larger proportion of fat provides less body water into which the alcohol can distribute, thus increasing BAC. For this reason, women generally have a higher BAC for a given number of drinks when compared to men.

(SEE ALSO: *Blood Alcohol Concentration, Measures of*)

BIBLIOGRAPHY

FISHER, H., SIMPSON, R., & KAPUR, B. (1987). Calculation of blood alcohol concentration (BAC) by sex, weight, number of drinks and time. *Canadian Journal of Public Health, 78,* 300–304.

MYROSLAVA ROMACH
KAREN PARKER

BNDD *See* U.S. Government Agencies: Bureau of Narcotics and Dangerous Drugs.

BOLIVIA, DRUG USE IN

Bolivia is a land of gaunt mountains, cold desolate plains, and semitropical lowlands situated in the central part of South America. Straddling the Andes mountains, Bolivia's 424,165 square miles occupy an area about the size of Texas and California combined. It is a big country, but with a population of only 6.7 million. About 14 percent are of European heritage; 25 percent are Aymara Indians, 30 percent are Quechua Indians, and 31 percent are mestizos (mixed Indian and European ancestry). Although Bolivia is rich in mineral resources—petroleum, natural gas, tin, lead, zinc, copper, and gold—it is an economically depressed country. Most of the population works in agriculture, which is generally unrewarding, while a small number work in the mines. At the beginning of the 1990s Bolivia had a national debt of some 4.1 billion in U.S. dollars, a large sum for a nation in which the annual gross domestic product (GDP) barely exceeds 4 billion dollars.

Much of the Bolivian population lives on the bleak, treeless, windswept Altiplano (high plain), a

plateau more than 13,000 feet above sea level. The Altiplano is an arid expanse of red earth, of about 40,000 square miles, with widely scattered llamas, sheep, cattle, and homesteads. However, the Altiplano is considered to be the most livable part of the country, with 70 percent of the population residing along its western quarter. Much of the rest of the people live in the Yungas, the Chapare, and the Beni—the tropical jungles of northeastern and central Bolivia, where *Erythroxylum coca* thrives. *Erythroxylum coca*, or simple "coca," is the shrub from which COCAINE is derived (Inciardi, 1992).

COCA PRODUCTION

Historically, the chewing of coca leaves was a cultural pattern among the Indian peasant laborers of the Andes. The mild stimulation received from the low cocaine-content leaves enabled workers to endure the burdens of their 12- to 14-hour days in the mines and in the fields, so both Bolivian and Peruvian laws have permitted controlled production of coca for domestic consumption—about 12,000 kilograms (kg) in Bolivia (which also includes production for international pharmaceutical use). A part of the Bolivian economy has therefore always depended on the cultivation, transport, and sale of coca leaves.

The growers of illegal coca in Bolivia are the thousands of farm families who have shifted away from the cultivation and harvest of more traditional crops. Coca accounts for as much as 40 percent of Bolivia's agricultural production, about 50 percent of its gross domestic product, and about 67 percent of its export earnings. In addition, between 40,000 and 70,000 peasants produce coca, and some 500,000 Bolivians (about 20% of the working population) depend on coca for a livelihood, either directly or through support industries (Healy, 1988; Burke, 1991).

COCA PASTE USE

Not surprisingly, drug use in Bolivia is related to the production of coca and cocaine. Many people in Bolivia have tried coca products in one form or another. However, in Bolivia abuse of cocaine generally involves neither the chewing of coca leaves nor the ingestion of either powder-cocaine or CRACK-cocaine, but rather, the smoking of COCA PASTE—an intermediate product in the transformation of the coca leaf into pure cocaine. In jungle refineries, coca leaves are treated with a wide variety of chemicals, including alcohol, benzol (a petroleum derivative used in the manufacture of motor fuels and insecticides), sulfuric acid, leaded gasoline, sodium carbonate, and kerosene. The process yields crude cocaine (coca paste). Whereas the cocaine content of leaves is relatively low, 0.5 percent to 1 percent by weight, paste has a cocaine concentration ranging up to 90% (Inciardi, 1992).

Known to most South Americans as *basuco*, *susuko*, *pasta basica de cocaina*, or just simply *pasta*, coca paste is typically smoked straight or in cigarettes mixed with either TOBACCO or MARIJUANA. The smoking of coca paste became popular in Bolivia and other parts of South America beginning in the early 1970s (Jeri, 1984). Readily available and inexpensive, it had a high cocaine content and was absorbed quickly. As the phenomenon was studied, however, it was quickly realized that paste smoking was far more serious than any other form of cocaine use. In addition to cocaine, paste contains traces of all the chemicals used to process the coca leaves initially, the oxidized products of these solvents, plus any number of other alkaloids present in the coca leaf.

When the smoking of paste was first noted in South America, the practice seemed to be restricted to the coca-processing regions of Bolivia, Colombia, Ecuador, and Peru, appealing primarily to low-income groups—it was a cheaper price than refined cocaine. By the early 1980s, however, it had spread to other South American nations and to the various segments of the social strata; throughout that decade, paste smoking further expanded to become a major drug problem for much of South America. Although there have been no systematic studies of coca paste use in Bolivia, most observers report that it is concentrated among the impoverished youths of the country's many rural and urban shantytowns (Farah, 1989; Germani, 1988; Noya, 1989), where it contributes to other health-compromising conditions such as poor nutrition, sniffing of gasoline or other INHALANT drugs, and excessive use of alcoholic beverages. New population surveys being completed in Bolivia should help people in that country understand the nature and magnitude of their coca-related problems and help them devise preventive strategies.

(SEE ALSO: *Coca Plant; Colombia As Drug Source*)

BIBLIOGRAPHY

BURKE, M. (1991). Bolivia: The politics of cocaine. *Current History, 90*, 65–68, 90.

FARAH, D. (1989). Bolivia's cocaine trade is consuming its children. *Washington Post*, September 18.

GERMANI, C. (1988). Coca addiction hits home among rural children of drug-producing Bolivia. *Christian Science Monitor*, September 29.

HEALY, K. (1988). Bolivia and cocaine: A developing country's dilemmas. *British Journal of Addiction, 83*, 19–23.

INCIARDI, J. A. (1992). *The war on drugs II: The continuing epic of heroin, cocaine, crack, crime, AIDS, and public policy*. Mountain View, CA: Mayfield.

JERI, F. R. (1984). Coca-paste smoking in some Latin American countries: A severe and unabated form of addiction. *Bulletin on Narcotics*, 15–31.

NOYA, N. (1989). Cocaine crisis in Bolivia. Paper presented at *What works: An international perspective on drug abuse, treatment, and prevention research*, October 22–25, New York.

JAMES A. INCIARDI

BOOZE   *See* Distilled Spirits; Slang and Jargon.

BORDER MANAGEMENT   In 1977, a U.S. government interagency team led by the Office of Drug Abuse Policy (ODAP) conducted a comprehensive review of border control and recommended consolidation of the principal border-control functions into a single border-management agency. Executive departments failed to agree on distribution of resources and organizational placement of the new agency. The border-management agency never materialized.

Border control in the United States was described in the review as an extremely complex problem involving vast distances, many modes of transportation, millions of arrivals and departures, and millions of tons of cargo. Laws to be enforced involved illegal drugs and other contraband, terrorists, public-health threats, agricultural pests and diseases, endangered species, entry visas, duties, and so forth. Nine federal agencies shared border-control responsibilities, contributing to overlap, duplication of effort, and duplicated management systems.

The ODAP report recommended consolidating the inspection and patrolling functions, including operational and administrative support. The potential for improved effectiveness in a consolidated border-management agency was widely recognized. A similar report by the U.S. Congress's General Accounting Office (GAO) also recommended single-agency management and responsibility for border control. Controversy over which activities to include and which executive department should control the new agency was, however, effective in blocking further action.

(SEE ALSO: *Drug Interdiction; Opereation Intercept*)

BIBLIOGRAPHY

HAVEMANN, J. (1978). Carter's reorganization plans—scrambling for turf. *National Journal, 10*, (20), 788–794.

RICHARD L. WILLIAMS

BRAIN STRUCTURES AND DRUGS

Drugs that alter behavior do so by modifying the actions of brain cells normally involved in the control of these behaviors. Therefore, to understand the actions of abused drugs on the brain, one must have an understanding of the functions that brain cells serve in the expression of behavior in general. This article focuses on information that will assist readers in understanding the biological basis of drug actions on the brain, and particularly the actions of commonly abused drugs. First, the general classification of brain cells will be discussed, followed by a discussion of brain structure as it relates to function and drug action. The classification of brain cells based on the chemical nature of communication between cells will then be discussed as it relates to the actions of drugs of abuse.

CLASSIFICATION OF BRAIN CELLS

The brain is a complex structure that has many different types of cells. Brain cells are subdivided into groups based on a number of criteria that include whether they serve as (1) structural support cells (glia) or (2) cells that receive and transmit information (called neurons or nerve cells). If the latter, then additional criteria are: (a) shape or size; (b) the distance over which they transmit information;

and (c) which chemicals are released to transmit information to other cells. Most of the effects that drugs produce, which are related to abuse potential, are situated on brain cells that process or transmit information. For that reason, the discussions to follow will consider only actions on neurons (nerve cells).

The actions of drugs on the brain are complex and seldom involve only one type of brain cell. Nerve cells have a high level of connectivity between one another; cells in one brain region send inputs to and receive outputs from other regions. These factors make the identification of cells in the brain responsible for a given drug effect difficult to distinguish. This is true of even the most simple behaviors, which involve complex interactions between millions of cells. For these reasons, the understanding of the processes underlying addiction is incomplete; however, significant progress has been made during the 1980s and 1990s. For example, it is generally believed that there are brain systems that are dedicated to the processes underlying euphoria and feelings of well-being that are stimulated by abused drugs.

ORGANIZATION OF THE BRAIN: BRAIN REGIONS

The Cerebral Cortex. A number of experimental approaches have developed to study the basis of behavior in the brain. One of these has been to study the role of brain regions in behavior. The brain is composed of distinct substructures. The most general categorization scheme separates the brain into five segments called *lobes* (Figure 1). From front to back these include the frontal, parietal and occipital lobes and the cerebellum. The temporal lobe is on the lateral surface of the brain. The outermost surface of the brain is called the *cortex*; this part of the brain has expanded in size the most in higher animals and is thought to be responsible for the high level of intelligence in primates (which includes humans). Areas of the cortex are specialized so that specific physiological functions are mediated by cells in defined cortical regions. For example, visual processes occur in cells located on the surface of the back of the brain in the occipital lobe. In front of this region, on the border between the parietal and frontal lobes, is the area that controls movement, which is called the motor cortex. The area of the

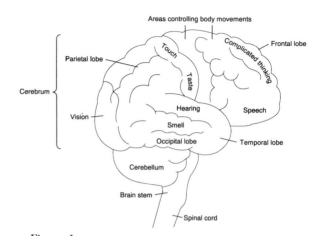

Figure 1
The Lobes of the Brain.
The brain consists of several major sections or lobes. These include the frontal, parietal, occipital, temporal, and cerebellum.

brain that controls sensation, the sensory cortex, is just in front of the motor cortex. The area in front of the sensory cortex in the most frontal portion of the brain, the frontal cortex, is involved in cognitive functions and thinking. It is the evolution of this region that is believed to be responsible for the superior cognitive functioning ability of humans.

The Thalamus. Information processing includes sensory information that comes in from sense organs (for example, eyes, ears, tongue) to the brain through the spinal cord or directly through cranial nerves (nerve cells connected directly to the brain). This incoming information from sense organs goes to a central relay station called the thalamus. The thalamus is specialized, much like the cerebral cortex, in that defined areas receive input that is specific to a sensory modality. For example, input from the eyes through the optic nerve goes to a region of the thalamus called the lateral geniculate body. The areas of the thalamus, in turn, send the information transmitted from sense organs to the appropriate area of the cortex. For example, the lateral geniculate sends visual information to the area of the cortex specialized for vision, which is located in the occipital lobe. Similarly, the cerebral cortex sends commands to the effector systems (usually muscles) that act on the environment through the same relay system. As one can easily see, the thalamus is a very important structure for the coordination of inputs and outputs from the brain. Thus, degenerative diseases of this

structure are very debilitating, as would be drugs that specifically altered the function of this structure.

The Brain Stem. Other areas of the brain are responsible for life processes of which we are not usually aware. These processes are generally controlled by the part of the brain called the brain stem, which is located between the spinal cord and the cerebral hemispheres of the brain. The brain stem contains the cell bodies for centers that maintain heart beat, blood pressure, breathing, and other life-sustaining processes that occur without our having to think about them. A number of psychoactive agents have actions on NEURONS located in the brain stem. For example, OPIATES such as morphine or heroin have a direct inhibitory effect on the brain stem respiration (breathing) centers. This is why heroin OVERDOSES are often fatal—since the breathing centers stop working. Morphine decreases pain by altering the sensitivity of brain cells involved in pain perception. The brain stem is important in the control of pain and also contains the cell bodies for some important nerve cells involved in the euphoric and behavioral actions of drugs.

BRAIN SYSTEMS

The Limbic System. Another important anatomical brain system through which abused drugs act is the LIMBIC SYSTEM. This system is a collection of structures that exist between the brain stem and the cerebral cortex. It includes the frontal and cingulate cortices, NUCLEUS ACCUMBENS, amygdala, hypothalamus, hippocampus, and septum—all of which have direct connections with one another. The limbic system is involved in the expression of emotional behaviors. Tumors or lesions of these structures often lead to abnormal emotional expression. Drugs that directly affect this system can produce changes in mood (euphoria) and emotions.

The Motor System. Motor function (movement) involves a number of brain structures that include the caudate nucleus-putamen, which sits above and in front of the thalamus, the premotor cortex, and the motor cortex as previously described. Drugs that increase (stimulants) or decrease (depressants such as alcohol) activity levels may do so by affecting the activity of these structures. Although the basic mechanisms may differ for drugs of different classes, the overall effect may be the same.

NEUROTRANSMITTER SUBSTANCES

Besides categorizing the parts of the brain by structure, the brain can also be separated into systems based on the distribution of the chemicals that nerve cells use to communicate with one another. Thus, cell bodies for some important nerve cells are localized in specific brain nuclei (collection of nerve cell bodies). Some drugs of abuse have specific actions on subsets of cells that use or release a specific chemical to communicate with other cells. For example, ALCOHOL (ethanol) is believed to act on three systems in the brain—the ones containing the nerve cells that release SEROTONIN, GLUTAMATE, and GAMMA-AMINOBUTYRIC ACID (GABA). The cell bodies of serotonin-releasing nerve cells are localized in the brain-stem region called the raphé nuclei, while glutamate-releasing and GABA-releasing cells are distributed widely throughout the brain (see Figure 2).

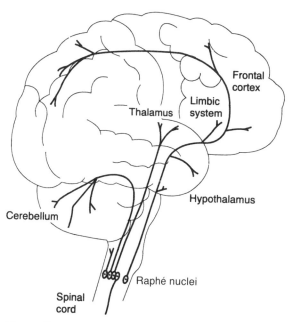

Figure 2
Serotonin Pathways.
The serotonergic neurons originate mainly from the raphé nuclei in the brain stem and project to the forebrain, the cerebellum, and the spinal cord. Very high concentrations of serotonin are also found in the pineal gland. Serotonin-containing neurons are involved in such functions as pain, temperature regulation, sensory perception, and sleep.

What about the different actions of drugs of abuse, where do they act in the brain? As stated previously, it is still not fully understood how the actions of drugs on the brain eventually affect behavior. Our knowledge as of the mid-1990s does, however, indicate some specific actions on some defined sites and cell systems. The so-called stimulant drugs (e.g., AMPHETAMINE, COCAINE, METHAMPHETAMINE) produce overall effects on the brain resulting in increased activity, faster speech and thought patterns, and euphoria. This overall effect results from changes in a number of specific behavioral patterns and represents a complex action of these drugs on several important neuronal systems in the brain. Neurochemical studies of the brain have shown that these stimulant drugs enhance and/or prolong the action of the neurotransmitters DOPAMINE, NOREPINEPHRINE, and SEROTONIN that are released by cells that produce these chemicals to communicate with other brain cells. The dopamine cells send inputs to only a few structures in the forebrain; these include the caudate nucleus that is involved in motor functions and areas of the limbic system that are involved in emotional behaviors and euphoria. The areas of this system to which dopamine cells send inputs include the amygdala, the nucleus accumbens, the olfactory tubercle, and the frontal and cingulate cortices (see Figure 3). Norepinephrine and serotonin cells send inputs more widely to most all forebrain regions, even though their cell bodies are localized in specific brain-stem nuclei. These drugs stimulate motor activity by increasing the function of the dopamine system, which sends inputs to the caudate nucleus. These drugs produce feelings of well-being and euphoria by enhancing dopaminergic activity in limbic areas. Serotonin also has a role in these effects of stimulants, but just how is not yet clear.

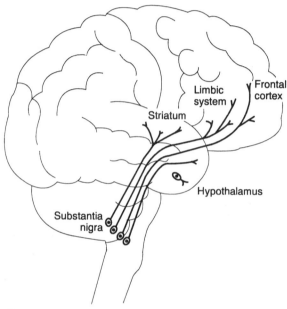

Figure 3
Dopamine Pathways.
Dopamine is not only a precursor of noradrenaline but also a transmitter of its own. Dopamine represents more than 50 percent of the total catecholamine content of the central nervous system. The highest levels of dopamine are found in the neostriatum, nucleus accumbens, and tuberculum olfactorium. There are four main dopaminergic systems in the brain: the nigrostriatal, the mesolimbic and mesocortical, and the tuberoinfundibular systems. The nigrostriatal system appears to be involved in motor function, while the tuberoinfundibular dopamine neurons are involved in hypothalamic-pituitary control. The functions of the mesolimbic and mesocortical systems are less well known, although it is conceivable that they play a role in psychotic disease and in the reinforcing effects of drugs.

BIOLOGICAL BASIS OF ADDICTION

Activation of Brain Hedonic Systems. One common property of drugs of abuse is that they produce feelings of euphoria and pleasantness or they decrease unpleasantness. This results from the activation of brain-cell systems that are naturally involved in these processes (eating desirable foods, listening to pleasing music, sex, leaving unpleasant circumstances, and so on). Chemical stimulation by drugs can however produce the activation of these systems far beyond that produced by normal behavior. Many researchers believe that the ability to modulate these systems by chemical agents is the factor that leads to abuse. Such euphoric effects appear to pose a particular problem for those adolescents who have underdeveloped inhibitory systems, have limited experience with socially accepted forms of personal gratification, and have higher than average levels of aggression. Even without such characteristics, adolescence is one of the most confusing and stressful periods of human development. Ready ac-

cess to simple chemical means of activating reward systems under these conditions can easily lead to abuse.

Drugs Do Not Have Intrinsic Hedonic Properties. The euphoria that occurs after chemical activation of these systems is not only the result of the direct actions of the drug on neurons but is also influenced by the expectations of the individual and the environment in which the drug is taken. Studies in laboratory animals have shown large differences in the effects of a drug on the brain to depend on whether the drug is self-administered or administered to the animal passively (not under its control). It has become clear that the act of drug taking and the control over when the drug is taken are perhaps the two most important factors in the pleasant feelings that follow drug intake. The drug itself has no consistent intrinsic hedonic properties.

Why is it important for an individual to control the onset of drug action through self-administration? Well, it suggests that the activation of these brain systems is also under behavioral influences. Drugs have behavioral effects that are not the same in everybody and can even change in the same individual. For example, alcohol can produce feelings of euphoria in a social situation or depression when one is alone. Another example is cocaine, a very potent stimulator of brain systems involved in euphoria and feelings of well-being—however, when animals are given simultaneous infusions of cocaine without control of delivery, cocaine becomes a stressor that will lead to the animal's death much faster than animals controlling and self-administering the drug.

Dopamine Hypothesis of Drug Abuse. It is widely accepted in the mid-1990s that the abuse potential of a wide variety of drugs, at least in part, is directly related to the direct actions of these chemical agents on brain mesolimbic and mesocortical dopamine cells (see Figure 3). These dopamine cells send inputs to the limbic system, including the limbic cortical regions. The dopamine hypothesis states that drugs that are abused directly activate these dopamine-releasing nerve cells, resulting in the production of reinforcement and/or feelings of euphoria and well-being. This may be correct for PSYCHOMOTOR STIMULANTS like amphetamine and cocaine, which have direct actions upon dopamine-releasing nerve cells, but convincing evidence for dopamine being primarily responsible for the abuse of alcohol (ethanol), opiates, and particularly for

BENZODIAZEPINES is lacking. Dopamine-releasing nerve cells clearly have an important function in the behavioral process and in the euphoria produced by psychomotor stimulants. To ascribe a universal role for these cells in all euphorogenic processes is however likely to be an oversimplification. Scientists have focused on studying dopamine nerve cells often at the expense of exploring the involvement of other brain neurochemicals. Complex brain neuronal networks are responsible for the complex behavioral effects that are called euphoria. It is likely that outputs of the cerebral cortex have a significant role in these processes. The role of these neuronal systems have not been explored.

A drug may be subject to abuse if it directly activates the neuronal networks that are responsible for feelings of well-being and euphoria (positive reinforcement) or if it decreases the unpleasant or aversive nature of the environment in which the individual exists (negative REINFORCEMENT). Most scientific studies of the biological basis of addiction have focused upon the positive reinforcing effects of drugs, a focus which has led some to emphasize the role of dopamine in addiction. However, drug self-administration in humans and in laboratory animal models likely involves both positive and negative reinforcement. The act of taking the drug itself may result in circumstances that produce strain and pressures upon one's normal patterns of living. In addition, the drug itself may activate body and brain systems that are involved in stress, thus, directly producing unpleasant circumstances. For these reasons, the research with animals that has implicated dopamine in the positive reinforcing effects of stimulant drugs is likely more the result of both positive and negative reinforcement. It could be that decreases in the unpleasant nature of one's existence may involve increases in the activity of dopamine-releasing nerve cells. It should also be noted that stressful situations can activate some of these same regions.

Neuronal Network Hypothesis of Drug Abuse. In addition to dopamine, the nerve cells that appear to be involved in the euphoric properties of drugs include acetylcholine, glutamate, opioid, and serotonin-releasing cells. As of the mid-1990s, there is significantly less research data supporting the involvement of these cells; however, it is clear that the brain's opioid receptors are necessary for the reinforcing effects of opiates and that dopamine may not be the exclusive mediator of the euphoric effects of

opiates. Serotonin and glutamate may have important roles in the euphoric properties of alcohol, while acetylcholine-releasing neurons may have a role in the general processes underlying euphoria.

Studies in laboratory animals have suggested that specific brain circuits are involved in the processes related to drug reinforcement. These include areas of the cortex, the midbrain, and the brain stem and involve acetylcholine, dopamine, glutamate, gamma-aminobutyric acid, norepinephrine, opioid, and serotonin-releasing neurons. The frontal and cingulate cortices (nucleus accumbens, lateral hypothalamus, and amygdala) that are included in the limbic system are part of these circuits, as are the ventral pallidum and thalamus. Brain-stem dopamine, norepinephrine, and serotonin nerve-cell nuclei send projections to these forebrain regions involved in such processes. In turn, these regions send output nerve cells to these structures that utilize acetylcholine and glutamate primarily. Some of these forebrain structures are in turn connected to the brain-stem cell nuclei for dopamine, norepinephrine, and serotonin-releasing nerve cells by GABA-releasing nerve cells.

CONCLUSION

This is a simplified description of complex neuronal networks that are believed by some scientists to play a major role in the production of euphoric effects or reinforcement in the brain. It is likely that this complexity will increase as research continues to define and delineate the basic biology of brain-behavior relationships. Investigations of drug self-administration are adding significantly to this field of study in the 1990s. This ongoing research will help us to understand the basic biology of drug abuse so that more efficient and effective forms of treatment and prevention can be developed.

ACKNOWLEDGMENTS

The production of this discourse was funded in part by USPHS Grants DA 00114, DA 01999, DA 03628, and DA 06634.

(SEE ALSO: *Reward Pathways and Drugs*)

BIBLIOGRAPHY

KOOB, G. F. (1992). Drugs of abuse: Anatomy, pharmacology and function of reward pathways. *Trends in Pharmacological Sciences, 13*, 177–184.

KOOB, G. F., & BLOOM, F. E. (1988). Cellular and molecular mechanisms of drug dependence. *Science, 242*, 715–723.

JAMES E. SMITH
STEVEN I. DWORKIN

BREATHALYZER Breath-analysis machines detect and measure the ALCOHOL present in deep lung air and convert this to an estimate of BLOOD ALCOHOL CONCENTRATION (BAC). The basis for this calculation is the relatively constant though small proportion of alcohol that the body excretes through the lungs. BAC is approximately 2,300 times breath alcohol concentration, although there is some variation among individuals. Breath analysis machines use methods such as thermal conductivity and infrared absorption to detect alcohol in lung air. Because breath alcohol analysis is quick and noninvasive, it is a useful tool in a variety of situations. The Breathalyzer has traditionally been associated with law enforcement agencies for monitoring drinking and driving. However, it is increasingly being used in clinical settings. A number of models—both portable and fixed ones—are available.

(SEE ALSO: *Driving Under the Influence (DUI); Drunk Driving*)

BIBLIOGRAPHY

GILES, H. G., ET AL. (1990). *Alcohol and the identification of alcoholics.* Lexington and Toronto: D. C. Heath.

MYROSLAVA ROMACH
KAREN PARKER

BRITAIN, DRUG USE IN The legal use of what we now term *illicit drugs* was widespread in nineteenth-century Britain. Opiates in various forms were used by all levels of society, both for self-medication and for what we now call recreational use. The differentiation between medical and nonmedical usage was not clearly drawn then. Concepts such as addiction were not then widely accepted. The story of drug use in Britain since the late nineteenth century is the story of how and why drugs became defined as a social problem and which factors

brought the establishment of certain forms of drug-control policy. These were, in fact, issues that often bore little relationship to the objective dangers of the drugs concerned.

In the early twentieth century, there was limited involvement either by doctors or by the state in the control of drug use and addiction. The supply of opiates and other drugs was controlled by the pharmaceutical chemist. As dispensers and sellers of drugs over the counter, they were de facto agents of control. A rudimentary medical system of treatment operated via the Inebriates Acts (codified in 1890), whereby some inebriates could be committed to a form of compulsory institutional treatment. Legislation covered only liquids that were drunk (e.g., LAUDANUM) not injectables. Users of hypodermic morphine or cocaine were therefore not included under this system.

Drug addiction was not perceived as a pressing social problem in early twentieth-century Britain, nor, indeed, was it one. Numbers of addicts decreased as overall consumption declined. No specific figures are available for that period, but various indicators, such as poisoning mortality statistics, indicate this conclusion. The twentieth century nevertheless brought increased controls and the classification of opiates and other drugs as *dangerous*. Dangerous drugs were regulated through a penal system of control rather than through the mechanisms of health policy.

Two factors brought regulation. The first was Britain's involvement in an international system of drug control; the second was the impact of World War I (1914–1918) and its aftermath. U.S. pressure on the international scene pushed an initially unwilling Britain into a system of control that rapidly extended from the 1909 Asian regulation discussed at Shanghai to the worldwide system envisaged in the 1912 Hague Convention.

Prior to World War I, however, only the United States, by way of the HARRISON NARCOTICS ACT of 1914, had put this system of drug control into operation. Britain favored a simple extension of the existing Pharmacy Acts. The influence of emergency wartime conditions, however, brought a differently located and more stringent form of control. The fear of a cocaine epidemic among British soldiers patronizing prostitutes in the West End of London—a fear which on later investigation proved to have been largely illusory—allowed the passage of drug regulation in 1916 under the Defence of the Realm Act.

International drug control in its turn became part of the postwar peace settlement at Versailles. The 1920 Dangerous Drugs Act therefore enshrined a primarily penal approach; control was located in the Home Office rather than in the newly established (1919) Ministry of Health.

British drug policy was henceforward marked by a tension between rival conceptualizations of the drug-addiction issue; drugs as a penal issue versus drugs as a health matter. The 1920s saw this conflict at its height. Britain seemed likely to follow a penal course similar to that of the United States, on whose 1914 act the British legislation was consciously modeled, but British doctors soon reasserted their professional control. By 1926, Britain's ROLLESTON REPORT legitimated a medical approach that could entail medical "maintenance prescribing" of opiates to a patient who would otherwise be unable to function. The Rolleston Report established what became known as the BRITISH SYSTEM of drug control—a liberal, medically based system—albeit one that operated within Home Office control.

This system remained in operation for nearly 40 years, until the rapid changes of the 1960s. The 1920s, 1930s, and 1940s were decades when the numbers of addicts were small and there were few nonmedical users (less than 500). It is generally recognized that the British System of medical control operated *because* of this situation rather than *as the cause* of it. This equilibrium began to break down after World War II (1939–1945), when more extensive recreational, or nonmedical, use of drugs (such as HEROIN and COCAINE) began to spread for a variety of reasons. These included the spread of cannabis (MARIJUANA)—from the new immigrant to the white population, overprescribing of heroin by a number of London doctors, thefts from pharmacies, and the arrival of Canadian heroin addicts. Other drugs—in particular, AMPHETAMINES—also became recreationally popular.

The official numbers of heroin addicts rose rapidly, from 94 persons in 1960 to 175 in 1962; and cocaine users increased from 30 in 1959 to 211 in 1964. Nearly all of these were nonmedical consumers. The average age of new addicts also dropped sharply. Initial government reaction, in the report of the first Brain Committee (1961), was muted; however, the second report (1965), produced when the committee was hastily reconvened, had an air of urgency. Controls were introduced on amphetamines in 1964. The report's proposals (implemented in the

Dangerous Drugs Act of 1967) took the prescribing of heroin and cocaine out of the hands of general practitioners and placed it in those of specialist hospital doctors working in drug-dependence units. A formal system was established that notified the Home Office about addicts.

The clinic system established in 1968 did not operate as originally intended. In the 1970s, as the rise in numbers of addicts appeared to stabilize, clinic doctors moved toward a more active concept of treatment, substituting orally administered METHADONE for injected heroin and often insisting on short-term treatment contracts rather than on maintenance prescribing. These clinic policies aided the emergence of a drug black market in Britain in the late 1970s. An influx of Iranian refugees from the Islamic revolution of 1979, bringing financial assets in the form of heroin, also stimulated the market.

The British elections of 1979 returned a Conservative government with a renewed emphasis on a penal response to illicit drugs. Britain participated enthusiastically in the U.S.-led international "war on drugs," but there were also strong forces inside Britain arguing for a more health-focused approach. In 1985, the discovery of acquired immunodeficiency syndrome (AIDS) among injecting drug users in Edinburgh, Scotland, was the trigger for policies that emphasized the reduction of harm from drug use rather than a prohibitionist stance. Nevertheless, in the early 1990s, the tension between penal and health concepts and the interdependence of the two approaches to policy still remained unresolved.

(SEE ALSO: *Anslinger, Harry J., and U.S. Drug Policy; British System of Drug-Addiction Treatment; Opioids and Opioid Control: History*)

BIBLIOGRAPHY

BERRIDGE, V. (1993). AIDS and British drug policy: Continuity or change? In V. Berridge & P. Strong (Eds.), *AIDS and contemporary history.* Cambridge: Cambridge University Press.

BERRIDGE, V. (1984). Drugs and social policy: The establishment of drug control in Britain, 1900–1930. *British Journal of Addiction, 79,* 17–29.

STIMSON, G., & OPPENHEIMER, E. (1992). *Heroin addiction: Treatment and control in Britain.* London: Tavistock.

VIRGINIA BERRIDGE

BRITISH JOURNAL OF ADDICTIONS
See Addiction.

BRITISH SYSTEM OF DRUG-ADDICTION TREATMENT

To many observers from outside the United Kingdom (U.K.), the British System is synonymous with heroin maintenance, with doctors supplying drugs on demand to addicts. To some it has been viewed as an approach of extreme folly; to others it is an effective policy of supreme pragmatism. To those who know and work within it, the British System is somewhat more complex. Indeed, the extent to which a clearly defined system can be identified has been the focus of debate. Here, we will demonstrate that a particular set of factors have combined in the U.K. to create an evolving system of care for drug takers, which has been responsive both to the changing drug scene and to the individual needs of the drug taker. This review will identify the key characteristics of the British System. Important historical milestones in the development of the system will be identified. Finally the effectiveness of the system will be discussed.

WHAT IS THE BRITISH SYSTEM?

Since a key characteristic of the system has been its evolution during the twentieth century, observers at different stages in this process have had different views as to its nature and purpose. This late twentieth-century view proposes the following five characteristics of the British System:

1. *An Evolving System of Health and Social Care for Drug Takers.* Since the 1920s, the British policy toward addiction has been principally in the form of treatment conducted by medical practitioners. This differed from most other jurisdictions, particularly the United States, where addiction was deemed a deviant and criminal activity under the HARRISON NARCOTICS ACT (1914) through and until the revisions of the late 1960s and early 1970s. While the burden of care for drug takers has expanded over time from the general practitioner to specialist psychiatrist and then back to the generalist (often the general practitioner), the British System has to a large extent been located in the public-health sector, latterly in the National

Health Service. Specialist drug-dependence clinics were established throughout the U.K. from the 1960s onward.

Two important consequences of the emphasis on health rather than correctional services have been (1) the attention to the health and social needs of the drug taker—particularly crucial in the wake of the recent human immunodeficiency virus (HIV) epidemic in injecting drug takers, and (2) the opportunity to influence the development of public-health strategies on a national scale through the existing system of health-care services. This second point has been important in the rapid development of new services, including injecting-equipment exchange schemes in the 1980s as a public-health measure to reduce the spread of HIV.

2. *Control and Monitoring of Drug Takers and Their Physicians.* The extent to which the British System has aimed to exercise control over drug takers has been variable. However, part of the purpose of prescribing drugs in the context of addiction treatment (see below) has been to attract drug takers to have contact with statutory services. An index of drug takers (mainly of HEROIN and COCAINE) known to physicians has been maintained by the Home Office since 1968. This has allowed some indication of the scale of the problem to be known as well as preventing individuals from attending more than one clinic. At the same time, in response to concerns over irresponsible prescribing by certain physicians, only those physicians in possession of a special license issued by the Home Office were entitled to prescribe heroin and cocaine to drug takers. Although the recent (early 1990s) policy has been to encourage the increased involvement of general practitioners in prescribing METHADONE, the prescribing behavior of all doctors in relation to addictive drugs remains closely monitored.

3. *Drug Prescribing.* The overall contribution of drug prescribing as a component of the British System has been important, but overestimated by some. Certainly, the right of physicians to prescribe, and drug takers to receive, addictive drugs in the context of treatment has been a key part of the British System since its inception. At various times, opiates (including injectable heroin and methadone), cocaine, AMPHETAMINES, and BARBITURATES have been prescribed in this context.

The aims of such prescribing, as well as the prescribing practices, have varied over time. The principal aim throughout has been to provide a method of detoxification that is as comfortable and medically safe as possible. It has also been accepted at a policy level, however, that some individuals who are either unable or unwilling to stop drugs may require long-term prescribing, with a view to stabilization or maintenance treatment—in some instances with injectable drugs. Since the 1970s, oral methadone (in the form of tablets or linctus—a syrup) has been the opiate prescription of choice in view of its greater safety and lower resale value on the black market. The prescription of other drugs had largely ceased in this context by the mid-1980s. Heroin prescribing and the prescribing of injectable methadone still have a few proponents, despite a lack of controlled scientific evidence to support their effectiveness (see below). Although there is international interest in the right of British doctors to prescribe injectable heroin, fewer than 200 patients receive such treatment (in the early 1990s). Of more interest is the prescribing of injectable methadone as a potential intermediate step in converting the heroin injector to oral methadone.

Drug prescribing within the British System has been characterized by a greater permissiveness and flexibility than is the case in most other jurisdictions, alongside a continued overall conservative approach by most medical practitioners to prescribing agonist drugs to the drug abuser.

4. *Competition with the Black Market.* A prominent aim at various times in the history of the British System has been that of attracting the drug taker away from the black market and into treatment. The putative benefits of such a scheme would be to remove demand for illicit drugs, leading to elimination of the black market; improve health benefits to the drug user for taking pharmaceutical rather than illicit drugs; and reduce criminality associated with purchasing illicit drugs. While there were relatively few heroin takers in Britain until the 1960s, the continued nonexistence of an imported black market meant that prescribed heroin was the main source of supply, but it actually contributed to a growth in the number of drug takers. Since no convincing evidence has emerged to support the value of this approach, the prescribing of drugs has become based more

on individual medical indications than on economic policy.

5. *Flexibility.* Perhaps the most striking feature of the British System has been its capacity to evolve in response to the changing drug problem. Whether this has been the result of deliberate policy or benign laissez faireism is open to debate. The result has allowed a flexible response without overt government intervention in medical practice (beyond the constraints described). Flexibility has been possible at two levels. At the system level, experiments in the provision of a range of services have been possible, including a wide range of drug prescribing-and-injecting equipment exchange schemes. At the individual level, treatments may be tailored to individual needs rather than having the imposition of tightly restricted, prescriptive, and blanket approaches.

MILESTONES IN THE HISTORY OF THE BRITISH SYSTEM

1920	Introduction of the Dangerous Drugs Act following the International Opium Convention at the Hague (1912). This restricts the dispensing of several drugs to physicians, including opiates and cocaine.
1926	The Rolleston Committee publishes its report (see ROLLESTON REPORT), which establishes the right of medical practitioners to prescribe drugs in the context of the treatment of addiction.
1920s–1960s	Small numbers (approx. 400–500) of mainly "therapeutic addicts" and addicted physicians receive opiate prescriptions as treatment for addiction.
1961	The first Brain Committee publishes its report. In reviewing the period since the Rolleston Committee's report it reaffirms the medical practitioner's role and recommends no change in the system.
1960s	A small number of physicians, mainly in London, are prescribing large quantities of heroin leading to a rapid increase in the number of heroin injectors (increasing to 2,000 in number) and growing public alarm.
1965	The Brain Committee is reconvened to consider the increasing problem and publishes its second report. It recommends several restrictions including (1) the establishment of specialized treatment clinics, (2) the licensing of medical practitioners for prescribing, and (3) the Home Office Addicts Index.
1967	The Dangerous Drugs Act implements the recommendations of the Brain Committee and prohibits physicians from prescribing heroin or cocaine to drug takers (except for the purpose of relieving pain caused by organic illness or injury) unless specially licensed by the Home Office. This practically restricts prescribing to the newly established specialist clinics.
1980s	An epidemic of heroin taking occurs on a scale not previously encountered in the U.K.—mainly as a result of new illicit trade routes opening up from the Golden Crescent (Iran, Pakistan, Afghanistan, etc.) to supplement the Far East's GOLDEN TRIANGLE trade. The epidemic is mainly among young Caucasian males in inner city areas throughout the U.K. and leads to a rapid increase in crime (to support the habit). The estimated number of heroin takers increases from around 20,000 in the early 1980s to as many as approximately 150,000 by the end of the 1980s.
1982	The Advisory Council on the Misuse of Drugs (ACMD) publishes its Treatment and Rehabilitation report. The specialist clinics are overwhelmed by the upsurge in demand; a new recognition emerges that drug takers represent a heterogeneous group, many of whom

do not require specialist treatment. The report recommends (1) an expanded role for the generalist, (2) the development of nonmedically based treatment approaches, and (3) a requirement for health authorities to monitor the scale of heroin problems in each community. This results in more restrictive opiate prescribing and a broadening of the range of treatment options.

1984 Guidelines for Good Clinical Practice in the Treatment of Drug Misuse are published. This is the first central guidance to physicians since the inception of the system, which indicates the flexibility of practice that physicians had been accorded.

1988/89 The ACMD publishes two reports on AIDS and Drug Misuse. This major policy review is prompted by the epidemic in HIV infection in injecting drug takers. The reports recommend greater emphasis on attracting and retaining in treatment drug takers unable or unwilling to change their behavior. This results in less restrictive opiate prescribing practices, the introduction of low threshold and user friendly services, and the further expansion of injecting-equipment exchange schemes through the late 1980s and early 1990s.

1991 New Guidelines for Good Clinical Practice is published. These are written by physicians for physicians and aimed at the generalist.

1992 Targets for reductions in drug injecting are introduced as part of the British Government's Health of the Nation white paper—and also formed a part of the AIDS response.

1993 The ACMD's third report on AIDS and drug misuse is published.

The constituency of concern is broadened from opiate injectors to those who inject amphetamines and BENZODIAZEPINES. Recommendations include a focus on the impact of hidden drug-taking populations, through outreach, and the introduction of oral methadone programs (along North American and Australasian lines).

HAS THE BRITISH SYSTEM BEEN EFFECTIVE?

Clearly, the question of effectiveness is difficult to answer in the context of a national problem subject to many external and internal influences, within a system that has evolved over many years. Further, it is difficult to compare the effects of policies toward drug problems in various countries: Drug problems are often culturally specific, and attempted solutions that may be acceptable in one setting may be unwelcome or unhelpful in another. On one level, it is clear that the U.K. has not been spared the epidemic rise in heroin taking experienced by other Western industrialized countries; nor has it avoided the spread of HIV in intravenous drug takers—however, the epidemic of HIV has been far less severe than in several other European countries and in the United States. Regional variation in the pattern of the HIV epidemic in the U.K. suggests that the areas where prescribing and specialist clinics were limited, such as in Edinburgh, Scotland, experienced a much more rapid spread—although closer examination reveals this to be insufficient as the sole explanation.

At times, there have been disadvantages to the British System. In particular, a situation where addictive drugs were overenthusiastically prescribed contributed to a worsening of the problem. Further, the U.K. experience with barbiturate and amphetamine prescribing was wholly negative and resulted in its complete discontinuation.

At an individual level, remarkably little controlled research has been carried out to evaluate different prescribing or other treatment approaches, given the opportunity available to do so in the British System. One controlled trial with ninety-six heroin takers involved the random assignment to either injectable heroin or oral methadone maintenance. The results suggested advantages and disadvantages in both

treatments. At one year follow-up, more in the methadone group were abstinent than in the heroin group, but more had also returned to illicit drug use in the methadone group.

With the increasing emphasis on treatment in the primary care setting in the 1980s, specialist Community Drug Teams were established with the brief of encouraging the increased involvement of general practitioners. The main advantage of such an approach is, in theory at least, that this should allow greater availability of services than could be provided by specialist clinics alone. During the 1990s, there has been an important but modest increase in the extent of general practitioner involvement, but this still falls short of the goal of universal availability of treatment for drug takers.

Overall, it can be said that the principal benefits of the British System have been (1) to ensure the humanitarian handling of drug takers, through treatment services, and (2) to allow the evolution of a system of care responsive to changing needs—which also has been relatively free from unnecessary governmental constraints.

(SEE ALSO: *Britain, Drug Use in; Injecting Drug Users and HIV; Needle and Syringe Exchanges and HIV/AIDS*)

BIBLIOGRAPHY

CONNELL, P. H. (1968). Drug dependence in Great Britain: A challenge to the practice of medicine. In H. Steinberg (Ed.), *Scientific basis of drug dependence*. London: Churchill.

DORN, N., & SOUTH, N. (EDS.). (1987). *A land fit for heroin?: Drug policies, prevention and practice*. Basingstoke: Macmillan.

JUDSON, H. F. (1974). *Heroin addiction in Britain*. New York: Harcourt Brace Jovanovic.

MACGREGOR, S. (ED.). (1989). *Drugs and British society: Responses to a social problem in the 1980s*. London: Routledge.

MINISTRY OF HEALTH (1926). *Report of the Departmental Committee on Morphine and Heroin Addiction* (Rolleston Report). London: HMSO.

STIMSON, G. V., & OPPENHEIMER, E. (1982). *Heroin addiction: Treatment and control in Britain*. London: Tavistock.

STRANG, J. (1989). 'The British System': Past, present, and future. *International Review of Psychiatry, 1,* 109–120.

STRANG, J., & GOSSOP, M. (1993). *Responding to drug abuse: The 'British System'*. Oxford: Oxford University Press.

STRANG, J., & STIMSON, G. V. (EDS.). (1990). *AIDS and drug misuse: Understanding and responding to the drug taker in the wake of HIV*. London: Routledge.

TREBACH, A. S. (1982). *The heroin solution*. New Haven: Yale University Press.

D. COLIN DRUMMOND
JOHN STRANG

BULIMIA   The term *bulimia* means "an extreme hunger," but the word is most commonly understood to refer to BULIMIA NERVOSA, a type of eating disorder. This disorder is characterized by a pattern of eating in which episodes of rapid, uncontrolled ingestion of large amounts of food occur over a short period of time (binges) and are followed by feelings of guilt, depressed mood, self-criticism, and physical discomfort (nausea, abdominal pain). The affected individual regularly engages in purging behavior (induced vomiting, use of diuretics or laxatives), strict dieting, or energetic exercise in an attempt to avert weight gain. Medical complications can ensue, including dehydration and electrolyte imbalances that can be life-threatening. Poisoning by ipecac (an agent used to induce vomiting) leading to death has been reported. Severe dental erosions can develop from vomitus destroying tooth enamel. A characteristic feature in the bulimic patient is a persistent concern with weight and body shape. Other psychiatric disorders can accompany bulimia, particularly major depression.

This disturbance in eating affects mostly young women—usually women of normal weight—and is often preceded by ANOREXIA nervosa (restricted eating). The bulimic symptoms may continue for many years with exacerbations and remissions. From the mid-1970s to the mid-1990s, the prevalence of eating disorders appeared to be increasing in industrialized countries. The etiology of bulimia is unknown, although psychological, sociocultural, and biological theories have been proposed. Many consider Western societies' increasing emphasis on thinness, especially among women, to be a contributing influence.

The treatment of bulimia depends on its severity. Many cases of the eating disturbance resolve on their

own. Specific interventions that may be tried include psychodynamic (individual, family, group) therapies as well as cognitive and behaviorally oriented therapies and pharmacological treatments. Modest improvements have been reported with the use of ANTIDEPRESSANT medication.

BIBLIOGRAPHY

GARFINKEL, P. E., & GARNER, D. M. (EDS.). (1987). *Role of drug treatments for eating disorders.* New York: Brunner/Mazel.

GARNER, D. M., & GARFINKEL, P. E. (EDS.). (1985). *Handbook of psychotherapy for anorexia nervosa and bulimia.* New York: Guilford Press.

HUDSON, J. I., & POPE, H. G., JR. (EDS.). (1987). *The psychobiology of bulimia.* Washington, DC: American Psychiatric Press.

MYROSLAVA ROMACH
KAREN PARKER

BULIMIA NERVOSA   Since 1980, bulimia nervosa has been recognized by the American Psychiatric Association as an autonomous eating disorder. It is characterized by recurrent episodes of binge eating followed by such regular activities as self-induced vomiting, excessive use of laxatives and/or diuretics, fasting or dieting, and vigorous exercise—all of which are directed at weight control. The full syndrome affects 1 to 3 percent of the adolescent and young adult female population, but many more experience subclinical variants of the disorder. Bulimia nervosa does occur in males, but such incidence is rare.

Parallels between bulimia nervosa and substance abuse have been drawn based on an ADDICTION model, a self-psychology model, and a psychobiological model. According to the *addiction model*, food is the "substance" that is abused in bulimia nervosa. Although there are superficial similarities in phenomenology between binge eating and substance abuse, these similarities are selective and rely on a loose definition of addiction. The *self-psychology perspective* is that both bulimia nervosa and substance abuse arise from a common deficit in psychological functioning. Difficulties regulating affect and tension generate a need for the external distraction provided by food or psychoactive substances, respec-

tively. This model may have some heuristic value but it has not, as of the mid-1990s, received empirical validation. The *psychobiological view* regards eating and drinking as consummatory behaviors with the potential for dysregulation. One possibility is a shared disturbance in the brain neurochemical functioning that regulates drives of appetite. There is some evidence that brain SEROTONIN function may be disrupted in both bulimia nervosa and ALCOHOL abuse; however, research in this area has just begun and the validity of this model is unknown as of the mid-1990s.

Although the cause of the link between bulimia nervosa and substance abuse is not known, there is substantial evidence that a connection does exist. Among women receiving treatment for substance abuse, estimates of the prevalence of bulimia nervosa range from 8 to 17 percent, and estimates of the prevalence of some eating disorder range from 26 to 47 percent. Similarly, estimates of alcohol abuse among women seeking treatment for bulimia nervosa range from 27 to 49 percent. Thus, substance abuse and bulimia nervosa occur together in young women much more frequently than would be expected for independent disorders. One potential source of this comorbidity lies in genetic risk. Several studies have indicated an overrepresentation of alcohol abuse in the families of women with eating disorders. Another possibility is that certain psychological factors place certain women at risk for the development of either bulimia nervosa or substance abuse. There is some limited evidence for an underlying ADDICTIVE PERSONALITY in both disorders. As well, women with both disorders seem to have more difficulties, generally, with impulsive behaviors.

BIBLIOGRAPHY

GOLDBLOOM, D. S. (1993). Alcohol abuse and eating disorders: Aspects of an association. *Alcohol and Alcoholism.*

MITCHELL, J. E. (1990). *Bulimia nervosa.* Minneapolis: University of Minnesota Press.

PEVELER, R., & FAIRBURN, C. (1990). Eating disorders in women who abuse alcohol. *British Journal of Addiction, 85,* 1633–1638.

VANDEREYCKEN, W. (1990). The addiction model in eating disorders: some critical remarks and a selected bibliography. *International Journal of Eating Disorders, 9,* 95–101.

WILSON, G. T. (1991). The addiction model of eating disorders: a critical analysis. *Advances in Behaviour Research and Therapy, 12,* 27–72.

MARION OLMSTED
DAVID GOLDBLOOM

BUPRENORPHINE Buprenorphine is a semisynthetic OPIATE which is produced from thebaine, a naturally occurring ALKALOID present in the ripe pods of the opium poppy (*Papaner somniferum*). Buprenorphine has an ANALGESIC potency twenty-five to fifty times greater than MORPHINE on a weight basis. However, the analgesic actions of buprenorphine are quite similar to those of morphine and the other opiates after taking into consideration its greater potency. It is assumed that these effects are dependent upon its ability to act at *mu* (morphine) receptors in the brain. Once bound to the receptor, however, buprenorphine only produces a limited effect, and thus it is termed a partial AGONIST. This ability to produce only a partial response may explain why buprenorphine lowers breathing (respiratory depression) less than drugs such as morphine. Because it is a partial agonist, buprenorphine administration to morphine-dependent patients does not elicit significant withdrawal symptoms and can therefore be used as a methadone-like opiate substitute in treatment programs. Another reason for the use of the agent in this respect is its particularly long duration of action. Single doses of buprenorphine can attenuate or prevent many of the actions of morphine for up to thirty hours. Thus, buprenorphine maintenance programs have been proposed to treat opiate addiction.

The interactions of buprenorphine with ANTAGONISTS are interesting. Buprenorphine actions can be readily prevented by antagonists such as NALOXONE, when the antagonist is administered prior to buprenorphine. However, antagonists given after buprenorphine do not readily reverse the opioid actions. This unique pharmacology distinguishes it from traditional opiates such as morphine. Many believe that this observation is due to the prolonged occupation of the receptor by buprenorphine. Once it is bound, other drugs can no longer get to the receptor.

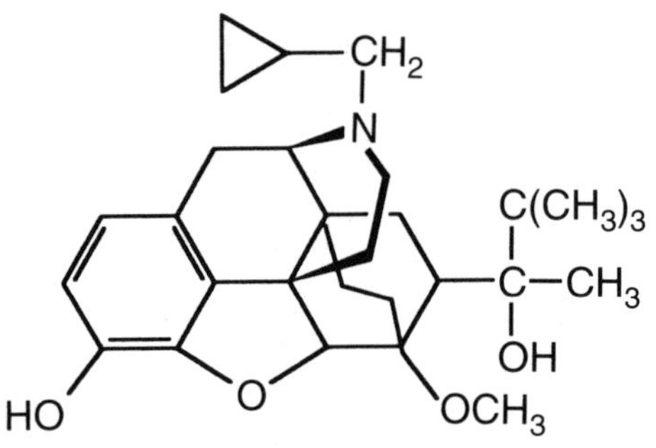

Figure 1
Buprenorphine

In the early 1990s, it was proposed that buprenorphine might also prove effective in lowering COCAINE use. Some studies in primates showed that buprenorphine lowered the amounts of cocaine taken. Although some small clinical studies in people also suggested a similar effect, more controlled studies did not show a special effect on cocaine use. More extensive work will be needed to determine whether buprenorphine can be useful in the treatment of cocaine abusers.

(SEE ALSO: *Heroin; Treatment/Treatment Types*)

BIBLIOGRAPHY

JAFFE, J. H., & MARTIN, W. R. (1990). Opioid analgesics and antagonists. In A. G. Gilman et al. (Eds.), *Goodman and Gilman's the pharmacological basis of therapeutics,* 8th ed. New York: Pergamon.

GAVRIL W. PASTERNAK

C

CAFFEINE Caffeine is the world's most widely used behaviorally active drug. More than 80 percent of adults in North America consume caffeine regularly. Average per capita caffeine intakes in the United States, Canada, Sweden, and the United Kingdom have been estimated at 211 milligrams, 238 milligrams, 425 milligrams, and 444 milligrams per day, respectively; the world's per capita caffeine consumption is about 70 milligrams per day. These dose levels are well within the range of caffeine doses that can alter human behavior: As little as 32 milligrams of caffeine, less than the amount of caffeine in most 12-ounce cola soft drinks, can improve vigilance performance and reaction time; and doses as low as 10 milligrams, less than the amount of caffeine in some chocolate bars, can alter self-reports of mood. These data suggest that a large number of people are daily consuming behaviorally active doses of caffeine.

Caffeine-containing foods and beverages are ubiquitously available in and widely accepted by most contemporary societies—yet dietary doses of caffeine can produce behavioral effects that share characteristics with prototypic drugs of abuse: physical dependence, self-administration, and TOLERANCE. Chronic administration of only 100 milligrams of caffeine per day, the amount of caffeine in a cup of coffee, can produce PHYSICAL DEPENDENCE, as evidenced by severe and pronounced withdrawal symptoms that can occur upon abrupt termination of daily caffeine. Under some circumstances, research volunteers reliably self-administer dietary doses of caffeine, even when they are not informed that caffeine is the drug under study; and some evidence indicates that daily use of caffeine produces tolerance to caffeine's behavioral and physiological effects.

CLASS AND CHEMICAL STRUCTURE

Caffeine is an ALKALOID that is often classified as a central nervous system stimulant. Caffeine is structurally related to xanthine, a purine molecule with two oxygen atoms (see Figure 1). Several important compounds, including caffeine, consist of the xanthine molecule with methyl groups attached. A methyl group consists of a carbon atom and three hydrogen atoms. These methylated xanthines, called methylxanthines, are differentiated by the number and location of methyl groups attached to the xanthine molecule. Caffeine is a 1, 3, 7-trimethylxanthine. The "tri" refers to the fact that caffeine has three methyl groups. The "1, 3, 7" refers to the position of the methyl groups on the purine molecule. Other important methylxanthines include theophylline, theobromine, and paraxanthine. All these methylxanthines are metabolites of caffeine. In addition, theophylline and theobromine are ingested directly in some foods and medications.

SALIENT FEATURES

Sources. Coffee and TEA are the world's primary dietary sources of caffeine. Other sources include soft drinks, cocoa products, and medications. Caf-

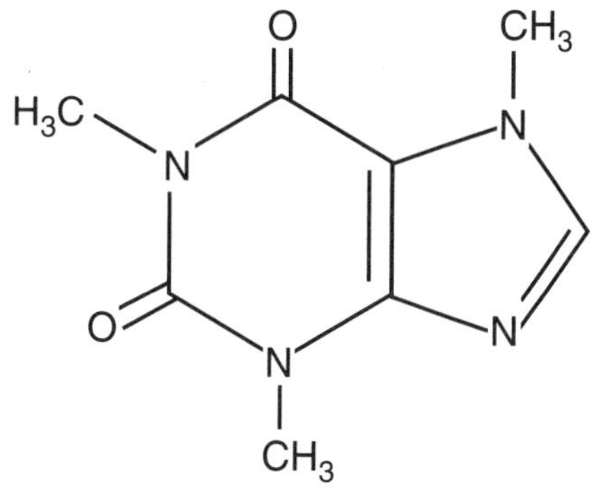

Figure 1
The Caffeine Molecule

feine is found in more than sixty species of plants. COFFEE is derived from the beans (seeds) of several species of *Coffea* plants, and the leaves of *Camellia sinensis* plants are used in caffeine-containing teas. CHOCOLATE comes from the seeds or beans of the caffeine-containing cocoa pods of *Theobroma cacao* trees. In developed countries, soft drinks, particularly COLAS, provide another common source of dietary caffeine. Only a portion of the caffeine in soft drinks comes from the kola nut (*Cola nitida*); most of the caffeine is added during manufacturing. Since the 1960s, a marked decrease in coffee consumption in the United States has been accompanied by a substantial increase in the consumption of soft drinks. Maté leaves (*Ilex paraguayensis*), guarana seeds, and yoco bark are other sources of caffeine for a variety of cultures. Table 1 shows the amounts of caffeine

found in common dietary and medicinal sources. As can be seen in the range of values for each source in this table, the caffeine content can vary widely depending on method of preparation or commercial brand.

Effects on Mood and Performance. It has long been believed that caffeine stimulates mood and behavior, decreasing fatigue and increasing energy, alertness, and activity. Although caffeine's effects in experimental studies have sometimes been subtle and variable, dietary doses of caffeine have a variety of effects on mood and performance. Doses below 200 milligrams have been shown to improve vigilance and reaction time, increase tapping speed, postpone sleep, and produce reports of increased alertness, energy, motivation to work, desire to talk to people, self-confidence, and well-being. Higher doses can both improve or disrupt performance of complex tasks, increase physical endurance, work output, hand tremor, and reports of nervousness, jitteriness, restlessness, and anxiousness.

DISCOVERY

Caffeine, derived from natural caffeine-containing plants, has been consumed for centuries by various cultures. Consumption of tea was first documented in China in 350 A.D., although there is some evidence that the Chinese first consumed tea as early as the third century B.C. Coffee cultivation began around 600 A.D., probably in what is now Ethiopia.

TABLE 1
Caffeine Content in Common Dietary and Medicinal Sources

Source	Standard Value (in milligrams)	Minimum (in milligrams)	Maximum (in milligrams)
Coffee (6 oz./180 ml)			
ground roasted	102	77	186
instant	72	35	211
decaffeinated	4	2	10
Tea (6 oz./180 ml)			
leaf or bag	48	34	58
instant	36	29	37
Cola Soft Drink (12 oz./360 ml)	43	2	58
Chocolate Milk (6 oz./180 ml)	4	2	5
Chocolate Bar (1.45–1.75 oz./40–50 g)	7	5	31
Caffeine-containing Over-the-Counter Medications			
analgesics and cold preparations	32	15	100
appetite suppressants and stimulants	100	50	350

Caffeine was first chemically isolated from coffee beans in 1820 in Germany. By 1865, caffeine had been identified in tea, maté (a drink made from the leaves of a South American holly), and kola nuts (the chestnut-sized seed of an African tree).

THERAPEUTIC USES

Caffeine is incorporated in a variety of over-the-counter preparations marketed as analgesic, stimulant, cold, decongestant, menstrual-pain, or appetite-suppression medications. As an ingredient in ANALGESICS (painkillers), caffeine is used widely in the treatment of ordinary types of headaches, although evidence for caffeine's analgesic effects is limited: Caffeine may only diminish headaches that result from caffeine withdrawal, but it is also combined with an ergot ALKALOID in the treatment of migraine. Caffeine may have some therapeutic effectiveness in its ability to constrict cerebral blood vessels. The use of caffeine as a central nervous system (CNS) stimulant does have an empirical basis, but there is little evidence that caffeine has appetite-suppressant effects.

Because of various effects of caffeine on the respiratory system, caffeine is used to treat asthma, chronic obstructive pulmonary disease, and neonatal apnea (transient cessation of breathing in newborns)—although other agents, including theophylline, are usually preferred for the treatment of asthma and chronic obstructive pulmonary disease.

Historically, caffeine has been used medically to treat overdoses with opioids and central depressants, but this use has decreased considerably with the development of alternative treatments.

ABUSE

Case reports have described individuals who consume large amounts of caffeine—exceeding one gram per day (1,000 milligrams). This excessive intake, observed particularly among psychiatric patients, drug and alcohol abusers, and anorectic patients, can produce a range of symptoms—muscle twitching, ANXIETY, restlessness, nervousness, insomnia, rambling speech, tachycardia (rapid heartbeat), cardiac arrhythmia (irregular heartbeat), psychomotor agitation, and sensory disturbances including ringing in the ears and flashes of light.

The disorder characterized by excessive caffeine intake has been referred to as caffeinism. There is some suggestion that excessive caffeine consumption can be linked to psychoses and anxiety disorders. Substantial amounts of caffeine are also used by a small percentage of competitive athletes, despite specific sanctions against such use.

Abused drugs are reliably self-administered under a range of environmental circumstances by humans and most are also self-administered by laboratory animals. Caffeine has been self-injected by laboratory nonhuman primates and self-administered orally and intravenously by rats, but there has been considerable variability across subjects and across studies.

Human self-administration of caffeine has been variable, as well; however it is clear that human subjects will self-administer caffeine, either in capsules or in coffee, and even when they are not informed that caffeine is the drug under study. For example, heavy coffee drinkers given repeated choices between capsules containing 100 milligrams caffeine or placebo under double-blind conditions showed clear preference for the caffeine capsules and, on average, consumed between 500 and 1,300 milligrams of caffeine per day. Experimental studies with low to moderate caffeine consumers have found that between 30 and 60 percent of those subjects reliably choose caffeine over placebo in blind-choice tests. Subjects tend to show less caffeine preference as the caffeine dose increases from 100 to 600 milligrams, and some subjects reliably avoid caffeine doses of 400 to 600 milligrams.

TOLERANCE

Chronic caffeine exposure can produce a decreased responsiveness to many of caffeine's effects (i.e., tolerance). This has been observed in both nonhumans and humans. Research with nonhumans has clearly demonstrated that chronic caffeine administration can produce partial tolerance to various effects of caffeine and can produce complete tolerance to caffeine's stimulating effect on locomotor activity in rats. A number of studies also suggest that tolerance to caffeine develops in humans: Daily doses of 250 milligrams of caffeine can increase systolic and diastolic blood pressure, however tolerance quickly develops to these effects within four days. The stimulating effects of caffeine on urinary and salivary output also diminish with chronic caffeine exposure. Although tolerance appears to develop to some of

the central nervous system effects of caffeine, this aspect of caffeine tolerance has not been well explored. Comparisons of the effects of caffeine between heavy and light caffeine consumers provide indirect evidence that repeated (regular) caffeine use diminishes the sleep-disturbing effects and alters the profile of self-reported mood effects. For example, 300 milligrams of caffeine may produce self-reports of jitteriness in people who normally abstain from caffeine but not in regular caffeine consumers. High chronic caffeine doses (900 mg per day) can eliminate the self-reported mood effects (tension, anxiety, nervousness and jitteriness) of 300 milligrams of caffeine given twice a day.

PHYSICAL DEPENDENCE

Evidence of physical dependence on caffeine is provided by the appearance of a withdrawal syndrome following abrupt termination of daily caffeine. Although there have been relatively few demonstrations of caffeine withdrawal in nonhumans, abrupt termination of chronic daily caffeine has been shown to clearly decrease locomotor behavior in rats. Considerably more is known about caffeine withdrawal in humans. Caffeine withdrawal is well documented in anecdotal case reports dating back to the 1800s and in experimental and survey studies from the 1930s to the present. Caffeine withdrawal is typically characterized by reports of headache, fatigue (e.g., reports of mental depression, weakness, lethargy, sleepiness, drowsiness, and decreased alertness), and possibly anxiousness. Descriptions of the withdrawal headache suggest that it develops gradually and can be throbbing and severe.

When caffeine withdrawal occurs, its intensity can vary from mild to severe. Anecdotal descriptions of severe withdrawal suggest that it can be incompatible with normal functioning and include flulike symptoms, fatigue, severe headache, nausea, and vomiting. In general, caffeine withdrawal begins twelve to twenty-four hours after terminating caffeine, peaks at twenty to forty-eight hours, and lasts from two to seven days. Caffeine withdrawal can occur following termination of caffeine doses as low as 100 milligrams per day, an amount equal to one strong cup of coffee, two strong cups of tea, or three soft drinks. Caffeine withdrawal effects can vary within an individual in that a given individual may not experience caffeine withdrawal during every pe-

riod of caffeine abstinence. The severity of the withdrawal symptoms usually appears to be an increasing function of the maintenance dose of caffeine. Caffeine suppresses caffeine withdrawal symptoms in a dose-dependent manner, so that the magnitude of suppression increases as a function of the administered caffeine dose.

The data described above indicate that the large majority of the adult population in the United States is at risk for periodically experiencing significant disruption of mood and behavior when there are interruptions of daily caffeine consumption.

CAFFEINE WITHDRAWAL

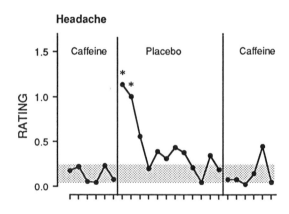

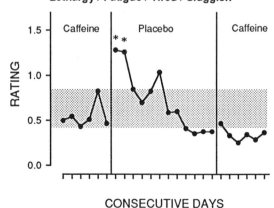

CONSECUTIVE DAYS

Figure 2

The Termination Effects of Daily Caffeine Consumption

SOURCE: Griffiths et al. (1990). *Low-dose caffeine physical dependence in humans.*

The nature and time course of effects of terminating daily caffeine consumption is illustrated in Figure 2, a recent experiment involving seven adult subjects. The subjects followed a caffeine-free diet throughout the study and received identically appearing capsules daily. Prior to the study, subjects had received 100 milligrams of caffeine daily for more than 100 days. Placebo capsules were substituted for caffeine without the subjects' knowledge, and subjects continued to receive placebo capsules for twelve days, after which caffeine administration was resumed. The top panel of the figure shows that substitution of placebo for caffeine produced statistically significant increases (asterisks) in the average ratings of headache during the first two days of placebo substitution. Headache ratings gradually decreased over the next twelve days and continued at low levels during the final caffeine condition. The bottom panel of the figure shows that substitution of placebo for caffeine produced similar time-limited increases in subjects' ratings of lethargy/fatigue/tired/sluggish.

ORGAN SYSTEMS

Caffeine affects the cardiovascular, respiratory, gastrointestinal and central nervous systems. Most notably, caffeine stimulates cardiac muscles, relaxes smooth muscles, produces diuresis by acting on the kidney, and stimulates the central nervous system. The potential of dietary doses of caffeine to stimulate the central nervous system is primarily inferred from caffeine's behavioral effects. Low to moderate caffeine doses can produce changes in mood (e.g., increased alertness) and performance (e.g., improvements in vigilance and reaction time). Higher doses produce reports of nervousness and anxiousness, measurable disturbances in sleep, and increases in tremor. Very high doses can produce convulsions.

Caffeine's cardiovascular effects are variable and depend on dose, route of administration, rate of administration, and history of caffeine consumption. Caffeine doses between 250 and 350 milligrams can produce small increases in blood pressure in caffeine-abstinent adults. Daily caffeine administration, however, produces tolerance to these cardiovascular effects within several days; thus comparable caffeine doses do not reliably affect blood pressure of regular caffeine consumers. High caffeine doses can produce a rapid heartbeat (tachycardia) and in rare cases ir-

regularities in heartbeat (cardiac arrhythmia). Caffeine's effects on peripheral blood flow and vascular resistance are variable. In contrast, caffeine appears to increase cerebrovascular resistance and decrease cerebral blood flow.

Moderate doses of caffeine can increase respiratory rate in caffeine-abstinent adults. Caffeine also relaxes the smooth muscles of the bronchi. Because of caffeine effects on respiration, it has been used to treat asthma, chronic obstructive pulmonary disease, and neonatal apnea (transient cessation of respiration in newborns).

Moderate doses of caffeine can act on the kidney to produce diuretic effects that diminish after chronic dosing. Caffeine has a variety of effects on the gastrointestinal system, particularly the stimulation of acid secretion. These effects can contribute to digestive upset and to ulcers of the gastrointestinal system.

Caffeine increases the concentration of free fatty acids in plasma and increases the basal metabolic rate.

TOXICITY

High doses of caffeine, typically doses above 300 milligrams, can produce restlessness, anxiousness, nervousness, excitement, flushed face, diuresis, gastrointestinal problems, and headache. Doses above 1,000 milligrams can produce rambling speech, muscle twitching, irregular heartbeat, rapid heartbeat, sleeping difficulties, ringing in the ears, motor disturbances, anxiety, vomiting, and convulsions. Adverse effects of high doses of caffeine have been referred to as caffeine intoxication, a condition recognized by the American Psychiatric Association. Extremely high doses of caffeine—between 5,000 and 10,000 mg—can produce convulsions and death.

Extremely high doses of caffeine, well above dietary amounts, have been shown to produce teratogenic effects (birth defects) in mammals. Although there is some evidence to the contrary, dietary doses of caffeine do not appear to affect the incidence of malformations or of low-birth-weight offspring. Although there has been some suggestion that caffeine consumption increases the incidence of benign fibrocystic disease and cancer of the pancreas, kidney, lower urinary tract, and breast, associations have not been clearly established between caffeine intake and any of these conditions. Similarly, dietary caffeine

has been associated with little, if any, increase in the incidence of heart disease.

Controversies continue over the medical risks of caffeine. Although research has not definitively resolved all the controversies, health-care professionals must make recommendations regarding safe and appropriate use of caffeine. In a recent survey of physician specialists, more than 65 percent recommended reductions in caffeine in patients with arrhythmias, palpitations, tachycardia, esophagitis/hiatal hernia, fibrocystic disease, or ulcers, as well as in patients who are pregnant.

PHARMACOKINETICS

Absorption and Distribution. Caffeine can be effectively administered orally, rectally, intramuscularly, or intravenously; however, it is usually administered orally. Orally consumed caffeine is rapidly and completely absorbed into the bloodstream through the gastrointestinal tract, producing effects in as little as fifteen minutes and reaching peak plasma levels within an hour. Food reduces the rate of absorption. Caffeine readily moves through all cells and tissue, largely by simple diffusion, and thus is distributed to all body organs, quickly reaching equilibrium between blood and all tissues, including brain. Caffeine crosses the placenta, and it passes into breast milk.

Metabolism and Excretion. The bloodstream delivers caffeine to the liver, where it is converted to a variety of metabolites. Most of an ingested dose of caffeine is converted to paraxanthine and then to several other metabolites. A smaller proportion of caffeine is converted to theophylline and theobromine; both of those compounds are also further metabolized. Some of these metabolites may contribute to caffeine's physiologic and behavioral effects.

The amount of time required for the body of an adult to remove half of an ingested dose of caffeine (i.e., the half-life) is 3 to 7 hours. On average, about 95 percent of a dose of caffeine is excreted within 15 to 35 hours. Cigarette smoking produces a twofold increase in the rate at which caffeine is eliminated from the body. There is a twofold decrease in the caffeine elimination rate in women using oral contraceptive steroids and during the later stages of pregnancy. Newborn infants eliminate caffeine at markedly slower rates, requiring over 10 days to eliminate about 95 percent of a dose of caffeine. By 1 year of age, caffeine elimination rates increase substantially, exceeding those of adults; school-aged children eliminated caffeine twice as fast as adults.

MECHANISMS OF ACTION

Three mechanisms by which caffeine might exert its behavioral and physiological effects have been proposed: (1) blockade of receptors for adenosine; (2) inhibition of phosphodiesterase activity resulting in accumulation of cyclic nucleotides; and (3) translocation of intracellular calcium. Only one of these, however, the blockade of adenosine receptors, occurs at caffeine concentrations in plasma produced by dietary consumption of caffeine. Adenosine (an autacoid—or cell-activity modifier), found throughout the body, has a variety of effects that are often *opposite* to caffeine's effects—although caffeine is structurally very similar to adenosine. As a result, caffeine can bind to the receptor sites normally occupied by adenosine, thereby blocking adenosine binding, and preventing adenosine's normal activity. Thus, caffeine's ability to stimulate the central nervous system, and increase urine output and gastric secretions, may be due to the blockade of adenosine's normal tendency to depress the central nervous system and decrease urine output and gastric secretions. The methylxanthine metabolites of caffeine (including paraxanthine, theophylline, and theobromine) are also structurally similar to adenosine and block adenosine binding.

(SEE ALSO: *Addiction: Concepts and Definitions; Tolerance and Physical Dependence*)

BIBLIOGRAPHY

DEWS, P. B. (ED.) (1984). *Caffeine*. New York: Springer-Verlag.

GRAHAM, D. M. (1978). Caffeine—Its identity, dietary sources, intake and biological effects. *Nutrition Reviews, 36*, 97–102.

GRIFFITHS, R. R., & WOODSON, P. P. (1988). Caffeine physical dependence: A review of human and laboratory animal studies. *Psychopharmacology, 94*, 437–51.

GRIFFITHS, R. R., ET AL. (1990). Low-dose caffeine physical dependence in humans. *Journal of Pharmacological and Experimental Therapy, 255*, 1123–1132.

HUGHES, J. R., AMORI, G., & HATSUKAMI, K. D. (1988). A survey of physician advice about caffeine. *Journal of Substance Abuse, 1*, 67–70.

RALL, T. W. (1990). Drugs used in the treatment of asthma. In A. G. Gilman et al. (Eds.), *Goodman and Gilman's the pharmacological basis of therapeutics*, 8th ed. New York: Pergamon.

SPILLER, G. A. (ED.) (1984). *The methylxanthine beverages and foods: Chemistry, consumption, and health effects.* New York: Alan R. Liss.

WEISS, B., & LATIES, V. G. (1962). Enhancement of human performance by caffeine and the amphetamines. *Pharmacological Review, 14,* 1–36.

KENNETH SILVERMAN
ROLAND R. GRIFFITHS

CALCIUM CARBIMIDE Citrated calcium carbimide is a mixture of two parts citric acid to one part calcium carbimide; it slows the metabolism of ALCOHOL (ethanol) from acetaldehyde to acetate, so it is used in the treatment of ALCOHOLISM. It is also known as calcium cyanamide. As an antidipsotropic—an antialcohol or alcohol-sensitizing medication (e.g., Temposil; Abstem), it is used for treatment in Canada, the United Kingdom, and Europe—but not in the United States. It was introduced for clinical use in 1956.

The PHARMACOKINETIC data on the absorption metabolism and elimination of carbimide in humans is incomplete. Since nausea, headache, and vomiting occur because of the rapid absorption of carbimide, for treatment purposes it is formulated as a slow-release tablet. Peak plasma concentrations of carbimide following oral administration in experimental animals occur at sixty minutes; the drug is then metabolized at a relatively rapid rate so that half disappears about every ninety minutes (i.e., an apparent elimination half-life of 92.4 minutes).

Alcohol (ethanol) is normally metabolized first to acetaldehyde, which is then quickly metabolized further so that levels of acetaldehyde are ordinarily quite low in the body (acetaldehyde is toxic). Carbimide produces competitive inhibition of hepatic (liver) aldehyde-NAD oxidoreductase dehydrogenase (ALDH), the enzyme from the liver responsible for oxidation of acetaldehyde into acetate and water. Within two hours of taking carbimide by mouth, ALDH inhibition occurs. If alcohol is then ingested, blood acetaldehyde levels are increased; also mild facial flushing, rapid heartbeat, shortness of breath, and nausea occur with just one drink. As more is drunk, the severity of the reaction increases, with ris-ing discomfort and apprehension. Severe reactions can pose a serious medical risk that requires immediate attention.

Repeated administration of carbimide produces few side effects. Carbimide does not cause drowsiness, lethargy, depression, or liver damage. Carbimide, however, exerts antithyroid activity, which can be clinically significant in patients with preexisting hypothyroid disease. The clinical significance of transient white blood cell increases remains unclear.

The rationale for use of carbimide in alcoholism treatment is as follows. The threat of an unpleasant reaction, which one may expect following drinking, is sufficient to deter drinking. For alcoholics in treatment who take a drink, the ensuing reaction is unpleasant enough to strengthen their overall conditioned aversion to alcohol. Their reduction of alcohol consumption during carbimide treatment is expected to result in a general bodily improvement. Carbimide is most often administered daily to provide continuous protection. Carbimide is not widely used in this manner, since its duration of action is short, and it must be taken twice a day. A second approach involves the use of carbimide as part of a RELAPSE-PREVENTION treatment, whereby an individual might take it in anticipation of a high-risk situation. As of the early 1990s, scientific evidence supporting the efficacy of carbimide in alcoholism treatment is inconclusive because of a lack of well-controlled clinical trials.

(SEE ALSO: *Causes of Drug Abuse: Learning; Disulfirum; Treatment Types: Aversion Therapy*)

BIBLIOGRAPHY

PEACHEY, J. E., & ANNIS, H. (1985). New strategies for using the alcohol-sensitizing drugs. In C. A. Naranjo & E. M. Sellers (Eds.). *Research advances in new psychopharmacological treatments for alcholism* (pp. 199–218). New York: Excerpta Medica.

PEACHEY, J. E. ET AL. (1989a). Calcium carbimide in alcoholism treatment. Part 1: A placebo-controlled, double-blind clinical trial of short-term efficacy. *British Journal of Addiction, 84,* 877–887.

PEACHEY, J. E. ET AL. (1989b). Calcium carbimide in alcoholism treatment. Part 2: Medical findings of a short-term, placebo-controlled, double-blind clinical trial. *British Journal of Addiction, 84,* 1359–1366.

JOHN E. PEACHEY

CALIFORNIA CIVIL COMMITMENT PROGRAM

The California Civil Addict Program (CAP) is considered the first true CIVIL COMMITMENT program implemented in the United States. It was initiated in 1961 following a recommendation by the Study Commission on Narcotics that there be a compulsory, yet nonpunitive, program for the treatment and control of narcotic addiction. The program was placed under the direction of the Department of Corrections and equipped with clear standards of commitment procedures.

PROGRAM DESCRIPTION

Addicts convicted of a felony or misdemeanor could be committed to CAP for seven years. Afterward, they would return to court for disposition of the original charges (which were typically dropped), or time served in CAP was credited toward the sentence. Addicts not charged with a criminal offense could be involuntarily committed for seven years or they could participate voluntarily for two and a half years. (These procedures are rarely used today.) Addiction was determined by two court-appointed physicians, and patients underwent a sixty-day evaluation period (McGlothlin, Anglin, & Wilson, 1977).

The program provided for an inpatient phase and an outpatient phase of treatment. The main inpatient facility was the 2,400-bed, moderate-security California Rehabilitation Center located 50 miles from Los Angeles. This institution was viewed as a modified therapeutic community, whose primary goal it was to cause the individuals to assume greater responsibility for their behavior. Academic and vocational training was also offered. Supervised community aftercare was provided by the Narcotic Addict Outpatient Program administered by 180 specially trained parole agents with average caseloads of 33 parolees; the former patients were also given regular objective antinarcotic testing (McGlothlin et al., 1977). During the initial 8 years of CAP (1961–1969), this outpatient program was very stringent and adhered to the parole agent's expression, "You use, you lose." During the 1970s, the program became more liberal. For example, some infrequent drug use might be tolerated if the overall behavioral pattern was deemed acceptable (Anglin, 1988).

Studies have found that the California civil commitment program, as originally implemented, had significant effects on suppressing daily heroin use by narcotic addicts (Anglin, 1988). Participants in CAP, moreover, exhibited sustained reductions in addiction use and in related behavior, particularly in property crime and other antisocial activities. Although nearly everyone in the study became readdicted at some point following their civil commitment, the treatment group had fewer multiple relapses and the relapses were of shorter duration and were separated by longer periods of nonaddiction (Anglin, 1988).

CURRENT STATUS

CAP has an important place in the history of compulsory substance-abuse treatment, but the program has been dramatically altered since the late 1970s, and today it is weak and underfunded. The length of the commitment period has been reduced from seven years to an average of three years. The community phase is disorganized and unfocused. Ancillary services have been substantially cut, and there is currently no treatment service beyond the minimal 120-hour Civil Commitment Education Program (Wexler, 1990). The efficacy and utility of compulsory treatment are being debated at all levels of government, but because long-term studies have shown that the earlier intensive programming in California was effective (Anglin, 1988), the continuation of a program is being considered. Recommendations for program modifications are under review (Wexler, 1990).

(SEE ALSO: *Coerced Treatment for Substance Offenders*)

BIBLIOGRAPHY

ANGLIN, D. (1988). The efficacy of civil commitment in treating narcotic addiction. *Compulsory treatment of drug abuse: Research and clinical practice* (NIDA Research Monograph 86). Rockville, MD: U.S. Department of Health and Human Services.

MCGLOTHLIN, W. H., ANGLIN, M. D., & WILSON, B. D. (1977). *An evaluation of the California Civil Addict Program* (Services Research Monograph Series). Rockville, MD: U.S. Department of Health, Education, and Welfare.

WEXLER, H. K. (1990). *Summary of findings and recommendations of the second invited review of California Department of Corrections substance abuse treatment*

efforts. Unpublished manuscript, Narcotics and Drug Research, Inc., New York.

HARRY K. WEXLER

CANADA, DRUG AND ALCOHOL USE IN

A 1990 survey found that 81 percent of Canadian adults aged 15 and over were current drinkers (who had consumed ALCOHOL at least once in the past year); an additional 11 percent said they were former drinkers, while only 8 percent said they never drank. Per-adult consumption was 10 quarts (9.5 l) of absolute alcohol (ethanol), mostly (54%) beer.

In 1990, 29 percent of adults were current smokers and another 35 percent were former smokers. Overall, the percentage of the population who smoke has been dropping since the 1970s. Practically all TOBACCO is consumed as cigarettes, with daily consumption per smoker estimated at 25 (more than one pack) in fiscal year 1985/86.

The 1990 survey found that 20 percent of adults had used MARIJUANA or HASHISH at some point, while 5 percent had used it the past year. One percent were current COCAINE or CRACK users, and 3 percent had used it at some time. A 1989 survey found that 0.4 percent of adults had used LYSERGIC ACID DIETHYLAMIDE (LSD), AMPHETAMINES (speed), or HEROIN in the past year, and 4.1 percent had used them at some point.

High school students and "street" kids reported the use of numerous illegal drugs, such as LSD, HALLUCINOGENS, speed, heroin, glue and other INHALANTS, or made nonprescription use of stimulants, BARBITURATES, and tranquilizers. Street kids also reported the use of synthetic drugs, such as "ecstasy" and "ice."

The 1990 survey found that 7 percent of adults used sleeping pills, 5 percent used tranquilizers such as Valium, 1 percent used diet pills or stimulants, 3 percent used antidepressants, and 11 percent used narcotic painkillers such as Demerol, morphine, or codeine. (In Canada, codeine of less than 8 milligrams per tablet is an over-the-counter drug.)

Alcohol, tobacco, and illegal drug use is generally greater in men than in women, but prescription psychoactive-drug use is greater in women. Among teenagers, there is little difference in alcohol use between the genders, although more adolescent girls and young women reported smoking tobacco than

did males of the same ages. By 1991, however, Ontario data no longer showed gender differences in tobacco use.

CONSEQUENCES

A 1989 survey found that 12 percent of drinkers had had a physical health problem due to their drinking at some point; in 7 percent it had affected their outlook on life. Almost 11 percent said it had interfered with their friendships or social life, and 6 percent said it had affected their home lives or marriages. Finally, 3.5 percent said it interfered with their work, studies, or employment opportunities, and 5 percent said it affected their financial positions.

In 1990, 23 percent of current drinkers reported drinking and DRIVING. Of fatally injured drivers who had been tested, 45 percent had positive BLOOD ALCOHOL CONCENTRATIONS (BAC), with 28 percent exceeding 150 milligrams. (The legal level in Canada for impairment is 80 milligrams.)

In 1991, there were 111,184 persons charged with 140,068 drunk driving offenses, consisting predominantly of impaired operation of a motor vehicle (91%), with the rest primarily consisting of the refusal to provide a breath sample. In fiscal 1991/92, there were 13,054 persons sentenced to jail for drunk driving offenses, representing 17 percent of all persons in jail.

In 1991, there were 145,239 adults charged with 221,185 provincial liquor act offenses, and 21,946 juvenile offenders were involved in liquor act offenses. In 1991/92, there were 5,766 admissions to provincial adult correctional facilities for liquor act offenses, forming 8 percent of all persons in jail.

In 1991, there were 56,123 drug offenses, with 40,864 adults and 3,249 juveniles charged for offenses under the Narcotic Control Act (96 percent of drug offenses) and the Food and Drugs Act. *Cannabis* (marijuana) accounted for 59 percent of drugs involved. Excluding *Cannabis*, there were 9,631 convictions for offenses under the federal drug acts in 1991, of whom 5,323 were sentenced to jail. Drug offenses in 1991/92 made up 15 percent of all admissions to federal penitentiaries (which hold prisoners sentenced for a term of 2 years or more) and 6 percent of sentenced admissions to provincial adult correctional facilities (which are used for prisoners sentenced for terms of under 2 years).

TREATMENT

In 1989, there were an estimated 486,100 alcoholics in Canada. In 1989/90, there were 39,357 cases treated as inpatients in general and psychiatric hospitals for primary alcohol diagnoses including alcoholic psychoses, alcohol dependence syndrome, nondependent abuse of alcohol, alcoholic pellagra, neuropathy, cardiomyopathy, gastritis, chronic liver disease and cirrhosis, portal hypertension, toxic effects of alcohol, excessive blood level of alcohol, and suspected damage to the fetus due to maternal alcohol addiction with noxious influences transmitted via the placenta or breast milk.

There were 21,507 cases treated in general and psychiatric hospitals for primary drug diagnoses including drug dependence, drug psychoses, nondependent abuse of drugs, suspected damage to the fetus from drugs, drug withdrawal syndrome in the newborn, or poisonings from analgesics, antipyretics and antirheumatics, sedatives and hypnotics, or psychotropic agents.

DEATHS

In 1990, there were 2,959 deaths directly due to alcohol diagnoses and another 15,665 deaths from cancer, cardiovascular and respiratory problems, accidents, suicide and homicide in which alcohol was a contributory cause; 422 deaths were due to drugs, and 35,717 were indirectly due to tobacco—that is, the deaths were caused by bronchitis, asthma and emphysema, cancer, and cardiovascular disease.

BIBLIOGRAPHY

ADRIAN, M., JULL, P., & WILLIAMS R. T. (B.) (1989). *Statistics on alcohol and drug use in Canada and other countries: Data available by 1988.* Toronto: Addiction Research Foundation.

SINGLE, E., WILLIAMS, B., & MCKENZIE, D. (1994). *Canadian profile: Alcohol, tobacco and other drugs 1994.* Ottawa and Toronto: The Canadian Centre on Substance Abuse and the Addiction Research Foundation of Ontario.

WILLIAMS, B., CHANG, K., & TRUONG, M. V. (1992). *Canadian profile: Alcohol and other drugs.* Toronto: Addiction Research Foundation.

MANUELLA ADRIAN

CANCER, DRUGS, AND ALCOHOL The likelihood of a substance to cause cancer is called carcinogenicity; this is determined in several ways. The first is to see if cells grown in vitro (in a test tube) with the potential carcinogen develop cell-structure abnormalities in the new-grown cells. The second is to see if the substance will result in cancers in animals. The third is to look at the clinical course of a disease in a human patient. Finally, the outcomes in a group of people exposed to the substance will be compared to outcomes in those not exposed—either by following the natural course of history in a population through time in cohort studies or in case-control.

ALCOHOL AND CANCER

Animal data does not show that administering ALCOHOL alone causes cancer, although there is sufficient evidence for the carcinogenicity of acetaldehyde (the major metabolite of alcohol). When alcohol was administered to animals who were also exposed to known carcinogens, the animals who were given the alcohol had a higher rate of tumors of pituitary and adrenal glands, pancreatic islet cells, esophagus, and lungs. They had higher levels of liver-cell (hepatocellular) carcinomas, liver angiosarcomas, and neoplastic nodules of the liver, as well as benign tumors of the nasal cavity and trachea.

Digestive-Tract Cancer. Epidemiological studies in humans have shown a causal relationship between alcohol consumption and cancer of the digestive tract, primarily cancer of the oral cavity, the pharynx (nasopharynx excluded), and the larynx—all two to five times more likely in alcoholics; the esophagus—two to four times more likely; and the liver—liver cancer was increased 50 percent, with primary liver cancer increasing twofold to threefold. These findings persisted, even after adjusting for the effects of smoking, with the relative risk for cancer increasing with the amount of alcohol consumed.

Colon Cancer. Evidence suggests there may be a possible causal link between alcohol, especially beer, and cancer of the large bowel. The risk for colon cancer was increased between 15 percent and threefold depending on the study; that for rectal cancer was increased up to twofold. Unfortunately, these studies did not control for differences in diet.

The mechanism of action appears to be that alcohol acts as a local irritant to the upper gastrointestinal tract, whereas chronic excessive drinking

affects the liver, because of the accumulation of the alcohol metabolite acetaldehyde.

Breast Cancer. A strong association exists between alcohol and breast cancer, and there appears to be a dose-response relationship with an apparent relative risk of 1.5 to 2. This relationship held even after controlling for a number of other factors known to affect breast cancer. As the full etiology (cause) of breast cancer is not yet known, an as-yet-unrecognized factor may account for some of these findings. Overall, it has been estimated that as many as 10 percent of all cancer deaths are due to alcohol.

OTHER DRUGS AND CANCER

The role of TOBACCO as a carcinogen is well established and is discussed elsewhere. The role of other drugs as a cause of cancer is more unclear. Some drugs may have a role in cancer development because of mode of administration or degree of carcinogenicity. In vitro studies have shown mutagenic properties in a number of drugs—LYSERGIC ACID DIETHYLAMIDE (LSD), OPIUM and its derivatives such as MORPHINE, synthetic narcotics such as METHADONE, and some compounds found in MARIJUANA. There have been clinical reports of cancers in the respiratory tract, primarily the lungs, in heavy marijuana users, of the nasal passages in cocaine users, and in a number of organs in LSD users. Higher rates of esophageal cancer have been reported in opium smokers.

Injection drug users who share needles are at risk of developing ACQUIRED IMMUNODEFICIENCY SYNDROME (AIDS); a high proportion of AIDS patients develop the otherwise rare Kaposi's sarcoma. This form of cancer is more likely a complication of AIDS than the result of drug carcinogenicity, however.

NARCOTIC and psychoactive drugs have an important role in cancer treatment. Narcotic ANALGESICS (painkillers) are widely used to control cancer pain. Cancer patients have used *Cannabis sativa* (marijuana) to reduce the nausea associated with chemotherapy. LSD has been used in treating psychological disturbances associated with cancer.

(SEE ALSO: *Complications: Immunological*)

BIBLIOGRAPHY

BRESLOW, N. E., & DAY, N. E. (1980/1987). *Statistical methods in cancer research, vol. 1, The analysis of case-control studies; vol. 2, The design and analysis of cohort studies.* Lyon, France: World Health Organization (IARC Scientific publications no. 32 and 82).

HANKS, G. W., & JUSTINS, D. M. (1992). Cancer pain: Management. *Lancet, 339*(8800), 1031–1036.

WORLD HEALTH ORGANIZATION. (1988). *Alcohol drinking—IARC monographs on the evaluation of carcinogenic risks to humans.* Lyon, France: World Health Organization International Agency for Research on Cancer, vol. 44.

MANUELLA ADRIAN

CANNABIS SATIVA This is the botanical name for the HEMP plant that originated in Asia. It is the basis of the hemp industry as well as the source of the widely used intoxicant TETRAHYDROCANNABINOL (THC), the active agent in MARIJUANA, HASHISH, GANJA, and BHANG.

HISTORY

The use of *Cannabis sativa* has been recorded for thousands of years, beginning in Asia. It was known to the ancient Greeks and later to the Arabs, who, during their spread of Islam from the seventh to the fifteenth centuries, also spread its use across the Levant and North Africa. Some 200–300 million people are estimated to use *Cannabis* in some form worldwide. Thus, it is not only one of the oldest known but also one of the most widely used of mind-altering drugs.

Since the 1960s, the rise in its use in the United States has been enormous and associated with the youth movement and countercultural revolution. Although the drug was in use before that time, it was popular only in some ethnic and specialized groups (e.g., jazz musicians). By the 1990s, some 30–40 million Americans are estimated to have used it and a substantial number use it regularly—although since 1979 the number of youngsters initiated into its use has been declining after a steep rise with an increasingly lower age of first use.

BOTANY

Cannabis sativa grows in the tropical, subtropical, and temperate regions. It is generally considered a single species of the mulberry family (Moraceae) with multiple morphological variants (e.g., *C. indica*

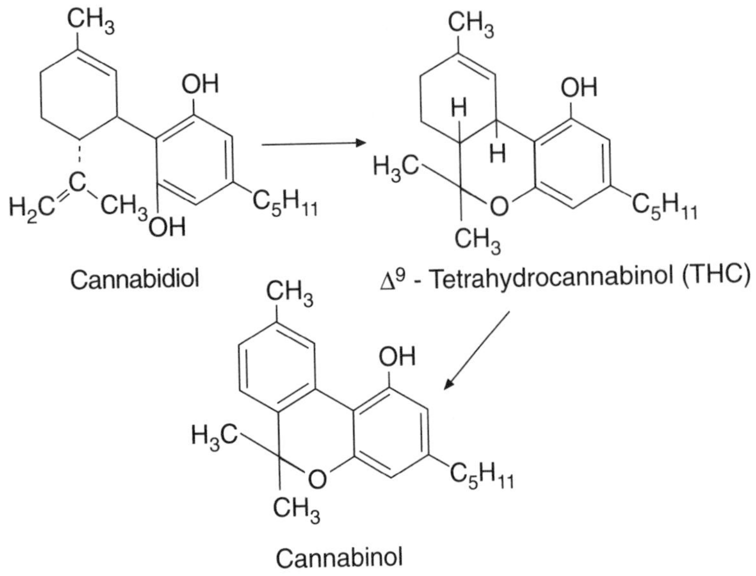

Figure 1
Biosynthetic Pathway of Cannabinoids

or *C. americana*). It is an herb of varying size; some are quite bushy and attain a height of 10 to 15 feet (3 to 4.6 m). Due to genetic differences, some plants produce strong fibers (but little THC) and others produce a substantial quantity of THC but weak fibers. The fiber-producer is grown commercially for cloth, rope, roofing materials, and floor coverings; this was cultivated as a cash crop in colonial America for such purposes (Hart, 1980). During World War II, when it appeared that the United States might be cut off from Southeast Asian hemp, necessary to the war effort, the plants were cultivated in the midwestern states. Some of them continue to grow wild today, but since they are of the fiber-producing variety, they contain little drug content.

The drug-producing variety is widely cultivated in societies where its use is condoned. Illegal crops are also planted, some in the United States. The choice parts are the fresh top leaves and flowers of the female plant. The leaves have a characteristic configuration of five deeply cut serrated lobes. When they are harvested, they often resemble lawn cuttings—which accounts for the slang term "grass."

CHEMISTRY

The collective name given to the terpenes found in *Cannabis* is cannabinoids. Most of the naturally oc-

curring cannabinoids have now been identified, and three are the most abundant—cannabidiol (CBD), tetrahydrocannabinol (THC), and cannabinol (CBN). The steps from CBD to THC to CBN represent the biosynthetic pathway in the plant. THC is an optically active resinous material that is very lipid-soluble but water-insoluble; these physical properties make pharmacological investigations difficult, since various nonpolar solvents must be used. Although many other materials have been found in this plant, the cannabinoids are unique to it and THC is the only one with appreciable mental affects. THC is believed to be largely, if not solely, responsible for the effects desired by those who use *Cannabis* socially. Virtually all the effects produced by smoking or eating some of the whole plant can be attained by using THC alone.

USE AS A SOCIAL DRUG

Cannabis grows so easily that it is called a weed. In the United States, where it remains illegal, it is possible for those who wish to use it as a social drug to grow their own supply. The ease of cultivation keeps the price of imported illicit marijuana down, which helps account for some of its widespread use. Such cultivation is, however, as illegal as possession of the drug obtained from illicit "street" sources.

(SEE ALSO: *Anslinger, Harry J., and U.S. Drug Policy; High School Senior Survey*)

BIBLIOGRAPHY

HART, R. H. (1980). *Bitter grass: The cruel truth about marihuana.* Shawnee Mission, KS: Psychoneurologia Press.

RAZDAN, R. K. (1986). Structure-activity relationships in cannabinoids. *Pharmacology Review, 38,* 75–148.

LEO E. HOLLISTER

CASUAL DRUG USE   *See* Addiction: Concepts and Definitions; Lifestyles and Drug Use Complications.

CATECHOLAMINES   The catecholamines are a series of structurally similar amines (e.g., DOPAMINE, epinephrine, NOREPINEPHRINE) that function as hormones, as NEUROTRANSMITTERS, or as both. Catecholamines are produced by enzymatic conversion of tyrosine, sharing the chemical root of 3,4 dihydroxyphenylethanolamine. The three major catecholamines (mentioned above) derive from sequential enzymatic reactions—tyrosine is converted to dihydroxyphenylacetic acid (dopa); dopa, which is not an end product but a common intermediate (and the medication of choice for Parkinson's disease), is converted to dopamine; dopamine to noradrenaline (also called norepinephrine); and noradrenaline to adrenaline (also called epinephrine). These substances are the transmitters for the sympathetic neurons (nerve cells) of the autonomic nervous system, as well as for three separate broad sets of brain neuropathways.

BIBLIOGRAPHY

SNYDER, S. H. (1980). *Biological aspects of mental disorder.* New York: Oxford University Press.

FLOYD BLOOM

CATHINONE   *See* Khat.

CAUSES OF SUBSTANCE ABUSE   This section contains articles on some of the many factors thought to contribute to substance use, abuse, and dependence. It includes discussions of *Drug Effects and Biological Responses, Genetic Learning,* and an article on the *Psychological (Psychoanalytic) Perspective. Sociocultural* causes and *Vulnerability,* are discussed in several articles throughout the Encyclopedia, for example, *Ethnicity and Drugs, Families and Drug Use, Lifestyle and Drug Use Complications, Poverty and Drugs.* See also the article *Disease Concept of Alcoholism and Drug Addiction* and the section on *Vulnerability.*

Drug Effects and Biological Responses
Although many indirect factors lead to an individual abusing drugs, a person's response to the effects of the drugs themselves contribute both to their use and abuse. These drug effects should be considered in relation to four phases of drug use: (1) initiation-consolidation, (2) maintenance, (3) repeated withdrawal and relapse, and (4) postwithdrawal. During the initiation-consolidation phase, behaviors that lead to the taking of a drug are gradually strengthened through operant and classical conditioning processes and by biochemical changes in brain. The drug effects include a cascade of discriminative or internally appreciated drug cues (i.e., subjective effects). The presence of these cues often leads to associated autonomic responses and reports of urges in humans. These responses and urges may result in an unfolding of a sequence of behavioral and physiological events leading to continued drug consumption.

After a pattern of chronic drug use is established, individuals may become tolerant to certain of a drug's effects. In addition, they may experience withdrawal effects when they stop taking a drug. Withdrawal effects are often opposite to the drug-induced state and usually involve some form of dysphoria—a state of illness and distress. Over time, withdrawal effects become associated with stimuli in the environment, as was the case for the euphoric and other direct effects of the drug. Because of operant and classical conditioning processes, these associated stimuli can then produce conditioned effects that are often characterized as urges or cravings and that may trigger relapse.

The underlying NEUROTRANSMITTER systems within the brain subserving these behavioral features

of drug effects are just beginning to be understood. Early RESEARCH on the neural substrates of reward in general used electrical brain stimulation as the reward. For example, Olds (1977) found that rats would press a lever to receive a brief electrical pulse to the hypothalamus; rats would press this lever to such an extent that they did not engage in consummatory reward activities such as eating and drinking. Subsequent research indicated that activation of certain systems in the brain, namely the mesolimbic and nigrostriatal dopaminergic systems, were most sensitive to brain stimulation reinforcement. Several theories have been suggested to explain the importance of the brain reward system for the survival of species (Conrad, 1950; Glickman & Schiff, 1967; O'Donahue & Hagmen, 1967; Roberts & Carey, 1965).

Further research demonstrated that most drugs of abuse lower the threshold for this brain stimulation reward, thus suggesting that such drugs may activate the same, or similar, reward pathways (see Koob & Bloom, 1988). As will be seen, furthermore, the reinforcing effects of the drugs themselves—that is, effects that lead individuals to take the drugs—are directly mediated by these REWARD systems. The fact that many drugs induce activation of these systems may indicate a mechanism underlying the addiction-related effects of drugs of abuse.

COCAINE AND OTHER STIMULANTS

COCAINE is an indirect catecholamine agonist that acts by blocking the reuptake of monoamines, including DOPAMINE (DA), NOREPINEPHRINE (NE), and serotonin (5-HT). During the process of reuptake, the previously released neurotransmitter is actively transported back from the synaptic cleft into the presynaptic terminal of the neuron where the neurotransmitter was produced and released (Pitts & Marwah, 1987). In contrast to cocaine, AMPHETAMINE acts not only by inhibiting uptake, but also by releasing catecholamines from newly synthesized storage pools from the presynaptic terminal of the neuron (e.g., Carlsson & Waldeck, 1966).

Amphetamine and cocaine are both potent PSYCHOMOTOR stimulants. They produce increased alertness and energy and lower ANXIETY and social inhibitions. The acute reinforcing actions of the stimulants are primarily determined by their augmentation of DA systems. With prolonged consumption: (1) acute TOLERANCE becomes substantial, and

(2) the individual starts to regularly consume higher and many more doses if the resources are available. Over time, in high-dose regimens, the behavioral pattern of use becomes stereotyped and restricted. In settings of low availability, the individual focuses on the acquisition and consumption of the drug. These effects of stimulants occur within weeks or months of continued use. The individual may also start "bingeing" during this period. A binge is characterized by the readministration of the drug approximately every ten to twenty minutes, resulting in frequent mood swings (i.e., alternations of highs and lows). Cocaine binges typically last twelve hours, but may last as long as seven days.

It has been proposed that cocaine abstinence consists of a three-phase pattern: crash, WITHDRAWAL, and extinction (Gawin & Kleber, 1986; Gawin & Ellinwood, 1988). The crash phase immediately follows the cessation of a binge and is characterized by initial depression, agitation, and anxiety. Over the first few hours, drug craving is replaced by an intense desire for sleep. During this time, the individual may use ALCOHOL, BENZODIAZEPINES, or OPIATES to induce sleep. Following the crash, hypersomnolence (excessive sleep) and hyperphagia (excessive appetite) develop. Following the first few days of hypersomnolence and hyperhagia, other symptoms emerge that are the opposite of the effects of cocaine—withdrawal symptoms. During this withdrawal period, which lasts three to ten days, individuals experience decreased energy, limited interest in their environment, and anhedonia. They are also strongly susceptible to RELAPSE and starting another binge cycle (Gawin & Ellinwood, 1988; Gawin & Kleber, 1986; Jaffe, 1985). This phase is followed in time by the extinction phase, in which relapse to cocaine use is prevented. During the extinction phase, brief periods of drug CRAVING also occur. These episodes of craving are thought to be triggered by conditioned stimuli that were previously associated with the drug. If the individual experiences these cues without the associated drug effects—that is, resists relapse—then the ability of these cues to elicit drug cravings should diminish over time, which in turn should lessen the probability of relapse (Gawin & Ellinwood, 1988).

As already noted, acute administration of cocaine produces profound inhibition of dopaminergic uptake (Fuxe, Hamberger, & Malmfors, 1967). The relation between cocaine dose and DA levels is linear; therefore, larger amounts of cocaine result in higher

TABLE 1
**Effects of Chronic Cocaine and Amphetamine Administration
on Dopaminergic Functioning**

	Amphetamine		Cocaine	
	Day 1	Day 7	Day 1	Day 7
Autoreceptor Sensitivity	sub	super	sub	super
Receptors	decreased	decreased	unclear	unchanged
Biosynthesis	reduced	reduced	unchanged	unchanged
Uptake Sites	decreased	decreased	unchanged	unchanged

extracellular DA levels. These levels of DA are thought to underlie the reinforcing effects of cocaine (Gawin & Ellinwood, 1988). Because both cocaine and amphetamine result in enhanced dopaminergic neurotransmission, thereby producing elevated extracellular levels of catecholamines, these elevated neurotransmitter levels would presumably have local time-dependent inhibitory effects on the enzyme tyrosine hydroxylase, which is responsible for controlling their rate of synthesis. Therefore, this substrate-inhibitory mechanism might compensate for the increased catecholamine levels and activity by decreasing their synthesis. Galloway (1990) found that cocaine, in a way that was consistent with this proposition, decreased DA synthesis in a dose-dependent manner in various brain regions.

Chronic, intermittent stimulant use (e.g., 1–2 injections per 24 hrs) produces other behavioral effects besides euphoria and increased energy: (1) stimulant psychosis, which is characterized by paranoia, anxiety, stereotyped compulsive behaviors, and HALLUCINATIONS, and (2) sensitization or "reverse tolerance." Sensitization refers to the fact that the effects of cocaine are progressively enhanced. Although sensitization has been demonstrated in animal studies, it is not clear whether it occurs in humans. There are nevertheless several possible explanations for sensitization. First, because cocaine blocks dopaminergic uptake, chronic cocaine use could somehow harm the functioning of the dopamine uptake mechanism; the evidence regarding this possibility is equivocal (Zahniser et al., 1988b). Second, sensitization could also be the result of enhanced dopaminergic release, similar to that found chronic after amphetamine administration (Castaneda, Becker, & Robinson, 1988). Akimoto, Hammamura, & Otsuki (1989) found enhanced DA release in the striatum one week following chronic

cocaine administration. Similar data have been obtained by others (Kalivas et al., 1988; King et al., 1993; Pettit et al., 1990). Cocaine levels in blood and cerebrospinal fluid have also been reported to the elevated in chronically treated subjects (Reith, Benuck, & Lajtha, 1987); however, these increases cannot account for most of the change in DA release (Pettit et al., 1990). Furthermore, some researchers report no consistent effects in this regard. Third, there could be changes in autoreceptor sensitivity following chronic cocaine administration. Autoreceptors for particular neurotransmitters are those receptors that reside on the same NEURON that releases the NEUROTRANSMITTER. The autoreceptors on the somatodendritic area of neurons regulate impulse flow along the neuron, whereas autoreceptors on the terminal regions of the neuron regulate the amount of neurotransmitter released per impulse and neurotransmitter synthesis (Cooper, Bloom, & Roth, 1986). Sensitization could, therefore, be the result of decreased autoreceptor sensitivity. Such subsensitivity would result in either increased impulse flow, if somatodendritic autoreceptors were altered, or increased neurotransmission/synthesis, if terminal autoreceptors were altered. The net effect, in either case, would be an increase in dopaminergic neurotransmission. There is some evidence of decreased somatodendritic autoreceptor sensitivity twenty-four hours after the cessation of chronic cocaine administration (Henry, Greene, & White, 1989). However, seven days after termination of daily cocaine injections, when cocaine-induced sensitization is still fully present, somatodendritic autoreceptors are no longer reduced in sensitivity (Zhang, Lee, & Ellinwood, 1992). Evidence regarding changes in terminal autoreceptor sensitivity is mixed. Dwoskin et al. (1988) found that terminal autoreceptors were supersensitive, not subsensitive, to a DA agonist

twenty-four hours following chronic cocaine use. Henry et al. (1989) also found that terminal autoreceptors were supersensitive to DA following chronic daily cocaine injections. Although autoreceptor supersensitivity cannot explain sensitization, it is a possible mechanism underlying the previously described anhedonia and anergy experienced by cocaine abusers during the withdrawal phase. Fourth and last, there could be an increase in the number or sensitivity of postsynaptic DA receptors. The evidence regarding this hypothesis is also mixed (Zahniser et al., 1988a). For example, Peris et al. (1990) found an increased number of postsynaptic D2 receptors in the NUCLEUS ACCUMBENS one day following cessation of chronic cocaine administration; however, after one week the number of receptors had returned to normal levels. In contrast, there is some evidence that postsynaptic DA receptors are decreased following chronic cocaine use. Volkow et al. (1990) found lower uptake values for [18 F]n-methylspiroperidol in human cocaine users who had been abstinent for one week, as compared with normal subjects. Uptake values were similar, however, for normal subjects and cocaine users who had been abstinent for one month.

In contrast with these results, Yi and Johnson (1990) have reported that chronic intermittent cocaine use impairs the regulation of synaptosomal 3[H]-DA release by DA autoreceptors, thus suggesting a subsensitivity or down-regulation of release-modulating DA autoreceptors seven days after chronic cocaine administration. The differences in the results of the Yi and Johnson (1990) and the Dwoskin et al. (1988) studies may be due to differences in the administration schedules or in the procedures used to measure autoreceptor sensitivity.

In contrast with the changes induced by intermittent but chronic drug administration, a regimen that involves the chronic administration of steady-state levels of drug results in decreased DA overflow when striatal brain slices are perfused with cocaine. This result may be due to the development of supersensitive autoreceptors. Autoreceptor supersensitivity would result in decreased dopaminergic activity. There is some support for this hypothesis from research involving the chronic administration of amphetamine. Lee and colleagues (Lee, Ellinwood, & Nishita, 1988; Lee & Ellinwood, 1989) found that twenty-four hours after withdrawal from a week of continuous administration of amphetamine, all indicators of autoreceptor activity demonstrated a pro-

nounced subsensitivity. Similar results have been found following the continuous infusion of cocaine (Zhang et al., 1992). However, by the seventh day of withdrawal (a period associated with anergia, irritability, and "urges" in human stimulant abusers), nigrostriatal somatodendritic autoreceptors progress from an initial subsensitivity to a supersensitive state, whereas terminal autoreceptors are normosensitive. The changes in sensitivity of receptors clearly depend on the way the drug is administered and which receptors are evaluated. The evidence, moreover, is not always consistent.

There is also evidence that chronic cocaine administration produces neurotoxicity—i.e., actual destruction of neural tissue—although there are conflicting results and the relationship of this neurotoxicity to the addiction process is unclear. For example, Trulson and colleagues (1986) demonstrated decreased tyrosine hydroxylase activity sixty days after chronic cocaine treatment (see also Trulson & Ulissey, 1987), thereby indicating decreased DA synthesis. (Tyrosine hydroxylase is the rate-limiting step in the biosynthesis of DA; Cooper et al., 1986.) Similarly, Taylor and Ho (1978) found that chronic administration of cocaine decreased tyrosine hydroxylase activity in the caudate, but Seiden and Kleven (1988) were unable to replicate the findings of Trulson. As contrasted with the inconclusive results on cocaine, research involving amphetamine is much clearer. First, chronic METHAMPHETAMINE administration reduces the number of DA uptake sites (Ricaurte, Schuster, & Seiden, 1980; Ricaurte, Seiden, & Schuster, 1984). Second, DA and tyrosine-beta-hydroxylase levels are reduced for extended periods following chronic amphetamine administration (Ricaurte et al., 1980, 1984). Third, there is evidence of neuronal degeneration, chromatolysis, and decreased catecholamine histofluorescence (Duarte-Escalante & Ellinwood, 1970).

As with cocaine's effects on DA reuptake, cocaine also blocks 5-HT reuptake. Since activation of 5-HT postsynaptic receptors affects neurotransmission in neurons that release DA, this blockade prolongs the inhibitory effects of 5-HT on dopaminergic neurotransmission (Taylor & Ho, 1978). However, cocaine also inhibits the firing rates of dorsal raphe 5-HT neurons (Cunningham & Lakoski, 1988, 1990). Thus, acutely the net effect of cocaine on 5-HT neurotransmission in the nucleus accumbens will depend on the relative contributions of uptake inhibition, which would increase synaptic 5-HT, and

inhibition of neuronal firing, which would decrease synaptic 5-HT. Broderick (1991) reported that acute, subcutaneous injections of cocaine resulted in a dose-dependent increase in DA levels, as measured by dialysis of the nucleus accumbens. This suggests a decrease in 5-HT levels that may result from activation of somatodendritic 5-HT autoreceptors located in the dorsal raphe nucleus. Acute cocaine administration has indeed been reported to almost completely inhibit the basal firing rate of dorsal raphe serotonergic neurons.

As with the effects of chronic amphetamine administration on the functioning of DA systems, chronic methamphetamine administration has been shown to induce pronounced long-term changes in tryptophan hydroxylase activity, as well as in 5-HT content and number of uptake sites (Ricaurte et al., 1980). The effects of chronic cocaine on serotonergic functioning are less well established. For example, Ho et al. (1977) found decreased levels of 5-HT following chronic cocaine administration. Seiden and Kleven (1988), however, failed to find any effects of chronic cocaine on the biosynthesis of serotonin.

Some of these discrepancies can be reconciled by the fact that different chronic dosing regimens produce different changes in 5-HT systems. For example, Cunningham and colleagues found that daily injections of cocaine resulted in an increased sensitivity of dorsal raphe somadendritic 5-HT autoreceptors to cocaine's inhibitory effects as measured by electrophysiological techniques (Cunningham & Lakoski, 1988, 1990). These results are consistent with the behavioral data of King and colleagues (1993a), who found that daily cocaine injections produced an enhanced inhibitory effect of NAN-190 on cocaine-induced locomotion and an enhanced excitatory effect of 8-OH-DPAT on locomotion. In contrast with these results, the continuous infusion of cocaine via an osmotic minipump results in a decreased sensitivity of dorsal raphe somadendritic 5-HT autoreceptors and a decreased excitatory effect of 8-OH-DPAT on locomotion (King, Joyner, & Ellinwood, 1993b; King et al., 1993b).

Interestingly, the depletion of 5-HT or reduction of 5-HT neurotransmission is associated with impulsive behavior. For example, Linnoila et al. (1983) found that violent offenders with a diagnosis of PERSONALITY DISORDER associated with impulsivity had lower levels of 5-hydroxyindoleacetic acid (5-HIAA, the metabolite of 5-HT) than other offenders.

Bouilliouc et al. (1978, 1980) reported suppressed 5-HIAA levels in the cerebrospinal fluid (CSF) of XXY-chromosome (aggressive) institutionalized criminals. Lithium has been successfully used to treat violent offenders (Sheard, 1971, 1975; Sheard et al., 1976; Tupin et al., 1973). Lithium treatment produces increases in CSF 5-HIAA in humans (Fyro et al., 1975; Sheard & Aghajanian, 1970), and central nervous system (CNS) 5-HT in nonhumans (Sheard & Aghajanian, 1970; Mandell & Knapp, 1977); this indicates that increases in 5-HT are associated with decreases in violent and aggressive behavior. After extensively reviewing the literature, Brown and Linnoila (1990) concluded that low levels of CSF 5-HIAA are related to disinhibition of aggressive/impulsive behavior and not to antisocial acts in and of themselves. The transition to high-dose cocaine use might be considered impulsive behavior because the individual is focusing on the immediate, short-term advantages of drug consumption while ignoring the long-term advantages of drug abstinence. Hence, the 5-HT receptor supersensitivity, and resulting inhibition of 5-HT neurotransmission, may be a contributing factor to the development of the high-dose, bingelike pattern of cocaine abuse.

OPIATES

The OPIATES are derived from the POPPY plant and have been used for centuries. A number of types of endogenous opiate RECEPTORS have been identified and their locations mapped. There are high concentrations of opiate receptors in the caudate nucleus, nucleus accumbens, periventricular gray region, and the nucleus arcuatus of medobasal hypothalamus (Pert, Kuhar, & Snyder, 1975, 1976). These areas may be differently involved in the reinforcing, aversive, and dependence-producing effects of the opiates. Furthermore, different receptor subtypes may mediate the different effects of the opiates.

The opiates produce ANALGESIA, changes in mood (e.g., euphoria and tranquility), drowsiness, respiratory depression, and nausea (Jaffe & Martin, 1990). These drugs also reduce motivated behavior; there is a decrease in appetite, sexual drive, and aggression. Intravenous administration of opioids results in initial effects of flushing of the skin and sensations in the abdominal regions that have been likened to a sexual orgasm (Jaffe, 1990).

With continuous use of opioids, marked tolerance develops to some, but not all, of the effects of these

drugs. Tolerance to opioids is generally characterized by a shorter duration of effect and attenuated analgesia, euphoria, and other CNS-depressant effects; however, there is less tolerance to the lethal effects of opiates. Therefore, if an individual administers ever larger doses to obtain the same effect (e.g., the rush or high), this may increase the probability of a lethal overdose (Jaffe, 1990).

Although the course and severity of withdrawal symptoms following opiate abstinence depend on which opiate was used, the dose and pattern of consumption, the duration of use, and the interdose interval, the opiate withdrawal syndrome follows the same general progression. Approximately eight to twelve hours after the last dose, individuals experience yawning, lacrimation, and rhinorrhea; twelve to fourteen hours after the final dose, they may fall into a fitful, restless sleep and awaken feeling worse than when they went to sleep. With the continuation of opiate withdrawal, they experience increasing dysphoria, anorexia, gooseflesh, irritability, agitation, and tremors. At the peak intensity of the withdrawal symptoms, they may experience exacerbated irritability, insomnia, intense anorexia, weakness, and profound depression. Common symptoms include alternating coldness and intense skin flushing and sweating, vomiting and diarrhea (Jaffe, 1990). This pattern of symptoms indicates that during the initial withdrawal phase there is a generalized CNS hyperexcitability. Thus, the addicted opiate abuser continues to recycle opiate use to both avoid or terminate the wtihdrawal symptoms, and to reexperience the euphoric effects. This powerful combination of euphoria, tolerance, and withdrawal can lead to profound levels of addiction.

Studies have found that rats and monkeys will self-administer opioids, thus indicating that these drugs serve as reinforcers (Koob & Bloom, 1988). Chronic opioid administration results in physical dependence, as demonstrated by the presence of a withdrawal syndrome following drug cessation. Most clinicians hold the classic position that PHYSICAL DEPENDENCE (i.e., avoidance of withdrawal symptoms) is a major motivating factor in opiate self-administration, but evidence indicates that reinforcement and withdrawal are separate processes. Bozarth and Wise (1984) demonstrated that rats will self-administer morphine into the ventral tegmental area without the presence or development of any apparent withdrawal symptoms. Chronic administration of morphine into the periaqueductal gray area, however, produces signs of a strong withdrawal syndrome.

Several lines of evidence indicate that dopaminergic neurotransmission may partially mediate the reinforcing effects of opiate administration. First, injection of met-enkephalin into the ventral tegmental area results in increases in DA release in the nucleus accumbens (Di Chiara & Imperato, 1988). Second, although opiates generally produce sedation, low doses of systemic morphine increase locomotor activity (Domino, Vasko, & Wilson, 1976). Third, injections of morphine into the ventral tegmental area produce circling behavior (Holmes, Bozarth, & Wise, 1983); injections of opiates into the ventral tegmental area produce increased locomotion, as with systemic injections of opiates, thereby suggesting increased dopaminergic transmission (Blaesig & Herz, 1980). Fourth, selective lesions of the dopaminergic system decrease opiate self-administration, although not to the extent of affecting cocaine self-administration (Bozarth & Wise, 1985). Fifth, rats learn to self-administer opiates directly into the ventral tegmental area (Bozarth & Wise, 1981), rats also inject opiates into the nucleus accumbens and the lateral hypothalamus (Goeders, Lane, & Smith, 1984). Sixth, administration of the D1 antagonist SCH 23390, but not the D2 antagonists sulpiride and spiperone, block the reinforcing effects of morphine.

Ettenberg et al. (1982) found no effect of alpha flupenthixol, primarily a D2 antagonist, on heroin self-administration, although the same doses decreased cocaine self-administration. Similar results have been reported by others using other dopaminergic antagonists (De Wit & Wise, 1977). Thus, both place preference and self-administration procedures indicate that opiates are not reinforcing through D2 receptors, which are vital to stimulant reinforcement. These results indicate that opiate reinforcement is at least partially independent of the D2 stimulant type of reinforcement, yet they do act through a dopaminergic mechanism to induce a significant part of their effects.

Chronic administration of opiates produces several behavioral and neurochemical effects that may be related to their reinforcing effects. First, chronic administration of MORPHINE results in the augmentation of the behavioral effects of low doses of morphine. In other words, subjects undergoing chronic opiate administration become sensitized to the be-

havioral effects of morphine (Ahtee, 1973, 1974). Second, chronic morphine administration results in decreased DA turnover in the striatum and limbic system during withdrawal (Ahtee & Atilla, 1980, 1987). Third, in mice withdrawn from morphine, the synthesis and release of DA are attenuated (Ahtee et al., 1987); similar results have been obtained with human heroin addicts in which CSF homovanillic acid concentrations were decreased (Bowers, Kleber, & Davis, 1971).

These results indicate that during chronic morphine administration there is a down-regulation of the dopaminergic system and a neuroadaptation to this depletion. During withdrawal from opiate administration there is an augmentation of dopaminergic mechanisms. Indeed, during withdrawal rats are sensitized to the behavioral effects of apomorphine (Ahtee & Atilla, 1987), and small doses of morphine increase striatal homovanillic acid levels more in withdrawn than in control rats, thereby indicating that the dopaminergic system is sensitized at this point (Ahtee, 1973, 1974). Thus, some of the withdrawal symptoms (e.g., irritability and dysphoria) may be mediated by changes in dopaminergic functioning.

Acute administration of opiates increases the synthesis of 5-HT and the formation of 5-HIAA, and these effects are eliminated by the administration of opiate antagonists (Ahtee & Carlsson, 1979), thus suggesting that opiate administration results in increased serotonergic functioning. Indeed, acute administration of dynorphin-(1-13), while it decreases striatal dopamine, actually increases striatal serotonin (Broderick, 1987). This increased serotonergic functioning may contribute to the "post-consummatory calm" produced by opiate drugs: Increasing serotonergic functioning would tend to inhibit incentive-motivated behaviors and produce a calm, tranquil state. Indeed, the atypical anxiolytic drug buspirone exerts its anxiety-reducing effects via serotonergic activation.

During withdrawal from chronic opioid administration, 5-HIAA levels are decreased (Ahtee, 1980; Ahtee et al., 1987). This pattern of serotonin results could well cause increased impulsivity and a higher probability of relapse, similar to that described earlier in relation to the psychomotor stimulants.

In summary, like cocaine, the opiates are consumed because of their reinforcing properties. These reinforcing properties are the result of activation of endogenous opiate receptors; furthermore, activation of the dopaminergic system modulates the reinforcing effects of opiates. During chronic opiate administration, subjects become physically dependent. There is an increase in dynorphin levels that may mediate some of the aversive aspects of the withdrawal syndrome (e.g., decreased dopaminergic functioning). Furthermore, during chronic administration, there is functional down-regulation of both the dopaminergic and serotonergic systems. Upon withdrawal from opiates, there is a subsequent supersensitivity of the dopaminergic system. This dopaminergic supersensitivity may be involved in opiate craving and general irritability during withdrawal.

(SEE ALSO: *Addiction: Concepts and Definitions; Brain Structures and Drugs; Opioids: Complications and Withdrawal; Tolerance and Physical Dependence; Research: Animal Model; Wikler's Pharmacologic Theory of Drug Dependence*)

BIBLIOGRAPHY

AHTEE, L. (1980). Chronic morphine administration decreases 5-hydroxytryptamine and 5-hydroxyindoleacetic acid content in the brain of rats. *Med Bio, 58,* 38–44.

AHTEE, L. (1974). Catalepsy and stereotypes in rats treated with methadone: Relation to striatal dopamine. *Euro J. Pharmacol, 27,* 221–230.

AHTEE, L. (1973). Catalepsy and stereotyped behaviour in rats treated chronically with methadone: Relation to brain homovanillic acid content. *J Pharm Pharmacol, 25,* 649–651.

AHTEE, L. & ATILLA, L. M. J. (1987). Cerebral monoamine neurotransmitters in opioid withdrawal and dependence. *Med Bio, 65,* 113–119.

AHTEE, L. & ATILLA, L. M. J. (1980). Opioid mechanisms in regulation of cerebral monoamines in vivo. In O. Eraenkoe, S. Soinila, & H. Paevaerinta (Eds.), Histochemistry and Cell Biology of Autonomic Neurons, SIF Cells, and Paraneurons. *Advances in Biochemical Psychopharmacology, 25,* 361–365. New York: Raven Press.

AHTEE, L., & CARLSSON, A. (1979). Dual action of methadone on 5-HT synthesis and metabolism. *Naunyn-Schmiedeberg's Arch Pharmacol, 307,* 51–56.

AHTEE, L., ET AL. (1987). The fall of homovanillic acid and 5-hydroxyindoleacetic acid concentrations in brain of

mice withdrawn from repeated morphine treatment and their restoration by acute morphine administration. *J Neural Trans, 68*, 63–78.

AKIMOTO, K., HAMMAMURA, T., & OTSUKI, S. (1989). Subchronic cocaine treatment enhances cocaine-induced dopamine efflux, studied by in vivo intracerebral dialysis. *Brain Research, 490*, 339–344.

BLAESIG, J., & HERZ, A. (1980). "Interactions of opiates and endorphins with cerebral catecholamines. In L. Szekeres (Ed.), *Handbook of experimental pharmacology: Adrenergic activators and inhibitors* (Vol. 54). Heidelberg: Springer-Verlag.

BOZARTH, M. A., & WISE, R. A. (1985). Involvement of the ventral tegmental dopamine system in opioid and psychomotor stimulant reinforcement. In L. S. Harris (Ed.), *Problems of Drug Dependence*, Washington, DC: U.S. Government Printing Office.

BOZARTH, M. A., & WISE, R. A. (1984). Anatomically distinct opiate receptor fields mediate reward and physical dependence. *Science, 244*, 516–517.

BRODERICK, P. A. (1987). Striatal neurochemistry of dynorphin-(1-13): In vivo electrochemical semidifferential analyses. *Neuropeptides, 10*, 369–386.

CARLSSON, A., & WALDECK, B. (1966). Effects of amphetamine, tyramine, and protriptyline on reserpine resistant amine-concentrating mechanisms of adrenergic nerves. *J Pharm Pharmacol, 18*, 252–253 (1966).

CASTANEDA, E., BECKER, J. B. & ROBINSON, T. W. (1988). The long-term effects of repeated amphetamine treatment in vivo on amphetamine, KCL and electrical stimulation evoked striatal dopamine release in vitro. *Life Sciences, 42*, 2447–2456.

CONRAD, L. (1950). The comparative method of studying innate behavior patterns. In *Symposia of Society for Experimental Biology* (Vol. 4; pp. 221–254).

COOPER, J. R., BLOOM, F. E., & ROTH, R. H. (1986). *The Biochemical Basis of Neuropharmacology* (5th ed.). New York: Oxford University Press.

CUNNINGHAM, K. A., & LAKOSKI, J. M. (1990). The interaction of cocaine with serotonin dorsal raphe neurons: Single-unit extracellular recording studies. *Neuropsychopharmacology, 3*, 41–50.

CUNNINGHAM, K. A., & LAKOSKI, J. M. (1988). Electrophysiological effects of cocaine and procaine on dorsal raphe serotonin neurons. *Euro J Pharmacol, 148*, 457–462.

DEWIT, H., & WISE, R. A. (1977). Blockade of cocaine reinforcement in rats with the dopamine receptor blocker pimozide, but not with the noradrenergic blockers phentolamine or phenoxybenzamine. *Canadian Journal of Psychology, 31*, 195–203.

DI CHIARA, G., & IMPERATO, A. (1988). Drugs abused by humans preferentially increase synaptic dopamine concentrations in the mesolimbic system of freely moving rats. *Proceedings of the National Academy of Sciences of the United States of America, 85*, 5274–5278.

DOMINO, E. F., VASKO, M. R., & WILSON, A. E. (1976). Mixed depressant and stimulant actions of morphine and their relationship to brain acetylcholine. *Life Sciences, 18*, 361–376.

DUARTE-ESCALANTE, O., & ELLINWOOD, E. H., JR. (1970). Central nervous system cytopathological changes in cat with chronic methedrine intoxication. *Brain Research, 21*, 151–155.

DWOSKIN, L. P., ET AL. (1988). Repeated cocaine administration results in supersensitivity of striatal D-2 DA autoreceptors to pergolide. *Life Sciences, 42*, 255–262.

ETTENBERG, A., ET AL. (1982). Heroin and cocaine intravenous self-administration in rats: Mediation by separate neural systems. *Psychopharmacology, 78*, 204–209.

FUXE, K. B., HAMBERGER, B., & MALMFORS, T. (1967). The effects of drugs on accumulation of monoamines in tubero-infundibular dopamine neurons. *Euro J Pharmacol, 1*, 334–341.

GALLOWAY, M. P. (1990). Regulation of dopamine and serotonin synthesis by acute administration of cocaine. *Synapse, 6*, 63–72.

GAWIN, F. H., & ELLINWOOD, E. H., JR. (1988). Cocaine and other stimulants: Actions, abuse and treatments. *New England Journal of Medicine, 318*, 1173–1182.

GAWIN, F. H., & KLEBER, H. D. (1986). Abstinence symptomatology and psychiatric diagnosis in cocaine abusers. *Archives of General Psychiatry, 43*, 107–113.

GLICKMAN, S. E., & SCHIFF, B. V. (1967). A biological theory of reinforcement. *Psychol Rev, 74*.

GOEDERS, N. E., LANE, J. D., & SMITH, J. E. (1984). Self-administration of methionine enkephalin into the nucleus accumbens. *Pharmacol Biochem Behav, 20*, 451–455.

HENRY, D. J., GREENE, M. A., & WHITE, F. J. (1989). Electrophysiological effects of cocaine in the mesoaccumbens dopamine system: Repeated administration. *Journal of Pharmacology and Experimental Therapeutics, 251*, 833–839.

HO, B. T., ET AL. (1977). Behavioral effects of cocaine-metabolic and neurochemical approach. In E. H. Ellinwood, Jr., M. M. Kilbey (Eds.), *Advances in Behavioral Biology: Cocaine and Other Stimulants*. New York: Plenum.

HOLMES, L. J., BOZARTH, M. A., & WISE, R. A. (1983). Circling from intracranial morphine applied to the ventral

tegmental area in rats. *Brain Research Bulletin, 11*, 295–298.

JAFFE, J. H. (1990). Drug addiction and drug abuse. In A. G. Gilman et al. (Eds.), *Goodman and Gilman's the pharmacological basis of therapeutics* (8th ed.). New York: Permagon.

JAFFE, J. H., & MARTIN, W. R. (1990). Opioid analgesics and antagonists. In A. G. Gilman et al. (Eds.), *Goodman and Gilman's the pharmacological basis of therapeutics* (8th ed.). New York: Permagon.

KALIVAS, P. W., et al. (1988). Behavioral and neurochemical effects of acute and daily cocaine administration in rats. *Journal of Pharmacology Experimental Therapeutics, 245,* 485–492.

KING, G. R., JOYNER, C., & ELLINWOOD, E. H., JR. (1993b). Withdrawal from continuous or intermittent cocaine: Behavioral responsivity to 5-HT₁ receptor agonists. *Pharmacology, Biochemistry, and Behavior, 45,* 577–587.

KING, G. R., KUHN, C., & ELLINWOOD, E. H. JR. (1993c). Dopamine efflux during withdrawal from continuous or intermittent cocaine. *Psychopharmacology, 111,* 179–184.

KING, G. R., ET AL. (1993a). Withdrawal from continuous or intermittent cocaine: Effects of NAN-190 on cocaine-induced locomotion. *Pharmacology, Biochemistry, and Behavior, 44,* 253–262.

KOOB, G. F., & BLOOM, F. E. (1988). Cellular and molecular mechanisms of drug dependence. *Science, 242,* 715–723.

LEE, T. H., & ELLINWOOD, E. H. JR. (1989). Time-dependent changes in the sensitivity of dopamine neurons to low doses of apomorphine following amphetamine infusion: Electrophysiological and biochemical studies. *Brain Research, 483,* 17–29.

LEE, T. H., ELLINWOOD, E. H., JR., & NISHITA, J. K. (1988). Dopamine receptor sensitivity changes with chronic stimulants. In W. Kalivas & C. B. Nemeroff (Eds.), *The mesocorticolimbic system.* New York: New York Academy of Sciences.

LINNOILA, M., ET AL. (1983). Low cerebrospinal fluid 5-hydroxyindolacetic acid concentrations differentiates impulsive from nonimpulsive violent behavior. *Life Sciences,* 2609–2614.

O'DONAHUE, N. F., & HAGMEN, W. D. (1967). A map of the cat brain for regions producing self-stimulation and unilateral attention. *Brain Research, 5,* 289.

PERT, C., KUHAR, M. J., & SNYDER, S. H. (1975). Autoradiographic localization of the opiate receptor in the rat brain. *Life Sciences, 16,* 1849–1854.

PERT, C., KUHAR, M. J., & SNYDER, S. H. (1976). The opiate receptor: Autoradiographic localization in the rat brain. *Proceedings of the National Academy of Sciences of the United States of America, 73,* 3729–3733.

PETTIT, H. O., ET AL. (1990). Extracellular concentrations of cocaine and dopamine are enhanced during chronic cocaine administration. *Journal of Neurochemistry, 55,* 798–804.

PITTS, D. K., & MARWAH, J. (1987). Neuropharmacology of Cocaine: Role of monoaminergic systems. *Monographs in Neural Science, 13,* 34–54.

REITH, M. E. A., BENUCK, M., & LAJTHA, A. (1987). Cocaine disposition in the brain after continuous or intermittent treatment and locomotor stimulation in mice. *Journal of Pharmacology and Experimental Therapeutics, 243,* 281–287.

RICAURTE, G. A., SCHUSTER, C. R., & SEIDEN, L. S. (1980). Long-term effects of repeated methylamphetamine administration on dopamine and serotonin neurons in the rat brain: A regional study. *Brain Research, 193,* 153–163.

RICAURTE, G. A., SEIDEN, L. S., & SCHUSTER, C. R. (1984). Further evidence that amphetamines produce long-lasting dopamine neurochemical deficits by destroying dopamine nerve fibers. *Brain Research, 303,* 359–364.

ROBERTS, W. W., & CAREY, R. J. (1965). Rewarding affects performance of gnawing aroused by hypothalamic stimulation in the rat. *J Comp Physio Psychol, 59,* 317.

SEIDEN, L. S., & KLEVEN, M. S. (1988). Lack of toxic effects of cocaine on dopamine or serotonin neurons in the rat brain. In D. Clouet, K. Asghar, & R. Brown (Eds.), *Mechanisms of cocaine abuse and toxicity* (National Institute on Drug Abuse Research Monograph No. 88). Washington DC: U.S. Government Printing Office.

TAYLOR, D., & HO, B. T. (1978). Comparison of inhibition of monoamine uptake by cocaine and methylphenidate and amphetamine, *Res Comm Clin Path Pharmacol, 21,* 67–75.

TRULSON, M. E., & ULISSEY, J. J. (1987). Chronic cocaine administration decreases dopamine synthesis rate and increases [³H]-spiroperidol binding in rat brain. *Brain Research Bulletin, 19,* 35–38.

TRULSON, M. E., ET AL. (1986). Chronic cocaine administration depletes tyrosine hydroxylase immunoreactivity in the rat brain nigral striatal system: Quantitative light microscopic studies. *Exp Neurology, 94,* 744–756.

VOLKOW, N. D., ET AL. (1990). Effects of chronic cocaine abuse on postsynaptic dopamine receptors. *American Journal of Psychiatry, 147,* 719–724.

YI, S-J., & JOHNSON, K. M. (1990). Chronic cocaine treatment impairs the regulation of synaptosomal ³H-DA re-

lease by D$_2$ autoreceptors. *Pharmacol Biochem Behav*, *36*, 457–461.

ZAHNISER, N. R., ET AL. (1988a). Repeated cocaine administration results in supersensitive nigrostriatral D-2 dopamine autoreceptors. In P. M. Beart, G. Woodruff, & D. M. Jackson (Eds.), *Pharmacology and Functional Regulation of Dopaminergic Neurons*, London: Macmillan.

ZAHNISER, N. R., ET AL. (1988b). Sensitization to cocaine in the nigrostriatal dopamine system. In D. Clouet, K. Asghar, & R. Brown (Eds.), *National Institute on Drug Abuse Research Monograph*, Washington, DC: U.S. Government Printing Office.

ZHANG, H., LEE, T. H., & ELLINWOOD, E. H. JR. (1992). The progressive changes of neuronal activities of the nigral dopaminergic neuron upon withdrawal from continuous infusion of cocaine. *Brain Research, 594*, 315–318.

EVERETT H. ELLINWOOD
G. R. KING

Genetic Factors Many drugs produce effects that can be described as pleasurable, novel, relaxing, or stimulating; these effects are not so powerful that repetitive drug use or abuse is inevitable. A significant amount of research has focused on why certain people cease drug use after experimentation; others continue drug use but do not become dependent; and still others become compulsive drug abusers. In research studies, investigators must distinguish between populations that do and do not abuse drugs, populations that abuse only a single drug, and populations that abuse multiple drugs. In addition, when attempting to identify genetic factors involved in drug abuse, researchers must also take into account sociological and environmental factors that may influence the pattern and degree of drug abuse.

The structure, function, and development of a biological system is determined by heredity, since all genetic information is transmitted from one generation to another generation through the genes. An inherited trait may be controlled by a single gene, a pair of genes, or many genes. If an inherited trait is determined by a single gene, there is no continuous variation in the population, and an individual would either possess the characteristic or lack it. If many genes are involved in the transmission of a specific trait, then the inherited characteristic will present a continuous variation within the population; this

transmission is termed multigenic or multifactorial inheritance.

Drug abuse is a complex phenomenon. It is improbable that a single gene determines whether an individual will abuse drugs. Most probably, drug abuse or dependence results from a combination of many genes that are influenced by multiple nongenetic factors. Genetic factors may contribute to drug abuse in that some individuals may experience a more intense response to drug use or, conversely, they may not experience intense adverse effects from drug use, such as the "flushing" reaction to alcohol, side effects like nausea from opioids, or other withdrawal phenomena. Such differences in drug response may be mediated by genetically influenced differences in the receptor proteins on which the drugs act, differences in the enzymes that metabolize these drugs, or differences in the proteins that remove these drugs from their sites of action.

GENETIC PREDISPOSITION

Increasing evidence exists for a genetic predisposition to ALCOHOLISM. Research on genetic predisposition has followed observations that some people easily become alcoholic while others seem to be resistant to alcohol abuse; that alcoholism occurs generation after generation in some families but not in others; that alcoholism is common in some ethnic groups but is rare in others. An impressive amount of evidence has been gathered since about 1980 to indicate that alcohol abuse can be genetically transmitted and that genetic factors are involved in a majority of alcoholism cases.

Identifying a single genetic factor that induces susceptibility to alcoholism has been difficult, however. The genetically determined biochemical and metabolic differences may be subtle changes at the molecular and subcellular levels, changes that in themselves are not pathologic and are manifested only when alcohol is consumed. Genetic transmission of alcoholism has been indicated by studies of twins, adopted children of alcoholics, and studies of half siblings from alcoholic parents. These studies indicate that alcoholism is more prominent in people with at least one biological parent who is alcoholic and that genetic predisposition is a more important factor than childhood environment in influencing the development of alcoholism. A potential source of genetic variation in certain people may be the ability

to metabolize alcohol. The major chemical pathway for the metabolism of alcohol requires two enzymes alcohol dehydrogenase, which converts alcohol to acetaldehyde, and acetaldehyde dehydrogenase, which converts acetaldehyde to acetic acid. Acetic acid is then converted by a series of enzymes to carbon dioxide and water. A prevalence of alcohol sensitivity exists among Asians as compared with Caucasians. Recent studies indicate that there is a genetic difference in the enzymes involved in the alcohol metabolism of Asians (including Native Americans who migrated from Asia thousands of years ago). In a significant proportion of Asians, alcohol dehydrogenase is more active than it is in Caucasians; it rapidly converts alcohol to acetaldehyde. In these same Asian populations, there is a genetic deficiency in acetaldehyde dehydrogenase, so that acetaldehyde accumulates and produces unpleasant symptoms including flushing, elevated skin temperature, increased pulse rate, headache, and often nausea. A potentially rapid and unpleasant accumulation of acetaldehyde is what alcoholic patients risk when they take the drug DISULFIRAM (Antabuse) to deter them from drinking. The acetaldehyde dehydrogenase deficiency may actually deter many Asians from abusing alcohol. Some studies have shown that acetaldehyde dehydrogenase deficiency is less common in Asian alcoholics than in the general Asian population.

Genetic alteration of neuronal (nerve)-cell membranes may also be a factor in alcoholism. Alcohol lowers the viscosity of the nerve-cell membrane, which may interrupt neural excitation or may interfere with the membrane proteins involved with neurotransmission. Studies performed on two genetic strains of mice demonstrated that mice that were sensitive to alcohol had nerve-cell membranes that were more susceptible to the viscosity-lowering effects of alcohol than were those of mice less susceptible to alcohol (Goldstein, 1982).

Another genetic factor may be alteration of membrane-bound neuronal enzymes, such as the enzyme sodium-potassium ATPase, which is an ion pump responsible for the movement of sodium-potassium ions through the cell membrane. Alcohol has been demonstrated to inhibit the sodium-potassium ATPase, leading to the speculation that there may be a genetic predisposition to alcoholism in those who show an atypical resistance to the alcohol-induced inhibition of this enzyme. Alcohol has been shown

to modulate the proteins that make up the RECEPTORS, including the NMDA receptor and the GABA receptor. The NMDA receptor mediates neuronal excitation following release of the NEUROTRANSMITTER glutamate. The NMDA receptor is also the major excitatory pathway in the brain, while the GABA pathway is the major inhibitory pathway in the central nervous system. Potential changes in either of these receptor proteins could predispose an individual to alcoholism. Additionally, alcohol interferes with the release of the neurotransmitters NOREPINEPHRINE, DOPAMINE, and GABA.

An advance in the field of genetic factors relating to alcoholism has been made by identification of a particular allele (a variation in form) of the dopamine D2 receptor protein that appears to be associated not only with alcohol abuse but also with polysubstance abuse (Uhl et al., 1992).

SAMPLE AND CONTROL VARIABLES

There are many confounds in this search for genetic factors that confer vulnerability for drug abuse; often there are genetic differences between ethnic populations, so a study must adequately match drug-abusing populations and control populations. Additionally, when comparing the genetic profile of any drug- or substance-abusing population with a control, there is a potential confound in the control. It is difficult to determine whether an individual in the control is resistent to drug abuse or vulnerable to drug abuse but not expressing the behavior during the study. The sample size must be adequate to detect small genotype changes from background. Variables such as severity of dependence need to be consistent between research groups in assessing the degree of drug abuse of their subjects.

PSYCHIATRIC DISORDERS

Certain psychiatric diagnostic categories are regularly overrepresented among those who seek treatment for alcoholism and drug dependency. These include depressive disorders, anxiety disorders, and ANTISOCIAL PERSONALITY. Personality has also been determined to be an inherited trait and can potentially confound observations made to determine genetic factors that are specific to drug abuse as distinct from personality factors that may be associated with risk. Although research in this area has

many confounding variables, significant advances in determining potential genetic factors that may predispose to substance abuse have been made since about 1980. It is conceivable that a series of genetic factors may predispose an individual to substance abuse and that these factors, when combined with sociological factors, will lead to drug dependence.

(SEE ALSO: *Conduct Disorder; Complications: Mental Disorders; Epidemiology of Drug Abuse; Women and Substance Abuse; Vulnerability As Cause of Substance Abuse*)

BIBLIOGRAPHY

GOLDSTEIN, D., CHIN, J., & LYON, R. (1982). Ethanol disordering of spin-labeled mouse brain membranes: Correlation with genetically determined ethanol sensitivity in mice. *Proceedings of the National Academy of Sciences, U.S.A., 79*, 4231–4233.

HESSELBROCK, M. M., & KEENER, J. (1985). Psychopathology in hospitalized alcoholics. *Archives of General Psychiatry, 42*, 1050–1055.

ROUNSAVILLE, B., WEISSMAN, M., KLEBER, H., & WILBER, C. (1982). Heterogeniety of psychiatric diagnosis in treated opiate addicts. *Archives of General Psychiatry, 39*, 161–166.

UHL, G. R., PERSICO, A. M., & SMITH, S. S. (1992). Current excitement with D_2 receptor gene alleles in substance abuse. *Archives of General Psychiatry*.

GEORGE R. UHL

Learning The role played by learning factors in drug and alcohol abuse has recently received much attention. Two basic learning mechanisms are thought to be activated when an organism repeatedly self-administers a psychoactive substance. First, classical conditioning processes are engaged when environmental stimuli signal the upcoming effects of the drug. Second, operant conditioning occurs as an organism learns that particular behaviors lead either to a drug reward or to punishment. The effects of these two processes presumably interact, and they are thought to influence repeated drug use and/or relapse to drug use following a period of abstinence.

Classical conditioning occurs when an organism learns about a contingency between two events in the external environment. The most common situation involves learning that a biologically neutral event (the conditioned stimulus, CS, such as a light or a tone) signals the upcoming occurrence of a biologically relevant event (the unconditioned stimulus, US, such as the effects of a drug or the WITHDRAWAL syndrome from absence of a drug). As a result of this signaling relationship, the CS produces conditioned responses (CRs), related to the US in use. In the area of drug use, a number of investigators have suggested that environmental events that signal upcoming withdrawal or drug use in humans elicit CRs—which motivate further drug taking (Baker, Morse, & Sherman, 1987).

Operant conditioning involves learning about contingencies between behaviors and their outcomes. A typical operant conditioning situation sets up contingencies between three different events—a response; the outcome of that response (the reward or reinforcer); and the stimulus situation in which that response–outcome relationship is established (the discriminative stimulus). Drugs of abuse function as potent reinforcers for human addicts, since a variety of behaviors are directed solely toward their attainment and use. Consequently, understanding the rules governing the acquisition of operant behaviors directed toward drug reinforcers may be critical to understanding addiction.

Classical and operant conditioning processes may be activated simultaneously during drug seeking and self-administration. Events that have consistently signaled drug use may eventually come to evoke CRs in the form of craving—urges to use the drug. In this way, signals of drug use may act as discriminative stimuli motivating the drug user to begin drug-seeking behavior. For example, walking past a known dealer might act as a CS for a heroin addict, evoking the CR of craving for HEROIN. This craving response might then increase the likelihood of behaviors that are rewarded by the desired drug effects—buying and preparing heroin.

OPERANT CONDITIONING WITH DRUG REINFORCERS

A large body of data shows that virtually all drugs of dependence in human beings act as reinforcers for animals in operant-conditioning situations. Typical studies on the reinforcing properties of drugs involve rats or monkeys fitted with venous catheters, through which a drug can be administered directly. Responses directed toward an object, such as a lever, result in infusions of the drug.

The basic finding of such studies has been that a wide variety of abused drugs—including COCAINE, MORPHINE, heroin, *d*-AMPHETAMINE, pentobarbital, and ALCOHOL—all serve to establish and maintain operant behaviors in animals. Other drugs with a lesser abuse potential in humans—such as aspirin, tricyclic antidepressants, hallucinogens, and opioid mixed agonist–antagonists—fail to support responding.

The degree to which a given drug of abuse reinforces behavior appears to depend more on the schedule of reinforcement of the drug than on its intrinsic properties. A schedule of REINFORCEMENT refers to the pattern of access provided to the reinforcing event. For example, ratio schedules require an animal to make some predetermined number of responses before a reinforcer is given. Yet interval schedules are set up so that responses are effective at producing a reinforcer only after a delay following the previous one. Reinforcers in ratio schedules depend solely on the number of responses made; therefore, these schedules typically result in higher response rates than interval schedules in which responses made too early are ineffective. Because reinforcement schedules largely determine the rate of responding in a given situation, the abuse potential of the various drugs cannot be reliably assessed by comparing how quickly animals respond for each substance.

Other techniques for making such comparisons are available, however, and one technique for comparing the reinforcing properties of various substances involves calculating for each a so-called breaking point under a fixed ratio schedule of reinforcement. A fixed ratio schedule requires that an animal make a fixed number of responses (the ratio) for each reinforcement received. For a given drug dose, the breaking point is reached when a ratio too high to support responding is required. The breaking point achieved with the highest tolerable dose of a drug is often taken to be an index of that drug's reinforcing properties. Drugs with the highest breaking point are considered to be the most reinforcing and hence to have the highest abuse potential. Of drugs studied with such a procedure, cocaine appears to have the highest breaking point. For example, in some experiments, animals have been willing to press a lever up to 12,000 times for a single dose of cocaine.

Choice experiments provide a second means for comparing the reinforcing properties of two different drugs. Animals in such designs are typically given a choice between two responses, each of which leads to infusions of a different drug. A preference for one response is taken to indicate a preference for the drug associated with that response. A finding of particular interest from such studies has been that cocaine appears to be preferred not only to a number of other drugs but also to nondrug rewards such as food and social contact (Johanson 1984)—but it is important to vary the other reinforcers as well. Animals will choose less cocaine when the amount of food provided is greater or tastier.

Far fewer systematic data on the reinforcing properties of drugs have been collected with human subjects. A number of experiments have shown that human subjects will work for tokens exchangeable for OPIOIDS, alcohol, pentobarbital, DIAZEPAM (Valium), and *d*-amphetamine. In addition, drug-abusing individuals will reliably produce arbitrary responses in a laboratory for immediate access to their drugs of choice. For example, heroin addicts will repeatedly push a button to receive heroin injections and cocaine users will choose to perform responses leading to cocaine injections over responses that yield injections of saline (Henningfield, Lukas, & Bigelow, 1986).

In sum, a body of both animal and human data now exists that documents the way drugs of abuse can act as potent reinforcing events. The pattern of drug use exhibited by an individual user, however, appears to depend as much on the schedule of drug availability as on the particular properties of the chosen drug. Therefore, predicting patterns of drug taking by humans will require a better understanding of the parameters of drug availability that exist in the real world.

CLASSICAL CONDITIONING OF DRUG-RELATED CUES

Conditioned Withdrawal Model. A number of investigators have advanced the idea that stimuli that reliably signal drug use elicit CRs that motivate further drug use. For example, Wikler (1980) noted that drug-free heroin addicts participating in discussions of drug use appeared to go through episodes of withdrawal as a result. Withdrawal refers to the unpleasant symptoms experienced by drug abusers following the abrupt cessation of drug use. Since the individuals discussed by Wikler had not used heroin for a long time, their withdrawal symptoms could not

have been the result of recent termination of the drug. Wikler proposed instead that events that reliably signal the onset of naturally occurring drug withdrawal become CSs capable of evoking withdrawal symptoms on their own. On future occasions, the mere presence of drug-related stimuli evoke conditioned withdrawal states that compel the person to counter these unpleasant feelings with drug use.

A second model that also invokes conditioning was put forth by Siegel (1979), who proposed that stimuli paired with drug use come to evoke conditioned compensatory responses, which oppose the direct effects of the drug. As these drug-opposite responses grow in size, over repeated conditioning experiences, they increasingly oppose the effects of the drug. Therefore, addicts should find that, over time, higher doses of the drug should be necessary to achieve a given effect. This pattern is indeed observed, and it is known as drug tolerance. According to Siegel's model, drug-related cues encountered in the absence of drug taking produce drug-opposite responses, which are not canceled by the direct effects of the drug. These drug-opposite responses are then thought to represent what the user experiences as withdrawal symptoms.

According to Siegel's model, conditioning can motivate drug use in two ways. First, the withdrawal symptoms generated by drug-related stimuli following a period of abstinence can lead to drug use aimed at relieving these unpleasant effects. Second, tolerance to the effects of a drug may motivate a user to increase his or her level of use in an attempt to maintain a fixed level of drug effect.

Siegel's model has not gone unchallenged. The primary problem appears to be that signals for drug use do not always produce drug-opposite responses. Instead, such signals sometimes produce responses that resemble the direct effects of the drug. The conditions determining whether CRs produced by drug-related stimuli are drug-like or drug-opposite have not been fully worked out. Yet drug-like responses (e.g., drug-induced euphoria) to environmental stimuli may act to motivate drug use as well. Some researchers have asserted that it is the memory of drug-induced euphoria that is the major factor contributing to continual drug use and relapse.

Conditioned Incentive Model. Stewart, deWit, and Eikelboom (1984) have proposed that conditioned drug stimuli provide the impetus for further drug use by producing mild drug-like effects, which

whet the appetite of the user (the priming effect). Thus drug-related events cause drug use by prompting the user to anticipate the pleasurable consequences. Like the models of Siegel and Wikler, this theory proposes that events signaling drug use become conditioned stimuli that encourage the drug user to initiate drug-seeking behaviors. The models differ only in their characterization of the CR elicited by the drug-related events.

Some evidence for this model lies in the observation that many animals show drug-like responses to stimuli paired with drug use, particularly when stimulant drugs such as cocaine or *d*-amphetamine are used. Furthermore, many researchers have found that animals that have stopped responding for a drug reinforcer may resume responding following a small unearned dose of the drug (a priming dose). Environmental signals for drug use may act in the same way as these priming doses. Some research suggests that the stimuli associated with rewards can elicit release of the neurotransmitter dopamine in the reward center of the brain, an event that is produced by the effects of most rewarding drugs.

Human Data. Since the 1970s, investigators have collected data from a number of sources to document that stimuli associated with drug use in humans acquire conditioned properties. Evidence for this statement has come from three primary sources:

- self-reports by addicts about the conditions under which they experience craving and withdrawal
- attempts to establish drug conditioning in the laboratory
- assessments of responding to cues thought to have become drug CSs in the natural environment

Self-reports of Conditioned Effects. Many drug-abuse patients report drug craving and withdrawal when faced with drug-related stimuli in their home environment. Wikler (1980) reported that drug-free heroin addicts who returned to their home (addiction) environment following a period of treatment experienced symptoms of heroin withdrawal. O'Brien (1976) took a more systematic approach, interviewing heroin addicts and constructing a hierarchy of real-world events that result in withdrawal feelings and craving for drug use. Such reports encouraged the idea that events that signal drug self-administration in the home environment come to evoke con-

ditioned responses, which motivate further drug use. Consequently, investigations have attempted to study this phenomenon under controlled laboratory conditions.

Laboratory Conditioning Studies. O'Brien and his colleagues (1986) found that neutral stimuli paired either with opiate administration or with opiate withdrawal appeared to elicit conditioned withdrawal reactions. A number of later studies using alcohol as the unconditioned stimulus have also tended to find drug-opposite responses elicited by the experimental CS, although drug-like effects have been observed as well.

Studies of conditioning in the laboratory permit the conclusion that such conditioning can occur as a consequence of drug-taking behavior by addicts. However, the contingency between potential CSs and drug effects in the natural environment is undoubtedly less precise than that programmed in the laboratory. Therefore, affirming the role of drug-related stimuli in drug use requires a direct assessment of the effects of such events.

Cue-Assessment Studies. To determine whether events associated with drug use in the natural environment acquire conditioned properties, many studies have recreated typical drug-related stimuli in the laboratory. In such studies, subjects are exposed to audiotapes, videotapes, slides, and paraphernalia with drug-related content, while physiological and self-report responses are obtained. Responses to such drug-related stimuli are then usually compared with the responses subjects make when they are exposed to comparable stimuli lacking a drug-specific content. So far, the majority of cue-assessment studies have been carried out with either opiate-abusing subjects or alcoholics, although there have also been some studies of cocaine abusers.

For opiate users, exposure to visual stimuli and paraphernalia typically associated with opioid use have been found to produce subjective reports of craving for opiates and withdrawal, as well as other physiological changes associated with withdrawal, including drops in skin temperature and skin resistance and increases in heart rate. Other studies have shown that opiate users given a medication that blocks the effects of the opiates (an opiate ANTAGONIST) initially report pleasurable sensations after only injection—even of placebo (saline solution). After several repetitions of the placebo, the injections begin to become aversive, including elements of

withdrawal. Opiate-related stimuli, then, appear to evoke both subjective and physiological responses related to drug use; it appears likely that such responses have a basis in prior conditioning.

Studies with alcoholics have yielded similar results. In a typical experiment, the reactions of alcoholics to the sight and smell of an alcoholic beverage are compared with responses evoked by a nonalcohol (placebo) drink. The results show greater craving/urges to drink induced by the alcohol cue than by the control stimulus. Such data suggest that alcohol-related cues acquire the ability to evoke conditioned responses.

Classical/Operant Conditioning Interaction. As mentioned earlier, much of the work on drug conditioning contains the implicit notion that classical and operant learning effects combine to motivate drug use (or craving, even if no drug use actually occurs). The most common idea is that drug CSs evoke craving and withdrawal states, which motivate the performance of drug-seeking behaviors. These behaviors are reinforced, in turn, by the effects of the drug. Although much evidence suggests that psychoactive drugs of abuse have powerful reinforcing properties—and that signals of those drugs elicit conditioned responses—the question remains as to whether these classically conditioned responses actually motivate drug-seeking behaviors.

One technique used in a study to examine this issue involved asking opiate abusers to describe the conditions under which they had relapsed to drug use. After some probing, many patients were able to describe specific events or stimuli that triggered craving and withdrawal leading to use. Such reports, however, suffer from the difficulty that retrospective self-reports from individuals reluctant to analyze their own behavior may be inaccurate.

Because of the ethical limitations imposed on providing drugs to drug abusers, little laboratory work has been done to examine whether drug-related cues increase drug self-administration. Some data have been collected by Ludwig, Wikler, and Stark (1974), however, who found that alcoholics in a barlike environment worked harder at an alcohol-rewarded task than did subjects in a laboratory setting. This result supports the idea that alcohol-related stimuli impact directly on the motivation to drink alcohol. Yet, no nonalcoholic subjects were studied, preventing the conclusion that alcoholics are uniquely susceptible to alcohol-related events.

Because the assumption that responses to drug-related stimuli motivate actual drug use is central to learning models, more studies using drug taking as a dependent measure are clearly needed.

Treatment Implications of Learning and Conditioning Theories. If classical and operant conditioning motivate drug use, then substance-abuse treatments should aim at reducing the impact of these learning effects. The most commonly discussed interventions include aversion training, extinction, and behavioral alternatives.

AVERSION THERAPY or training involves teaching subjects that stimuli and responses that once terminated in drug effects will lead to unpleasant outcomes. The most common technique has been to pair self-administrations of a drug with electric shock or with chemically induced sickness (e.g., Antabuse). Although there are some anecdotal reports of successful treatment using such therapy, systematic data are not abundant. In addition, such aversion training suffers from the disadvantage that patients are unlikely to continue to administer punishments to themselves once outside treatment. Since the treatment setting is clearly different from the home environment, subjects may simply learn that drug-taking behavior is reinforced at home but punished in the clinic.

Extinction training consists of exposing subjects repeatedly to drug-related stimuli and responses, to break the association between these events and the effects of drug use. Operant extinction procedures require subjects to perform repeatedly drug-use behaviors in the absence of a drug reinforcer. This can be accomplished by having subjects administer their drug of abuse in the usual way—while they are maintained on medication blocking the drug's effects. In this way, drug responses go unreinforced, because the drug effects are missing. Extinction of classically conditioned stimuli typically requires subjects to view repeatedly drug-related scenes and handle paraphernalia without using the abused substance. Such training has the advantage of not requiring subjects to accept punishment. Nevertheless, subjects might show extinction in the context of the clinic but continue to experience conditioned effects that lead to drug use in the unaffected home environment.

Behavioral alternatives training represents a third approach to reducing the impact of conditioning on drug abuse. This technique involves teaching subjects to avoid drug-related situations or to make alternative responses in the presence of drug-related stimuli. These new responses are designed to compete with the drug-seeking behaviors usually elicited by drug-related cues. Rather than try to eliminate craving produced by drug cues, this treatment attempts to give the patient ways to avoid cues plus alternatives to drug use. Behavioral alternatives to drug use range from simple time-out periods, to forming images inconsistent with drug use, to acting in ways that reduce the chances of use or cue exposure (e.g., going out to eat). These approaches are now commonly designated as COGNITIVE THERAPIES or RELAPSE-PREVENTION approaches. The advantage of these procedures over aversion therapy and extinction lies in the greater potential for patients to use their training in the clinic to deal with high-risk situations in the real world.

Whether these particular conditioning interventions provide lasting help to substance abusers remains to be seen—but data exist to suggest that in alcoholics, opiate addicts, and cocaine users, these cognitive therapies and relapse-prevention techniques do have value. They reduce the probability of relapse.

(SEE ALSO: *Addiction: Concepts and Definitions; Conditioned Tolerance; Naltrexone in Treatment of Drug Dependence; Wikler's Pharmacologic Theory of Drug Addiction*)

BIBLIOGRAPHY

BAKER, T. B., MORSE, E., & SHERMAN, J. E. (1987). The motivation to use drugs: A psychobiological analysis of urges. In C. Rivers (Ed.), *The Nebraska symposium on motivation: Alcohol use and abuse.* Lincoln: University of Nebraska Press.

CHILDRESS, A. R., MCLELLAN, A. T., & O'BRIEN, C. P. (1985). Behavioral therapies for substance abuse. *International Journal of Addiction, 20,* 947–969.

DRUMMOND, D. C., COOPER, T., & GLAUTIER, S. P. (1990). Conditioned learning in alcohol dependence: Implications for cue exposure treatment. *British Journal of Addiction, 85,* 725–743.

GOUDIE, A. J., & DEMELLWEEK, C. (1986). Conditioning factors in drug tolerance. In S. R. Goldberg & I. P. Stolerman (Eds.), *Behavioral analysis of drug dependence.* Orlando, FL: Academic Press.

HENNINGFIELD, J. E., LUKAS, S. E., & BIGELOW, G. E. (1986). Human studies of drugs as reinforcers. In S. R. Goldberg & I. P. Stolerman (Eds.), *Behavioral analysis of drug dependence*. Orlando, FL: Academic Press.

JOHANSON, C. E. (1984). Assessment of the dependence potential of cocaine in animals. National Institute on Drug Abuse Research Monograph, No. 50, 54–71. Rockville, MD.

JOHANSON, C. E., & FISCHMAN, M. W. (1989). The pharmacology of cocaine related to its abuse. *Pharmacological Review, 41*, 3–52.

LUDWIG, A. M., WIKLER, A., & STARK, L. H. (1974). The first drink: Psychobiological aspects of craving. *Archives of General Psychiatry, 30*, 539–547.

O'BRIEN, C. P. (1976). Experimental analysis of conditioning factors in human narcotic addiction. *Pharmocological Review, 27*, 533–543.

O'BRIEN, C. P., EHRMAN, R. N., & TERNES, J. W. (1986). Classical conditioning in human opioid dependence. In S. R. Goldberg & I. P. Stolerman (Eds.), *Behavioral analysis of drug dependence*. Orlando, FL: Academic Press.

SHERMAN, J. E., JORENBY, M. S., & BAKER, T. B. (1988). Classical conditioning with alcohol: Acquired preferences and aversions, tolerance and urges/craving. In D. A. Wilkinson & D. Chaudron (Eds.), *Theories of alcoholism*. Toronto: Addiction Research Foundation.

SIEGEL, S. (1979). The role of conditioning in drug tolerance and addiction. In J. D. Keehn (Ed.), *Psychopathology in animals: Research and treatment implications*. New York: Academic Press.

STEWART, J., DEWIT, H., & EIKELBOOM, R. (1984). The role of unconditioned and conditioned drug effects in the self-administration of opiates and stimulants. *Psychological Review, 91*, 251–268.

YOUNG, A. M., & HERLING, S. (1986). Drugs as reinforcers: Studies in laboratory animals. In S. R. Goldberg & I. P. Stolerman (Eds.), *Behavioral analysis of drug dependence*. Orlando, FL: Academic Press.

WIKLER, A. (1980). Opioid dependence: Mechanisms and treatment. New York: Plenum.

STEVEN J. ROBBINS

Psychological (Psychoanalytic) Perspective The psychological study and understanding of substance abusers has tended to be difficult, controversial, and complicated. Part of this derives from the nature of addictive illness; the acute (short-term) and the chronic (long-term) use of drugs and ALCOHOL cause individuals to seem pleasure oriented, self-centered, and/or destructive to self and others, thus making them difficult to approach, understand, or treat. In other respects, the controversy or lack of understanding derives from competing ideas or schools of thought that debate (if not hotly contend) whether substance abuse is a disease or a symptom, whether biological and genetic factors are more important than environmental or psychological ones, and/or whether substance abuse causes or is the result of human psychological suffering. Furthermore, in recent years psychological factors were minimized, because we entered the era of biological psychiatry/psychology, in which empirical interest in brain structure and function (down to the microscopic and molecular level) has predominated over interest in the person, the person's mind, and subjective aspects of human psychological life that govern our emotions and behavior. Although one cannot ignore that substances of abuse are PSYCHOACTIVE—powerful chemicals that act on the brain—there is a tendency to lose sight of the total person whose ways of thinking, feeling, and behaving (including subjective feelings about self and others) are equally and profoundly affected both by that chemistry and by the subjective effects produced by those psychoactive substances.

Clearly, biological, genetic (i.e., hereditary), and sociological factors are important in the development of drug abuse and dependence. Such factors are best studied by empirical methods, and modern technology—including the computer—has yielded new and valuable data since the late 1960s to explain aspects of addictive behavior. It is also noteworthy that during this period (a time when substance abuse has been most prevalent, studied, and treated), clinical work with substance abusers has yielded data and findings of equal importance and validity, and this work has focused on some of the important subjective psychological factors that also explain aspects of addictive behavior—some which empirical methods alone do not adequately fathom or explain.

I will present here a psychological understanding of drug abuse and dependence based on the perspective gained from clinical work with alcoholic and drug-dependent individuals. In psychology and clinical psychiatry, it is referred to as the *case method* of study of human psychological problems. Guided

by psychodynamic principles (the assumptions of which will be explained), this article will review what three decades of clinical work and case study with substance abusers has yielded on some of the main psychological influences that make likely or compelling the dependence on, and continued use and relapse to, drugs and alcohol.

ASSUMPTIONS

A psychodynamic perspective of human psychological life problems rests on the principle that we are all more or less susceptible to various forms of human psychological vulnerabilities—at the same time, we are also more or less endowed with human psychological strengths or capacities to protect against these vulnerabilities. Without ignoring hereditary factors, especially those that affect temperament, a psychological, and in this case psychodynamic perspective attempts to understand psychological forces at work (for example, drives and feelings) that operate within the individual at the same time that there is a corresponding interest in the psychological structures and functions that observably (and just as often, less obviously) operate to regulate or control our drives, feelings, and behavior.

A psychodynamic approach to human psychology greatly depends on a developmental perspective or an appreciation of the psychological forces, structures, and functions as they develop and change over one's lifetime. Psychodynamic clinicians are especially interested in the way individuals are influenced in the earliest phases of development by parents (and other caregivers), and then in the development of relationships with other children and peers, and later in the life cycle in relationships with adults and small and large groups—all of which shape our life views and experiences, as well as our attitudes, values, and characteristic ways of reacting and behaving.

Based on these assumptions, clinicians have the opportunity, most usually in the context of treating patients, to study and understand how the degree of developmental impairments (or strengths) has predisposed toward (or protected against) psychological and psychiatric dysfunction, including addictive vulnerability. In my experience, and that of my associates, we believe that modern psychodynamic-clinical approaches are as relevant and useful for studying and treating substance-dependent individuals as

they are for the many other patients who benefit from this perspective.

The psychological study and understanding of addictive illness necessarily requires the condition of abstinence (being free of drug/alcohol use). Again, there is considerable debate about the duration of abstinence required before meaningful or valid psychological inferences can be made about individuals with addictive disorders. In my experience, however, the confounding effects of acute and chronic drug/alcohol use are variable, and it is often surprising that within days or weeks—but certainly within several months of abstinence—how much can be learned about a person's makeup and psychology that predisposed him or her to use and become dependent on substances. This point about the requirement for a period of abstinence from drugs and alcohol is important to emphasize, otherwise it can be and is rightfully argued that what appear to be the psychological causes of dependence on psychoactive substances are actually the result of such a dependence. Fortunately, in recent years, the combination of modern detoxification approaches, psychoeducational/rehabilitation/RELAPSE PREVENTION programs, TWELVE-STEP groups, and individual and group psychotherapeutic approaches, have been increasingly successful in establishing and maintaining abstinence. This, in turn, has made psychological treatments and understanding increasingly possible.

PSYCHOLOGICAL SUFFERING AND SELF-CONTROL

A clinical-psychodynamic perspective suggests that human psychological suffering and problems with self-control are at the heart of addictive disorders. In fact, it is probably safe to say that to understand the psychology of addictive behavior is to understand a great deal about human psychological problems of suffering and control in general. The suffering that influences addictive behavior occurs at many levels, but it principally evolves out of susceptibilities involving people's self-esteem, relationships, emotions, and capacities to take care of themselves. Individuals who find various or particular drugs appealing (including alcohol) or who become dependent on them, discover that, short-term, the drug action or effect relieves or controls their distress—that is, such drugs are used to self-medicate distress. Although problems with self-esteem

and relationships are important parts in the equation of addictive behavior, it is mainly the problems with how substance-dependent individuals experience, tolerate, and express their feelings and their problems with self-care that makes addictive behavior so malignantly likely and compelling.

Problems with emotions and self-care painfully and repetitiously become involved with attempts to control suffering and behavior. This process includes such self-defeating coping patterns as action, activity, psychological defensiveness (e.g., denial, boastful or arrogant postures, attitudes of invulnerability and toughness), and, ultimately, the use of drugs and alcohol. What originally was a "solution" for suffering and self-regulation—where substances were used for relief or control—turns into a problem where there is a progressive loss of control of one's self, the drugs or alcohol employed to combat one's difficulties, and possibly life itself.

THE SELF-MEDICATION HYPOTHESIS

The self-medication hypothesis specifically refers to some individuals who, by dint of temperament or developmental factors, experience and find that certain painful feelings (or affects) are intense and unbearable and that the specific action or effect of one of the various classes of abused drugs (e.g., analgesics, depressants, or stimulants) relieves their psychological pain and suffering. The self-medication hypothesis also implies that the particular drug or class of drugs preferred is not random. Rather, it is determined by how that class of drugs with its specific actions interact with emotional states or particular painful feelings unique to the individuals who use or select their "drug-of-choice."

This is only one aspect of addictive suffering—namely that emotions are experienced in the extreme and that addictive-prone individuals feel too much pain, so resort to particular drugs to *relieve* their suffering. Another aspect of addictive suffering, to be covered subsequently, is that emotions are just as often absent, nameless, and confusing and that such individuals experience pain of a different type; they consciously feel too little of their distress and do not know when or why they are bothered (e.g., feeling empty, void, or cut off from emotions), and drugs or alcohol in these instances are used to change or *control* their emotions or suffering. In the

first instance the operative motive is the relief of suffering; in the second, it is the control of suffering.

The self-medication hypothesis rests on the observation that patients, if asked, will indicate that they prefer or discover that one class of drugs has more appeal than another. (It should be added parenthetically that the drugs preferred by an individual are not the ones that are always used. Drugs that are actually used are just as often the result of other factors, such as cost and availability.)

The three main classes of drugs that have been studied are the OPIOID analgesics (pain relievers), depressants or SEDATIVE-HYPNOTICS (soothing, relaxing, or sleep-inducing drugs), and STIMULANTS (activating or energizing drugs). The main appeal of opioids (e.g., HEROIN, MORPHINE, oxycodone) is that they are powerful subduing or calming agents. Besides calming or subduing physical pain for which they were originally intended, opioids are also very effective in reducing or alleviating distressing or disruptive emotions. Beyond its calming influence on physical and emotional pain in general, however, I have found that the main and specific action of opioids, namely as an anti-rage or anti-aggression drug, makes them especially appealing and compelling for those who struggle within, and with others, with feelings of intense anger, AGGRESSION, and hostility. Such a state of affairs is not uncommon for people who, in their early life development or in later life experiences, have suffered major trauma, neglect, or abuse. Such individuals, when they first use opioids, discover the extraordinary calming and soothing effects of these drugs on their intense anger and rage—and thus they become powerfully drawn or attached to them.

Whereas opioid-dependent people have much difficulty controlling their feelings, especially anger and rage, those who prefer or who are dependent on depressants generally have the opposite problem—namely they are too controlled or too "tightly wrapped" around their feelings. As is the case with other substance abusers, developmental life experiences, in this case often involving distrust and traumatic disappointment, have had a special influence on their experience of emotions. In the case of those who prefer depressants, they are the ones who have special difficulties experiencing emotions involving loving or caring feelings, interpersonal dependency, and closeness; in psychological terms, they are defensive and repressed around these emotions and

have difficulty in experiencing or expressing them. Depressants (e.g., alcohol, Seconal, Xanax) have appeal for these people, because such drugs help them to relax their defenses and release them from their repressions. Mainly, such drugs briefly (the short- or quick-acting depressants) produce a sense of safety and an inner sense of warmth, affection, or closeness that otherwise these people cannot experience or allow.

Finally, stimulants (i.e., AMPHETAMINES and COCAINE are the most popular and widely used) have appeal for those who suffer with overt and/or subtle states of depression, mania, and hyperactivity—in which problems with activation, activity, and energy are common. For example, ambitious driven types—for whom performance, prowess, and achievement are essential—find such drugs especially appealing on two counts: (1) stimulants are uplifting when they become depressed if their goals and ambitions, often unrealistic, fail them; (2) stimulants are facilitating and make action and activity easier when such people are on the upswing, making it easier for them to be the way they like to be when they are performing at their best. Stimulants cast a wide net of appeal because, in addition, they counter feelings of low energy, low activity, and low self-esteem in those suffering with overt or less overt (unrecognized or atypical) depression. Finally, those individuals suffering with attention deficit-hyperactivity disorder (ADHD), often subclinical or not recognized, are also drawn to and become dependent on stimulants, because of the paradoxically opposite calming effect that stimulants have for people with this disorder—much like hyperactive children who respond well to the prescribed stimulant Ritalin.

SELF-REGULATION VULNERABILITIES

To explain why people became addicted, early psychodynamic theory placed great emphasis on subconscious and unconscious factors, pleasure and aggressive instincts or drives, and the symbolic meaning of drugs. To some extent, the stereotype of substance abusers as pleasure-seeking destructive characters (to self and others), in part, still persists and derives from these early formulations. Albeit useful and innovative at the time, much of this early perspective is now outdated, counterempathic, and does disservice to understanding the motives of addicted and alcoholic individuals.

In contrast, the self-medication hypothesis has evolved from contemporary psychodynamic theory, which has placed the centrality of feelings (or affects) ahead of drives or instincts and has emphasized the importance of self-regulation, involving self-development or self-esteem (i.e., self-psychology), relationship with others (i.e., object-relations theory), and self-care (i.e., ego or structural psychology/theory). These contemporary psychodynamic findings have evolved since the 1950s, based on the works of investigators such as Weider and Kaplan, Milkman and Frosch, Wurmser, Krystal, Woody and associates, Blatt and associates, Wilson, Dodes, and Khantzian.

Although the self-medication hypothesis has gained wide acceptance as an explanation for drug/alcohol dependency, it is not without its critics and it fails to deal with at least two fundamental problems or observations that it does not explain or address. First, many individuals suffer with the painful feelings and emotions that substance abusers experience, but they do not become addicted or alcoholic. Secondly, the self-medication hypothesis fails to take into account that addicted and alcoholic individuals suffer as much if not more as a result of their drug/alcohol use, and this might appear to contradict the hypothesis that substances are used to relieve suffering.

Many of these criticisms, inconsistencies, and apparent contradictions are better understood or resolved when addictive problems are considered more broadly—in terms of self-regulation vulnerabilities or as a self-regulation disorder. For humans, life is the constant challenge of self-regulation, as opposed to the release, relief, or control of instincts and drives as early theory suggested. What is in need of regulation involves feelings, the sense of self (or self-esteem), relationships with others, and behavior. Those prone to addictive problems are predisposed to be so, because they suffer with a range of self-regulation vulnerabilities. Their sense of self, including self-regard, is often shaky or defective from the outset of their lives. A basic sense of well-being and a capacity for self-comfort and self-soothing is very often lacking or underdeveloped from the earliest phases of development. Subsequent development of self-esteem and self-love, if it develops at all, remains shaky and inconsistent, given the compromised sense of self from which self-regard evolves. Needless to say, a poor sense of self or low self-esteem (which usually originates in a compromised or deficient self–other parenting relationship), ultimately

affects subsequent self–other relationships and profoundly affects one's capacity to trust or to be dependent upon or to become involved with others.

It should not be surprising, then, that for some the energizing and activating properties of stimulants help self-doubting reticent individuals to overcome their depressive slumps and withdrawal; or that the soothing, relaxing effects of depressants help individuals who are restricted and cut off from others to break through their inhibitions and briefly experience the warmth and comfort of human contact that they otherwise do not allow or trust; or yet still, that those whose lives are racked by anger and related agitation would find a drug like heroin (an opioid analgesic) to be a powerful containing calming antidote to their intense and threatening emotions, which disrupt them from within and threaten most of their relationships with others. These examples, and those previously covered in relation to self-medication motives that govern drug use and dependency, help demonstrate the how and why of specific drug effects—which often become so compelling that they may consume the lives of some users.

In my experience, the regulation of *feelings* (or affects) and *self-care* become the two most compelling self-regulatory problems; they combine to make dependence on substances more likely than any other self-regulation factors. Focus on these two factors explains clearly why most people who suffer subjective painful emotions do not necessarily become addicted as well as why so many substance abusers persist in using debilitating substances despite the great suffering that ensues from their abuse.

It should be noted that in this article I have stressed psychological factors and have not pursued how the regular use of addictive drugs causes TOLERANCE AND PHYSICAL DEPENDENCE; the drugs, then, are used to remain "normal." It is not insignificant however, that the emotional pain involved in physical WITHDRAWAL just as often is an exaggerated form of the pain that the drug-of-choice originally relieved. This aspect of drug use and relapse are covered elsewhere in this encyclopedia.

As I have indicated, substance abusers suffer in the extreme with their emotions—they feel too much or they feel too little. When there is too much, we have described how drugs can relieve the intense unbearable feelings that addicts and others experience. Where there is too little and people are (or seem to be) devoid of, cut off from, or confused by their feelings (e.g., alexithymia, disaffected, or affect deficits), addicts prefer to counter the helplessness and loss of control caused by their lack of feelings. Instead, they choose to use drugs to change and control their feelings, even if it causes them more distress. They exchange feelings that are vague, confusing, and out-of-control, for drug-induced feelings that they recognize, understand, and control, even if such are painful and uncomfortable. Therefore, the factors of *relief* and *control* dominate people's motives for depending on drugs—even if these people have to endure the pain that their dependence on drugs also entails.

Finally, *deficits in self-care* (again deriving from early-life developmental problems) make it likely that certain individuals will become involved with hazardous activities and relationships that lead to drug experimentation, use, and dependence. Self-care deficits refer to a major self-regulation problem, wherein individuals feel and think differently around potential or actually dangerous situations and activities, including those that involve drug/alcohol experimentation and use. Where most of us would be apprehensive or frightened or would anticipate some guilt and shame, addictive and alcoholic-prone individuals show little or no such worry. Studying these patients' pre- and postaddictive behavior patterns very often reveals similar unfeeling, unthinking, fearless behavior in conducting other aspects of their lives—for example, preventable accidents, health-care problems, and financial difficulties seem evident and common. Being out of touch with, or not feeling, their feelings (that is, their "affect deficits" or "dis-affected state") contributes to their self-care problems and thus makes it more likely that they would engage in the dangerous pursuit of drug/alcohol abuse, where others with better self-care functions would not (even in those instances where the unbearable psychological suffering and states of distress are like those experienced by addicts). In this respect, painful or unbearable feelings, alone, are not sufficient to cause substance abuse or dependence. Rather, it is when individuals lack adequate self-care capacities *and* experience intense suffering that conditions exist for addictive behavior to develop or be likely.

(SEE ALSO: *Addiction: Concepts and Definitions; Causes of Substance Abuse: Learning; Comorbidity and Vulnerability; Complications: Mental Disorders; Conduct Disorder and Drug Use; Coping and Drug*

Use; Disease Concept of Alcohol and Drug Abuse; Vulnerability: Psychoanalytic)

BIBLIOGRAPHY

BLATT, S. J., ET AL. (1984). Psychological assessment of psychotherapy in opiate addicts. *Journal of Nervous and Mental Disease, 172*, 156–165.

DODES, L. M. (1990). Addiction, helplessness and narcissistic rage. *Psychoanalytical Queries, 59*, 398–419.

KHANTZIAN, E. J. (1990). Self-regulation and self-medication factors in alcoholism and the addictions. In M. Galanter (Ed.), *Recent developments in alcoholism*, vol. 8. New York: Plenum.

KHANTZIAN, E. J. (1985). The self-medication hypothesis of addictive disorders. *American Journal of Psychiatry, 142*, 1259–1264.

KHANTZIAN, E. J., & MACK, J. E. (1983). Self-preservation and the care of the self-ego instincts reconsidered. *Psychoanalytical Study of Childhood, 38*, 209–232.

KHANTZIAN, E. J., HALLIDAY, K. S., & MCAULIFFE, W. E. (1990). *Addiction and the vulnerable self.* New York: Guilford Press.

KRYSTAL, H. (1982). Alexithymia and the effectiveness of psychoanalytic treatment. *International Journal of Psychoanalysis and Psychotherapy, 9*, 353–378.

MILKMAN, H., & FROSCH, W. A. (1973). On the preferential abuse of heroin and amphetamine. *Journal of Nervous and Mental Disease, 56*, 242–248.

WEIDER, H., & KAPLAN, E. (1969). Drug use in adolescents. *Psychoanalytical Study of Childhood, 24*, 399–431.

WILSON, A., ET AL. (1989). A hierarchical model of opiate addiction: Failures of self-regulation as a central aspect of substance abuse. *Journal of Nervous and Mental Disease, 177*, 390–399.

WOODY, G. E., ET AL. (1986). Psychotherapy for substance abuse. *Psychiat. Clin. N. Am. 9*, 547–562.

WURMSER, L. (1974). Psychoanalytic considerations of the etiology of compulsive drug use. *Journal of the American Psychoanalytical Association, 22*, 820–843.

E. J. KHANTZIAN

CENTER FOR SUBSTANCE ABUSE PREVENTION (CSAP)/CENTER FOR SUBSTANCE ABUSE TREATMENT (CSA)

See U.S. Government Agencies: Substance Abuse and Mental Health Services Administration (SAMHSA).

CENTER ON ADDICTION AND SUBSTANCE ABUSE (CASA)

The Center on Addiction and Substance Abuse at Columbia University (152 W. 57th St., New York, NY 10019) was established in 1992 as a not-for-profit entity affiliated with the university. Organized by Joseph A. Califano, Jr., a former secretary of health, education, and welfare in the Carter administration, CASA has been funded by major grants from the Robert Wood Johnson Foundation. Califano, who has had a long-term interest in substance-abuse problems, was a vigorous advocate of SMOKING-CESSATION programs when he was a member of the Carter cabinet. In organizing CASA, he assembled a board of directors that includes many prominent people from politics, industry, academia, ADVERTISING, and the media, and he recruited a prominent drug-abuse researcher, Dr. Herbert Kleber, as executive vice-president and medical director.

The work of the center has emphasized analysis of available data on social and ECONOMIC COSTS of substance abuse (ALCOHOL, TOBACCO, and illicit drugs) and developing alternative strategies for preventing drug use and treating drug dependence. CASA has been called on by the Clinton administration to develop proposals for providing coverage for substance-abuse treatment and prevention services under the president's health-care reform proposal. CASA has also developed strategies for international demand-reduction programs. Through articles in the popular press, conferences, and testimony before congressional committees, CASA conducts a continuing campaign to raise public awareness about the pervasiveness and SOCIAL COSTS of substance abuse. Its board of directors has set the following priorities for the organization: To explain to the American people the social and economic cost of substance abuse and its impact on their lives; to identify what can be done—which PREVENTION and TREATMENT programs work and for whom; to encourage individuals and institutions to take responsibility to prevent and combat substance abuse.

(SEE ALSO: *Economic Costs of Substance Abuse; Social Costs of Substance Abuse*)

JEROME H. JAFFE

CENTRAL NERVOUS SYSTEM (CNS)

See Brain Structures and Drugs; Limbic System; Neurotransmission.

CHAIN OF CUSTODY   *See* Drug Testing.

CHEMICAL DEPENDENCE   *See* Addiction: Concepts and Definitions.

CHEWING TOBACCO   *See* Tobacco, History of; Tobacco, Smokeless.

CHILD ABUSE AND DRUGS   In the United States, on average, a child is abused every 13 seconds. Because of social awareness, reported child abuse has increased dramatically since the 1980s. Some states have experienced a 20 percent increase in reported child abuse between 1990 and 1991. The American Public Health Association (APHA) estimates that 1.7 million children are abused or neglected annually in the United States. This means that by 1992 a total of about 63.5 million (2.8%) of children under 18 years of age were abused.

Reported cases in the 1990 National Child Abuse and Neglect Data System totaled about half the APHA estimate, or only 893,856. This total comprised 227,057 victims of physical abuse; 403,430 victims of neglect; 138,357 victims of sexual abuse; 59,974 victims of emotional maltreatment; and 68,207 other. These figures represent only the reported cases—the proverbial tip of the iceberg. Research suggests that as many as 10 percent of children may be sexually abused and even more children physically abused or neglected. In addition, each year a higher percentage of U.S. children are being raised in poverty, often by overstressed and drug-abusing parents.

INCIDENCE AND PREVALENCE OF DRUG ABUSE

Although the casual use of drugs is decreasing in the United States, reported drug abuse is increasing in women of child-bearing age. While it is believed that more children are being raised by alcohol, tobacco, or other drug (ATOD)-abusing parents, the scope of the problem is undetermined. Longitudinal studies of children of ATOD-abusing parents are currently underway by the U.S. Centers for Disease Control (CDC), the NATIONAL INSTITUTE ON DRUG ABUSE (NIDA), and the NATIONAL INSTITUTE ON ALCOHOLISM AND ALCOHOL ABUSE (NIAAA). The Children of Alcoholics Foundation estimates that in the U.S. population about one in eight were raised in homes with one alcoholic parent. Studies suggest that as many as 11 percent of newborns are drug-exposed in utero. About 6 million women of childbearing age are marijuana users and 10,000 children per year are born to women using opiates. Polydrug use and frequent use of alcohol and other drugs by parents increases the difficulty of researching any causal relationships between a specific drug and child abuse.

The National Committee for the Prevention of Child Abuse (NCPCA) estimates that 10 million U.S. children are raised by ATOD-abusing parents or caretakers and at least 675,000 children every year are seriously abused by ATOD caretakers. ATOD-abusing women have higher fertility rates and more multiple births than non-ATOD-abusing women. Reasons for these repeated pregnancies in drug-abusing women may include lack of sex education and birth control, irregular menstruation, carelessness when using drugs, peer pressure and cultural norms, the desire to replace lost children, the need for increased welfare payments, the enjoyment of being pregnant (decreased depression), and having an infant to love them. These interesting findings should be studied further to determine their validity and related psychological, sociological, and biological causal mechanisms.

RELATIONSHIP OF CHILD MALTREATMENT AND ATOD ABUSE

Do addicted parents abuse their children more? Do both addicted and nonaddicted parents abuse their children more when using alcohol or drugs? Unfortunately, clear empirical research is lacking on the relationship of child abuse and alcohol abuse or child abuse and drug abuse. The validity of existing research is threatened by problems, such as unclear definitions of ATOD abuse, lack of control groups or longitudinal causal studies, and inappropriate statistical and research design techniques for separating causation and coincidence (Bays, 1990). Nevertheless, a relationship does exist between child abuse and ATOD abuse—and definitely between ATOD abuse and child neglect. Similar risk factors for child abuse exist for both child-abusing parents and substance-abusing parents, such as poor parenting

skills, family disorganization, involvement in criminal activity, and a disproportionately high incidence of mental and physical illness. Several types of child abuse and neglect involving children of drug abusers are reviewed below.

Prenatal Drug Exposure. A number of states legally define *in utero exposure* to alcohol and other drugs as child abuse. States with excessively punitive laws requiring child removal are rapidly changing these laws. For example, in California, the law mandating that infants be removed from detected drug-abusing mothers at birth has been modified; in San Francisco, a positive urine toxicology alone cannot be the only reason for removal. In many states, medical or social-services personnel are mandated to report such cases to protective-services workers; this can result in the avoidance of prenatal care by ATOD-abusing pregnant women. For this reason, social-services employees may hesitate to notify authorities. Additionally, notification of authorities can appear to be (and may in some cases be) racially biased and discriminatory if more poor women of color are referred. In one Florida study only 1 percent of white but 10 percent of African-American drug-abusing pregnant women were reported to child-protective services.

Alcohol and other drugs can cause teratogenic effects—resulting in abnormalities in the fetus. Isolating the specific effects of individual drugs has been complicated by the large proportion of women who are polydrug abusers and by additional factors of poor nutrition, disease, stress, and lack of prenatal care. Alcohol and tobacco are the drugs most commonly used by women during pregnancy; as many as 1 percent of births may be affected by FETAL ALCOHOL SYNDROME (FAS). Some researchers assert that FAS may be the major cause of mental retardation. Characteristics of FAS include facial anomalies, retarded growth, and abnormalities of the heart, kidneys, ears, and skeletal system. The long-term effects of FAS are still being studied but appear to include reduced intelligence, attention deficits, learning disorders, hyperactivity, impulsivity, and more antisocial behaviors than the norm.

The perinatal and long-term effects of other drugs have been studied—such as COCAINE, METHAMPHETAMINE, MARIJUANA, OPIATES, and PHENCYCLIDINE (PCP). Although a number of immediate problems are apparent, including drug withdrawal and developmental delays, with good postnatal environments

many of these children can overcome their in utero exposure if structural damage is not severe. Some researchers have reported that even when cocaine-exposed infants were reared in adoptive homes from birth, some showed neurological deficits. Significant central nervous system (CNS) damage occurs with cocaine exposure. The major effects at birth of most drugs, however, including alcohol and tobacco, are preterm deliveries of low-birthweight infants, indicative of growth retardation that may affect both brain and physical development. Sudden infant death syndrome (SIDS) is also two to twenty times higher in infants exposed to cocaine and opiates.

Few longitudinal studies have tracked the impact of drug exposure on children. The longest follow-up study is of prenatal opiate-exposed children evaluated at ten years of age, and it is very difficult to separate the impact of a poor postnatal environment from prenatal drug exposure (unless the children are adopted). The few longitudinal studies conducted of prenatal drug-exposed infants have found almost no long-term developmental problems directly related to their drug exposure. A few cross-sectional studies of children of drug abusers have found clinically significant negative impacts on their emotional, academic, and behavioral status. These studies suggest that the greater the degree of maternal drug abuse, the greater the negative impact on the child's mental and behavioral status as measured by standardized clinical measures.

The quality of the infant's postnatal environment as actively constructed by the mother (or caregiver) appears to be the most significant factor in determining the impact of drugs on the drug-exposed or non-drug-exposed infant. Studies find that children born to drug-abusing mothers can look normal or be resilient to their in utero exposure to drugs if they are provided a nurturing environment that includes responsiveness to their needs, stimulation, and early childhood education.

Postnatal Exposure to Drugs. Children can be hurt by ingesting or inhaling alcohol, tobacco, and other drugs. In 1978, PCP was the second most common cause of poisoning in young children at Los Angeles Children's Hospital. Four major ways exist for children to become intoxicated: passive inhalation, accidental self-ingestion, being given drugs by a minor, and deliberate poisoning by an adult. In addition, infants can ingest alcohol, nicotine, and other drugs through breast milk. Passive inhalation of to-

bacco is recognized as a health hazard to children; however, passive inhalation of CRACK (freebase cocaine), PCP, marijuana, or hashish also has negative effects. Children living with parents who manufacture synthetic DESIGNER DRUGS in their homes, such as methamphetamine, are exposed to hazardous toxic chemicals. Some ATOD-abusing parents allow their children to drink alcohol or use the drugs they find lying around the house. Some parents deliberately give their children alcohol or other drugs (tincture of opium) to reduce their crying, sedate them, or to induce intoxication to amuse the parents. Any relatively healthy child with unexplained neurological symptoms, seizures, or death may have been exposed to drugs.

Physical Abuse. Until recently, child-welfare agencies did not routinely screen for alcohol and drug abuse in caregivers of abused children. Because only about 40 percent of public child-welfare agencies and 71 percent of private child-welfare agencies even inquire about caregiver substance use, little is known about the incidence of substance abuse in child abuse cases. The Child Welfare League of America (CWLA) reported, after a 1990 survey of its 547 member agencies, that 37 percent of the children served by state agencies and 57 percent of children served by private agencies were estimated to be affected by ATOD family problems. A review of the literature found five studies suggesting a strong overlap between physical abuse and parental alcoholism. Physical and sexual abuse has been reported in 27 percent of alcoholic families and 19 percent of opiate-addicted families. Serious neglect was even more common (30.5%). Overall, 41 percent of children of addicts were found to be physically abused or neglected. In 1987, 50 percent of all reported child abuse and neglect cases in New York City were associated with parental drug abuse and 64 percent of cases were associated with parental alcohol and drug abuse. Of all child fatalities, 25 percent had related to a positive child drug toxicology (an overdose, OD).

ATOD abuse is frequently implicated when the courts remove a child from the home. A 1986 Illinois study indicated that 50 percent of all outplacements were from substance-abusing families and 68 percent of these parents refused ATOD treatment. Children growing up in abusive environments have increased, unfulfilled dependency needs, low self-esteem, distrust of others, and problems with aggression and anxiety.

Child Sexual Abuse. A high percentage of drug abusers report that they were sexually abused as children. Child molesters are often intoxicated when the abuse occurs. Alcohol's influence on the brain allows a disinhibition of socially proscribed behaviors, including incest and the sexual molestation of children. A 1986 review of the research suggests that alcohol is involved in about 30 to 40 percent of child sexual-abuse cases, particularly when girls are abused. A 1988 study found 48 percent of fathers who had committed incest were alcoholic but 63 percent of fathers were drinking at the time of the abuse. Because of the high heritability rate of male-limited alcoholism (the most severe type of alcoholism associated with early drinking and antisocial behavior in males), sexually molested children may be more genetically vulnerable to ATOD-abused antisocial behavior. Thus, the cycle of childhood sexual abuse is perpetuated over generations, because of the overlap between the two types of abuse.

Childhood sexual abuse is a major risk factor for greater psychological distress, dissociative experiences, depression, eating disorders, relationship, trust, and intimacy difficulties, posttraumatic stress response, psychotic disorder, and heavy drug abuse. A very high percentage of drug abusers at inpatient and residential treatment programs report being sexually abused as children. When direct questions were asked of the clients, men's reports of childhood sexual abuse increased from 4 percent to 16 percent for adult males, and to 42 percent for adolescent males. Reports by women increased from 20 percent to 75 percent of adult women, and to a high of 90 percent for adolescent women. Other studies indicate that between 25 and 44 percent of female drug abusers report childhood sexual abuse compared to 15 percent of nonaddicted women.

Psychological conflicts arising from childhood sexual abuse are often a hidden factor contributing to drug abuse and relapse. Sexually molested children are reported to experience boundary inadequacy, resulting in difficulty establishing and enforcing the personal, psychological, or social boundaries necessary to maintain a sense of the self that is separate from other people. Hence, survivors of childhood sexual assault often do not see themselves as individuals separate from the desires or demands of others. The concept of refusing another person access to their bodies (and in later life, to their privacy, time, physical space, and possessions)

has not been incorporated into their sense of identity. This leaves the survivors vulnerable to subsequent violations or coercive tactics throughout their lives. It could also lead adult survivors to become perpetrators who abuse their own or other children because of their own boundary inadequacies.

Risk Factors for Child Abuse by Substance Abusers. Child-welfare authorities consider parental substance abuse to be a major risk factor for child abuse. Under the influence of alcohol and other drugs, adults are less inhibited and have reduced judgement and emotional control. Uppers (stimulants such as cocaine, methamphetamine, PCP, and amphetamines) can cause anxiety, irritability, paranoia, and aggressiveness. Downers (depressants such as alcohol, opiates, sedatives, and barbiturates) have also been related to depression, irritability, and loss of control while disciplining children. It has been suggested that organic brain damage, hypoglycemia, and sleep disturbances caused by alcohol exacerbates child abuse by alcoholics. ATOD-abusing parents are often irritable and angry because of neurochemical imbalances caused by persistent drug abuse. Some researchers attest that these neurochemical imbalances can last for several years after detoxification. Furthermore, neurotransmitter imbalances, which can be either biologically inherited or lifestyle-induced, may precede parental drug abuse and lead to self-medication with drugs. For example, excessively aggressive adolescent human and monkey males have been found to have lower levels of serotonin. Alcohol and carbohydrates increase brain levels of serotonin. Low levels of serotonin are associated with depression and eating disorders. Doctors prescribe serotonin-uptake inhibitors, such as fluoxetine (Prozac), to reduce mental disorder like depression and bulimia.

Psychosocial risk factors for child abuse include the following:

1. *Modeling Physical and Sexual Abuse and Violence* as seen in the child's home as enacted by adults or the abuser's friendship groups, or as portrayed in popular media (movies, television, radio). Drug abusers often belong to subcultures where violence is common. Children raised in violent homes are more likely to become abusers as adults, thus perpetuating the cycle of violence.

2. *Family Violence and Conflict.* High levels of family conflict found in drug-abusing families can lead to family violence. Absence of empathy and support among family members in the home environment increases the risk of child abuse and family violence. Ironically, women who are victimized by their spouses have pregnancy rates 2.3 times higher than national averages. Children growing up in abusive homes experience increased anxiety, powerlessness, and self-deprecation, which may lead to ATOD abuse and, in turn, to aggression, conflict, and physical/sexual abuse of others.

3. *Poor Parenting Skills.* Drug-abusing parents or caretakers have been found to have less adequate parenting skills, spend less time with their children, have unrealistic developmental expectations that can lead to excessive punishment, and lax, overly severe, or inconsistent discipline. Verbal abuse in the form of threatening, chastising, belittling, and criticizing are common. Studies have found that drug-abusing parents, whether in recovery or not, are able to increase their parenting skills after participating in a 14-week parent-training program (The Strengthening Families Program).

4. *Poverty and Stress.* Many children of drug-abusing parents or caretakers are raised in poverty. Money that would normally be available for food, clothing, transportation, medical and dental care, and to provide social and educational opportunities for the children is often diverted into purchases of tobacco, alcohol, and drugs. Crack-addicted parents sometimes use food stamps and welfare checks to purchase crack. Lack of money to handle daily crises elevates the usual level of life stressors and increases parental anger and irritability. Unemployment, which frequently results in low self-esteem, can lead to increased child abuse—the "kick the kid" syndrome.

5. *Mental Disorders.* Approximately 90 percent of drug abusers have other mental disorders, such as depression, bipolar-affective disorder, narcissism, ANTISOCIAL PERSONALITY, organic brain disease, and psychosis. Mental disorders of this nature can have a severe impact on a person's ability to parent and can lead to child abuse. Parents suffering from antisocial personality and narcissism are less empathic toward their children's suffering. It is harder to decenter from their own perspective, needs, and emotions in order to consider the child's feelings. Depression, bipolar disorder, and psychosis can cause parents to become angry, irrational, and abusive. Parents with per-

sonality disorders are less likely to internalize so-cietal taboos against child abuse and sexual abuse.

6. *Physical Illness and Handicaps.* Physical illness and physical handicaps can reduce the patience parents need to handle the stress inherent in dealing with children. Physical illness is more common in ATOD-abusing families because of their lifestyle and lack of preventive health care. Intravenous drug abusers and their children have higher rates of common infections, as well as in-creased exposure to diseases transmitted through the blood (HIV/AIDS and hepatitis), sexually transmitted diseases (syphilis, gonorrhea, and herpes), and tuberculosis.

7. *Criminal Involvement.* Drug-abusing parents are at high risk for criminal involvement by nature of their use alone or by the need to obtain consid-erable sums of money to support their habit. Prostitution, theft, and drug dealing are reported in about half of all drug-abusing parents. Arrest and incarceration may increase the stress on the family and can reduce inhibitions to sexual abuse upon reunification of the family.

Children of ATOD abusers frequently are more difficult to parent because of the increased preva-lence of ATTENTION DEFICIT DISORDER (ADD), hyperactivity, CONDUCT DISORDERS, and learning dis-orders. Some of these difficult temperament charac-teristics are caused by in utero exposure to drugs, some by genetic inheritance, and others by lack of nurturing and inconsistent parenting. Regardless of the cause, children of ATOD-abusing parents are fre-quently the most difficult to raise, even if they are raised by unstressed, happy, healthy, non-ATOD–abusing parents.

PROPOSED RESPONSES

Reasonable evidence exists to indicate that chil-dren who are raised by ATOD-abusing parents are at increased risk of abuse and neglect, as well as of sub-sequent addiction and delinquent behaviors. Addi-tional research is needed on the long-term effects of reported physical or sexual abuse. Because of the high overlap between child abuse and drug abuse, ATOD treatment agencies should routinely ask their clients if they have been or are being physically or sexually abused. It is also important that child-wel-fare agencies routinely determine whether caregiver or family member ATOD abuse is contributing to the maltreatment of children.

Because it is not possible to remove all children from risky family environments, additional research is needed on ways to protect children. Caregivers and professionals can help maltreated children to avoid abuse or become more psychologically resilient to future ATOD abuse. Some children are resilient to negative outcomes, even though they were exposed to drugs in utero or lived with drug-abusing parents. Some of these children were really never exposed to the same degree of negative influences because they were sheltered by a caring adult who addressed their needs. The emerging literature on resiliency pro-cesses and mechanisms should be reviewed and used to inform resiliency research with children of drug abusers and to make prevention interventions more effective.

Negative outcomes primarily appear to be related to the physical and emotional abuse and neglect typ-ically endured by children of drug-abusing parents. Even children of drug-addicted mothers can be re-silient to their high-risk environments if their moth-ers realize the negative impact of their chaotic street lives on their infants and work to improve their parenting skills. This may include finding external supports to learn parenting skills, such as parent-and-family-skills training programs, locating good early-childhood education for the child and outside child care, and possibly even considering foster care or adoption. Research has shown more positive out-comes for drug-exposed infants if the mothers were willing to use whatever external social supports were necessary to provide the best opportunities for learn-ing and emotional growth for the child. Such moth-ers clearly were able to understand and empathize with their children's needs and were willing to sep-arate from their infants for short or long periods, if necessary, for the welfare of their children.

(SEE ALSO: *Childhood Behavior and Later Drug Use; Coping and Drug Use; Family Violence and Sub-stance Abuse; Poverty and Drug Use*)

BIBLIOGRAPHY

ADAMS, E. H., GFROERER, J. C., & ROUSE, B. A. (1989). Ep-idemiology of substance abuse including alcohol and cigarette smoking. *Annals of the New York Academy of Science, 562,* 14.

BURGESS, A. (ED.). (1985). *Rape and sexual assault: A research handbook.* New York: Garland.

CWLA NORTH AMERICAN COMMISSION ON CHEMICAL DEPENDENCY AND CHILD WELFARE. (1992). *Children at the front: A different view of the war on alcohol and drugs.* Washington, DC: Child Welfare League of America.

CONTE, J. R., & BERLINER, L. (1988). The impact of sexual abuse on children: Empirical findings. In L. E. A. Walker (Ed.), *Handbook of sexual abuse of children.* New York: Springer.

FINKELHOR, D. (1984). *Child sexual abuse: New theory and research.* New York: Free Press.

FINKELHOR, D. (1986). *A Sourcebook on child sexual abuse.* Beverly Hills: Sage.

RUTTER, M. (1990). Psychosocial resilience and protective mechanisms. In J. E. Rolf et al. (Eds.), *Risk and protective factors in the development of psychopathology.* New York: Cambridge University Press.

KAROL L. KUMPFER
JAN BAYS

CHILDHOOD BEHAVIOR AND LATER DRUG USE

Social scientists can point with confidence to risk factors from childhood that predict drug use and deviance in adolescence. In general, the findings indicate that certain childhood personality traits, family experiences, and ecological factors strongly affect adolescent drug-using behavior.

A child who is irritable and easily distracted, who throws temper tantrums, fights often with siblings, and engages in predelinquent behavior is more likely than others to use drugs in adolescence. Other investigators have also found childhood AGGRESSION to be a most powerful predictor of adolescent drug use and deviance.

Poor childhood impulse control and a difficult temperament have been related to adolescent marijuana use. When problematic factors continue into adolescence, both the use of illicit drugs and the psychopharmacological effects of some drugs may then actually serve to exacerbate and enlarge the adolescent's feelings of irritability and aggressiveness—as evidenced by continuing temper tantrums, aggression, and delinquent behavior.

In addition to childhood personality traits that predict future drug use, family experiences can also serve as predictors. In a longitudinal study dealing with the early childhood precursors of adolescent drug use, researchers found that greater mother in-

volvement with the child protected against later drug use. It has also been found that children who became frequent drug users had mothers who were cold and who gave them little encouragement. The connection between peer factors during childhood and later drug use has received little empirical study, although peer factors during adolescence are of demonstrable importance and have long been known to be so. Childhood ecological factors, such as relatively low socioeconomic status, are related to greater adolescent drug use.

Factors from childhood do not directly affect adolescent drug use; rather, they are mediated by the factors of adolescence. More specifically, the risk factors of childhood are associated with the risk factors of adolescence—and these, in turn, become related to drug use.

The risk factors of adolescence include aspects of unconventionality (such as rebelliousness); difficulty in the parent–child mutual attachment relation (such as low parental affection and identification); and the adolescent's associating with deviant peers. More specifically, findings indicate that nonachieving aggressive children—those with difficulty in emotional control and those who have received little economic and psychosocial support—are most likely to be rebellious adolescents, to have a conflictual, nonaffectionate relationship with their parents, and to be associated with deviant peers. Adolescent rebelliousness, difficulty in the parent–child attachment, and associating with deviant peers are the factors related to greater drug use in adolescence.

PREVENTION AND TREATMENT

Because the risk factors of both childhood and adolescence come from different domains, a multifactorial approach to drug-use prevention and treatment is essential. Moreover, since childhood drug-prone personality characteristics and adverse childhood experiences seem to determine adolescent risks for drug use, logic and social responsibility dictate that one intervene early in the development of a child at risk. Where needed or warranted, early intervention may facilitate the development of later drug-resistant personality traits—a positive parent–child bond and association with nondeviant peers—and, consequently, the result should be lower drug use in the adolescent.

As children grow up, there are also several later points and distinct psychological domains at which

it is possible to ameliorate drug use. During adolescence, for example, a decrease in the risk factors that relate to personality, family, or peers may also result in less to no drug use.

(SEE ALSO: *Child Abuse and Drugs; Conduct Disorders; Coping and Drug Use; Families and Drug Use; Family Violence and Substance Abuse; Poverty and Drug Use*)

BIBLIOGRAPHY

BLOCK, J., BLOCK, J. H., & KEYES, S. (1988). Longitudinally foretelling drug usage in adolescence: Early childhood personality and environmental precursors. *Child Development, 59,* 336–355.

BROOK, J. S., WHITEMAN, M., & FINCH, S. (1992). Childhood aggression, adolescent delinquency and drug use: A longitudinal study. *Journal of Genetic Psychology, 153*(4), 369–383.

BROOK, J. S., ET AL. (1992). African-American and Puerto Rican drug use: Personality, familial, and other environmental risk factors. *Genetic, Social, and General Psychology Monographs, 118*(4), 417–438.

COHEN, P., COHEN, J., & BROOK, J. S. (1993). An epidemiological study of disorders in late childhood and adolescence—II. Persistence of disorders. *Journal of Child Psychology and Psychiatry, 34*(6), 869–877.

LOEBER, R. T. (1991). Antisocial behavior: More enduring than changeable? *Journal of the American Academy of Child and Adolescent Psychiatry, 30,* 393–397.

McCORD, J. (1981). Alcoholism and criminality. *Journal of Studies on Alcohol, 42,* 739–748.

SHEDLER, J., & BLOCK, J. (1990). Adolescent drug use and psychological health: A longitudinal inquiry. *American Psychologist, 45,* 612–630.

JUDITH S. BROOK
PATRICIA COHEN

CHINA *See* Asia, Drug Use in; Golden Triangle.

CHINESE AMERICANS, ALCOHOL AND DRUG USE AMONG

In 1980, the Chinese-American community, with a population of 812,178, comprised the largest subpopulation of Asian/Pacific Islanders in the United States. During the 1980s, the population of Chinese Americans nearly doubled—

1,618,973 according to the 1990 data from the U.S. Bureau of the Census (although the Filipino-American community had by then become the largest Asian subgroup). The largest numbers of Chinese Americans reported in the 1990 census are in the states of California (704,850), New York (284,144), Hawaii (68,804), Texas (63,232), New Jersey (59,084), Massachusetts (53,792), and Illinois (49,936). The Chinese-American ethnic community actually consists of people from many countries, and recent waves of immigration especially contribute to the heterogeneity of this ethnic group. Chinese immigrants have come to the United States from British Hong Kong, the People's Republic of China, the Republic of China (Taiwan), and from various countries in Southeast Asia, Latin America, and the Caribbean. Approximately 63.3 percent of the Chinese-American respondents to the 1980 census had been foreign (non-U.S.) born.

ALCOHOL

In China, historically, alcohol was sanctioned for religious ceremonies, especially ancestor worship. Today, in China and among Chinese immigrants, alcohol is commonly served at celebrations and banquets, and some people consume alcohol at meals—beer, wine, brandy, or whiskey. Drinking-centered institutions, however, are absent (Hsu, 1955; Singer, 1972; Wang, 1968). In Chinese tradition, moderate drinking is believed to have medicinal effects, but excessive use is believed to bring on "nine-fold harm" (Yu & Liu, 1986/87) and is condemned in folk culture as one of the four vices. Many hypothesize that cultural influences are important in shaping drug-use patterns as well as beliefs about drug use. Some research ties cultural beliefs to differences in drinking patterns, despite similarities in availability (Glassner & Berg, 1980; Mizruchi & Perrucci, 1962).

Chinese cultural beliefs regarding the religious and medicinal benefits of moderate drinking and the harm associated with excessive use may control drinking patterns in China, but when people move into a new cultural setting, their alcohol use may be influenced by the extent to which they adopt the values of the surrounding culture. Sue (1987) states that alcohol abuse is more congruent with American than Chinese values, since Chinese values are antithetical to alcohol abuse. This "acculturation hypothesis" (Austin & Lee, 1989) has received mixed support with respect to the experience of Chinese

Americans. This suggests that more investigation is necessary to help determine which influences result in the retention of cultural values and which result in adaptation to the new culture.

OPIUM

OPIUM is thought to have been introduced to China by Arab traders during the ninth century. Initially, it was taken internally as medicine (Singer, 1974). Not until the mid-seventeenth century was the practice of smoking opium (usually in pipes) introduced by the Portuguese. Little of the opium poppy (*Papaver somniferum*) was actually grown or used in China before the sixteenth century. By the eighteenth century, however, opium had become a profitable cash cargo—from British India to China's ports, where foreigners were allowed only confined access to trade—for the Portuguese, Dutch, and English—and then after 1810 for the Americans (Goodie, 1963). Smoking opium had become so widespread and so debilitating in China that its sale was forbidden by imperial decree as early as 1729 and its importation was prohibited in 1800. The emperor's declarations were not universally honored, however, and much disagreement existed on how to deal with opium addictions, the drain of silver to foreigners, and the tribute system of then-developing foreign relations (Fairbank, Reischauer, & Craig, 1965).

Meanwhile, an illicit opium trade continued to grow—for example, from approximately 5,000 chests imported to Canton in 1821 by British traders to approximately 30,000 chests by the late 1830s (Fairbank, Reischauer, & Craig, 1965). Efforts in an anti-opium campaign were stepped up, and hostilities between China and Britain eventually led to the Opium Wars. Britain had asserted that it was not bound by the trade restrictions imposed on Canton, and Britain won the wars. As a result, Hong Kong, a major port and center for all kinds of trade, was ceded to Britain in 1842. Illicit opium remained an important export until 1911, at which time the British Parliament forbade its shipment to China. By this time, however, cultivation of the opium poppy was flourishing in China, and markets for MORPHINE, HEROIN, and other narcotic concentrates were growing. Although opium dens provided an atmosphere and opportunity for drug use by individuals or as a social activity, in China opium smoking remained one of the four vices.

Much of the research on ALCOHOL and other drug use has grouped all Asians and Pacific Islanders together. Only two studies have compared Asian groups, and they have suggested significant differences among them. In a 1981 study conducted in Los Angeles, Kitano and Chi (1986–1987) found differences in alcohol consumption patterns among respondents from four groups of Asians: Chinese, Japanese, Korean, and Filipino. Most of the respondents were from thirty to sixty-one years old. Except in the Japanese sample, the majority were foreign born and most had an average annual income of 20,000 to 30,000 dollars. Among these four groups, the following identified themselves as abstainers: 31.2 percent of Chinese males and 68.8 percent of females; 32.8 percent of Japanese males and 33.8 percent of females; 34.5 percent of Filipino males and 80.0 percent of females; and 45.8 percent of Korean males and 81.6 percent of females.

The lowest prevalence of heavy drinking was reported by Chinese Americans (14% male, 0% female), followed by Koreans (25.8% male, 0.8% female), Filipinos (29.0% male, 3.5% female), and Japanese (28.9% male, 11.7% female). Most of the male heavy drinkers were in the age category 26–35 among Chinese, in the age category 36–45 among Koreans, and evenly divided among age categories for Japanese and Filipinos.

Kitano and Chi found that among Chinese Americans in their Los Angeles sample those most likely to drink at any level were men, under the age of forty-five, and of relatively high social and educational background. They found that parental drinking and going to or giving parties were the most important variables distinguishing drinkers from abstainers among their Chinese adult male sample (Chi, Kitano, & Lubben, 1988). Going to bars and having friends who drank were also significant factors.

CONCLUSION

More rigorous surveys are still needed to obtain an accurate picture of alcohol- and other drug-use patterns among Chinese Americans. Since this is a heterogeneous group, future studies should take into account whether people in the sample are U.S. or foreign born, their country of origin, their degree of acculturation, and other demographic characteristics that will provide a better basis for comparison with other groups.

Although it has long been asserted that responses to drug problems should be sensitive to cultural diversity, until recently little research has focused on drug use among people other than blacks or whites in the United States. Such research would be useful for developing culturally appropriate interventions.

(SEE ALSO: *Ethnic Issues and Cultural Relevance in Treatment; Ethnicity and Drugs; Papaver somniferum*)

BIBLIOGRAPHY

AUSTIN, G., & LEE, H. (1989). *Substance abuse among Asian-American youth* (Prevention Research Update 5, Winter). Portland, OR: Western Center for Drug-Free Schools and Communities.

CAHALAN, D., & CISIN, I. H. (1976). Drinking behavior and drinking problems in the United States. In B. Kissin & H. Begleiter (Eds.), *Social aspects of alcoholism.* New York: Plenum.

CHI, I., KITANO, H. H. L., & LUBBEN, J. E. (1988). Male Chinese drinking behavior in Los Angeles. *Journal of Studies on Alcohol, 49*(1), 21–25.

FAIRBANK, J. K., REISCHAUER, E. O., & CRAIG, A. M. (1965). *East Asia: The modern transformation.* Boston: Houghton Mifflin.

GLASSNER, B., & BERG, B. (1980). How Jews avoid alcohol problems. *American Sociological Review, 45,* 647–664.

GOODRICH, L. C. (1963). *A short history of the Chinese people.* New York: Harper & Row.

HSU, F. L. K. (1955). *American and Chinese.* London: Cresset Press.

KITANO, H. H. L., & CHI, I. (1986–1987). Asian-Americans and alcohol use. *Alcohol Health and Research World, 11*(2), 42–47.

MIZRUCHI, E. H., & PERRUCCI, R. (1962). Norm qualities and differential effects of deviant behavior: An exploratory analysis. *American Sociological Review, 27,* 391–399.

SINGER, K. (1974). The choice of intoxicant among the Chinese. *British Journal of Addiction, 69,* 257–268.

SINGER, K. (1972). Drinking patterns and alcoholism in the Chinese. *British Journal of Addiction, 67,* 3–14.

SUE, D. (1987). Use and abuse of alcohol by Asian Americans. *Journal of Psychoactive Drugs, 19*(1), 57–66.

WANG, R. P. (1968). A study of alcoholism in Chinatown. *International Journal of Social Psychiatry, 14,* 260–267.

YU, E. S. H., & LIU, W. T. (1986/87). Alcohol use and abuse among Chinese-Americans. *Alcohol Health and Research World, 11*(2), 14–17, 60.

RICHARD F. CATALANO
TRACY W. HARACHI

CHLORAL HYDRATE   Chloral hydrate is one of the oldest sedative agents still in use. It was made by the German chemist Liebig in 1832 and introduced into general use in 1869 as a substitute for LAUDANUM, an alcoholic solution of OPIUM. Chloral hydrate differs from the BARBITURATES in that it is a simple molecule composed of two carbon atoms, three hydrogen atoms, two oxygen atoms, and three chloride atoms. It is the famous (or infamous) substance added to alcohol to make a *Mickey Finn,* a drink known to cause those who drank it to become unconscious. Because it shares many effects of other central nervous system depressants, it can be used to treat the alcohol withdrawal syndrome. Chloral hydrate was a popular sedative for elderly patients because its effects occur quickly, last only a short time, and leave no nagging hangover effect. However, it is inconvenient to use (up to 2 grams must be taken by mouth) and, after the introduction of the BENZODIAZEPINES, its use has decreased.

BIBLIOGRAPHY

HARVEY, STEWART C. (1980). Hypnotics and sedatives. In A. G. Gilman, L. S. Goodman, & A. Gilman (Eds.), *Goodman and Gilman's the pharmacological basis of therapeutics,* 6th ed. New York: Macmillan.

S. E. LUKAS

CHLORDIAZEPOXIDE   Chlordiazepoxide (Librium) is a member of the BENZODIAZEPINE family of drugs currently used to treat insomnia, anxiety, muscle spasms, alcoholism, and some forms of epilepsy. It was the first benzodiazepine to be used in clinical practice in the 1960s as an alternative to PHENOBARBITAL or MEPROBAMATE in treating psychoneuroses, anxiety, and tension. It is sometimes used to treat insomnia and to reduce anxiety before surgery, but in addition to its use as an antianxiety agent it is most frequently used to treat the seizures or DELIRIUM TREMENS (DTs) that appear during alcohol withdrawal. Its advantage over BARBITURATES

Figure 1
Chlordiazepoxide

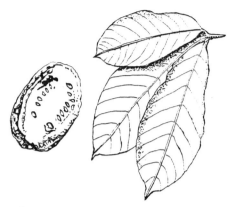

Figure 1
Cacao Leaves and Pod

and other central nervous system depressants is that it is less toxic, especially after an overdose.

SCOTT E. LUKAS

CHLOROFORM   *See* Inhalants.

CHOCOLATE   An ingredient of many popular treats—candies, sweets, baked goods, soft drinks, hot drinks, ice cream, and other frozen desserts. It is prepared, often as a paste, from the roasted crushed seeds (called cocoa beans) of the small South American cacao tree called *Theobroma cacao* (this is not the shrub known as the COCA PLANT, which produces COCAINE, *Erythroxylon coca*).

The cacao tree has small yellowish flowers, followed by fleshy yellow pods with many seeds. The dried, partly fermented fatty seeds are used to make the paste, which is mixed with sugar to produce the chocolate flavor loved throughout the world. Cocoa butter and cocoa powder are other important extracts from the bean. Cocoa beans were introduced to Europe by the Spanish, who brought them back from the New World in the sixteenth century. They had first been used by the civilizations of the New World—Mexicans, Aztecs, and Mayan royalty—in a ceremonial unsweetened drink and as a spice in special festive foods, such as molé. They were first used in Europe by the privileged classes to create a hot, sweet drink. By the seventeenth century, cocoa shops and COFFEE shops (cafés) became part of European life, serving free TOBACCO with drinks and thereby increasing trade with the New World colonies.

Chocolate produces a mild stimulating effect caused by the THEOBROMINE and CAFFEINE it contains. Both are ALKALOIDS of the chemical class called xanthines. Theobromine in high doses has many effects on the body, and it is possible to become addicted to some xanthines, such as caffeine. Nevertheless, some people are so attracted to the flavor that compulsive or obsessive use has resulted in the newly coined term *chocoholic*. Some scientists are researching the phenylethylamine in chocolate as the factor that encourages compulsive chocolate ingestion.

BIBLIOGRAPHY

RALL, T. W. (1990). Drugs used in the treatment of asthma: The methylxanthines, cromolyn sodium, and other agents. In A. G. Gilman et al. (Eds.), *Goodman and Gilman's the pharmacological basis of therapeutics*, 8th ed. New York: Pergamon.

MICHAEL J. KUHAR

CIGARETTE COMPANIES   *See* Advertising and Tobacco Use.

CIGARETTES AND SMOKING   *See* Nicotine; Tobacco, History of; Treatment: Tobacco, An Overview.

CIRRHOSIS   *See* Complications: Liver; Complications: Liver (Alcohol).

CIVIL COMMITMENT

CIVIL COMMITMENT The term commonly used for compulsory treatment is *civil commitment*. Typically, civil commitment serves as an alternative to incarceration (prison) by providing compulsory, court-ordered treatment for chronic drug abusers, especially antisocial addicts responsible for committing a large number of criminal acts. It is generally believed that narcotic addicts must be brought into a supervised environment for an extended period of time for any treatment to be meaningful. Civil commitment is a useful strategy for diverting into treatment those who ordinarily would not seek assistance voluntarily, and it has been shown to suppress daily narcotic use and criminal involvement (Leukefeld & Tims, 1988).

HISTORICAL CONTEXT

The concept of compulsory treatment for drug abusers in the United States was proposed shortly after Congress passed the Harrison Act of 1914. By 1919, the Narcotics Unit of the U.S. Treasury Department had convinced Congress to establish a chain of federal "narcotics farms," where heroin users convicted of federal law violations could be incarcerated and treated for addiction (Inciardi, 1988). The first of such farms, established in 1935, was the U.S. PUBLIC HEALTH SERVICE HOSPITAL in Lexington, Kentucky. Three years later, a sister hospital was established in Fort Worth, Texas. The goal of the facilities was to use vocational and psychiatric therapy to help free the addicts of their psychological dependence on drugs, to treat withdrawal illness, and to correct mental and social problems. Follow-up studies from Lexington in the 1940s indicated that addicts treated under legal coercion with posthospital supervision had better outcomes than voluntary patients, primarily because voluntary patients rarely completed the treatment program (Maddux, 1988). However, later studies failed to support these early positive findings. During the 1950s, hospital staff members recommended the enactment of a federal civil commitment law for narcotic addicts, but legal counsel in the Department of Health, Education, and Welfare considered such a law unconstitutional.

Then in 1962, President John F. Kennedy convened a White House Conference on Narcotic and Drug Abuse where nearly all members approved the civil commitment of narcotic addicts. Civil commitment was advocated as protection for society and rehabilitation for the individual. A compulsory treatment program aimed at the federal offender was enacted by the NARCOTIC ADDICT REHABILITATION ACT (NARA) of 1966. By that time, about twenty-five states had laws permitting civil commitment, and a few major programs were enacted in response to public fears of growing drug-related street crime (Inciardi, 1988). California, in 1961, launched its Civil Addict Program (CAP), the first large-scale civil commitment program to be implemented in the United States. Because of its relative success, in 1966, New York's Narcotic Addiction Control Commission (NACC) established the largest and costliest civil commitment program in history—the NEW YORK STATE CIVIL COMMITMENT PROGRAM.

CIVIL COMMITMENT PROGRAMS

The federal NARA and the California and New York compulsory treatment programs had a similar intent: They made it possible for the necessary legislation to be enacted and for commitment procedures to be carried out. They served to control and rehabilitate the compulsive drug abuser by providing secure treatment environments as an alternative to regular incarceration in correctional facilities (Leukefeld & Tims, 1988). Eligible addicts convicted of a crime could be committed by the court or could choose commitment over incarceration. Addicts not involved in criminal proceedings could commit themselves voluntarily or could be involuntarily institutionalized upon the petition of another individual (such as a peace officer) (McGlothlin, Anglin, & Wilson, 1977). Integral to each of these programs was supervised aftercare with antinarcotics testing. Length of commitment terms ranged from three years in NARA to seven years in CAP.

Although nearly all of NARA's civil commitment programs were generally considered unsuccessful, the funding for community programs contained within the same legislation did provide seed money for the nationwide establishment of some of today's basic drug-treatment programs in the community (Maddux, 1988). It was also believed that the New York State Civil Commitment Program failed, and the process of dismantling it began in 1971 (Inciardi, 1988). The failure of this program is partly attributable to the fact that it was administered by the social welfare agency of New York State, which had little experience controlling addicts. In contrast, CAP, California's program, was deemed at least moderately effective in modifying behavior, because it was

implemented through the California Department of Corrections, which had trained personnel who were familiar with treating substance abusers (Anglin, 1988).

Follow-up studies of the California program found that participants exhibited reductions in daily heroin use as well as in property crime and antisocial activities. Although many patients did become readdicted at some point, their relapses were typically of shorter duration and less frequent than those not involved in treatment (Anglin, 1988). The general conclusion drawn from these studies was that civil commitment, when adequately implemented, might be an effective means of reducing narcotic addiction and of minimizing adverse, addiction-related behavior. Repeated interventions are typically required, however, because drug dependence is a chronic condition marked by relapses (Leukefeld & Tims, 1988).

LIMITATIONS OF CIVIL COMMITMENT

Civil commitment helps get drug abusers into treatment and keeps the abusers in treatment for an extended length of time. Outcome studies have generally shown that the success of treatment is directly related to the amount of time spent in treatment and that long-term supervision upon return to the community, with objective monitoring (DRUG TESTING), is an essential component of a successful program. Furthermore, civil commitment often makes treatment available before a crime is committed, and it provides clear treatment goals rather than only providing punishment (Leukefeld & Tims, 1988). Still, such a program as civil commitment has serious limitations. It is costly and can overwhelm facilities unless adequate funding, facilities, and staff have been made available. Many addicts are considered unwilling or unsuitable for participation. External coercion can bring drug users into treatment, but it cannot assure that as patients they will participate in treatment. Even with the advent of intensive interventions designed to engage the patients, some patients simply participate passively (Maddux, 1988). Finally, the scope of civil commitment is restricted by constitutional guarantees of individual liberty. The question remains: Within a free society, to what extent should the government curtail the civil liberties of a compulsive drug user?

Today, there is an increasing tendency to see civil commitment as control rather than treatment and it serves only a limited number of addicts who are sufficiently problematic in their behavior to warrant commitment. The use of such measures as civil commitment in a better coordinated and expanded fashion, however, could produce significant individual and social benefits (Anglin, 1988). Civil commitment may also gain more popular support as a means for dealing with intravenous drug users who are at risk for contracting or transmitting AIDS.

(SEE ALSO: *Coerced Treatment for Substance Offenders; Treatment Alternatives to Street Crime [TASC]*)

BIBLIOGRAPHY

ANGLIN, D. (1988). The efficacy of civil commitment in treating narcotic addiction. *Compulsory treatment of drug abuse: Research and clinical practice* (NIDA Research Monograph 86). Rockville, MD: U.S. Department of Health and Human Services.

INCIARDI, J. A. (1988). Compulsory treatment in New York: A brief narrative history of misjudgment, mismanagement, and misrepresentation. *The Journal of Drug Issues, 18*(4), 547–560.

LEUKEFELD, C. G., & TIMS, F. M. (1988). An introduction to compulsory treatment for drug abuse: Clinical practice and research. *Compulsory treatment of drug abuse: Research and clinical practice* (NIDA Research Monograph 86). Rockville, MD: U.S. Department of Health and Human Services.

LEUKEFELD, C. G., & TIMS, F. M. (1988). Compulsory treatment: A review of findings. *Compulsory treatment of drug abuse: Research and clinical practice* (NIDA Research Monograph 86). Rockville, MD: U.S. Department of Health and Human Services.

MADDUX, J. F. (1988). Clinical experience with civil commitment. *Compulsory treatment of drug abuse: Research and clinical practice* (NIDA Research Monograph 86). Rockville, MD: U.S. Department of Health and Human Services.

MCGLOTHLIN, W. H., ANGLIN, M. D., & WILSON, B. D. (1977). *An evaluation of the California Civil Addict Program* (Services Research Monograph Series). Rockville, MD: U.S. Department of Health, Education, and Welfare.

HARRY K. WEXLER

CLANDESTINE LABORATORIES See
Amphetamine Epidemics; Colombia As Drug Source.

CLASSIFICATION OF DRUG TYPES   *See* Drug Types.

CLONIDINE   While not itself life-threatening, the opioid WITHDRAWAL syndrome is extremely unpleasant and contributes to further opioid use and relapse. HEROIN addicts report that the acute withdrawal syndrome begins in approximately eight hours after their last injection and includes the following: craving for the drug, anxiety, perspiration, perspiration with hot and cold flashes, tearing of the eyes and nose, restlessness, problems sleeping, problems falling asleep, goose bumps, aching bones and muscles, loss of appetite, nausea, vomiting, diarrhea, abdominal cramps, spontaneous yawning, and a group of symptoms called flu-like.

During the later years of the nineteenth century and early years of the twentieth, some cures for this opioid withdrawal syndrome have been far worse than the withdrawal itself—with some actually causing death. Soon after it became available in the mid-nineteenth century, injectable MORPHINE was proposed as a treatment for opium eating; then heroin or COCAINE were, in the late nineteenth century, proposed as cures for morphine addiction. From the mid-twentieth century until the 1970s, most medical treatment of the opioid withdrawal syndrome involved either gradual reduction of the dose of opioid or the substitution of METHADONE, followed by its gradual reduction. In 1978, Gold and coworkers proposed that the nonopiate antihypertensive clonidine could be an effective nonopiate treatment for opiate withdrawal distress. The scientific basis for the proposition that clonidine would be useful was based on the hypothesis that the opiate withdrawal syndrome was caused by hyperactivity or hyperexcitability of a specific brain nucleus composed of noradrenergic neurons, called the locus coeruleus (LC). There was considerable neuroscientific research to support this withdrawal hypothesis and the rationale for the efficacy of clonidine.

Since 1977/78, clonidine has been tried in numerous inpatient and outpatient opioid addict populations worldwide and studied by researchers in numerous well-controlled studies. In virtually all studies, clonidine has been shown to be a safe and effective nonopioid treatment that could control several aspects of opioid withdrawal. Clonidine, while having opiate-like effects in reversing several aspects of opiate withdrawal, is not an opiate and is therefore not subject to the burdensome regulatory CONTROLS that have been placed on the use of opioids. Clonidine has its most demonstrable effects on autonomic elements of opioid withdrawal: sweating, gastrointestinal complaints (cramps, diarrhea, nausea), and elevated blood pressure. It does not have substantial capacity to alleviate muscle aches, insomnia, or craving for opioids.

Research and clinical experiences since the original discoveries have (1) supported the notion of LC hyperactivity as one of the neural substrates for the opioid withdrawal syndrome; (2) supported the efficacy of clonidine—establishing clonidine detoxification as one of the standard treatments for adult opioid addicts—and extended it to neonates (newborns) and to alcohol, nicotine, and other drug withdrawals that share a preponderance of behaviours with opioid withdrawal; (3) demonstrated that abstinence could be maintained by some opioid addicts and that others could benefit from antagonist therapy with NALTREXONE, thanks to clonidine or accelerated clonidine—naltrexone detoxification; (4) led to considerable progress in the understanding of the critical cellular event causing LC hyperactivity in opioid withdrawal and hypoactivity in the presence of clonidine or opioid agonists; and (5) led to the rapidly expanding clinical armamentarium available to treat addicts on the basis of rodent and primate studies.

CLONIDINE SHORTENS DETOXIFICATION

Detoxification of opioid addicts with clonidine has been used to facilitate the transition from chronic opioid administration to naltrexone (a long-acting opioid ANTAGONIST) or to drug-free status. Naltrexone possesses opioid-blocking action at all opioid-receptor sites in the body and brain, rather than having an affinity for a specific type of opioid receptor. It is a useful medication for those patients willing to take it to prevent relapse. When recovering addicts take naltrexone, they make opioid effects unavailable to themselves. The affinity of naltrexone for the receptors is such that they are unable to feel the effects of heroin, methadone, or other exogenous opioids. While the original clonidine treatment protocol of Gold and his colleagues (1978a, b) facilitated the initiation of naltrexone by avoiding the extra ten-day wait required after the last methadone, an accel-

erated detoxification protocol has been developed using naltrexone and clonidine simultaneously. Since clonidine reduces precipitated withdrawal as well as the withdrawal that results from simply discontinuing chronic opioid administration, total withdrawal and naltrexone induction time has been shortened to six days with little loss in success rate (Charney, Heninger, & Kleber, 1986).

OTHER NEW USES

Clonidine has been tried with varying success in a number of medical problems where the behaviors, signs, and/or symptoms resemble those seen in opiate withdrawal or following LC electrical or chemical stimulation to a certain degree. Clonidine has also been tried in humans on the basis of noradrenergic hyperactivity in generalized and panic ANXIETY; obsessive-compulsive symptomatology; Gilles de La Tourette's syndrome; mania; ATTENTION DEFICIT DISORDER; narcolepsy; neuroleptic-induced akathisia; ALCOHOL withdrawal and NICOTINE withdrawal; and phaeochromocytoma. Clonidine's ANALGESIC effects have been rediscovered and given orally, transdermally (skin patches), epidurally (into the area around the spinal canal), and parenterally (by injection)—to decrease anesthetic requirements and to effect less respiratory depression than OPIOIDS alone.

CONCLUSION

Using clonidine for withdrawal distress allows the brain to reestablish normal homeostatic patterns when given as part of a long-term recovery program. It allows patients sufficient motivation to achieve and sustain drug-free existence.

(SEE ALSO: Opioid Complications and Withdrawal)

BIBLIOGRAPHY

CHARNEY, D. S., HENINGER, G. R., & KLEBER, H. D. (1986). The combined use of clonidine and naltrexone as a rapid, safe and effective treatment of abrupt withdrawal from methadone. *American Journal of Psychiatry, 143,* 817–831.

GOLD, M. S. (1991). *The good news about drugs and alcohol.* New York: Villard Books.

GOLD, M. S., REDMOND, D. E., & KLEBER, H. D. (1978a). Clonidine in opiate withdrawal. *Lancet* I (8070), 929–930.

GOLD, M. S., REDMOND, D. E., & KLEBER, H. D. (1978b). Clonidine blocks acute opiate symptoms. *Lancet* II, 599–602.

JAFFE, J. H. (1990). Drug addiction and drug abuse. In A. G. Gilman, et al. (Eds.), *Goodman and Gilman's the pharmacological basis of therapeutics*, 8th ed. New York: Pergamon.

MARK S. GOLD

COAST GUARD *See* Drug Interdiction; U.S. Government: The Organization of U.S. Drug Policy.

COCA-COLA *See* Cola/Cola Drinks.

COCAETHYLENE In 1990, a U. S. survey reported that 5.3 million Americans used cocaine during the previous month concurrently with (during the same month) alcohol. COCAINE was used simultaneously (on the same occasion) with ALCOHOL by 4.6 million people during the month before the survey. Of cocaine abusers, 59 percent also have lifetime alcohol abuse dependence diagnoses, and 30 percent of cocaine users drink alcohol (ethanol) on almost every occasion of cocaine use.

The co-abuse of cocaine and ethanol represents a serious health hazard. This combination was found to be the most common among individuals reporting to hospital emergency rooms with substance-abuse problems. Concurrent use of cocaine and alcohol produces a high death rate and a high incidence of cases requiring emergency medical care.

Cocaine as an agent with probable toxic effects on the liver in men was first reported in 1967, but very little is known about the long-term consequences of chronic cocaine and ethanol co-abuse on hepatic (liver) function. Cocaine-induced liver damage can be potentiated by pretreatment with drugs that activate cocaine metabolism by the microsomal cytochrome P-450 system, producing more toxic products. Alcohol is one such agent. Chronic or acute alcohol consumption significantly increases the susceptibility of hepatocytes (liver cells) to cocaine-medicated toxicity. Ingestion of alcohol may modify the rate and/or pattern of cocaine metabolism and may thereby enhance its hepatotoxicity.

A novel (new) cocaine metabolite, cocaethylene, was found in the postmortem blood, liver, and brain of people who had used cocaine concurrently with alcohol shortly before death. Cocaethylene is a benzolegonine ethyl ester. It is formed by transesterification of cocaine in the presence of ethanol in a reaction catalyzed by a nonspecific liver enzyme, acetylesterase. In the absence of ethanol this enzyme hydrolyzes cocaine to benzoylecgonine. Another isoform of liver acetylesterase catalyzes hydrolysis of cocaine to ecgonine methylester and has no activity in the synthesis of cocaethylene.

Cocaethylene formation does not occur in brain homogenates or whole blood. Cocaethylene is as potent as cocaine in binding to the DOPAMINE transporter in tissues taken from the striatal area of the human brain, but it is much less potent in binding to norepinephrine binding sites of the occipital cortex membranes or to the serotonin transporter in the frontal cortex of the brain.

The pharmacological activity of cocaethylene formed during the simultaneous use of cocaine and ethanol can be additive or synergistic to that of cocaine. Little data are so far available on the hepatotoxic potential of cocaethylene.

(SEE ALSO: *Complications: Liver Disease*)

BIBLIOGRAPHY

PIROZHKOV, S. V., WATSON, R. R., & ESKELSON, C. D. (1993). Ethanol enhances hepatotoxicity of cocaine and cocaethylene. *Advances in the Biosciences, 86,* 609–611.

RONALD ROSS WATSON

COCAINE The abuse of cocaine has become a major public-health problem in the United States since the 1970s. During that period it emerged from relative obscurity, described by experts as a harmless recreational drug with minimal toxicity. By the mid-1980s, cocaine use had increased substantially and its ability to lead to drug taking at levels that caused severe medical and psychological problems was obvious. Cocaine (also known as "coke," "snow," "lady," "CRACK" and "ready rock"), is an ALKALOID with both local anesthetic and PSYCHOMOTOR STIMULANT properties. It is generally taken in binge cycles, with periods of hours to days in which users take the drug repeatedly, alternating with periods of days to weeks when no cocaine is used. Many users are recalcitrant to treatment, and the introduction of substantial criminal penalties associated with its possession and sale have not yet been effective in reducing its prevalence of heavy use. In fact, although occasional use of cocaine diminished somewhat by the early 1990s, heavier use did not.

HISTORY

Cocaine is extracted from the COCA PLANT (*Erythroxylon coca*), a shrub now found mainly in the Andean highlands and the northwestern parts of the Amazon in South America. The history of coca plant use by the cultures and civilizations who lived in these areas (including the Inca) goes back more than a thousand years, with evidence of use found archeologically in their burial sites. The Inca called the plant a "gift of the Sun god" and believed that the leaf had supernatural powers. They used the leaves much as the highland Indians of South America do today. A wad of leaves, along with some ash, is placed in the mouth and both chewed and sucked. The ash helps in the extraction of the cocaine from the coca leaf—and the cocaine is efficiently absorbed through the mucous membranes of the mouth.

During the height of the Inca Empire (11th–15th centuries) coca leaves were reserved for the nobility and for religious ceremonies, since it was believed that coca was of divine origin. With the conquest of the Inca Empire by the Spanish in the 1500s, coca use was banned. The Conquistadors soon discovered, however, that their Indian slaves worked harder and required less food if they were allowed to chew coca. The Catholic church began to cultivate coca plants, and in many cases the Indians were paid in coca leaves.

Although glowing reports of the stimulant effects of coca reached Europe, coca use did not achieve popularity. This was no doubt related to the fact that coca plants could not be grown in Europe and the active ingredient in the coca leaves did not survive the long ocean voyage from South America. After the isolation of cocaine from coca leaves by the German chemist Albert Niemann in 1860 and the subsequent purification of the drug, it became more popular. It was aided in this regard by commercial endeavors in which cocaine was combined with wine (e.g., Vin de

Coca), products for which there appeared many enthusiastic and uncritical endorsements by notables of the time.

Both interest in and use of cocaine spread to the United States, where extracts of coca leaves were added to many patent medicines. Physicians began prescribing it for a variety of ills including dyspepsia, gastrointestinal disorders, headache, neuralgia, toothache, and more—and use increased dramatically. By the beginning of the twentieth century, cocaine's harmful effects were noted and caused a reassessment of its utility. As part of a broader regulatory effort, the U.S. government began to control its manufacture and sale. In 1914, the HARRISON NARCOTIC ACT forbade use of cocaine in over-the-counter medications and required the registration of those involved in the importation, manufacture, and sale of either coca or opium products. This had the effect of substantially reducing cocaine use in the United States, which remained relatively low until the late 1960s, when it moved into the spotlight once again.

MEDICAL UTILITY

Cocaine is a drug with both anesthetic and stimulant properties. Its local anesthetic and vasoconstriction effects remain its major medical use. The local anesthetic effect was established by Carl Koller in the mid-1880s, in experiments on the eye, but because it has been found to cause sloughing of the cornea, it is no longer used in eye surgery. Because it is the only local anesthetic capable of causing intense vasoconstriction, cocaine is beneficial in surgeries where shrinking of the mucous membranes and the associated increased visualization and decreased bleeding are necessary. Therefore, it remains useful for topical administration in the upper respiratory tract. When used in clinically appropriate doses, and with medical safeguards in place, cocaine appears to be a useful and safe local anesthetic.

PHARMACOKINETICS

Cocaine can be taken by a number of routes of administration—oral, intranasal, intravenous, and smoked. Although the effects of cocaine are similar no matter what the route, route clearly contributes to the likelihood that the drug will be abused. The likelihood that cocaine will be taken for nonmedical purposes is assumed to be related to the rate of increase in cocaine brain level (as measured by blood levels) associated with those routes that provide the largest and most rapid changes in brain level being associated with greater self-administration. The oral route of administration, not a route used by cocaine abusers, is characterized by relatively slow absorption and peak levels that do not appear until approximately an hour after ingestion. Cocaine, however, is quickly absorbed from the nasal mucosa when it is inhaled into the nose as a powder (cocaine hydrochloride). Because of its local anesthetic properties, cocaine numbs or "freezes" the mucous membranes, a quality used by those purchasing the drug on the street to test for purity. When cocaine is used intranasally ("snorting"), cocaine blood levels, as well as subjective and physiological effects, peak at about 20 to 30 minutes, and reports of a "rush" are minimal. Intranasal users report that they are ready to take a second dose of the drug within 30 to 40 minutes after the first dose. Although this route was the most common way for people to use cocaine in the mid-1980s, it is not as efficient in getting the drug to the brain as either smoking or intravenous injection, and it has declined in popularity.

When taken intravenously, venous blood levels peak virtually immediately and subjects report a substantial, dose-related rush. This route was, until the mid-1980s, traditionally the choice of the experienced user, since it provided a rapid increase in brain levels of cocaine with a parallel increase in subjective effects. Blood levels of cocaine dissipate in parallel with subjective effects, and subjects report that they are ready for another intravenous dose within about 30 to 40 minutes. Users of intravenous cocaine are also more likely to combine their cocaine with HEROIN (e.g., a "speedball") than are users by other routes.

In the mid-1980s, smoked cocaine began to achieve popularity. FREEBASE, or "crack," is cocaine base, which is not destroyed at temperatures required to volatilize it. As with intravenous cocaine, blood levels peak almost immediately and, as with intravenous cocaine, a substantial rush ensues after smoking it. Users can prepare their own freebase from the powdered form they purchase on the street, or they can purchase it in the form of crack, or "ready-rock." The development of a smokable form of cocaine provided a more socially acceptable route of drug administration (both NICOTINE and MARIJUANA cigarettes provided the model for smoking cocaine), resulting in a drug that was both easy to use and highly toxic, since the route allowed for frequent

repeated dosing with a readily available and relatively inexpensive drug. The use of intravenous cocaine, in contrast, was limited to those able to acquire the paraphernalia and willing to put a needle in a vein. The toxicity of the smoked route of administration is in part related to the fact that a potent dose of cocaine is available to anyone who can afford it.

Cocaine is frequently taken in combination with other drugs such as alcohol, marijuana, and OPIATES. In fact, almost 75 percent of cocaine deaths reported in 1989 involved co-ingestion of other drugs. When taken in combination with alcohol, a metabolite—COCAETHYLENE—is formed, which appears to be only slightly less potent than cocaine in its behavioral effects. It is possible that some of the toxicity reported after relatively low doses of cocaine might well be due to the combination of cocaine and alcohol.

Cocaine is broken down rapidly by enzymes (esterases) in the blood and liver. The major metabolites of this action (all relatively inactive) are BENZOYLECGONINE, ecgonine, and ecgonine methyl ester, all of which are excreted in the urine. Cocaethylene is an additional metabolite when cocaine and alcohol are ingested in combination. People with deficient plasma cholinesterase activity—fetuses, infants, pregnant women, patients with liver disease, and the elderly—are all likely to be sensitive to cocaine and therefore at higher risk for adverse effects than are others.

PHARMACOLOGY

Research has been focused on the neurochemical and neuroanatomical substrates that mediate cocaine's reinforcing effects. Although a number of NEUROTRANSMITTER systems are involved, there is growing evidence that cocaine's effects on dopaminergic neurons in the mesolimbic and/or mesocortical neuronal systems of the brain are most closely associated with its reinforcing and other behavioral effects. The initial site of action in the brain for its reinforcing effects has been hypothesized to be the dopamine transporter of mesolimbocortical neurons. Cocaine action at the DOPAMINE transporter has the effect of inhibiting dopamine re-uptake, resulting in higher levels of dopamine at the synapse. These dopaminergic pathways may mediate the reinforcing effects of other stimulants and opiates as well. A substantial body of evidence suggests that dopamine plays a major role in mediating cocaine's reinforcing

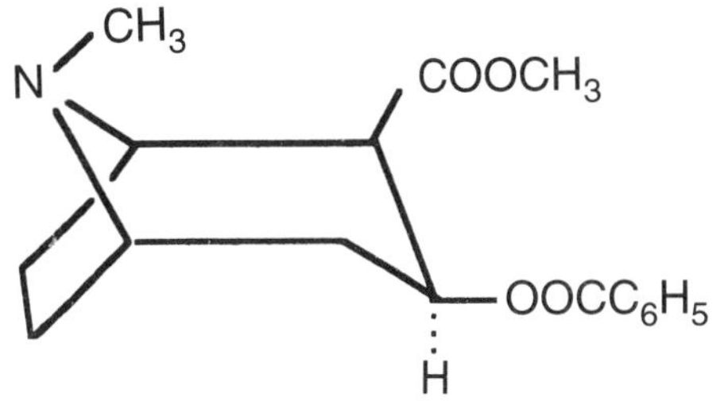

Figure 1
*Chemical Structure
of Cocaine*

effects, although it is clear that cocaine affects not only the dopamine but also the SEROTONIN and noradrenaline systems.

TOXICITY

In addition to blocking the re-uptake of several neurotransmitters, cocaine use results in central nervous system stimulation and local anesthesia. This latter effect may be responsible for the neural and myocardial depression seen after taking large doses. Cocaine use has been implicated in a broad range of medical complications covering virtually every one of the body's organ systems. At low doses, cocaine causes increases in heart rate, blood pressure, respiration, and body temperature. There have been suggestions that cocaine's cardiovascular effects can interact with ongoing behavior, resulting in increased toxicity. Cocaine intoxication has been associated with cardiovascular toxicity, related to both its local anesthetic effects and its inhibition of neuronal uptake of catecholamines, including heart attacks, stroke, vasospasm, and cardiac arrhythmias.

Cocaine is generally taken in binges, repeatedly, for several hours or days, followed by a period in which none is taken. When taken repeatedly, chronic cocaine intoxication can cause a psychosis, characterized by paranoia, anxiety, a stereotyped repetitive behavior pattern, and vivid visual, auditory, and tactile hallucinations. Less severe behavioral reactions to repeated cocaine use include irritability, hypervigilance, paranoid thinking, hyperactivity, and eating and sleep disturbances. In addition, when a cocaine binge ceases, there appears to be a crash response, characterized by depression, fatigue, and eating and sleep disturbances. Initially, the crash is accompanied by little cocaine craving, but as time increases since the last dose of cocaine, compulsive drug seek-

ing can occur in which users think of little else but the next dose.

BEHAVIORAL EFFECTS

Nonhuman Research Subjects. One of cocaine's characteristics, as a PSYCHOMOTOR STIMULANT, is its ability to elicit increases in the motor behavior of animals. Single low doses produce increases in exploration, locomotion, and grooming. With increasing doses, locomotor activity decreases and stereotyped behavior patterns emerge (continuous repetitious chains of behavior). When administered repeatedly, cocaine produces increased levels of locomotor activity, increases in stereotyped behavior, and increases in susceptibility to drug-induced seizures (i.e., "kindling"). This sensitization occurs in a number of different species and has been suggested as a model for psychosis or schizophrenia in humans. Although sensitization to cocaine's unconditioned behavioral effects generally occurs, such effects are related to dose, environmental context, and schedule of cocaine administration. For example, sensitization occurs more readily when dosing is intermittent rather than continuous and when dosing occurs in the same environment as testing.

Learned behaviors, typically generated in the laboratory using operant schedules of reinforcement in which animals make responses that have consequences (e.g., press a lever to get food), generally show a rate-dependent effect of cocaine. As with AMPHETAMINE, cocaine engenders increases in low rates of responding and decreases in high rates of responding. Environmental variables and behavioral context can modify this effect. For example, responding maintained by food delivery was decreased by doses of cocaine that either had no effect or increased comparable rates of responding maintained by shock avoidance. Cocaine's effects can also be modified by drug history. Although repeated administration can result in the development of sensitization to cocaine's effects on unlearned behaviors, repeated administration generally results in tolerance to cocaine's effects on schedule-controlled responding. This decrease in effect of the same dose after repeated dosing is influenced by behavioral as well as pharmacological factors.

Human Research Subjects. A major behavioral effect of cocaine in humans is its mood-altering effect, generally believed related to its potential for abuse. Traditionally, subjective effects have provided the basis for classifying a substance as having abuse potential—and the cocaine-engendered profile of subjective effects is prototypic of stimulant drugs of abuse. Thus, cocaine produces dose-related reports of "high," "liking," and "euphoria"; increases in stimulant-related factors, such as increases on Vigor and Friendliness scale scores; ratings of "stimulated"; and decreases in various sedation scores. Subjective effects correlate well with single intravenous or smoked doses of cocaine, peaking soon after administration and dissipating in parallel with decreasing plasma concentrations. When cocaine is administered repeatedly, tolerance develops rapidly to many of its subjective effects and the same dose no longer exerts much of an effect. This means that the user must take increasingly larger amounts of cocaine to achieve the same effect. Tolerance to the cardiovascular effects of cocaine is less complete; the result here is a potential for drug-induced toxicity, since more and more drug is taken when the subjective effects are not present but the disruptions in cardiovascular function are still present.

Although users of stimulant drugs claim that their performance of many activities is improved by cocaine use, the data do not support their assertions. In general, cocaine has little effect on performance except under conditions in which performance has deteriorated from fatigue. Under those conditions, cocaine can bring it back to nonfatigue levels. This effect, however, is relatively short-lived, since cocaine has a half-life of less than one hour.

TREATMENT

Despite substantial efforts directed toward treatment of cocaine abuse, in the mid-1990s we are still unable to treat successfully many of the cocaine abusers who seek treatment. For many years the only approach to treating these people was psychological or behavioral. As of 1994, the most promising of these include BEHAVIORAL THERAPY, RELAPSE PREVENTION, rehabilitation (e.g., vocational, educational, and social-skills training) and supportive PSYCHOTHERAPY. A major problem with these treatment approaches is related to their lack of selectivity. Rather than tailoring programs to an individual's background, drug-use history, psychiatric state, and socioeconomic level, individuals receive the treatment being delivered by the particular program they

happen to attend. Treatment programs that focus on specific target populations will be far more successful than those which cover all who apply. For example, patients with relatively mild symptoms might do quite well in a behavioral intervention with some relapse-prevention instructions but those with more severe problems might require the addition of PHARMACOTHERAPY.

Pharmacological approaches to treating cocaine abusers have focused on potential neurophysiological changes related to chronic cocaine use. Thus, because dopamine appears to mediate cocaine's reinforcing effects, dopamine agonists such as AMANTADINE and bromocriptine have been tried. METHYLPHENIDATE, a stimulant, has been suggested as a possible substitution medication, and ANTIDEPRESSANTS such as desipramine have been studied because of their actions on the dopaminergic system. In addition, because cocaine blocks re-uptake of SEROTONIN at nerve terminals, serotonin-uptake blockers, such as fluoxetine, have also been tested. Although most of the potential medications have been shown to be successful in some patients under open label conditions, none have been clearly successful in double blind placebo-controlled clinical trials.

Clearly, no medication yet exists for the treatment of cocaine abuse. It may well be that different medications may be effective for the various target populations and that variations in dosages and durations of treatment might be required, depending on a variety of patient characteristics. In fact, several medications have been shown to be effective only for small and carefully delineated populations (e.g., lithium for cocaine abusers diagnosed with concurrent bipolar manic-depressive or cyclothymic disorders). An artificial enzyme has been developed that inactivates cocaine as soon as it enters the blood-stream by binding the cocaine and breaking it into two inactive metabolites, and this has the potential for destroying much of the cocaine before it reaches the brain. As of 1994, this technique is unavailable for human use. In addition, and most importantly, cocaine abuse (and drug abuse in general) is a behavioral problem, and it is unlikely that any medication will be effective unless it is combined with an appropriate behavioral intervention.

(SEE ALSO: *Cocaine, Treatment Strategies; Colombia As Drug Source; Epidemics of Drug Abuse; Epidemiology of Drug Abuse; National Household Survey on Drug Abuse; Treatment: Cocaine*)

BIBLIOGRAPHY

BOCK, G., & WHELAN, J. (1992). *Cocaine: Scientific and social dimensions.* Ciba Foundation Symposium 166. Chichester: Wiley.

JOHANSON, C. E., & FISCHMAN, M. W. (1989). Pharmacology of cocaine related to its abuse. *Pharmacological Reviews, 41,* 3–52.

KLEBER, H. D. (1989). Treatment of drug dependence: What works. *International Review of Psychiatry, 1,* 81–100.

LANDRY, D. W., ET AL. (1993). Antibody-catalyzed degradation of cocaine. *Science, 259,* 1899–1901.

MARIAN W. FISCHMAN

COCAINE ANONYMOUS   More than 200 mutual-help groups have been modeled after the TWELVE STEPS of ALCOHOLICS ANONYMOUS (AA), but only Cocaine Anonymous (CA) deals with as seductive a drug as COCAINE. "Laboratory animals will give up both food and sex for self-administered doses of cocaine and will even starve to death to continue receiving cocaine instead of food" (Goode, 1984: 103). Most studies of users agree that the cocaine high is unusually pleasant and that these effects reinforce each other until the user quickly wants to continue its use again and again. Thus, it is easy to lose control, and recreational users run great risks of addiction. Members of CA tend to believe that the unique features of the drug call for a twelve-step group that focuses exclusively on cocaine and its capacity to produce addiction.

Prior to 1985, cocaine was so expensive that only the affluent or those adroit at lucrative crimes could afford it. For the most part, it was inhaled or rubbed on mucous membranes; sometimes it was made into a solution and injected. The advent of an inexpensive special form of cocaine in 1985/86—CRACK or rock cocaine—reduced the cost. Crack is usually smoked in pipes. Injecting or smoking cocaine (coke) "can lead to almost continual consumption and drug-seeking behavior, destructive to personal competence and productivity" (Van Dyke & Byck, 1982:140). Moreover, the risk of dying from a cocaine overdose due to injecting or smoking it has increased significantly since the mid-1970s, and the side effects of abuse are extremely unpleasant (e.g., insomnia, gross paranoia, and severe anxiety attacks).

Cocaine Anonymous is very closely patterned after Alcoholics Anonymous. For example, the focus is only on cocaine, much as AA focuses exclusively on ALCOHOL. In this regard, both AA and CA differ from NARCOTICS ANONYMOUS (NA), whose focus is more inclusive—a wide variety of drugs. Members of CA insist, and with good reason, that they had rapidly lost control of their lives and were largely powerless over just cocaine. Some say they underscore step one of the Twelve Steps, because they believe that of all the drugs they claimed to have experienced, they most easily lost control of cocaine. At the same time, they point to recreational users they know who seem to use modest amounts without harmful effects; but they stress that they themselves cannot use the drug in this fashion. Like members of AA, they believe they must be "clean"—there is no moderate use for them. Furthermore, they are certain that those who believe they are recreational users will someday become addicted. Somewhat unlike AA, however, when using the twelve-step program, CA's literature tends to emphasize the first three steps, with the twelfth step in the background. Although some AA groups do much the same, in CA this tendency seems to be more pronounced.

One of the first points CA makes with newcomers is that its members at one time uniformly denied addiction. Members admit that it may be difficult to determine if one is addicted with certainty, but "one thing is sure, though: every single one of us *denied* being an addict for months, for years, who now freely admit that we are cocaine addicts who thought that we could control cocaine when in fact it was controlling us" (Cocaine Anonymous, 1990:1). Also, CA uses the AA notion of "hitting bottom" from at least three standpoints: a physical bottom such as sexual impotency; an emotional bottom when no matter how much cocaine they used they never again experienced the former elation, only a brief, brief respite from severe depression; and a bottom made up of a gridlock web of spending, lying, dealing, and credit manipulations. "Most of us were brought down by a medley of financial, physical, social, and spiritual problems."

The CA notion of being clean is a total one, without any exceptions. Even though its focus is largely on cocaine, it discourages the use of all mind- and mood-altering drugs, including alcohol and MARIJUANA. In this it goes beyond AA, where the essential focus is on alcohol, although some attention has been given to urging members to avoid other drugs. Thus, CA cautions against "substitute addictions" as a real threat. It insists that the admonition comes from the bitter experiences of its members.

Although CA is directed toward only one drug, it nevertheless appears to have many groups. No census of all groups in the United States is publicly available, but more than seventy meetings occur weekly in the New York metropolitan area.

(SEE ALSO: *Cocaine, Treatment Strategies; Rational Recovery; Treatment; Women for Sobriety*)

BIBLIOGRAPHY

Cocaine Anonymous World Service. (1990). *To the newcomer.* Culver City, CA: Author.

GOODE, E. (1984). *Deviant behavior* (2nd ed.). Englewood Cliffs, NJ: Prentice-Hall.

VAN DYKE, C., & BYCK, R. (1982). Cocaine. *Scientific American* (March), 128–141.

HARRISON M. TRICE

COCAINE, EPIDEMICS *See* Epidemics of Drug Abuse.

COCAINE, TREATMENT STRATEGIES

Strategies for treating COCAINE abuse include PSYCHOTHERAPIES, pharmacotherapies for attaining abstinence and RELAPSE PREVENTION, and treatments for "dual diagnosis" conditions of either multiple-drug abuse or substance abuse with major psychiatric disorders. Multiple-drug abuse in cocaine abusers involves problems with ALCOHOL, OPIOIDS, and BENZODIAZEPINES. The medical and psychosocial consequences of using these drugs in various combinations are often more severe than using each drug alone, and combinations of treatment options may be needed for many of these drugs. Specific treatments may include relapse-prevention psychotherapies targeted toward drug-related cues, and pharmacotherapies targeted toward cocaine as well as other drugs of abuse, such as NALTREXONE for opioid abuse and DISULFIRAM for alcohol abuse. Few controlled clinical trials are as yet available with cocaine abusers, so that what are currently considered successful treatments must be viewed tentatively un-

til confirmed by adequate controlled and independently replicated trials.

PSYCHOTHERAPEUTIC APPROACHES

The psychotherapeutic approaches for cocaine abusers are generally similar to the approaches for abusers of other drugs, although treatments for ALCOHOLISM and drug abuse have evolved somewhat differently and the models used may conflict at certain points. A great deal of discussion has been generated about these conflicts in combined treatment for alcohol- and drug-dependent patients, but, overall, the literature is positive about the merits of combining approaches. The distinctions in program staffing and procedures for treating cocaine abuse, moreover, may depend on distinctions that are not directly related to the type of drug abused. For example, "streetwise" and "nonstreetwise" abusers may require different types of treatment programming. In developing psychotherapies, researchers need to consider two issues in particular: increased comorbid psychopathology and insistence on complete abstinence, including abstinence from alcohol.

Several diagnostic studies have found high rates of psychopathology among cocaine abusers. In a recent survey of cocaine abusers, the rates of affective disorders and childhood ATTENTION-DEFICIT DISORDERS were quite high. In a large survey of opiate addicts, the cocaine-abusing opiate addict had high rates of DEPRESSION and ANTISOCIAL PERSONALITY DISORDER. Specific psychotherapeutic approaches have been developed for depression, for example, and the resources to provide these professional services may frequently be required.

A second issue in the treatment of cocaine abusers who also abuse alcohol is controlled alcohol use once cocaine consumption is controlled. While complete abstinence from HEROIN, cocaine, and SEDATIVES is generally considered the treatment goal, some serious consideration has been given to controlled drinking among "recovered" alcoholics and may be considered among alcoholic cocaine abusers. This issue of abstinence versus controlled drinking has been examined in a study of alcoholic methadone-maintained patients. The study compared thirty-six control patients with forty-two entering an abstinence-oriented ALCOHOLICS ANONYMOUS (AA) treatment and with forty-two entering a controlled-drinking, behavior-modification program. During a

6-week educational period for the two treatment groups, dropout was quite high (58% of patients lost), making later comparisons difficult. Furthermore, after 12 more weeks of treatment, the only significant difference was in total alcohol consumption, and the abstinence-oriented group had done worse than the control group. Overall, this study showed no efficacy for additional psychotherapies aimed at alcoholism among alcoholic methadone-treated patients. This finding has interesting implications for the cocaine-abusing alcoholic, but it has not been specifically examined.

Psychotherapeutic approaches for substance abuse may involve SELF-HELP, behavioral, COGNITIVE, interpersonal, or family treatments. One study examined both a cognitive and a psychodynamic form of therapy for methadone-maintained patients, and it does not appear that any specific changes would be needed in order to use these therapies with methadone patients who were also abusing cocaine. Family approaches to therapy also have been described for cocaine abusers. Behavioral therapy has been described in conjunction with outpatient treatment programs and can be an important part of residential treatment programs using a therapeutic community model. Two other types of substance-abuse psychotherapy have also been described: interpersonal psychotherapy and relapse prevention.

Relapse-prevention therapy requires specific interventions based on precipitants that have been identified as associated with risk of returning to abuse of a specific drug. These precipitants, which include negative emotional states, interpersonal conflict, social pressure, and specific drug-related cues, may be quite different for different drugs of abuse. For example, in a methadone-maintained patient, the precipitants for using heroin or cocaine may be closely related to being with particular "friends" and then "getting high." This "getting high" on heroin can be pharmacologically blocked by large doses of METHADONE; large methadone doses will not have a similar effect on cocaine use. Self-monitoring is used to identify risk situations for the specific drug, and then coping strategies are developed using rehearsal of coping behaviors such as anger management and social skills. Preventing relapse focuses on ensuring that brief lapses to cocaine use do not become full relapses. A lapse may be seen as a discreet isolated event—not uncommon in recovery and not nullifying all progress. Reduction of this ABSTINENCE VIO-

LATION EFFECT by reframing the concept in this way may work with all drugs of abuse, although in multiple-drug abusers, sequential lapses in *each* drug must be prevented—by carefully emphasizing the importance of abstinence and not giving "permission" for experimenting with isolated use of the various abused drugs.

Interpersonal psychotherapy (IPT) was first developed as a treatment for depression and was adapted for opiate addicts and, later, cocaine abusers. This psychotherapy for substance abusers is based on the premise that drug abuse is one way in which an individual attempts to cope with problems in interpersonal functioning. An exploratory stance focuses on interpersonal relationships and the impact of drug abuse on these relationships. In helping the patient stop his or her substance abuse, the practitioner selects the important component of treatment; they may include documenting the adverse effects of the drugs compared with their perceived benefits, identifying the thoughts and behaviors that precede drug use, and developing strategies to deal with drug-related cues and high-risk situations. Only after attaining abstinence are interpersonal difficulties directly addressed, including the roles of drug use in these relationships.

A key strategy with IPT is to develop more productive means for achieving the desired social gratification or tension reduction for which the drug abuse substitutes. This substitution may differ markedly for various drugs that a multiple-drug abuser may be using. For example, the abuser may be using cocaine to reduce social isolation and to "meet exciting new people," but may be abusing alcohol because the cocaine "crash" is reduced by the alcohol. Since only the cocaine, and not the alcohol, is directly related to the social deficit, only the cocaine abuse will directly benefit from interpersonal therapy. In general, the interpersonal impact will be somewhat different for the abuse of licit drugs such as alcohol, illicit drugs such as heroin and cocaine, and "doctor shopping" drugs such as benzodiazepines. Among cocaine addicts, for example, the licit drugs such as alcohol are often used in response to interpersonal tension, while the illicit drugs such as heroin lead to consequences of increased interpersonal tension, rather than being used in response to tension. In summary, interpersonal therapy must identify the relationship of each particular drug to the interpersonal setting as either primary associa-tion or secondary to other drug effects and as either a tension reliever or inducer.

PHARMACOTHERAPIES FOR COCAINE DEPENDENCE

The pharmacological treatment of cocaine abuse is defined as the use of medication to facilitate initial abstinence from cocaine abuse and to reduce subsequent relapse to cocaine abuse. The initiation of abstinence from cocaine abuse involves reduction in the withdrawal symptoms associated with cessation of cocaine. This withdrawal syndrome resembles depression but also includes a great deal of anxiety and craving for use of cocaine. Craving for cocaine often persists for several weeks after abstinence has been attained, and cues—places or things associated with cocaine use in the past—can continue to stimulate cocaine craving for many months after cocaine use has been stopped. Because of the persistence of what is called *conditioned craving*, relapse to cocaine abuse can occur after the patient has become abstinent. Preventing relapse is a second important function of medication treatment.

An objective of the use of medications in cocaine dependence is to reverse changes that are caused in the brain after chronic cocaine use. These brain changes, called *neuroadaptation*, are easily demonstrated in animal models. Chemical analyses of animal brains chronically exposed to cocaine show abnormalities in the neurotransmitter receptors on brain cells, including the DOPAMINE receptors and the SEROTONIN receptors.

Direct and indirect evidence that there are changes in these brain receptors can also be found in human studies. Prolactin is a hormone that is controlled by the neurotransmitters dopamine and serotonin. In some heavy cocaine abusers, prolactin levels become abnormally high after the abuse has stopped, and they remain elevated for a month or more. According to this evidence, both dopamine and serotonin brain systems are perturbed by cocaine and the abnormality persists for some time. Other evidence of persistent abnormalities in the dopamine systems come from brain imaging studies directly examining dopamine receptors. Human imaging studies using positron emission tomography (PET) have shown a marked reduction in dopamine receptors on brain cells that are ordinarily very rich in dopamine receptors. The abnormally low amount

of dopamine receptors persists for at least two weeks after a patient stops using cocaine. That several medications may reverse these neurochemical receptor changes has been an important rationale for their use.

These brain changes are thought to occur after chronic heavy cocaine use and not after single, small amounts of cocaine use. These brain changes are also most likely to occur when cocaine abusers are using intensive routes of administration, such as smoking or injecting. Much larger dosages of cocaine get to the brain than with other routes of administration, such as orally or intranasally ingested (snorted) cocaine. In addition to these direct biological indicators of neuroadaptation, neuropsychological tests have documented sustained deficits in thinking, concentration, and learning among chronic cocaine abusers. These deficits may persist for months after cocaine use has been stopped.

The biological abnormalities induced by cocaine in the brains of abusers may be manifest by a characteristic WITHDRAWAL syndrome. The very early phases of this syndrome, commonly called the "crash," may involve serious psychiatric complications such as paranoia with agitation and depression with suicide. These complications require medications for symptomatic management, including ANTIPSYCHOTIC agents, such as chlorpromazine and haloperidol, or large dosages of benzodiazepines to calm highly agitated patients. Many patients self-medicate these crashes using such sedating substances as benzodiazepines or alcohol. Because this crash phase is usually relatively brief, rarely lasting more than several days, sustained medication is generally not necessary. The more important role for medications occurs during the later phase of withdrawal, which may persist for several weeks. This phase resembles a depressive syndrome, with substantial anxiety and craving to use cocaine. The neurobiological changes noted in both human and animal studies after chronic cocaine use occur at about the same time as the depressive syndrome. This temporal correspondence has provided a further rationale for the use of ANTIDEPRESSANT medications in the treatment of cocaine dependence and withdrawal.

A wide range of pharmacological agents besides antidepressants have been used to treat cocaine withdrawal and to reduce relapses to dependence. These other agents include dopamine agonists, dopamine blockers, lithium, monoamine oxidase inhibitors, amino acids, and opioid-derived medications. The use of opioid-derived medications to treat cocaine dependence has an ironic twist, because Sigmund FREUD had suggested that cocaine might be an appropriate treatment for morphine (an opioid) addiction. Clearly, substituting one drug of abuse for another drug of abuse is a risky treatment approach, but new ideas are emerging on the use of opioids with lower abuse potential than morphine, such as BUPRENORPHINE for patients dependent on both opioids and cocaine. Evaluation of these proposed medications must use controlled studies involving a double-blind procedure and random assignment to either an active medication or an inactive placebo.

In controlled studies, several antidepressants have been found superior to placebo. Desipramine was felt to promote cocaine abstinence by reducing craving, but the initial study has not been successfully replicated. Fluoxetine is another antidepressant that an initial controlled study found efficacious in reducing cocaine abuse, but it has also not been independently replicated. Pilot studies with other antidepressants have suggested their efficacy, but controlled studies are usually lacking. Likely agents include imipramine, maprotiline, phenelzine, and trazadone.

In theory, dopaminergic agents may be particularly useful in ameliorating early withdrawal symptoms after cocaine binges, and a variety of agents including AMANTADINE, bromocriptine, L-dopa, METHYLPHENIDATE, mazindol, and pergolide have been applied to this purpose with few adequate, placebo-controlled, double-blind studies. Several trials have examined amantadine at 200 and 300 milligrams (mg) daily and found that it reduces craving and use for several days to a month, but side effects have limited the utility of several other dopaminergic agents. A number of other agents recently have been utilized to treat different aspects of cocaine abuse and dependence, including a dopamine blocker flupenthixol, a mixed opiate agonist-antagonist, buprenorphine, and the anticonvulsant carbamazepine, although their utility has not been confirmed in double-blind studies.

MULTIPLE-DRUG USE

During the last few years, multiple-drug use has become a major problem among cocaine abusers.

Cocaine abusers in a recent study of 300 treatment-seeking patients reported alcohol abuse in 70 percent of the cases and sedative abuse in 43 percent. In this study, opiate abusers were specifically excluded. The alcohol-abuse rate among these cocaine abusers was quite interesting in that only 20 percent of the cocaine abusers were primary alcoholics who had become alcoholic before becoming cocaine abusers. For the rest of the alcohol abusers, the alcohol was used to cope with the dysphoria that followed cocaine use (the crash), and when they abstained from cocaine, they also did not abuse alcohol. For sedative abuse, the pattern was primarily sporadic use with only 11 percent of these cocaine abusers reporting weekly use. These findings suggest that although multiple-drug abuse is quite prevalent, treatment of the cocaine abuse alone may be sufficient to control the abuse of the alcohol or sedatives. Thus the major treatment efforts need only focus on the relatively easier task of detoxification for alcohol or sedatives, rather than simultaneous maintenance treatments for several abused drugs.

Unfortunately, the situation with combined cocaine and opiate abusers appears more complex, with concurrent abuse of cocaine occurring commonly within opioid treatment programs such as METHADONE MAINTENANCE. Cocaine abuse has been relatively common among heroin addicts, with a 50 percent lifetime rate. Current multiple-drug use rates are somewhat lower, but still high, with current cocaine abuse reported by 29 percent of addicts, both in and out of opioid-maintenance treatment programs. The reasons for cocaine abuse by heroin addicts are apparently to "improve" the euphoria from heroin. These findings suggest that control of heroin abuse in many patients may directly reduce cocaine abuse, and the reduction in cocaine abuse reported by several surveys of methadone-maintenance programs support this assertion.

Within treatment programs, multiple-drug use is a problem for initial retention of patients and for rehabilitation in those programs with good retention such as methadone maintenance. Initial retention is reduced by the need for prolonged multiple detoxifications, since patients tend to leave the hospital or drop out of outpatient detoxification programs. In programs with better outpatient retention, such as methadone maintenance, multiple-drug use undermines efforts at social rehabilitation. The cocaine-abusing methadone patient will be unable to have

sustained employment or education and may continue to engage in criminal activity in order to obtain the drug.

Detoxification. While detoxification is not a critical issue with the "pure" cocaine abusers, detoxification from the more typical multiple-drug abuse pattern of contemporary cocaine abusers can be a complex procedure requiring inpatient treatment. Inpatient detoxification may not be required for dependence on drugs such as cocaine or for treatments in which patients do not have to be drug-free (such as initiation of methadone maintenance). However, the combination of sedatives and sometimes of alcohol with cocaine dependence may require carefully monitored inpatient treatment protocols, because of the potential for seizures, organic psychotic states, and death. Substantial medical interventions may be needed, as when an alcoholic develops delerium tremens. Following detoxification, DISULFIRAM may be appropriate, but liver functioning must be monitored, since alcohol is a liver toxin.

In cocaine abusers, sedative and alcohol detoxification may be feasible as an outpatient procedure. While the interruption of binges of cocaine abuse may require hospitalization for some patients, many cocaine abusers are treated as outpatients, sometimes with the use of adjunctive medications to reduce cocaine craving and maintain abstinence. For those cocaine abusers who are also dependent on alcohol or sedatives, outpatient detoxication from either of these two nonstimulants might be considered, using either carbamazepine or clonazepam given in tapering dosages over several days, in order to treat alcohol (5 days of tapering) or other sedatives (10–12 days of tapering). The cocaine could be discontinued abruptly without needing detoxification. The major issue after stopping cocaine is careful observation of the patient during the cocaine "crash." The risks during a crash can be substantial, because a severe postcocaine depression may precipitate suicide attempts or be associated with a paranoid psychosis, but this degree of severity is unusual. Thus, hospitalization might be required for management of the cocaine abuse independent of other concurrent drug dependence.

Detoxification from the combination of cocaine and opiate dependence usually is managed in the outpatient setting. Detoxification from the opiate is needed, if long-term residential or outpatient NAL-

TREXONE treatment is being considered, but opiate detoxification is not needed for methadone maintenance or for the investigational medication buprenorphine. As indicated above, cocaine dependence usually does not require inpatient treatment for detoxification; the most common treatment for these dually addicted patients is methadone maintenance. For those entering residential treatment, opiate detoxification using clonidine or clonidine with naltrexone precipitation of withdrawal might be considered. Among those being detoxified using clonidine, it is important that tricyclic antidepressants, a treatment for cocaine abuse, not be started until the opiate detoxification is complete; the tricyclics may interfere with the withdrawal-suppressing effects of clonidine.

Maintenance Pharmacotherapies for Poly-drug Abuse. Maintenance pharmacotherapies have been developed for opiate dependence and, to a lesser extent, for alcoholism in combination with cocaine dependence. Maintenance pharmacotherapies have developed from somewhat different rationales for each of the abused drugs: AGONISTS (methadone) or ANTAGONISTS (naltrexone) for opiates, aversive agents (disulfiram) for alcohol, and "anticraving" agents (desipramine, amantadine) for cocaine, as previously reviewed.

In general, the preferred approach in using PHARMACOTHERAPY for the multiple-drug abuser would be to select a single medication to manage all abused drugs or at least the most problematic drug and to use nonpharmacological approaches for the other abused drugs. The closest approximation to a single medication for opiate, cocaine, and alcohol dependence is naltrexone, a long-acting opioid antagonist used primarily in the treatment of opiate addicts. Recent data have suggested that it may also decrease relapse to alcohol abuse by preventing "slips," or relatively brief lapses into alcohol use, from developing into full alcoholic relapses, and preliminary data have shown significantly less cocaine abuse in naltrexone-maintained than in methadone-maintained former opiate addicts. Another medication that may be useful for more than one abused drug is buprenorphine, because in some preliminary studies it reduced not only opiate abuse but also cocaine abuse. Finally, for cocaine-abusing opiate addicts in methadone maintenance, desipramine, amantadine, and other agents have been examined in combination with methadone.

SUMMARY

The treatment of cocaine abuse typically involves a mixture of psychotherapy and pharmacotherapies. A variety of psychotherapies have been developed for cocaine abusers, including relapse-prevention therapy, interpersonal psychotherapy, family therapy, and therapies targeted at secondary psychopathology, particularly depression. Pharmacotherapies for primary cocaine abuse have included antidepressants, dopamine agonists, and several miscellaneous agents such as buprenorphine and carbamazepine. These agents may also complement the treatment of multiple-drug abusers such as the cocaine-abusing alcoholic or heroin addict.

Combination pharmacotherapies of these cocaine anticraving agents with methadone or naltrexone for heroin addiction and with disulfiram or naltrexone for alcoholism have also been tried with some success. An important clinical need with patients dependent on opiates, alcohol, or sedatives in addition to cocaine is for detoxification. While cocaine withdrawal is not associated with major medical complications, withdrawal from these other drugs can be medically significant and often needs specific pharmacological interventions. The high rates of major psychiatric disorders in cocaine abusers has also led to the use of various psychotherapies and pharmacotherapies targeted toward these disorders rather than specifically toward reduction in cocaine abuse. The efficacy of these various treatments has received limited evaluation, but many promising leads have developed.

(SEE ALSO: *Treatment: Cocaine, Pharmacotherapy; Withdrawal: Cocaine*)

BIBLIOGRAPHY

GAWIN, F. H., & ELLINWOOD, E. H. (1988). Cocaine and other stimulants: Actions, abuse, and treatment. *New England Journal of Medicine, 318,* 1173–1182.

KOSTEN, T. R. (1989). Pharmacotherapeutic interventions for cocaine abuse: Matching patients to treatment. *Journal of Nervous and Mental Disease, 177*(7), 379–389.

KOSTEN, T. R., & KLEBER, H. D. (1992). *Clinician's guide to cocaine addiction.* New York: Guilford Press.

ROUNSAVILLE, B. J., GAWIN, F. H., & KLEBER, H. D. (1985). Interpersonal psychotherapy (IPT) adapted for ambu-

latory cocaine abusers. *American Journal of Drug and Alcohol Abuse, 11,* 171–191.

THOMAS R. KOSTEN

COCA PASTE Coca paste is the first crude extraction product of coca leaves from the COCA PLANT; it is obtained in the process of extracting COCAINE from these leaves. The leaves are mashed with alkali and kerosene and then sulfuric acid (and sometimes also potassium permanganate). The result is an off-white or light-brown paste containing 40 to 70 percent cocaine, as well as other ALKALOIDS, benzoic acid, kerosene residue, and sulfuric acid (ElSohly, Brenneisen, & Jones, 1991). Peruvian and Bolivian paste is illegally exported to Ecuador or Colombia, where it is purified into cocaine hydrochloride and then illicitly shipped to markets throughout the world. Although cocaine is the major component of coca paste, the paste is chemically complex, reflecting additives used by the clandestine laboratories performing the extraction from the coca leaves.

Coca paste, also called cocaine paste or pasta, is smoked, primarily in Latin American countries, by mixing about 0.2 ounces (0.5 g) of it with TOBACCO (called "tabacazo") or with MARIJUANA (called "mixto") in a cigarette. When this dab of coca paste is smoked with tobacco, only 6 percent of the cocaine reaches the smoker—but since most the paste samples contain significant amounts of manganese as well as several gasoline residues, the inhaled condensate is an extremely toxic substance. Despite the low bioavailability of cocaine from coca paste when it is smoked, use of this illegal substance by the smoking route reached epidemic proportions in Latin America in the late 1970s. More recently, coca paste smoking has been reported in the NETHERLANDS, the Antilles, Panama, and the United States, although the level of use remains very low.

The effects of coca-paste smoking have been reported to be as toxic as those seen after intravenous or smoked cocaine (i.e., CRACK) in the United States. In fact, coca-paste smokers can achieve cocaine blood levels comparable to those seen in users injecting or smoking cocaine (Paly et al., 1980). Smoking the paste leads to an almost immediate euphoric response, and users smoke it repeatedly. As with smoking cocaine (FREEBASING), large quantities of the paste are taken repeatedly within a single smoking session, which is terminated only when the drug supply is depleted. Users report a dysphoric response (unease, illness) within about thirty minutes after smoking, so more paste is generally smoked at this time if available.

Substantial toxicity has been reported for chronic use of the coca-paste–tobacco combination, with users smoking it repeatedly, and progressing from stimulant-related effects and euphoria to HALLUCINATIONS and paranoid psychoses. In fact, studies carried out in Peru defined a mental disorder of coca-paste smoking, made up of four distinct phases—euphoria, dysphoria, hallucinosis, and paranoid psychosis (Jeri et al., 1980). Since substantial amounts of paste are smoked at one time, the paranoid psychosis seen after chronic stimulant use has also been reported for paste use. As with cocaine abusers, experienced users of coca paste usually turn to criminal activities to support their illicit drug use.

(SEE ALSO: *Bolivia, Drug Use in; Complications; Crime and Drugs; Pharmacokinetics; Psychomotor Stimulants*)

BIBLIOGRAPHY

ELSOHLY, M. A., BRENNEISEN, R., & JONES, A. B. (1991). Coca paste: Chemical analysis and smoking experiments. *Journal of Forensic Sciences, 36,* 93–103.

JERI, F. R., ET AL. (1980). Further experience with the syndromes produced by coca paste smoking. In F. R. Jeri (Ed.), *Cocaine 1980.* Lima: Pacific Press.

PALY, D., ET AL. (1980). Cocaine: Plasma levels after cocaine paste smoking. In F. R. Jeri (Ed.), *Cocaine 1980.* Lima: Pacific Press.

VAN DYCK, C., & BYCK, R. (1982). Cocaine. *Scientific American, 246,* 128–141.

MARIAN W. FISCHMAN

COCA PLANT The coca plant is a cultivated shrub, generally found in the Andean Highlands and the northwestern areas of the Amazon in South America. The plant, however, can be grown in many parts of the world and in the early part of the twentieth century much of the cocaine used in medicine was obtained from plants grown in Asia. Of the more than 200 species of the genus *Erythroxylon*, only *E. coca* variety *ipadu*, *E. novogranatense*, and *E. novogranatense* variety *truxillense* contain significant amounts of COCAINE, ranging from 0.6 to 0.8 percent.

Figure 1
Coca Leaf

In addition to cocaine, the leaves of the coca plant contain eleven other ALKALOIDS, although no others are extracted for their euphorogenic effects.

Coca plants have long histories of use for both their medicinal and stimulant effects. Coca leaves are believed to have been used for well over a millenium, since archeological evidence from Peruvian burial sites of the 6th century A.D. suggests coca use. In fact, ancient Indian legends describe its origin and supernatural powers. The Inca called the coca plant a "gift of the Sun God," and attributed to it many magical functions. The Inca and the other civilizations of the Andes used coca leaves for social ceremonies, religious rites, and medicinal purposes. Because of their energizing property, coca leaves were also used by soldiers during military campaigns or by messengers who traveled long distances in the mountains. Under the Spanish conquest of the sixteenth century, coca plants were systematically cultivated and the custom of chewing coca leaves or drinking coca tea was widely adopted as part of the Indian's daily life in South America. Use of coca leaves as both a medicinal and a psychoactive substance continues to be an integral part of the daily life of the Indians living in the Andean highlands. Substantial societal controls have existed concerning the use of these leaves, and minimal problematic behavior related to use of the coca leaves has been reported.

In the highland areas of Peru and Bolivia, and less frequently, in Ecuador and Colombia, the dried leaves are mixed with lime or ash (called "tocra") and both chewed and sucked. A wad containing 0.4 to 1 ounce (10 to 30 g) of leaf is formed, and daily consumption by coca-leaf chewers is between 1 and 2 ounces (30 and 60 g). The Indian populations in the Amazonean areas, however, crush the dried leaves, mix the powder with an alkaline substance, and chew it. Coca leaves are chewed today in much the same way that they were chewed hundreds of years ago.

Substantial cocaine plasma levels can be attained when coca leaves are chewed along with an alkaline substance, which increases the bioavailability of the drug by changing its pH. Volunteers allowed to chew either the leaf or the powdered form of coca mixed with an alkaline substance reported numbing in the mouth and a generally stimulating effect which lasted an average of approximately an hour after the coca chewing was begun (Holmstedt et al., 1979). This time-course corresponded to the ascending limb of the cocaine plasma-level curve, suggesting that the effect was cocaine-induced. Absorption of cocaine occurs from the buccal mucosa (inner cheek wall) as well as from the gastrointestinal tract after saliva-containing coca juice is swallowed. In fact, plasma concentrations in coca chewers are about what would be predicted if a dose of cocaine equivalent to that in the leaves was administered in a capsule (Paly et al., 1980).

(SEE ALSO: *Bolivia, Drug Use in; Coca Paste; Colombia As Drug Source*)

BIBLIOGRAPHY

HOLMSTEDT, B., ET AL. (1979). Cocaine in the blood of coca chewers. *Journal of Ethnopharmacology, 1*, 69–78.
PALY, D., ET AL. (1980). Plasma levels of cocaine in native Peruvian coca chewers. In F. R. Jeri (Ed.), *Cocaine 1980*. Lima: Pacific Press.

MARIAN W. FISCHMAN

CODEINE Codeine is a natural product found in the opium poppy (*Papaver somniferum*). An alkaloid of OPIUM, codeine can be separated from the other opium ALKALOIDS, purified, and used alone as an ANALGESIC (painkiller). It is however most often used along with mild nonopioid analgesics, such as aspirin, acetaminophen, and ibuprofen. These combinations are particularly effective; the presence of the mild analgesics permits far lower codeine doses. Using lower doses of codeine has the advantage of

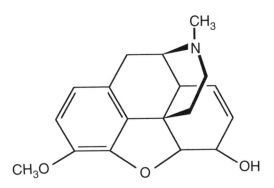

Figure 1
Codeine

reducing side effects, such as constipation. Codeine is one of the most widely used analgesics for mild to moderate pain.

Structurally, codeine is very similar to MORPHINE, differing only by the presence of a methoxy (-OCH$_3$) group at position 3, instead of morphine's hydroxy (-OH) group. The major advantage of codeine is its excellent activity when taken by mouth, unlike many opioid analgesics. Codeine itself has very low affinity for opioid receptors, yet it has significant analgesic potency. In the body, it is metabolized into morphine, and it is believed that the morphine generated from codeine is actually the active agent. Codeine has also been widely used as a cough suppressant. Codeine can be abused, and problems of abuse have often been linked to codeine-containing cough medicines, since they were once easily obtained over the counter. Chronic dosing with high codeine doses will produce TOLERANCE AND PHYSICAL DEPENDENCE, much like morphine.

(SEE ALSO: *Papaver somniferum*)

BIBLIOGRAPHY

GILMAN, ALFRED G., ET AL. (EDS.). (1990). *Goodman and Gilman's the pharmacological basis of therapeutics*, 8th ed. New York: Pergamon.

GAVRIL W. PASTERNAK

CODEPENDENCE The term *codependence* replaced an earlier term, *coalcoholism*, in the early 1970s and achieved widespread acceptance among the general public during the 1980s. Both terms point to problematic beliefs and behaviors that family members of chemically dependent people tend to have in common, although the term codependence broadens the concept to cover a wider range of family dysfunctions than chemical dependence alone.

A rather large nonscientific literature has developed on the topic of codependence. Much of it is couched in terms of the need to deal with injuries to emotions sustained during childhood—that is, to heal the wounds of the "*inner child*," a term popularized by *John Bradshaw*.

Despite the current popularity of codependence, awareness that one person's alcoholism affects everyone in the family is not new. *The Big Book of Alcoholics Anonymous* (1939; 1976) described the experience of family members of alcoholics in the following manner:

> We have had a long rendezvous with hurt pride, frustration, misunderstanding and fear. These are not pleasant companions. We have been driven to maudlin sympathy, to bitter resentment. Some of us veered from extreme to extreme, ever hoping that our loved one would be themselves once more.
>
> We have been unselfish and sacrificing. We have told innumerable lies to protect our pride and our husband's reputations. We have prayed, we have begged, we have been patient. We have struck out viciously. We have run away. We have been hysterical. We have been terror stricken. We have sought sympathy. We have had retaliatory love affairs with other men.
>
> Usually we did not leave. We stayed on and on [pp. 104–106].

In his book *I'll Quit Tomorrow* (1973), *Vernon Johnson* described the same experiences when he wrote that the *ism* of alcoholism is shared by other family members. In his words,

> While there may be only one alcoholic in a family, the whole family suffers from the alcoholism. For every harmfully dependent person, most often there are two, three, or even more people immediately around him who are just as surely victims of the disease. They too need real help and should be included in any thoroughgoing model of therapy.... With every drunk there is a sick dry who is almost a *mirror image.* [italics added]
>
> The people around the alcoholic person have predictable experiences that are psychologically damaging. As they meet failure after failure, their feelings of fear, frustration, shame, inadequacy, guilt, resentment, self-

pity, and anger mount, and so do their defenses. They too use rationalization as a defense against these feelings because they are threatened with a growing sense of self-worthlessness. They too begin to project these masses of free-floating negative feelings about themselves upon the children, back on the spouse, on other family members, on employees, and everybody else at hand. Their defenses have begun to operate in the same way as the alcoholic's, although they are unconscious of this, and they are victimized by their own defenses rather than helped. Out of touch with reality, just like the alcoholic, they say, "*I* don't need help. It's his problem, not mine!" The chemically dependent and those around him all have impaired judgment; they differ only in the degree of impairment [p. 30].

DEFINITION

Although considerable debate still remains among professionals regarding the definition and meaning of codependence, most addiction specialists agree that the concept has successfully ushered huge numbers of people into recovery. Perhaps the best general definition of codependence is called the Scottsdale definition, after the conference location where several lecturers met to achieve consensus:

Co-dependence is a pattern of painful dependence on compulsive behaviors and on approval from others in an attempt to find safety, self-worth and identity.

CHARACTERISTICS

The following five characteristics form the common thread weaving through the lives of many, if not most, family members of alcoholics and other drug addicts:

1. *Codependents change who they are, and what they are feeling, to please others.* Codependents are chameleons who sacrifice their own identity in an effort to get others to love them. They do this for two reasons. First, they fear being abandoned if people know how they really feel or who they really are. Second, they have so little sense of who they are that they need to be in relationships in order to organize their lives and feel complete. Unless they are in a relationship, and can take their cues from another person, they feel desperately lonely and worthless. As a result, codependents are split between two worlds. One world is the facade they show other people—the

false version of themselves. The other world is how chaotic, fearful, and empty their life feels underneath.

2. *Codependents feel responsible for meeting other peoples' needs, even at the expense of their own needs.* Codependents are so afraid of rejection that they will do anything to keep other people happy, including sacrificing their own needs to keep people from leaving them. They actually get more upset if others are disappointed or hurt than if their own problems go unsolved. This habit of focusing more on others often leads to codependents' enabling a family member's drinking. *Enabling* means that the codependent protects the chemical dependent from the negative consequences of their drinking and other drug usage to keep the other person from having to feel any pain or embarrassment.

3. *Codependents have low self-esteem.* Most people who are chemically dependent feel ashamed of themselves and are inwardly very self-critical. So perhaps it is not strange that other family members also begin to feel bad about themselves. For codependents, low self-esteem comes from two main places:
   (a) It comes from having very little sense of self to esteem. By always pleasing others and taking their whole identity from others, codependents end up not knowing who they are apart from the relationships they are in. It's hard to respect people who are afraid to be themselves, even when it's you!
   (b) Low self-esteem also comes from believing that they truly are responsible for someone's alcohol/drug use. Once they believe this, they will always feel inadequate when they fail to control the chemical dependent's behavior. This mistaken sense of what should be under their control is at the very core of *both* codependence and chemical dependence.

4. *Codependents are driven by compulsions.* Codependents feel they do not have any real choices about what is happening to them. They typically feel *compelled* to keep the family together, to stop the drinking or other drug use, to save the family from shame, to work, to eat or diet, to take physical risks, to spend or gamble, to have affairs, to be religious, to keep the house clean, and on and on. The driven quality of compulsions accomplishes two things:
   (a) Compulsions create excitement and drama.

As people battle their compulsions, the adrenaline begins to flow, and simple decisions, such as what to eat or how much to work, are turned into life and death struggles. This drama temporarily gives a feeling of purpose and vitality.

(b) Compulsions also occupy a lot of time and block people from their deeper feelings. Codependents often get locked into compulsive behaviors to avoid more painful feelings of fear, sadness, anger, and abandonment caused by a family member's chemical dependence.

5. *Codependents have the same use of denial and distorted relationship to willpower that is typical of active alcoholics and other drug addicts.* Denial and an unwillingness to accept human limitations are the two most destructive parts of the *ism* of alcoholism described by Vernon Johnson. In their own way, codependent family members fall into the same distorted relationship to reality and willpower as the chemical dependent. Both deny reality and think they can control alcoholism (their own or another's) if they just use enough willpower. For example, if chemical dependents deny that they are abusing alcohol or other drugs and remain unaware of its impact on their lives and their relationships with family members, friends, and coworkers, then codependents show exactly the same denial. They often refuse to see that a family member is chemically dependent, or they refuse to acknowledge that their children are being hurt. Shame and the compulsion to keep things under control cause codependents to deny the problem. Denial is a universal human trait, but it is overused by every member of a chemically dependent family.

Codependents are driven by the firm belief that their coping strategies fail because of personal inadequacy. When they cannot control the drinking or other drug use of someone they love, they blame themselves for not trying hard enough—or for not trying the right way. When codependents take too much responsibility for another person's recovery, it keeps the chemical dependent from seeing that only they can be responsible for their own recovery.

PSYCHIATRIC PERSPECTIVE

In many ways, codependence is the mirror image of a chemical dependent's self-centeredness and grandiosity. Another term for such self-centeredness is *narcissism*. Codependence is the complement of narcissism, just as a glove complements the hand it is shaped to fit.

In the Greek's myth that gives us the prototype for self-centeredness, Narcissus had relationships only with people who shared his values and interests. He was unable to feel a sense of human connection with people who were separate from him, just as chemical dependents may break off relationships with people who do not support their denial. The myth of Narcissus also gives us the prototype for *other-centeredness* in Echo—who is the perfect reflection of Narcissus. The two fit together and seemed to complete each other. Their relationship had intense chemistry.

In the eternal struggle within each individual between the need to be nurtured and the need to nurture others, Narcissus and Echo (and chemical dependents and codependents) strike a balance between two extreme positions. Rather than balancing the two needs within each of themselves, they allot the need to be validated and appreciated to Narcissus and the need to nurture and be in a relationship to Echo. Neither is capable of a truly mutual relationship—but, together, they create an intense experience of connectedness.

In healthy families, children remain comfortable with the competing, normal childhood needs to be unconditionally loved and validated as worthwhile (i.e., to be the center) and the opposite need to be completely dependent upon all powerful and good parents (i.e., to have others be the center). When parents are unable to tolerate not being the center of relationships, even with their children (which often happens with a chemically dependent parent), children often renounce their own need to be focused on. They become the opposite of narcissistic; they become codependent.

(SEE ALSO: *Adult Children of Alcoholics; Al-Anon; Alateen; Families and Drug Use*)

BIBLIOGRAPHY

ALCOHOLICS ANONYMOUS WORLD SERVICES. (1939; 1976). *The Big Book of Alcoholics Anonymous.* New York: Author.

BRADSHAW, J. J. (1988). *Healing the shame that binds you.* Deerfield Beach, Fl: Health Communications.

CERMAK, T. L. (1986). *Diagnosing and treating co-dependence.* Minnesota: Johnson Institute.

CERMAK, T. L.. (1990–1991). *Evaluating and treating adult children of alcoholics.* Minnesota: Johnson Institute.

JOHNSON, V. (1973). *I'll quit tomorrow.* New York: Harper & Row.

TIMMEN L. CERMAK

COERCED TREATMENT FOR SUBSTANCE OFFENDERS

Efforts in the nineteenth century to reform alcoholics ("drunkards") in the United States were noted sometime before the Lincoln presidency. Perhaps the earliest record of the concept of compulsory treatment for addiction was a proposal made in 1835 by Samuel Woodward, the superintendent of a Massachusetts asylum. Woodward suggested that alcoholics were not amenable to voluntary treatment but that mandated treatment might achieve the desired ends. His ideas were well received by some and ushered in the "inebriate asylum movement." In 1864, New York became the first state to establish such a hospital, and within the next several decades, other states—Massachusetts, Iowa, and Minnesota—followed New York's lead. The movement to establish these asylums did not take hold nationally, and it ended before the days of Prohibition (1919–1933).

Toward the end of the 1800s, officials and the general public alike were concerned with an even more threatening spectacle—the opiate addict. By the early 1900s, public indignation and concern about the opiate addict became entwined in ever-increasing debates on how to curb the use of alcohol and *Cannabis* (marijuana).

COERCED TREATMENT BEGINNING IN THE 1930s

With the repeal of Prohibition, as alcohol consumption once again spread across the United States, renewed attention was given to the problem of opiate addiction. Addicts were seen as contributing to the criminal element, and the federal government sought a solution. In 1929, in response to the growing problem of narcotics addiction and violations of federal law Congress passed an act establishing "farms" for "compulsory treatment" of addicts. The first, in Lexington, Kentucky, became operational in 1935. Soon after, in 1938, a second farm was established in Fort Worth, Texas, also for the purpose of treating opiate addicts. Voluntary clients were admitted along with mandated clients, however, which made it difficult to evaluate the success of the compulsory treatment. The suggested term of treatment was five months, but most voluntary patients terminated treatment well before then. It was reported that a majority (70%) of patients failed to complete treatment, and, as a result, both farms were widely critcized.

Proponents of compulsory treatment, however, were encouraged by subsequent research in the form of a twelve-year follow-up study in New York. According to the findings of George Vaillant, a physician, researcher, and author, although 90 percent of narcotic addicts surveyed had relapsed at one time or another following treatment, 46 percent were drug free at the time of follow-up.

Disagreement and confusion over the rate of relapse have often centered on the question of whether return to drug use following treatment constitutes failure. Returning to drug use once or twice is measured by some as failure, but in long-term studies prolonged abstinence is viewed as success, even though the addict may have relapsed to use at some time during the follow-up period. The question of whether a brief relapse (or even a couple of relapses) constitutes treatment failure is crucial to long-term follow-up studies, especially since few addicts overcome addiction permanently after one episode of treatment. In the view of many, it is the long-term follow-up that provides the most realistic evaluation of both treatment and subsequent prolonged abstinence. (Of course, long-term follow-up is both difficult and expensive, and attrition rates may be high.)

In Vaillant's 12-year follow-up, the 30 percent who were rated as the best adjusted were found to have an average length of abstinence of 7 years. According to Vaillant, those who underwent compulsory aftercare were judged to have the best outcomes, a finding of critical importance to advocates of compulsory treatment.

COERCED TREATMENT THROUGH CIVIL COMMITMENT IN THE 1960s

In 1961, the California Rehabilitation Center was established by the California Youth and Adults Corrections Agency. The Center's clients were offender-addicts ordered into treatment by the court. After

completion of inpatient treatment, clients continued treatment on an outpatient basis and were eligible for discharge only after maintaining three years of abstinence from drugs. Drug testing was performed both regularly and on a random basis, and a failed drug test resulted in a return to inpatient status.

Also in the early 1960s, there were isolated reports of successful outcomes of court-ordered referrals into treatment for alcoholism in Minnesota. Nevertheless, the concept of coercion into treatment was resisted by most treatment professionals, who believed that addicts could not truly be helped unless they *wanted* help.

Those in the legal community, however, were becoming increasingly receptive to the concept of compulsory treatment. A few federal and state programs initiated CIVIL COMMITMENT policies for addicts in the 1960s, most notably in California and New York. The program in California was known as the Civil Addict Program (CAP) and was the result of the Civil Addicts Commitment Law, passed in 1961. The New York program resulted from the passage of the Metcalf-Volker Act of 1962. Also, in 1966, a federal commitment law, the NARCOTIC ADDICT REHABILITATION ACT, was passed. The programs resulting from the enactment of these laws focused on opiate addiction; they produced mixed results. The New York program was the largest project of its kind ever undertaken, and also the most expensive.

A distinction must be made between court-ordered treatment for those charged with criminal offenses and court-ordered treatment for those not charged with violating the law. In fact, the civil commitment process is often part of a criminal proceeding that results in the addict being ordered into treatment by the court in lieu of incarceration. But noncriminal addicts may be ordered into treatment for addiction if there is a finding by the court that they are a threat to themselves or to others.

COMPULSORY TREATMENT FOR CRIMINAL ADDICTS SINCE THE 1970s

Since addition is frequently associated with a criminal lifestyle, it was perhaps inevitable that the criminal-justice system would become increasingly attracted to the concept of compulsory treatment. Vaillant was an early advocate of compulsory parole supervision for drug and alcohol abusers. In 1966, he reported that parolees who were required to ab-

stain from using were more likely to succeed than those who were not required to abstain.

In the early 1970s, the federal government, in conjunction with local jurisdictions around the United States, began an initiative to provide treatment for drug abuse to the growing number of offenders with a history of drug abuse. The program, designated TREATMENT ALTERNATIVES TO STREET CRIME (TASC), was an effort to channel more offenders into treatment and fewer into jails and prisons. The TASC program was made possible by a confluence of events, including the drug epidemic of the 1960s and 1970s, which increased public awareness of the amount of drug-related crime, and the appointment by President Richard M. Nixon of Dr. Jerome H. Jaffe as the first director of the Special Action Office for Drug Abuse Prevention (SAODAP). Jaffe saw the need for more drug treatment and also realized that large numbers of criminal-justice clients were actually addicts in need of treatment.

Ironically, at about this time, criminal-justice officials had become overwhelmed by mounting crime, burgeoning prisons, and recidivism. The prevailing view in most corrections circles was that "nothing worked" in terms of treatment and rehabilitation of offenders. In short, within the criminal-justice community there was growing resistance to the concept of treatment at the very moment it was most needed. Politically, too, there was a call for longer sentences and the abolishment of parole (also pronounced a failure).

The necessary collaboration between the criminal justice and treatment communities was not an intuitive one, and long-standing hostilities existed between the two fields. Prison and jail officials viewed treatment professionals as outsiders and do-gooders who could easily be fooled by offenders. In addition, prosecutors, judges, and the police were largely uninformed about treatment, but their cooperation was essential to the success of TASC.

The SAODAP officials took the lead in formulating policies and procedures for the identification and assessment of drug-abusing offenders and for their referral into community-based treatment programs—rather than sending them to jail or prison. If treatment was completed successfully, defendants might avoid prosecution.

A key to the eventual success of TASC was the cooperation of Jeris Leonard, director of the Law Enforcement Assistance Administration (LEAA). Leon-

ard's support was vital to the program and contributed to its fruiton. Although LEAA internal staff were never completely committed to the concept of treatment, they gave the TASC program begrudging cooperation because they saw that it was well funded. Over time, LEAA officials became increasingly supportive as they saw the opportunity to obtain additional funding for TASC projects and thus enhance their own budgets.

Although the original focus of TASC was opiate addiction, the program expanded to include non-opiate abusers, including alcoholics. In 1974, SAO-DAP was replaced by the National Institute on Drug Abuse (NIDA), which continued the mission of attempting to facilitate the cooperation of the criminal-justice and treatment communities.

The "drug courts" of the late 1980s and early 1990s owe their surge of growth to the successes of TASC. One of the best known of the drug-court initiatives is the Diversion and Treatment Program of Dade County, Florida, known locally as the Miami drug court. The Miami drug court uses the power of the judiciary to foster a partnership between the Dade County criminal-justice and treatment systems to provide drug services to court-ordered clients. The court is regularly informed of the progress of the arrestee in treatment, and it uses its judicial powers to authorize treatment and reincarceration for erring offenders (those who continue to abuse drugs or otherwise violate the conditions of the court).

Preliminary results, published in 1993, indicate that the drug-court participants had lower incarceration rates, less frequent rearrests, and longer times before rearrest. Both criminal justice professionals and treatment professionals acknowledge, however, that there have not yet been rigorous studies of the mushrooming drug courts. Furthermore, it is acknowledged that positive findings may be partly the result of the selection process, since violent and serious offenders are not eligible. Other jurisdictions have attempted to implement similar programs by building on the Miami model and adding features designed to respond to known problems encountered in the early stages of the Dade County project.

By 1994, an emphasis on drug treatment for offenders at all levels—pretrial, in prison, on probation or parole—was clearly established. No longer does there appear to be much disagreement among criminal-justice officials about whether a large percentage of offenders may be rehabilitated through drug treatment. Instead, the debate has moved to questions of how to provide treatment most effectively and to identify evaluation strategies for successful outcomes and aftercare. The proponents of coerced drug treatment have become legion and include Douglas Lipton, a New York drug researcher who questioned the efficacy of correctional treatment during the 1950s and 1960s. Other advocates of coercion into treatment include Douglas Anglin, whose research includes the finding that clients who have been legally coerced into treatment do as well as clients who have been voluntarily admitted. Anglin favors expansion of treatment for offenders, because he maintains that legally mandated clients may stay in treatment longer. Since length of time in treatment has been associated with reduced drug use, the advantages of legal coercion into treatment for drug-abusing offenders are obvious. The client who is ordered into treatment by the court is more likely to stay the required length of time.

Like other profesionals in the field, George DeLeon, an expert and proponent of the effectiveness of THERAPEUTIC COMMUNITIES, has written that "there is little evidence for differential outcomes between legally referred and nonlegally referred clients" (DeLeon, 1990, p. 133). DeLeon also reported on the relationship between length of time in treatment and favorable treatment outcomes.

Ironically, although many treatment professionals support the concept of coercion into treatment, there continues to be a vocal minority who oppose the involvement of the criminal justice community. Antagonism exists among treatment staff who staunchly defend the concept of confidentiality of records and who resist reporting to criminal-justice staff their clients' "dirty urines" or relapse to drug use. Such tensions are not surprising, but they need to be resolved. The sheer magnitudfe of the problem of addiction ensures that there will be a continued, energetic search for solutions.

(SEE ALSO: *Shock Incarceration and Boot Camp Prisons; Treatment, History of; U.S. Government Agencies: U.S. Public Health Service Hospitals*)

BIBLIOGRAPHY

ANGLIN, M. D., & HSER, Y. (1990). Treatment of drug abuse. In M. Tonry & S. Q. Wilson (Eds.), *Drugs and crime*. Chicago: University of Chicago Press.

DeLeon, G. (1990). Treatment strategies. In J. A. Inciardi (Ed.), *Handbook of drug control in the United States.* Westport, Conn.: Greenwood Press.

National Institute on Drug Abuse. (1988). In C. Leukefeld & F. Tims, *Compulsory treatment of drug abuse: Research and clinical practice.* (Research Monograph Series No. 86). Rockville, MD: U.S. Government Printing Office.

President's Commission on Law Enforcement and Administration of Justice. (1967). *Task force report: Narcotics and drug abuse.* Washington, DC: U.S. Government Printing Office.

Vaillant, G. (1984). *The natural history of alcoholism: Causes, patterns, and paths to recovery.* Cambridge, MA: Harvard University Press.

Wellisch, J., Prendergast, M., & Anglin, M. D. (1993). Criminal justice and drug treatment systems linkage: Federal promotion of interagency collaboration in the 1970s. *Contemporary Drug Problems* (Winter), 614–650.

MARIE RAGGHIANTI

COFFEE   Coffee is the world's most common source of CAFFEINE, providing a little more than half of all caffeine consumed daily. In the United States, coffee is usually a beverage made by percolation or infusion from the roasted and ground or pounded seeds of the coffee tree (genus *Coffea*), a large evergreen shrub or small tree, which was native to Africa but now is grown widely in warm regions for commercial crops. Caffeine from coffee accounts for an estimated 125 milligrams of the 211 milligrams of U.S. caffeine consumed per capita per day. Recent estimates suggest that more than 50 percent of the adolescents and adults in the United States consume some type of coffee beverage. Coffee is one of the main natural commodities in international trade, ranking second only to petroleum in dollar value. Approximately fifty countries export coffee and virtually all of those countries rely on it as a major source of foreign exchange. An estimated 25 million people make their living in the production and distribution of coffee products.

In addition to caffeine, roasted coffee contains at least 610 other chemical substances, which may contribute to its smell, taste, and physiological effects. Nevertheless, coffee's primary psychoactive ingredient is caffeine. The amount of caffeine in an individual cup of coffee varies considerably, depending on the type and amount of coffee used, the form of the final coffee product (e.g., ground roasted or instant), and the method and length of brewing. On average, a 6-ounce (177 milliliters) cup of ground roasted coffee contains about 100 milligrams caffeine; the same amount of instant coffee typically contains about 70 milligrams caffeine. However, the caffeine content of any given 6-ounce cup of coffee can vary considerably and can reach as much as 210 milligrams. Drip coffee typically contains more caffeine than percolated; decaffeinated coffee contains a small amount of caffeine, approximately 4 milligrams in a 6-ounce cup. Individual servings of caffeinated coffee contain amounts of caffeine that have been shown experimentally to produce a range of effects in humans including the alteration of mood and performance and the development of physical dependence with chronic daily use.

Coffee cultivation probably began around 600 A.D. in Ethiopia, but the drink was spread into the Middle East and Europe. Today, much of the world's coffee is grown in South and Central America, particularly Brazil and Colombia, and in several African countries. Coffee beverages derive primarily from the seeds of two species of *Coffea* plants, *Coffea arabica* and *Coffea canephora* var. *robusta*. Robusta coffees contain approximately twice as much caffeine as Arabicas. Arabica beans are used in the majority of the coffee consumed today, particularly in the higher quality coffees. Since processing for instant and decaffeinated coffee extracts flavor components from the bean, the stronger flavored beans, typically Robusta beans, are used for these coffee products. Caffeine extracted in the decaffeination process is sold for use in soft drinks and medications.

The coffee bean, covered with several layers of skin and pulp, occupies the center of the coffee berry. During the first part of coffee production, the outer layers of the coffee berry are removed, leaving a green coffee bean. The beans are then roasted, removing between 14 and 20 percent of their water and changing their color from green to various shades of brown; generally, the beans get darker as more water is extracted. The beans are then ground and ready for use. To produce instant coffee, roasted and ground coffee is percolated to produce an aqueous coffee extract. That extract is dehydrated by spray or freeze-drying to produce water-soluble coffee extract solids. Since this process removes flavor

Figure 1
Coffee

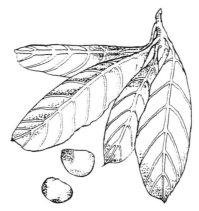

Figure 1
Kola

and aroma from the coffee, compounds are added to the extracts at the completion of the process to restore the lost characteristics.

BIBLIOGRAPHY

BARONE, J. J., & ROBERTS, H. (1984). Human consumption of caffeine. In P. B. Dews (Ed.), *Caffeine.* New York: Springer-Verlag.

SPILLER, G. A. (ED.). (1984). *The methylxanthine beverages and foods: Chemistry, consumption, and health effects.* New York: Alan R. Liss.

KENNETH SILVERMAN
ROLAND R. GRIFFITHS

COGNITIVE THERAPY OF ADDICTIONS *See* Treatment Types.

COLA/COLA DRINKS Cola drinks are carbonated soft drinks, sodas, that contain some extract of the kola nut in their syrup. Kola nuts are the chestnut-sized and -colored seeds of the African kola tree (*Cola nitida* or *Cola acuminata*). For the soft-drink industry, the trees are now grown on plantations throughout the tropics. Historically, kola nuts were valued highly among African societies for their stimulating properties. Kola nuts were cracked into small pieces and chewed for the effect—which increased energy and elevated mood in extremes of heat, hunger, exhaustion, and the like. The European colonists in Africa learned of the effect and

some chewed it. In the 1800s, Europeans brought kola nuts to various strenuous endeavors in Africa and in other regions, and they began to increase the areas under cultivation. Kola nuts were soon finely powdered and made into syrups for ease of use—with no loss of effect, it was claimed.

The active ingredient responsible for these stimulatory properties is CAFFEINE, a powerful brain stimulant, which is also present in other plants such as COFFEE, cocoa, TEA, maté, and others. Besides reversing drowsiness and fatigue, a heightened awareness of stimuli and surroundings may occur. Studies have shown that less energy may be expended by the musculature with equal or greater results—in animals as well as humans—but excess use causes TOLERANCE and dependence, often unrealized until deprivation results in severe headaches. Large doses can cause nervous irritation, shaking, sleep disturbances, insomnia, and aggravation of stomach ulcers or high blood pressure.

In the late 1800s, in the United States, cola drinks came onto the market with other carbonated or phosphated (fizzy) drinks. Coca-cola™, one of the first and most popular, contained extracts of both the COCA PLANT (cocaine) and the kola nut (caffeine)—but by the early 1900s, with the realization of COCAINE's dangers, this was removed and replaced by additional caffeine. Drinking cola is part of American culture, emulated and enjoyed worldwide, with many brands competing for a huge and growing consumer market. Colas are now available with sugar or artificial sweeteners, with or without caffeine, with or without caramel coloring (clear)—thus indicating

that people seem to like the flavor regardless of the specific ingredients or the "effect."

BIBLIOGRAPHY

GRAHAM, D. M. (1978). Caffeine—its identity, dietary sources, intake and biological effects. *Nutritional Review, 36,* 97–102.

LEWIN, L. (1964). *Phantastica: Narcotic and stimulating drugs, their use and abuse.* New York: Dutton.

MICHAEL J. KUHAR

COLLEGE ON PROBLEMS OF DRUG DEPENDENCE (CPDD), INC.

The College on Problems of Drug Dependence is an incorporated, non-profit, scientific organization that acts independently of both the U.S. government and the pharmaceutical industry. It is now a World Health Organization Collaborating Center for research and training in the field of drug dependence. Among the goals of the CPDD are the following: (1) to support, promote and carry out ABUSE-LIABILITY research and testing, both at the preclinical and clinical levels; (2) to serve as advisor to both the public and private sectors, nationally and internationally; (3) to sponsor an annual scientific meeting in fields related to drug abuse and chemical dependence. The annual scientific meeting of the CPDD has become one of the few forums where scientists from diverse disciplines can discuss problems of drug abuse and drug dependence at a rigorous academic and scientific level.

A primary goal of the CPDD is the publication of data on the physical-dependence potential and abuse liability of OPIOIDS, stimulants, and depressants, as well as the development of new methodology for drug evaluation. These data provide an independent scientific evaluation of drugs that might have abuse liability. A number of scientists from various medical schools work collaboratively to assess these drugs. The data are collated in the Laboratory of Medicinal Chemistry, National Institute of Diabetes, Digestive and Kidney Diseases (NIDDK), National Institutes of Health (NIH), Bethesda, Maryland. They are discussed by the Drug Evaluation Committee of the CPDD before publication. Government agencies can use the data to help determine whether a medically useful drug should be scheduled under the CONTROLLED SUBSTANCES ACT, to restrict access and thus reduce possible abuse.

The contemporary CPDD originated in 1913, as the Committee on Drug Addiction of the Bureau of Social Hygiene in New York City. In 1928, funds were provided by the Bureau of Social Hygiene to the Division of Medical Sciences, National Research Council (NRC), of the National Academy of Sciences (NAS), for the support of a chemical, pharmacological, and clinical investigation of narcotic drugs by the Committee on Drug Addiction, NRC, NAS. This research continued until World War II. From 1939 to 1947 the Committee on Drug Addiction served as an advisory group to the U.S. Public Health Service (Eddy, 1973).

The Committee on Drug Addiction was reestablished in 1947 as the Committee on Drug Addiction and Narcotics (CDAN), in the Division of Medical Sciences of the NRC, NAS. In 1965, CDAN's name was changed to the Committee on Problems of Drug Dependence (CPDD). The CPDD remained as an NRC, NAS committee until 1976, when it became an independent scientific organization, the Committee on Problems of Drug Dependence (CPDD), Inc. It was guided by a Board of Directors with the sponsorship of nine major scientific organizations, including such diverse groups as the American Chemical Society and the American Medical Association (May & Jacobson, 1989).

In 1991, the CPDD underwent its most recent reorganization and its name was modified to reflect its contemporary role. Now known as the College on Problems of Drug Dependence, Inc., the CPDD has become a scientific membership organization, which will enable its members to have a voice on issues relating to drug abuse. The members of the CPDD are involved in all the aspects of the effects of drugs subject to abuse—encompassing the enormous range from social issues through basic research in molecular biology and the study of the interaction of these drugs with specific RECEPTORS in the central nervous system.

(SEE ALSO: *Drug Types; World Health Organization Expert Committee on Drug Dependence*)

BIBLIOGRAPHY

EDDY, N. B. (1973). *The National Research Council involvement in the opiate problem, 1928–1971.* Washington, DC: National Academy of Sciences.

MAY, E. L., & JACOBSON, A. E. (1989). The Committee on Problems of Drug Dependence: A legacy of the Na-

tional Academy of Sciences. A historical account. *Drug and Alcohol Dependence, 23,* 183–218.

ARTHUR E. JACOBSON

COLOMBIA AS DRUG SOURCE

Smuggling and the commerce in contraband have been a way of life in Colombia for nearly 500 years. Approximately 1,000 miles (1609 km) of largely unpatrolled Pacific and Caribbean coastline and vast tracts of mostly uninhabitable territory—ranging from tropical jungles in the south, to rugged Andean mountain slopes in the east, to sparsely populated deserts in the north—have made Colombia a haven for smugglers of illegal COCAINE, MARIJUANA, and, most recently, HEROIN. Violence, corruption, inadequate control by the central government over much of its territory, and an ineffective judicial system have hampered Colombia's drug-control efforts. Consequently, Colombian cocaine is the single largest supply of that illicit drug to be smuggled into the United States.

During the 1980s, despite positive law enforcement and crop-control programs, Colombian laboratories processed large volumes of COCA PLANT (*Erythroxylon coca*) into cocaine; by the 1990s, their sophisticated trafficking infrastructure had diversified into heroin production and distribution, adding to the already large Asian and Mexican supply in the United States. To reduce the level of violence and achieve peaceful coexistence throughout the country, the Colombian government offered a type of amnesty, or plea bargain, to major drug traffickers willing to surrender and cease their trafficking operations.

Colombia has powerful, often violent, trafficking organizations based in the major urban areas of Medellin and Cali, as well as the oldest continuous political insurgency in the Western Hemisphere, the Revolutionary Armed Forces of Colombia (FARC). Although the extradition treaty with the United States is no longer constitutional, the Colombian government has implemented the strongest drug law-enforcement efforts of any Andean country. In 1991, Colombia began eradicating its burgeoning crop of opium poppy (*Papaver somniferum*), increased its cocaine seizures by almost 70 percent, and improved and expanded its ability to investigate drug crimes through the development of a new code of criminal procedure; tougher chemical control; an-

timoney-laundering, asset-seizure, and evidence-sharing procedures.

Colombia was a signatory to the 1961 SINGLE CONVENTION ON NARCOTIC DRUGS and the 1971 Convention on Psychotropic Substances. It has signed but not yet ratified the 1988 UN Convention Against Illicit Traffic in Narcotic Drugs and Psychotropic Substances. Although money laundering is not illegal in Colombia, the government is actively tracking narcotics proceeds (e.g., exchanging tax information and writing tougher financial disclosure laws) to improve its drug-investigative capacity.

ROLE AS COCAINE SUPPLIER

Proximity to the U.S. marketplace, remote and vast tracts of unpatrollable land, powerful criminal organizations, indigenous entrepreneurial spirit, and a long tradition of violence and smuggling make Colombia an ideal source for illegal drugs. The U.S. government estimated that in 1991 Colombia cultivated approximately 92,000 acres (37,500 ha) of the world's 533,000 acres (213,000 ha) of coca, mostly in the Llanos (plains) region, which encompasses almost half of eastern Colombia. Heavy growth also occurs in the Caqueta, Guaviare, Putumayo, and Vaupes departments (counties or provinces), with evidence of crop expansion in the Bolivar department and in south and southwest Colombia. Despite some new cultivation patterns, aerial surveys showed a 7 percent decline in coca plant cultivation from the 1990 estimate of 96,000 acres (40,000 ha).

Colombia's importance to the U.S. government's international narcotics control strategy lies in its role as the world's leading processor and distributor of cocaine hydrochloride (HCl)—the cocaine salt (powder) that is sniffed or snorted—and cocaine base, or CRACK cocaine—the rock or crystal that has been converted from cocaine HCl. Colombian cocaine-trafficking organizations are sophisticated and well-organized industries; they derive their strength from longtime control of cocaine laboratories and the smuggling routes to North America. Colombian traffickers, after financing the cultivation of coca plants in neighboring BOLIVIA and Peru, often oversee the processing of the leaves into COCA PASTE (and, sometimes, base), which may then be shipped to laboratories in Colombia to be refined into cocaine HCl by the ton. Recently, growers in Peru and Bolivia have stopped shipping some of their coca products to Colombia and have begun to refine them into cocaine

HCl in laboratories near their coca fields. Colombian traffickers still operate the greatest number of base and HCl labs—ranging from small simple operations to large sophisticated complexes.

HISTORICAL AND INSTITUTIONAL FACTORS

In the mid-to-late 1970s, the United States directed its international drug-control attention to eliminating the heroin and marijuana crossing our border at Mexico. As the U.S.-Mexican crackdown began to achieve positive results and the number of U.S. smokers of Mexican marijuana diminished, Colombian traffickers seized the opportunity to break into the lucrative U.S. drug market by smuggling large amounts of marijuana and small packages of cocaine. In the early 1980s, Florida became the destination of choice for smugglers because of its long coastlines, access to boats and planes, location in the Caribbean, and large Hispanic population; by 1986, Colombia supplied an estimated 80 percent of the cocaine HCl.

More than 150 Colombian drug groups organized loosely into autonomous business cartels in the two principal urban areas of Colombia—Medellin and Cali—represent one of the most important power centers. The government of Colombia and the roughly 10,000 members of the two antigovernment guerilla groups called the Revolutionary Armed Forces of Colombia (FARC) and the National Liberation Army (ELN) are also political forces there (Lee, 1989).

In controlling all stages of cocaine production, from cultivation to sale, Colombian traffickers have developed sometimes competitive, sometimes symbiotic, relationships with each other and with insurgent groups. The guerillas benefit from the illicit drug trade by providing protection for the coca fields, laboratories, and storage facilities; they carry out kidnapping and terrorism to support the traffickers' aims. In return, the political insurgents "tax" the profits of the drug trade, thereby earning hard currency and occasionally extracting payment in weapons.

Ironically, the Medellin cocaine-trafficking cartel grew out of a guerilla kidnapping of the sister of two traffickers. In 1982, afraid that their enormous wealth made them likely kidnapping targets, over 200 traffickers banded together in a group called MAS, the Spanish acronym for *death to kidnappers*.

Through bribes and threats (the Spanish equivalent is *plata o plomo*, silver or lead), this group became immune from both the authorities and the insurgents.

Elected in 1982, Colombia's President Belisario Betancur appointed a strong counternarcotics minister of justice, Rodrigo Lara Bonilla. In 1984, Lara Bonilla ordered a police raid on a Medellin cartel laboratory complex in a remote area called Tranquilandia, which produced more than 3 tons of cocaine per month. The police seized some 15 tons (13.8 metric tons) of cocaine, several airplanes, a variety of arms, and chemicals essential to the processing of cocaine. One month later, the minister of justice was assassinated in the capital, Bogota. In retaliation, President Betancur signed the first extradition order for Carlos Lehder, the Medellin cartel's transportation czar. Lehder was extradited in 1987, convicted in a Florida federal court, and is serving a sentence of life plus 135 years in the United States.

The assassination of Lara Bonilla marked a turning point in the history of drug control in Colombia, but all too sadly, it represented only one of many such violent acts. Another incident was the attack on Colombia's Palace of Justice, in 1985, in which 95 people died. It was engineered by the traffickers to prevent the justices from discussing the extradition treaty—the "weapon" feared most by the traffickers. Between 1984 and 1987, a total of 15 drug traffickers were extradited to the United States. In 1984, 1988, and 1990, major traffickers met privately with senior Colombian politicians to discuss the possibility of a general amnesty and an end to extradition—in return for ceasing the violence, abandoning the cocaine business, and relinquishing the assets used in the industry (e.g., planes, laboratories, airstrips). Although the first three offers were rejected, in September 1990, the Gaviria Administration gave the *Extraditables* the option of accepting reduced sentences and guarantees against extradition if they surrendered to the authorities, confessed their crimes, and offered up their assets.

Under President Vergilio Barco, 1986–1990, the violence throughout Colombia's countryside increased in response to the continuing extraditions. In the 1980s, the murders in Medellin, a city of 2 million inhabitants, increased tenfold—from 700 in 1980 to 7,000 by 1991—due in part to a bounty of $300 offered by Medellin cartel boss Pablo Escobar in 1990 for every policeman killed. In August 1990, the populist politician Luis Carlos Galan, a presiden-

tial contender, was assassinated at a political rally because of his strong support for the extradition treaty and antidrugs position. His murder mobilized the civilian population against the cartels and incensed Colombia's national police and military. Through a series of raids on cartel laboratories, the government began an unparalleled crackdown, culminating in the shooting of Medellin strongman Rodrigo Gacha and the destruction of one of their largest cocaine-processing centers.

The Cali cartel has always been less violent and less obtrusive than its Medellin counterpart. As a principal supplier of cocaine to New York, this cartel has all too quickly begun to move into the position vacated by the Medellin leaders, many of whom are either in prison or have been killed.

In 1991, the Cali cartel diversified into the more lucrative heroin trade by hiring Cauca-region Indians. Some 40 miles (64 km) southeast of Cali, they planted opium poppies on the slopes of the Andes. The U.S. government estimates that in 1991 about 6,200 acres (2,500 ha) of opium poppy was grown, with more terrain being prepared for future plantings. Cultivation appears to be concentrated in Huila and Cauca, but police have located sparse plantings in Antioquia, Boyaca, Tolima, and Cundinamarca. Only five small MORPHINE labs were found, indicating that most of the opium gum produced may be exported for processing into heroin in foreign labs.

DRUG-REDUCTION EFFORTS

Among the principal Andean SOURCE COUNTRIES for coca, Colombia has become most committed to defeating the cocaine cartels, since they threaten to undermine its society. Colombians recognize and fear that the violence and corruption endemic to drug trafficking are harming their economy, their political system, and their society.

Colombia's national police, the military, and the security forces have conducted major law-enforcement operations against the Medellin and Cali cartels. In 1989, with the assistance of U.S. technical and material support, they attacked and killed several mid-level Medellin traffickers, including Rodrigo Gacha. Colombia has also used the extradition to incarcerate or immobilize major traffickers. In late 1990, President Cesar Gaviria's offer of amnesty for major traffickers resulted in a decrease in violence and the surrender and imprisonment of five traffickers and one terrorist—including Pablo Escobar and

the three Ochoa brothers (Jorge Luis, Juan David, and Fabio). Colombia's government has however paid a heavy price for its action—suffering 429 deaths by assassination or during enforcement operations.

In 1991, President Gaviria's government security forces, spearheaded by the narcotics control units in the national police and the military, seized almost 100 tons of cocaine HCl and 10 tons of cocaine base (a 66% increase over 1990). They destroyed 220 cocaine-processing labs and numerous airstrips, seized almost 1 million gallons of essential chemicals, and made 1,400 arrests. Colombia's police have also eradicated virtually all *Cannabis sativa* (marijuana) cultivation in the traditional growing areas of the north coast (Guajira peninsula). Most importantly, the government of Colombia damaged the leadership structure of the Medellin cartel by incarcerating its leader, Pablo Escobar—who escaped and was killed in a shoot-out, December 1993.

As a signatory of the 1961, 1971, and 1988 United Nations International Narcotics Control Conventions, Colombia demonstrates its political commitment to attempt to immobilize drug traffickers and eradicate coca, cannabis, and opium. Significant among the new Colombian initiatives is the creation of the public-order courts, which have tried 700 persons and convicted 420 since the beginning of 1992. Colombia has also begun to share evidence, reform its judiciary, and to track substantial money flows—requiring banks to keep records on cash transactions of over 10,000 U.S. dollars. Despite widespread testing of various coca herbicides, Colombia has not yet begun a major coca-eradication effort—largely because coca fields are located in terrorist-controlled land and coca eradication is not a focus of the antidrug efforts.

In January 1992, fearing a new and burgeoning heroin production and distribution problem, the Colombian government agreed to spray the common garden herbicide glyphosate (Roundup) in an effort to eradicate the growing number of opium poppy fields. In 1991, almost 3,000 acres (1,100 ha) of poppy had been destroyed manually. Marijuana production continues to be minimal, since the herbicidal campaigns of the mid-1980s were relatively effective.

(SEE ALSO: *Crop-Control Policies; Drug Interdiction; Foreign Policy and Drugs; International Drug Supply Systems; Source Countries for Illicit Drugs*)

BIBLIOGRAPHY

BAGLEY, B. M. (1988). Colombia and the war on drugs. *Foreign Affairs* (Fall), 70–92.

BUREAU OF INTERNATIONAL NARCOTICS MATTERS, U.S. DEPARTMENT OF STATE. (1991). *International narcotics control strategy report (INCSR)*. Washington, DC: Author.

CRAIG, R. B. (1987). Illicit drug traffic: Implications for South American source countries. *Journal of Interamerican and World Affairs, 29* (2), 1–34.

GUGLIOTTA, G., & LEEN, J. (1989). *Kings of cocaine: Inside the Medellin cartel.* New York: Simon & Schuster.

LEE, R. W., III. (1991). Making the most of Colombia's drug negotiations. *Orbis 35*(2), 235–252.

LEE, R. W., III. (1989). *The white labyrinth.* New Brunswick, NJ: Transaction.

JAMES VAN WERT

COMMISSIONS ON DRUGS

For hundreds of years, governments have been using commissions of inquiry to help them investigate pressing social problems and to formulate plans for organized social and governmental responses to these problems. One of the first government commissions on problems associated with drug use, appointed in 1893, was the Indian Hemp Drugs Commission— whose purpose was to investigate the extent to which the hemp plant was cultivated, the preparation of drugs from it, the trade in those drugs, the extent of their use, and the effects of their consumption on the social, physical, and moral conditions of the people. In addition, they were asked to investigate certain economic aspects of the use of hemp and also the potential political, social or religious results of the prohibition of hemp. After one year of investigation, the commission published a report summarizing its conclusion that the occasional use of hemp in moderate doses was not harmful but that excessive use did cause injury.

Commissions often can create a firm basis for long-lasting social policy. One of the best examples relating to drugs is provided by the Shanghai Commission and the Smoking Opium Act of 1909. Several laws directed at the traffic in narcotics had been introduced into the U.S. Congress before 1908, but only after President Theodore Roosevelt convened the Shanghai Opium Commission in 1909 was federal enactment of any legislation accomplished. The Shanghai meeting had been called to aid China in its attempt to eliminate opium addiction. Then the first U.S. federal legislation against American narcotic abuse, the Smoking Opium Exclusion Act of 1909, outlawed the importation into the United States of opium prepared for smoking.

Within Great Britain, the Rolleston Commission served the important function of establishing basic principles for long-lasting social policy. For more than sixty years, the recommendations of the Rolleston Commission, called the ROLLESTON REPORT, have guided British social and governmental policy toward the prevention and control of nonmedical usage of heroin and other opioid drugs.

In North America, there have been a series of important commissions directed toward nonmedical drug use. In 1939, President Franklin D. Roosevelt asked Mayor of New York Fiorello LaGuardia to chair a scientific commission to investigate the effects of marijuana and other drugs in communities, especially within urban areas of the United States. After looking into these issues for five years, the LaGuardia commission members published their final report entitled *The Marijuana Problem in the City of New York: Sociological, Medical, Psychological and Pharmacological Studies.* Based on interviews with hundreds of users, and after sociological studies and laboratory investigations, the commission concluded that marijuana was not an addictive drug as compared to morphine and that tolerance to marijuana was developed to only a limited degree. Although the LaGuardia report is a significant contribution to the marijuana literature, its conclusions must be qualified because of various weaknesses in the experimental methodologies available at that time.

In 1962, President John F. Kennedy appointed the President's Advisory Commission on Narcotics and Drug Abuse. This commission considered how best to begin reexamination of the problem of drug abuse in the United States, as well as what specific recommendations to make regarding the control of the problem of addiction by law enforcement. The commission's final report in 1963 suggested that psychological treatment might be useful against addiction. The report also marked a shift away from the trend to consider all NARCOTICS the same under law.

During the late 1960s and early 1970s, both the United States and Canada launched commissions of inquiry into a growing problem of nonmedical drug use among the young people of their countries. The Canadian LeDain Commission was appointed in

1969 with the goals of referencing the existing literature and data regarding the nonmedical use of drugs; reporting on the current state of medical knowledge of these drugs; studying the motivation underlying such nonmedical use of drugs; investigating social, economic, educational and philosophical factors related to such use; and finally, recommending ways the Canadian government could act to reduce the problems associated with such use.

Shortly afterward, in the United States, the NATIONAL COMMISSION ON MARIHUANA AND DRUG ABUSE issued two important reports, the first entitled *Marihuana: A Signal of Misunderstanding* and the second called *Drug Use in America: Problem in Perspective*. For the most part, the commission's recommendations about MARIJUANA fell on deaf ears. A most important recommendation of the commission was realized in the form of a well-funded national program of periodic epidemiologic surveillance concerning nonmedical drug use and the consequences of such use, now apparent in the MONITORING THE FUTURE studies and in the NATIONAL HOUSEHOLD SURVEY ON DRUG ABUSE.

(SEE ALSO: *Marihuana Commission; Opioids and Opioid Control; Shanghai Opium Conference of 1909; U.S. Government: The Organization of U.S. Drug Policy*)

JENNIFER K. LIN

COMMITTEES OF CORRESPONDENCE

This organization works toward a drug-free America. It is a nonprofit organization that does not accept government funding and is headed by Otto and Connie Moulton, 24 Adams Street, Danvers, Massachusetts 01923, (508) 774-2641. Drug Watch International and its subsidiary, the International Drug Strategy Institute, were founded to expand the information gathering and dissemination efforts of the Committees of Correspondence.

In 1977, Otto had been coaching little-league baseball and youth hockey and was uninformed on the youth drug culture. After four of his players changed in attitude and ability, he discovered the cause—MARIJUANA. The Moultons began learning about the health effects of marijuana, which was not an easy task. PRIDE and The American Council on Marijuana provided research reports and, armed

with facts, the Moultons shared them in their local communities, alerting parents and students to marijuana's effects.

At the 1980 PRIDE conference, they joined with other groups to found a grass-roots PARENTS MOVEMENT called The National Federation of Parents for Drug-Free Youth. The Moultons also revived the Committees of Correspondence, an organization originally founded in 1772 by Samuel Adams—to exchange ideas on building colonial unity. The modern version was formed to build national unity by exchanging facts and ideas on drug prevention, with a newsletter and letter campaign to government favoring antidrug legislation.

The Moultons have served Massachusetts state government and the U.S. government as advisors in the 1980s and 1990s. The Committees of Correspondence maintains a large library on drug-culture history, with books, videotapes, and publications to provide information for global requests; it also provides the public, policymakers, and the media with current research data in an ongoing effort to counter drug advocacy.

OTTO MOULTON

COMMUNITY PARTNERSHIPS *See* Parents Movement; Prevention Movement.

COMORBIDITY AND VULNERABILITY

For more than 100 years, clinicians who treat people with drug problems have observed that these patients often have a history of psychiatric disturbances or personality traits that are likely to make them more vulnerable to drug problems. The addictive disorders of these patients follow after and so are *secondary* to their psychiatric disorders or personality problems.

Psychiatrists also have noted the emergence of mental disorders that seem to have been caused by drug use. In this case, the psychiatric disorders follow after and so are secondary to the drug use; without the drug use, perhaps no mental disorder would have occurred.

A third group of patients have both a psychiatric disorder and an addictive disorder, but it is not at all clear which came first and which came second. In addition, sometimes the patient's history is compli-

cated by some very harmful early life experiences, such as having been abused as a child, or some other underlying condition that might very well have caused or might account both for the psychiatric disorder and also for the addictive disorder.

No matter which came first—the psychiatric condition or the drug problem—and even when some underlying condition might lead a patient to have both disorders, the co-occurrence of a psychiatric disorder with alcohol and other drug (AOD) problems is known as *comorbidity*. Comorbidity has become an important common theme in contemporary biomedical research, with implications for neuroscientists and other laboratory scientists, but also for clinical researchers and for epidemiologists.

One reason for an increased interest in psychiatric comorbidity is a set of population studies within the United States that helped clarify how often alcohol and drug problems co-occur with psychiatric disorders like major depression. Known as the Epidemiologic Catchment Area surveys, these studies identified cases of mental disorders like major DEPRESSION and SCHIZOPHRENIA in the community and found that 22 percent of the people with these disorders had a history of currently active or former alcohol abuse or dependence, and 15 percent had a history of abuse or dependence involving other drugs, such as cocaine and marijuana. These figures were substantially higher than the proportions found when people without mental disorders were studied. At the population level, this is important evidence on the strength of the association between alcohol and drug disorders and the other mental disorders. This evidence, however, does not tell us which came first and whether there might not be some other predispositions or experiential factors that link them together.

Other important evidence on psychiatric comorbidity has come from studies of public hospital patients. For example, among public mental hospital patients admitted for inpatient care, 55 to 60 percent were found to have a currently active alcohol or drug disorder. When alcohol or drug disorders were found together with other mental disorders, about one-half of the patients had substance-induced mental disorders. In addition to revealing the scope or extent of alcohol and other drug problems among public mental hospital patients, these statistics indicate that psychiatric comorbidities are present much more often among patients who are receiving care than among people who are in community samples like those of the Epidemiologic Catchment Area survey, where large but lower levels of comorbidity were found.

One of the most thoroughly studied links between psychiatric disturbances and drug use involves early conduct problems, antisocial behavior, and ANTISOCIAL PERSONALITY disorders. Particularly among boys, it has been possible to link rule breaking, fighting, and other misbehavior in early elementary school with later heavy drug use. Evidence even links earlier childhood misbehavior and the later risk of becoming an intravenous (IV) drug user.

In the Epidemiologic Catchment Area studies, the association between antisocial personality disorder and alcohol or drug dependence was very strong: People with an antisocial personality disorder were about thirty times more likely to have developed an alcohol or drug disorder, when compared with people without an antisocial personality disorder. On the basis of evidence like this, scientists now are looking into the biological and behavioral substrata that might be shared by early conduct problems, antisocial personality disorder, and drug use. Developmental scientists are carrying out prevention research to see whether elementary school programs to reduce rule breaking might have benefits in the form of reduced drug use later in life.

Clinicians also have identified patients who do not get into trouble except when they are intoxicated. For these patients, the rule breaking and misbehavior seem to be promoted by drinking or drug use. In addition, among many adults and sometimes among younger drug users, drug use or drinking is followed by a period of depressed mood and associated symptoms. When adults with alcohol dependence have been studied, a large proportion of them have been found with intensive depressed mood during the period of withdrawal from alcohol, and for many of them, the depression persisted for weeks after withdrawal; in a smaller proportion, the depressed mood disappeared within two weeks. These secondary complications of drinking and drug use are examples of the hazards of psychoactive-drug use that can appear in the form of psychiatric comorbidity. If these complications are to be prevented, more research and projects to prevent intoxication or reduce (AOD) use will be needed.

In the late 1980s and early 1990s, studies disclosed what might prove to be an important causal

pathway leading from depressed mood to AOD problems. For example, adults with a history of depression were found to be more vulnerable than adults without such a history to developing alcohol dependence or alcohol abuse in adulthood. The link between depression and later AOD dependence or abuse does not seem to be as strong as that between conduct disorders and AOD problems, although some studies show low self-esteem to be a predictor of later illicit drug use.

It is quite clear that the extended use of stimulants, such as AMPHETAMINES, can produce signs and symptoms of psychoses that resemble an episode of schizophrenia. These signs and symptoms include extreme suspiciousness, paranoia, and firmly held false beliefs about people or experiences (e.g., the delusion that one's thoughts are being completely controlled by alien beings), hallucinations, apathy, and neglect of one's personal appearance. In addition, studies have highlighted the problem of heavy drinking and drug use among people with schizophrenia who were discharged from mental hospitals. That is, these patients were released into the community, their adaptation to the community environment outside the hospital all too often included heavy drinking or drug use—which interferes with their outpatient treatment and rehabilitation. Thus, in some way, schizophrenia seems to signal a vulnerability to maladaptive drug taking or alcohol use.

It is possible to think about these vulnerabilities in relation to self-medication. Mentally ill persons might be resorting to alcohol or illicit drug taking to help themselves feel better or even to relieve some symptoms of their psychiatric disturbances. The illicit drug use or alcohol use might thus be serving the function that prescribed medications are intended to serve. The idea that mentally ill patients self-medicate with alcohol or other drugs also applies to people who have anxiety disorders such as phobias (irrational fears) or panic attacks. Although this view has intuitive appeal, conclusive experimental evidence is lacking that self-medication is a reason for such individuals' abuse of alcohol and other drugs. There is evidence, moreover, that their use can actually produce a worsening of psychiatric problems.

Some especially interesting scientific problems are surfacing in relation to a pattern of comorbidity that involves tobacco dependence and depression. Some people say they smoke tobacco in an effort to help themselves deal with a low mood or feelings of depression; in these cases, tobacco smoking is a form of self-medication. Other smokers complain that they get depressed when they try to stop smoking; this is an example of depression as a secondary medical complication of smoking or, specifically in this case, a complication due to cutting back on smoking. Evidence from twin studies suggests that there is an especially complex relationship between AOD dependence and depression—and that there might be an underlying genetic predisposition that can help account for the co-occurrence of tobacco dependence and depression, besides any stressful life events or other life experiences that might have caused these two conditions to co-occur within a single individual or family. Teasing apart the separate or interacting genetic and experiential causes of this linkage is an important task for future studies.

SUMMARY

Psychiatric comorbidity refers to the co-occurrence of two or more psychiatric disorders within a single individual. Studies of community populations and studies of patients in treatment have shown that many people with mental disorders also are affected by alcohol or other drug problems; among patients of public mental hospitals, more than 50 percent were so affected. By themselves, these statistics do not indicate which came first or what was causing these disorders to co-occur within individuals. Greater attention to comorbidity in studies of vulnerability for drug and alcohol dependence will lead to more certain knowledge of the underlying explanations for such high levels of comorbidity. At the same time, the linkage of alcohol and drug use with secondary medical complications in the form of psychoses, depression, or other mental disorders cannot be neglected. Research now under way should help determine how often a psychiatric comorbidity is the result of an underlying genetic predisposition to acquiring two or more mental disorders—the result of early life stressors or other experiences, or some combination of both genetic predispositions and life experiences.

(SEE ALSO: *Causes of Drug Abuse: Psychological; Complications: Mental Disorders; Conduct Disorder; Coping and Drugs; Epidemiology of Drug Abuse*)

BIBLIOGRAPHY

ANTHONY, J. C., & PETRONIS, K. (1989). Epidemiologic evidence on suspected causal associations between cocaine use and psychiatric disturbances. In C. Schade & S. Schober (Eds.), *The epidemiology of cocaine use and abuse* (National Institute on Drug Abuse Research Monograph). Washington, DC: U.S. Government Printing Office.

CRUM, R. M., HELZER, J. E., & ANTHONY, J. C. (1993). Level of education and alcohol abuse and dependence in adulthood: A further inquiry. *American Journal of Public Health, 83,* 830–837.

KHANTZIAN, E. J. (1985). The self-medication hypothesis of addictive disorders: Focus on heroin and cocaine dependence. *American Journal of Psychiatry, 142,* 1259–1264.

LEHMAN, A. F., ET AL. (1994). Prevalence and patterns of "dual diagnosis" among psychiatric inpatients. *Comprehensive Psychiatry, 35,* 106–112.

REGIER, D. A., ET AL. (1990). Comorbidity of mental disorders with alcohol and other drug abuse: Results from the Epidemiologic Catchment Area (ECA) study. *Journal of the American Medical Association, 264,* 2511–2518.

SCHUCKIT, M. A. (1985). The clinical implications of primary diagnostic groups among alcoholics. *Archives of General Psychiatry, 42,* 1043–1049.

JAMES C. ANTHONY

COMPLICATIONS: This section has articles on some aspects of the physical and psychological complications of substance abuse. It contains an overview of *Medical and Behavioral Toxicity* and individual articles on the following: *Cardiovascular System; Cognition; Dermatological; Endocrine and Reproductive Systems; Immunologic; Liver (Alcohol); Liver Damage; Mental Disorders; Neurological; Nutritional;* and those *Due to Route of Administration.* Each article is extensively cross-referenced and will refer the reader to other articles throughout the Encyclopedia that will either expand or simplify concepts introduced here, and to articles on the many other behavioral and nonmedical complications that arise as a result of alcohol and drug use.

Cardiovascular System (Alcohol and Cocaine) Since the 1960s, the effects of ALCOHOL (ethanol) on the heart and blood vessels have been extensively studied. Clearly, the toxic effects of both acute and chronic ingestion are independent of nutritional and cardiovascular risk factors. In 1964, a relationship was established between the duration and quantity of alcohol use and the degree of heart disease in patients without nutritional or liver disease. Alcoholism, once felt to be coincidental with heart muscle damage, is now among the most frequently identified causes, according to two recent studies that report a 32 percent and a 45 percent incidence, respectively.

Attention was also paid to COCAINE abuse as the popularity of this drug skyrocketed in the 1980s (an estimated 30 million Americans had used it and some 6 million were users in 1985). Here too came the recognition that acute and chronic-use users were associated with cardiopulmonary manifestations. A 1990 survey estimated that nearly 12 million Americans use cocaine and alcohol together, and researchers have found a unique metabolite of this combination—COCAETHYLENE.

ALCOHOL

Effects of Acute Administration in Those With and Without Heart Disease. Even mildly intoxicating levels of alcohol affect the cardiovascular system. The magnitude of effects may depend on the chronicity of use. When six ounces of scotch were given to both noncardiac alcoholics and noncardiac nonalcoholics over a two-hour period, only the nonalcoholics demonstrated evidence of diminished heart muscle contractility (pumping less blood per contraction). This depressant effect is enhanced by increasing blood levels 75 milligrams/100 milliliters, but remains for only a few hours after ingestion ceases. However, the amount of blood the heart can pump per minute may actually increase in normal subjects acutely exposed to alcohol, because of an acceleration of the heart rate (increase of contractions per minute).

During the late-intoxication/early-withdrawal stage of acute alcohol consumption, blood pressure may be affected in the noncardiac alcoholic. Blood-pressure elevation is the rule. Blood levels of certain hormones, as well as urinary levels of certain breakdown products, correlate directly with blood-pressure response, which appears to vary with the degree of alcohol intake.

Abnormal heart rhythms are also commonly described in noncardiac patients during acute alcohol intoxication. The so-called holiday heart syndrome represents an acute transient-rhythm disturbance in persons without overt heart disease who are examined following heavy drinking. Atrial fibrillation—a very rapid and irregular but generally not life-threatening heart rhythm—was the most common arrhythmia. Normal rhythm was restored in all cases, but recurrence of the syndrome is common.

Cardiac patients who have already had at least one prior episode of decompensated heart failure, usually with symptoms of severe fatigue and shortness of breath, may exhibit even greater sensitivity to acute alcohol consumption. Such individuals given six ounces of scotch over two hours exhibited a substantial increase in measured internal heart pressures, suggesting poor heart function and reserve.

EFFECTS OF CHRONIC ALCOHOL USE

Subclinical Dysfunction. Results of studies on alcoholic subjects without evidence of heart disease suggest that a subclinical disease state may exist. In one study, those with at least a ten-year history of heavy alcohol consumption were compared to controls. These patients had biopsy-proven fatty liver disease (a common sequel to chronic alcohol use) without prior history of heart disease. The response of the heart's main pumping chamber—the left ventricle—to storehouses was assessed. Alcoholic hearts were found to have abnormally high internal pressures and could not appropriately compensate by increasing forward blood blow. A contractility index was correspondingly low in the noncardiac alcoholics compared to control. Doppler echocardiography, an ultrasound technique for assessing heart function, has also demonstrated an inability of the heart muscle to relax and its chambers to fill properly with blood. Therefore, the pumping chamber of the heart cannot fill or expel blood in a proper manner. Not surprisingly, autopsy specimens from these patients demonstrate fibrosis and scarring of the heart muscle. Consideration must also be given to cardiac status in cirrhotics, a group once thought relatively resistant to heart muscle failure. A group of thirty-seven patients with cirrhosis of the liver but no evidence of heart disease were studied. One subset with poor heart function at rest had an abnormal response to left ventricle stressors. The other subset, which had abnormally vigorous function at rest, also failed to respond appropriately to stressors, suggesting cardiac dysfunction in this group of patients despite a lack of symptoms.

MECHANISMS OF ACTION

The exact mechanism by which alcohol exerts its effects on the heart muscle, at the cellular level, remains speculative. It is believed that altered movement of calcium ions within these cells may be of major importance.

The Picture of Clinical Heart Failure. A full-blown cardiomyopathic picture, although relatively uncommon among alcoholics, is not unlike that seen with other causes of this syndrome. Most commonly, complaints of weakness and fatigue are present before a history of exertional and nocturnal shortness of breath. Engorgement of the veins of the neck, an enlarged, tender liver, and swelling of the legs and feet are other peripheral signs of heart failure frequently seen in these patients. Heart size is variable. The most significant cardiomegaly (enlargement of the heart) is seen in those patients who develop atrio-ventricular valve regurgitation, a leaking of blood backward from the large to the small heart chambers, as a result of muscle weakness. This abnormal flow creates heart murmurs and may give a clue to the severity of disease. Blood clots are also a common feature of this syndrome. They often form in the vein of the leg or along the walls of the heart chambers and can travel to the lungs and brain causing near instantaneous death or stroke.

Studies performed in the 1960s clearly documented this clinical decompensation. In one, a patient was fed twelve to sixteen ounces of Scotch daily. Gradually, over a four-month period, he developed the signs and symptoms of heart failure. These reversed when the alcohol was discontinued. This sensitivity to alcohol has individual variability and, unfortunately, the factors contributing to this are unknown. There may be some difference in susceptibility between the sexes. In a study of matched subjects with alcoholic cirrhosis under age forty-five, measures of heart muscle performance were significantly worse in men than in women. Experience also shows that heart failure is rare in alcoholic women prior to menopause.

In either gender, heart failure is generally found in those who continuously (chronically) ingest intoxicating amounts for a minimum of ten years. Isolated

studies do, however, suggest the potential for the beneficial effects of cessation of alcohol ingestion. One study of thirty-one alcoholics matched for degree of heart size and symptoms found that all twelve abstainers survived over several years, while all nineteen who continued drinking died. A study of sixty-four patients found a 9 percent mortality rate among abstainers over four years and a 57 percent mortality rate in persistent drinkers.

Coronary Artery Disease. Chest pain in patients with alcoholic hearts is not uncommon, although it had long been held that chronic alcoholics have less severe atherosclerotic coronary artery disease than nonalcoholics. Yet, it has been postulated that many heavy drinkers may have a higher incidence of heart attacks (myocardial infarction). Heart attacks have even been reported in a small number of patients without significant coronary disease—possibly related to the scarring of coronary arteries or the spontaneous formation of blood clots within them.

Alcohol may exert at least an indirect effect on coronary anatomy via interaction with fat metabolism. Three major components of fat metabolism, low density lipoprotein (LDL) cholesterol, high density lipoprotein (HDL) cholesterol, and triglycerides all impact on coronary atherosclerosis or plaquing. Levels of LDL cholesterol (the so-called bad cholesterol responsible for atherosclerosis) appear to be lowered by heavy alcohol use. Moderate use (1–3 drinks per day) does not seem to affect levels of either triglycerides or LDL cholesterol.

HDL cholesterol, the so-called good cholesterol (believed to be protective against atherosclerosis), has shown a consistent relationship with even moderate alcohol consumption. Alcohol is believed to enhance the activity of the enzymes lipoprotein lipase and hepatic triglyceride lipase—thereby favoring HDL production. Autopsy studies have shown a lower prevalence of atherosclerosis in alcoholics and cirrhotics. Whether this is due to malnutrition or direct effects on fat metabolism remains unclear.

A reduced degree of coronary disease has also been documented in light as compared to nondrinkers. Two recent studies both support the hypothesis that light or moderate alcohol intake appears to reduce the risk of fatal and nonfatal heart attacks and the need for coronary angioplasty and bypass operations in both men and women. Neither cohort included many heavy drinkers, and multiple coronary-risk factors (including cigarette SMOKING) were

accounted for. Although these studies supported the positive association between increased HDL cholesterol levels and alcohol intake, no consensus exists to promote alcohol use for the primary prevention of coronary artery disease, owing to its ease of addiction and other multiple deleterious effects.

COCAINE

Cocaine abuse has dramatically increased in the United States since the 1980s. One consequence has been the recognition of cocaine's unpredictable medical side effects—the most dramatic and devastating of which are cardiovascular. These include acute impairment of blood supply to the heart muscle, and heart attacks, as well as hypertension, accelerated arteriosclerosis, ruptured aorta, inflammation of heart muscle, cardiomyopathy, arrhythmias, and sudden death.

Cocaine has two separate primary pharmacologic effects on the heart and vascular system. First, it causes an accumulation of CATECHOLAMINES—accomplished by increasing their release (which include epinephrine [also called adrenaline], NOREPINEPHRINE, and DOPAMINE) from both brain and spinal cord stores and by blocking their reuptake at nerve endings. The result is a more pronounced and long lasting stimulation of the sympathetic nervous system, heart muscle, and vascular smooth muscle, represented clinically by an increase in heart rate, blood pressure, myocardial contractility, vasoconstriction, and coronary vascular resistance. Cocaine may also induce vasoconstriction via direct stimulation of calcium into smooth muscle cells.

Cocaine's second pharmacologic property is a local anesthetic effect on cardiac tissue. Cocaine can paralyze the movement of the ions (like sodium and potassium) required for the inherent electrical stimulation of heart-muscle function. Therefore, severe toxicity may result—with acute electromechanical dysfunction manifest as an abnormally slow heart rate or acute pump failure.

Several controlled trials have assessed the acute effects of cocaine on humans. Investigators have demonstrated a dose-related increase in heart rate and blood pressure. Others have induced a significant reduction in coronary blood flow in patients receiving intranasal cocaine at 2 milligrams per kilogram, with coronary angiography revealing a diffuse reduction in vessel caliber. Coronary vasospasm has also been documented with intranasal adminis-

tration. The effects of chronic use have been studied in animal models, with two experiments demonstrating that cocaine fed to rabbits on high cholesterol diets significantly increased aortic atherosclerosis as compared to rabbits fed similar diets minus the cocaine.

Since all the mechanisms discussed above either increase myocardial oxygen demand or decrease myocardial oxygen supply, heart attacks are natural consequences of cocaine toxicity. Many case reports have temporarily related cocaine use to myocardial infarction, frequently in patients without coronary artery narrowing. Thrombosis, the acute formation of an occlusive clot in a blood vessel, has been documented in nearly one-third of cases. This is consistent with evidence that cocaine causes an increase in the aggregation of platelets, blood elements that cling together to initiate blood clotting. Cocaine toxicity may also cause rupture of pre-existing small lipid-containing bulges on the arterial walls, as well as coronary spasm, both of which trigger thrombus formation.

Case reports have also suggested that cocaine has a primary depressant effect on cardiac muscle. Otherwise healthy patients have developed acute cardiac dilation and pump failure associated with acute or chronic cocaine use. With abstinence, such signs and symptoms resolved over days to weeks. Animal studies have confirmed this phenomenon. Autopsy studies on patients with cocaine in the bloodstream have shown inflammation and scarring of heart muscle cells in up to 20 percent, as compared with 4 percent of controls, suggesting a pathologic link.

Ultimately, most cocaine-related deaths are caused by cardiac arrhythmias. Abnormally fast and abnormally slow heart rates have been reported with cocaine use. Low and moderate doses often trigger the fast and massive doses the slow, including the complete cessation of heart beats (termed *asystole*). The combination of alcohol and cocaine may be even more dangerous than cocaine alone. In the presence of alcohol (ethanol), in humans, cocaine is metabolized to the compound COCAETHYLENE. This chemical renders the combination of cocaine and alcohol more lethal than either alone, and a twenty-one times greater risk of sudden death exists in people with associated coronary artery disease.

Cerebrovascular Disease. The effects of alcohol and cocaine on vascular physiology do not bypass the brain. Epidemiologic studies indicate a passive association between the amount consumed and the risk of cerebral vascular accidents, generally presenting as intracranial hemorrhage. In one study, heavy drinkers were twice as common among men and seven times as common among women who had sustained intracranial hemorrhage than in the general population. Furthermore, heavy drinkers were more likely to have been intoxicated in the twenty-four hours prior to their event. Young adults and women, generally unlikely candidates for intracranial hemorrhage, are not immune when subjected to acute, heavy intoxication.

The mechanism for this remains unclear. Hypertension is a known risk factor for stroke in general and alcohol-induced hypertension may be a causative factor. The same may be postulated with cocaine use. As with alcohol-induced cardiomyopathy, individuals who reduce their alcohol intake have a significantly lower risk of developing hemorrhagic stroke than those who continue its use.

(SEE ALSO: *Alcohol; Complications*)

BIBLIOGRAPHY

AHMED, S. S., & REGAN, T. J. (1992). Cardiotoxicity of acute and chronic ingestion of various alcohols in cardiovascular toxicology. In D. Acosta, Jr. (Ed.), *Cardiovascular Toxicology* (2nd. ed.). New York: Raven Press.

REGAN, T. J. (1991). Alcohol and the heart. In K. Chatterjie & W. Parmley (Eds.), *Cardiology*. New York: Gower Medical Publishing.

THOMAS, B. A., & REGAN, T. J. (1990). Interactions between alcohol and cardiovascular medication. *Alcohol, Health, and Resort World, 14,* 333–339.

TIMOTHY REGAN
ROBERT SAPORITO

Cognition PSYCHOACTIVE DRUGS of abuse, used for their perceived mind-altering effects, often have additional cognitive effects of which the drug user may not be aware. A cognitive effect is an impact on mental functions—including processes of learning, perceiving, imagining, remembering, feeling, thinking, reasoning, knowing, and judging.

Psychoactive drugs produce cognitive effects by causing chemical changes in the brain. These effects are mostly short-lived and correspond to the duration and intensity of the chemical changes in the brain. However, cognitive effects can persist after the drug has been eliminated from the body, and some

can be irreversible. The common cognitive effects of some psychoactive drugs of abuse are summarized below.

ALCOHOL

Ethanol (also called ethyl alcohol) is the drinking ALCOHOL of BEERS AND BREWS, wine, distilled spirits, or medicinal compounds; it acts by depressing or reducing cognitions. Initially, alcohol reduces inhibitions, and this results in more spontaneity or impulsivity and a feeling of relaxation. As the amount of alcohol acting on the brain increases, the ability to perceive, remember, reason, and judge is progressively impaired. Further increases in the amount of alcohol can depress the brain and cognitions to the point of loss of consciousness. Due to cognitive impairment, the person may not perceive the impairment (e.g., "I'm not drunk") and take undue risks (e.g., DRUNK DRIVING, indiscretions).

Alcoholic blackouts are impairments of memory for events that occurred while one was conscious but under the influence of alcohol. Such black-outs are not limited to chronic alcoholics. Long-term use of alcohol can lead to subtle impairment of perceiving, responding, and remembering that may not be detectable without special psychometric tests. A particular form of impairment of memory, called the amnestic syndrome, has been seen in alcoholics; they are unable to remember recent events although memories from long ago remain reasonably intact. By contrast, in alcoholic dementia, deficits in all cognitive functions are seen. Some deficits may persist for life even if the person stops drinking.

Paranoid states of unfounded suspicion or jealousy may manifest or be aggravated under the influence of alcohol. In alcoholic HALLUCINATIONS people can have vivid but unreal perceptions while awake; these typically occur as a result of neurochemical changes in the brain when alcohol use is abruptly discontinued after periods of excessive drinking. Even after months have elapsed since their last drink, alcoholics can have cognitive deficits, especially in visual-spatial abilities, hand-eye coordination, abstract reasoning, and new learning.

TRANQUILIZERS, SEDATIVES, AND HYPNOTICS

These drugs are often collectively referred to as "downers." Persons taking them are at risk for the

cognitive impairments discussed above, under "Alcohol." The ELDERLY are particularly at risk for confusion.

STIMULANTS

Stimulant drugs have effects that are the reverse of depressant drugs—they arouse the nervous system. They include such drugs as COCAINE, AMPHETAMINES (speed), and CAFFEINE. In low doses, perception is heightened, attention is increased, and thought processes are speeded up, resulting in a feeling of greater alertness. MEMORY, however, may be affected, resulting in impaired recall of material learned while under the influence of stimulants. Higher doses intensify the above effects, leading to restlessness and rapidity of thoughts, which reduce attention. Vulnerable persons may become paranoid or even psychotic. Higher effective doses of stimulants affecting the brain rapidly, as in the use of cocaine, result in an abrupt "rush" or "high." These effects are typically short-lived but are so intense (pleasurable) that individuals may repeat doses. Discontinuation of stimulants after a long period of use often leads to a temporary period of DEPRESSION.

MARIJUANA

Cannabis sativa is often used for the subjective effects of relaxation and a decreased awareness of conflicts. It is also known to distort perception of time and to reduce responsiveness. Long-term use of *Cannabis* has been associated with apathy, underachievement, and lack of motivation.

HALLUCINOGENS

Hallucinogenic drugs distort perceptions and cause hallucinations. They include LYSERGIC ACID DIETHYLAMIDE (LSD), PHENCYCLIDINE (PCP), MESCALINE, PSILOCYBIN mushrooms, and several newer drugs with hallucinatory and stimulant effects. Apart from profound effects on perceptions, responsiveness, learning, and judgment are affected. Some users experience flashbacks—spontaneous vivid recollections of experiences that occurred while under the influence of hallucinogens.

USE DURING PREGNANCY

Psychoactive drugs used during PREGNANCY affect the developing fetus. Prenatal exposure to alcohol,

Cannabis, or stimulants has been associated with cognitive impairments detectable early in the child's life.

(SEE ALSO: *Imaging Techniques*)

PETER MARTIN
GEORGE MATHEWS

Dermatological With the exceptions of rashes and other skin conditions resulting from idiosyncratic or allergic reactions to drugs, most drug use complications involving skin damage result from the use of hypodermic needles or other means of drug injection that involve breaking the skin surface. There are three primary types of injection: (1) *subcutaneous,* also known as "skin-popping," wherein the needle is injected into or directly under the skin surface; (2) *intramuscular* (IM), wherein the needle is injected into muscle mass, often in the shoulder or buttock; and (3) *intravenous* (IV), direct injection into a blood vessel. Skin damage can result from repeated injection in the same area, failure to clean the injection site, nonsterile needles, and/or impurities or insoluble materials in the substance injected. Adulterants in the drugs, liquids used to dilute the drugs and contaminated injection paraphernalia, and the surface of the injection site all provide sources of viruses, bacteria, and fungi.

The most common skin damage from repeated injection is needle-track scars. These are usually caused by unsterile injection techniques or by the injection of fibrogenic particulate matter—material often used by dealers to add bulk and weight to the drug or buffers in tablets that have been ground up and liquified for injection. Carbon deposited on needles by users who try to sterilize by heating the needle tip with a match may produce a "tattoo" discoloration, accompanied by a mild inflammatory reaction under the skin at the point of entry. Such scars are found mostly on the arms, but they can occur anywhere on the user's body that has been used as an injection site, including thighs, ankles, neck, and penile veins.

Needle abscesses, characterized by redness; a stinging, itching sensation; and swelling at the site, often result from repeated injection without cleaning the injection site. Such skin flora as staphylococci and streptococci are driven beneath the skin surface to infect the site, often with pus formation. Forms of

contact dermatitis can also result from allergic reactions, especially to fluids used to sterilize the skin at injection sites. Infections and allergic reactions increase as an individual's resistance decreases with a drug-compromised immune system.

Subcutaneous injection of SEDATIVE-HYPNOTIC drugs, such as BARBITURATES, can cause cellulitis—where the tissue becomes reddened, hot, painful, and swollen at the injection site. If not treated, the cellulitis may last for a long time. In extreme cases, the cellulitis may eventually cover most of the user's body as new needle sites are used to avoid painful areas. Superficial cellulitis, septic thrombophlebitis, and simple needle abscesses can usually be treated with local heat, incision, and drainage, followed by culture and sensitivity testing and appropriate antimicrobial therapy.

Repeated intravenous injection may produce anaerobic infections or abscesses that produce a foul-smelling discharge, sometimes gas formation, and a cellulitis that is characterized by a rapidly progressing stony or wooden-hard tenseness, often some distance from the original needle site. Although the mechanism of these infections is unclear, it is thought to involve a disruption of blood supply to the area from edema (fluid collection) resulting from the cellulitis. Treatment involves wide incision and pressure reduction in the affected area.

(SEE ALSO: *Allergies to Alcohol and Drugs; Complications: Route of Administration*)

BIBLIOGRAPHY

COHEN, S., & GALLANT, D. M. (1981). *Diagnosis of drug and alcohol abuse.* Brooklyn: Career Teacher Center, State University of New York.

SENAY, E. C., & RAYNES, A. E. (1977). *Treatment of the drug abusing patient for treatment staff physicians.* Arlington, TX: National Drug Abuse Center.

SEYMOUR, R. B., & SMITH, D. E. (1987). *The physician's guide to psychoactive drugs.* New York: Hayworth Press.

DAVID E. SMITH
RICHARD B. SEYMOUR

Endocrine and Reproductive Systems Many fundamental challenges remain in understanding the impact of ALCOHOL and drugs on en-

docrine and reproductive function. This article presents what is currently known; before beginning, a few caveats deserve attention.

Many factors may influence the degree to which illegal drug or alcohol abuse may cause an abnormality of endocrine or reproductive function. These factors include (1) the amount and duration of consumption; (2) the route of illegal drug administration; (3) whether there is preexisting or concurrent damage to an endocrine/reproductive organ; (4) concurrent use of another drug; and (5) the genetic predilection for an endocrine disorder. Often, our knowledge about these factors and how they interact with one another is more limited than what is known about the range of endocrine and reproductive dysfunction associated with the chronic consumption of alcohol and the abuse of illicit drugs.

Knowledge is also limited because some endocrine or reproductive consequences may be manifested only by an abnormal result from a laboratory (biochemical) test. The absence of a physical sign or a clinical symptom may lead to the false impression that there is no endocrine/reproductive consequence. In addition, there are challenges in ascertaining whether the alcohol- or drug-abuse related endocrine/reproductive dysfunction is due to the drug itself or to the social context in which the drug is used. Finally, endocrine or reproductive disturbances may also occur from the consequences of WITHDRAWAL syndromes when the drugs or alcohol ingestion is stopped or reduced. To the extent to which these issues have been clarified, we will note them here.

HYPOTHALAMIC/PITUITARY

Most endocrine and reproductive function is influenced directly or indirectly by the BRAIN—specifically by the functional interactions of the brain's hypothalamus and pituitary with the target endocrine organs. The hypothalamus produces pituitary-regulating hormones; all are peptides except one (DOPAMINE). In response to each of these hypothalamic hormones, the pituitary releases a hormone, which influences the function of an endocrine or reproductive organ.

Alcohol. The anecdotal reports of changes in sexual function following alcohol consumption was the stimulus for much of the research targeting hypothalamic-pituitary relationships, since impairments here may often result in sexual dysfunction.

Although acute alcohol use has been reported in public surveys to be associated with increased sexual drive and functioning, clinical and animal research have revealed major hormonal dysfunctions in chronic or heavy alcohol users.

Prolactin (PRL)—the pituitary hormone associated with preparation during pregnancy for breast-milk secretion—is increased with heavy alcohol use; however, chronic alcohol use inhibits the pituitary release of luteinizing hormone (LH) and follicle-stimulating hormone (FSH). Both LH and FSH are important in regulating the sex hormones produced by the testes in males and the ovaries in females. Yet, when alcohol is administered acutely, there are no significant changes in PRL, LH, or FSH serum levels.

Heavy alcohol consumption is associated with an increase in pituitary-secreted adrenocorticotropic hormone (ACTH), partly explaining the "pseudo"-Cushing's syndrome (moon faced appearance, central obesity, muscle weakness) and the increased melanocyte-stimulating hormone (MSH), which possibly leads to darkening skin pigmentation. Although there is no consistent effect of heavy alcohol use on the pituitary's release of thyroid-stimulating hormone (TSH) or growth hormone (GH), a rise in the blood alcohol level is associated with the inhibition of antidiuretic hormone (ADH) release from the posterior pituitary, resulting in increased urination.

Drugs. Complaints of derangements of libido (sex drive) and sexual functioning in OPIOID addicts (HEROIN) were among the first lines of clinical evidence to suggest the possible role of such narcotics in altering hypothalamic-pituitary functioning. Although most of what is known about drug-abuse related hypothalamic-pituitary abnormalities focuses on heroin use, the epidemic proportions of COCAINE abuse and dependency in the 1980s have brought renewed scientific interest to this area.

Studies have shown that opioid use has been associated with increased serum PRL without disturbances in serum GH or TSH levels; and cocaine use has been associated with both high and low PRL levels. The contradictory findings in the case of cocaine use might be attributable to the variations in patterns of cocaine use. Animal studies have shown that gonadotropin-releasing hormone (GnRH), released from the hypothalamus, did not stimulate PRL following acute cocaine administration, and it did not prevent acute cocaine-associated PRL suppression.

Elevated levels of dopamine have been observed during acute cocaine administration, but chronic cocaine use may deplete dopamine.

In patients receiving METHADONE therapy for opioid addiction, some investigators have reported a normal rise in TSH released by the pituitary, in response to stimulation by the hypothalamic hormone called thyrotropin-releasing hormone (TRH). Others have observed a blunted TSH and PRL response following TRH administration in active heroin users.

Although normal basal LH secretion has been observed in cocaine abuse, opiate use is associated with decreased basal FSH and LH levels in males. In female heroin addicts, these low levels of the pituitary gonadotropins have clinical relevance, resulting in a consistently normal FSH response and a variable LH response following a GnRH challenge.

Some researchers have demonstrated normal functioning of the hypothalamic-pituitary-adrenal (HPA) axis in former heroin addicts who were maintained on methadone both long-term and only for a number of months. However, there is also evidence suggesting alteration of the normal biological rhythm of hormonal secretion.

SEX HORMONES

Diminished sexual drive and performance in opioid users have raised questions about the relationship between such narcotic drug use and disturbances in the levels of sex hormones. Although some reports show no significant differences in serum-testosterone levels between heroin addicts, METHADONE-MAINTAINED patients, and normal controls, other studies have not confirmed these results. Some researchers have reported plasma levels of testosterone to be consistently lower in active heroin addicts, in addicts who self-administer heroin in controlled research settings, and to be within normal range in long-term methadone-maintained patients. Additionally, some evidence shows that plasma testosterone levels that are depressed under circumstances of heroin administration followed by methadone maintenance and then withdrawal gradually returned to preheroin-use levels.

Opioid effects on the estrogens of both males and females may be responsible for the clinical observations of sexual dysfunction. In the male heroin addicts studied, the plasma estradiol concentrations were either low or within normal ranges; in the female, the plasma estrogens are low. A clear explanation of these observed derangements in plasma testosterone and estrogens is unknown. Female heroin addicts frequently experience cessation of or irregular menses. However, most regain normal menstrual funcion when stabilized on methadone and under these circumstances fertility seems unaffected. The anecdotal reports and, in limited cases, experimental evidence of the influence of MARIJUANA on sexual function and sex-hormone levels are also inconsistent and confusing.

The illicit drug-related disturbances discussed above suggests that the narcotic-related depressions in sex-hormone production of the ovaries and testes may occur because they reduce the pituitary's stimulation of these sex organs. Still, this has not been a consistent finding.

REPRODUCTION AND PREGNANCY

Impotence, atrophy of the testes, infertility, and decreased libido are not uncommon complaints in male alcoholics. These observations are thought to be secondary to the direct effects of alcohol on testicular tissue, to an alcohol-associated decrease in sperm motility, and to an alcohol-related decrease in Vitamin A and zinc. Both Vitamin A and zinc are important in maintaining testicular tissue growth. In young females, alcohol abuse is associated with amenorrhea and anovulation; in chronic users, with early menopause. There is evidence that vaginal blood flow decreases as the alcohol serum level increases.

Despite these clinical observations, when rigorously investigated, there were no consistent changes in estradiol, progesterone, or testosterone. Consequently, it is difficult to determine whether these observations were due to alcohol-related liver disease, malnutrition, or the direct toxic effects of chronic alcohol use.

During PREGNANCY, alcoholism is associated with increased risk of spontaneous abortion and the development of FETAL ALCOHOL SYNDROME (FAS). FAS is comprised of a characteristic pattern of skin/facial abnormalities with growth and development impairments, which are believed to be related to alcohol's suppression of the sex hormone progesterone. While the features of FAS may vary, fetal abnormalities associated with alcohol can be divided into the following four categories: (1) growth deficiency; (2) central

nervous system dysfunction; (3) head and facial abnormalities, and (4) other major and minor malformations.

ADRENALS

Our understanding of the relationship between opioid drug use and the functioning of the adrenal gland is based on incomplete and often contradicting information. Some scientists have published reports of normal plasma cortisol levels (from the adrenals) during heroin use and withdrawal, under research conditions of heroin self-administration, and during methadone-maintenance treatment. In methadone-treated patients, ACTH produced by the pituitary stimulates the adrenal gland to produce cortisol. In another study, there was a decreased plasma-cortisol response to intravenous cosyntropin (an ACTH-like substance) stimulation in methadone-treated patients. There are also reports of low normal or subnormal plasma-cortisol levels in heroin users and disturbances in the daytime cortisol secretion from the adrenal gland in methadone-maintained patients.

The variable findings from the several studies may be attributed to differences in the types of drugs used, in the state of stress-associated drug withdrawal, in patterns of drug use, in study design, or to a combination of these and other as yet unknown factors. There is also the well-known problem of ACTH measurement, often resulting in falsely low values.

CARBOHYDRATE METABOLISM

The opioids are virtually the only class of *illegal* drugs for which there is information about the pharmacologic effects on serum-glucose levels. We have long-standing reports of opiate-associated hyperglycemia, but the mechanisms explaining these empirical observations are incompletely understood. In association with chronic opioid use, there are reports of both low levels of serum glucose and high levels of insulin. The conflicting results of some investigations may be due, in part, to differences in study design (e.g., nutritional state of the research subjects, amount of the glucose used in clinical studies, or the time(s) administered).

To briefly review the regulation of glucose control: The pancreas, an endocrine organ located in the upper abdomen, plays a central role by secreting glucagon to raise serum-glucose levels and by secreting insulin to lower serum-glucose levels. After the discovery of endogenous opioid peptides in the human pancreas, subsequent research provided information that one such endogenous opioid, beta-endorphin, stimulates the secretion of glucagon and a biphasic rise of insulin. This may, in part, explain the observations of both elevated and reduced serum-glucose levels in heroin users. Whatever the nature of the exact mechanism, glucose metabolism is deranged in both heroin and methadone use by some direct or indirect parameter of serum-glucose regulation.

The alcohol-related aberrations of carbohydrate metabolism are also quite complex. Some investigators have demonstrated that acutely administered alcohol may result in a reversible and mild resistance to the glucose-lowering effects of insulin, perhaps explaining the observations of a rise in glucose following alcohol use. In fasting individuals, alcohol administration can lead to severe depressions of serum glucose, primarily by reducing the liver's production of glucose. Serum-glucose levels are also lower in chronic alcohol users with concurrent alcohol-related liver disease. Nevertheless, serum levels are elevated in alcoholics with concurrent alcohol-related destruction of the pancreas. Even without other concurrent diseases, alcohol consumption may result in either no changes or in minimal to mild elevations or reductions in serum glucose.

THE THYROID

Located in the anterior aspect of the neck, the thyroid gland secretes thyroxine (T_4) and other hormones whose principal purpose is to regulate the metabolism of other tissues in the body. The production of T_4 by the thyroid is under the control of the TSH produced by the pituitary. Therefore alterations in thyroid function can be the result of problems directly involving the gland or disruptions in the TSH-mediated control of the thyroid gland.

Despite the frequency, duration, or amount of use, it appears that there are no clinical signs or symptoms of thyroid dysfunction in chronic heroin or alcohol users. Disturbances in biochemical indices (laboratory tests) of thyroid function are, however, not uncommon in opiate or alcohol use. The total T_4 is decreased while the amount of biologically available T_4 (free T_4) and other indices of thyroid function are normal in heavy alcohol users.

In active heroin users or during heroin withdrawal, total T_4 levels are increased in association with normal, subnormal, or high levels of other parameters of thyroid function. In methadone-maintained persons, there are reports of normal and slight-to-significant increases in total T_4 in conjunction with increased levels of thyroxine-binding globulin (TBG), a protein that binds thyroid hormones in blood. Interestingly, successful methadone maintenance is associated with a correction of these biochemical disturbances.

There are a number of possible explanations for the biochemical derangements observed during opiate use. To maintain an adequate range of biologically active T_4, the total T_4 is increased whenever there is an increase in TBG. Perhaps the increase in total T_4 is the result of a direct opiate-induced elevation of TBG. It is also possible that the altered liver function seen in chronic heroin and alcohol users may be responsible TBG abnormalities, leading to disturbances in T_4 levels. Finally, it is possible that opiate-related or alcohol-related disturbances may be due to a combination of the above mechanisms as well as to some other still undefined processes.

BONE METABOLISM

The observations of increased fractures sustained by alcoholics has prompted investigations about the role alcohol may play in disturbances of the structure and mechanical properties of bone. Some studies have shown reduced bone mass in alcoholics, while others have reported decreases in compact and trabecular bone mass—a condition called osteoporosis. Some of these disturbances in new bone formation may be mediated by alcohol's impairment of calcium and Vitamin D metabolism, both of which are crucial to bone metabolism. Nevertheless, there does remain considerable doubt as to whether the bone complications are due to alcohol itself or due to alcohol-related liver disease, to malnutrition, or to a potential host of other factors. Chronic liver disease unrelated to alcohol has also been a cause of osteoporosis and other bone diseases.

CONCLUSION

The endocrine and reproductive consequences of illicit drug and alcohol abuse are extensive and profound. Both drug and alcohol abuse result in

clinically significant multiglandular derangements. Although knowledge about the dimensions of such disturbances to endocrine and reproductive function slowly increases, explanations of the scientific mechanisms accounting for these observations remain to be elucidated. Given the role of alcohol and illicit drugs in society, however, the spectrum of related endocrine and reproductive complications can be expected to expand and, thereby, increase in public-health significance.

(SEE ALSO: *Complications: Liver Disease*)

BIBLIOGRAPHY

FELIG, P., ET AL. (1987). *Endocrinology and metabolism*, 2nd ed. New York: McGraw-Hill.

HAMID, A., ET AL. (1987). Alcohol and reproductive function: A review. *Obstetrical and Gynecological Survey*, 42(2), 69–74.

LALOR, B. C., ET AL. (1986). Bone and mineral metabolism and chronic alcohol abuse. *Quarterly Journal of Medicine*, 59, 497–511.

SMITH, C. G., & ASCH, R. H. (1987). Drug abuse and reproduction: Fertility and sterility. *The American Fertility Society*, 48(3), 355–373.

LAWRENCE S. BROWN, JR.

Immunologic This article describes the basic and clinical immunologic aspects of alcohol and drug abuse.

ALCOHOL

The physiological characteristics of ALCOHOL (ethanol) allow it to interfere intensively with the functions of immune cells. Alcohol is completely miscible with water and, to some degree, is fat soluble. It crosses membranes by diffusion across a concentration gradient. Historically, alcohol has been associated with lower host resistance and increased infectious disease. For example, ALCOHOLISM has been closely associated with lung abscesses, bacteremia, peritonitis, and tuberculosis. Although these infections might be a result of malnutrition or poor living conditions, prolonged consumption of alcohol also results in alterations of immune responses, ul-

timately manifested by increasing susceptibility to infectious agents. Overwhelming evidence is showing that alcohol abuse broadly suppresses the various immune responses, seriously impairing the body's normal host defense not only to invading microbes but also to its defenses against CANCER cells.

These disruptions are the combined result of direct toxic effects on the immune system and indirect effects such as malnutrition, oxidative stress, endocrine changes, and the complications of liver disease. The alcoholic's predisposition to extracellular and intracellular infection indicates the effects of alcohol consumption at the local, humoral, and cellular levels, inhibiting immune response and host defense. Recent evidence suggests that aberrant regulation of the neuroimmune-endocrine networks may be a major risk factor for the development of alcohol-induced immunosuppression, leading to the collapse of host defense. Bidirectional communication can occur between the immune and neuroendocrine systems. Accordingly, stimulated lymphoid cells send signals mediated by cytokines and other immune products to inform the central nervous system about the activity of the immune system. Neuroendocrine molecules, in turn, may complete a feedback loop by modulating the immune response via the pituitary-endocrine axis as well as the autonomic neural output. Thus, effective feedback communications between the endocrine and the immune system may be crucial to the host's defense responses.

Clinical and experimental studies indicate a relationship between excessive alcohol use and compromised immune responses. Human studies have shown that chronic alcohol ingestion is associated with abnormalities of both humoral and cellular immunity. These abnormalities include a depression of serum bacteriocidal activity, alterations of immunoglobulin production, leukopenia, defects in chemotaxis, decreased antigen trapping and processing, and decreased T-cell mitogenesis. The clear association of alcoholism and infections such as tuberculosis and listeriosis indicates defective functioning of cell-mediated immunity. A study has linked alcohol abuse and deficient T-cell responsiveness. Skin-test reactivity using purified protein derivative and dinitrochlorobenzene have also demonstrated poor responses in alcoholics with liver disease. Natural killer (NK) cell activity is impaired in acute alcohol intoxication and in chronic alcoholic liver disease; NK cells are programmed to recognize and destroy abnormal cells, such as virus-infected or tumor cells. Some researchers have speculated that decreased NK cell activity may be intimately involved in the increased incidence of tumors in alcoholics.

Giving alcohol to animals also has a profound effect on decreasing the weight of their peripheral lymphoid organs as measured by a decreased number of thymocytes and splenocytes. In mice, alcohol use produces thymic and splenic atrophy and alterations in the circulating lymphocytes and lymphocyte subpopulations, as well as alterations in cellular and humoral immunity and impaired cytokine production. Also impaired by dietary alcohol are antibody-dependent cellular cytotoxicity, lymphocyte proliferation, B-lymphocyte functions, and cytokine production by lymphoid cells (the lymph and lymph nodes). Thus, alcohol-induced immunosuppression may render alcoholics more susceptible to tumorigenesis and infection.

Alcoholics are susceptible to infections by bacteria such as *Listeria monocytogenes*, *Vibrio vulnificus*, *Pasteurella multocida*, and *Aeromonas hydrophilia*. The severity of these infections has raised the possibility of a neutrophil (white blood cell) abnormality in these patients. The proper functioning of neutrophils is critical for host defense against microorganisms. Neutrophils are the chief phagocytic leukocyte of the blood; they are short-lived cells having a life span of approximately four days. Their production is a tightly regulated process centered in the bone marrow. Chronic alcoholics have often been noted to be leukopenic (abnormally low in leukocytes). The toxic effect of alcohol is now believed to be caused by the depression of the T-cell-derived colony-stimulating factor rather than to direct suppression of myeloid (bone-marrow) precursors secondary to bone-marrow toxicity.

Neutrophils must recognize the invading pathogens, engulf them, and destroy them using a number of killing mechanisms, which include adherence, chemotaxis, locomotion, phagocytosis, and intracellular killing. Several functions of neutrophils are affected by alcohol in vitro, including impairment of chemotaxis, decreased migration of neutrophils within vessels, altered adherence to nylon fibers in vitro, impaired phagocytosis, and decreased intracellular killing of bacteria. In human with advanced cirrhosis from chronic (prolonged and excessive) ingestion of ethanol and impaired phagocytic capacity, decreased metabolic activity was observed in the

liver's reticuloendothelial system; there also were impairments of neutrophil chemotaxis, bacterial phagocytosis and killing, and alterations of neutrophil-antigen expression. Neutrophil dysfunction is therefore responsible for aggravating the susceptibility to secondary infections seen in alcoholics.

The balance of cellular and humoral immune response to antigens is controlled by communication between immunocompetent cells. They are regulated to a great extent by soluble mediators (termed *cytokines*) produced mostly by T-helper cells and macrophages. Cytokines are biologically active polypeptide intercellular messengers that regulate growth, mobility, and differentiation of leukocytes. Thus, cytokines have extremely important roles in the communicating network that links inducer and effector cells to immune and inflammatory cells.

Since any perturbation in the tightly controlled cytokine regulatory system can result in immune alterations modifying host resistance to infectious disease and cancer, the influence of alcohol consumption on cytokine secretion has been considerably investigated. Several studies have indicated a correlation between circulating levels of macrophage-derived cytokines and disease progression during chronic alcohol consumption. Increased plasma concentrations of tumor necrosis factor have been observed in cases of alcoholic liver disease and, interestingly, relate significantly to decreased long-term survival; plasma Interleukin-1 is also significantly increased in these patients (relative to healthy controls) but does not correlate with increased mortality.

In a model of alcohol-fed mice, we found that, compared to controls, production of all cytokines was suppressed by chronic alcohol consumption, suggesting general immunosuppression. The elevated levels of cytokines in some animals with murine (mice and rats) AIDS were, however, increased further by alcohol ingestion as compared to controls that indicated alcohol-induced aggravation of some AIDS symptoms. Similarly those cytokines suppressed by murine AIDS were further suppressed by alcohol. Thus alcohol exacerbated their immune dysfunction. Several pathways may be involved in mediating the interaction between the endocrine system and the immune system. Recent findings indicate that pituitary peptide hormones can directly influence immune response. In addition, when a neurotransmitter is released in lymphoid tissues, it may locally modify functional properties of lymphocytes and release of cytokines.

In human studies regarding alcohol, all parameters, such as hormone levels and immune responses to monitor changes of immune response and neurotransmitter, are usually detected in serum. Since the serum levels of these parameters cannot accurately reflect the real situation in the lymphoid organs or tissues, some results from them, therefore, could be misleading. No animal model for alcohol studies can mimic the complications of alcoholic liver disease seen often in human alcoholics. Furthermore, the facts that different hormonal status in individual animals, even within same strain of animal, and difficulty in defining hormonal status in animals indicate that some results from animal studies could also be misleading. Therefore, the research on the mechanism of alcohol's effects on the neurological system at cellular and systemic levels and the interaction between endocrine and immune system should continue if we are to understand the complex changes caused by direct and indirect effects of alcohol consumption.

COCAINE

COCAINE acts directly on lymphoid cells (the lymph and lymph nodes) and indirectly modulates the immune response by affecting the level of neuroendocrine hormones. The first studies about the impact of cocaine use on the immune response were initiated because epidemiological data demonstrated a high prevalence of ACQUIRED IMMUNODEFICIENCY SYNDROME (AIDS) in polydrug users. Depending on the different administration routes, the plasma levels of cocaine in humans appear to be in the range of 0.1 to 1 micrograms per milliliter ($\mu g/ml$). Such concentrations last only for thirty to sixty minutes at these levels and then decline because of cocaine's short biological half-life of about one hour. The direct effects of cocaine and its metabolites on immune cells should occur only during a short time, except in heavy cocaine users who use the drug several times a day every day. Besides the direct effects on immune cells, cocaine could indirectly affect the immune response via its impact on the neuroendocrine system—and both have been shown.

Short-term exposure of mice to cocaine by daily intraperitoneal injection for fourteen days reduced

body, spleen, and thymus weight in the animal. Cocaine increased the responsiveness of lymphocytes to mitogens (cell proliferation initiators) and the delayed hypersensitivity responsiveness, but it suppressed the antibody response. All the animal studies, however, suggest that the immune system requires continuous exposure to cocaine to demonstrates its suppressing or stimulating effects. After a single dose of cocaine (0.6 mg/kg), nonhabitual cocaine users showed a significant stimulation of natural killer cell activity, which is vital to defend against cancers. The levels of natural killer cells were also increased, but the levels of T-helper and suppressor cytotoxic cells, B cells, and monocytes were not elevated.

Cocaine causes neuroendocrine-mediated effects on the immune response. Cocaine stimulates the brain's hypothalamus to increase secretion, producing potentiated secretion of beta-endorphin. As a result of the cocaine administration, beta-endorphin binds to opioid receptors on monocytes and lymphocytes and exerts multiple stimulating and suppressing effects on these cells, including secretion of immunoregulatory cytokines. The net outcome of the reactions related to the immune response of the host is difficult to assess, because other determinants of these interactions (such as the psychological and social situation of the cocaine users) are also possible.

There are other mechanisms that might be operating to mediate cocaine-induced immunomodulation, including nutritional deficiencies and their impact on lymphoid cells. As early as 1870, the French physician Charles Gazeau suggested that coca leaves might be used to suppress the appetite. With food deprivation, which is common under conditions of habitual drug use, the self-administration of cocaine by rats increased. Although data indicate a poor nutritional status for cocaine users, no study has yet assessed the nutritional status of drug users as it contributes to a compromised immune competence. Cocaine clearly modifies hormones with immunoregulatory properties via neurological effects. In addition, malnutrition could be a factor on cocaine use resulting in altered disease and tumor resistance. Intravenous use of drugs, including cocaine, is associated with the transmission of HUMAN IMMUNO-DEFICIENCY VIRUS (HIV), hence ultimately the development of AIDS. Immunomodulation by cocaine after HIV infection could accelerate disease development as well as overall resistance to a variety of pathogens found frequently in intravenous drug users.

TOBACCO

While it is now well known that the use of TO-BACCO is a major health hazard, millions of Americans still continue to smoke and the popularity of smokeless tobacco is on the rise. Tobacco use is the chief cause of lung cancer in smokers and is strongly linked with the oral cancers of those who use chewing tobacco or SNUFF.

The pulmonary alveolar macrophage (PAM) is the cellular component of the immune system comprising the first line defense of the lung, offering protection against inhaled particles, including irritants and microbial invaders. Because the PAM has exposure to both the bloodstream and to the atmosphere, it is uniquely suited to perform its protective functions, which include clearance, immune modulation, and modulation of surrounding tissue. There is general agreement that the number of PAMs in smokers' lungs is increased 2 to 20 times above that found in the lungs of nonsmokers. It also appears that there is a difference in the morphology and certain aspects of the function of alveolar macrophages between the two groups. In general, PAMs from smokers are larger, contain more lysosomes and lysosomal enzymes, and are more metabolically active than those nonsmokers, suggesting that they may be in a chronically stimulated, more active state. This might lead to the inference that there would be greater phagocytic capacity in the lungs of smokers, resulting in increased clearance of foreign matter. However, the responsiveness of smokers' macrophages to foreign bodies or bacteria was equal to or less than that of nonsmokers, leading researchers to conclude that chronic stimulation of PAMs by cigarette smoke may be harmful rather than beneficial to the immunocompetence of the lung.

There is some disagreement as to whether smoking affects the phagocytic and bactericidal activity of PAMs. The question of whether tobacco smoke alters the tumoricidal ability of PAMs has not yet been answered. Thus the relationship between cigarette smoking, neutrophil accumulation in the lung, and lung destruction continues to be researched. It is known that particles from cigarette smoke are present in the PAMs of smokers, and researchers have

found that the PAMs of cigarette smokers released a potent chemotactic factor for neutrophils, whereas those of nonsmokers did not. Therefore, the cigarette smokers had an increased number of neutrophils in the lavage fluid and in lung biopsy tissue as compared to nonsmokers. Neutrophils store and release elastase, a substance implicated in the development of certain lung diseases. Smokers' lungs are exposed to a large chronic burden of elastase from neutrophils, which may predispose them to lung destruction.

A number of animal and human studies comparing peripheral blood samples of smokers and nonsmokers have indicated that smokers have altered immunoglobulin levels. Elevated levels of immunoglobulin E (IgE) were present in a high proportion of the smoke-exposed animals but in none of the controls. Studies in human subjects have also revealed that IgE levels were higher in smokers than in nonsmokers. A study of coal workers indicated that both mining and nonmining smokers had depressed serum IgA and IgM levels as compared with similar groups of nonsmokers. A disturbing finding in relationship to increased immunoglobulin levels in smokers is the effect that maternal smoking may have on the fetus. In newborn infants of mothers who smoked during pregnancy, IgE was elevated threefold. Tobacco smoke affects fetal immunoglobulin synthesis, stresses the fetal immune system, and can predispose the infant to subsequent sensitization. Thus 34 percent of the reported asthma in childhood may be caused by maternal smoking.

Natural killer cells, thought to serve important antitumor and antiviral functions in the body, have been found to be decreased in smokers. Studies of the white blood cells called basophils in the peripheral blood indicated that there are alterations linked to tobacco smoking as well.

When considering the effect that tobacco use has on immunocompetence, other confounding variables must also be accounted for, including genetic factors, preexisting disease, and nutritional status. Smoking has been observed to cause deficiencies of Vitamin C, beta-carotene (Vitamin A), and other nutrients which have important functions in protecting immunity.

Tobacco smoking causes deleterious effects on the pulmonary and systemic immune systems of experimental animals and in humans. Aspects of both cell-mediated and humoral immunity are affected. It is often difficult to compare studies directly because of the variability in smoking behaviors and the differences among tobacco products. Although it is expected that heavy smoking causes the most amount of immune system damage, that does not mean that light to moderate smoking is safe. Thus, if some alterations due to smoking are reversible, it is not yet known whether long-term smoking may cause the impairment of the immune system to become permanent. Further, simultaneous exposure to other air contaminants or air pollution may exert damaging synergistic effects upon local or systemic immune defenses.

MORPHINE AND OTHER OPIOIDS

Several studies have drawn a parallelism between MORPHINE abuse and immune inhibition. In vitro studies have shown that polymorphonuclear cells and monocytes form in patients subjected to morphine treatment but that they were severely depressed in their phagocytic and killing properties as well as in their ability to generate superoxide. OPIOID addiction also caused alterations in the frequencies of T-cells and null lymphocytes in human peripheral blood.

There is convincing evidence of the presence of opioid receptors on various types of human immune cells. The presence of opioid receptors on immune cells may allow for modulation of specific immune functions in the presence of exogenous opiates. Various administration schedules for opioids were shown to potentiate infections by *Klebsiella pneumoniae* and *Candida albicans*. The increased susceptibility was partly due to a decrease in reticuloendothelial-system activity as well as a reduction in the number of phagocytes, not by a direct cytotoxic effect of the opioid.

Chronic administration of morphine has also inhibited a primary antibody response of mice. These effects were antagonized by naloxone (a nonaddictive analog of morphine), indicating that morphine inhibits the immune system in a specific manner—via its interaction with opioid receptors. Other studies in animals have shown that morphine can affect NK cell activity, perhaps yielding reduced resistance to tumors.

Such changes, which can also include morphine suppression of spleen and body weight, show evidence of significant changes in immune functions.

MARIJUANA

Several approaches have been used to study the effects of MARIJUANA or its active component, TETRAHYDROCANNABINOL (THC), on human immune systems. These include using cells isolated from chronic marijuana smokers, using cells from volunteers who have been only exposed to marijuana smoke, or using cells from nonexposed donors but exposing their cells to THC in the laboratory. A survey of chronic marijuana smokers showed that the response of their cells was depressed to stimulation with mitogens (substances that cause cell division).

Several studies have shown that neither marijuana smoking nor THC is immunosuppressive. Nevertheless, other immune alterations have been associated with marijuana or THC, including significantly reduced serum IgG levels in chronic smokers; inhibition of natural killer cell activity; inhibition of phagocytic activity; elevation of serum IgD levels; and reduced T-cell numbers. THC also inhibited DNA-, RNA-, and protein synthesis in stimulated human lymphocytes.

Studies performed in animals have produced more consistent findings than those in humans. In most cases, THC is associated with immunosuppression of various immune parameters. The greater consistency observed in animal studies probably reflects the influence of genetic factors, of consistent dosage levels, and of controlled diets and other conditions.

Animal studies have thus provided strong evidence of the immunosuppressive effects of THC. Such effects were clearly demonstrated when animals exposed to THC were more susceptible to infections than were others. THC has also exacerbated viral infection, as has been shown in mice and guinea pigs, and it has reduced resistance to bacterial pathogens.

Obvious differences in the susceptibility of humans versus animals to the effects of marijuana and THC will be resolved when other, more regulated research studies are carried out in immunosuppression and decreased disease resistance.

BIBLIOGRAPHY

MacGregor, R. R. (1986). Alcohol and immune defense. *Journal of the American Medical Association, 256,* 1474–1479.

Wallace, C. L., & Watson, R. R. (1990). Immunomodulation by tobacco. In *Drugs of Abuse and Immune Function.* Boca Raton, FL: CRC Press.

Watzl, B., & Watson, R. R. (1990). Immunomodulation by cocaine—A neuroendocrine mediated response. *Life Sciences, 46,* 1319–1329.

Yahya, M. D., & Watson, R. R. (1987). Minireview: Immunomodulation by morphine and marijuana. *Life Sciences, 41,* 2503–2510.

RONALD R. WATSON

Liver (Alcohol) For all the attention being directed toward HEROIN and COCAINE, the favorite mood-altering drug in most human societies is ALCOHOL. Alcohol, in different quantities for different people, is also a toxic drug—its overconsumption taxes the body's economy, produces pathological changes in liver and other tissues, and can cause disease and death. In urban areas of the United States, just one of the complications—namely scarring or cirrhosis of the liver—is the fourth to fifth most frequent cause of death for people between ages of twenty-five and sixty-five. In recent years, changes in liver and other tissues have been directly associated with specific steps in the metabolism of alcohol (also called ethanol or ethyl alcohol), giving some hope that rational methods can be developed for prevention and treatment.

PATHOLOGY OF ALCOHOL ABUSE

Alcohol abuse affects all organs of the body (Lieber, 1992a). It atrophies many tissues, including the brain and the endocrine glands. Indeed, altered hepatic (liver) metabolism plays a key role in a variety of endocrinological imbalances (such as gonadal dysfunctions and reproductive problems). Alcohol also exerts toxic effects on the bone marrow and alters hematological status (e.g., macrocytic anemias), and it scars the heart and other muscles. This article focuses mainly on the liver and gastrointestinal tract, since this is where alcohol penetrates into the body and has its most vicious effects; this focus will also allow exemplification of the insights and possible benefits that can be derived from the application of newly acquired knowledge in biochemistry, pathology, and molecular biology.

Liver disease, one of the most devastating complications of alcoholism, was formerly attributed exclusively to the malnutrition associated with ALCOHOLISM. Indeed, nutritional deficiencies are common in the alcoholic for various reasons, some socioeconomic, but also because alcohol is a unique compound. Alcohol is a drug, a psychoactive drug, but unlike other drugs, which have negligible energy value, alcohol has a high energy content—each gram of alcohol contributes 7.1 kilocalories, which means that a cocktail or a glass of wine will provide 100 to 150 kilocalories. Thus, alcoholic beverages are similar to food in energy terms, but, unlike food, they are virtually devoid of vitamins, proteins, and other nutrients; they act as a provider of empty calories.

As shown in Figure 1, because of its large energy load, alcohol decreases the appetite for food and displaces other nutrients in the diet, thereby promoting primary malnutrition (Lieber, 1991a). Nutrition is also impaired because alcohol affects the gastrointestinal tract. Alcohol-induced intestinal lesions, including pancreatitis, are associated with maldigestion and malabsorption, causing secondary malnutrition. Moreover, malnutrition itself will create functional impairment of the gut. Finally, alcohol (ethanol or its metabolite acetaldehyde) also ad-

versely affects nutritional status by altering the hepatic activation or degradation of essential nutrients.

Indeed, in experimental animals, malnutrition may produce a variety of liver alterations, including fatty liver and fibrosis; however, the extent to which malnutrition contributes to the development of liver disease in the alcoholic remains unclear. Furthermore, studies conducted in the past three decades have shown that either the initial liver lesion—the fatty liver—or the ultimate stage of cirrhosis can be produced by excess alcohol, even in the absence of dietary deficiencies (Lieber & DeCarli, 1991), because ethanol (via its metabolism and/or its metabolite acetaldehyde) exerts direct hepatotoxic effects. Thus, malnutrition plays a permissive, but not an obligatory, role in alcohol-related somatic pathology.

METABOLISM OF ETHANOL AND SOME INTERACTIONS

Ethanol is readily absorbed from the gastrointestinal tract. Only 2 percent to 10 percent of that absorbed is eliminated through the kidneys and lungs; the rest is oxidized in the body, principally in the liver. Except for the stomach, extrahepatic (outside the liver) metabolism of ethanol is small. This relative organ specificity probably explains why, despite the existence of intracellular mechanisms to maintain homeostasis (equilibrium), ethanol disposal produces striking metabolic imbalances in the liver (Lieber, 1991b). These effects are aggravated by the lack of a feedback mechanism to adjust the rate of ethanol oxidation to the metabolic state of the hepatocyte (liver cell) and the inability of ethanol, unlike other major sources of calories, to be stored in the liver or to be metabolized or stored to a significant degree in peripheral tissues. As summarized here, the displacement by ethanol of the liver's normal substrates and the metabolic disturbance produced by the oxidation of ethanol and its products explain many of the hepatic and metabolic complications of alcoholism.

A major pathway for ethanol disposition involves alcohol dehydrogenase (ADH), an enzyme of the cell sap (cytosol) that catalyzes the conversion of ethanol to acetaldehyde. Liver ADH exists in multiple molecular forms that arise from the association in various permutations of different types of subunits. Extrahepatic tissues contain isozymes of ADH with a much lower affinity for ethanol than the hepatic

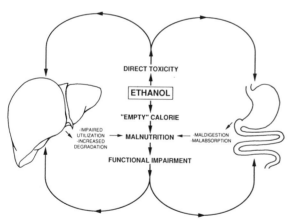

Figure 1
Interaction of Direct Toxicity of Ethanol on Liver and Gut with Malnutrition Secondary to Dietary Deficiencies, Maldigestion and Malabsorption.
SOURCE: *Lieber, C. S. (1991a). Alcohol, liver, and nutrition.* Journal of the American College of Nutrition, 10, *602–632.*

isozymes; as a consequence, at the levels of ethanol achieved in the blood, these extrahepatic enzymes are inactive, and therefore, extrahepatic metabolism of ethanol is negligible, with the exception of gastric metabolism. Because of the extraordinary high gastric ethanol concentration after alcohol consumption, even the gastric ADH with low affinity for ethanol becomes active, and significant gastric ethanol metabolism ensues. This decreases the bioavailability of ethanol and represents a kind of protective barrier against systemic effects, at least when ethanol is consumed in small social-drinking amounts. This gastric barrier disappears after gastrectomy (Caballeria et al., 1989) and may be lost, in part, in the alcoholic, because of a decrease in gastric ADH (Di Padova et al., 1987).

Similar effects may also result from gastric ADH inhibition by some commonly used drugs. For example, aspirin, or some H_2-blockers such as those used in treatment of ulcers (Di Padova et al., 1992) were found to inhibit gastric ADH activity and to result in increased blood levels of ethanol when alcohol was consumed in amounts equivalent to social drinking. Women also have a lower gastric ADH activity than do men (Frezza et al., 1990); as a consequence, for a given intake, women's blood ethanol levels are higher, an increase that is compounded by their body composition (more fat, less water than men) and, on average, a lower body weight. Their higher blood ethanol levels, in turn, may therefore contribute to women's greater susceptibility to alcohol.

Alcohol dehydrogenase converts ethanol to acetaldehyde and hydrogen. Hydrogen is a form of fuel that can be burned (oxidized). Normally, the liver burns fat to produce the energy required for its own functioning but, when alcohol is present, its hydrogen displaces fat as the preferred fuel. When the liver stops burning fat and instead burns the hydrogen from the ethanol, however, fat accumulates, and a fatty liver develops, which is the first stage of alcoholic liver disease (Lieber, 1992a). Once a fatty liver has developed, fat accumulation does not increase indefinitely, even though alcohol consumption may be continued (Salaspuro et al., 1981). Fat deposition is offset, at least in part, by lipoprotein secretion, resulting in hyperlipemia—elevated amounts of fat in blood. Hyperlipemia of a moderate degree is commonly associated with early stages of alcoholic liver injury, but it wanes with the progression of liver disease (Lieber & Pignon, 1989). In some individuals,

marked hyperlipemia may develop, sometimes associated with Zieve's syndrome—hemolytic anemia, fatty liver, and jaundice. This represents the potentiation, by alcohol, of an underlying abnormality in the metabolism of either lipids (essential hyperlipemia) or carbohydrates (prediabetes, pancreatitis). In addition, the degree of hyperlipemia is also influenced by the duration of alcohol intake. The capacity for a hyperlipemic response develops progressively and is accompanied by an increased activity of enzymes of the endoplasmic reticulum (within the living cells) engaged in lipoprotein production. This hyperlipemia involves all lipoprotein classes, including high-density lipoproteins (HDL), which have been said to be involved in the protection against atherosclerosis and in the lesser incidence of coronary complications in moderate drinkers (compared to total abstainers). However, factors other than alcohol may also contribute to this apparent protection. The ability of the liver to respond with hyperlipemia reflects the integrity of the hepatocytes; its capacity decreases with the development of more severe liver injury.

Elucidation of the hepatic redox (contraction of oxidation reduction) associated with ethanol oxidation via the alcohol dehydrogenase pathway has also furthered our understanding of associated disorders in carbohydrate, purine, and protein metabolism—including hypoglycemia (low blood sugar), hyperlactacidemia (excessive levels of lactic acid in the blood) and acidosis, as well as hyperuricemia (elevated uric acid levels in blood) (Lieber, 1992a).

In addition to the enzyme ADH, alcohol is also oxidized in the liver by the enzyme system referred to as the microsomal ethanol-oxidizing system (MEOS), which involves a specific cytochrome P-450 (P450IIE1) (Lieber, 1987). Contrary to ADH, this pathway is inducible by chronic alcohol consumption. In rat livers (Lieber et al., 1988) and in human liver biopsies of heavy drinkers (Tsutsumi et al., 1989), a five to tenfold increase of this alcohol-inducible form was found. This induction represents one of the most striking biochemical differences between heavy drinkers and normals and provides an explanation for the metabolic tolerance to ethanol—a more rapid metabolism—that develops after alcohol abuse. The induction spills over to microsomal systems that metabolize other substrates, resulting in cross-tolerance to other drugs—not only sedatives and tranquilizers but also many commonly used medications such as anticoagulants and hypo-

glycemic agents. Thus, heavy drinkers require an increased dosage of many commonly used medications, at least at the initial stage, prior to the development of severe liver disease, which, when it develops, it will offset the enzyme induction, at which time drug dosage may have to be decreased. What complicates treatment of heavy drinkers even further is the fact that the microsomal enzymes (especially P450IIE1) also activate many xenobiotic agents (substances from outside the body) to highly toxic compounds. This explains the increased vulnerability of heavy drinkers to the hepatotoxicity of industrial solvents, anesthetics, analgesics (painkillers), and chemical carcinogens. The latter contribute to the increased incidence of various cancers in the alcoholic.

Alcohol has a major impact on gastrointestinal cancers, with a significant increase in the incidence of neoplasms of the oropharynx, the esophagus, the stomach, the liver, and the colon (Garro & Lieber, 1990). There is also activation to toxic metabolites of commonly used drugs and even over-the-counter analgesics (acetaminophen or paracetomal) (Sato et al., 1981) and vitamins, such as Vitamin A. In the heavy drinker, there are both increased breakdown and hepatic depletion of Vitamin A (Leo & Lieber, 1982), with adverse consequences. In addition, alcohol potentiates the toxicity of Vitamin A (Leo et al., 1982), which complicates supplementation with the vitamin in the presence of alcohol abuse. Alcohol abuse also promotes the microsomal breakdown of testosterone and its conversion to estrogens, which, together with testicular toxicity and decreased testosterone production, results in hypoandrogenism—the loss of masculinity (Lieber, 1992a).

In addition to environmental factors, there are individual differences in rates of ethanol metabolism that appear to be genetically controlled, and the possible role of heredity in the development of alcoholism in humans has been emphasized. The induction of the MEOS pathway also leads to increased conversion of alcohol to acetaldehyde, a highly reactive and thus potentially toxic compound.

TOXICITY OF ACETALDEHYDE

Acetaldehyde causes injury through the formation of adducts with proteins, resulting in antibody formation, inactivation of many key enzymes, decreased deoxyribonucleic acid (DNA) repair, and al-

terations in cell structures such as microtubules, mitochondria, and plasma membranes (Lieber, 1988, 1992a). Acetaldehyde also promotes synthesis of hepatic collagen—the key protein of scar tissue; furthermore, it causes glutathione depletion, thereby exacerbating the toxicity mediated by free radicals, which results in lipid peroxidation and other tissue damage (Lieber, 1991b). Because of the far-reaching toxicity of this metabolite of ethanol, some of the liver cells die; this attracts inflammatory cells, which results in the more severe stage of alcoholic hepatitis, one of the precursors to the ultimate scarring or cirrhosis.

Once there is cirrhosis, a number of complications ensue, including obstruction of blood flow—with portal hypertension (elevated pressure in the veins leading from the intestine to the liver) and internal, life-threatening bleeding of distended veins, so-called varices. There is also a buildup of water in the abdominal cavity, so-called ascites (Lieber, 1992a).

Acetaldehyde is particularly elevated if drinking occurs in pregnancy; it crosses the placenta (Karl et al., 1988) and has been incriminated in the pathogenesis of the FETAL ALCOHOL SYNDROME (FAS), the most common preventable cause of cogenital abnormalities.

The bulk of acetaldehyde is oxidized to acetate by an acetaldehyde dehydrogenase of the liver mitochondria. Lack of the active form of the enzyme in some Asians explains their high blood acetaldehyde and flushing reaction after alcohol. DISULFIRAM (Antabuse—a drug used in recovering alcoholics) is an inhibitor of acetaldehyde dehydrogenase. It raises the acetaldehyde levels after drinking and thereby causes flushing and several adverse effects that can be utilized to sustain abstinence in patients motivated to take the compound.

TREATMENT AND CONCLUSION

Alcoholics suffer commonly from malnutrition. Therefore, nutritional deficiencies, when present, should be corrected—although such efforts were found to be ineffective in fully preventing liver disease in view of the intrinsic toxicity of ethanol (Lieber, 1991b; Lieber & DeCarli, 1991).

Although progress is being made at offsetting the direct toxicity of ethanol through chemical means (Lieber, 1992b), at present, the single fully effective

way of preventing somatic alcoholic injury remains control of the toxic agent—ethanol—through control of consumption. Full abstinence is required in those who are genetically (or otherwise) prone to develop craving or to exhibit dependence, or those who are predisposed to develop the major somatic complication with chronic use of alcohol.

For the others, moderation is recommended. What is considered "moderate" or "excessive" has been the subject of debate. One view is that on the average, moderate drinking represents no more than one drink a day in women and no more than two drinks a day in men—with a drink being 12 ounces of regular beer, 5 ounces of wine, or 1.5 ounces of distilled spirits (80 proof) (Dietary Guidelines, 1990). It is important, however, that "excess" be defined individually, taking into account not only gender, but also heredity and personal idiosyncrasies.

(SEE ALSO: *Addiction: Concepts and Definitions; Alcohol: Complications; Cancer, Drugs, and Alcohol; Complications: Liver Damage; Social Costs of Alcohol and Drug Abuse*)

BIBLIOGRAPHY

CABALLERIA, J. ET AL. (1989). The gastric origin of the first pass metabolism of ethanol in man: Effect of gastrectomy. *Gastroenterology, 97,* 1205–1209. Dietary Guidelines. (1990). *Nutrition and your health: Dietary guidelines for Americans* (3rd ed.). Washington, DC: U.S. Department of Agriculture.

DI PADOVA, C. ET AL. (1987). Effects of fasting and chronic alcohol consumption on the first pass metabolism of ethanol. *Gastroenterology, 92,* 1169–1173.

DI PADOVA, C. ET AL. (1992). Effects of ranitidine on blood alcohol levels after ethanol ingestion: Comparison with other H_2-receptor antagonists. *Journal of the American Medical Association, 267,* 83–86.

FREZZA, M. ET AL. (1990). High blood alcohol levels in women: Role of decreased gastric alcohol dehydrogenase activity and first pass metabolism. *New England Journal of Medicine, 322,* 95–99.

GARRO, A. J., & LIEBER C. S. (1990). Alcohol and cancer. *Annual Review of Pharmacology and Toxicology, 30,* 219–249.

KARL, P. I. ET AL. (1988). Acetaldehyde production and transfer by the perfused human placental cotyledon. *Science, 242,* 273–275.

LEO, M. A., & LIEBER, C. S. (1982). Hepatic vitamin A depletion in alcoholic liver injury. *New England Journal of Medicine, 307,* 597–601.

LEO, M. A., ARAI, M., SATO, M., & LIEBER, C. S. (1982). Hepatotoxicity of vitamin A and ethanol in the rat. *Gastroenterology, 82,* 194–205.

LIEBER, C. S. (1987). Microsomal ethanol oxidizing system (MEOS), *Enzyme, 37,* 45–56.

LIEBER, C. S. (1988). Metabolic effects of acetaldehyde, *Biochemical Society Transactions, 16,* 241–247.

LIEBER, C. S. (1991a). Alcohol, liver and nutrition. *Journal of the American College of Nutrition, 10,* 602–632.

LIEBER, C. S. (1991b). Hepatic, metabolic and toxic effects of ethanol: 1991 update. *Alcoholism: Clinical and Experimental Research, 15,* 573–592.

LIEBER, C. S. (ED.). (1992a). *Medical and nutritional complications of alcoholism: Mechanisms and management.* New York: Plenum Press.

LIEBER, C. S. (1992b). Hepatotoxicity of alcohol and implications for the therapy of alcoholic liver disease. In J. Rodes & V. Arroyo (Eds.), *Therapy in liver disease* (pp. 348–367). Barcelona: Ediciones Doyma.

LIEBER, C. S., & DECARLI, L. M. (1991). Hepatotoxicity of ethanol. *Journal of Hepatology, 12,* 394–401.

LIEBER, C. S., & PIGNON, J-P. (1989). Ethanol and lipids. In J. C. Fruchart & J. Shepherd (Eds.), *Human plasma lipoproteins: Chemistry, physiology and pathology* (pp. 245–280). New York: Walter De Gruyter.

LIEBER, C. S. ET AL. (1988). Role of acetone, dietary fat and total energy intake in induction of hepatic microsomal ethanol oxidizing system. *Journal of Pharmacology and Experiment Therapeutics, 247,* 791–795.

SALASPURO M. P. ET AL. (1981). Attenuation of the ethanol induced hepatic redox change after chronic alcohol consumption in baboons: Metabolic consequences in vivo and in vitro. *Hepatology, 1,* 33–38.

SATO, C., MATSUDA, Y., & LIEBER, C. S. (1981). Increased hepatotoxicity of acetaminophen after chronic ethanol consumption in the rat. *Gastroenterology, 80,* 140–148.

TSUTSUMI, R. ET AL. (1989). The intralobular distribution of ethanol-inducible P450IIE1 in rat and human liver. *Hepatology, 10,* 437–446.

CHARLES S. LIEBER

Liver Damage The liver is the largest organ of the human body, normally weighing about 3.3

pounds (1.5 kg). It occupies the right upper quadrant of the abdominal cavity just below the diaphragm.

As befitting its anatomical prominence, its function is essential to maintain life. If we surgically removed the entire liver from any animal (including humans), it will fall into a coma shortly and die. The absence of a certain critical mass of functioning liver tissue is incompatible with life. While the human liver has a remarkable resilience and regenerative capacity after injury or illness, this is true only up to a certain point. If illness pushes the liver beyond the "point of no return," the person dies.

The liver has a multitude of complex functions and is justly called the "laboratory" of the human body. It secretes a digestive juice into the intestine, called bile; it produces a number of essential proteins, clotting factors, and fatty substances; it stores and conserves energy-producing sugars; it detoxifies both internally produced and external toxins and drugs that would otherwise be poisonous to the human organism—just to name some of its important functions.

What can seriously jeopardize this very important organ and consequently the well-being and survival of the individual? For one, there are diseases—both congenital and acquired—over which we have little or no control, such as some genetically determined and developmental abnormalities, circulatory liver problems, certain tumors, and infections.

A very large part of hepatology (the technical term to describe the study and treatment of liver diseases) is, however, devoted to liver problems created by a peculiar human behavior—the abuse of ALCOHOL and drugs. Whereas discussions as to whether ALCOHOLISM and DRUG ABUSE are truly self-inflicted problems elicit a variety of opinions, the liver disease that results from substance abuse in a given individual could have been avoided if the substance-abusing behavior had not occurred. Beyond the psychosocial consequences of substance abuse, diseases of the liver (and brain) represent the major COMPLICATIONS of alcohol and drugs.

The morbidity (disease incidence) and mortality (death incidence) from alcoholic and drug-induced liver injury are very high. In the scientific literature, it is well established that the mortality from alcoholic liver disease is correlated with the per capita alcohol consumption; in fact, the prevalence of alcoholism in a given society has been calculated from liver mortality statistics. While alcohol is a direct liver toxin, most of the other commonly abused psychoactive substances are generally not known to affect the liver directly to a great extent; their major contribution to liver morbidity and mortality is via exposing people to viral hepatitis—a potentially fatal disease.

ALCOHOLIC LIVER DISEASES

Another article in this encyclopedia discusses the relationship between alcohol and the liver in great detail. In any article dealing with the effects of drugs on the liver, however, alcohol must be addressed.

The gamut of alcoholic liver diseases and their interrelationship is illustrated in Figure 1.

Alcoholic Fatty Liver. Fat accumulation in the liver is an almost universal response to excessive alcohol consumption. It occurs in the majority of heavy drinkers. How and why fat accumulates in liver cells is complicated and not completely understood; but we know for sure that it happens. If you examine a piece of biopsied liver tissue from an alcoholic under the microscope, you see that many liver cells are loaded with big bubbles consisting of fat, almost totally occupying the cell. In most cases, this fatty change does not matter too much as far as the patient's health is concerned. It is an almost invariable response to too much alcohol consumption and an early warning. The person who has nothing worse than an alcoholic fatty liver may not feel sick at all, and only if a biopsy is done can the fatty liver be diagnosed. The doctor may feel an enlarged liver by palpation, which may be a bit tender. The laboratory test may show a slight elevation in the blood of some liver enzymes, best known by their initials: SGOT (or AST) and SGPT (or ALT). These enzymes are elevated because some of them tend to leak out of the fatty liver cells into the blood.

If a person stops drinking, the fat disappears from the liver cells, the swelling subsides and the AST and ALT levels become normal. The two-way arrow in the diagram of Figure 1 indicates that fatty liver is reversible with abstinence, and the condition may fluctuate back and forth between normal and fatty liver with abstinence and drinking, respectively. Thus, this, per se is not likely a serious situation; it is an early warning that "your liver does not like alcohol" and that possibly worse things might yet come. There was a time when fatty liver was regarded as the precursor of the end-stage liver disease called cirrhosis (indicated by the broken arrow and

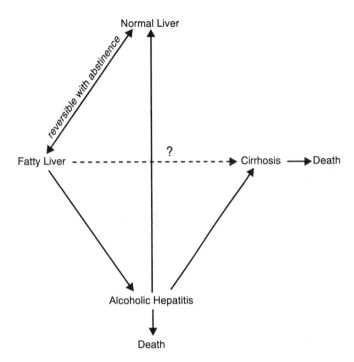

Figure 1

Interrelationships between Various Forms of Alcoholic Liver Disease

question mark on Figure 1), but in the 1990s, most physicians do not believe that this direct connection exists.

Alcoholic Hepatitis. This is a potentially more serious form of alcoholic liver disease. A certain proportion of alcoholics, in addition to accumulating fats in their livers when drinking, will develop inflammation (hepatitis means liver inflammation)—consisting of an accumulation of white blood cells, the death (necrosis) of some of the liver cells, and the presence of some very characteristic material (called Mallory bodies). Again, all this can be seen under the microscope in a biopsied piece of tissue.

The clinical picture of alcoholic hepatitis can be very variable. At one extreme is the person who feels perfectly well and only the biopsy could tell that something is wrong. At the other extreme is the patient with a swollen and painful liver, yellow jaundice (a yellowing of the entire body from bile pigment leaking into the blood), fever, and disturbed consciousness—who dies. Between these extremes are people with varying degrees of seriousness of the illness; for example, with or without some jaundice, with or without pain and fever, etc. The blood's white cell count is usually elevated. The bilirubin (bile pigment) level may be elevated in patients who are yellow (a pale to deep mustard). The liver enzymes are

higher than normal in the blood, because they leak out of the inflamed liver cells. However, these values are not as high as in viral hepatitis and, characteristically, in alcoholic hepatitis AST (SGOT) is higher than ALT (SGPT), which helps to distinguish alcoholic hepatitis from viral hepatitis (difficult to do at times). In viral hepatitis not only are the absolute enzyme values higher, but the ratio is reversed: ALT is higher than AST.

Thus, the outcome of alcoholic hepatitis can be death (worst scenario) or recovery (best scenario)—as shown on Figure 1. Repeated episodes of drinking and alcoholic hepatitis, however, even if the patient does not die in a given episode, can lead to the end-stage of alcoholic liver disease: cirrhosis.

Alcoholic Cirrhosis. In terms of histology (tissue damage) this indeed is an end-stage disease: a cirrhotic liver *cannot* become normal; in Figure 1, there is no arrow between cirrhosis and normal liver. Clinically, cirrhosis is a serious disease, potentially fatal, but not inevitably so. Alcoholism is not its only cause, but it is by far the most common.

Under the microscope a cirrhotic liver shows a disorganized architecture: the dead (necrotic) liver cells have been replaced by scar tissue. The liver tries to repair itself: In a somewhat haphazard fashion it attempts to produce new liver tissue in the

form of nodules, which are separated from each other by scar. These newly formed liver nodules may indeed sustain liver function and thus life for a time, but at a price: the liver's blood circulation is mechanically compressed. Thus, the pressure increases in the blood vessels leading to the liver. Some of these overloaded blood vessels, especially those on the border of the stomach and esophagus (called esophageal varices), can rupture any time, causing a major hemorrhage.

Those who develop the cirrhotic stage of alcoholic liver disease present their symptoms in various ways. Some of them look quite normal and only the biopsy will reveal the presence of cirrhosis. Others are jaundiced, the yellow color coming from bile pigment leaking out of the damaged liver into the blood, thus staining the skin and the whites of the eyes. Still others have large fluid accumulations in their extremities (edema) or in their abdominal cavity (ascites); the latter may make these patients—men or women—look like they are nine months pregnant. Some may vomit blood, because of the hemorrhaging. In most advanced cases, there is just not enough functioning liver tissue left; the liver no longer can perform its "laboratory" function, and the person slips into a coma and may die.

When cirrhotic patients are examined by doctors, their livers do not feel smooth on palpation, but bumpy from all those nodules that formed. At first the liver may be swollen and enlarged, but at the later stages it shrinks. The ultrasound picture suggests a patchy, disorganized architecture of the liver. The spleen may enlarge from the increased pressure in the blood vessels. The liver enzymes (AST and ALT) may be moderately elevated as in other forms of alcoholic liver disease, but this has no prognostic importance. More ominous signs pointing toward severely compromised liver functions are the following: a decrease in blood level of albumin (an important protein manufactured by the liver), deficiency in blood-clotting factors that are also made in the liver, and the presence of anaemia (low hemoglobin and red blood cell count).

What Can Kill a Cirrhotic Patient? Ascites (fluid accumulation in the abdomen) is very uncomfortable and unsightly but, by itself, usually does not kill—unless it gets spontaneously infected, which is always a threat. Generally, cirrhosis compromises the immune system, rendering cirrhotic alcoholics susceptible to all sorts of potentially overwhelming infections. Bleeding from the ruptured esophageal varices can cause death. Finally, total decompensation of liver-cell function may cause coma and death.

The good news is that even when there is irreversible cirrhosis at the tissue level, death may not be inevitable. Survival depends mainly on two factors: luck and alcohol abstinence. Abstaining alcoholics with cirrhosis can stabilize and survive on what's left of their liver tissue without necessarily and relentlessly progressing to one of the fatal outcomes. A famous Yale University study many years ago showed clearly the correlation between abstinence and survival in cirrhosis.

Who Gets Which Alcoholic Liver Disease? There are still no certain answers to this question. Fatty liver is an almost universally predictable response to heavy alcohol consumption, but this by itself is seldom a serious problem. A smaller number of people develop alcoholic hepatitis and still fewer (variously estimated in different populations between 5 to 25% of alcoholics) end up with cirrhosis. Considering the large number of alcoholics in our society, the minority who develop cirrhosis still represents huge numbers; cirrhosis *is* one of the leading causes of all deaths.

Still, why do some alcoholics develop alcoholic hepatitis and cirrhosis, while others who drink equally heavily do not? The amount of alcohol consumption and the length of time of heavy drinking is certainly one risk factor. Gender may be another: Women's livers generally are more vulnerable to the effects of alcohol than those of men, given equal alcohol exposures. Finally, there may be a genetically determined (but still unclarified) individual susceptibility, which may explain why some people never get cirrhosis, why some do after many years of alcoholism, and why still others get cirrhosis at a young age or after a relatively short drinking career.

Prognosis and Treatment. The issues of prognosis and treatment cannot be separated from each other. The cornerstone of treatment is complete abstinence from alcohol; achieving abstinence can arrest the progression of liver disease, even in established cirrhosis. Continued drinking leads to deterioration and death.

One therapeutic issue relating to alcoholism itself, should be addressed here because it is relevant to liver disease. The drug DISULFIRAM (Antabuse) is sometimes prescribed to reinforce abstinence: its unpleasant, sometimes severe interaction with alcohol is used as a deterrent against drinking. Since disulfiram (as so many other drugs) has been occasionally

reported to produce liver toxicity of its own, the presence of alcoholic liver disease is sometimes regarded as a relative contraindication against the prescription of disulfiram. The liver toxicity caused by alcohol far outweighs any risk that may be caused by disulfiram.

Are there any other treatment techniques available beyond abstinence that can help the recovery from alcoholic liver damage? In the late 1980s, a Toronto research group reported the beneficial effect of propylthiouracil (PTU). This is a drug normally used for the treatment of thyroid disease, but by reducing oxygen demand in the body (including in the liver), it might help to repair the damage caused by alcohol. The early results were promising but it is still not a widely accepted treatment. Other drugs, such as corticosteroids (to decrease inflammation) or colchicine (to decrease scar formation) have dubious value.

There are relatively effective treatments available for some of the complications of alcoholic liver disease so that the patient may survive and thus begin his or her abstinence program. The fluid accumulation in the extremities (edema) or in the abdomen (ascites) can be helped by diet modifications (salt restriction), water removing drugs (diuretics), albumin infusion, or tapping the abdomen. Infections can be treated with antibiotics. The brain syndrome of liver failure (so-called hepatic encephalopathy or, in severe cases, hepatic coma) can improve with dietary means (protein restriction) or some drugs (e.g., neomycin, lactulose). The potentially or actually bleeding esophageal varicose veins can be obliterated by certain injections through a gastroscope (so-called sclerotherapy), and the bleeding risk can be lessened by beta-blocking drugs or some surgical procedures to decrease pressure.

Finally, in the 1990s we have the possibility of liver transplantation. If all else fails, a successful liver transplant cures alcoholic liver disease. Apart from the general problems of donor matching and supply, some people have raised objections on ethical grounds to offering transplantation for alcoholic (i.e., "self-inflicted") liver disease. This is not an acceptable objection and goes against medical ethics. Well-motivated recovering alcoholics are entitled, as much as anybody else, to a life-saving procedure. In fact, studies have shown that the very dramatic and heroic nature of this operation may be an extremely powerful motivator for future abstinence by liver recipients. Numerous successful transplants have been carried out on alcoholics.

DRUGS AND THE LIVER

There are many drugs in medicinal use that can have direct liver toxicity. Peculiarly, most of the psychoactive drugs that people tend to abuse are not known to be particularly harmful to the liver. Occasional liver damage has been reported with SOLVENT sniffing and COCAINE use, but this is not a common problem. Narcotics (opioids), anti-anxiety, and other sedative drugs (such as BARBITURATES), MARIJUANA, and HALLUCINOGENS do not usually cause liver injury.

There are, however, several relevant secondary issues concerning drug abuse and the liver. For one, a damaged liver (for example from alcohol or hepatitis) results in poor tolerance of SEDATIVES, because good liver function is necessary to eliminate sedatives properly; impaired liver function can therefore result in exaggerated sedative effect. Conversely, some sedatives, notably BARBITURATES (which were often abused in the past and sometimes still are), actually stimulate ("induce") certain liver enzymes, which can result in increased elimination (i.e., decreased effect) of another therapeutically necessary drug. For example, a barbiturate user (or abuser) may have poor effect from a clotting preventative (anti-coagulant) drug that is necessary in heart disease or after a stroke. Some drugs do the opposite—they inhibit liver enzymes. For example, the anti-ulcer drug cimetidine (Tagamet), which per se has no PSYCHOACTIVE effect, can cause such enzyme inhibition; if a person at the same time also happens to use or abuse a sedative, the sedative can have an exaggerated effect.

Generally speaking, the normal liver transforms or inactivates drugs (detoxification) to less active or harmless forms. A notable and important exception is acetaminophen, one of the most commonly used medications against PAIN and fever (e.g., the various Tylenol preparations). The liver can transform acetaminophen into a toxic metabolic derivative that might cause a potentially lethal, acute liver injury. Generally, this does not happen at ordinary therapeutic acetaminophen dose levels. In the case of an acetaminophen overdose, however, such severe liver toxicity can occur that a person will die within days. Most of such overdoses are, of course, suicidal attempts.

Acetaminophen itself does not have any psychoactive (mind-altering) properties; thus people do not abuse it for such reasons. Many marketed acet-

aminophen preparations, however, are combined with CODEINE (a NARCOTIC). People seeking narcotic "highs" from such preparations might ingest them in large enough quantities to subject themselves to potentially severe liver injury. It is the codeine they are after, but it is the bystander acetaminophen that may kill them. There is an antidote against acetaminophen poisoning (called acetyl-cysteine), but it is effective only if it is given within a few hours (less than a day) after the ingestion of the drug. The person who is overdosing with a suicidal intent is more likely to be discovered and brought to quick medical attention than an unintentionally overdosing drug abuser. An additional issue in the acetaminophen story is that there is strong evidence of increased risk when alcohol and acetaminophen are combined. In alcoholics, relatively low, even therapeutic, doses of acetaminophen can cause severe and potentially fatal liver damage.

Apart from acetaminophen, direct liver toxicity is not a major feature of substance abuse except in the case of alcohol. The liver disease that is *very* commonly associated with drug use is viral hepatitis (liver inflammation) which is not caused by the drugs themselves but by infection with a virus. It is then transmitted from person to person through contaminated needles and syringes. The problem of viral hepatitis, then, is largely that of injecting drug users (IDUs).

VIRAL HEPATITIS IN DRUG ABUSERS

In the mid-1990s, at least five types of disease-causing hepatitis viruses have been identified, and they are designated by the letters of the alphabet,

A–E. Table I summarizes some of their important characteristics. Of the five, hepatitis A and E are not particularly associated with injecting drug abuse; but the other three very much are and they will be discussed in some detail in that context.

Hepatitis B. This virus (which used to be called "serum hepatitis") is endemic to some parts of the world, such as Southeast Asia, where as much as 10 percent of the population may be infected. In the Western world, IDUs represent the greatest reservoir for hepatitis B virus. It is transmitted through a direct blood-borne route, such as (1) contaminated needles and syringes (which drug addicts notoriously did not sterilize in the past); (2) from an infected mother across the placenta and through the umbilical cord of a developing FETUS; (3) from blood contaminating accidental needle-stick injuries in health-care workers; and (4) from any blood to blood contact occurring during sexual intercourse. At one time blood transfusions were a common source of infection, but since the 1970s we have had a reliable test to screen out infected donors.

The symptoms of hepatitis B infection vary. In its severest form, it can cause general unwellness, fever, jaundice, coma, and death. The majority of patients, even with marked jaundice and fever, do not die. Many infected people do not even have an overt illness; they may not feel sick at all or may just have transient "flu-like" symptoms. There may be a tender enlargement of the liver. If such people are tested in the laboratory, they have elevated enzymes, such as AST (also known as SGOT) and ALT (also known as SGPT), which are usually much higher than the values found in alcoholic liver disease (in contrast to alcohol, viral hepatitis tends to cause

TABLE 1
Viral Hepatitis: Major Features of Different Types

Hepatitis Virus	Transmission	Course	Vaccine
A	mouth stool	acute	yes
B	blood	acute chronic	yes
C	blood	acute chronic	no
D	blood	acute chronic	hepatitis B only
E	mouth stool	acute	no

more elevation in ALT than in AST). The bilirubin (bile pigment) level will be high if the person has yellow jaundice.

There are now quite good serological tests for hepatitis B. A virus particle (hepatitis B's antigen) can be identified in infected people. Those who recover from the illness and clear the virus out of their body will develop a protective antibody that will prevent their reinfection. The antibody can be detected in a laboratory test.

The majority of people who get infected with hepatitis B do recover and acquire protective antibodies. A sizable minority of those who survive, however, perhaps 10 percent, will continue to carry the virus (remain "antigen positive"); and some of these will have a chronic liver inflammation that can end up in cirrhosis. The cirrhosis caused by hepatitis B is essentially similar to alcoholic cirrhosis, with the same consequences and potential complications described above. Moreover, hepatitis B has the potential to cause liver cancer in some of those who develop cirrhosis. Not only is hepatitis B in such chronically infected individuals a threat to their own survival, but it is also a source of infection to others, particularly to their needle-sharing partners, to their sexual partners, and to their developing fetuses and newborn babies.

Hepatitis C. Until about 1990, this was called "non-A-non-B hepatitis," because we knew that there were viral hepatitis cases that were caused by neither of the two identifiable viruses, A or B. An antibody test can now identify this virus, which is called hepatitis C. The antibody detected is not a protective antibody, but it is similar to the AIDS (HIV) antibody in that it indicates the presence of the virus. A lot of the viral hepatitis caused by blood transfusions in the past was due to hepatitis C infection; the antibody test can eliminate this source of transmission, since it is used to screen the donor blood supply.

Injecting drug users, however, remain a major reservoir and source for the spread of this virus. Hepatitis C is transmitted similarly to hepatitis B—and, for that matter, to HIV—primarily through direct blood to blood contact (by contaminated injection PARAPHERNALIA) and to a lesser extent, but still possibly, via sex and from mother to fetus. The primary infection goes very often unnoticed. The laboratory tests, in addition to hepatitis C antibodies, will show elevated ALT and AST levels. Since this is a newly

identified virus, the natural history of hepatitis C is not yet clear. A fair amount of evidence suggests that chronic hepatitis, eventual cirrhosis, and liver cancer may be an even greater risk with hepatitis C than it is with hepatitis B. Some studies in the current medical literature indicate that 50 to 80 percent of intravenous drug addicts may be positive for hepatitis C, so we are not talking about a trivial problem here.

Hepatitis D. This is a very peculiar virus, which was originally called "delta agent" and later renamed hepatitis D. It is an incomplete virus that can exist only in the presence of hepatitis B. When the two organisms combine, the outcome is a particularly nasty, potentially lethal hepatitis, both in terms of acute mortality and chronic consequences. Discovered in Italy about 1990, in North America hepatitis D is known to be primarily harbored by the injection drug-using population.

Prevention and Treatment of Viral Hepatitis. Obviously the best prevention for injection drug users would be to stop injecting drugs. Other, often more realistic prophylactic measures—which are now all familiar from the HIV scene—are the use of sterile (or at least bleached) needles and syringes, needle-exchange programs, and condoms for sexual activities.

Immediately after a known or suspected exposure to hepatitis B, the injection of an antibody preparation (known as "hepatitis B immune globulin") can prevent illness. A more permanent prophylaxis in high-risk populations is provided by the hepatitis B vaccine, which gives long-term immunity in previously uninfected individuals. IDUs certainly represent one of these high risk populations, although the widespread use of the hepatitis B vaccine in this group raises some obvious ethical and logistic dilemmas. At the present time, there is no passive or active immunization available for hepatitis C.

The acute phase of any form of viral hepatitis cannot be treated effectively. Chronic hepatitis B and C infection may respond, to a certain extent, to some antiviral drugs known as interferons, which are currently widely studied. Finally, as mentioned under alcoholic liver disease, the most radical form of therapy in the end-stages is liver transplantation.

(SEE ALSO: *Complications: Liver (Alcohol); Needle and Syringe Exchanges; Social Costs of Alcohol and Drug Use; Vulnerability As Cause of Substance Abuse*)

BIBLIOGRAPHY

GOLD, M. S. (1991). *The Good News about Drugs and Alcohol.* New York: Villard.

LIEBER, C. S. (1992). *Medical disorders of alcoholism.* New York: Plenum.

MEZEY, E. (1982). Alcoholic liver disease. In H. Popper & F. Shaffner (Eds.), *Progress in liver diseases.* New York: Grune & Stratton.

SHAPIRO, C. N. (1994). Transmission of hepatitis viruses. *Annals of Internal Medicine, 120,* 82.

WOOLF, G. M., & LEVY, G. A. (1988). Chronic viral hepatitis. *Medicine in North America, 21,* 3379.

PAUL DEVENYI

Medical and Behavioral Toxicity Overview Alcohol and other drugs of abuse have caused and continue to cause considerable adverse health effects to both the individual and to society. Both legal and illegal drugs (substances) of abuse are taken to modify mood, feeling, thinking, and perception. As with most drugs (medications), both acute and chronic toxicities occur. In general, the term *acute* refers to the short period of time when the drug is present in the body, exerting its main effects. The term *chronic* refers to a longer term, usually years.

Acute toxicity results in the impairment of behavior leading to other complications (e.g., trauma) and, in the case of some drugs, high doses can decrease breathing (respiratory depression) or change the rhythm of the heart, leading to accidental or intentional death. Chronic use can result in organ damage, which may lead to chronic illness or death (as with alcoholic cirrhosis of the liver). Persistent use of many classes of drugs also leads to TOLERANCE (an increased amount is required to produce the same effects) and physiologic (physical) dependence, so that a WITHDRAWAL syndrome is associated with sudden cessation of drug use. Drug users who employ hypodermic needles and syringes (injecting drug users [IDUs]) are at risk for blood-borne diseases associated with the use of unsterile equipment, such as hepatitis and human immunodeficiency virus (HIV 1 and 2—the viruses responsible for AIDS; see ACQUIRED IMMUNODEFICIENCY SYNDROME).

This article focuses on ALCOHOL as the representative drug, but other drugs of abuse will be referred to where appropriate. In North America, diagnosis of alcohol and other psychoactive substance abuse/dependence is usually made according to the DIAGNOSTIC AND STATISTICAL MANUAL (DSM) of the American Psychiatric Association (APA). The fourth edition, called DSM-IV, defines psychoactive substance *dependence* as at least three of the following (occurring in the same 12-month period):

1. tolerance, as defined by either of the following:
    a. need for markedly increased amounts of the substance to achieve intoxication or desired effect
    b. markedly diminished effect with continued use of the same amount of the substance
2. withdrawal, as manifested by either of the following:
    a. the characteristic withdrawal syndrome for the substance
    b. the same (or closely related) substance is taken to relieve or avoid withdrawal symptoms
3. the substance is often taken in larger amounts or over a longer period than was intended
4. a persistent desire for or unsuccessful efforts in cutting down or controlling substance use
5. a great deal of time is spent in activities necessary in obtaining the substance (e.g., visiting multiple doctors or driving long distances), using the substance (e.g., chain-smoking), or recovering from its effects
6. important social, occupational, or recreational activities are given up or reduced because of substance use
7. continued substance use despite knowledge of having had a persistent or recurrent physical or psychological problem that was likely to have been caused or exacerbated by the substance (e.g., recurrent cocaine use despite recognition of cocaine-induced depression; continued drinking despite recognition that an ulcer was made worse by alcohol consumption)

The diagnosis of alcohol and other substance *abuse* (as opposed to dependence) relies on:

A. A maladaptive pattern of substance use leading to clinically significant impairment or distress as manifested by one or more of the following occurring at any time during the same twelve-month period:
    1. recurrent substance use resulting in a failure to fulfill major obligations at work, school, or

home (e.g., repeated absences or poor work performance related to substance use; substance-related absences, suspensions, or expulsions from school; neglect of children or household)

2. recurrent use in situations in which it is physically hazardous (e.g., driving an automobile or operating a machine when impaired by substance use)

3. recurrent substance-related legal problems (e.g., arrests for substance-related disorderly conduct)

4. continued substance use despite having persistent or recurrent social or interpersonal problems caused or exacerbated by the effects of the substance (e.g., arguments with family members about consequences of intoxication; physical fights)

B. Has never met criteria for Substance Dependence for this class of substance.

These criteria continue to evolve and are likely to be somewhat changed in the future. Clearly the lack of one of the above diagnoses does not preclude a given person from being at risk for complications of alcohol or drug use (e.g., trauma as a result of intoxication).

THE ACUTE EFFECTS OF ALCOHOL

At the level of the cell, very high doses of alcohol (ethanol) seem to act by disrupting fat (lipid) structure in the central nervous system (anesthetic effect). Lower doses are thought to interact with various proteins and NEUROTRANSMITTERS (which act as RECEPTORS), such as GLUTAMATE, GABA (GAMMA-AMINO BUTYRIC ACID), cyclic AMP (adenosine mono-phosphate), and G proteins. Other actions may involve ion (calcium) channels. The reinforcing (rewarding) effects of alcohol may be mediated via DOPAMINE (a neurotransmitter) in specific brain regions; dopamine acts as an intermediary compound in the reinforcement process. The reinforcement of responses to other drugs of abuse, such as COCAINE, are also thought to be mediated via dopamine.

For most persons at least some of the acute effects of alcohol are well known on the basis of personal experience. Low doses cause blood vessels to dilate. The skin becomes flushed and warm. There is relaxation and mild sedation. Persons become talkative with loss of inhibitory control of emotions. Small doses (one to two drinks) do not impair complex intellectual ability; however, as the dose increases (two or more drinks or as the blood alcohol concentration approaches and exceeds the legal limit) impairment at multiple levels of the nervous system occurs. All types of motor performance are eventually affected, including maintenance of posture, control of speech, and eye movements. These movements become slower and more inaccurate. There is a decrease in mental functioning, such that there is impairment in attention and concentration, and a diminishing ability to make mental associations. There is a decreased ability to attend to incoming sensory information. Night and color vision are impaired. Judgment and discrimination and the ability to think and reason clearly are adversely affected. Even higher doses result in a stuporous condition associated with sleeping, vomiting, and little appreciation of surroundings. This is followed by coma and sometimes death from decreases in the functioning of the brain centers that control respiration.

ACUTE EFFECTS OF OTHER DRUGS OF ABUSE

Other drugs of abuse can be classified into stimulants, depressants, OPIOIDS, and drugs that alter perception (including HALLUCINOGENS). The effects of any drug depends on the dose taken at any one time, the previous drug experience of the user, the circumstances in which the drug is taken, and the manner (route of administration) in which the drug is taken.

Stimulants such as cocaine and AMPHETAMINE produce euphoria, increased confidence, increased sensory awareness, increased ANXIETY and suspiciousness, decreased appetite, and a decreased need for sleep. Physiological effects include increases in heart rate, blood pressure, and pupil size, and decreases in skin temperature.

Depressants such as the minor tranquillizers (including the BENZODIAZEPINES, BARBITURATES, and other SEDATIVE-HYPNOTICS) produce acute effects of a similar nature to alcohol—also a depressant. Actual effects vary according to drug, so that benzodiazepines (such as diazepam/Valium) produce less drunkenness compared to alcohol or barbiturates.

The term opioid refers to both drugs derived from OPIUM (opiates) and other synthetic drugs with similar actions—those acting on the same receptor sys-

tem. The term NARCOTIC is usually synonymous with opioid, but it can technically also include other drugs included in the HARRISON NARCOTIC ACT (e.g., cocaine). Opioids produce euphoria, sedation (to which rapid tolerance develops), itching, increased talkativeness, increased or decreased activity, a sensation of stomach turning, nausea, and vomiting. There are minor changes in blood pressure and the pupils become constricted (made smaller).

Drugs that alter perception include those above as well as MARIJUANA, PHENCYCLIDINE (PCP), and LYSERGIC ACID DIETHYLAMIDE (LSD). In general, most drugs of abuse can cause hallucinations under some circumstances. The drugs which more specifically affect perception (hallucinogens) produce a combination of depersonalization, altered time perception, body-image distortion, perceptual distortions (usually visual), and sometimes feelings of insight. Physiological effects such as changes in heart rate and blood pressure may also occur.

HARMFUL EFFECTS

Accidents. Alcohol is a significant factor in accident-related deaths. The main causes are motor-vehicle ACCIDENTS, falls, drownings, and fires and burns. Approximately 50 percent of motor-vehicle fatalities (driver, pedestrian, or cyclist) in the United States are alcohol-related, with the incidence having fallen a little in recent years. These alcohol-related accidents are more common at nights and on weekends. Falls are the most frequent cause of nonfatal accidents and the second most frequent cause of fatal accidents. According to various surveys of fatal falls, those that are alcohol-related range from 17 to 53 percent; for nonfatal falls, from 21 to 77 percent (Hingson & Howland, 1987). The higher the blood alcohol content (BAC), the higher is the risk for falls. The third leading cause of accidental death in the United States is drowning. About half (47–65%) of adult deaths by drowning are alcohol-related (Eighth Special Report to U.S. Congress, 1993). Fires and burns are the fourth leading cause of accidental death in the United States. Studies on burn victims show that alcohol intoxication is common. Cigarette smoking while drinking is an additional cause of fires and burn injuries. Estimates of the rates of intoxication range from 37 to 64 percent.

Users of other drugs of abuse (e.g., cocaine and opioids) also have higher rates of accidents in comparison to the nondrug-abusing population. The combination of cocaine and alcohol has been reported to be commonly associated with motor-vehicle deaths. Between 1984 and 1987 in New York City, 18 percent of motor-vehicle deaths showed evidence of cocaine use at autopsy. Cigarette smokers have higher rates of accidents than do non-smokers. Drugs that alter perception, such as PCP, are also associated with accidents mostly related to an impaired sense of judgment.

Crime. Associations between criminal activity and alcohol use have been established; however, methodological inadequacies of studies in this area preclude a clear causal relationship between alcohol use and crime. The strongest association between crime and alcohol use occurs in young males. Other forms of drug abuse (e.g., HEROIN and cocaine) have much higher associations with criminality. For example the majority of persons enrolled in METHADONE programs have extensive criminal careers. Those involved in drug dealing are at a high risk of being both a perpetrator or a victim of homicide.

Family Violence. Several studies indicate an association between alcohol use/abuse and spousal abuse; however, the nature of this interaction is not well understood. Intoxication is associated with an increase in negative behavior for episodic drinkers while less negative behavior is seen in steady drinkers, suggesting that drinking may be a short-term solution to problems for regular drinkers. Clearly, alcohol use is associated with physical VIOLENCE in some families, and there also appears to be a link between alcohol and child abuse. Female caregivers with a diagnosis of alcohol abuse, alcohol dependence, recurrent depression, or ANTISOCIAL PERSONALITY are more likely to report physical abuse of their children than those without these diagnoses (Bland & Orn, 1986).

Suicide. From 20 to 36 percent of SUICIDE victims have a history of alcohol abuse or had been drinking prior to death. Alcohol use is linked more to impulsive than to premeditated suicides, and to the use of firearms, rather than to other modes of killing. Death from OVERDOSE of illicit drugs is common; most of these are thought to be accidental but some are intentional.

Trauma or Severe Injuries. A history of trauma has been found to be a marker for (sign of) alcohol abuse. Emergency room trauma victims often have high rates of intoxication. Furthermore, heavy alcohol use both interferes with recovery from

serious injuries and increases rates of mortality for a given injury. Users of illicit drugs have a higher age-adjusted rate of mortality than do nonusers. Many of these deaths result from trauma.

Fetal Alcohol Syndrome (FAS). Since the 1970s, alcohol has become firmly established as a teratogen (an agent that produces defects in the developing fetus). It is considered the most common known cause of mental retardation. FAS defects range from specific structural bodily changes to growth retardation and subtle cognitive-behavioral abnormalities. The diagnostic criteria for FETAL ALCOHOL SYNDROME are the following: prenatal (before birth) and postnatal (after birth) growth retardation; characteristic craniofacial defects; central nervous system dysfunction; organ system malformations. When only some of these criteria are met, the diagnosis is termed *fetal alcohol effects* (Eighth Special Report to U.S. Congress, 1993). The abnormalities in physical appearance seem to decrease with age whereas the cognitive deficiencies tend to persist. There is no clear dose-response relationship between alcohol use and abnormalities. The safe amount of drinking during pregnancy (if it exists at all) is unknown. The peak level of blood (or brain) alcohol attained and the timing in relation to gestation (and particular organ development) are probably more important than the total amount of alcohol consumed during pregnancy. Genetic and maternal variables also seem to be important. Native-American and African-American children seem to be at high risk. While the public is generally aware of the relationship between alcohol consumption and fetal abnormality, surveys reveal that there is a need for greater public education in this area.

Smoking is associated with low birthweight. Cocaine use in PREGNANCY has been associated with complications (e.g., placental separation and in utero bleeding), and it appears to be associated with congenital abnormalities. Heroin use in pregnancy is associated with premature delivery and low birthweight; often there is a withdrawal syndrome in the baby at birth. Methadone (a long-acting opioid) usually reduces rates of prematurity and low birthweight but still causes as much or more opioid withdrawal in the newborn.

Cancer. There is clear epidemiologic evidence for an increased risk of certain types of CANCER in association with alcohol consumption. These include cancer of the esophagus, oropharynx, and liver. Other cancers possibly associated with alcohol

consumption include cancer of the breast, stomach, prostate, and colon (Geokas, 1986). Alcohol plays a synergistic (additive) role with smoking TOBACCO in the development of cancer, particularly with respect to the head, neck and esophagus. There are several possible mechanisms through which alcohol enhances the onset of cancer. Alcohol appears to modify the immune response to cancers, facilitate delivery of carcinogens (substances which enhance cancer onset), and impair protective responses. Overall, alcohol is considered to act as a co-carcinogen; for example, it increases the likelihood of certain smoking-induced cancers.

Smoking is, of course, well established as a cause of lung as well as other cancers. Smoking is responsible for 85 percent of lung cancers and has been associated with cancers of the mouth, pharynx, larynx, esophagus, stomach, pancreas, uterine cervix, kidney, ureter, and bladder (Bartecchi et al., 1994). Chewing tobacco (SMOKELESS TOBACCO) is associated with mouth cancer. The chewing of BETEL NUTS with lime is common in Asia and results in absorption of arecoline (a mild stimulant). This practice also causes cancer of the mouth. It has been suggested that MARIJUANA smoking also causes lung cancer, since high tar levels are present in the smoked products.

THE EFFECTS OF ALCOHOL ON BODILY SYSTEMS

Neurologic. Acute alcohol consumption causes impairment as described above. Alcohol potentiates the action of many drugs that produce acute effects on the brain. High blood-alcohol levels can result in "blackouts." This condition is due to acute loss of memory associated with intoxication, although the person usually behaves in apparently normal fashion during this period. Blackouts are also seen with the taking of other central nervous system depressants, such as the barbiturates and the benzodiazepines.

The main adverse consequences of chronic alcohol consumption with respect to the nervous system are the following: brain damage (manifested by dementia and alcohol amnestic syndrome); complications of the withdrawal syndrome (seizures, HALLUCINATIONS); and peripheral neuropathy. Chronic alcohol consumption results in tolerance, followed by an increased long-term consumption that likely leads to tissue damage. PHYSICAL DEPENDENCE may also develop (i.e., a withdrawal syndrome occurs on

sudden cessation of drinking). The brain damage, when severe, is usually classified as one of two main disorders. The first is a type of global (general) dementia. It is estimated that 20 percent of admissions to state mental hospitals suffer from alcohol-induced dementia (Freund & Ballinger, 1988). The second is an alcohol-induced amnestic (memory-loss) syndrome, more commonly known as Wernicke-Korsakoff syndrome. This is related to thiamine (Vitamin B_1) deficiency. The Wernicke component refers to the acute neurologic signs, which consist of ocular (eye) problems such as a sixth cranial nerve palsy (disturbed lateral gaze), and ataxia (impaired balance); the Korsakoff component refers to the memory impairment, which tends to be selective for short-term memory and is usually not as amenable to treatment once it has become manifest.

Milder forms of these disorders are also detectable with neuropsychologic testing or brain IMAGING TECHNIQUES (CAT scans; MRI). Studies of detoxified alcoholics (without other evidence of organic brain damage) reveal that 50 to 70 percent have impairments in neuropsychologic assessment (Eckardt & Martin, 1986). In most of these cases there is reversibility with abstinence from alcohol. Severe liver disease (e.g., advanced cirrhosis or acute hepatitis) may also contribute to this neurologic impairment. Computerized tomography (CT) scans reveal that many alcoholics have cerebral atrophy—this consists of decreased brain weight, an increase in spaces (sulci) between various regions of the brain, and an increase in size of ventricles (spaces filled with cerebrospinal fluid). In a minority of cases, these structural changes are reversible with abstinence. Seizures are associated with heavy alcohol consumption and usually occur in association with alcohol withdrawal. Abstinence from alcohol is usually the only treatment needed for this type of seizure. The hallucinations that are mostly associated with alcohol withdrawal are usually treated with drugs—benzodiazepines and phenothiazines.

Peripheral neuropathy is seen in association with chronic alcoholism. Peripheral neuropathy usually refers to toxic damage to peripheral nerves. Concurrent nutritional deficiencies often contribute to this damage. This neuropathy results in changes in sensation and occasionally motor function, usually in the legs. Sometimes this can occur acutely with intoxication. For example, the abnormal posture in association with a drunken stupor can result in radial nerve ("Saturday night") palsy. Alcoholics are also at

increased risk of subdural hematomas (blood clots due to ruptured intracranial veins secondary to trauma) and of stroke.

The neurologic complications associated with the acute use of other drugs of abuse include seizures (convulsions) and strokes in association with cocaine. High doses of some opioids, such as propoxyphene (Darvon) or MEPERIDINE (Demerol) can also cause seizures. Substances which can cause delirium (reversible disorientation) include *Cannabis* (MARIJUANA), phencyclidine (PCP), lysergic acid diethylamide (LSD), and atropine. Sudden cessation of use of central nervous system depressants (benzodiazepines, barbiturates, and alcohol) can result in seizures and hallucinations. Chronic use of other substances of abuse can also result in neurologic complications. Tobacco use is associated with increased rates of stroke (but it appears to be associated with lower rates of Parkinson's disease—a progressive disorder affecting control of movement). Solvent abuse (inhaling) can cause damage to the cerebellum (the part of the brain controlling movement) and to peripheral nerves. A form of "synthetic heroin," MPTP (1-methyl-4-phenyl-1,2,5,6-tetrahydropyridine), an analogue of meperidine (Demerol), has been demonstrated to cause a severe form of Parkinson's disease.

Psychiatric. Alcohol-related diagnoses are common among psychiatric patients. For example, a recent study (Moore et al., 1989) showed that 30 percent of those admitted to a psychiatric unit had a concurrent alcohol-related diagnosis. Alcohol alone may produce symptoms and signs that mimic psychiatric disorders. Examples include depression, anxiety disorder, psychosis, and antisocial personality disorder. Alternatively, an alcohol-related disorder may co-exist with one of these or may aggravate the psychiatric disorder.

Alcohol as a central nervous system (CNS) depressant tends to *cause* low mood states (hypophoria) with chronic use. It does not commonly cause long-lasting significant clinical depression, but it may aggravate it. If alcohol is the primary cause of a low mood state, then abstinence from alcohol, as the sole treatment, rapidly improves the disorder. Hallucinations may occur during alcohol withdrawal, mimicking a psychotic disorder. Similarly, the anxiety associated with alcohol withdrawal may mimic an anxiety disorder. Anxiety and hallucinations may also be seen during withdrawal from sedative-hypnotics. Behavior associated with alcoholism may lead

to an erroneous diagnosis of antisocial personality disorder.

When alcohol is used for self-medication in some psychiatric conditions, such as anxiety disorders, it tends only to be of short-term help and leads to more long-term problems. Other drugs of abuse, such as the stimulants cocaine and amphetamine, also produce anxiety and occasionally may produce a psychotic state associated with acute intoxication. This usually disappears rapidly as the drug effects wear off. Withdrawal following chronic use of stimulants may be associated with depression, excessive fatigue, and somnolence (a "crash"). Tobacco smoking also appears to be somewhat associated with depression. (Individuals with a history of depression are more likely to smoke, and may develop depression when they try to stop.) The nature of this relationship is unclear, but patients with psychiatric diagnoses have higher rates of smoking than the general population. Hallucinogens (such as LSD and PCP) may also cause an acute psychotic disorder which typically disappears as drug effects wear off; however, in some cases there may be longer lasting effects. Antisocial personality disorder is a common diagnosis in those who abuse drugs.

Endocrine and Reproductive. Alcohol produces both acute and chronic effects on virtually all endocrine organs (glands). Acutely, alcohol raises plasma catecholamines, which are chemicals released from nerve endings that are responsible for certain emotional reactions—"fear, flight, and fight". Epinephrine (adrenaline) is released from the inside of the adrenal gland (medulla) and norepinephrine (noradrenaline) from sympathetic NEURONS (nerve cells) and the adrenal glands. Alcohol also causes release of cortisol from the outside (cortex) of the adrenal gland both acutely and chronically. Cortisol is a hormone (chemical messenger) responsible for multiple effects on the body, including changes in the immune response, glucose regulation, fat breakdown, blood pressure, and mood. Alcohol-induced cortisol excess can mimic Cushing's disease (a condition associated with excess cortisol production, often caused by a tumor on the adrenals) and is known as pseudo-Cushing's disease. Alcohol affects the hypothalamus (an area of the brain), where it modifies chemical-releasing factors, which in turn control release of hormones from the pituitary (a gland in the brain linked to the hypothalamus by a special blood supply), which in turn affect endocrine

organs throughout the body. Acutely, alcohol also inhibits the release antidiuretic hormone (ADH) from the posterior pituitary, which results in an increase in urine production.

The best documented chronic endocrine effect of alcohol is male hypogonadism. This is a condition resulting from low sex-hormone function. Signs of this are small testes and decreased body hair. Symptoms include loss of libido and impotence. Hypogonadism can occur as a result of alcohol lowering testosterone levels. Alcohol acts both directly on the testes and indirectly via the hypothalamus. Alcoholic liver disease may also produce feminization in men, as a result of impaired metabolism (breakdown) of female sex hormones such as estrogen. Signs of such feminization in men include gynecomastia (enlarged breasts) and female fat distribution. In women who drink alcohol excessively, there is a high prevalence of gynecologic disorders (missed periods and problems in functioning of ovaries) and a possibly earlier onset of menopause than in nondrinkers. In women, also, alcohol is metabolized at different rates according to the particular phase of the menstrual cycle.

Abnormalities of both growth hormone (impaired release) and prolactin (increased release) have been described in association with acute alcohol ingestion. Thyroid function (which controls rate of body metabolism) can be indirectly affected as a result of alcoholic liver disease. This results in impaired conversion of T4 (one version of thyroid hormone) to T3 (a more active form of thyroid hormone). Furthermore, in alcoholism there are abnormalities in the proteins to which thyroid hormone binds. This results in making thyroid function tests difficult to interpret. Overall thyroid function is usually normal despite mild abnormalities in the tests.

Other drugs, particularly the opioids, also have multiple effects on the endocrine system. Opioids produce a degree of hypogonadism as a result of lowered testosterone in males and disturbed menstrual function in females. This results from opioid inhibition of gonadotropin releasing hormone (GRH) in the hypothalamus, which in turn inhibits release of LH (lutenizing hormone) and FSH (follicle stimulating hormone) from the pituitary. Opioids also inhibit corticotropin releasing factor (CRF), which results in decreased adrenocorticotrophic hormone (ACTH) and decreased cortisol. Nicotine causes release of epinephrine and norepinephrine, which in turn increase blood pressure and heart rate. Nicotine also

enhances the release of ADH from the hypothalamus, which decreases urine output (i.e., counteracts alcohol's effects).

Cardiovascular. Alcohol has direct effects on both cardiac muscle and cardiac electrophysiology (electrical functioning). These effects are also dependent on prior history of alcohol use (i.e., whether there have been underlying changes due to chronic use) and whether there is any evidence of underlying heart disease. Acutely, alcohol is a myocardial depressant (decreases muscle function) and, chronically, it may cause a degeneration of cardiac muscle (known as cardiomyopathy), which can lead to heart failure (condition due to excess body fluids because of inadequate pumping function of the heart). Abstinence from alcohol leads to improvement in function in some cases.

Both acute alcohol intoxication and acute withdrawal can lead to cardiac arrhythmias (abnormal heart beats). The most frequent association is with atrial fibrillation (frequent uneven and uncoordinated contraction of the atria). This is usually not life-threatening and mostly disappears without specific treatment. High levels of alcohol consumption are associated with increased rates of coronary (blood vessels which supply heart muscle) heart disease, while low levels of consumption (in comparison to complete abstinence) may be associated with a mild protective effect (the so-called U-curve relationship). However, low levels of consumption are not recommended as a preventive measure against coronary heart disease. Cigarette smoking is a much greater risk factor for coronary heart disease than is alcoholism. It should be noted however, that 80 to 90 percent of alcoholics are also cigarette smokers.

Multiple epidemiologic studies have established a relationship between alcohol and high blood pressure (hypertension). Somewhere between 5 and 24 percent of hypertension is considered to be alcohol related (Klatsky, 1987). The relationship seems to hold most strongly for white males over the age of 55 consuming at least 3 standard drinks per day on a chronic basis. Many cases resolve with abstinence. Acute alcohol withdrawal has also been associated with hypertension, but this usually lasts for only a few days.

The acute use of cocaine (a stimulant) results in increases in heart rate and blood pressure and causes narrowing of peripheral and coronary artery blood vessels. Repeated use of cocaine has been associated with abnormal heart beats, myocardial infarction (heart attack), and possibly myocardial fibrosis (an increase of scar tissue within the heart).

Acute tobacco use also results in constriction (narrowing) of blood vessels and an increase in heart rate because of the nicotine. Chronic tobacco use is the most important of the preventable causes of coronary heart disease. The coronary arteries supply the heart muscle. Long-term tobacco use results in an increase in atherosclerosis (build up of fat and other products inside the walls of blood vessels) in most of the arteries throughout the body and increases coagulation (clotting). This has important effects on the following blood vessels: the coronary (causes angina [chest pain] and infarction [heart attack]; the aorta (causes aneurysms, the ballooning effect on the arterial wall, which can be fatal); the carotid (cause of strokes); the femoral (causes intermittent claudication, pain on walking); and the kidney (cause of kidney failure and some hypertension). Acute use of opioids have minor effects on blood pressure. There are not thought to be important chronic adverse effects of opioids on the cardiovascular system. Marijuana acutely causes increases in heart rate and blood flow.

Respiratory System. Acutely, alcohol does not usually interfere with lung function; however a decrease in cough reflexes, a predisposition to reflux of stomach fluids, and the impairment of bacterial clearance in the respiratory tract occur after intoxication. For some persons with asthma, alcoholic beverages can induce bronchospasm (airway narrowing). This is thought to be related to nonalcoholic components in the beverage. Acute alcohol consumption also has a direct depressant effect on the respiratory center located in the brainstem. Accordingly, an overdose (intentional or unintentional) can result in death from respiratory failure (decreased ability to breathe). Alcohol also contributes to respiratory depression when taken with other central nervous system depressants such as barbiturates and benzodiazepines (minor tranquillizers). Acute alcohol intake increases sleep apnea (period of time not breathing) in those who suffer from this disorder.

Chronic alcohol consumption is associated with several pulmonary infectious diseases (in addition to risks associated with tobacco smoking). These include pneumonia, lung abscess, and tuberculosis. Aspiration pneumonia occurs in association with high levels of alcohol intoxication; it is thought to be

caused by the inhalation of bacteria caused by the impairment of the usual reflexes, such as coughing. Pancreatitis and alcoholic cirrhosis are associated with pulmonary effusions (build-up of fluid on the lung).

Among the other drugs, cigarette smoking causes emphysema, chronic bronchitis, and lung cancer. The smoking of marijuana on a frequent long-term basis may also increase the likelihood of these disorders. Acute use of opiate drugs intravenously may cause pulmonary edema (accumulation of fluid in the lungs). which can be life-threatening. Chronic use of intravenous drugs may cause pulmonary fibrosis (increased scar tissue). This is probably related to impurities, such as talc, associated with the cutting of the drug (diluting the dose with fillers) prior to its sale and eventual injection.

Gastrointestinal Tract and Pancreas. Acutely, alcohol alters motor function of the esophagus. Chronic use of alcohol increases gastro-esophageal reflux. Alcohol alone does not appear to cause peptic ulcers (cigarettes do), but alcohol interferes with healing. Alcohol disrupts the mucosal barrier in the stomach and causes gastritis (inflammation of the stomach) which can lead to hemorrhage, especially when combined with aspirin. Alcohol also interferes with the cellular junctions within the small intestine, which can result in the disturbance of fluid and nutrient absorption, producing, diarrhea and malabsorption. Any resulting nutritional deficiencies can further aggravate this process.

Heavy alcohol use interferes significantly with pancreatic structure and function. Alcohol abuse and gallstone disease are the major causes of pancreatitis, and alcoholism alone is responsible for most cases of chronic pancreatitis. Alcohol changes cellular membranes, resulting in changes in transport mechanisms and the permeability of vital ions and nutrients essential for normal cellular function. Acetaldehyde, which is a breakdown product of alcohol (and also present in cigarette smoke), is toxic to cells and has been proposed as a causative agent in the development of this disorder (Geokas, 1984). Acute pancreatitis is life-threatening; patients have abdominal pain, nausea, and vomiting. Increased levels of pancreatic enzymes, such as amylase and lipase, accompany this disorder. Treatment is usually by conservative measures, such as replacement of fluids and pain relief. Chronic pancreatitis can be without symptoms; can become evident with chronic abdominal pain and evidence of malabsorption (weight loss, fatty stools, nutritional deficiencies); or, uncommonly, with diabetes mellitus as a result of the destruction of the endocrine as well as the exocrine function of the pancreas.

Liver. Alcoholic liver disease is a major cause of morbidity and mortality in the United States; in 1986, cirrhosis of the liver was the ninth leading cause of death. Alcohol causes three progressive pathological (abnormal) changes in the liver—fatty liver, alcoholic hepatitis, and cirrhosis. These changes are useful in a prognostic sense but can only be diagnosed with a liver biopsy, which is not always feasible or practical. More than one pathological condition may exist at any one time in a given patient. Fatty liver is the most benign of the three conditions, and is usually completely reversible with abstinence from alcohol; it occurs at a lower threshold of drinking compared to alcoholic hepatitis and cirrhosis. Alcoholic hepatitis ranges in severity from no symptoms at all to severe liver failure with a fatal outcome; it can be followed by complete recovery, chronic hepatitis, or cirrhosis. Treatment is primarily supportive. Similarly, the symptoms and signs of cirrhosis range from none at all to coma and death. Cirrhosis consists of irreversible changes in liver structure resulting from an increase in scar tissue. A consequence of this is an abnormal flow of blood through the liver (shunts), which can result in the adverse health consequences of bleeding and presentation of toxic substances to the brain. This, in turn, may result in effects ranging from impaired thinking to coma and death. Abstinence from alcohol can prevent progression of cirrhosis and reduces mortality and morbidity (illness) from this condition. Medications may also help to reduce mortality from alcoholic liver disease. These include propothiouracil (an antithyroid drug) and prednisone (a steroid). The former reduces the oxygen requirements for areas of the liver that are poorly perfused. The latter reduces inflammation. Women appear to be at higher risk for liver damage than are men.

Opioid use alone has not been associated with liver disease, but some opioids such as morphine can cause spasm of the bile duct, which results in acute abdominal pain. Tobacco use is associated with a more rapid metabolism (breakdown) of certain drugs in the liver. This means that sometimes higher or more frequent dosing of medicatons is required for smokers. This effect is thought to relate to the tars

in tobacco rather than to the nicotine. High doses of cocaine have been associated with acute liver failure.

Acute and chronic hepatitis (types B, C, and D) is common in users of intravenous drugs. It is not usually the drug itself that causes hepatitis (inflammation of the liver) but rather the introduction of disease-producing organisms associated with the sharing of needles. Viruses and bacteria introduced by injecting drugs cause other problems, such as HIV infection and AIDS, endocarditis (infection of heart valves), cellulitis (skin infection), and abscesses.

Immune System. Alcohol affects the immune system both directly and indirectly. It is often difficult to discern the direct effects of alcohol from concurrent conditions, such as malnutrition and liver disease. Alcohol affects host defense factors in a general way; it also seems to predispose those who drink heavily to specific types of infection. With respect to host factors, alcohol alone can reduce both the number and function of white blood cells (both polymorphonuclear leucocytes and lymphocytes). This both predisposes toward infection while it interferes with the ability to counteract infection. Mechanical factors are also of importance. For example intoxication with alcohol and a depressed level of consciousness (and depressed cough reflex) predisposes toward aspiration pneumonia. Specific infections that alcoholics are at higher risk for, compared to the population at large, include pneumococcal pneumonia (the most common form of pneumonia), other lung infections (Hemophilus influenzae, Klebsiella), abscesses (anaerobic infections,), and pulmonary tuberculosis.

Alcoholics with liver disease are at increased risk of spontaneous bacterial peritonitis (inflammation of the lining of the abdominal cavity). Other infections possibly associated with alcoholism include bacterial endocarditis (infection of the heart valves), bacterial meningitis, pancreatitic abscess, and diphtheria. HIV infected drug abusers are at increased risk of tuberculosis as well as a multitude of other infections. As mentioned above, injecting drug users are also susceptible to a variety of infections associated with use of unsterile equipment.

Changes in immune function have been reported to occur in users of other drugs of abuse, including heroin, cocaine, and marijuana. The precise relationship of the immune function change to the drug of abuse is not yet understood. Lifestyle factors such as poor nutrition are also likely to contribute to this.

Nutrition. In heavy alcohol consumers, malnutrition as a result of poor dietary habits is common. In women, heavy alcohol consumption is associated with lower than usual body weight to a degree similar to that also associated with tobacco smoking. There is less of a weight-lowering effect in men. Specific nutritional disorders associated with alcoholism include anemia (due to iron or folate deficiency); thiamine (Vitamin B_1) deficiency—causing beri-beri or Wernicke's encephalopathy or neuropathy; malabsorption; and defective immune and hormonal responses. Alcohol also interferes with the absorption of vitamins (such as pyridoxine and Vitamin A), minerals (such as zinc), and other nutrients (such as glucose and amino acids) (Mezey, 1985).

Metabolic. Alcohol is metabolized (broken down) in the liver to acetaldehyde and hydrogen, and then to carbon dioxide and water. Acetaldehyde is toxic to many different cellular functions. Alcohol affects carbohydrate, lipid (fat), and protein metabolism. Alcohol can cause low blood glucose (hypoglycemia) due to inhibition of glycogen (liver stores of carbohydrate) metabolism. Alcohol also raises blood sugar and acids (alcoholic ketacidosis). By interfering with the elimination of uric acid, alcohol may precipitate acute attacks of gout. Increased urinary excretion of magnesium can result in muscle weakness. Alcohol causes disturbances in blood lipids, with mostly an increase in triglycerides and high density lipoprotein (HDL) cholesterol.

Acute alcohol consumption can decrease, whereas chronic consumption can increase, the metabolism of certain drugs. Tobacco smoking also increases the metabolism of some drugs, such as theophylline and caffeine. This results from the increased activity of various liver enzymes as discussed above.

Hematologic (Blood) System. The effects of alcohol on the hematologic system can either be direct, or it can be indirect (as a result of liver disease or nutritional deficiencies). Uncommonly acute alcohol consumption (a very large dose in a short span of time) has direct effects on the bone marrow, resulting in decreased production of red cells, white cells, and platelets.

The most frequent effect seen in alcoholics following chronic consumption is an increase in the size of the red blood cells (macrocytosis). This is mainly due to direct toxic effects on the red cell membrane rather than to folate (a vitamin found in

green vegetables) deficiency, which also causes macrocytosis. A folate deficiency, however, is sometimes seen in alcoholics caused mainly by impaired intake and absorption of folate. Iron deficiency anemia is also seen because of impaired intake of iron and because of frequent bleeding (due to a variety of factors, such as coagulation defects, gastritis, and the impaired healing of peptic ulcer). Iron-overload syndromes are also diagnosed in alcoholics and are due to a multiplicity of causes. Chronic alcohol consumption can also lead to hemolytic (excess breakdown of red blood cells) anemia, which is mainly seen in association with liver disease. Platelet production and function can be suppressed by alcohol, resulting in prolonged bleeding times.

Other drugs also exert hematologic effects. Experimental addiction to opioids results in a reversible anemia and a reversible increase in erythrocyte sedimentation rate (a nonspecific indicator of disease process). Smoking allows carbon monoxide to enter the body and bind to hemoglobin (carboxyhemoglobinemia), which consequently causes an increase in red cell production (erythrocytosis). The hematocrit value and the plasma fibronogen (a clotting factor) rise and increase blood viscosity; platelets (sticky constituents of blood important in wound healing) aggregate more in smokers. These thickening factors, together with damage to the insides of blood vessels, increase the probability of both stroke and heart attack (myocardial infarction) in smokers. White cells are also at increased levels in smokers (leucocytosis).

Skeletal Muscle. Chronic alcohol consumption can result in muscle cell necrosis (death). Two main patterns are seen: (1) An acute alcoholic myopathy (disturbance of muscle function) occurs in the setting of binge drinking, sometimes associated with stupor and immobilization. This results in severe muscle pain, swelling elevated creatine kinase (a muscle enzyme), and myoglobinuria (muscle protein in the urine which can cause kidney failure). (2) This pattern consists of a more slowly evolving syndrome of proximal muscle (those closest to the trunk) weakness and atrophy (decreased size). Milder degrees of muscle injury are quite common and consist of elevated levels of the muscle enzyme creatine kinase.

Cocaine use can also cause muscle damage (rhabdomyolysis), resulting in abnormalities of creatine kinase. Most drugs of abuse (especially depressants) may indirectly cause muscle damage as a result of

prolonged abnormal posture, for example, sleeping in an intoxicated state on a hard surface.

Kidney. Alcohol abuse causes a variety of electrolyte and acid-base (blood chemistry) disorders, which include decreases in the levels of phosphate, magnesium, calcium, and potassium. These abnormalities relate to disorders within the functioning kidney tubules (involved in secretion and reabsorption of minerals). The abnormalities usually disappear with abstinence from alcohol.

Chronic use of other abused substances is also associated with kidney (renal) damage and failure. Long-term consumption of pain-relieving medicines (daily use over many years) has been associated with kidney failure (analgesic nephropathy). This is especially associated with the combination products—those that include two or more of CODEINE, CAFFEINE, aspirin, and phenacetin. The rewarding effects that perpetuate this form of drug use most probably relate to the codeine (an opioid) and the caffeine (a stimulant). Heroin use has been associated with a form of kidney failure known as heroin nephropathy. Its precise relationship to heroin use is unclear. Secondary effects on the kidneys from drug and alcohol abuse also occur (for example, from effects of trauma or muscle damage as described above).

(SEE ALSO: *Crime and Alcohol; Crime and Drugs; Family Violence and Substance Abuse; Inhalants; Social Costs of Alcohol and Drug Use; Substance Abuse and AIDS; Tobacco: Complications*)

BIBLIOGRAPHY

AMERICAN PSYCHIATRIC ASSOCIATION (1994). *Diagnostic and statistical manual of mental disorders-fourth edition.* Washington, DC: Author.

BARTECCHI, C. E., MACKENZIE, T. D., & SCHRIER, R. W. 1994. The human costs of tobacco use. *New England Journal of Medicine, 330,* 907–912.

BLAND, R., & ORN, H. (1986). Family violence and psychiatric disorder. *Canadian Journal of Psychiatry, 31,* 129–137.

ECKARDT, M. J., & MARTIN, P. R. (1986). Clinical assessment of cognition in alcoholism. *Alcoholism (NY) 10*(2), 123–127.

FREUND, G., & BALLINGER, W. E. (1988). Loss of cholinergic muscarinic receptors in the frontal cortex of alcohol abusers. *Alcoholism (NY), 12*(5), 630–638.

GEOKAS, M. C. (ED.). (1984). Ethyl alcohol and disease. *Med Clin N America, 68*(1), 1–246.

HINGSON, R., & HOWLAND, J. (1987). Alcohol as a risk factor for injury or death resulting from accidental falls: A review of the literature. *Journal of Studies in Alcohol, 48,* 212–219.

KLATSKY, A. L. (1987). The cardiovascular effects of alcohol. *Alcohol and Alcoholism, 22*(suppl. 1), 117–124.

MEZEY, E. (1985). Effect of ethanol on intestinal morphology, metabolism, and function. In H. K. Seitz & B. Kommerell (Eds.), *Alcohol related diseases in gastroenterology.* Berlin: Springer-Verlag.

MOORE, R. D., ET AL. (1989). Prevalence, detection, and treatment of alcoholism in hospitalized patients. *Journal of the American Medical Association, 261,* 403–407.

U.S. DEPARTMENT OF HEALTH AND HUMAN SERVICES. (1993). *Eighth special report to the U.S. Congress on alcohol and health.* Washington, DC: U.S. Government Printing Office.

JOHN T. SULLIVAN

Mental Disorders Psychiatric disorders have long been recognized as being associated with psychoactive-substance-use disorders (commonly referred to as drug or alcohol abuse). The term *dual diagnosis* is frequently used to describe people with substance-use disorders combined with other psychiatric disorders. The term COMORBIDITY is also used to describe the situation in which an individual has two or more distinct disorders. Anxiety disorders and mood disorders are generally the disorders thought to occur in individuals with substance abuse, but other psychiatric disorders also demonstrate high rates of psychoactive-substance use also, including eating disorders (particularly BULIMIA), posttraumatic stress disorder (PTSD), personality disorders, somatization disorder, and SCHIZOPHRENIA. Children with ATTENTION DEFICIT hyperactivity disorder (ADHD) may also be at increased risk of substance abuse as adults. Various relationships may exist between drug and alcohol use and the development of these psychiatric disturbances. In understanding the relationships between substance abuse and psychiatric disorders, the concepts of *primary* and *secondary* are of critical importance. The primary–secondary dichotomy is based on the time sequence in which each disorder developed. When a disorder is referred to as primary, this would indicate that it presented first. The rationale for using the primary–secondary concept involves improved prediction of familial clustering of the psychiatric disorder, implications for treatment, and improved outcome prediction.

In addition to the primary–secondary distinction, the approach to the individual with both a substance use and a nonsubstance use psychiatric disorder should incorporate a similar but slightly more encompassing approach. The drug or alcohol use in such individuals may be involved as a form of self-medication for the psychiatric disturbance; it may itself induce psychiatric symptoms in an otherwise unaffected individual; or the individual may be affected by both disorders (substance abuse and other psychiatric disorders) through separate routes of VULNERABILITY.

As can be anticipated from this introduction to dual diagnosis issues, the relationship of psychiatric disorders and substance abuse is complex. Despite this complexity, the extent of such problems underscores the need for attention to this area. Studies have shown higher prevalence rates of substance abuse in individuals with psychiatric disorders than in the general population, and conversely patients seeking care for substance abuse have shown high rates of other psychiatric disorders. Large epidemiologic studies of community samples in the United States reveal that greater than 50 percent of substance abusers have at least one other mental illness (Regier, 1990). Data from this same study indicate that approximately one third of those identified as having a mental disorder also have a substance-abuse disorder.

ALCOHOLISM AND MOOD DISORDERS

Depression and Alcoholism. The rate of depression in individuals with alcoholism and rate of alcoholism in individuals affected with mood disorders (depression and mania) varies greatly according to different studies. The reason for the lack of agreement regarding such rates involves problems shared by all combinations of dual diagnosis. Two such problems include the means of assessment (both of substance abuse and psychiatric symptoms), and the timing of the assessment of psychiatric symptoms (i.e., in relationship to the last occurrence of substance use).

The effect that the means of assessment has upon psychiatric comorbidity is well illustrated by depression. Different rates of depression in alcoholics are seen if one uses standard clinical interviews, structured research interviews or self-report measures. Such methodological differences lead to widely differing conclusions regarding both comorbidity rates and comorbid influence. For example, rates of depression in alcoholism are reported to range from 8.6 percent to 44 percent, depending on the type of assessment (Keeler, 1979).

The critical importance of the timing of the psychiatric assessment and its relationship to comorbidity is demonstrated by studies from the Alcohol Research Center in San Diego (Brown, 1988; Schuckit, 1990). Symptoms of depression in 191 alcoholics were recorded within 48 hours of admission for alcohol detoxification and again after 4 weeks of abstinence (Brown, 1988). Significant levels of depression were noted in 42 percent of alcoholics in the first assessment, but in only 6 percent in the follow-up evaluation irrespective of any specific antidepressant therapy. Other studies of depression in alcoholics would suggest that the rate of primary depression in alcoholics is similar to that seen in the general population, which is approximately 3 to 10 percent (Winokur, 1971).

These studies demonstrate that for a number of alcoholics, psychiatric symptoms are directly induced by the intake of alcohol, and that these symptoms should be regarded as secondary to the alcoholism. This is important for two reasons. First, if the psychiatric symptoms are secondary to the alcoholism, treatment of the psychiatric symptoms alone will not treat the main disorder (alcoholism). Second, risk for relapse of the alcoholism is high.

Another complication of alcoholism in regard to depression is poor treatment response to standard ANTIDEPRESSANT therapy, both pharmacologic and nonpharmacologic in type. The reason for this is unknown, but may be related to adverse social complications of the alcoholic behavior (e.g., legal problems, job difficulties, marital separation, and divorce) (Cook, 1991).

Mania and Alcoholism. Individuals with bipolar affective disorder (manic-depressive disorder) have been noted to have increased use of alcohol during their manic episodes. Two studies have suggested that alcoholic bipolar patients have high rates of alcoholism in their families as compared to nonalcoholic bipolar patients (Morrison, 1974; Dunner,

1979). This fact suggests that the risk for alcoholism in bipolar disorder may occur due to a familial predisposition (e.g., genetic predisposition, behavior modeling, etc.), and not necessarily from a complication of the manic episode itself. Regardless, impulsive behavior during manic episodes clearly includes risk for excessive alcohol use.

SUBSTANCE ABUSE AND MOOD DISORDERS

Nearly all substances of abuse have the potential to alter mood symptoms. Classically, PSYCHOSTIMULANTS, such as AMPHETAMINES and COCAINE, may induce an appearance of elevated mood, racing thoughts, increased energy, and sense of well-being. Individuals who have developed tolerance to stimulants will experience, upon their discontinuation, withdrawal. These withdrawal symptoms will overlap characteristic depressive symptoms, including severe dysphoria, insomnia followed by hypersomnia, irritability, and fatigue. OPIATES induce a sense of elevated mood, and increased self-esteem. A sense of decreased anxiety is also frequently reported. Upon withdrawal, depressive symptoms are accompanied by characteristic physical symptoms such as muscle aches, drug CRAVING, lacrimation (secretion of tears), and piloerection (goose flesh).

SUBSTANCE ABUSE AND PERSONALITY DISORDERS

Personality disorders by definition involve maladaptive patterns of relating to one's environment and self that lead to conflict. To meet the definition of a disorder, these patterns should be enduring qualities, and the onset of such disorders is late adolescence. Behavior induced by substance abuse should be carefully separated from behavior demonstrated during periods of abstinence. This is important, since maladaptive behavior associated with personality disorders will persist through adulthood, while maladaptive behavior that is induced by substance abuse should subside during abstinence from the substance. Two personality disorders that are closely associated with substance abuse are ANTISOCIAL PERSONALITY DISORDER and borderline personality disorder.

Antisocial Personality and Alcoholism. A great deal is known regarding the relationship of antisocial personality disorder and alcoholism. This di-

agnostic combination is estimated to involve as many as 2 percent of the male population of the United States. Most studies of this combination of illnesses indicate that the antisocial alcoholic has an earlier onset of drinking difficulties, more family history of alcoholism, more social complications of alcoholism, and a greater number of symptoms of other psychiatric disturbances, e.g., drug abuse, depression, mania, schizophrenia, and psychotic symptoms. Antisocial alcoholics have also been reported to attempt suicide more frequently. In addition to these more severe symptoms at the time of initial evaluation, antisocial personality disorder influences the natural history of the substance use disorders and alcoholism. This change in course is demonstrated by the following studies.

Schuckit (1985) utilized standardized research criteria to divide a group of 541 alcoholics into those who were primary alcoholics, primary drug abusers, primary antisocials, and primary affective disorders. Intake and one-year outcome were then evaluated. The primary antisocial, along with the primary drug abusers, had a poorer one-year outcome in terms of drug use, police and social problems, and higher scores (worse outcome) on a clinical-outcome scale. Schuckit concluded from this study that antisocial personality predicted a poor prognosis in terms of continued alcohol problems.

In a carefully designed study, Rounsaville and co-workers (1987) evaluated 266 alcoholics one year after treatment. Multiple-outcome measures were utilized in this study and over 84 percent of the original cohort were reevaluated. In this study, it was found that in males, an additional diagnosis of major depression, antisocial personality disorder, or drug abuse were associated with poor prognosis at one year. Further analysis in this study also supported the conclusion that the diagnosis was the factor that conveyed the poor prognosis, not general severity of psychopathology or degree of alcohol dependence.

A recent study that looked at outcome of alcohol problems in subtypes of antisocials was conducted by Liskow (1991). In this study, antisocial alcoholics were subtyped as to presence of additional diagnoses of drug abuse and depression. An alcoholism-only group was included as a control group. In this study, the alcoholism-only group had the best outcome on a number of measures, while the antisocial alcoholic with drug abuse had the worst outcome. The antisocial alcoholic with no other diagnosis and the antisocial alcoholic with depression were similar in

outcome—and intermediate in outcome. Overall, the differences in the alcoholism-only compared to the antisocial-only and the antisocial depressed alcoholics were small compared to the differences between the antisocial plus drug group and all other groups. This study suggests that the poor prognosis in antisocial alcoholics may depend in part on other additional psychopathology (i.e., drug abuse).

Antisocial Personality Disorder and Substance Abuse. Individuals with antisocial personality disorder have high rates of drug use. Conversely, the dysfunctional lifestyle of an individual actively involved with substance abuse frequently involves lying, joblessness, and the inability to comply with social norms concerning issues such as child care, finances, and the legal system. This makes disentangling these disorders difficult. If one looks for evidence of conduct disturbance, particularly during the late adolescence that predates the substance abuse, then the diagnosis of antisocial personality disorder is much more reliable. The implication for this distinction involves improved ability to predict changes in the individual should long-term abstinence be achieved. The abstinent antisocial will likely continue to demonstrate behavior problems in a variety of areas, whereas someone with an intact personality would be expected to have a better prognosis.

Borderline Personality Disorder. Less is known about borderline personality disorder and substance abuse or alcoholism. Individuals with borderline personality disorder clearly have high rates of substance abuse, and the criteria published in the DI-AGNOSTIC AND STATISTICAL MANUAL OF MENTAL DISORDERS, 3rd edition, revised (DSM-III-R) (American Psychiatric Association, 1987) for this personality disorder include self-damaging behavior—such as substance abuse—as one of the five symptoms required to make the diagnosis. Dulit (1990) has suggested that drug abuse may be an important factor in the development of this disorder.

POSTTRAUMATIC STRESS DISORDER (PTSD) AND SUBSTANCE ABUSE

Following extreme stresses beyond the realm of normal human experience, symptoms of anxiety including intrusive recollections of the trauma, autonomic hyperactivity, and nightmares have long been observed, but PTSD as a psychiatric diagnosis is much newer. Following recognition of this disorder,

the link with substance abuse has been the subject of a number of studies. Rates of alcoholism in PTSD range from 40 to 80 percent, while other forms of substance abuse may range from 20 to 50 percent. This high rate of substance abuse has led to the hypothesis that the drug use may be explained by a self-medication theory. Jelinek (1984) has proposed that in the treatment of PTSD, those with substance abuse be divided into groups with abuse that preceded the trauma and those whose abuse followed the trauma. This latter group is considered as a "self-medication group," and in this group treatment of the PTSD is felt to be the primary goal. Following treatment for the PTSD, it is believed that the substance abuse in this group will then decrease or end. In the former group, detoxification from the substance abuse and abstinence is felt to be the primary goal, and that following this the PTSD symptoms will improve.

SUBSTANCE ABUSE AND OTHER ANXIETY DISORDERS

Alcohol and Anxiety. Individuals with anxiety disorders (e.g., generalized anxiety disorder, panic disorder, phobic disorders, obsessive-compulsive disorder) find that alcohol provides temporary relief from some of their anxiety symptoms. Large community studies of individuals with phobias suggest over a twofold increase in alcoholism risk. Panic disorder patients have rates of alcoholism approaching 20 percent, and male relatives of individuals with panic disorder have a two to three times increased rate of alcoholism when compared to controls, further suggesting a relationship between alcoholism and anxiety disorders. Another known fact is that anxiety symptoms are experienced during withdrawal. Schuckit (1990), in a study of anxiety symptoms during withdrawal, evaluated 171 alcoholics for anxiety and panic symptoms. Nearly all subjects had at least one anxiety symptom during heavy drinking, or upon abrupt discontinuation of drinking, but only 4 percent fulfilled DSM-III-R criteria for generalized anxiety disorder when three or more months of abstinence were achieved.

Anxiety and Substance Abuse. Panic attacks have been shown to be induced by psychostimulants, particularly COCAINE. The rate of panic attacks among users of cocaine has been reported to be as high as 64 percent. Anxiety symptoms during the withdrawal phase from cocaine also increases the

risk for alcohol abuse and/or benzodiazepine abuse. These substances are frequently used to ease the "crash" phase.

SUBSTANCE ABUSE AND SCHIZOPHRENIA

Only recently has the high prevalence of alcoholism in schizophrenia been noted. Likewise, the recognition of high rates of other substance abuse in the schizophrenic population was not appreciated until the 1980s. A review of published estimates of the prevalence of alcohol abuse in schizophrenia reported a range of 8.4 to 47 percent (Mueser, 1990). Stimulant abuse in this review was reported between 4 and 15 percent. The question of whether substance abuse induces a chronic schizophrenic-like psychosis even after the drugs are stopped is still open to debate. It is generally held, however, that individuals who develop schizophrenia coupled with drug abuse would most likely have developed schizophrenia regardless, but the abuse may have caused an earlier onset. The early drug use may represent efforts at self-treatment. Treatment of the schizophrenic with drug abuse presents a major clinical challenge. Such patients tend to be disruptive, prone to frequent relapse of psychosis and drug use, and do not easily fit into conventional treatment settings. Optimal care is thus difficult, and improved strategies for treatment are needed.

SUBSTANCE ABUSE AND EATING DISORDERS

Individuals with eating disorders (ANOREXIA and BULIMIA) abuse a number of drugs and alcohol. During the course of their lives, they often use agents to reduce weight, such as laxatives, emetics, diet pills, and diuretics. Of those individuals with eating disorders who seek psychiatric treatment, as many as 35 percent have a significant substance-abuse history. Alcoholism, particularly in bulimia and bulimic anorectic patients, appears to be common. Substance abuse in eating disorders is generally thought to convey a poor prognosis for recovery.

SUBSTANCE ABUSE AND ATTENTION DEFICIT DISORDER

Children with attention deficit hyperactivity disorder (ADHD) have been noted to be at risk for development of alcoholism and cocaine abuse as they

grow into adolescence and adulthood. Family studies of children with ADHD and alcoholism have demonstrated higher rates of alcoholism in family members than that seen in the general population. Goodwin (1975) compared previously hyperactive adult adoptees with and without alcoholism. As children, these alcoholics were hyperactive, truant, shy, aggressive, disobedient, and friendless. In these adoptees, those with alcoholism clearly had an excess of alcoholism in their biological parents. No alcoholism was found among the biological parents of the nonalcoholic hyperactive adoptees. These findings suggest that in the case of alcoholism and hyperactivity, the risk for alcoholism comes from a genetic basis and not necessarily from just having ADHD.

SUBSTANCE ABUSE AND OTHER COMPLICATIONS

Suicide. Alcoholics have a 15 percent lifetime risk of SUICIDE. Alcohol is involved in at least 50 percent of successful suicides. Substance abusers are also recognized to have an elevated risk of suicide, and it has been reported that 70 percent of suicides in young people are associated with substance abuse. Studies of successful suicides demonstrate the ambivalent nature of this act. Alcohol or substance abuse may act as the weight that tips the scale toward suicide, or may induce the psychiatric symptoms that elicit the suicidal urge. Regardless, substance abuse is among the strongest risk factors for suicide.

Organic Brain Syndromes. A variety of organic brain syndromes, including DELIRIUM and dementia are associated with acute and chronic use of drugs and alcohol. Abrupt WITHDRAWAL from alcohol or sedative-hypnotic drugs can cause withdrawal delirium (DTs). These organic effects from drug use must be carefully separated from the psychiatric conditions discussed earlier, and from neurologic conditions which can overlap their symptoms. The impact of chronic drug use and personality is an area in need of further study.

Acquired Immunodeficiency Syndrome (AIDS). Intravenous drug use, needle sharing, and high-risk sexual practices among drug users are major risk factors for AIDS. Psychiatric manifestations of AIDS may present in a number of ways, including mood disorders, dementia, psychosis, and behavioral impairment. Suicide risk among AIDS victims is high. In evaluating the substance abuser with neuropsychiatric changes, HIV testing should be completed and treatment for AIDS should incorporate educating the patient about these risks.

SUMMARY

Substance abuse of all kinds and many psychiatric disorders have been shown very conclusively to be associated one with the other. The combination of these disorders, as generally agreed, make such individuals more difficult to treat from the standpoint of both their psychiatric and their substance-abuse problems. Research is being conducted to determine better ways of understanding the origins of these associations.

(SEE ALSO: *Comorbidity and Vulnerability; Conduct Disorder; Research: Mood and Drugs; Social Costs of Alcohol and Drug Abuse; Structured Clinical Interview for DSM-III-R; Vulnerability*)

BIBLIOGRAPHY

AMERICAN PSYCHIATRIC ASSOCIATION. (1987). *Diagnostic and Statistical Manual of Mental Disorders-3rd ed.-rev.* Washington, DC: Author.

BROWN, S. A., & SCHUCKIT, M. A. (1988). Changes in depression among abstinent alcoholics. *Journal of the Study of Alcohol, 49,* 412–417.

COOK, B. L., ET AL. (1991). Depression and previous alcoholism in the elderly. *British Journal of Psychiatry, 158,* 72–75.

DULIT, R. A., ET AL. (1990). Substance abuse in borderline personality disorder. *Psychiatry, 147,* 1002–1004.

DUNNER, D. L., HENSEL, B. M., & FIEVE, R. R. (1979). Bipolar illness: Factors in drinking behavior. *American Journal of Psychiatry, 136,* 583–585.

GOLD, M. S., & SLABY, A. E. (1991). *Dual diagnosis in substance abuse.* New York: Marcel Dekker.

GOODWIN, D., ET AL. (1975). Alcoholism and the hyperactive child syndrome. *Journal of Nervous and Mental Disorders, 160,* 349–353.

JELINEK, J. M., & WILLIAMS, T. (1984). Post-traumatic stress disorder and substance abuse in Vietnam combat veterans: Treatment problems, strategies and recommendations. *Journal of Substance Abuse Treatment, 1,* 87–97.

KEELER, M. H., TAYLOR, C. I., & MILLER, W. C. (1979). Are all recently detoxified alcoholics depressed? *American Journal of Psychiatry, 136,* 586–588.

Liskow, B., Powell, B. J., & Nickel, E. (1991). Diagnostic subgroups of antisocial alcoholics: Outcome at 1 year. *Comparative Psychiatry, 31,* 549–556.

Lowinson, J. H., et al. (1992). *Substance abuse: A comprehensive textbook,* 2nd ed. Baltimore, MD: Williams & Wilkins.

Morrison, J. R. (1974). Bipolar affective disorder and alcoholism. *American Journal of Psychiatry, 131,* 1130–1133.

Mueser, K. T., Yarnold, P. R., & Levinson, D. F. (1990). Prevalence of substance abuse in schizophrenia: Demographic and clinical correlates. *Schizophrenia Bulletin, 16,* 31–56.

Regier, D. A., et al. (1990). Comorbidity of mental disorders and other drug abuse: Results from the epidemiology catchment area (ECA) study. *Journal of the American Medical Association, 264*(19), 2511–2518.

Rounsaville, B. J., et al. (1987). Psychopathology as a predictor of treatment outcome in alcoholics. *Archives of General Psychiatry, 44,* 505–513.

Schuckit, M. A. (1985). The clinical implications of primary diagnostic groups among alcoholics. *Archives of General Psychiatry, 42,* 1043–1049.

Schuckit, M. A., Irwin, M., & Brown, S. A. (1990). The history of anxiety symptoms among 171 primary alcoholics. *Journal of the Study of Alcohol, 51,* 24–51.

Winokur, G., Rimmer, J., & Reich, T. (1971). Alcoholism IV: Is there more than one type of alcoholism? *British Journal of Psychiatry, 118,* 525–531.

BRIAN L. COOK
GEORGE WINOKUR

Neurological Alcohol (ethanol, also called ethyl alcohol) and other psychotropic drugs are taken because of their ability to affect the central nervous system (CNS) and thereby alter mental functioning. However, the possible reinforcing effects are offset by a cost: it is now well established that CNS structural and functional integrity can be compromised by heavy or prolonged intake of many abused substances. This article addresses the effects of alcohol and other psychotropic drugs on nervous system structure and function. It will briefly review and synthesize information from studies using various methods and technologies, including neurological examination, postmortem examination of the brain, neuropsychological tests, and neuroradiological techniques such as magnetic resonance imaging (MRI), which shows the living brain in fine detail, and positron-emission tomography (PET), which indicates the level of functioning of particular brain regions while the individual is at rest or is engaged in a cognitive task. Both acute and chronic effects of substances on brain and behavior and the reversibility of drug-related impairments are addressed. Because most of the relevant research on this issue has been conducted with ALCOHOL, more limited information is presented regarding the effects of other drugs of abuse on CNS.

ALCOHOL: ACUTE EFFECTS

Blood alcohol concentrations (BAC) above the legal limit (0.08%) typically impair the operation of complex machinery, as should be obvious from public information and programs regarding DRIVING while intoxicated. The signs of intoxication, such as impaired judgment, slurred speech, and motor incoordination, are due to CNS depression. Sensitive testing also reveals impairments in a number of specific cognitive operations, including selective attention, decision making, and hand-eye coordination, at lower blood alcohol concentrations. Intoxication can increase risk-taking, aggressive, or dangerous behaviors because of diminished inhibitory control coupled with the person's inability to evaluate the consequences of his or her actions. Therefore, it is not surprising that intoxication is frequently associated with traumatic injuries, including traumatic brain injuries, and is a common factor in fatal motor vehicle accidents and violent incidents.

A binge of heavy drinking can lead to MEMORY lapses or alcoholic blackouts, in which the individual is unable to recollect events that took place during the period of intoxication even though he or she may have seemed "normal" to observers at the time. Although the pathogenesis of these episodes is not yet defined, it appears that the mechanisms underlying memory storage are temporarily disrupted during the blackout. Less severe difficulties with storage of new information can be seen even when drinking is below the legal limit of intoxication.

Very high doses of alcohol depress consciousness, leading to sleepiness, coma, respiratory depression, and death. The acute effects outlined here are clearly dose-dependent and are due to depression of successively more regions of the nervous system with increasing dose.

ALCOHOL: TOLERANCE AND WITHDRAWAL

Dependence on alcohol, and many other drugs, is characterized by TOLERANCE and WITHDRAWAL. Tolerance refers to the fact that with chronic use, increasing doses of the drug are needed to achieve the same behavioral effects. Thus, the degree of acute impairment outlined above will vary with the individual's tolerance. People who have developed alcohol tolerance also show cross-tolerance to other CNS depressants, including general anesthetics. Loss of tolerance appears to occur in the ELDERLY and in alcoholics who have developed organic brain impairments due to alcohol use or other factors, such as head injury. However, tolerance does not appear to develop to the direct neurotoxic effects of long-term alcohol abuse.

Following heavy drinking, many alcoholics experience a tremulous-hyperexcitable withdrawal syndrome, which is characterized by postural tremor, agitation, confusion, and ataxia. Generalized seizures can also appear in withdrawal, typically 10 to 48 hours after cessation of drinking. It has been hypothesized that long-term alcohol use may establish an epileptogenic state of the brain that becomes manifest upon alcohol withdrawal. For this reason, it has become common practice in many treatment facilities to guard against withdrawal seizures in patients with known susceptibility by giving prophylactic anticonvulsants or tranquilizers. Long-term treatment is usually not indicated because the withdrawal syndrome is self-limiting. In some patients, the acute withdrawal syndrome can progress to DELIRIUM TREMENS (DTs). This more severe form of withdrawal is characterized by delirium, HALLUCINATIONS, and a hyperautonomic state manifested by sweating and tachycardia. DTs are associated with approximately 15 percent mortality rate, possibly due to cardiac toxicity caused by the hyperadrenergic state. Treatment of the disorder involves rehydration and haloperidol (a neuroleptic drug) as well as medication to control withdrawal.

ALCOHOL: CHRONIC EFFECTS

Alcohol has direct toxic effects on neurons and, in association with other medical consequences of alcohol abuse, such as liver damage and inadequate nutrition, can result in significant and lasting cog-

nitive deficits. There is no clear indication of the level of consumption that might put one at risk for such consequences, but "safe" drinking guidelines of no more than twelve to fourteen drinks per week probably represent a minimum level. Although there are no precise data on the incidence of neurological or cognitive impairment in alcohol abusers, it is estimated that 50 to 70 percent of individuals seeking treatment may present with some form of neurocognitive impairment. Most of these report drinking more than thirty standard drinks (each containing 13.6 grams of ethanol) per week and at least a five-year history of such use.

Victor and his colleagues (1989) have made the definitive studies of the best-known disorder associated with alcohol abuse, the Wernicke-Korsakoff syndrome (WKS). There are three major symptoms in the acute phase, known as Wernicke's encephalopathy: abnormalities of eye movements; ataxia; and a confusional state that includes poor responsivity, disorientation, and deficits in attention and memory. The disorder has been demonstrated to be caused by a thiamine (Vitamin B_1) deficiency that is probably due to decreased B_1 in the diet and decreased absorption or utilization of B_1 induced by alcohol-related gastrointestinal disorders or other mechanisms. These symptoms usually improve substantially when the patient is immediately treated with thiamine. The chronic phase, known as Korsakoff's syndrome, is marked by a profound memory deficit that includes both retrograde amnesia (an inability to recall information from the remote past) and anterograde amnesia (an inability to learn and retain new information). It should be noted that there is frequently no prior Wernicke's encephalopathy recognized in Korsakoff patients. Although there may be difficulties in other cognitive capacities, the levels of general intellectual functioning, verbal abilities, and many other specific skills remain intact in these patients. Although partial or complete recovery from the amnesia is seen in some individuals, at least 50 percent of cases show slight or no recovery.

Postmortem analysis indicates that the lesions in Korsakoff patients generally involve diencephalic areas known to be important to memory functioning. These include the mammillary bodies and the dorsomedial nuclei of the thalamus. Neuronal loss is also prominent in other areas surrounding the cerebral ventricles, such as the periaqueductal gray of

the mesencephalon, hippocampus, and basal fore-brain. Modern imaging techniques that permit in vivo examination of the neuropathology of WKS are consistent with the neuropathological data. Analysis of MRI scans reveals small or absent mammillary bodies, as well as more general cerebral atrophy.

Estimates of the prevalence of WKS, based on hospital records, suggest that it is relatively rare. However, it appears that the diagnosis is often missed during life, despite its seemingly dramatic presentation. Autopsy series by Harper and his colleagues (1987) have indicated that less than 20 percent of patients with the characteristic brain lesions of WKS had been correctly identified antemortem.

A second profile of alcohol-related brain dysfunction, which is much more common than WKS, has been described by many investigators since the 1970s. These individuals may not show overt neurological symptoms, but selective impairments are seen in cognitive functions when sensitive neuropsychological tests are used. Extensive reviews of these effects are available in a book edited by Parsons, Butters, and Nathan (1987). The most prominent deficits are in complex visual-motor functions, particularly when speed of response is important. Thus, visual search, manual tracking, symbol copying, and other psychomotor functions are marked by imprecision and slowness. Problem-solving abilities, such as abstraction, hypothesis generation, and mental flexibility are also deficient. Mild deficits are apparent in new learning and memory, especially for nonverbal material. The memory difficulties are increased when the task requires the patient to use strategies for organizing and retrieving the information.

Studies of BRAIN STRUCTURES in these chronic alcoholics reveal apparent atrophy of the cortex, with enlargement of ventricles and sulci. For example, autopsy studies demonstrate that cerebral atrophy and low brain weight are associated with alcoholism, and some studies even show loss of cerebral tissue in "moderate" drinkers. Although such damage may be relatively widespread, several lines of research implicate a predominant involvement of frontal cortical regions in alcohol-related cerebral dysfunction. Autopsy studies show significant reduction in neuronal counts in the superior frontal cortex but not the motor cortex. Research with MRI and PET scanning in vivo reveals a consistent decrease in brain volume and functioning at rest in frontal regions, although most studies have implicated other brain areas as well. This evidence is generally consistent with studies using neuropsychological techniques, which also suggest impairments through tests sensitive to frontal-lobe dysfunction. However, further research is necessary to refine and correlate these different sources of evidence before strong conclusions can be made regarding selective effects of alcohol on localized cortical regions.

Many investigators recognize a more severe and global impairment in mental functioning that is different from both the "typical" picture of chronic alcoholic brain dysfunction and from WKS. This is generally referred to as alcoholic dementia, to underscore the severe and global nature of the cognitive deficits. However, it is not known whether alcoholic dementia exists as a separate pathological entity, which represents the end point of chronic alcoholism in some older individuals, or is an extension of WKS to other brain regions and cognitive domains. There is no clear set of clinical diagnostic or neuropathological criteria for this disorder. We have found a relatively high prevalence of alcohol-related dementias among residents of several long-term care institutions in northern Ontario. In that study, 24 percent of cognitively impaired residents fit this diagnostic profile, a figure substantially higher than had been reported previously. Given that the proportion of elderly individuals in North America is growing, we might expect an increase in research activity associated with this disorder during the 1990s.

ALCOHOL: RECOVERY OF FUNCTION

When alcoholics stop drinking, their presenting neurocognitive impairment can often show marked recovery over weeks to months with maintained abstinence. Substantial recovery has been demonstrated in only a minority of patients with WKS, given appropriate thiamine treatment and abstinence from alcohol use. In 1978, Carlen and colleagues were the first to report reversibility in measured cerebral atrophy on computerized tomography (CT) scans in chronic alcoholics after several months of abstinence. This reversibility has been replicated by several other research groups. Many studies have also reported recovery in the cognitive performance deficits for a majority of patients, with improvements depending critically on abstinence. In general, the more novel, complex, and rapid the in-

formation-processing requirements of the task, the longer the time for recovery to normal levels of function. As of the mid-1990s, only modest correlations between the measures of brain atrophy and cognitive functioning have been shown. The mechanisms underlying the pathogenesis and reversibility of cerebral atrophy are still under study.

ALCOHOL: SUMMARY AND FURTHER QUESTIONS

Chronic alcohol ingestion has potentially devastating effects on neurocognitive functioning. Impairments associated with alcohol use range from transient deficits observed in acute intoxication to potentially permanent and severe disorders, such as alcoholic dementia. There are also various other neurological conditions associated with chronic alcohol abuse. Some of these are relatively common in alcoholics, including cerebellar degeneration, peripheral neuropathies, movement disorders, and hepatic encephalopathy. Alcohol-related conditions that are relatively rare include central pontine myelinolysis, pellegra encephalopathy, and Marchiafava Bignami disease. Although some important relationships between alcohol misuse and neurocognitive functioning have been discerned since the 1970s, many important questions remain. Outstanding issues include the prevalence of impairment in the alcohol-dependent population; individual risk factors that mediate the expression of deficits; the relation between levels and patterns of consumption and resulting impairments; the rate and extent of recovery of function and treatments that may enhance recovery; precise specification of the profiles of cognitive impairment in different clinical syndromes and their relation to measures of brain damage; and implications of cognitive dysfunction for prevention and treatment of substance abuse.

OTHER CNS DEPRESSANTS

Several other classes of drugs act as depressants on the central nervous system. The profile of impairment with barbiturate intoxication largely resembles the acute effects of alcohol. Because of the relatively high abuse potential and severe withdrawal associated with these drugs, the BENZODIAZEPINE-shave largely displaced BARBITURATES in prescriptions for SEDATIVE-HYPNOTIC drugs. Currently the most prescribed class of PSYCHOACTIVE drugs in Western industrialized countries, they are typically used for muscle relaxation, sedation, and reduction of ANXIETY. It has become clear that benzodiazepines (e.g., Valium) can be associated with adverse behavioral changes, particularly in older individuals. In acute administration, they can cause impaired memory, slowing of reaction time and decision making, and disrupted attention. These effects are similar to those produced by drinking alcohol, and the effects these two drugs taken together can be additive. Although patients appear to develop some tolerance to the sedating effects of benzodiazepines when they are administered for long periods, new evidence suggests that memory and cognitive impairments can remain or even increase with chronic administration. At present, it does not appear that these drugs have direct toxic effects on brain structures, so that their effects on behavior are likely mediated by a temporary and reversible pharmacological blocking of normal routes of neural information processing.

CANNABIS

Cannabis (MARIJUANA) intoxication leads to widespread changes in cognitive functioning, including disrupted attention, memory, and perceptual-motor abilities. For example, individuals may be unable to remember information learned while intoxicated, even when tested in drug-free conditions. Like alcohol, there is also impairment in the complex visual, motor, and decision-making skills needed to operate complex machinery. Numerous reports from the 1970s indicated a lack of enduring cognitive impairment associated with chronic cannabis use. However, later work reported poor learning and memory in newly abstinent users, with recovery of function documented over a six-week period. Although further research is needed, there is clearly not the same degree of brain and behavioral dysfunction that has been associated with alcohol.

OPIOIDS

Primarily used for therapeutic management of severe pain, the OPIOIDS can have a profound mood-altering euphoric effect that can lead to dependence. Administration of opioids (HEROIN, MORPHINE, or Demerol) to relatively naive users leads to a generalized depression of cognition that can be referred to as "mental clouding." Impairments in perception, learning and memory, and reasoning accompany the

drowsiness and mood changes induced by the drug. Generally, tolerance develops to the depression in mentation with chronic administration. Thus, there appears to be little long-term performance or brain dysfunction associated with this drug class.

AMPHETAMINES AND COCAINE

AMPHETAMINES and COCAINE act as CNS stimulants, which means that they generally increase arousal and psychomotor activity. As might be expected, acute administration can improve performance on many tasks, particularly when vigilance or speed of response is important. It can also reverse the effects of fatigue, which suggests that attentional resources are enhanced. However, this increase in arousal can be coupled with a dysfunction in higher-order control processes used to monitor or inhibit ongoing behavior, such that there is a corresponding increase in errors, impulsivity, and hyperactivity. Neurological consequences of a single dose of cocaine can include intracerebral hemorrhages, seizures, and strokes. These complications appear in only a small proportion of cocaine users, but the factors that place one at risk are not yet known. Although there is little research on the existence or nature of cumulative effects of chronic stimulant use, some recent evidence indicates mild impairments in memory and attention. A few studies have documented the recovery of function with abstinence and a correlation between the level of consumption and impairment, suggesting a direct relationship between drug use and behavioral deficiency. Convergent evidence for a transient disruption in brain function is provided by the work of Volkow and colleagues (1988). Their PET studies indicated decreased blood flow in the prefrontal cortex of cocaine users and an apparent return to normal levels with abstinence. In animal models, high doses of amphetamines, particularly METHAMPETAMINE, can produce damage to serotonergic and dopaminergic neurons.

SOLVENTS

Solvents are chemical compounds, such as benzene and toluene, typically used to dissolve oils or resins. Although a small proportion of young people voluntarily abuse or inhale these substances (e.g., sniffing glue or gasoline), many more individuals are exposed to them as workers in an industrial setting.

These chemicals are unique in their ability to cause damage to the CNS after fairly limited exposure. Clinical observation of acute effects of these drugs shows that users experience euphoria, dizziness, and "drunkenness," which are usually accompanied by fatigue, muscle weakness, and impairments in concentration, memory, and reasoning. This can progress through loss of self-control, disorientation, and coma. Chronic neuropsychological impairments are seen in a variety of domains, including motor coordination, memory, and attention, and can resemble the symptoms of dementia. Neurological impairments include diminished sensitivity to pain and touch, shrinkage of the cortex, and lesions in the cerebellum. Although there is not yet sufficient evidence to be conclusive, it is likely that much of the damage and disruption to function are permanent.

SUMMARY

Although much is known about the effects of drugs on the brain and behavior, many important questions remain to be answered. Even given the intensive research on alcohol, there is still lack of consensus regarding the durability of impairments, the risk factors that determine individual susceptibility, and the relationship between consumption and brain damage. In particular, the separation of any neurocognitive dysfunction that may precede drug abuse from that which is consequent to chronic drug use remains an important issue that is difficult to address without prospective studies. Research attempting to discover the relationships among a given drug's effects on various indices of brain integrity is also a relatively new area and requires further elaboration. For example, only a few studies attempting to relate brain atrophy in specific regions to particular cognitive impairments in chronic alcoholics have shown significant and reliable correlations. It is also increasingly common to find that individuals will use and abuse several different drugs, yet research on the interacting effects of various drug combinations is in its infancy. In the final analysis, much of the work summarized here prepares us to ask better questions regarding the consequences of drug use on neurocognitive functioning.

(SEE ALSO: *Accidents and Injuries from Alcohol; Imaging Techniques; Inhalants; Inhalant Abuse and Its Dangers*)

BIBLIOGRAPHY

Carlen, P. L., et al. (1978). Reversible cerebral atrophy in recently abstinent chronic alcoholics measured by computed tomography scans. *Science, 200,* 1076–1078.

Chesher, G. B. (1989). Understanding the opioid analgesics and their effects on skills performance. *Alcohol, Drugs and Driving, 5,* 111–138.

Fornazzari, L., Wilkinson, D. A., Kapur, B. M., & Carlen, P. L. (1983). Cerebellar, cortical, and functional impairment in toluene abusers. *Acta Neurologica Scandinavica, 67,* 319–329.

Harper, C. G., Kril, J., & Daly, J. (1987). Are we drinking our neurones away? *British Medical Journal, 294,* 534–536.

Hartman, D. E. (1988). *Neuropsychological toxicology: Identification and assessment of human neurotoxic syndromes.* New York: Pergamon.

Lishman, W. A. (1990). Alcohol and the brain. *British Journal of Psychiatry, 156,* 635–644.

Mody, C. K., et al. (1988). Neurologic complications of cocaine abuse. *Neurology, 28,* 1189–1193.

O'Malley, S., Adamse, M., Heaton, R. K., & Gawin, F. H. (1992). Neuropsychological impairment in chronic cocaine abusers. *American Journal of Drug and Alcohol Abuse, 18,* 131–144.

Parsons, O. A., Butters, N., & Nathan, P. E. (1987). *Neuropsychology of alcoholism: Implications for diagnosis and treatment.* New York: Guilford Press.

Thomas, P. K. (1986). Alcohol and disease: Central nervous system. *Acta Medica Scandinavica* (suppl.), *703,* 251–264.

Victor, M., Adams, R. D., & Collins, G. H. (1989). *The Wernicke-Korsakoff syndrome and related neurological disorders due to alcoholism and malnutrition* (2nd ed.). Philadelphia: F. A. Davis.

Volkow, N. D., et al. (1988). Cerebral blood flow in chronic cocaine users: A study with positron emission tomography. *British Journal of Psychiatry, 152,* 641–648.

Woods, J. H., Katz, J. L., & Winger, G. (1992). Benzodiazepines: Use, abuse and consequences. *Pharmacological Reviews, 44,* 151–347.

Peter L. Carlen
Mary Pat McAndrews

Nutritional This entry discusses nutritional complications in alcoholics, smokers, and abusers of other addictive drugs.

ALCOHOL

Alcoholic beverages were long used as a source of nourishment for the sick, as a means of promoting appetite, and as a treatment for pain and infection—all before other means and medications were developed for these situations. Traditionally, wine and BEER were foods, used ceremonially and as part of ceremonial healing for the ailing (and pregnant) who refused or could not tolerate a solid diet. Eventually, alcoholic beverages moved from purely ceremonial occasions to a reason for social occasions in some cultures, among some classes, and for some individuals. Alcoholic beverages have a habit-forming or addictive element for some people that may become life threatening to fatal.

Use of Alcohol in Medicine: Recent History. In 1900, Atwater and Benedict reported on their experiments at Wesleyan University, which attempted to define whether ALCOHOL could actually be considered a food; they showed that alcohol is oxidized in the body and that the energy so derived can be used as a fuel for metabolic purposes. Before that, F. E. Anstie (1877) had written his treatise *On the Uses of Wine in Health and Disease,* and, in fact, the long tradition of using alcoholic beverages within the medical profession persisted into the twentieth century. Sir Robert Hutchison, a noted British physician, wrote in 1905 that there was reason *to believe* (not that there was evidence) that alcohol increases disease resistance. Alcohol was actually used to treat serious infectious disease, such as typhus, into the late 1920s—until it was shown that patients treated with milk and beef tea had greater survival rates.

The use of alcohol to treat such disease was linked to a supposition that debility would somehow be overcome and strength regained. Other than this, the major indication for alcohol was for analgesia (pain suppression). The basic analgesic properties of alcohol and alcoholic beverages were utilized for hundreds of years in the management of the injured and those requiring surgery. For example, prior to the time of anesthetics, patients were offered brandy to reduce the agonizing pain of amputations. Decline and cessation of these medical uses of alcohol came about with the development of inhalation anesthetics and more efficient analgesics.

Alcohol, Obesity, and Wasting. In nonalcoholics, calories from alcohol are utilized as efficiently as calories from carbohydrates or fats (and

alcohol provides more calories per gram than does carbohydrate). Indeed, while carbohydrate yields 4 kilocalories per gram (kcal/gr) on combustion, when alcohol is combusted in a bomb calorimeter it yields 7.1 kcal/gm. This suggests that when alcohol is consumed in addition to a diet that maintains body weight, weight gain occurs. Fictitious characters such as Shakespeare's Sir John Falstaff provide evidence that obesity was already in the 1600s considered a characteristic of heavy drinkers.

The realization came about gradually that in fact chronic heavy drinking leads not to obesity but to weight loss and an inability to sustain adequate nutritional status. Wasted alcoholics were first portrayed by artists such as William Hogarth (1697–1764) who were intent on showing both the social and medical evils of drinking gin—a recent import from Holland that became a fad. All ages and classes indulged in the new drink at all hours of the day and night. In eighteenth and nineteenth century England, when artists were portraying the physical deterioration associated with heavy gin drinking, it was assumed that drinking eventually led to wasting only because the drunkard was disinterested in food. This idea persisted into the twentieth century. By the 1940s, it was also well recognized that chronic alcoholics are malnourished because of impaired utilization of nutrients.

It is well-known today that, whereas obesity may occur in heavy eaters who consume alcohol, chronic alcoholics are undernourished. Furthermore, recent studies have shown that long-term, heavy consumption of alcohol in addition to food is not associated with the gain in body weight that would be expected from the calorie intake (Lieber, 1991). In addition, if dietary carbohydrate is replaced by alcohol, weight loss occurs (as in the so-called Drinking Man's Diet of the 1960s and 1970s). This energy deficit has been attributed to induction of the system that metabolizes alcohol and at the same time uses chemical energy and generates heat. Lieber, in reviewing current knowledge of the question, notes that this does not explain the fact that there is little or no weight deficit when alcohol is consumed with a very low-fat diet.

Alcohol and Malnutrition. Diet-related causes of malnutrition in alcoholics include low dietary intake of calories and nutrients—because of poor appetite, inebriation, and diversion of food dollars into support of the alcohol habit. In addition, malnutri-

tion may be caused by impaired absorption of nutrients, poor nutrient utilization, and increased nutrient losses in body wastes. In 1940, it was suggested that ALCOHOLISM is the major cause of malnutrition in the industrialized world (Jolliffe, 1940). Malnutrition in alcoholics may be caused by impaired absorption of nutrients because of the reduced absorptive capacity of the alcohol-damaged gut. Nutrients that are poorly absorbed by alcoholics include the B vitamins—folic acid, thiamin (Vitamin B_1), and riboflavin (Vitamin B_2). Folic acid deficiency, which causes an anemia, is particularly common in heavy drinkers. Multiple nutritional deficiencies, including deficiencies of water and fat-soluble vitamins, are also common in those alcoholics who have pancreatic and liver disease. Chronic alcoholic pancreatitis (inflammation of the pancreas) develops commonly in people who consume 150 grams or more of alcohol per day for at least ten years and at the same time eat a high-fat diet. The digestive functions of the pancreas become impaired, and therefore food is not broken down into nutrients that can be absorbed. This type of pancreatitis is a major cause of malabsorption of nutrients in alcoholics. Alcoholic cirrhosis is a condition in which liver cells that are responsible for the conversion of nutrients to active forms are replaced by fibrous tissue. Cirrhosis develops slowly in heavy drinkers and is a special risk in those who consume about 35 percent or more of their total caloric intake as alcohol. Cirrhosis is the chief cause of impaired nutrient utilization (Morgan, 1982); however, cirrhosis is caused not by a nutritional deficiency but by the toxic effects of alcohol on the liver (Lieber, 1988).

Mineral and trace-element deficiencies, particularly zinc deficiency, are common in alcoholics. Contributory causes are low intake and increased losses in the urine.

Alcohol, Nutrition, and Brain Damage. Alcoholics are at risk for brain damage when they go on drinking sprees without food. Evidence exists for this condition only in Caucasians who are genetically predisposed. An acute confusional state may occur, called Wernicke's encephalopathy; this condition can be rapidly reversed if the patient is given massive doses of thiamin, intravenously, within a period of forty-eight hours from the onset of the symptoms. If this acute condition is not treated with thiamin, a chronic state of irreversible brain damage develops,

in which there is moderate to severe dementia (Victor et al., 1957).

Alcohol and Heart Disease. As of the early 1990s, evidence indicates that while moderate drinking may reduce the risk of heart disease, alcohol abuse is associated with an increased risk of heart disease. Alcohol has the effect of increasing blood (plasma) levels of high-density lipoproteins (HDL)—and elevation of these blood lipids is associated with a lower risk of heart disease. In a British study (Razay et al., 1992), it was shown that women consuming a moderate amount of alcohol (1–20 gm/day) have lower fat (triglyceride) levels in their blood and higher HDL levels. The authors consider this strong evidence for supporting a lower risk of heart disease. It is important to note that in this study the women who were the moderate drinkers were slimmer than the non-drinking group. Lower body weight, found in this study among the moderate drinkers, is also a known factor in reduced risk of heart disease. Heart disease in alcoholics is due to the direct toxic effects of alcohol on heart muscle (Brigden and Robinson, 1964).

Alcohol and Osteoporosis. The formation of new bone tissue is reduced in heavy drinkers, and this causes a marked decrease in bone mass and strength, leading to severe osteoporosis. Alcohol abuse is recognized as a risk factor for osteoporosis in both men and women. Because inebriation is also associated with a high risk of falls, alcoholics who have osteoporosis are likely to sustain hip fractures. Low intake of calcium in foods is an additional risk factor for osteoporosis in alcoholics (Bikle et al., 1985).

Methods for Assessing Nutritional Status in Alcoholics. The methods required for the assessment of caloric and nutrient intake in actively drinking alcoholics include direct observation (seldom feasible outside a treatment facility) and so-called tray weigh back (also feasible only in the detoxification section of a rehabilitation facility, hospital, or nursing home). The term *tray weigh back* means weighing the food served to a patient, weighing the uneaten food, then computing intake from the difference.

When alcoholics are asked to recount what they have eaten, they tend to confabulate: When asked leading questions, they provide answers that the question indicates are correct or ideal. They may provide the questioner with an account of a make-believe diet, or they exaggerate the amounts of food they have eaten. These responses, which are worthless for the purpose of assessing the amount of calories consumed from food or for assessing the nutrients consumed, are given by alcoholics who may not remember what was eaten and also because they may want to please the dietitian, physician, or nurse seeking information. Not only do alcoholics confabulate, they may also exaggerate the amount they eat, reporting what is served to them rather than what was consumed. (This is also the type of overreporting of food intake frequently found in people consuming other drugs that suppress appetite.)

The presence of malnutrition is assessed in alcoholics (as well as nonalcoholics) by using anthropometric (body) measurements—including weight-for-height measurements, calculation of the body–mass index (the weight/height squared), and the circumference of the upper arm and the thickness of the fat on the back of the arm. Alcoholics show muscle wasting in the upper arms, which may suggest malnutrition even when body weight is not markedly decreased. Although alcoholics with advanced liver disease are frequently wasted, weight loss may not register in numerical terms because of fluid retention within the abdominal cavity (ascites).

Biochemical measurements are valuable for assessing the nutritional status of alcoholics. The measurement of plasma albumin levels is particularly important—a value of less than 3.5 grams per 100 milliliters of plasma indicates that protein–energy malnutrition exists.

Nutrient Intolerance in Alcoholics with Liver Disease. Alcoholics with liver disease are very intolerant of high-protein diets. If high-protein diets are provided during periods of nutritional rehabilitation, such alcoholics may develop signs of liver failure. Such alcoholics are also intolerant of Vitamin A if this vitamin is taken in amounts that exceed 10,000 international units (IU) per day. Continued intake of Vitamin A at a high daily dosage level leads to further liver damage and may also precipitate liver failure (Roe, 1992).

Nutritional Rehabilitation of Alcoholics. Nutritional rehabilitation of alcoholics can be carried out successfully only when abstinence is enforced or the alcoholic voluntarily stops drinking. If the alcoholic has advanced liver disease or impairment of pancreatic function such that digestion and absorption of nutrients is impaired, optimal nutritional sta-

tus cannot be maintained. The goal of nutritional rehabilitation is the treatment of existing protein-energy malnutrition by increasing caloric intake from carbohydrates and the treatment of existing vitamin, mineral, and trace-element deficiencies. Appetite returns after alcohol withdrawal symptoms have abated; however, recovery of efficient absorption of vitamins may not occur until ten to fourteen days after drinking ceases. Initially, intolerance of milk and other dairy foods is common during rehabilitation, because of lactose intolerance, and extreme caution has to be exercised in diet prescription because of protein intolerance (Roe, 1979).

TOBACCO

Smoking diminishes appetite and on average, smokers have lower body weights than nonsmokers. Nevertheless, on average, smokers have greater waist-to-hip circumference ratios than nonsmokers. This suggests that smoking may have an effect on body-fat distribution. Central (torso) adiposity, reflected by this change in circumferential measurements, has been shown to worsen the risk of cardiovascular disease. Cessation of smoking is usually associated with moderate weight gain, caused at least in part by increased food intake (Troisi et al., 1991).

OTHER ADDICTIVE DRUGS

Multiple Substance Abuse and Nutrition. Drug abuse includes the experimental use of various addictive drugs as well as chronic addiction to one or more of these social drugs. The term *addiction* here refers both to PHYSICAL DEPENDENCE on the drug, such that when the drug is withdrawn specific physical withdrawal symptoms occur, and to PSYCHOLOGICAL DEPENDENCE on the drug—even without physical dependence. Alcohol has been called *the* GATEWAY DRUG, because its early use is frequently accompanied by and/or followed by use of other drugs.

Effects of multiple-drug use on nutrition depend on the properties and toxic characteristics of the drug most used, as well as on doses, frequencies, and duration of use/abuse, and the time in life when the drug or drugs are abused. NARCOTIC drugs, such as HEROIN, impair appetite—so food intake is often diminished. If the drug is injected intravenously, malnutrition may be secondary to blood-borne bacterial

infection or ACQUIRED IMMUNODEFICIENCY SYNDROME (AIDS). AMPHETAMINE ("speed") is the stimulant drug that has the most inhibitory effect on appetite; if taken in large doses, it also prevents sleep and stimulates activity—therefore energy expenditure may be high and weight loss is common. COCAINE and CRACK are also stimulants, they reduce appetite and may in addition induce gastrointestinal symptoms such as nausea, which further lessen food intake (Brody et al., 1990).

Substance Abuse and Nutrition in Pregnancy. Relationships between substance abuse and impaired nutrition of the fetus and newborn have been summarized in a 1990 report by the National Academy of Sciences, *Nutrition during Pregnancy*.

Alcohol use during PREGNANCY has led to poor birth outcomes. One condition is infants with specific defects in neuronal and cranial development, designated FETAL ALCOHOL SYNDROME. Even the daily drinking of more than two glasses of wine or a daily mixed drink has led to fetal alcohol syndrome, but this condition is most common among the offspring of mothers who are chronic drinkers or binge drinkers.

Alcohol use during pregnancy is also known to be associated with prenatal and postnatal growth retardation. After birth, infants of heavy drinkers may fail to suck, either because of the presence of withdrawal symptoms or because of cleft palate (which may be part of the fetal alcohol syndrome).

Cigarette smoking during pregnancy can affect both maternal and fetal nutrition (Werler et al., 1985). Effects are due to increased metabolic rate in smokers and to toxic effects from tobacco that impair the mother's utilization of certain nutrients, including iron, Vitamin C, folic acid (part of the B complex), and zinc. Low-birthweight infants are more likely to be the offspring of smokers than of nonsmokers—because their caloric intake is likely to be less and because the transfer of nutrients from the mother to the fetus via the placenta may be reduced in smokers.

Cocaine and amphetamine use in pregnancy also lead to increased numbers of low-birthweight infants. This may be caused by low food intake by the mother, since these drugs reduce appetite. The risk of malnutrition in the newborns of women who have used cocaine during pregnancy is caused by the abnormal development of the infant's small intestine. These intestinal disorders in the infant may be extremely severe and may be associated with entero-

colitis or bowel perforation, which may be fatal. If these infants survive, special methods of feeding via a vein are required. Although drugs other than cocaine are known to cause constriction of blood vessels in the pregnant woman, none other than cocaine have been shown to produce these bowel disorders in infants (Telsey et al., 1988; Spinazzola et al., 1992).

(SEE ALSO: *Alcohol History of*; *Complications: Liver [Alcohol]*; *Overeating and Other Excessive Behaviors*)

BIBLIOGRAPHY

ANSTIE, F. E. (1877). *On the uses of wine in health and disease.* London: Macmillan.

ATWATER, W. O., & BENEDICT, F. B. Experiments on the metabolism of matter and energy in the human body. (Bulletin No. 69). Washington, DC: U.S. Department of Agriculture.

BIKLE, D. D. ET AL. (1985). Bone disease in alcohol abuse. *Annals of Internal Medicine, 103*, 42–48.

BRIGDEN, W., & ROBINSON, J. (1964). Alcoholic heart disease. *British Medical Journal, 2*, 1283–1289.

BRODY, S. L., SLOVIS, C. M., & WRENN, K. D. (1990). Cocaine-related medical problems: Consecutive series of 233 patients. *American Journal of Medicine, 88*, 325–330.

HUTCHISON, R. (1905). *Food and the principles of dietetics.* New York: W. Wood.

JOLLIFFE, N. (1940). The influence of alcohol on the adequacy of B vitamins in the American diet. *Quarterly Journal of the Study of Alcohol, 1*, 74–84.

LIEBER, C. S. (1991). Perspectives: Do alcohol calories count? *American Journal of Clinical Nutrition, 54*, 976–982.

LIEBER, C. S. (1988). The influence of alcohol on nutritional status. *Nutrition Review*, 241–251.

MORGAN, M. Y. (1982). Alcohol and nutrition. *British Medical Bulletin, 38*, 21–29.

NUTRITION DURING PREGNANCY. (1990). Washington, DC: National Academy Press.

RAZAY, G., ET AL. (1992). Alcohol consumption and its relation to cardiovascular risk factors in British women. *British Medical Journal, 304*, 80–83.

ROE, D. A. (1992). *Geriatric nutrition.* Englewood Cliffs, NJ: Prentice Hall.

ROE, D. A. (1979). *Alcohol and the diet.* Westport, CT: AVI.

SPINNAZOLA, R., ET AL. (1992). Neonatal gastrointestinal complications of maternal cocaine abuse. *New York State Journal of Medicine, 92*, 22–23.

TELSEY, A. M., MERRIT, A., & DIXON, S. D. (1988). Cocaine exposure in a term neonate: Necrotizing enterocolitis as a complication. *Clinical Pediatrics, 27*, 547–550.

TROISI, R. J., ET AL. (1991). Cigarette smoking, dietary intake and physical activity: Effects on body fat distribution—the Normative Aging Study. *American Journal of Clinical Nutrition, 53*, 1104–1111.

VICTOR, M., ADAMS, R. D., & COLLINS, G. H. (1971). *The Wernicke-Korsakoff syndrome: A clinical and pathological study of 245 patients, 82 with post-mortem examination.* Philadelphia: F. A. Davis.

WERLER, M. M., POBER, B. R., & HOLMES, L. B. (1985). Smoking and pregnancy. *Teratology, 32*, 473–481.

DAPHNE ROE

Route of Administration

The mode of drug administration—ingestion (by mouth), insufflation (snorting), inhalation (smoking), or injection (intravenous, subcutaneous, or intramuscular)—can be responsible for a number of medical complications to alcohol and other drug use. In the following, these complications are discussed as direct and indirect results of the various modes (route) of administration in the above order.

COMPLICATIONS DUE TO INGESTION

Ingestion is the way ALCOHOL, liquid medicines, pills and capsules are usually taken. Ingested drugs enter the gastrointestinal (GI) system, undergo some digestive processing, and enter the bloodstream through the walls of the stomach and intestines. Most medical complications from drug ingestion are a result of the corrosive and irritant effects of the drugs on the GI system. Alcohol and a variety of medicines, including aspirin, can cause intense, localized irritation to the GI mucous membranes, leading to ulceration and GI bleeding. Pharmaceutical manufacturers attempt to decrease the danger of GI irritation by adding buffers to their pills and capsules. Buffers are inert or nonactive ingredients that cushion the corrosive effect of the active ingredients. However, if drug users attempt to dissolve pills intended for oral use and inject them, these buffers will often cause problems, such as abscesses or embolisms.

COMPLICATIONS DUE TO INSUFFLATION (SNORTING)

Medical complications from insufflation (snorting) are usually caused by stimulant drugs, such as the AMPHETAMINES or COCAINE. These drugs are breathed into the nose and absorbed into the bloodstream through the capillaries in the nasal mucous membrane. While these drugs cause a certain amount of surface irritation, the major damage is caused by their action as vasoconstrictors—they reduce the diameter of blood vessels, and with chronic use can severely limit the delivery of blood through the capillaries to the inner membranes of the nose. The result of this is that tissue damaged by contact with the drugs is unable to repair itself, and progressive necrosis (tissue death) follows. With chronic cocaine use, this process can result in actual holes through the septum (the dividing tissue) between the nostrils. When tobacco is insufflated as snuff, the risk of cancer of the nasal passages is increased.

COMPLICATIONS DUE TO SMOKING (INHALING)

The fastest delivery of large amounts of drug directly to the brain is through smoking (inhaling). Drugs taken in this way go directly to the lungs and are absorbed along with oxygen directly into the blood heading for the brain. The two terms, *smoking* and *inhaling*, as a means of drug intake, are clearly differentiated when, on the one hand, material is actually burned and the resulting smoke is taken into the lungs—as with TOBACCO or MARIJUANA—or on the other hand, when fumes from volatile substances are inhaled, such as glue or gasoline. They may be confused or used interchangeably, however, when material is vaporized through heat and the vapor is inhaled—as with cocaine FREEBASE (crack).

Smoking. Smoke from any material will act as an irritant to the lungs and bronchial system, eventually causing problems that can range from chronic bronchitis to emphysema or cancer of the mouth, throat and/or lungs. Both tobacco and marijuana contain a number of tars and potential carcinogens (cancer-triggering substances) and both produce potentially toxic concentrations of carbon monoxide. While it has been argued that tobacco is the worst danger because it is smoked very frequently, it has also been pointed out that the use mode of marijuana is worse—holding the smoke in the lungs for a long time. The argument is moot, since both can produce profound damage. As a vasoconstrictor, nicotine in tobacco promotes mouth ulceration and gum disease. It can be said that people who smoke lose their teeth, while those who don't, don't. Besides its irritant effects, the smoking of tobacco may also promote respiratory disease by weakening the immune system and by paralyzing the cilia (the tiny hairlike organs) in the lungs that push out foreign matter.

Inhalation (Sniffing). The inhalation (sniffing) of volatile hydrocarbons, such as solvents, can cause death by asphyxiation or suffocation, can impair judgment, and may produce irrational, reckless behavior. Abnormalities also have occurred in liver and kidney functions, and bone-marrow damage has occurred. These may be due to hypersensitivity to the substances or chronic heavy exposure. Chromosome damage and blood abnormalities have been reported, and solvents have been cited as a cause of gastritis, hepatitis, jaundice, and peptic ulcers—such effects are due more to the actions of the drugs than to the route of administration. Chronic users have developed slow-healing ulcers around the mouth and nose, loss of appetite, weight loss, and nutritional disorders. Irreversible brain damage has been reported, too. Many deaths attributed to solvent inhalants are caused by suffocation when users pass out with the plastic bags containing the substance still glued to their noses and mouths. There is also a very real danger of death from acute solvent poisoning or aerosol inhalation. The mere provision of adequate ventilation and the avoidance of sticking one's head in a plastic bag are by no means sufficient safeguards against aerosol dangers.

Other hazards may include freezing the larynx or other parts of the airway when refrigerants are inhaled, and potential spasms as these areas defrost. Blockage of the pulmonary membrane, through which oxygen is absorbed into the lungs, can occur. Death may also result from the ingestion of toxic ingredients along with the aerosol substance. The possibility is made more likely by the fact that commercial products not produced for human consumption are not required to list their ingredients on the label. Individual substances may produce a spectrum of toxic reactions depending on their contents. These have included gastric pain, headaches, drowsiness, irritability, nausea, mucous-membrane irrita-

tion, confusion, tremors, nerve paralysis, optic-nerve damage, vomiting, lead poisoning, anemia, and so on. The inhaling of aerosol fluorocarbons can cause "sudden-sniffing death" (SSD), wherein the heart is hypersensitized to the body's own hormone epinephrine (adrenaline), leading to a very erratic heartbeat, increased pulse rate, and cardiac arrest.

The inhaling of amyl, butyl or isobutyl nitrites can cause intense headaches, an abrupt drop in blood pressure, and loss of consciousness through orthostatic hypotension (increased heart rate and palpitations), with a threat of myocardial infarction (heart attack).

COMPLICATIONS DUE TO INJECTION

The injection of drugs generally involves the use of the hypodermic needle, first invented in the early nineteenth century and used initially for the medical delivery of the opiate painkiller MORPHINE, for the rapid control of intense PAIN. This combination was first used extensively for battlefield wounds during the Crimean War (1853–1856) and the American Civil War (1861–1865). As its name implies, the hypodermic needle pierces the skin—the dermis. Hypodermic injections may be subcutaneous, directly beneath the skin surface; intramuscular, into the muscle tissue; or intravenous, into a blood vessel. (*Note:* Although a number of injection-related medical complications are directly skin-related, these are discussed in the article *Complications: Dermatological.*

While the hypodermic needle is the primary means of drug injection, drug addicts who do not have access to hypodermics have made use of a number of ingenious, and often very dangerous, substitutes. Nonhypodermic-needle means of injection may involve such paraphernalia as lancets or scalpels, or any small sharp blade to make an opening, and the insertion of an eyedropper, tubing and bulb, or any means of squirting the drug into the resultant wound. In extremes, addicts have used such implements as a pencil, ballpoint or fountain pen, or the sharpened end of a spoon.

Intra-arterial Injection. Injections are never made intentionally into arteries. Accidental intra-arterial injection will produce intense pain, swelling, cyanosis (blueness), and coldness of the body extremity injected. Intra-arterial injection resulting in

these symptoms is a medical emergency and, if untreated, may produce gangrene of the fingers, hands, toes, or feet and result in loss of these parts.

Transmittal of Disease through Injection. The greatest number and variety of medical complications of drug use caused by the mode of administration occur as a result of injection. Among the highest risk, and that with the most frequent fatal and disabling consequences is the transmittal of disease through the use of unsterile needles and the sharing of such needles.

Human Immunodeficiency Virus (HIV). Needle-using drug abusers comprise one of the primary high-risk populations for contracting human immunodeficiency virus (HIV). The primary recognized routes of transmission for HIV are (1) sexual contact through unprotected anal or vaginal intercourse—particularly if there are damaged tissue or sores present that provide direct access to the bloodstream; (2) contact with infected blood through needle sharing or through transfusions of blood or blood products; and (3) in utero or at-birth transmission from a mother to her baby. ACQUIRED IMMUNODEFICIENCY SYNDROME (AIDS), the most severe and life-threatening result of HIV infection, involves the destruction of a person's immune system and the development of cancers and infections that can no longer be fought off.

The incidence of HIV infection among needle-using drug abusers is closely related to local use traditions, habits, and the prevalence of HIV infection among other addicts. The highest incidence is in areas such as New York City, where there is a tradition of needle sharing or where "shooting galleries"—places where users can rent or share "works"—are commonly utilized and where there was a high prevalence of HIV among the homosexual population. Users in other geographical locations, such as San Francisco, seem to be more conservative in their social-usage patterns, and when they do share needles, tend to keep the same "shooting partners" over a longer period of time. HIV-prevention efforts in some areas have focused on NEEDLE AND SYRINGE EXCHANGE, while others, particularly where needle exchange is not legalized, have community-outreach workers teaching users how to sterilize their needles between each use with household bleach. The gist of both campaigns is that users who share their needles or who use dirty needles are at risk for contracting HIV through their

drug use. Those who use sterile needles are not. Both approaches are considered stopgap, however, and are apt to be condemned as "encouraging of drug abuse."

All needle-using drug abusers are considered at extremely high risk for HIV infection, and HIV screening is performed routinely at most drug-treatment centers. The virus has a very long incubation period and may be present for seven or more years before active symptoms of opportunistic disease appear. Early symptoms may include: a persistent rash or lesion; unexplained weight loss; persistent night sweats or low-grade fever; persistent diarrhea or fatigue; swollen lymph glands; DEPRESSION or states of mental confusion.

Hepatitis and Other Liver Disorders. Hepatitis B, and related strains, often referred to as serum (fluid-related) hepatitis, are the most common medical complication of needle drug use. Like HIV, hepatitis can spread in other ways than needle use, such as sexual intercourse or other direct sharing of blood and bodily fluids. Several strains, however, can be spread by contaminated foods, particularly shellfish, or by unhygienic practices in food handling. Current research indicates that some forms of hepatitis spread via an anal/oral progression—so it is recommended that hands are washed thoroughly after all bowel movements or any other anal-area or fecal-matter handling, as a means of prophylaxis.

Unlike AIDS, hepatitis is often not fatal if it is detected and treated at an early stage. Symptoms of all forms of hepatitis include fatigue, loss of appetite, pain in the upper abdomen, jaundice—yellow skin and a yellowish-to-chartreuse tinge to the sclerae (white of the eye), general itching, dark urine reaching the color of cola drinks with light-tan to cream-colored feces, and mental depression. Gamma globulin injection can provide short-term immunity to all forms of hepatitis and can reduce the symptoms of serum hepatitis if it is given during the gestation period. Treatment includes bed rest, nutritional support, and avoidance of alcohol or any other substance that may further irritate the liver. Caregivers should wear rubber gloves for handling patients. Patients with any form of hepatitis should avoid preparing food for others and use separate towels, bedlinens, and eating utensils until symptoms disappear. Toilet seats and any spilled bedpan matter should be disinfected and hands should then be washed thoroughly with soap. Condoms should be used for any genital contact.

Hepatitis can cause hepatic fibrosis—the development of fibrous tissue in the liver. It can also cause or exacerbate cirrhosis (scarring of the liver), although this is most often a result of chronic alcohol abuse. Symptoms of cirrhosis include jaundice (yellowish skin and eye whites), fatigue, ankle swelling, enlargement of the abdomen, and a full feeling in the right upper abdomen.

Tetanus and Malaria. According to Senay and Raynes, the first case of tetanus associated with needle-using substance abuse was reported in England in 1886. By the 1990s, between 70 and 90 percent of tetanus cases have occurred to drug abusers. As a medical complication to drug injection, tetanus most often occurs from "skin-popping"—which is cutaneous injection. A majority of cases occur in women, and this is attributed to less-substantial venous development than in men and a smaller population with tetanus immunization.

Malaria (caused by the *Plasmodium* parasite) was first reported among drug users in the United States in 1926. It affects intravenous drug abusers and was brought to this country by needle-sharing sailors who had been exposed to malaria in Africa. The initial outbreak in New Orleans spread to New York City in the 1930s and resulted in several hundred deaths from tertian malaria among drug abusers. A second outbreak occurred in the 1970s, as a result of malaria-infected veterans returning from Vietnam.

The spread of both these diseases among needle-sharing drug abusers has been kept somewhat in check, particularly on the East Coast and in Chicago, by the inclusion of 15 to 30 percent quinine (a natural antimalarial), as filler, to stretch profits in illicit opioid drug mixtures in those areas. Quinine (an alkaloid from chinchona bark) is a protoplasmic poison that prevents the germination of the fastidious tetanus anaerobe, *Clostridium tetani*, under the skin and in adjacent muscle tissue. Although the quinine amount is not sufficient to eradicate malaria once it has taken hold in the body, it does help prevent the disease by killing the malarial parasites in the hypodermic syringe.

COMPLICATIONS TO HEART AND BLOOD VESSELS

Drug abuse is related to a number of heart and blood vessel medical complications. Some of these, such as alcohol cardiomyopathy, are a direct result

of the drug's toxic effects. Others are at least partially related to needle use.

Endocarditis, an infection of the tissues in the heart, usually a heart valve, is a progressive disease characterized by frequent embolization (obstruction of blood vessels) and severe heart-valve destruction that can be fatal if not treated. This disease can result from repeated injection of the infective agents into the blood system, usually from nonsterile needles and/or unusual methods of injection. Infective endocarditis is highly prevalent among drug abusers and should be suspected in any needle-using abuser who shows such symptoms as the following: fever of unknown origin; heart murmur; pneumonia; embolic phenomena; blood cultures that are positive for *Candida*, *Staphylococcus aureus* or *enterococcus*, or Gram-negative organisms.

MISCELLANEOUS COMPLICATIONS

Blood-vessel changes caused by necrotizing angiitis (polyarteritis—the inflammation of a number of arteries) or a swelling that leads to tissue loss have been demonstrated in intravenous amphetamine abusers, resulting in cerebrovascular occlusion (blockage in brain blood vessels) and intracranial hemorrhage or stroke.

Problems in the lungs often develop from inert materials that are included as cutting agents or as buffers and binding agents in drugs that come in pill form but are liquified and injected. These substances do not dissolve, so their particles may become lodged in the lungs, causing chronic pulmonary fibrosis and foreign-body granulomas. These same buffers and binding agents may as well become lodged in various capillary systems, including the tiny blood vessels in the eye.

Finally, injection-induced infections reaching the skeleton can be responsible for such bone diseases as septic arthritides and osteomyelitis. Gangrene can develop from cutting off circulation to the extremities and may necessitate amputation or be fatal.

(SEE ALSO: *Inhalant Abuse and Its Dangers; Needle and Syringe Exchange and HIV/AIDS*)

BIBLIOGRAPHY

COHEN, S., & GALLANT, D. M. (1981). *Diagnosis of drug and alcohol abuse.* Brooklyn: Career Teacher Center, State University of New York.

SENAY, E. C., & RAYNES, A. E. (1977). *Treatment of the drug abusing patient for treatment staff physicians.* Arlington, VA: National Drug Abuse Center.

SEYMOUR, R. B., & SMITH, D. E. (1990). Identifying and responding to drug abuse in the workplace. *Journal of Psychoactive Drugs, 22*(4), 383–406.

SEYMOUR, R. B., & SMITH, D. E. (1987). *The physician's guide to psychoactive drugs.* New York: Hayworth.

WILFORD, B. B. (1981). *Drug abuse: A guide for the primary care physician.* Chicago: American Medical Association.

DAVID E. SMITH
RICHARD B. SEYMOUR

COMPULSORY TREATMENT *See* Civil Commitment; Coerced Treatment for Substance Offenders; Narcotic Addict Rehabilitation Act (NARA).

CONDITIONED TOLERANCE *Tolerance* refers to the diminishment or the loss of a drug effect over the course of repeated administrations. Some researchers have postulated that an important factor in the development of tolerance is Pavlovian conditioning of drug-compensatory responses. The administration of a drug may be viewed as a Pavlovian conditioning trial. The stimuli present at the time of drug administration are the conditional stimulus (CS), while the effect produced by the drug is the unconditional stimulus (UCS). Many drug effects involve disruption of the homeostatic level of physiological systems (e.g., alcohol lowers body temperature), and these disruptions elicit compensatory responses that tend to restore functioning to normal levels. The compensatory, restorative response to a drug effect is the unconditional response (UCR). Repeated administrations of a drug in the context of the same set of stimuli can result in the usual predrug cues coming to elicit as a conditional response (CR) the compensatory, restorative response. The conditional drug-compensatory CR would tend to reduce the drug effect when the drug is administered with the usual predrug cues—thus accounting for tolerance, or at least some aspects of tolerance.

One test of the Pavlovian conditioning model of tolerance is whether conditional drug-compensatory responses are elicited by predrug cues. In one experiment with rats (Crowell, Hinson, & Siegel, 1981), injections of alcohol in the context of one set of stim-

uli were alternated with injections of saline solution in the context of a different set of stimuli for several days. Each day, the rats' body temperatures were measured. Alcohol lowered body temperatures the first time it was given, but this effect diminished over the course of the repeated alcohol administrations—that is, tolerance developed to the hypothermic effect of alcohol. To determine if a drug-compensatory CR was elicited by the usual predrug cues, the rats were given a placebo CR test. In a placebo CR test, saline solution is administered instead of the drug. The placebo CR test was given to some rats under conditions where they were expecting alcohol; that is, saline was administered with the usual predrug cues. For the remaining rats, the placebo CR test was given under conditions where there should have been no expectancy of alcohol, that is saline was administered with cues that had previously signaled only saline. Rats given saline with the usual predrug cues had elevated body temperatures, while rats given saline without the usual predrug cues showed little temperature change. Thus, it was possible to directly observe the drug-compensatory CR, in this case hyperthermia opposed to the hypothermic effect of alcohol. Other experiments similar to the one just described have found drug-compensatory CRs following the development of tolerance to various effects of OPIATES, BARBITURATES, and BENZODIAZEPINES (Siegel, 1983).

Conditioned responses occur only when the conditional stimulus is presented. If drug-compensatory CRs contribute to tolerance, then tolerance should only be evident in the presence of the usual predrug cues that are the CS. This expectation was tested in the experiment by Crowell, Hinson, and Siegel (1981), involving tolerance to the hypothermic effect of alcohol. After all rats had developed tolerance to the hypothermic effect of alcohol, a test was given in which some rats received alcohol with the usual predrug cues, while other rats received alcohol when the usual predrug cues were not present. Although all rats had displayed tolerance prior to the test, only those rats given alcohol in the presence of the usual predrug cues (i.e., with the CS) showed tolerance during the test. The explanation of this "situational specificity" of tolerance is that when alcohol is given with the usual predrug cues, the drug-compensatory CR occurs and reduces the drug effect—but when alcohol is given without the usual predrug cues, the drug-compensatory CR does not occur and the drug effect is not reduced. Other research has demon-

strated situational specificity with regard to tolerance to opiates, barbiturates, and benzodiazepines (for a complete review see Siegel, 1983).

In order to eliminate a CR, it is necessary to present the CS not followed by the UCS, a procedure termed *extinction*. Research indicates that the loss of tolerance occurs as a result of extinction of drug-compensatory CRs. Again referring to the experiment of Crowell, Hinson, and Siegel (1981), rats were given alcohol in the presence of a consistent set of cues until tolerance developed. Then, all drug injections were stopped for several days. During this period some animals were given extinction trials, in which the usual predrug cues were presented but only saline was injected. The other animals did not receive extinction trials and were left undisturbed during this time. Subsequently, all animals were given a test in which the drug was given with the usual predrug cues. The animals that had received extinction trials were no longer tolerant, whereas animals that had not been given extinction trials retained their tolerance. Similar results—in which tolerance is retained unless extinction trials are given—occur for tolerance to opiates, barbiturates, and benzodiazepines (Siegel, 1983).

The drug-compensatory CRs that contribute to tolerance may also be involved in withdrawal-like symptoms that occur in detoxified drug addicts. Detoxified addicts often report experiencing withdrawallike symptoms when they return to places where they formerly used drugs, although they are now drug free. The places where the addict formerly used drugs act as CSs and still elicit drug-compensatory CRs; even when the addict is drug free, the drug-compensatory CRs achieve expression. Thus, it is postulated that the drug-compensatory CRs elicited by the usual predrug cues in the drug-free postaddict result in a withdrawal-like syndrome (Hinson & Siegel, 1980). This conditional postdetoxification withdrawal syndrome may motivate the postaddict to resume drug taking (to alleviate the symptoms).

(SEE ALSO: *Addiction: Concepts and Definitions; Causes of Drug Abuse: Learning; Tolerance and Physical Dependence; Wikler's Pharmacologic Theory of Drug Addiction*)

BIBLIOGRAPHY

CROWELL, C., HINSON, R. E., & SIEGEL, S. (1981). The role of conditional drug responses in tolerance to the hy-

pothermic effects of alcohol. *Psychopharmacology, 73,* 51–54.

HINSON, R. E., & SIEGEL, S. (1980). The contribution of Pavlovian conditioning to ethanol tolerance and dependence. In H. Rigter & J. C. Crabbe (Eds.), *Alcohol tolerance and dependence.* Amsterdam: Elsvier/North Holland Biomedical Press.

SIEGEL, S. (1983). Classical conditioning, drug tolerance, and drug dependence. In F. B. Glaser et al. (Eds.), *Research advances in alcohol and drug problems,* vol. 7. New York: Plenum.

RILEY HINSON

CONDUCT DISORDER AND DRUG USE

A behavior pattern characterized by stealing, violence, running away from home, and truancy occurs in about 10 percent of children under 16 years of age. Within the framework of the DIAGNOSTIC AND STATISTICAL MANUAL *of Mental Disorders,* 3rd edition, revised (DSM-III-R), this serious and persistent pattern of antisocial behavior is diagnostically labeled *conduct disorder* (CD); it is the most common psychiatric disorder in emotionally disturbed youth, present in about 75 percent of cases. Boys outnumber girls in ratios of 4:1, but 8:1 for property and violent crimes. Emerging evidence suggests, however, that the gender gap is narrowing; and by adolescence, commonly associated problems include alcoholism, drug addiction, criminality, incarceration, sexually transmitted diseases (STDs), pregnancies, prostitution, traumatic injuries, dropping out of school, and comorbid psychiatric disorders.

DIAGNOSIS

In the American Psychiatric Association classification system for diagnosing mental disorder (DSM-III-R), conduct disorder is defined as "a persistent pattern of conduct in which the basic rights of others and major age-appropriate societal norms or rule are violated" (American Psychiatric Association, 1987). In the 1990s, conduct disorder is one of the most valid and reliably diagnosed psychiatric disturbances. The problem behavior is transsituational—it is manifested in the home, at school, and in daily social functioning. Often, CD youth are suspicious of others and, consequently, they misinterpret the intentions and actions of others. By adolescence, ag-

gression may become so severe that violent assault, rape, and homicide are committed. Precocious sexual behavior and sexual misbehavior, especially among females, are also common. Denial and minimization generally occur when the youngsters are confronted about their behavior. Typically, feelings of guilt are not experienced.

Other, less severe, types of behavior disorders are also known. The most common that resemble CD are (1) adjustment disorder with disturbances of conduct; (2) childhood (or adolescent) antisocial behavior; and (3) oppositional defiant disorder.

Substantial differences in the behavioral manifestations of CD have prompted efforts to develop subtypes. The most well-known subtyping criteria are (1) socialized versus unsocialized; (2) aggressive versus nonaggressive; and (3) overt versus covert. Just one variant of CD, the solitary aggressive type, characterizes approximately 50 percent of incarcerated youth; they are usually socioeconomically disadvantaged and typically derive from broken homes or dysfunctional families. Moral development is arrested, cognitive abilities are low, and behavior is often dangerous both to self and others. This CD variant should not be confused with *adaptive delinquency,* in which the behavior is an attempt to adjust to the manifold disadvantages of inner-city living.

NATURAL HISTORY

Other psychiatric disorders frequently occur in conjunction with conduct disorder. The most prevalent comorbid (coexisting) psychiatric disorder is *attention deficit disorder.* By adolescence, the comorbid conditions of psychoactive substance use disorder with depressive disorders often emerge; however, virtually any type of psychiatric disorder can be present concurrently with CD (Rutter, 1984). By adulthood, an ANTISOCIAL PERSONALITY disorder is the most common outcome of CD; this disorder may also be accompanied by any other psychiatric disorder.

Among those who have CD with attention deficit disorder, the onset age of behavior problems tends to be earlier and more severe than in cases with either disorder alone. In the situation where both are present, children are also at greater risk for developing criminal behavior and substance abuse by adolescence or young adulthood.

The coexistence of CD and substance abuse has been frequently observed. It is estimated that as

many as 50 percent of serious offenders are substance abusers. In these cases, CD usually preceded the onset of substance abuse. Some evidence has been marshaled to suggest that, for many individuals, substance abuse and CD are the overt expressions of a common underlying predisposition. Only in some cases, does the onset of CD follow the onset of substance abuse. Drug use during adolescence, by virtue of its pattern of illegal behavior plus association with nonnormative peers, increases the risk for violent assault as well as getting arrested and convicted for drug possession or distribution. In effect, the use of drugs in this circumstance socializes a person to a deviant lifestyle by early to mid-adolescence.

Approximately 30 percent of boys with CD also qualify for a diagnosis of DEPRESSION. In this comorbid condition, there appears to be a lower risk of depression in adulthood compared to cases of depression in childhood without CD. Since the outcome of depressed children with comorbid CD is similar to nondepressed CD children, this suggests that the affective disturbance is a secondary condition.

CD in childhood is associated with an increased risk for antisocial personality disorder in adulthood. Compared to other psychiatric disorders of childhood, CD is the most likely to remain stable. Persistence of conduct problems into adulthood is most likely if the behavior problems are serious, are generalized across multiple environments or situations, have an early age onset, and lead the person into the criminal-justice system (Loeber, 1991).

ETIOLOGY

Adoption and family studies implicate a genetic predisposition for the development of antisocial behavior in many. A genetic propensity does not, however, appear to invariably ensure this adverse outcome. Other complicating factors include being a child in a dysfunctional family where the parents are abusive, neglectful, or absent or where there are poor parenting skills. Alcoholic and physically abusive parents have been frequently linked to CD in their own childhoods. Neurologic injuries (e.g., trauma) and neurodevelopmental disability (e.g., dyslexia) can exacerbate the expression of CD. Socioeconomic and ethnic factors (e.g., POVERTY, street GANGS) also influence the development of CD.

TREATMENT

The following are generally inadequate for the treatment of youth with CD: individual psychotherapy, behavior modification, group counseling, family therapy, milieu therapy, and immersion in a long-term therapeutic community. The most promising approaches emphasize training parents in the skills necessary to promote normal socialization in their children, accompanied by training children in the use of problem-solving strategies (Kazdin et al., 1987). The complexity and severity of CD-associated problems dictate the need for multimodal treatment. Primary consideration should be given to containment and limit setting—which create the conditions for treatment, to provide safety, and to delineate a comprehensive program of intervention encompassing behavioral and social-skill building, family therapy, and educational assistance.

(SEE ALSO: *Adolescents and Drugs; Comorbidity and Vulnerability; Crime and Drugs; Families and Drug Use; Family Violence and Substance Abuse; Vulnerability As Cause of Substance Abuse*)

BIBLIOGRAPHY

AMERICAN PSYCHIATRIC ASSOCIATION. (1987). *Diagnostic and statistical manual of mental disorders-3rd ed.-rev.* (DSM-III-R). Washington, DC: Author.

KAZDIN, A. E., ET AL. (1987). Effects of parent management training and problem-solving skills combined in the treatment of antisocial child behavior. *Journal of the American Academy of Child and Adolescent Psychiatry, 26*, 416–424.

LOEBER, R. (1991). Antisocial behavior: More enduring than changeable? *Journal of the American Academy of Child and Adolescent Psychiatry, 30*, 393–397.

RUTTER, M. (1984). Psychopathology and development. I. Childhood antecedents of adult psychiatric disorder. *Australian and New Zealand Journal of Psychiatry, 18*, 225–234.

ADA C. MEZZICH
RALPH E. TARTER

CONDUCT DISORDER IN CHILDREN

One of several childhood behavioral disturbances, CONDUCT DISORDER refers to repeated patterns of conduct by a child that violate the basic rights of

others or transgress age-appropriate societal rules. The behavior is socially disruptive and generally more serious in its consequences than typical childhood mischief. The duration of the behavior, its severity, and the kinds of actions involved distinguish conduct disorder from general misbehavior. Conduct disorder is the most common behavioral problem seen in child psychiatric settings in North America.

The behaviors that characterize this disorder include theft, vandalism, physical fights—sometimes with weapons—fire setting, running away from home, truancy, repetitive lying, forcing sexual activity on others, physical cruelty to animals and to people, and substance abuse. Legal involvement may ensue. Different children may manifest different combinations of these behaviors, and these in turn may change at different points of child development. Conduct disorder appears to be more common in boys than in girls.

The etiology of conduct disorder is considered multifactorial. Psychological and social factors believed to contribute to its development include the child's particular temperament, a family history of ANTISOCIAL PERSONALITY disorder or alcohol dependence (or both), poor parenting skills, a chaotic home environment, and lower socioeconomic status. Mild central nervous system abnormalities have been found in children with a history of violent behavior, and they are thought to contribute to the children's impulsivity. ATTENTION-DEFICIT HYPER-ACTIVITY DISORDER and specific developmental disorders are common associated diagnoses. Children who display significant antisocial behavior have a poorer long-term prognosis with greater psychiatric impairment in adulthood (including antisocial personality disorder), poorer educational achievement, overt criminal behavior, higher rates of unemployment, impaired social functioning—and considerably higher rates of smoking and alcohol abuse, illicit drug use and dependence.

Treatment of conduct disorder in children and adolescents can include family therapy, parent management training, behavioral and cognitive therapies, residential treatment programs, and, less frequently, pharmacotherapy.

(SEE ALSO: *Comorbidity and Vulnerability; Crime and Drugs*)

BIBLIOGRAPHY

CANTWELL, D. P. (1989). Conduct disorder. In H. I. Kaplan & B. J. Sadock (Eds.), *Comprehensive textbook of psychiatry*, 5th ed., vol. 2. Baltimore, MD: Williams & Wilkins.

CHAMBERLAIN, C., & STEINHAUER, P. D. (1983). Conduct disorders and delinquency. In P. D. Steinhauer & Q. R. Grant, (Eds.), *Psychological problems of the child in the family*, 2nd ed. rev. New York: Basic Books.

MYROSLAVA ROMACH
KAREN PARKER

CONTINGENCY CONTRACTS These have been used extensively and effectively in the treatment of a wide variety of behavioral disorders, including drug dependence. They have been used effectively in the treatment of a variety of forms of drug dependence, including ALCOHOL (Hunt & Azrin, 1973; Miller, 1972), AMPHETAMINE (Boudin, 1972), COCAINE (Anker & Crowley, 1982; Higgins et al., 1993, 1994), MARIJUANA (Budney et al., 1991), OPIOIDS (Hall & Burmaster, 1977), and SEDATIVES (Pickens & Thompson, 1984).

A contingency contract is usually a written document that carefully specifies in objective detail the desired behavior change, the consequences to follow success or failure in making that change, and the time required for completing the contract. Details on how to write an effective contingency contract are available (e.g., DeRisi & Butz, 1975).

Contingency contracts are appropriately considered a major subset of *contingency-management procedures*. That is, contingency contracts always include some form of contingency management, although the converse is not necessarily true. The empirical evidence supporting the efficacy of contingency-management procedures in the treatment of drug dependence is compelling. Stitzer and colleagues, for example, have demonstrated in numerous controlled studies that explicit use of positive reinforcement contingencies is effective in reducing the use of illicit drugs and in improving compliance with treatment regimens (see Stitzer, Bigelow, & McCaul, 1985). More recent research illustrates their utility in the treatment of cocaine dependence (Budney et al., 1991; Higgins et al., 1993, 1994).

Most typically, but not always, contingency contracts are used as a component of a more comprehensive treatment for drug dependence (e.g., Anker & Crowley, 1982; Budney et al., 1991; Higgins et al., 1991). The first example of directly reducing drug use via a contingency contract comes from a case study conducted with an amphetamine abuser (Boudin, 1972). The patient deposited 500 dollars into a joint account with a therapist. The contract stipulated that evidence of amphetamine use would result in a 50 dollar check being sent as a contribution to one of the patient's most disliked organizations. Any money remaining in the account at treatment termination would be returned to the patient. Only one check had to be mailed; the patient discontinued amphetamine use and remained abstinent through a two-year follow-up.

As was noted, contingency contracts can also be used to improve compliance with treatment regimens. For example, contingency contracts have been used effectively to increase compliance with antabuse therapy in alcoholics (Bigelow et al., 1976; Keane et al., 1984). As another example, Stitzer and colleagues used reinforcement contingencies to improve attendance at counseling sessions among patients enrolled in a methadone clinic for treatment of heroin dependence (Stitzer et al., 1977). Patients generally visit the clinic daily to take their medication under supervision and attend regularly scheduled counseling sessions approximately once a week. Unfortunately, patients often take their medication but fail to attend the required counseling sessions. To counter this problem, patients in this study were allowed to take methadone doses home from the clinic (i.e., a break from the grind of daily clinic attendance) provided that they attend the counseling sessions. Implementation of this contingency increased counseling attendance significantly.

As illustrated in the preceding, the use of contingency contracts can be effective in directly increasing drug abstinence and in improving compliance with treatment regimens for various types of drug dependence. The most notable shortcoming associated with their use is that treatment gains are often lost at some point after the contract expires. This is actually a problem shared by all interventions for drug dependence. Nevertheless, it is an important obstacle to be surmounted. Systematic use of multimodal interventions designed to address the many changes likely to be necessary to achieve long-term absti-

nence appears to offer a reasonable approach to this important challenge (e.g., Higgens et al., 1993, 1994; Hunt & Azrin, 1973).

(SEE ALSO: *Treatment Types: Behavior Modification*)

BIBLIOGRAPHY

ANKER, A. A., & CROWLEY, T. J. (1982). Use of contingency contracts in specialty clinics for cocaine abuse. In L. Harris (Ed.), *Problems of drug dependence 1981, NIDA research monograph, No. 41*. Washington, D.C.: U.S. Government Printing Office.

BIGELOW, G., ET AL. (1976). Maintaining disulfiram ingestion among outpatient alcoholics: A security deposit contingency contracting procedure. *Behavior Research and Therapy, 14*, 378–381.

BOUDIN, H. M. (1972). Contingency contracting as a therapeutic tool in the reduction of amphetamine use. *Behavior Therapy, 3*, 604–608.

BUDNEY, A. J., ET AL. (1991). Contingent reinforcement of abstinence with individuals abusing cocaine and marijuana. *Journal of Applied Behavior Analysis*. In press.

DeRISI, W. J., & BUTZ, G. (1975). *Writing behavioral contracts: A case simulation practice manual*. Champaign, IL: Research Press.

HALL, S. M., & BURMASTER, S. (1977). Contingency contracting as a therapeutic tool with methadone maintenance clients: Six single-subject studies. *Behavior Research & Therapy, 15*, 438–441.

HIGGINS, S. T., ET AL. (1994). Incentives improve outcome in outpatient behavioral treatment of cocaine dependence. *Archives of General Psychiatry, 51*, 568–576.

HIGGINS, S. T., ET AL. (1993). Achieving cocaine abstinence with a behavioral approach. *The American Journal of Psychiatry, 150*, 763–769.

HUNT, G. M., & AZRIN, N. H. (1973). A community-reinforcement approach to alcoholism. *Behavior Research and Therapy, 11*, 91–104.

KEANE, T., ET AL. (1984). Spouse contracting to increase antabuse compliance in alcoholic veterans. *Journal of Clinical Psychology, 40*, 340–344.

MILLER, P. M. (1972). The use of behavioral contracting in the treatment of alcoholism: A case report. *Behavior Therapy, 3*, 593–596.

STITZER, M. L., BIGELOW, G. E., & McCAUL, M. E. (1985). Behavior therapy in drug abuse treatment: Review and evaluation. In R. S. Ashery (Ed.), *Progress in the development of cost-effective treatment for drug abusers*.

NIDA monograph No. 58, 85-1401. Washington, D.C.: U.S. Government Printing Office.

STITZER, M. L., ET AL. (1977). Medication take-home as a reinforcer in a methadone maintenance program. *Addictive Behaviors, 2*, 9–14.

STEPHEN T. HIGGINS
ALAN J. BUDNEY

CONTROLLED SUBSTANCES ACT OF 1970

Until 1970, psychoactive drugs were regulated at the federal level by a patchwork of statutes enacted since the turn of the century. These statutes were shaped by an evolving conception of congressional power under the U.S. Constitution. The first federal law on the subject was the Pure Food and Drug Act of 1906, which required the labeling of substances such as patent medicines if they included designated NARCOTICS (e.g., OPIATES and COCAINE) and were shipped in interstate commerce. In 1909, Congress banned the importation of smoking opium. Then in 1914, in the HARRISON NARCOTICS ACT, Congress deployed its taxing power as a device for prohibiting the distribution and use of narcotics for nonmedical purposes. (The taxing power was used because U.S. Supreme Court decisions implied that Congress would not be permitted to use its power to regulate interstate commerce in banning "local" activities, such as the production and distribution of narcotics.) The scheme established by the Harrison Act required the registration and payment of an occupational tax by all persons who imported, produced, or distributed narcotics; it imposed a tax on each transaction; and it made it a crime to engage in a transaction without paying the tax. Mere possession of narcotics without a prescription was presumptive evidence of a violation of the act. The Marihuana Tax Act of 1937 utilized the same model.

In 1965, Congress prohibited the manufacture and distribution of "dangerous drugs" (stimulants, depressants, and hallucinogens) for nonmedical purposes. By this time, Congress's constitutional authority to enact such legislation under the commerce clause was no longer in doubt. (In 1968, Congress made simple possession of the drugs a misdemeanor.) An important feature of the 1965 "dangerous drug" legislation was its delegation of authority to the secretary of Health, Education and Welfare (HEW) to control previously uncontrolled drugs if they had a "potential for abuse" due to their depressant, stimulant, or hallucinogenic properties. (In 1968, this scheduling authority was transferred to the U.S. attorney general.)

All this legislation was replaced by a comprehensive regulatory structure in the 1970 Controlled Substances Act (CSA). Under the new statutory scheme, all previously controlled substances were classified—in five schedules—according to their potential for abuse and accepted medical utility; an administrative process was then established for scheduling new substances, building on the model of the 1965 act. Schedule 1 lists drugs that have no traditional recognized medical use, such as HEROIN, LSD, and cannabis (MARIJUANA). Schedule 2 lists the drugs with medical uses that have the greatest potential for abuse and dependence, such as MORPHINE and cocaine. The remaining schedules use a sliding scale that balances each drug's ABUSE POTENTIAL and its legitimate medical uses.

Different degrees of control are applied to manufacturers, distributors, and prescribers—depending on the schedule in which the drug has been placed. The regulatory structure of the Controlled Substances Act is predicated on the assumption that tighter controls on legitimate transactions will prevent diversion of these substances and will thereby reduce the availability of these substances for nonmedical use.

The drafting of the Controlled Substances Act reflected a continuing controversy regarding the locus of administrative authority for scheduling new drugs and for rescheduling previously controlled drugs. Under the bill passed by the Senate, this responsibility would have rested with the U.S. attorney general, who was required only to "request the advice" of the secretary of HEW (now Health and Human Services, HHS) and of a scientific advisory committee; the attorney general was not required to follow this advice although the various criteria in the act require primarily scientific and medical judgments. The Senate rejected an amendment that would have made the recommendations of the "advisor" binding on the attorney general. Under the bill passed in the House of Representatives, however, the secretary's decision declining to schedule a new drug was binding on the attorney general, and the secretary's recommendation concerning rescheduling was binding as to its medical and scientific aspects. The House version prevailed in the 1970 law as it was finally adopted.

After enactment of the federal Controlled Substances Act, the National Conference of Commissioners on Uniform State Laws promulgated a Uniform Controlled Substances Act, which was modeled after the federal act. (Earlier state laws were modeled on the 1934 Uniform Narcotic Drug Act, which had also been promulgated by the National Conference.) Every state has enacted the Uniform Controlled Substances Act.

(SEE ALSO: *Anslinger, Harry J., and U.S. Drug Policy; Controls: Scheduled Drugs; Legal Regulation of Drugs and Alcohol*)

BIBLIOGRAPHY

BONNIE, R. J., & WHITEBREAD, C. H., II. (1974). *The marihuana conviction.* Charlottesville, VA: University Press of Virginia.

MUSTO, D. F. (1987). *The American disease.* New York: Oxford University Press.

SONNENREICH, M. R., ROCCOGRANDI, A. J., & BOGOMOLNY, R. L. (1975). *Handbook on the 1970 federal drug act.* Springfield, IL: Charles C. Thomas.

RICHARD BONNIE

CONTROLS: SCHEDULED DRUGS/DRUG SCHEDULES, U.S.

The Comprehensive Drug Abuse Prevention and Control Act of 1970, commonly known as the CONTROLLED SUBSTANCE ACT (CSA) establishes the procedures that must be followed by drug manufacturers, researchers, physicians, pharmacists, and others involved in the legal manufacturing of, distributing of, prescribing of, and dispensing of controlled drugs. These procedures provide for accountability of a drug from its initial production through distribution to the patient and are intended to reduce widespread diversion of controlled drugs from legitimate medical or scientific use.

CRITERIA FOR CONTROLLING AND SCHEDULING DRUGS

Several factors are considered before a drug is controlled under this act. These factors include the potential for abuse (i.e., history, magnitude, duration, and significance), risk to public health, and potential of physical or psychological dependence.

TABLE 1
Criteria for U.S. Drug Scheduling

Schedule	Potential for:		Medical Use & Safety
	Abuse	Dependence	
I	+ + + +	+ + + +	No
II	+ + + +	+ + + +	Yes
III	+ + +	+ + +	Yes
IV	+ +	+ +	Yes
V	+	+	Yes

Drugs controlled under this act are divided into five Schedules (I–V) according to their potential for abuse, ability to produce dependence, and medical utility. Drugs in Schedule I have a high potential for abuse and/or dependence with no accepted medical use or they lack demonstrated clinical safety. Those in Schedules II–V may have a high potential for abuse or ability to produce dependence but also have an accepted medical use. (However, some substances which have no accepted medical use but which are precursors to clinically useful substances may also be found in Schedules II–V. For example, thebaine, found naturally in OPIUM, has no medical use but it is a substance used in the manufacture of CODEINE and a series of potent OPIOID compounds as well as opioid ANTAGONISTS.) The potential for abuse and the ability to produce dependence is considered to be the greatest for Schedule I and II drugs and progressively less for Schedule III, IV, and V (see Table 1).

The amount of controlled drug in a product can also determine the schedule in which it is placed. For example, AMPHETAMINE, METHAMPHETAMINE, and codeine, as pure substances, are placed in Schedule II; however, these same drugs in limited quantities and in combination with a noncontrolled drug are placed in Schedules III and V. Drugs in Schedule V generally contain limited quantities of certain narcotic drugs used for cough and antidiarrheal purposes and can only be distributed or dispensed for medical purposes.

LISTS OF SCHEDULED DRUGS

Drugs controlled under the CSA are listed by schedule and drug class in Table 2 (Schedules I and II) and Table 3 (Schedules III, IV, and V). A listing

TABLE 2
List of Controlled Drugs

SCHEDULE I

Opiates		Opium Derivatives	Hallucinogens	Depressants	Stimulants
Acetyl-alpha-methylfentanyl	Hydroxpethidine	Actorphine	4-bromo-2,5-DMA	Mecloqualone	Fenethylline
Acetylmethadol	Ketobemidone	Acetyldihydrocodeine	2,5-DMA	Methaqualone	(±) cis-4-methylam-
Allylprodine	Levomoramide	Benzylmorphine	PMA		inorex
Alphameprodine	Levophenacylmorphan	Codeine methylbromide	MMDA		N-ethylamphetamine
Alphamethadol	3-methylfentanyl	Codeine-N-Oxide	DOM, STP		N,N-dimethyl-am-
Alpha-methylfentanyl	3-methylthiofentanyl	Cyprenorphine	MDA		phetamine
Alpha-methylthiofentanyl	Morpheridine	Desomorphine	MDMA		
Benzethidine	MPPP	Dihydropmorphine	MDEA		
Betacetylmethadol	Noracymethadol	Drotebanol	N-hydroxy MDA		
Beta-hydroxyfentanyl	Norlevorphanol	Etorphine (except HCl salt)	3,4,5-trimethoxy		
Beta-hydroxy-3-methylfentanyl	Normethadone	Heroin	amphetamine		
Betameprodine	Norpipanone	Hydromorphinol	Bufotenine		
Betamethadol	Para-fluorofentanyl	Methyldesorphine	DET		
Betaprodine	PEPAP	Methyldihydromorphine	DMT		
Clonitazene	Phenadoxone	Morphine methylbromide	Ibogaine		
Dextromoramide	Phenampromide	Morphine methylsulfonate	LSD		
Diampromide	Phenomorphan	Morphine-N-Oxide	Marihuana		
Diethylthiambutene	Phenoperidine	Myrophine	Mescaline		
Difenoxin	Piritramide	Nicocodeine	N-ethyl-3-piperidyl		
Dimenoxadol	Proheptazine	Nicomorphine	benzilate		
Dimepheptanol	Properidine	Normorphine	N-methyl-3-piperidyl		
Dimethylthiambutene	Propiram	Pholcodine	benzilate		
Dioxaphetyl butyrate	Racemoramide	Thebacon	Peyote		
Dipipanone	Thiofentanyl		Phencyclidine analogs		
Ethylmethylthiambutene	Tilidine		PCE, PCPy, TCP,		
Etonitazene	Trimeperidine		TCPy		
Etoxeridine			Psilocybine		
Furethidine			Psilocyn		
			Synhexyl		
			Tetrahydrocannabinols		

SCHEDULE II

Opiates	Opium & Derivatives	Hallucinogens	Depressants	Stimulants	Others
Alfentanil	Raw opium	Dronabinol	Amobarbital	Amphetamine	Opium poppy
Alphaprodine	Opium extracts	Nabilone	Glutethimide	Methamphetamine	Poppy straw
Anileridine	Opium fluid		Pentobarbital	Phenmetrazine	Coca leaves
Bezitramide	Powdered opium		Phencyclidine	Methylphenidate	Intermediate precursors
Bulk dextropropoxy-	Granulated opium		Secobarbital		Amphetamine
phene	Tincture of opium				Methamphetamine
Carfentanil	Codeine				Phencyclidine
Dihydrocodeine	Ethylmorphine				
Diphenoxylate	Etorphine hydrochloride				
Fentanyl	Hydrocodone				
Isomethadone	Hydromorphone				
Levo-alphacetylmethadol	Metopon				
Levomethorphan	Morphine				
Levorphanol	Oxycodone				
Metazocine	Oxymorphone				
Methadone	Thebaine				
Metadone-Intermediate					
Moramide-Intermediate					
Pethidine					
Pethidine-Intermediate-A					
Pethidine-Intermediate-B					
Pethidine-Intermediate-C					
Phenazocine					
Piminodine					
Racemorphan					
Racemthrophan					
Sufentanil					

TABLE 3
List of Controlled Drugs

SCHEDULE III

Narcotics	Depressants	Stimulants	Anabolic Steroids		Others
Limited quantities of:	Mixtures of	Limited mixture of	Boldenone	Methandrostenolone	Nalorphine
Codeine,	Amobarbital	Schedule II	Chlorotestosterone	Methenolone	
Dihydrocodeinone,	Secobarbital	amphetamines	Clostebol	Methyltestosterone	
Dihydrocodeine,	Pentobarbital	Benzphetamine	Dehydrochlor-	Mibolerone	
Ethylmorphine,	Derivative of	Chlorphentermine	methyltestosterone	Nandrolone	
Opium, and	barbituric acid	Clortermine	Dihydrotestosterone	Norethandrolone	
Morphine	Chlorhexadol	Phendimetrazine	Drostanolone	Oxandrolone	
in combination	Lysergic acid		Ethylestrenol	Oxymesterone	
with non-narcotics.	Lysergic acid amide		Fluoxymesterone	Oxymetholone	
	Methyprylon		Formebulone	Stanolone	
	Sulfondiethylmethane		Mesterolone	Stanozolol	
	Sulfonethylmethane		Methandienone	Testolacton	
	Sulfonmethane		Methandranone	Testosterone	
	Tiletamine		Methandriol	Trenbolone	
	Zolazepam				

SCHEDULE IV

Narcotics	Depressants		Stimulants	Others
Limited quantity of diferoin	Alprazolam	Loprazolam	(+)-norpseudoephedrine	Fenfluramine
in combination with atropine sulfate.	Barbital	Lorazepam	Diethylpropion	
	Bromazepam	Lormetazepam	Fencamfamin	Pentazocine
Dextropropoxyphene	Camazepam	Mebutamate	Fenproporex	
	Chloral betaine	Medazepam	Mazindol	
	Chloral hydrate	Meprobamate	Mefenorex	
	Chlordiazepoxide	Methohexital	Pemoline	
	Clobazam	Methylphenobarbital	Phentermine	
	Clonazepam	Midazolam	Pipradrol	
	Clorazepate	Nimetazepam	SPA	
	Clotiazepam	Nitrazepam		
	Cloxazolam	Nordiazepam		
	Delorazepam	Oxazepam		
	Diazepam	Oxazolam		
	Estazolam	Paraldehyde		
	Ethchlorvynol	Petrichloral		
	Ethinamate	Phenobarbital		
	Ethyl loflazepate	Pinazepam		
	Fludiazepam	Prazepam		
	Flunitrazepam	Quazepam		
	Flurazepam	Temazepam		
	Halazepam	Tetrazepam		
	Haloxazolam	Triazolam		
	Ketazolam			

SCHEDULE V

Narcotics	Stimulants
Buprenorphine	Pyrovalerone
Limited quantities (less than Schedules III & IV) of:	
Codeine,	
Dihydrocodeine,	
Ethylmorphine,	
Diphenoxylate, Opium, and Difenoxin in combination with non-narcotics.	

of controlled chemical derivatives, immediate precursors (chemical which precedes the active drug), and chemicals essential for making a controlled drug, along with drugs exempt from control can be found in the most current edition of the *Controlled Substances Handbook.* Brand names for drugs in Schedules II–V are not included in the Tables but can also be found in the latest edition of the *Controlled Substances Handbook.*

PRESCRIBING AND DISPENSING OF CONTROLLED DRUGS

Medical practitioners have to follow specific rules for each schedule when prescribing or dispensing controlled drugs. Drugs in Schedule I can only be obtained, prescribed, and dispensed to an individual after special approval is obtained from the Food and Drug Administration (FDA). Drugs in Schedule II cannot be refilled or dispensed without a written prescription from a practitioner, except in an emergency. When they are dispensed in an emergency, a written prescription must be obtained within 72 hours. Drugs in Schedule III and in Schedule IV may not be dispensed without a written or an oral prescription. Prescriptions for these drugs may not be filled or refilled more than six months after their issue date or refilled more than five times unless authorized by a licensed practitioner. Drugs in Schedule V can be refilled, with a practitioner's authorization, without the limitation on number of refills or time. Certain Schedule V drugs may be purchased directly from a pharmacist, in limited quantities, without a prescription. The purchaser must be at least 18 years of age and furnish appropriate identification and the transactions must be recorded by the dispensing pharmacist.

When drugs in Schedule II, III, and IV are dispensed, a warning label stating "Caution: Federal law prohibits the transfer of this drug to any person other than the patient for whom it was prescribed," must be affixed to the dispensing container. The warning label regarding transfer does not apply to Schedule V drugs.

REGISTRATION, ORDERING, QUOTAS, AND RECORDS OF CONTROLLED DRUGS

Each individual or institution engaged in manufacturing, distributing, or dispensing of any con-

trolled drug must be authorized by and register annually with federal and state drug-enforcement agencies, unless specifically exempted. A unique registration number is assigned to each individual or institution registered under the act. A separate registration is required for practitioners who dispense narcotic drugs to individuals for the purpose of addiction treatment (i.e., METHADONE and LAAM [L-ALPHA-ACETYLMETHADOL] for opioid detoxification or maintenance).

All orders for Schedule I and II drugs must be made using a special narcotic order form. Proof of registration is required when ordering Schedule III–V drugs.

Annual production quotas are established for drugs in Schedule I and Schedule II. Everyone registered to handle controlled drugs must maintain records, conduct inventories, and file periodic reports specific to their business or professional activity.

SPECIAL ISSUES

State and local laws either parallel the federal regulations as described by the CSA, or impose additional restrictions. Individuals registered to handle controlled drugs must abide by the law (state or federal) that is most stringent in governing their business or professional activity. Examples where state law may be more stringent than federal law include (1) the requirement for TRIPLICATE PRESCRIPTION forms or (2) the placing of a drug in a higher schedule.

(SEE ALSO: *Addiction: Concepts and Definitions; Abuse Potential; Legal Regulation of Drugs and Alcohol*)

BIBLIOGRAPHY

BAUMGARTNER, K., & HOFFMAN, D. (EDS.) (1993). Schedules of controlled substances. In *Controlled substances handbook.* Arlington, VA: Government Information Services, J.J. Marshall Publisher.

CODE OF FEDERAL REGULATIONS (21 CFR Parts 1301–1308). (1992). *Food and drugs—Drug Enforcement Administration, Department of Justice.* Washington, DC: U.S. Government Printing Office.

SIMONSMEIER, L. M., & FINK, J. L. (1990). The comprehensive drug abuse prevention and control act of 1970. In

A. R. Gennaro (Ed.), *Remington's pharmaceutical sciences*, 18th ed. Easton, PA: Mack Publishing Company.

ROLLEY E. JOHNSON
ANASTASIA E. NASIS

COPING AND DRUG USE Coping is the capacity to surmount negative emotional states, including ANXIETY, DEPRESSION, anger, loneliness, and alienation. These aversive states are induced by internal psychological conflict or by external STRESS. Effectiveness in appraising and overcoming emotional distress that results from predisposing or triggering stressors determines, to a large extent, psychological well-being. In contrast, ineffectiveness in coping, as well as a subjective perception of ineffectiveness, exacerbates emotional distress, which comprises for some people an important factor in promoting ALCOHOL, TOBACCO, and other drug (ATOD) consumption.

The association between ATOD use and coping is complex. In some individuals there is a direct connection. In effect, PSYCHOACTIVE DRUGS are consumed to reduce tension and associated negative emotions. The consumption of drugs is motivated by their palliative effects. In most individuals, however, the connection between drug consumption and coping is more complicated. Numerous factors such as psychiatric illness, low self-esteem, deviant social values, maladaptive learned behaviors, inadequate social support, poor social skills, and personality disposition moderate and mediate the relationship between ATOD use and coping. No specific association has been established between coping style and VULNERABILITY to drug use or abuse. Thus, whereas it is generally recognized that a substantial proportion of the ATOD-using population is deficient in coping capacity, it is important to understand that many factors influence this association.

Coping and substance use and abuse become so intertwined over time that cause–effect relationships cannot always be discerned. Deficient coping capacity initially may, directly or indirectly, lead to ATOD consumption. Neurobehavioral, psychopathological, and social adjustment disturbances that occur along with chronic ATOD consumption may also diminish coping ability.

Substantial variation among individuals occurs with respect to both coping capacity and drug-use behavior across the life span. Drug consumption among youth is most frequently related to negative feelings such as depression and anxiety, social deviancy, and interpersonal problems—whereas substance use among the ELDERLY is more commonly associated with life crises, psychiatric disorder, bereavement, sleep disturbances, and unremitting pain.

Drug-abusing youth and adults, as a group, exhibit less ability to cope than the general population (Peele, 1985). It is essential to emphasize, however, that ATOD use and abuse may also be motivated by reasons other than the need to cope. In this context, ATOD consumption often stems from the desire for a euphoric effect or some other desirable state, a desire that may reflect accurate as well as inaccurate beliefs about the pharmacological effects of the chosen drug. For example, ATOD consumption may be motivated by perceived APHRODISIAC effects, energy or alertness enhancement, or social facilitation.

Among those whose ATOD consumption is motivated by deficient coping skills, it appears that augmenting competency improves the likelihood of successful treatment. In other words, treatments designed to enhance their coping skills are superior to treatments that emphasize their exploration of feelings (Getter et al., 1992). Furthermore, active coping strategies present 2 years after treatment are associated with a superior outcome at 10-year posttreatment follow-up (Finney & Moos, 1992).

The role of coping in ATOD use needs to be evaluated on a case-by-case basis. Assessment can be conducted using the Ways of Coping scale (Lazarus & Folkman, 1984) or the more comprehensive Constructive Thinking Inventory (Katz & Epstein, 1989). Severity of ATOD-use disorder can be efficiently quantified by employing the Drug Use Screening Inventory (Tarter, 1990). This brief self-report evaluates the severity of the disorder in ten key domains: (1) substance use, (2) psychiatric disorder, (3) behavior patterns, (4) health status, (5) family system, (6) work adjustment, (7) social competence, (8) peer relationships, (9) school adjustment, and (10) leisure/recreation. A treatment protocol to enhance coping has also been developed for alcoholics (Kadden et al., 1992); this practical approach to intervention is also applicable for treating individuals with other types of drug abuse.

(SEE ALSO: *Prevention: Life Skills Training; Relapse Prevention; Treatment Types: Cognitive Therapy of Addictions*)

BIBLIOGRAPHY

FINNEY, J. W., & MOOS, R. H. (1992). The long-term course of treated alcoholism: II. Predictors and correlates of 10-year functioning and mortality. *Journal of Studies on Alcohol, 53,* 142–153.

GETTER, H., ET AL. (1992). Measuring treatment process in coping skills and interactional group therapies for alcoholism. *International Journal of Group Psychotherapy, 42,* 419–430.

KADDEN, R., ET AL. (1992). *Cognitive behavioral coping skills therapy manual.* Project MATCH Monograph Series, Vol. 3, DHHS Publication No. (ADM) 92–1895.

KATZ, L., & EPSTEIN, S. (1989). Constructive thinking and coping with laboratory induced stress. *Journal of Personality and Social Psychology, 61,* 789–800.

LAZARUS, R., & FOLKMAN, S. (1984). *Stress, appraisal and coping.* New York: Springer.

PEELE, S. (1985). *The meaning of addiction.* Lexington, MA: Lexington Books.

TARTER, R. (1990). Evaluation and treatment of adolescent substance abuse. A decision tree method. *American Journal of Drug and Alcohol Abuse, 16,* 1–46.

RALPH E. TARTER

CRACK Crack (sometimes called crack-cocaine) is an illicit drug, the smokable form of CO-CAINE, made by adding the bases ammonia or baking soda and water to cocaine hydrochloride. The white powder illicitly purchased as cocaine is in the hydrochloride form; it cannot be smoked, because it is destroyed at the temperatures required for smoking. Therefore, in order to be used by the smoked route, cocaine must be converted to the base state. A mixture is made and heated to remove the hydrochloride, resulting in a pellet-sized cakelike solid substance that can be smoked. This form of cocaine is inexpensive, available for purchase "on the street," and is called "crack," because of the cracks formed in the solid as it dries.

Although crack can be smoked in tobacco cigarettes or marijuana cigarettes, it is generally smoked in a special crack pipe. In its simplest form, this is a glass tube with a hole at the top of one end and a hole at the other end through which the smoke is inhaled. The crack pellet is placed on fine wire mesh screens that cover the hole distal to the smoker and a flame is applied directly to the pellet. Soda bottles, small liquor bottles, etc. are all used to manufacture crack pipes. They have in common the use of fine mesh screens so that the crack is not lost as it melts. Temperatures of approximately 200°F (93°C) are most efficient in providing the largest amount of cocaine to the user. Higher temperatures destroy more of the cocaine.

Smoking cocaine began with the use of FREEBASE cocaine, prepared by its users from the cocaine hydrochloride illicitly purchased by them. Soon after this form of cocaine had achieved its popularity, single doses of cocaine already prepared for smoking (i.e., crack), became available through the illicit drug market. Unlike the process for forming freebase cocaine, the crack manufacturing process does not rid the cocaine of its adulterants. Smoking cocaine rapidly became a popular route of administration once crack became readily available, since it was so convenient to use. Blood levels peak rapidly when cocaine is smoked, because of efficient respiratory absorption, and the smoked route of cocaine administration yields effects (peak, duration of effect, half-life) comparable to the intravenous route of administration. This means that the smoker of cocaine can achieve rapid onset of effect, including a cocaine "rush" and substantial cocaine blood levels, and can do this repeatedly using a more socially acceptable route of administration—one that requires none of the PARAPHERNALIA associated with hardcore illicit drug use (e.g., syringes, needles, etc.).

The more rapid the onset of the drug effect, the more likely it is that the drug will be abused. Thus, although the effects of smoking crack are no different than the effects of cocaine by any other route, the ease with which the drug can be taken, combined with its toxicity make this an extremely dangerous substance.

From a financial perspective, crack is more desirable for both the buyer and the seller. A gram of cocaine hydrochloride costs approximately 50 to 60 dollars. This gram can be turned into 10 to 25 crack pellets, each selling for 2 to 20 dollars. Thus, a gram of cocaine can generate a substantial profit for the seller, and, as well, is available in single-dose units to anyone with only a few dollars to spend.

(SEE ALSO: *Coca Paste; Freebasing; Pharmacokinetics; Street Value*)

BIBLIOGRAPHY

INCIARDI, J. A. (1991). Crack-cocaine in Miami. In S. Schober & C. Schade (Eds.), *The epidemiology of co-*

caine use and abuse. NIDA Research Monograph No. 110. Rockville, MD: National Institute on Drug Abuse.

SIEGEL, R. (1982). Cocaine smoking. *Journal of Psychoactive Drugs, 14,* 271–359.

MARIAN W. FISCHMAN

CRAVING The term *craving* is generally defined as a state of desire, longing, or urge for a drug that is responsible for ongoing drug-use behavior in drug-dependent individuals. Craving is also viewed by many drug-abuse researchers and clinicians as the main cause of relapse among drug users attempting to remain abstinent. During periods of abstinence, drug-dependent individuals often complain of intense craving for their drug. Several systems for diagnosing drug abuse include persistent desire or craving for a drug as a major symptom of drug-dependence disorders.

The belief that an addict's inability to control drug use is caused by craving and irresistible desire was a prominent feature of descriptions of addictive disorders provided by many nineteenth-century writers. Craving continued to be important in many models of addiction developed in the twentieth century. The use of craving as a key mechanism in theories of addiction peaked in the 1950s, supported largely by E. M. Jellinek's writings on the causes of alcoholism.

Jellinek contended that sober alcoholics who consumed a small amount of alcohol would experience overwhelming craving that would compel them to continue drinking. The proposal that craving and loss of control over drinking were equivalent concepts was adopted by many clinicians and addiction researchers. Equally popular was the position, also supported by Jellinek, that craving was a direct sign of drug withdrawal. WITHDRAWAL-based craving was often described as physical craving, distinguishing it from craving that led to relapse during long periods of abstinence after withdrawal had subsided. Craving that occurred after an addict no longer was experiencing withdrawal was typically viewed as the result of psychological factors. The craving concept was sufficiently controversial that a committee of alcoholism experts brought together by the World Health Organization in 1954 (WHO Expert Committees on Mental Health and on Alcohol, 1955) recommended that the term craving not be used to describe various aspects of drinking behavior seen in alcoholics.

The use of craving as a key process in theories of addiction decreased during the 1960s and early 1970s as a result of several factors. During this period, many studies showed that alcoholics did not necessarily engage in loss of control drinking when they drank small doses of alcohol. The failure to confirm Jellinek's conceptualization of alcoholic drinking cast doubt on the idea that craving was synonymous with loss of control over drug intake. Furthermore, withdrawal models of craving could not account for the common observation that many addicts experienced craving and relapsed long after their withdrawal had disappeared. Finally, addiction research was increasingly dominated by behavioral approaches that focused on the influence of environmental variables in the control of drug taking and avoided the use of subjective concepts, such as craving, to explain addictive behavior.

Even though many researchers questioned the value of using craving to explain addictive behavior, it persisted as an important clinical issue, as many addicts complained that craving was a major barrier to their attempts to stop using drugs. Craving continued to be cited as a major symptom of drug dependence in formal diagnostic systems of behavioral disorders, and the notion that craving was responsible for compulsive drug use remained at the core of several popular conceptualizations of drug addiction. Scientific interest on the role of craving in addictive disorders reemerged in the middle 1970s as a result of two developments. First, behavioral theories of addiction were increasingly influenced by social-cognitive models of behavior that were more sympathetic to the possibility that hypothetical entities such as craving might be useful in explaining addictive processes. Second, animal research on the contribution of learning processes to drug tolerance and drug withdrawal provided support for the hypothesis that learned withdrawal effects might produce craving and relapse in abstinent addicts.

THEORIES OF CRAVING

Although there is considerable disagreement across current theories regarding the processes that supposedly control craving, nearly all models describe craving as, fundamentally, a subjective state and agree

on the impact of craving on drug use. With few exceptions, modern theories of craving assume that craving is a necessary, but probably not sufficient, condition for drug taking among addicts. These theories suppose that addicts are driven to use drugs because of their craving, and craving is generally described as the principal cause of relapse in addicts trying to remain abstinent. Moreover, all the comprehensive models of craving invoke some sort of learning or cognitive process in their descriptions of the mechanisms controlling craving, and these models make little distinction between physical and psychological forms of drug craving. It is important to note that, at the present time, research on craving is not sufficiently advanced to fully evaluate the validity of any of the major models of craving.

Many modern theories associate craving with drug withdrawal and suggest that craving may be merely a part of drug withdrawal. For example, the diagnostic system published by the American Psychiatric Association in 1987 listed craving as one of the symptoms of withdrawal for nicotine and opiates. Other approaches assume that cravings are distinct from withdrawal, but represent an addict's anticipation of, and desire for, relief from withdrawal. To explain the presence of craving following long periods of abstinence, it has been posited that learning processes are responsible for the maintenance of withdrawal effects. For example, Wikler's conditioning model of drug withdrawal (see WIKLER'S PHARMACOLOGIC THEORY) hypothesizes that situations reliably paired with episodes of drug withdrawal become conditioned stimuli that can produce conditioned withdrawal responses. An addict who has been abstinent for an extended period may reexperience withdrawal if faced with these conditioned stimuli. This learned-withdrawal reaction will trigger drug craving, that, in turn, may lead to relapse. A similar theory is based on the suggestion that drug-tolerance processes can become conditioned to environmental stimuli. Some have hypothesized that conditioned drug-tolerance effects will produce withdrawal-like reactions that, as in Wikler's theory, should promote craving and relapse to drug use (Poulos, Hinson, & Siegel, 1981).

Another perspective on craving is that it is strongly associated with the positively reinforcing, or stimulating, effects of drugs. For example, Marlatt (1985) has suggested that craving is a subjective state produced by the expectation that use of a drug will produce euphoria, excitation, or stimulation. Similarly, Wise (1988) proposed that craving represents memories for the pleasurable or positively reinforcing effects of drugs. There are also multiprocess models, in which expectancies of positive reinforcement and anticipation of withdrawal relief, as well as other factors, including mood states and access to drugs, generate craving (Baker, Morse, & Sherman, 1987; Gawin, 1990).

In contrast to models that contend that craving is responsible for all addictive drug use, a recent cognitive theory suggests that drug use may operate independently of craving (Tiffany, 1990). According to this theory, as a result of a long history of repeated practice, most of an addict's drug-use behavior becomes automatic. That is, drug use may be easily triggered by certain cues, difficult to stop once triggered, and carried out effortlessly with little awareness. Addicts attempting to withdraw from drug use will experience craving as they try to stop these automatized actions from going through to completion.

MEASURES OF CRAVING

Craving is generally measured through three types of behaviors—self-reports of craving, drug-use behavior, and physiological responding. In the most frequently used measure, self-report, addicts are simply asked to rate or describe their level of craving for a drug. Recently, questionnaires have been developed that ask addicts to rate a variety of questions related to craving. These questionnaires produce results that are considerably more reliable than a single rating of craving and tend to show that an addict's description of craving may have multiple dimensions. Measures of drug-use behavior have also been used to assess drug craving. This is entirely consistent with the common assumption that craving is responsible for drug use in addicts. Finally, as several theories posit that craving should be represented by particular patterns of physiological changes, physiological measures, primarily those controlled by the autonomic nervous system, have been included in several studies as an index of craving. These measures have included changes in heart rate, sweat gland activity, and salivation. In general, withdrawal-based theories predict that the physiology of craving should look like the physiology of drug withdrawal. In contrast, models that emphasize positive

reinforcement in the production of craving would associate drug desire with physiology characteristic of the excitatory effects of drugs.

RESEARCH ON CRAVING

Two kinds of studies have been used to investigate drug craving. The first, naturalistic studies, examine changes in addicts' descriptions of craving as they are attempting to stop using drugs. These studies generally have shown that cravings are especially strong in the first several weeks of abstinence, but decline over time as addicts stay off drugs. They also reveal that craving rarely remains at a constant level throughout the day, but grows stronger or weaker depending on the situations the addict encounters. These situations tend to be strongly associated with previous use of drugs, such as meeting drug-using friends or going to locations where the addict used drugs in the past.

Laboratory studies attempt to manipulate craving by presenting addicts with stimuli or cues that have been associated with their previous drug use. For example, a heroin addict may watch a videotape of someone injecting heroin, or smokers may be asked to imagine a situation in which they would want to smoke. These cue-reactivity studies allow the measurement of self-reports of craving, drug-use behavior, and physiological reactions under controlled conditions. Results from these studies indicate that abstinence from drugs, drug-related stimuli, and negative moods can influence craving measures.

Many of the results of naturalistic and laboratory studies have presented a challenge to the dominant assumption that craving is directly responsible for drug use in addicts (Kassel & Shiffman, 1992; Tiffany, 1990). For example, across many cue-reactivity studies, there is not a very strong correlation between addicts' reported levels of craving and their level of drug consumption in the laboratory. Correlations between self-reported craving and physiological reactions also tend to be weak. Other studies reveal that, although addicts frequently complain that cravings are a major difficulty they face as they try to stay off their drugs, few addicts who relapse say that they experienced craving just before their relapse episode. These findings show that the exact function of craving in drug dependence remains a controversial issue.

Despite these negative indications, millions of dollars are spent each year to develop pharmacological agents that might be capable of blocking, preventing, or reducing craving for various drugs.

(SEE ALSO: *Addiction: Concepts and Definitions; Causes of Drug Abuse: Learning; Research: Conditioned Drug Effects; Research, Animal Model: Conditioned Drug Effects*)

BIBLIOGRAPHY

AMERICAN PSYCHIATRIC ASSOCIATION. (1987). *Diagnostic and statistical manual of mental disorders-3rd ed.-rev.* Washington, DC: Author.

BAKER, T. B., MORSE, E., & SHERMAN, J. E. (1987). The motivation to use drugs: A psychobiological analysis of urges. In P. C. Rivers (Ed.), *The Nebraska symposium on motivation: Alcohol use and abuse* (pp. 257–323). Lincoln: University of Nebraska Press.

JELLINEK, E. M. (1955). The "craving" for alcohol. *Quarterly Journal of Studies on Alcohol, 16,* 35–38.

KASSEL, J. D., & SHIFFMAN, S. (1992). What can hunger teach us about drug craving? A comparative analysis of the two constructs. *Advances in Behaviour Research and Therapy, 14,* 141–167.

MARLATT, G. A. (1985). Cognitive factors in the relapse process. In G. A. Marlatt & J. R. Gordon (Eds.), *Relapse prevention* (pp. 128–200). New York: Guilford Press.

POULOUS, C. W., HINSON, R., & SIEGEL, S. (1981). The role of Pavlovian processes in drug tolerance and dependence: Implications for treatment. *Addictive Behaviors, 6,* 205–211.

TIFFANY, S. T. (1990). A cognitive model of drug urges and drug-use behavior: Role of automatic and nonautomatic behavior. *Psychological Review, 97,* 147–168.

WIKLER, A. (1948). Recent progress on neurophysiological basis of morphine addiction. *American Journal of Psychiatry, 105,* 329–338.

WISE, R. A. (1988). The neurobiology of craving: Implications for understanding and treatment of addiction. *Journal of Abnormal Psychology, 97,* 118–132.

WORLD HEALTH ORGANIZATION EXPERT COMMITTEES ON MENTAL HEALTH AND ALCOHOL. (1955). Craving for alcohol: Formulation of the Joint Expert Committees on Mental Health and Alcohol. *Quarterly Journal of Studies on Alcohol, 16,* 63–64.

STEPHEN T. TIFFANY

CREATIVITY AND DRUGS   Accounts of alcohol and drug use to stimulate creativity are apocryphal and anecdotal. For example, Samuel Taylor Coleridge reportedly composed much of his unfinished poem *Kubla Khan* while in an opium dream. In ancient Greece, however, the Pythian priestesses of the oracle at Delphi inhaled medicinal fumes to facilitate revelatory trances—as did the priests and peoples of most ancient societies. The institutionalized twentieth-century Native American Church continues to use the PEYOTE of their ancestors to promote profound religious experiences.

Psychedelic drugs, such as LYSERGIC ACID DIETHYLAMIDE (LSD), MESCALINE, PSILOCYBIN, and methylene dioxyamphetamine (MDA) have been used—both legally and illicitly—to increase aesthetic appreciation, improve artistic techniques, and enhance creativity. MARIJUANA has been used to heighten the sense of meaning, foster creativity, and heighten perceptions (also both legally and illicitly); and ALCOHOL has been employed by countless people worldwide to relieve inhibitions, increase spontaneity, and stimulate innovation and originality.

In the industrial West, the common belief in, and positive association between, alcohol or drug use and creativity is strengthened by the popular stereotypes of artists, writers, actors, and others in the creative and performing arts as heavy users or abusers of such substances. Despite these anecdotal claims, little scientific evidence supports the notion that alcohol and drug use actually increase creativity.

Part of the reason that creativity is attributed to drug use involves the actions of many psychoactive substances in producing altered states of consciousness. These altered states are characterized by some or all of the following features: (1) *alterations in thinking*, in which distinctions between cause and effect become blurred and in which logical incongruities may coexist; (2) *disturbances in time sense*, whereby the sense of time and chronology may become greatly altered; (3) a sense of *loss of control*, during which the person becomes less inhibited and self-possessed; (4) a *change in emotional expression;* (5) *body image change*, with a dissolution of boundaries between one's self and the world, resulting in transcendental or mystical experiences of "oneness" or "oceanic feelings"; (6) *perceptual distortions*, including illusions, pseudohallucinations, heightened acuity, and increased visual imagery; (7) *hypersuggestibility*, representing a decrease in the use of critical faculties; (8) a heightened *sense of meaning and significance*; (9) *sense of the ineffable*, in which the experience cannot be expressed in words; and (10) feelings of *rebirth and rejuvenation*. When people experience such features as these, it is understandable that they attribute creativity to certain drug experiences.

The immediate problem, however, in evaluating whether this is really so depends on the definition of creativity. At the outset, three dimensions of creativity need to be distinguished: those pertaining to the creative person, those pertaining to the creative process, and those pertaining to the creative product. If creativity pertains to an attribute of the person (e.g., original thinking), then any unusual or extraordinary experiences should qualify as "creative," even if nothing of social value emerges. If creativity pertains to a process (e.g., discovery, insight), then the testing and validation of the insights must take place as well. If creativity pertains to a product, it not only should possess some measure of social utility but should embody such qualities as novelty, surprise, uniqueness, originality, beauty, simplicity, value, and/or coherence. For both the creative process and the creative product, there is no substantive evidence to indicate that alcohol or drugs have benefit, despite the ongoing belief of many that they do. The experience of the *sense of meaning or significance* produced by drugs may have no bearing on whether that experience has *true* meaning or significance. The American philosopher and psychologist William James's claim that alcohol makes things seem more "utterly utter" is especially apt. This also happens with PSYCHEDELIC DRUGS, which have the capacity to induce a sense of profundity and epiphany (intuitive grasp of reality), but usually without any substantive or lasting benefit or practical value.

What, then, is the actual state of knowledge about the relationship between substance use and creative achievement? What few studies exist, in fact, indicate mostly detrimental effects of drugs on creativity, especially when these substances are taken in large amounts and over an extended period of time. The results of studies on the actions of alcohol typify this. As early as 1962, for example, Nash demonstrated that small doses (about equal to two martinis) of alcohol, in normal volunteers, tended to facilitate mental associations, while large doses (about equal to four martinis) had adverse effects. With the large doses, they had more trouble in discriminating and

assimilating details and performing complex tasks. In another study, Hajcak (1975) found that male undergraduates permitted alcohol on an ad-lib basis (without limits or restraints) showed greater initial productivity than when not allowed to drink but showed decreased appropriateness and decreased creative problem solving when intoxicated.

In an anecdotal study with seventeen artists who drank, Roe (1946) found that all but one regarded the short-term effects of alcohol as deleterious to their work, but they sometimes used it to overcome various technical difficulties. The general sentiment was that alcohol provided the freedom for painting but impaired the discipline. In a more extensive study of thirty-four eminent writers, Ludwig (1990) found that more than 75 percent of artists or performers who drank heavily experienced negative effects from alcohol—either directly or indirectly, on creative activity, particularly when they did not refrain from drinking when they were working. More positive effects of alcohol were found in a small number of cases, among those who used it in moderate amounts early in their careers to remove certain roadblocks, to lessen depression or mania, or to modulate the effects of other drugs.

With many anecdotal accounts to the contrary, the weight of scientific and clinical evidence suggests that long-term alcohol and drug use exert mostly negative effects on creativity. That drugs and alcohol are used so widely within the creative arts professions seems to have less to do with creativity than with social expectations and other extraneous factors. In fact, people use pharmacological substances for many reasons other than the stimulation of their imaginations. These reasons include relaxation, the facilitation of sleep, self-medication, social rituals, pleasure, or simply habituation or addiction.

Because writers, artists, actors, or musicians may write about, portray, or act out certain aspects of their pharmacological experiences does not logically or necessarily mean that these experiences are essential for the creative process. Creative people often exploit all aspects of their experiences—whether pathological or healthy and whether drug-induced or not—in a creative way; they try to translate personal visions and insights within their own fields of expression into socially acceptable, useful, or scientifically testable truths. Without some measure of social utility, unique drug-induced experiences represent little more than idiosyncratic to quasi-psychotic produc-

tions, having value and meaning only, perhaps, to the substance user.

(SEE ALSO: *Lifestyle and Drug Use Complications*)

BIBLIOGRAPHY

HARMON, W. W., ET AL. (1966). Psychedelic agents in creative problem-solving. *Psychological Reports, 19,* 211–227.

KOSKI-JANNES, A. (1985). Alcohol and literary creativity. *Journal of Creative Behavior, 19,* 120–136.

KRIPPNER, S. (1969). The psychedelic state, the hypnotic trance and the creative act. In C. Tart (Ed.), *Altered states of consciousness.* New York: Wiley.

LUDWIG, A. M. (1990). Alcohol input and creative output. *British Journal of Addiction, 85,* 953–963.

LUDWIG, A. M. (1966). Altered states of consciousness. *Archives of General Psychiatry, 15,* 223–234.

LUDWIG, A. M. (1989). Reflections on creativity and madness. *American Journal of Psychotherapy, 43,* 4–14.

MCGLOTHLIN, W., COHEN, S., & MCGLOTHLIN, M. S. (1967). Long-lasting effects of LSD on normals. *Archives of General Psychiatry, 17,* 521–532.

NASH, H. (1962). *Alcohol and caffeine: A study of their psychological effects.* Springfield, IL: Charles C. Thomas.

ROE, A. (1946). Alcohol and creative work. *Quarterly Journal of the Study of Alcohol,* 415–467.

TART, C. T. (1971). *On being stoned.* Palo Alto, CA: Science and Behavior Books.

ARNOLD M. LUDWIG

CRIME AND ALCOHOL The relationship between ALCOHOL and involvement in crime is not a simple one. Drinking is a very common activity, and most drinking is not followed by criminal behavior. Understanding the alcohol-crime relationship requires an identification of those drinking effects and circumstances that are related to crime. Alcohol's relationship to crime also varies by the type of crime. The major crime-type distinction is between *violent* personal crime (such as homicide, forcible rape, and assault) and *property* crime (such as burglary and larceny). Alcohol's effects differ with respect to violent crime and property crime. Individual characteristics are also implicated in the alcohol-crime relationship. Age and gender, for example, affect whether drinking leads to criminal behavior. Young

adult males are more likely than older adult males and females of all ages to engage in alcohol-related offenses.

According to the available evidence, drinking is more likely to be implicated in violent than in property crime. Moreover, violent offenses are often thought of as expressive or instrumental. Expressive violent offenses are typically those that result from interpersonal conflict that escalates from verbal abuse to physical AGGRESSION. Such violence often involves a drinking offender or drinking by both (or multiple) parties to the violent conflict. Instrumental offenses have rational goals, typified by stealing to realize the value of the stolen money or goods. Alcohol is not thought to be an important causal factor in acquisitive crimes such as theft.

This article covers the following topics:

How often drinking precedes the commission of crimes

Whether alcohol contributes to the drinker's being a crime victim

Whether drinking is involved in FAMILY VIO-LENCE

How or why alcohol contributes to crime (and the tendency to use alcohol as an excuse for committing crimes)

Research has shown that alcohol is an important factor in the occurrence of expressive interpersonal violence, that alcohol use increases the risk of being a crime victim, that the alcohol-crime relationship is complex (involving multiple factors in addition to alcohol), and that alcohol is often blamed without justification for criminal offenses.

HOW OFTEN DOES DRINKING PRECEDE THE COMMISSION OF CRIMES?

In 1991, Beck and other statisticians at the Bureau of Justice Statistics (BJS) surveyed 13,986 state prison inmates (Beck et al., 1993). These inmates represented a sampling of the more than 700,000 adults who were held in state prisons in the United States at the time of the survey. During face-to-face interviews, the inmates were asked whether they had been under the influence of alcohol and/or drugs at the time of the offense that resulted in their incarceration. Table 1 shows the results of this question, organized by type of offense.

Overall, almost a third of the inmates (18% + 14%) said they were under the influence of alcohol alone or alcohol *and* drugs when they committed the offense. The percentage under the influence of al-

TABLE 1
Substance Abuse at the Time of Criminal Involvement

Current Offense	Inmates Under the Influence (%)		
	Alcohol Only	Drugs Only	Both
All offenses	18	17	14
Violent offenses	21	12	16
Homicide	25	10	17
Sexual assault	22	5	14
Robbery	15	19	18
Assault	27	8	14
Property offenses	18	21	14
Drug offenses	8	26	10
Public-order offenses	31	10	9
Driving while intoxicated (DWI)	70	3	8
Other	20	11	10

Courtesy of A. Beck et al. (1993). *Survey of State Prison Inmates, 1991* (NCJ-136949, p. 26). Washington, DC: U.S. Department of Justice, Office of Justice Programs, Bureau of Justice Statistics.

cohol alone or alcohol and drugs was somewhat higher for violent offenses (37%) and public-order offenses (40%).

Surveys of crime victims also indicate that offenders often had been drinking. Since 1973, the National Crime Victimization Survey (NCVS) has presented findings from interviews with individuals age 12 and older living in U.S. households. Zawitz and other BJS researchers (1993) reported that during the period 1973 to 1992, victims of violent offenses reported on the offender's drug or alcohol use. When reported, the victim thought the offender was under the influence of alcohol alone or alcohol and drugs in 54 percent of cases. In 40 percent of the victimizations, the victims thought the offenders were under the influence of alcohol only.

Most studies of alcohol and crime focus on offenses known to the police or on offenders serving sentences for crimes that resulted in conviction. A notable exception is a community study in Thunder Bay, located in the province of Ontario, Canada. Pernanen (1976, 1981, 1991) collected information from a representative sample of 1,100 community residents. Among those who had been victimized, the assailant had been drinking in 51 percent of the occasions when violence occurred; two-thirds of the time (68%), the assailant was judged to have been "drunk." Pernanen also noted that the findings for the Thunder Bay study are consistent with many other North American studies using police records. It is usually found that half of all violent offenders had been drinking at the time of the offense.

The most common pattern found in studies of violent crimes is that 60 percent or more of the events involve drinking by the offender, both the offender and the victim, or the victim only. The results of Wolfgang's classic study (1958) of criminal homicides in Philadelphia are typical (see table 2). The most common pattern when alcohol was present was for both victim and offender to have been drinking.

If the foregoing findings indicated the extent to which drinking was *causally* implicated in violent crime, it would be remarkable. It could then be argued that alcohol accounts for a majority of violent offenses. But neither the presence of alcohol in a crime nor the intoxication of an offender is necessarily an indication that alcohol influenced the occurrence of the crime. Because drinking is such a common activity, it is likely that alcohol is sometimes simply present but not causally relevant.

TABLE 2
Alcohol Presence at the Time of Criminal Homicide

Alcohol Presence	Criminal Homicides (%)
In homicide situation	63.6
In both victim and offender	43.5
In victim only	9.2
In offender only	10.9

Courtesy of M. E. Wolfgang (1958). *Patterns of Criminal Homicide*. (Table 14, p. 13). Philadelphia: University of Pennsylvania Press.

Drinking is also sometimes offered by offenders as an excuse for the crime, as a way of avoiding being held accountable. (In later sections of this article, we discuss the causal importance of alcohol to crime and alcohol as an excuse for crime.)

ALCOHOL USE AND CRIME VICTIMIZATION

Alcohol use raises the likelihood that the drinker will be a victim of violent crime. Substantial percentages of homicide, assault, and robbery victims were drinking just before their victimization. Medical examiners have done a significant number of homicide studies by running toxicological tests of the body fluids of homicide victims. Separate reviews by Greenberg (1981) and by Murdoch, Pihl, and Ross (1990) found that the percentage of homicide victims who had been drinking ranges widely, but is usually about 50 percent. Goodman et al. (1986) tested the alcohol levels of several thousand homicide victims; they found that 46 percent of the victims had consumed alcohol in the period before being killed, and three of ten victims had alcohol levels beyond the legal intoxication level.

Roizen (1993) examined studies of alcohol use by robbery and rape victims. The percentage who had been drinking before their victimization ranged widely—from 12 to 16 percent for robbery victims and from 6 to 36 percent for rape victims. Abbey (1991) and Muehlenhard and Linton (1987) also found in their studies of date rape that both offenders and victims had commonly been drinking. Abbey suggested that drinking by the offender and/ or by the victim contributes to rape by the impaired

communication and misperception that results from alcohol's effects on cognitive ability (among other contributing factors). Males who have been drinking, for example, may mistakenly attribute sexual intent to their date.

Alcohol may increase the risk that the drinker will be a crime victim because of effects that alcohol has on judgment and demeanor. Someone who has been drinking may take risks that might not be taken when sober, such as walking in a dangerous area of a city at night. Alcohol also causes some individuals to be loud and verbally aggressive. Such demeanor can be offensive and might sometimes precipitate physical attack.

DOES DRINKING GENERATE FAMILY VIOLENCE?

Unfortunately, violence is common in American households, and alcohol is a contributing factor, according to research done by Kantor and Straus (1989) and by Straus, Gelles, and Steinmetz (1980), among others. Hotaling and Sugarman (1986) found that alcohol appears to be most relevant to the occurrence of husband-against-wife violence. Hamilton and Collins (1981) reviewed about 25 studies that examined the role of alcohol in spouse and CHILD ABUSE. They found alcohol to be most relevant to wife beating, where it was present in one-quarter to a half of all such events. (Alcohol was present in less than one in five incidents of child abuse.) The most common pattern was for only the husband to be drinking or for both parties to have consumed alcohol. It was uncommon for only the wife to have been drinking. Studies also indicate that husbands or partners with alcohol problems were more likely to be violent against their wives/partners.

Some recent work by Jones and Schecter (1992) and by Barnett and Fagan (1993) on family violence suggests that violence against women may lead to their own use of alcohol and drugs as a coping mechanism. Drinking and/or drug use may be a response to the physical and emotional pain and fear that result from living in a violent relationship. Miller, Downs, and Testa (1993) found that women in alcohol-treatment programs had higher rates of father-to-daughter violence than did the women in the comparison group. These findings underline the importance of interpreting the meaning of alcohol's association with family (and other forms of) violence carefully. As previously noted, alcohol is often present but irrelevant to the occurrence of violence. Some recent literature on family violence indicates that alcohol use may sometimes be a *response* to violent victimization.

HOW OR WHY DOES ALCOHOL CONTRIBUTE TO CRIME?

There are a number of possible explanations offered for alcohol's role in crime:

- The need for money to support drinking may cause some individuals to commit crimes to generate cash to support their habit.
- The pharmacological effects of alcohol can compromise drinkers' cognitive ability and judgment and raise the likelihood of physical aggression.
- Expectations that alcohol makes drinkers aggressive may increase the chance of violence.
- Standards of conduct and accountability for behavior may differ for sober and drunken activities (these differences can result in an increase in the likelihood of criminal behavior after drinking).

These possible explanations are not mutually exclusive. All may sometimes accurately describe how drinking causes crime. Two or more of the explanations may even apply to the same incident.

Committing "income crimes"—to obtain money for drinking—is not thought to be an important explanation. Although the cost of maintaining an addiction to relatively more expensive drugs (e.g., HEROIN and COCAINE) is high, the price tag for supporting heavy drinking is usually modest. In most of the United States, for example, one could support a habit of daily heavy drinking for 10 dollars or less. The majority of individuals could maintain such a habit without resorting to crime, although many heavy drinkers spend more than this minimal amount on alcohol. There is virtually no information in the research literature about the likelihood or frequency of involvement in income crime to support drinking, but alcohol is not thought to be a major factor in income crimes. This does not mean it never happens, only that it is uncommon.

If alcohol is not an important factor in the occurrence of income-generating crime, why do so many property offenders (18% alcohol only, 14% alcohol and drugs) report they were under the influence of alcohol at the time they committed such offenses (see table 1)? At least two explanations are possible for the high correlation between drinking and property crime. The first suggests that the correlation is simply coincidental, not causal. A second reason (put forward by both Collins, 1988, and Cordilia, 1985), is that a property offender who has been drinking is more likely to be caught than is a sober one. This reason makes sense based on the known impairment effects of alcohol. A drinking offender may not be as competent or careful as a sober one, so drinking offenders may be overrepresented among offenders who are caught and thus known to criminal-justice officials.

Alcohol impairs cognitive ability, including one's own capacity to communicate clearly and the capacity to understand the verbal and behavioral cues of others. In addition, a person whose abilities have been impaired by alcohol is less able to make decisions and carry out appropriate and effective actions. Pernanen, in his early work (1976), discussed how alcohol-impaired cognitive ability can lead to violence. When one or both parties who are interacting have been drinking, there is an increased potential for misunderstanding that can lead to conflict and that may in turn escalate to violence. One factor in such a scenario is what may be called a "reduced behavioral response repertoire." Alcohol impairs a drinker's capacity to conceive and utilize the wide range of verbal and other behavioral options that are available to sober individuals. Alcohol-induced cognitive impairment may also diminish the drinker's capacity to foresee the negative implications of violent actions. In summary, one way that alcohol increases the likelihood of violence is its negative effect on cognitive capacities, and these effects lead to an increased risk of violence.

It has been demonstrated in laboratory experiments that both actual alcohol use and the *belief* that alcohol has been consumed can raise levels of aggression. In laboratory experiments using competitive encounters between opponents in which the winner can apply an electrical shock to the loser, subjects who have been given alcohol behave more aggressively. Evidence gathered by Bushman and Cooper (1990) and by Hull and Bond (1986) also indicates that subjects who have been told they have received alcohol, but who actually have been given a placebo, are more aggressive in their administration of electrical shocks. These findings suggest that beliefs about alcohol's behavioral effects can themselves affect behavior.

Expectations that alcohol use leads to aggressive behavior probably have sociocultural roots. Anthropologists such as Heath (1976a, 1976b) and MacAndrew and Edgerton (1969), for example, noted that societies differ in the behavior that occurs after drinking. Some of these differences may be attributable to racial or ethnic differences in physiological reactions to alcohol, but it is also clear that there are normative variations in what behaviors are expected or acceptable after drinking. In fact, behavioral norms after drinking may vary within societies. MacAndrew and Edgerton noted that during certain "time-out" periods, usual standards of behavior are suspended. For example, festivals or Mardi Gras celebrations are often characterized by high levels of drinking and behavior that is considered deviant or criminal during normal times.

Alcohol appears to be implicated in violence in another indirect way. Drinking is sometimes used as an excuse for crime or as a way to avoid accountability after the fact. McCaghy (1968) has referred to this phenomenon as "deviance disavowal." The deviance disavowal potential of alcohol can account for a drinker's involvement in crime in two ways: (1) individuals may drink or say that they have been drinking as an advance excuse for their conduct, (2) drinking may be offered as an excuse after the fact.

SUMMARY

Drinking alcohol and involvement in criminal behavior occur together frequently. Some of the time alcohol has a causal role in crime, but often it is merely present. Drinking is most likely to be relevant causally to expressive interpersonal violence—including family violence. Drinking can increase the risk of being victimized as well. Drinking may also sometimes help account for the commission of crimes to obtain money to support the habit, but alcohol is not a major factor in the occurrence of income crime. Drinking leads to criminal behavior in a number of ways, including the effects that alcohol use has on cognition and the rules that govern behavior and accountability for behavior. The alcohol-

crime relationship is complex. It is clear that drinking is rarely the only cause of criminal behavior, and that when it does contribute, it is usually only one of a number of relevant factors.

(SEE ALSO: *Complications: Cognition; Crime and Drugs; Drunk Driving; Economic Costs of Alcohol Abuse and Alcohol Dependence; Expectancies; Family Violence and Substance Abuse; Social Costs of Alcohol and Drug Abuse*)

BIBLIOGRAPHY

ABBEY, A. (1991). Acquaintance rape and alcohol consumption on college campuses: How are they linked? *Journal of American College Health, 39*(4), 165–169.

BARNETT, O. W., & FAGAN, R. W. (1993). Alcohol use in male spouse abusers and their female partners. *Journal of Family Violence, 8*(1), 1–25.

BECK, A., ET AL. (1993). *Survey of state prison inmates, 1991,* (NCJ-136949). Washington, DC: U.S. Department of Justice, Office of Justice Programs, Bureau of Justice Statistics.

BUSHMAN, B. J., & COOPER, H. M. (1990). Effects of alcohol on human aggression: An integrative research review. *Psychological Bulletin, 107*(3), 341–354.

COLLINS, J. J. (1988). Alcohol and interpersonal violence: Less than meets the eye. In N. A. Weiner & M. E. Wolfgang (Eds.), *Pathways to criminal violence.* Newbury Park, CA: Sage Publications.

CORDILIA, A. (1985). Alcohol and property crime: Exploring the causal nexus. *Journal of Studies on Alcohol, 46*(2), 161–171.

GOODMAN, R. A., ET AL. (1986). Alcohol use and interpersonal violence: Alcohol detected in homicide victims. *American Journal of Public Health, 76*(2), 144–149.

GREENBERG, S. W. (1981). Alcohol and crime: A methodological critique of the literature. In J. J. Collins Jr. (Ed.), *Drinking and crime: Perspectives on the relationships between alcohol consumption and criminal behavior.* New York: Guilford Press.

HAMILTON, C. J., & COLLINS, J. J., JR. (1981). The role of alcohol in wife beating and child abuse: A review of the literature. In J. J. Collins Jr. (Ed.), *Drinking and crime: Perspectives on the relationship between alcohol consumption and criminal behavior.* New York: Guilford Press.

HEATH, D. B. (1976a). Anthropological perspectives on alcohol: An historical review. In M. W. Everett, J. O. Waddell, & D. B. Heath (Eds.), *Cross-cultural approaches to the study of alcohol.* The Hague: Mouton.

HEATH, D. B. (1976b). Anthropological perspectives on the social biology of alcohol: An introduction to the literature. In B. Kissin & H. Begleiter (Eds.), *The biology of alcoholism:* Vol. 4. *Social aspects of alcoholism.* New York: Plenum.

HOTALING, G., & SUGARMAN, D. (1986). An analysis of risk markers in husband to wife violence: The current state of knowledge. *Violence and Victims, 1,* 101–124.

HULL, J. G., & BOND, C. F. (1986). Social and behavioral consequences of alcohol consumption and expectancy: A meta-analysis. *Psychological Bulletin, 99,* 347–360.

JONES, A., & SCHECHTER, S. (1992). *When love goes wrong: What to do when you can't do anything right.* New York: HarperCollins.

KANTOR, G.-K., & STRAUS, M. A. (1989). Substance abuse as a precipitant of wife abuse victimizations. *American Journal of Drug and Alcohol Abuse, 15*(2), 173–189.

MACANDREW, C., & EDGERTON, R. B. (1969). *Drunken comportment: A social explanation.* Chicago: Aldine.

MCCAGHY, C. (1968). Drinking and deviance disavowal: The case of child molesters. *Social Problems, 16,* 43–49.

MILLER, B. A., DOWNS, W. R., & TESTA, M. (1993). Interrelationships between victimization experiences and women's alcohol/drug use. *Journal of Studies on Alcohol,* Suppl. 11, 109–117.

MUEHLENHARD, C. L., & LINTON, M. A. (1987). Date rape and sexual aggression in dating situations: Incidence and risk factors. *Journal of Counseling Psychology, 34,* 186–196.

MURDOCH, D., PIHL, R. O., & ROSS, D. (1990). Alcohol and crimes of violence: Present issues. *International Journal of the Addictions, 25*(9), 1065–1081.

PERNANEN, K. (1991). *Alcohol in human violence.* New York: Guilford Press.

PERNANEN, K. (1981). Theoretical aspects of the relationship between alcohol use and crime. In J. J. Collins Jr. (Ed.), *Drinking and crime: Perspectives on the relationships between alcohol consumption and criminal behavior.* New York: Guilford Press.

PERNANEN, K. (1976). Alcohol and crimes of violence. In B. Kissin & H. Begleiter (Eds.), *The biology of alcoholism:* Vol. 4. *Social aspects of alcoholism.* New York: Plenum.

ROIZEN, J. (1993). Issues in the epidemiology of alcohol and violence. In S. E. Martin (Ed.), *Alcohol and interpersonal violence: Fostering multidisciplinary perspectives.* (NIH Publication No. 93-3496, Research

Monograph 24). Rockville, MD: U.S. Department of Health and Human Services.

STRAUS, M. A., GELLES, R. J., & STEINMETZ, S. K. (1980). *Behind closed doors: Violence in the American family.* Garden City, NY: Anchor-Doubleday.

WOLFGANG, M. E. (1958). *Patterns of criminal homicide.* Philadelphia: University of Pennsylvania Press.

ZAWITZ, M. W., ET AL. (1993). *Highlights from 20 years of surveying crime victims: The National Crime Victimization Survey, 1973–92,* NCJ-144525. Washington, DC: U.S. Department of Justice, Office of Justice Programs, Bureau of Justice Statistics.

JAMES J. COLLINS

CRIME AND DRUGS Because of widespread public and political concern over drug-related crime, there has been an urgent need to understand the drugs–crime relationship. However, despite numerous studies on this topic, only recently have significant empirical advances in our understanding emerged.

Authors of a comprehensive literature review published in 1980 concluded that the drugs–crime relationship was far more complex than originally believed (Gandossy et al.). While acknowledging significant contributions of previous research, the authors felt that methodological problems in the studies they reviewed had obscured an understanding of the linkage between drugs and crime. As these and other reviewers have observed, perhaps the most serious of these weaknesses was the use of official arrest records as indicators of criminal activity. Studies using confidential self-report methods in settings in which there is immunity from prosecution have consistently documented that less than 1 percent of offenses committed by drug abusers result in arrest. With a subsequent emphasis on confidential self-report data, studies conducted since 1980 have permitted more realistic estimates of the extent of criminality among drug abusers.

This article focuses on research findings since 1980 on the drugs–crime relationship. It concentrates on the relationship between the use of illicit drugs, including HEROIN, COCAINE, MARIJUANA, AMPHETAMINES, and PSYCHEDELIC substances, and two kinds of crime: *violent offenses*, including assault, robbery, rape, and homicide; and *property offenses*,

including vehicle theft, shoplifting, burglary, other theft, and "fencing," or selling or receiving stolen goods. (Offenses involving drug sales or possession and those associated with alcohol use are discussed elsewhere in this encyclopedia). The report also presents theories attempting to explain associations between drug use and crime. Finally, findings on the relationship between drug abuse and the commission of frequent, persistent, and dangerous crimes are discussed.

THE CRIMINALITY OF DRUG ABUSERS

In examining the criminality of drug abusers, it is important to note that the onset of illicit drug use typically does not result in the onset of criminal behavior. Rather, it is the *frequency*, not the *onset*, of drug use that increases criminal activity. Further, the positive relationship between drug-use frequency and crime frequency is not consistent across all types of drug use and all types of crime. Such a relationship has been observed with respect to only three types of drug abuse: heroin addiction, cocaine abuse, and multiple-drug use. In addition, such associations are more common for property crime than for violent crime.

Narcotic Drug Use. Much of our knowledge about the relationship between drugs and crime comes from detailed self-report information on the type, extent, and severity of criminal activity of NARCOTIC (mainly heroin) addicts. Recent large-scale, independently conducted studies have convincingly shown that increases in property crime and robbery, which has components of both property crime and violence, are associated with increased heroin use. Such a relationship, however, is less clear for violent crimes other than robbery.

Prevalence and Scope of Property and Violent Crime. Several key studies reveal an exceptionally high prevalence of property crime among narcotic addicts. Anglin and Speckart (1988) found that 82 percent of a sample of 386 California male narcotic addicts reported involvement in property crime over an average five-year period of daily narcotic use. Anglin and Hser (1987) reported that 77 percent of a sample of 196 female narcotic addicts from California admitted to involvement in property crime during an average six-year narcotic addiction period. Inciardi (1986) noted that almost all of a sample of

573 male and female narcotic abusers from Miami had reportedly engaged in theft during the year prior to interview. Inciardi also found that these individuals reported involvement in more than 77,000 property crimes (on average, 135 per subject) over a 12-month period while at large in the community. This figure included 6,669 burglaries, 841 vehicle thefts, 25,045 instances of shoplifting, and 17,240 instances of fencing. While these studies varied in sampling methods and definitions of property crime (e.g., including and not including robbery), all provided evidence that a substantial majority of narcotic abusers routinely engage in property crime.

Property crime comprises a considerable portion of the crime, other than drug distribution, committed by narcotic addicts. For instance, Nurco et al. (1991a) found that of the nondistribution crimes committed by a sample of 250 male narcotic addicts during an average 7.5-year addiction period, approximately 48 percent were property crimes.

Research has also consistently documented that among heroin addicts, violent crime is less prevalent and occurs with less frequency than property crime. Earlier studies noted that addicts tended to prefer property crime over violent crime and appeared to be less violent than other offenders. While findings from later studies have continued to show that violence accounts for only a small proportion of all addict crime (approximately 1% to 3%, a rate that is much smaller than the property-crime figure), the actual number of violent crimes is still quite large because addicts commit so many crimes. For example, in Inciardi's 1986 sample of 573 Miami narcotic abusers, violent crime comprised only 2.8 percent of all offenses (5% of nondistribution offenses) committed by the subjects in the year prior to interview. However, this relatively small percentage amounted to 6,000 incidents of violent crime (on average, 10.4 per subject), since a total of 215,105 offenses were committed.

Researchers have also suggested that heavy heroin use and, more recently, heavy cocaine abuse have contributed to record numbers of homicides in large cities in the United States. The ways in which drugs can contribute to violence is the basis for a prominent theory in the drugs-crime field, discussed later in this article.

Crime and Frequency of Heroin Use. Recent studies have provided consistent evidence of a direct, functional relationship between the frequency of narcotic drug (primarily heroin) use and the frequency of property crime. These investigations have employed a unique longitudinal design in which crime data are obtained for each subject over periods during which the frequency of narcotic use may vary. These studies of addiction careers reveal that property-crime rates are significantly higher during narcotic addiction periods than during periods of nonaddiction. Such a relationship tends to be linear, with the highest property-crime rates occurring at the highest levels of narcotic use (three or more times per day). In addition, although most addicts commit property crime prior to addiction, the frequency of such crime increases significantly from preaddiction to addiction, remaining high over subsequent addiction periods and low during intervening nonaddiction periods. While other factors also influence property-crime rates, the simplest explanation for these results is that property crime is functionally related to narcotic addiction—since addicts need cash to support their habits.

Evidence of a similar relationship between heroin use and violent crime is less conclusive. Studies have consistently shown that rates for robbery, in which there are property-crime features, are considerably higher during addiction periods than during either preaddiction or nonaddiction periods. However, when rates for composite measures of violence and rates of assault alone are examined, the relationship appears less clear.

In compiling composite measures of violence, Ball et al. (1983) found that for a sample of 243 male Baltimore addicts, the number of days on which violent crime was committed was considerably higher during the first addiction period than during the first nonaddiction period. However, in subsequent studies of 250 male addicts from Baltimore and New York City, most of whom had multiple periods of addiction, more complex relationships were observed. Over an addiction career, violent-crime rates for the total sample were significantly higher for combined addiction periods than for combined nonaddiction periods (Nurco et al., 1986; Nurco et al., 1988a). This result stemmed largely from high levels of crime committed during the first addiction period; violent crime actually decreased over subsequent periods of addiction, a finding that appeared to be age-related. The fact that mean rates for violence were found to be even higher for preaddiction (10 days per year) than for addiction periods (8 days per year) also re-

flected an inverse relationship between age and violent criminal activity.

Nonnarcotic Drug Use. Investigation of the nonnarcotic drugs–crime relationship has only recently emerged as a major research question. In the 1980 literature review, Gandossy and associates found that, of the few studies conducted on the nonnarcotic drug-crime relationship, evidence linking the use of various nonnarcotic substances to either property crime or violent crime was weak. Another reason for the unclear relationship between nonnarcotic drug use and criminal behavior is that various narcotic and nonnarcotic drugs are often used in combination. Thus, disentangling their separate relationships to criminal activity, let alone determining cause and effect, is especially problematic. Despite these difficulties, significant advances have been made in understanding the nonnarcotic drugs–crime relationship since 1980.

Cocaine. Data analyses on a nationwide probability sample of 1,725 adolescents strongly supported a cocaine-crime connection (Johnson et al., 1991). Adolescents who reported using cocaine in the year preceding the interview (comprising only 1.3% of the sample) were responsible for a disproportionately large share of the property and violent crime committed by the sample during this period. The cocaine users accounted for 60 percent of all minor thefts, 57 percent of felony thefts, 41 percent of all robberies, and 28 percent of felony assaults committed by the entire sample.

Typological studies involving seriously delinquent youth and female crack-cocaine abusers also revealed that subjects who reported the heaviest levels of cocaine use also had substantially higher rates of property and violent crime than subjects who used crack less frequently. Among a sample of 254 youth identified by Inciardi and coworkers (1993a) as serious delinquents, the 184 CRACK dealers (86% of whom were daily crack users) were responsible for 45,563 property crimes (an average of 231 per user) during the year preceding the interview. In contrast, the 70 subjects who were not crack dealers and who used crack less frequently (approximately three times per week) averaged 135 property crimes per year. In addition, the heavy cocaine users averaged 10 robberies per year, compared with 1 per year for the remaining subjects. Similar results were reported for a sample of 197 female crack abusers (Inciardi et al., 1993b). The average adjusted annual rates for

the 58 subjects classified as heavy cocaine users (8 or more doses per day) were 12, 14, and 320 for violent crime, major property offenses, and minor property crimes, respectively. These rates were substantially higher than rates for the 90 subjects classified as "typical" users (4–7.99 doses per day). For those 49 users who took less than 4 doses per day, the average adjusted annual rates for violence and major property crime were less than 1, and the rate for minor property offenses was 24.

Increased cocaine use among narcotic addicts has also been associated with increased property and violent-crime rates. Both Nurco et al. (1988b) and Shaffer et al. (1985) found that male narcotic addicts who had higher rates of cocaine use tended to have higher rates of property and violent crime than addicts who did not abuse cocaine.

Other Nonnarcotic Drugs. The use of other nonnarcotic drugs appears to be unrelated to increased criminal activity. While there is considerable evidence that frequent users of *multiple* nonnarcotic substances, including amphetamines, BARBITURATES, marijuana, and PCP, typically have high crime rates (although somewhat lower than rates for heroin addicts), such is not the case for users of *single* nonnarcotic drugs. Although single use may be related to offenses like disorderly conduct or DRIVING while impaired, it is not generally associated with predatory crime.

Marijuana. Research on the relationship between marijuana use and crime has found that, with the possible exception of the sale of the drug and disorderly conduct or driving while impaired, the use of marijuana is not associated with an increase in crime. Some studies have reported that marijuana use may actually reduce inclinations toward violent crime.

A major problem in studying the association between marijuana use and criminal behavior is that the exclusive use of marijuana is generally short-lived. Further, like other illicit nonnarcotic substances, marijuana is often used in combination with other drugs. Under such circumstances, it is difficult to isolate the effects of heavy marijuana use from those associated with the use of various drug combinations.

Amphetamines. Literature reviews published during the late 1970s and early 1980s (Gandossy et al., 1980; Greenberg, 1976) reported that the association between amphetamine use and crime was difficult to

determine because, among other factors, of the diversity of amphetamine users. More recent ethnographic studies of drug abusers (Goldstein, 1986) have reported that amphetamine use is related to violent crime in some individuals. In the general population, however, the association between amphetamine use and crime is not readily apparent. Despite assertions of the media in the 1960s and 1970s, the prevalence of amphetamine-related violence among American youth is likely to be quite low.

Psychedelic Substances. Most studies investigating the relationship between psychedelic-substance abuse and crime have involved PHENCYCLIDINE (PCP). Much of this research has examined the relationship between PCP and violence. As in the coverage of many other nonnarcotic drugs, media reports, principally in the 1970s and early 1980s, emphasized a perceived link between PCP use and violent behavior. However, the actual extent of this link is greatly exaggerated. In his report on the subject, Kinlock (1991) noted that serious methodological problems in some recent studies and contradictory findings in others disallowed a conclusive answer to the question of whether PCP use increased violent crime. Researchers have suggested, nevertheless, that the inconsistency of study findings may indicate that PCP use facilitates violence in a small proportion of users (Inciardi, 1986; Kinlock, 1991). There is agreement that biological, psychological, situational, and other factors underlying seemingly drug-related aggressive behavior should be examined in future research.

THEORIES ON THE DRUGS–CRIME RELATIONSHIP

Inciardi (1986) has noted that numerous theories have been posited to explain the drugs-crime relationship. Many of these theories have dealt with the etiology of drug use and crime. Early etiological theories tended to be overly simplistic, focusing on what Inciardi termed the "chicken-egg" question: Which came first, drugs or crime? This question polarized the drugs-crime field for over fifty years. It typically reflected two mutually exclusive positions: that addicts were criminals to begin with, and addiction was simply another manifestation of a deviant lifestyle; or that addicts were not criminals but, rather, were

forced into committing crime to support their drug habits.

Reflecting a middle-ground position, more recent theories argue for a diversity among narcotic addicts with regard to the predispositional characteristics and motives underlying drug-related criminal behavior. For example, on the basis of their research with narcotic addicts, Nurco and his associates (1991b) concluded that there is considerable variation among addicts in their propensity toward criminal activity. Some addicts had been heavily involved in crime prior to addiction, whereas others were extensively involved in crime only when addicted.

In the late 1970s, drugs-crime theories became increasingly complex, partly because studies tended to have fewer methodological problems that interfered with the measurement of both drug use and crime. With improvements in techniques, researchers gradually become more aware of heterogeneity among drug abusers on many dimensions, including the type and severity of drug-use patterns and related criminal activity. Also, more recent studies have found that drug use and crime, in most instances, do not initially have a causal relationship but, rather, are often the joint result of multiple influences. Among the many factors contributing to drug use and/or crime are those involving the family, such as lack of parental supervision, parental rejection, family conflict, lack of discipline, and parental deviance; association with deviant peers; school dropout, failure, and discipline problems; and early antisocial behavior. Consistent with the notion that all drug abusers are not alike, varying combinations of factors probably contribute to different patterns of deviant behavior in individuals at risk.

However, as Inciardi and his associates (1993a) have noted, some limitations of these theories remain. Most theories discuss drug abuse only as one of several manifestations of delinquency. Further, as in earlier years, the primary concern has been with the etiology of deviant behavior. Very little attention has been paid to explaining events that occur after the onset of drug use and criminal behavior, specifically how certain types of drug abuse increase the frequency of criminal activity. Finally, theories have typically focused on adolescence, without incorporating attributes and events that influence behavior during childhood and adulthood.

Among the most prominent theories in the drugs-crime field is that of Paul Goldstein (1986, 1989)

regarding the relationship between drugs and violence. Goldstein's theory is based on his numerous ethnographic accounts of violent drug-related acts obtained from both perpetrators and victims in New York City. According to this theory, drugs and violence can be related in three separate ways: psychopharmacologically, economic-compulsively, and systemically. Within the psychopharmacological model, violent crime results from the short- or long-term effects of the ingestion of particular substances, most notably crack-cocaine and heroin. According to the economic-compulsive model, violent crime is committed as a means to obtain money to purchase drugs, primarily expensive addictive drugs such as heroin and cocaine. The systemic model posits that drug-related violence results from the traditionally aggressive patterns of interaction found at various levels within systems of illicit-drug distribution. Examples include killing or assaulting someone for failure to pay debts; for selling "bad," or adulterated, drugs; or for transgression on one's drug-dealing "turf."

Several key studies have analyzed data in the light of Goldstein's concepts. In a study of 578 homicides in Manhattan in 1981, 38 percent of the male and 14 percent of the female victims were murdered as a result of drug-related activity (Tardiff et al., 1986). The investigators contended that these percentages were higher than those previously reported in the United States. In a subsequent study by Goldstein and his coworkers (1989) involving 414 homicides in New York City that occurred over an eight-month period, 53 percent were classified by the police and researchers as being drug-related. In both studies, most of the drug-related homicides were attributed to systemic causes. Interestingly, in the former study, most of the homicides involved heroin, whereas in the latter study, most involved crack-cocaine.

Drug Use and High-Rate, Serious Criminality. As indicated earlier, the onset of illicit drug use typically does not result in the onset of criminal behavior. In most cases, both drug use and crime begin in the early teens. Generally, the less serious the drug or crime, the earlier the age at onset of involvement. For example, among illicit drugs, marijuana is more commonly used at a younger age than are sedatives or tranquilizers, and these drugs, in turn, are typically used at a younger age than are "hard" drugs, such as heroin and cocaine. Similarly, minor forms of crime (e.g., shoplifting, vandalism) have an earlier

onset than more serious types of crimes, such as assault, robbery, and drug dealing.

Most marijuana users do not become heroin addicts, and most youths who commit minor property crimes do not subsequently become involved in more serious offenses. In both instances, the salient variable appears to be age of onset—the younger the individual is when first using a "soft" drug or committing a minor crime, the more likely he or she will move on to more serious forms of deviance. In general, the more deviant the environment (family, peers, community), the earlier the onset of deviance.

Since 1980, independent studies have identified several core characteristics of high-rate, serious offenders. According to Chaiken and Chaiken (1990), these studies have consistently found that predatory individuals tend to commit many different types of crime, including violent crime, at high rates, and to abuse many types of drugs, including heroin and cocaine. Also, research findings have consistently reported that among heroin addicts, prisoners, and seriously delinquent youth, the younger one is at onset of heroin and/or cocaine addiction, the more frequent, persistent, and severe one's criminal activity tends to be. In these studies, individuals with early onsets of addiction (typically before age 16) tended to abuse several types of drugs and to have disproportionately high rates of several types of crime, regardless of addiction status. Such findings have been observed in various geographic locations and are independent of ethnic group. These results are similar for both males and females, with one notable exception: females with early onsets of addiction are more likely to commit prostitution, shoplifting, and other property crimes at high rates, whereas males with early onsets are more likely to commit violent acts.

Chaiken and Chaiken's 1982 study of over two thousand male prisoners in three states was significant for at least two reasons. First, it challenged the long-held perception that drug abusers were less violent than other arrestees. While 65 percent of Chaiken and Chaiken's sample reported having used illicit drugs during the one- to two-year period preceding the arrest leading to the most recent incarceration, an even higher proportion (83%) of high-rate, serious offenders, identified as "violent predators," had used drugs during the same period. Among the offenders studied, violent predators were also most likely to have had histories of "hard" drug use (including heavy multiple-drug use and heroin

addiction) and to have had early onsets of several types of drug use and criminal activity. Second, and perhaps more important, the information on an offender's drug history was more likely than official arrest records to be related to the amount and seriousness of self-reported criminal activity. As in the results of drug-crime studies discussed earlier, official arrest data were poor indicators of the type, amount, and severity of crime committed by these respondents.

These findings suggest a potential for using an individual's history of illicit drug use, including age of onset, in identifying high-rate, dangerous offenders. However, this approach has several limitations. First, a general caution is in order whenever findings based on aggregate data are applied to the individual case. Second, although self-reports of drug use and crime are generally valid when obtained from individuals who are either at large in the community, entering a drug-abuse treatment program, or already incarcerated, they tend to be less accurate for individuals being evaluated for initial disposition in the criminal-justice system. Approximately one out of every two new arrestees identified as drug users by urine testing conceal their recent drug use, even in a voluntary, confidential interview having no bearing on their correctional status.

(SEE ALSO: *Antisocial Personality; Conduct Disorder; Crime and Alcohol; Families and Drug Use; Family Violence and Substance Abuse*)

BIBLIOGRAPHY

ANGLIN, M. D., & HSER, Y. (1987). Addicted women and crime. *Criminology, 25,* 359–397.

ANGLIN, M. D., & SPECKART, G. (1988). Narcotics use and crime: A multisample, multimethod analysis. *Criminology, 26,* 197–233.

BALL, J. C., ET AL. (1983). The day-to-day criminality of heroin addicts in Baltimore: A study in the continuity of offense rates. *Drug and Alcohol Dependence, 12,* 119–142.

CHAIKEN, J. M., & CHAIKEN, M. R. (1990). Drugs and predatory crime. In M. Tonry & J. Q. Wilson (Eds.), *Drugs and crime.* Crime and Justice: A Review of Research, Vol. 13. Chicago: University of Chicago Press.

CHAIKEN, J. M., & CHAIKEN, M. R. (1982). *Varieties of criminal behavior.* Santa Monica, CA: Rand.

ELLIOTT, D. S., ET AL. (1989). *Multiple problem youth: Delinquency, substance use, and mental health problems.* New York: Springer-Verlag.

GANDOSSY, R. P., ET AL. (1980). *Drugs and crime: A survey and analysis of the literature.* National Institute of Justice. Washington, DC: U.S. Department of Justice.

GOLDSTEIN, P. J. (1989a). Drugs and violent crime. In N. A. Weiner & M. E. Wolfgang (Eds.), *Pathways to criminal violence.* Newbury Park, CA: Sage.

GOLDSTEIN, P. J., ET AL. (1989b). Crack and homocide in New York City, 1988: A conceptually-based event analysis. *Contempory Drug Problems, 16,* 651–687.

GOLDSTEIN, P. J. (1986). Homicide related to drug traffic. *Bulletin of the New York Academy of Medicine, 62,* 509–516.

GREENBERG, S. W. (1976). The relationship between crime and amphetamine abuse: An empirical review of the literature. *Contemporary Drug Problems, 5,* 101–103.

INCIARDI, J. A., ET AL. (1993a). *Street kids, street drugs, street crime: An examination of drug use and serious delinquency in Miami.* Belmont, CA: Wadsworth.

INCIARDI, J. A., ET AL. (1993b). Women and crack-cocaine. New Yokr: Macmillan.

INCIARDI, J. A. (1986). *The war on drugs: Heroin, cocaine, and public policy.* Palo Alto, CA: Mayfield.

INCIARDI, J. A. (Ed.). (1981). *The drugs-crime connection.* Sage Annual Reviews of Drug and Alcohol Abuse, Vol. 5. Beverly Hills, CA: Sage.

JOHNSON, B. D., ET AL. (1991). Concentration of delinquent offending: Serious drug involvement and high delinquency rates. *Journal of Drug Issues, 21,* 205–229.

KINLOCK, T. W. (1991). Does phencyclidine (PCP) use increase violent crime? *Journal of Drug Issues, 21,* 795–816.

NURCO, D. N., ET AL. (1991a). A classification of narcotic addicts based on type, amount, and severity of crime. *Journal of Drug Issues, 21,* 429–448.

NURCO, D. N., ET AL. (1991b). Recent research on the relationship between illicit drug use and crime. *Behavioral Sciences and the Law, 9,* 221–242.

NURCO, D. N., ET AL. (1988a). Differential criminal patterns of narcotic addicts over an addiction career. *Criminology, 26,* 407–423.

NURCO, D. N., ET AL. (1988b). Nonnarcotic drug use over an addiction career: A study of heroin addicts in Baltimore and New York City. *Comprehensive Psychiatry, 26,* 450–459.

NURCO, D. N., ET AL. (1986). A comparison by ethnic group and city of the criminal activities of narcotic addicts. *Journal of Nervous and Mental Disease, 12,* 297–307.

SHAFFER, J. W., ET AL. (1985). The frequency of nonnarcotic drug use and its relationship to criminal activity among narcotic addicts. *Comprehensive Psychiatry, 26,* 558–566.

TARDIFF, K., ET AL. (1986). A study of homicide in Manhattan, 1981. *American Journal of Public Health, 76,* 139–143.

DAVID N. NURCO
TIMOTHY W. KINLOCK
THOMAS E. HANLON

CROP-CONTROL POLICIES (DRUGS)

Eliminating drug crops at the source through crop eradication and/or crop substitution has been a central, or at least an integral, part of U.S. international narcotics-control policy for the past twenty years. U.S. government policy officials maintain that eradication of illicit narcotics closest to the source of the raw material represents the most cost-effective and efficient approach to narcotics control within the overall supply-reduction strategy. The source of the illicit crop is believed to be the most commercially vulnerable point in the chain from grower to user. Since 1990, however, U.S. government policy officials have shifted away from crop control in favor of enhanced interdiction and targeting major trafficking organizations. Despite the best efforts of the United States and cooperating drug-SOURCE COUNTRIES, controlling the crop has been a difficult, if not impossible, task. Several HEROIN and MARIJUANA crop-control successes have occurred, most notably in MEXICO and COLOMBIA, but these programs had their problems. To date, notwithstanding minor, short-term successes in BOLIVIA, coca crop-control has remained an elusive goal in the Andes. Undertaking a drug crop-control program involves political as well as economic costs for both the source country and the United States.

CROP-ESTIMATING METHODOLOGIES

For more than a decade, the U.S. government has estimated the total acreage under illicit-drug cultivation at home and abroad, applying proven methods similar to those used to estimate the size of legal crops. The government knows with less certainty, however, actual crop yields (the amount of coca leaf or OPIUM gum produced per acre). Soil fertility, weather, farming techniques, and plant diseases can produce wide variations in crop yields. Given the clandestine nature of the drug business and variations from year to year and place to place, the government cannot estimate accurately the quantities harvested and available for processing. Furthermore, wide variations in processing efficiencies (depending on the orgin and quality of raw material, technical processing method, size and sophistication of laboratories, and the skill and experience of workers and chemists) make cocaine and heroin production estimates extremely complicated. Using commonly believed processing efficiencies, the government estimates a range for COCAINE and heroin production. Estimating the amount of this production that enters the United States poses still more difficult challenges.

The government uses two principal methods of estimating illicit-drug acreage under cultivation: (1) photographic-based aerial surveys; and, (2) remote sensing from satellite surveillance. Both methods have validity and reliability problems, but aerial surveys matched by ground truth (the verification of cultivation in the areas photographed) produce the best estimates. Satellite surveillance data are problematic because of weather, instrument calibration, cultivation under foliage, the small size of fields, false positives and negatives that result from color spectrum (signature) inaccuracies, and lack of ground truth.

EFFECTIVE CROP-CONTROL STRATEGIES

Many believe the illicit-drug trade is most susceptible to disruption at the organizational center of gravity—the traffickers' home country of production. Once the product leaves the production area and enters the distribution networks, it becomes more difficult to locate and control. Consequently, the U.S. government's drug-supply reduction programs have historically focused major attention on the drug source, which represents the smallest, most localized point in the grower-to-user chain.

International supply-reduction efforts close to the source of the drug also complement domestic supply-and-demand reduction efforts and give them a better chance of success. The U.S. 1991 *National Drug Control Strategy* states that when it is judged to be feasible politically, particularly when the market price of the raw product has been depressed below the cost of production, cooperative efforts can

and should be taken to reduce the net cultivation of the NARCOTIC crops. Such crop control, or eradication, would then occur through manual or herbicidal means, crop substitution, income replacement, and area-development projects that provide income and raise the standard of living. Additionally, efforts would be made to convince the cultivators to plough under or cut down their drug crops voluntarily or not to plant them in the first place.

Crop-control strategies are important for cocaine, heroin, and marijuana reduction, although neither the United States nor the source governments have much control over, or much access to, the largest opium-producing reigons of Southwest Asia (the Middle East) and Southeast Asia (especially Myanmar [formerly Burma] and Afghanistan, the world's largest producers of illicit opium). The Peruvian government has exercised only limited sovereignty over the world's largest area of coca cultivation (for cocaine), the insurgent-controlled Upper Huallaga Valley (UHV). When government commitment and ability exists for controlling the source of the illicit product, crop eradication and/or income substitution can be effective ways to achieve a net reduction in the production of the illegal crop.

COCA

Both a legal and an illegal commodity, the COCA PLANT is mainly grown in Andean South America—in Peru, Bolivia, and Colombia; it belongs to the genus *Erythroxylon* within the family Erythroxylaceae. Most cocaine comes from the leaves of 2 of the 250 identified species: *E. coca Lam* and *E. novograntense.* Each of these species, in turn, has two varieties. Agriculturally speaking, coca is a hardy, relatively labor-free, subtropical perennial plant that thrives at high to somewhat lower elevations and in dry to slightly humid climates, depending on variety. Coca plants are shallow-rooted, broad-leaved woody shrubs that grow to heights of 3 to 10 feet (1 to 3 m) and live about 30 years (White, 1989; LaBatt-Anderson, 1990). It takes 30 to 36 months before the coca bush is mature enough to produce leaves that can be used in the production of cocaine. Once the plant is mature, three to four crops of coca leaves may be harvested per year for an estimated yield of 1 to 3 tons per acre (0.8 to 2.7 metric tons [MT] per hectare) of dry coca leaf cultivated per year. Actual yield depends on microclimate, plant maturity, and species. After harvesting and drying, the leaves are soaked in a mixture of solvents, and the resulting COCA PASTE is precipitated, further refined to coca base, and finally refined to cocaine hydrochloride (HCl), the salt or white powder form of cocaine. Leaves weighing 1.1 tons (1MT) produce approximately 6.6 pounds (3 kg) of paste. These 6.6 pounds (3 kg) of paste, called *pasta*, are then converted to about 3.3 pounds (1.5 kg) of coca base, which is equivalent to 3.3 pounds (1.5 kg) of cocaine HCl. The U.S. government estimates that the Andean countries produced approximately 900 tons (816 MT) of cocaine for world consumption in 1991 (INCSR, 1992).

OPIUM POPPY

Unlike the perennial coca bush, the opium poppy flower is an annual that requires the planting of seeds for each crop. Poppies of many species grow throughout the world, but only *Papaver somniferum* yields opium and its derivatives—the medicinal analgesics CODEINE and MORPHINE as well as the now illicit addictive heroin. Once planted, the life cycle of the labor-intensive poppy lasts for 120 to 150 days, from planting until the petals fall off. Day and night, certain nitrogen-containing compounds, alkaloids, are produced by the plant and stored in its cells. After the petals fall, the seed capsule swells and is incised while still green. A milky alkaloid-rich sap seeps from tiny tubes in the capsule wall, which dries, darkens, and turns gummy—becoming a substance called opium gum. Raw opium gum is converted by crude refineries into morphine base; a few pots, simple chemicals, and a source of fresh water are all that is needed to create the morphine base from which illicit heroin is made. About 22 pounds (10 kg) of opium make 2.2 pounds (1 kg) of morphine base; treating the base with acetic anhydride creates heroin. *Papaver somniferum* is cultivated in dozens of varieties adapted to do well in various soils and climates, ranging from southern Sweden to the Equator. Depending on the soil and climatic conditions, the growers can harvest at least two crops per year (White, 1985). Although the opium poppy seems to flourish at about 3,000 feet (915 m) in low humidity, it also grows and survives in humid lowlands, under foliage, or in full sunlight. With the aid of irrigation and pesticides, the poppy growers have expanded the conditions and acreage in which the plant will produce. The U.S. government estimated that 4,187 tons (3,800 MT) of opium, or approxi-

mately 418 tons (380 MT) of heroin—if the total were converted—were produced throughout the source countries of Southeast Asia, the Middle East, Mexico, Guatemala, and Colombia.

CANNABIS

Marijuana, a by-product of the plant *Cannabis sativa*, remains the most commonly used illicit substance in the United States, although its use has been decreasing steadily for the past several years. Both the plant and its psychoactive ingredient TETRAHYDROCANNABINOL (THC) are controlled substances. The U.S. government estimates that Mexico still supplies the majority of the marijuana available in the United States, perhaps as much as 63 percent. Domestic supply accounts for another 18 percent, Colombia for 5 percent, Jamaica for 3 percent, and the remaining 11 percent comes form Belize, Laos, the Philippines, Thailand, Lebanon, Pakistan, and Afghanistan. Brazil and Paraguay also cultivate cannabis, but the majority is consumed locally or exported to neighboring countries, with very little, if any, finding its way to the United States (NNICC, 1990).

The flowering tops and the leaves of the marijuana plant are collected, dried, and used for psychoactive effect—usually smoked as a cigarette or in a pipe but also ingested as an ingredient of food. The plant is an annual; it is planted from seed and harvested traditionally in two seasons of five- to six-month cycles each year. Cannabis can grow almost anywhere outside of the ice-bound frigid zones, if provided with adequate sunshine and water. As a chlorophyll-based green plant, it requires photosynthesis to grow and mature. Coating the plant's leaves with a contact herbicide such as paraquat or glyphosate has been very effective in killing the plant in Mexico, Belize, and Colombia. Unlike the coca plant, which has a perennial root system, most types of marijuana can be destroyed easily by digging, spraying, or cutting.

CROP ERADICATION

Illicit crops probably constitute the least expense in the narcotics chain. Producers devote few economic resources to prevent detection, although it is easier to locate and destroy crops in the field than to locate the processed drugs once they enter the smuggling routes or on U.S. city streets. Despite this be-

lief, however, few effective crop-control policies have been implemented, and crop control remained a secondary approach, at best, in the U.S. government's 1991 and 1992 *National Drug Control Strategy*. It was easier for U.S. agencies to function at the border or within the United States than to get compliance on command from source countries. DRUG INTERDICTION and immobilizing the trafficking organizations were the preferred policy approaches.

Crop eradication can be effected forcibly or voluntarily (1) by manual plant removal, (2) by biological control through the use of pathogens or predators, or (3) by the use of herbicides. Of the three methods, herbicidal eradication has been the most effective and efficient, though not the most neutral politically. Payment to the growers for the labor of uprooting the plants voluntarily is an important short-term element, while development assistance is a longer-term component of the successful implementation of any eradication effort that invites voluntary reduction.

Because the coca bush is a perennial, destroying the plant can have a devastating effect on the productive capacity of the trafficking organizations. After the plant dies, it would take nearly three years for the grower to be reemployed on that land productively. For the most part, the coca fields of Peru, Bolivia, and Colombia are remote relatively small plots. There is little or no intercropping, where the grower mixes his coca fields with yucca or other agricultural produce, therefore, aerial herbicidal eradication is an efficient option—but one that requires significant amounts of herbicide to kill the hardy coca bush. The opium poppy and the marijuana plant are easier to eradicate than coca bushes but, because they are annuals and planted from seed, with several harvests a year, they require year-round crop-control efforts.

Manual Eradication. Manual eradication involves physical removal or cutting. For coca, removing the plant is more effective than cutting, because coca can sprout new shoots from the base of the stump. The plants, however, are difficult to remove after three years of growth because of their root systems. Coca, poppy, and marijuana plants can be removed only if grown in areas that are easily accessible. Many of the illicit growing areas are in remote corners of the countryside, so transporting personnel and equipment may be expensive and hazardous. More importantly, the manual eradicators place themselves at great personal risk, as for example in coca-growing areas of Peru and Colom-

bia—where violence directed at eradication personnel have forced the suspension of such efforts.

Biological Control. Numerous biological control agents—parasites, predators, pathogens— have been identified that may destroy coca and poppy plants. Too little is known, however, about the effect of these agents on the ecology of the growing regions. Another issue is the possible negative impact on legitimate crops, which, if inadvertently destroyed, could cause famine and/or severe economic losses. Use of biological agents may be a future possibility, but the current state of research on such pests and their potential impact does not make this option feasible yet.

Herbicides. Control of coca by foliar application of herbicides was attempted in Colombia in 1985, with inconsistent results. In 1987 and 1988, further small-scale testing was conducted in Peru, involving both foliar- and soil-applied herbicides, which confirmed the efficacy of soil-applied herbicides. Two herbicides were chosen for further testing: tebuthiuron and hexazinone. Based on further tests in Peru in 1989, researchers learned that aerial application of either pellet tebuthiuron or granular hexazinone, at lower rates than used typically in the United States, can kill a significant percentage of coca plants in a field and force its abandonment. Environmental tests were conducted in Peru for more than two years to measure such ecological effects as translocation of the herbicide into the soil, water, and air; water solubility; effect on the flora and fauna; and ability of the herbicide to leech on the clay molecules in the soil. Tebuthiuron and hexazinone were judged environmentally safe and effective. Other herbicides such as dicamba, imazapyr, picloram, and triclopyr were also tested and found to be more toxic but less effective. Use of herbicides on poppy and marijuana has been tested and thoroughly documented. Concise environmental reviews (measuring impact on environment) and environmental-impact statements (measuring health consequences for consumers who use the drug product after it has been sprayed with herbicide) have been filed for 2,4-D, glyphosate, and paraquat use on poppy and marijuana fields.

CROP-ERADICATION SUCCESSES

In Mexico, Colombia, Belize, Myanmar (formerly Burma), Bolivia, Jamaica, and Thailand, crop-eradication efforts continue to have varying degrees of success in reducing illicit crop cultivation. In the mid-1970s, Mexico began an aerial herbicidal-eradication program on both opium and marijuana and reduced the cultivation of these illicit crops significantly. In 1991, Mexico reportedly destroyed some 16,000 of its 25,000 acres (6,500 of its 10,000 hectares) of opium and 27,000 of its 71,500 acres (11,000 of its 29,000 hectares) of cannabis. In the early 1980s, the Colombian government used glyphosate in the north to eradicate most of its marijuana there. In the early 1990s, Colombia planned to use the same herbicide on newly discovered opium in the Cauca and Huila departments. In 1987, the U.S. government supported the government of Belize in an aerial marijuana-eradication program, which resulted in a 90 percent decline in cannabis production.

In the 1987–1988 growing season, Myanmar sprayed the herbicide 2,4-D from fixed-wing agricultural-spray aricraft to destroy about 31,000 acres (12,500 ha) of opium poppy. This program came to a halt late in 1988 when the government moved its limited military resources to attack mounting antigovernment protests. The political balance also changed abruptly when Myanmar's ruling military eliminated the opium-eradication program, abolished human rights, and accommodated certain trafficking insurgents (the Wa and Kokang Chinese components of the now-defunct Burmese Communist Party, who were fighting one of Burma's principal enemies, Khung Sa and his Shan United Army).

In addition to these aerial herbicidal-eradication efforts, manual eradication of crops continued in 1991 in a number of countries, to include destruction of nearly 30 percent of Thailand's 10,500 acres (4,200 ha) of opium, about half of Colombia's 6,000 acres (2,500 ha) of opium, approximately 10 percent of Bolivia's 131,000 acres (53,000 ha) of coca, and a little less than half of Jamaica's 4,500 acres (1,800 ha) of cannabis.

CROP-ERADICATION DIFFICULTIES

Conceptual, political, and technical arguments are often raised against drug-crop eradication. Opponents of eradication believe that the reduction of foreign supplies of illicit drugs is probably not achievable, or short-term at best; they say that even if eradication had a longer-term impact in the source country, it would not have a meaningful effect on levels of illicit-drug consumption in the United

States, where the consumer would simply switch to other available drugs. Moreover, some fear that inordinate environmental damage will result from herbicide use. Others question whether a global policy of crop control is feasible politically, because many growing areas are far beyond government control, and even when there is government jurisdiction, crop eradication becomes impractical because the grower can continually shift areas of cultivation. Instituting effective eradication efforts in some source countries, such as Peru, might also drive political insurgents (who co-locate with the drug traffickers) into threatening alliances that would undermine the central govenment even further. Finally, some question the value of supply-reduction efforts at the source altogether, since world production and supply of illicit drugs vastly exceed world demand. If the worldwide supply were reduced dramatically, it would not be felt in the United States until the supply had dried up throughout the rest of the world, because U.S. consumers often pay higher prices than those in any other market; moreover, U.S. dollars are the preferred narco-currency (Perl, 1988).

CROP-SUBSTITUTION EFFORTS

Crop substitution—the replacement of opium, coca, or marijuana production with a legal agricultural commodity—can never be successful by itself, because of the immense profits from illicit-drug cultivation. A more broadly defined income-replacement approach (which may include an agricultural crop-substitution component), however, coupled with strong law enforcement, may succeed in convincing drug growers to stop planting the illicit crop.

In the Malakand District of Pakistan's North-West Frontier Province, the U.S.–Pakistani efforts in the early 1980s to implement a fully integrated, rural development project, which provided roads, water, electrification, and agricultural substitutes (e.g., peanuts, apples), resulted in a net reduction in opium poppy production. Providing project support for the 300,000 inhabitants of the district enabled local residents to earn an income from an extensive road-building and infrastructure development program, thus making cultivation of the opium poppy unnecessary. A side benefit of the development efforts was the creation of valuable infrastructures to raise the standard of living throughout the region and encourage the nomadic populations to establish roots and achieve more stable living conditions.

The Highland Village Project in northern Thailand has provided similar benefits to the culturally diverse, opium-producing hill tribes. This resulted in decreased opium cultivation consistently in the early 1990s. In Laos, the Houaphanh project near the remote border region with Vietnam began in 1990 to provide area development incentives (of improved water and roads and medical and educational benefits) for growers who opted to cease planting opium. In the Western Hemisphere, a principal component of the Andean Strategy to eliminate illicit coca in Peru and Bolivia has an economic-assistance element that provides hard-currency earnings, trade incentives, and local project assistance to entice the growers away from coca cultivation. Some funds have already been expended on agro-research (discovering viable crops), infrastructure development, extension training, and rudimentary marketing.

In the broadest sense of the term, crop substitution worked rather effectively in Turkey in the early 1970s when a government cash subsidy permitted the farmers to harvest the poppy *before* the plant ripened to produce the opium gum. In this way, the traditional cooking and ceremonial uses of poppy could be maintained through the poppy straw process, as it is called, but the seed pod would not be available for the illicit opium gum.

CROP-SUBSTITUTION DIFFICULTIES

Several inherent difficulties exist with crop-substitution approaches.

1. Many of the growing regions are remote inhospitable areas, outside central government control.
2. In a free-market economy, no legitimate crop can compete with either coca or opium as an income-producing agricultural commodity. Even if there were competitive substitutes, with the immense profits from the drug trade, the drug traffickers could continue to raise the price to compete for willing cultivators.
3. Much of the land in the growing zones cannot produce legitimate agricultural products sufficient to support the farming population.
4. The presence of political insurgents and threat of violence in some of the growing areas (Peru, Myanmar, Afghanistan) create an unfavorable climate for crop substitution.
5. There are difficulties in finding international markets to accept the substitute crop; for exam-

ple, in Bolivia's Chapare region, oranges and coffee are viable agricultural products, but the international coffee cartel and U.S. citrus growers do not allow Bolivian products to compete for shares of existing markets.

6. In some regions, such as Peru's Upper Huallaga Valley and Bolivia's Chapare, the vast majority of coca cultivators are not farmers and know nothing about agriculture. Many were unemployed urban dwellers and laborers who moved to the coca-growing regions to seek a viable living after the collapse of the tin market in the late 1970s and early 1980s in Bolivia. In the long run, regional development efforts in the urban areas may be required to attract the cultivators back to their places of origin.

7. Successful crop substitution takes years of agro-research, infrastructure development, training, and marketing; it may take too long for subsistence-crop and/or cash-crop producers to make a living. In Pakistan's Malakand project, it took more than five years to develop the agricultural component. Some argue that in the absence of strong law enforcement and control, crop substitution becomes only an *additional* income generator, not a true substitute. The growers will accept the substitute and continue to cultivate the illicit-drug crop.

8. Corruption and powerful interest groups in the growing areas pose serious impediments to any crop-control efforts.

(SEE ALSO: *Foreign Policy and Drugs; Golden Triangle; International Drug Supply Systems*)

BIBLIOGRAPHY

BUREAU OF INTERNATIONAL NARCOTICS MATTERS, U.S. DEPARTMENT OF STATE. (1990). International cocaine strategy: Report to the Congress. Washington, DC: Author.

BUREAU OF INTERNATIONAL NARCOTICS MATTERS, U.S. DEPARTMENT OF STATE. (1990). *International narcotics control strategy report (INCSR)*. Washington, DC: Author.

LABATT-ANDERSON (1990). Environmental assessment of the use of herbicides to eradicate illicit coca overseas (Revised) Washington, DC: U.S. Department of State.

MORALES, E. (1989). *Cocaine: white gold rush in Peru*. Tucson: University of Arizona Press.

MUSTO, D. (1973). *The American disease*. New Haven & London: Yale University Press.

NADELMANN, E. A. (1988). U.S. drug policy: A bad export. *Foreign Policy, 70*, 83–108.

OFFICE OF NATIONAL DRUG CONTROL POLICY. (1992). *National Drug Control Strategy, 1992*. Washington, DC: Author.

PERL, R. F. (1989). Congress and international narcotics control. *CRS Report for Congress*. Washington, DC: The Library of Congress.

REUTER, P. (1985). Eternal hope: America's quest for narcotics control. *The Public Interest, 79*, 79–95.

VAN WERT, J. (1988). The U.S. State Department's narcotics control policy in the Americas. *Journal of Interamerican Studies and World Affairs, 30*(2 & 3), 1–18.

WALKER, W. O., III. (1981, 1989). *Drug control in the Americas*. Albuquerque: University of New Mexico Press.

WHITE, P. T. (1985). The poppy—for good and evil. *National Geographic, 167* (2), 143–189.

WHITE, P. T. (1989). Coca—an ancient herb turns deadly. *National Geographic, 175* (1), 6–51.

JAMES VAN WERT

CROSS-DEPENDENCE *See:* Addiction: Concepts and Definitions; Tolerance and Physical Dependence.

CROSS-TOLERANCE *See* Addiction: Concepts and Definitions; Tolerance and Physical Dependence.

CULTS AND DRUG USE The interface between cults and drug use is one filled with complexity and contradiction. Traditionally, cults are groups that diverge from major religions or that form new philosophical/religious systems, often around a charismatic leader. Consequently, at any given time it may be difficult to distinguish a cult from a newly formed religious sect. Some cults last and become sects or new religions; some remain cults, or die. The line is hard to draw and open to interpretation, even by social scientists and the clergy who specialize in this field.

BACKGROUND

Historically, some cults and cultlike groups have sponsored the use of drugs as an integral aspect of ritual. In ancient Greece, for example, the use of ergot (genus *Claviceps*, a fungus that grows on grains and causes hallucinations) appears to have been a significant aspect of the rituals performed within the Eleusinian mysteries, celebrated in the worship of the goddesses Demeter and Persephone. As poets noted, "I have seen the truth within the kernel of wheat"—which presented an enigmatic foreshadowing of the Countercultural Revolution/Love Generation, when a purified ergot derivative (LYSERGIC ACID DIETHYLAMIDE, LSD) afforded adepts a similar opportunity.

In Islam, alcohol is forbidden, but medieval Islamic sects were formed to use HASHISH (a form of *Cannabis sativa*, MARIJUANA, which came into use in the Middle East after the basic tenets of the religion were established in the seventh century by the Prophet Mohammed and his followers); the hashish was used to offer a taste of the paradise to come. In pre-Columbian America, a wide variety of drugs were utilized within religious rituals; today, the Native American Church continues to use the HALLUCINOGENS peyote and mescaline (both derived from the small cactus *Lophophora williamsii*) as an integral aspect of their practices. Recent judicial decisions have protected and reaffirmed their right to use them within church ceremonies. As Preston and Hammerschlag (1983) have noted, the use of hallucinogens within this context is part of a rigidly proscribed pattern—the substance is used clearly as part of a transcendent experience, accompanied by rituals of purification, and not lending itself to use on a promiscuous basis.

TWENTIETH CENTURY

The 1960s and 1970s were characterized by an extraordinary youth movement of baby-boomers with an efflorescence of interest in the cultic and the occult—and by a popularization of drug use within mainstream American society. Some of this was fueled by the philosophies and practices of Asia, especially Southeast Asia, where the Vietnam war was being fought; some was inspired by the Shangri-la nature of the lands of the Himalayas, where Buddhism was practiced in secluded monasteries and *nirvana* was sought. As the "Greening of America"

proceeded through the two decades, mind-altering substances in addition to ALCOHOL and NICOTINE became available, "on the street," and were no longer confined to the disenfranchised or marginal. There was an increasing juxtaposition of the so-called transcendent religious experience (the mind-expanding experience) with drug use that often became drug abuse.

This juxtaposition had been anticipated by some nineteenth-century poets, such as Rimbaud and Baudelaire, and by cult figures such as Alaister Crowley, Aubrey Beardsley, and William Blake. By a dual embrace of Transcendental Meditation/Mahareshi (Buddhistic-based movements) and "Lucy in the Sky with Diamonds," the much adored singing group called the Beatles both mirrored and promoted the use of hallucinogens as providing a readily accessible transcendental experience—although in Buddhism the goal of all existence is the state of complete redemption (*nirvana*, a state achieved by righteous living, *not* by drugs). Unlike Aldous Huxley who combined an interest in Vedanta (an orthodox system of Hindu philosophy) and the use of mescaline, the Beatles and their putative mentor, the Mahareshi Mohesh Yogi, proclaimed the desirability of enlightening the masses rather than restricting enlightenment to a righteous educated elite.

In literary monuments of that era, such as Armistead Maupin's *Tales of the City* (1977), his characters routinely advocate and use mild-altering substances (especially marijuana) without any apparent appreciation of their darker potential—but that was the general attitude toward TOBACCO and alcohol use at that time as well. In addition, there was no special appreciation that drug use, in and of itself, might encourage cult affiliation, yet this was very much the time of the burgeoning Hare Krishnas and the Moonies and the Nation of Islam.

The relationship of such cults and drug use is paradoxical. Deutsch (1983) has noted that prolonged drug use may encourage this type of cult affiliation, and many cult groups offer themselves to the public and to vulnerable individuals as quasi-therapeutic contexts where the individual will be able to transcend the need for drugs. This aspect of cult-appeal turned thousands of lost and confused free spirits and flower children into vacant-eyed smiling cultists who signed over to the cult all their worldly goods— to spend their days wandering the streets, airports, and bus or train stations shaking bells and tambourines or offering flowers to passing strangers. Rigor-

ous training programs, called "brainwashing" by parents of the lost children and other skeptics, were fashioned to strip cultists of free will and substitute nodding acquiescence.

The Moonies. The charismatic leaders of such cults include the Korean businessman Sun Myung Moon, who in 1954 established the Unification Church (Tong Il in Korea), which is a messianic millennial movement dedicated to restoring the kingdom of heaven on Earth. In the late 1950s, his missionaries went to Japan and to the West; in the 1970s he moved to the United States, where he purchased and started businesses and bought property. He has created worldwide communities and opportunities for his followers, many of whom toil on farms and ginseng plantations. Thousands at one time have been married in spectacular mass weddings (known as The Blessing) to partners chosen for them by the church. During the 1970s and 1980s, the Moonies were especially notorious for members whose parents "kidnapped" back their own children to have them "un"-brainwashed.

The People's Temple. Another charismatic cult leader was the Reverend Jim Jones, leader of the People's Temple. His claim of curing drug abuse was only one of the lures. After moving around the United States for awhile, he brought his people to an isolated spot in South America, where one of the former substance abusers mixed for them a massive batch of poisoned Kool-Aid for the cult's final event—a basically unexplained mass suicide.

Scientology. Nor was the People's Temple unique—organizations such as Narcanon have stated that their treatment of substance abusers reflects the dianetics-based teachings of L. Ron Hubbard, a U.S. science-fiction writer, whose Scientology movement expanded in the 1950s when he moved to England. Scientology is a quasi-philosophical system that claims to improve the mental and physical well-being as followers advance within the cult, by undertaking (and paying well for) a series of course levels.

The Nation of Islam. Also known as the American Black Muslim movement, this is another organization that stresses a self-help "do for self" policy that attracts substance abusers seeking a means to give up drugs. Led for forty-one years by Elijah Muhammad (died 1975), the Nation of Islam became centered in Chicago, and its founder Wallace D. Fard, now known as the Prophet Fard, is worshipped as Allah; Elijah Muhammad is the Messenger of Al-

lah, and he built on the teachings of Fard with aspects of Christianity, Islam, and black nationalism.

Hare Krishnas. The familiar saffron-robed youth who chanted on streetcorners during the 1970s were called Hare Krishnas; they were converts to the International Society for Kirshna Consciousness (ISKCON), the missionary form of devotional Hinduism brought to the United States in 1965 by the guru A. C. Bhaktivedanta Swami Prabhupada. The movement was known for the public chanting of the Hare Krishna mantra and its distribution of *Back to Godhead* and other literature specially translated into English for converts. It was a religious alternative to and an embodiment of the counterculture's values and attitudes. It attracted thousands of young people at lectures and devotional services, and preaching centers were founded in Los Angeles, Berkeley, Boston, and Montreal. In less than twenty years, it became an international movement with over 60 centers in the United States and another 160 in other countries. Devotees followed a daily regimen of rising at 3:30 A.M. for services and honored the four monastic regulative principles: no meat eating, no intoxicants of any kind, no sexual activity of any kind, and no gambling.

Twelve-Step Programs. Clearly, intense religious and quasi-religious feelings form an integral part of the enormous and expanding SELF-HELP recovery movement. Founded in the 1930s, ALCOHOLICS ANONYMOUS (AA) is the oldest and most prominent—and the prototype—for many similar organizations. At their base is the AA TWELVE-STEP program, which emphasizes a submission to "God *as we understand Him*" (italics theirs), as a major factor in achieving SOBRIETY. This underlines the fact that chronic substance abusers are often the product of a chaotic and unstructured environment. Their problem may reflect a need for continuity which they have not heretofore received. Their recovery may then be predicated on a sense of continuity which may be based on the religious or quasi-religious but explicit commitment. (Some twelve-step organizations have varied or discarded the supernatural commitment for those uncomfortable with such a concept.)

Synanon and Other Treatment Communities. If intense religious commitment is a significant aspect of the twelve-step recovery movement, of concern is the potential for this commitment to transmute itself into a species of cult affiliation. Rebhun (1983) is one among many who has noted the

potential for cultic evolution in drug-treatment programs such as SYNANON. Synanon was not unique; the history of residential-treatment center/drug-treatment center includes a number of organizations that were initially formed on the basis of an intense commitment to change but which mutated into authoritarian and hierarchical organizations. The very difficulty that recovered substance-abusers experience in leaving the world of the THERAPEUTIC COMMUNITY to become independent members of mainstream society is a testimony to the intensity of feelings mobilized within these organizations—the need of the staff and members alike to find magical transcendent solutions to serve problems and the false resolution of these problems through fusion with an authoritarian and charismatic leader who will ostensibly provide the continuity and structure for which the substance abuser hungers.

SUMMARY

Drugs and other mind-altering substances have formed an integral part of some cultic/religious ritual from very ancient times. In the mid-to-late twentieth century, the structure provided by groups that mobilize intense religious or quasi-religious feelings have sometimes enabled vulnerable individuals to transcend their personal difficulties. On occasion, the very intensity of the substance user's object hunger has lent itself to the transformation of otherwise viable or valuable organizations—thereby turning them into cultlike formations.

(SEE ALSO: *Religion and Drug Use*)

BIBLIOGRAPHY

BARKER, E. (1984). *The making of a Moonie: Choice or brainwashing?* Oxford: Oxford University Press.

DEUTSCH, A. (1983). Psychiatric perspectives on an Eastern-style cult. In D. A. Halperin (Ed.), *Psychodynamic perspectives on religion, sect, and cult.* Littleton, MA: James Wright-PSG.

MAMIYA, L. H. (1987). Elijah Muhammad. In M. Eliade (Ed.), *The encyclopedia of religion.* New York: Macmillan.

PRESTON, R., & HAMMERSCHLAG, C. (1983). The Native American Church. In D. A. Halperin (Ed.), *Psychodynamic perspectives on religion, sect, and cult.* Littleton, MA: James Wright-PSG.

REBHUN, J. (1983). The drug rehabilitation program: Cults in formation? In D. A. Halperin (Ed.), *Psychodynamic perspectives on religion, sect, and cult.* Littleton, MA: James Wright-PSG.

SHINN, L. D. (1987). International Society for Kirshna Consciousness. In M. Eliade (Ed.), *The encyclopedia of religion.* New York: Macmillan.

DAVID A. HALPERIN

CUSTOMS SERVICE, U.S. *See* U.S. Government/U.S. Government Agencies: U.S. Customs Service.

DARE *See* Appendix II, Volume 4, Drug Abuse Resistance Education (DARE) Regional Training Centers.

DARP *See* Drug Abuse Reporting Program.

DAWN *See* Drug Abuse Warning Network.

DAYTOP VILLAGE (*also* Daytop for a Drug Free World) Daytop Village, Inc., which began in 1964, had its roots in a research project conducted by Alex Bassin and Joseph Shelly of the Probation Department of the Second Judicial District of the Supreme Court of New York. They were awarded a grant from the National Institute of Mental Health to initiate a new approach for treating drug-addicted convicted felons. This new approach would offer an alternative to incarceration, in the form of a residential treatment center modeled roughly after a pioneering drug treatment effort in California known as SYNANON. The founders of Daytop Village included Dr. Daniel Casriel—a New York psychiatrist who had observed the Synanon community—David Deitch, a former Synanon director, and Monsignor William B. O'Brien, a Roman Catholic priest who continued to guide Daytop into organizational maturity and expansion of the THERAPEUTIC COMMUNITY approach.

In the early years, Daytop's primary effort was long-term residential treatment, generally lasting eighteen months to two years. Beginning in the 1970s, as an outgrowth of community-based intake and reentry centers, Daytop also created a series of day-care models. In the mid-1970s, as the age of onset for drug use dropped in the United States, Daytop began to separate adults and adolescents into discrete residential and day-care treatment units. During the mid-1980s, Daytop expanded its program to include working adults—both after work and during special employer-contracted daytime hours. In the late 1980s Daytop instituted special programs for pregnant women; by late 1991, over 150 nonaddicted well baby births had occurred.

Within each of these components on a continuum of care model, Daytop uses drug-free therapeutic community methodologies. The essential tenets of this philosophy are SELF-HELP and acceptance of personal responsibility in a "must function" demanding environment that is couched in copious amounts of love and caring discipline. Interventions to identify problems are accomplished through behavioral reality-oriented group, individual, and family approaches. The basic assumption underlying the treatment system is that drug dependence is a mix of educational, biomedical, emotional, spiritual, and psychosocial factors—and the treatment environment must attend to all of these. Intra- and interpersonal dysfunctions force the individual, despite a desire to change, into patterns of behavior most compatible with continued drug usage. The Daytop

therapeutic community treatment environment is designed to uncover and expose these dysfunctional behaviors and search for alternatives that will serve the client more positively.

Currently, Daytop operates a major diagnostic and assessment unit, a series of outreach centers with both day and evening programs, adolescent and adult residential services, and a residential unit that exclusively serves youth from correctional facilities. The corporate offices and the majority of these services are located in New York, but some are also in Texas, California, and Florida.

Inasmuch as Daytop has acted as a prototype (and often direct coach) for most of the drug-treatment therapeutic communities in the United States and overseas, outcome studies conducted on therapeutic communities in general attest to Daytop's relative effectiveness. Studies reveal a success rate among graduates (program completers) of over 75 percent for as long as seven years after graduation; even for dropouts, success rates average in the 33 percent range. Those in treatment a year or more show the best outcome. Retention in treatment is the key variable for success.

(SEE ALSO: *Amity, Inc.; Gateway Foundation; Marathon House; Operation PAR; Phoenix House; Treatment Types; Appendix III, Volume 4: State-by-State Treatment and Prevention Programs*)

BIBLIOGRAPHY

DEITCH, D. A. (1973). Treatment of drug abuse in the therapeutic community: Historical influences, current considerations and future outlook. In *National Commission on Marihuana and Drug Abuse. Report to Congress and the President*, vol. 5. Washington, DC: U.S. Government Printing Office.

DAVID A. DEITCH
ROBIN SOLIT

DEA *See* U.S. Government: The Organization of U.S. Drug Policy.

DECRIMINALIZATION *See* Marihuana Commission; Policy Alternatives.

DELINQUENCY AND DRUG ABUSE
See Conduct Disorder and Drug Use; Crime and Drugs; Gangs and Drugs.

DELIRIUM Delirium has been defined in many ways. Some use the term to refer to an acute, hyperactive, confusional state. Psychiatrists define it more broadly to describe clinical states characterized by a reduced level of consciousness, an inability by the affected individual to sustain or shift attention appropriately, disorganized thinking, disorientation to time, place, or person, and memory impairment. In addressing the affected individual, questions need to be repeated, the individual may perseverate in responses, and speech may be rambling or incoherent. Additional features include an altered sleep-wake cycle, sensory misperceptions, disturbances in the pace of psychological and motor activity, and varying mood states (e.g., apathy, euphoria). Sensory misperceptions—usually visual ones—may include illusions (e.g., specks on the floor are thought to be insects) or hallucinations (one "sees" a relative in the room when there is actually no one there). Delusions may be present (e.g., the person is convinced that medical staff are secret government agents). The individual may respond emotionally (e.g., with anxiety) and behaviorally (e.g., attack those viewed as threatening) to the context of the delusion. There may be elevated blood pressure, a rapid heartbeat, and sweating and dilated pupils. The onset of such a clinical state is relatively rapid (taking an hour to days), the symptoms fluctuate throughout the course of illness, and the duration is usually brief (about one week). It is important to note that the altered level of consciousness exists on a continuum. Hypervigilance can progress to confusion and drowsiness.

The factors that may cause delirium are numerous. They can include head trauma, infections (e.g., meningitis), metabolic disorders, liver and kidney disease, postsurgical states, and psychoactive substance intoxication and withdrawal. The common underlying functional disturbance in delirium is diffuse impairment of brain-cell metabolism and stability. These changes can frequently be seen on an electroencephalogram (EEG). Delirium can occur at any age but is more common in the very young and the very old. It is most often seen in hospital settings. The treatment of delirium consists of maintaining critical bodily functions (i.e., cardiac and respiratory

functions and hydration), correcting the precipitating problem, and managing the psychological and behavioral symptoms.

(SEE ALSO: *Delirium Tremens; Withdrawal: Alcohol*)

BIBLIOGRAPHY

AMERICAN PSYCHIATRIC ASSOCIATION. (1987). *Diagnostic and statistical manual of mental disorders-3rd edition-revised.* Washington, DC: Author.
HORVATH, T. B., ET AL. (1989). Organic mental syndromes & disorders. In H. I. Kaplan & B. J. Sadock (Eds.), *Comprehensive textbook of psychiatry,* 5th ed., vol. 1. Baltimore: Williams & Wilkins.
LISHMAN, W. A. (1987). *Organic psychiatry,* 2nd ed. London: Blackwell.
PLUM, F., & POSNER, J. B. (1980). *The diagnosis of stupor and coma,* 3rd ed. Philadelphia: Davis.

MYROSLAVA ROMACH
KAREN PARKER

DELIRIUM TREMENS (DTs) This clinical disorder is a DELIRIUM that occurs after abrupt cessation of, or reduction in, ALCOHOL consumption in an individual who has been a heavy drinker for many years. It represents the most severe form of the alcohol WITHDRAWAL state and is not very common. It is associated, however, with a significant mortality rate of those who develop it (10–15%), if left untreated.

Typically, the delirium sets in 48 to 72 hours after the last drink or after reduction in drinking. The course of illness is generally short, lasting, in most cases, 2 to 3 days. The disorder becomes significantly more life threatening if there is concurrent physical illness, such as liver failure, infection, or trauma.

Clinical signs and symptoms are the same ones that are characteristic of a delirium and include disorientation, fluctuating levels of consciousness, vivid hallucinations, delusions, agitation, fever, elevated blood pressure, rapid pulse, sweating, and tremor. The delirium may at times be preceded by a withdrawal seizure. Close monitoring and medical treatment in a hospital setting are required.

(SEE ALSO: *Withdrawal: Alcohol*)

BIBLIOGRAPHY

GOODWIN, D. W. (1989). Alcoholism. In H. I. Kaplan & B. J. Sadock (Eds.), *Comprehensive textbook of psychiatry,* 5th ed., vol. 1. Baltimore: Williams & Wilkins.
PLUM, F., & POSNER, J. B. (1980). *The diagnosis of stupor and coma,* 3rd ed. Philadelphia: Davis.

MYROSLAVA ROMACH
KAREN PARKER

DEMEROL *See* Meperidine.

DEPENDENCE *See* Addiction: Concepts and Definitions; Disease Concept of Alcoholism and Drug Abuse.

DEPRESSANTS *See* Drug Types.

DEPRESSION The term *depression* has been used to refer both to an emotional state and a group of psychiatric disorders. As an emotional state, it is also known by various comparable terms: dejection, despair, sadness, despondency, lowering of spirits. Cognitions (perceptions and judgments) of a negative nature often accompany depressed mood.

Most people experience brief periods of depressed or despondent mood, often in response to a disappointing life event. Each individual utilizes different COPING skills and relies on available social supports to deal with such episodes, which generally pass within hours to days.

When a dysphoric mood becomes more severe, is persistent, and impairs functioning, a major depression as a clinical syndrome has developed. Concurrent clinical features include a loss of interest or pleasure in usual activities, a sense of hopelessness, poor or alternatively increased sleep, loss of appetite or overeating with resultant changes in weight, fatigue, anxiety, restlessness, obsessive thinking, difficulty concentrating, irritability, feelings of worthlessness, recurring thoughts of death, and sui-

cidal ideation or an actual attempt to end one's life. Suicidal disturbances are of serious concern; approximately 66 percent of depressed patients contemplate suicide, and it is estimated that 10 to 15 percent succeed. In some cases, psychotic features such as hallucinations and delusions may develop.

Depression is one of the most common psychiatric disorders seen in adults, with a lifetime prevalence of approximately 5 percent. Some individuals suffer from chronically depressed mood of a less intense nature than that experienced in a major depressive episode; this is referred to as dysthymia. A depressive syndrome may occur as part of manic-depressive illness, and depression as a symptom (i.e., a depressed mood) can be found in many other psychiatric disorders.

Depression should be distinguished from the normal despair of bereavement and from the various medical disorders (e.g., Parkinson's disease) and chemical agents (e.g., alcohol or drugs for heart conditions) that can produce symptoms of depressed mood. The cause of depression is unknown. Biological factors (e.g., dysregulation of neurotransmitter systems), genetic factors, and psychosocial factors (e.g., life events, learned behaviors, and cognitions) have been proposed, and it is likely that all interact to varying extents. Depression is a treatable (but not really curable) illness in the vast majority of people. Treatment consists of a number of modalities, depending on the type and severity of the depression. PSYCHOTHERAPY, antidepressant medications, and electroconvulsive therapy are the main interventions used.

(SEE ALSO: *Causes of Substance Abuse; Comorbidity and Vulnerability; Complications: Mental Disorders*)

BIBLIOGRAPHY

KELLER, M. B. (ED.) (1988). Unipolar Depression. In A. J. Frances & R. E. Hales (Eds.), *American Psychiatric Press review of psychiatry* (Vol. 7). Washington DC: American Psychiatric Press.

MOLLICA, R. F., ET AL. (1989). Mood Disorders. In H. I. Kaplan & B. J. Sadock (Eds.), *Comprehensive textbook of psychiatry*, 5th ed., vol. 1. Baltimore: Williams & Wilkins.

MYROSLAVA ROMACH
KAREN PARKER

DESIGNER DRUGS   Designer drugs are synthesized chemical analogues of known, dangerous drugs; they are designed to produce pharmacological effects similar to the drugs they mimic. In the pharmaceutical industry, the development of new drugs often utilizes principles of basic chemistry, so that the structure of a drug molecule may be slightly altered to change its pharmacological activity. For therapeutic purposes, these strategies have had a long and successful history; for medical pharmaceutics, many useful new drugs or modifications of older drugs have resulted in improved health care. The principle of structure-activity relationships has been applied to many medically approved drugs in the marketplace, especially in the search for painkillers—nonaddicting opioid analgesics.

The clandestine production of new street drugs is, however, intended to avoid federal regulation and control. This practice can often result in the appearance of unknown substances, with wide-ranging degrees of purity, which have the potential to cause dangerous toxicity and serious health consequences for the unwitting drug user (the quality of personnel involved in clandestine drug synthesis can range from cookbook amateurs to highly skilled chemists). The most publicized case regarding the tragic consequences associated with the manufacture and use of designer drugs on the street involves MPTP (1-methyl, 4-phenyl, 1,2,3,6-tetrahydropyridine), a substance that was later found to cause a Parkinsonian syndrome in humans.

A controlled substance that has served as a template for the design of new look-alike OPIOID drugs is MEPERIDINE (Demerol). A slight change in its chemical structure yields the drug known as MPPP (1-methyl-4-propionoxy-4-phenylpyridine), a meperidine look-alike drug, which is known on the streets as synthetic heroin. In California in 1982, four young drug abusers developed Parkinsonian symptoms after the illicit intravenous use of street HEROIN. The analysis of their remaining drug samples revealed the presence of both MPPP and MPTP. The dealer involved in this illicit synthesis and sale of MPPP was a bad chemist, since MPTP represents a side product formed through the inadequate control of the temperature and/or acidity of the chemical reaction.

Another opioid that has resulted in serious health hazards on the street is fentanyl (Sublimaze), a potent and extremely fast-acting NARCOTIC ANALGESIC

(painkiller) with a high ABUSE LIABILITY. This drug has also served as a template for many look-alike drugs that come out of clandestine chemical laboratories. Very slight modifications in the chemical structure of fentanyl can result in analogues such as para-flouro-, 3-methyl-, or alpha-methyl-fentanyl—with, respectively relative potencies 100, 900, and 1,100 times that of MORPHINE. During the 1980s, the Drug Enforcement Administration (DEA) has reported a steady increase in deaths from drug overdoses associated with fentanyl-like designer drugs. Not every "designer" drug is actually thought up by chemists in illegal labs; some were actually synthesized for legitimate medical uses but were never marketed. HALLUCINOGENIC drugs, such as LSD or MESCALINE, rarely cause death—except as ACCIDENTS related to drug-induced mental aberrations. Adverse reactions to typical hallucinogens are usually treated by support, reassurance, and a quiet environment. Hallucinogenic designer drugs, however, include such substituted AMPHETAMINE ("speed") analogues as methylenedioxyamphetamine (MDA), methylenedioxymethamphetamine (MDMA or "Ecstasy"), and methylenedioxyethamphetamine (MDEA or "Eve"). Both acute and chronic toxicity have been reported following the administration of these drugs. Acute toxicity is usually manifested as restlessness, agitation, sweating, high blood pressure, tachycardia, and other cardiovascular effects, all of which are suggestive of excessive central nervous system stimulation. Following chronic administration in animals, MDA has been demonstrated to produce a degeneration of serotonergic nerve terminals in rats, implying that MDA might induce chronic neurological damage in humans as well. This also suggests that extreme caution should be exercised regarding the manufacture and use of MDA, MDMA, and related drugs—although a few psychotherapists claim that MDMA is a useful adjunct in the treatment of some patients.

The widespread illicit manufacture and use of designer drugs with unknown chronic toxicity could result in millions of people experimenting with the drug before the toxic effect was recognized; this could potentially produce an epidemic of neurodegenerative disorders.

(SEE ALSO: *Complications; Controlled Substances Act; MDMA; MPTP*)

BIBLIOGRAPHY

BARNETT, G., & RAPAKA, R. S. (1989). Designer drugs: An overview. In K. K. Redda, C. A. Walker, & G. Barnett (Eds.), *Cocaine, marijuana, designer drugs: Chemistry, pharmacology, and behavior* (pp. 163–174). Boca Raton, FL: CRC Press.

DOWLING, G. P., McDONOUGH, E. T., III, & BOST, R. O. (1987). "Eve" and "ecstasy": A report of five deaths associated with the use of MDEA and MDMA. *Journal of the American Medical Association, 257,* 1615.

NICHOLS, D. E. (1989). Substituted amphetamine controlled substance analogues. In K. K. Redda, C. A. Walker, & G. Barnett (Eds.), *Cocaine, marijuana, designer drugs: Chemistry, pharmacology, and behavior* (pp. 175–185). Boca Raton, FL: CRC Press.

NICK E. GOEDERS

DESOXYN *See* Methamphetamine.

DETOXIFICATION: AS ASPECT OF TREATMENT

Detoxification is the term commonly used to describe the process or set of procedures involved in readjusting a drug- or alcohol-dependent person to a lower or absent tissue level of the substance (drug) of dependence. With chronic (long-term) use of many drugs, there is adaptation within the nervous system. Readaptation of the nervous system to the absence of a particular substance can cause a WITHDRAWAL syndrome (as a manifestation of PHYSICAL DEPENDENCE). The patient would reasonably be expected to have symptoms (what they tell the health-care provider) and exhibit signs (what the observer sees) of a withdrawal syndrome.

The detoxification process usually occurs in a supportive environment, which might be a hospital or clinic, but not always; it might also involve the use of medications (other drugs) in order to control or suppress symptoms and signs of withdrawal, but not always. The level of care and the use of medications depends on the substance of abuse and the level of physical dependence (severity of withdrawal syndrome), complications, or potential for complications. The more severe complications are seen most frequently in association with alcohol or sedative-hypnotic withdrawal. The goal of detoxification is to provide a safe and comfortable transition to a

drug-free state. Detoxification is generally the first step in the process of treatment for rehabilitation.

(SEE ALSO: *Addiction: Concepts and Definitions; Clonidine; Treatment Types: Overview; Treatment Types: Nonmedical Detoxification; Withdrawal: Alcohol*)

JOHN T. SULLIVAN

DETOXIFICATION: NONMEDICAL   *See* Treatment Types.

DEXTROAMPHETAMINE   This is the *d*-isomer of AMPHETAMINE. It is classified as a PSYCHOMOTOR STIMULANT drug and is three to four times as potent as the *l*-isomer in eliciting central nervous system (CNS) excitatory effects. It is also more potent than the *l*-isomer in its ANORECTIC (appetite suppressant) activity, but slightly less potent in its cardiovascular actions. It is prescribed in the treatment of narcolepsy and OBESITY, although care must be taken in such prescribing because of the substantial ABUSE LIABILITY.

High-dose chronic use of dextroamphetamine can lead to the development of a toxic psychosis as well as to other physiological and behavioral problems. This toxicity became a problem in the United States in the 1960s, when substantial amounts of the drug were being taken for nonmedical reasons. Although still abused by some, dextroamphetamine is no longer the stimulant of choice for most psychomotor stimulant abusers.

(SEE ALSO: *Amphetamine Epidemics; Cocaine*)

MARIAN W. FISCHMAN

DIAGNOSIS OF DRUG ABUSE: DIAGNOSTIC CRITERIA   Diagnosis is the process of identifying and labeling specific disease conditions. The signs and symptoms used to classify a sick person as having a disease are called diagnostic criteria. Diagnostic criteria and classification systems are useful for making clinical decisions, estimating disease prevalence, understanding the causes of disease, and facilitating scientific communication.

Diagnostic classification provides the treating clinician with a basis for retrieving information about a patient's probable symptoms, the likely course of an illness, and the biological or psychological process that underlies the disorder. For example, the DIAGNOSTIC AND STATISTICAL MANUAL (DSM) of the American Psychiatric Association (1987) is a classification of mental disorders that provides the clinician with a systematic description of each disorder in terms of essential features, age of onset, probable course, predisposing factors, associated features and differential diagnosis. Mental health professionals can use this system to diagnose substance use disorders in terms of the following categories: acute INTOXICATION, ABUSE, DEPENDENCE, WITHDRAWAL, DELIRIUM, and other disorders. In contrast to screening, diagnosis typically involves a broader evaluation of signs, symptoms, and laboratory data as these relate to the patient's illness. The purpose of diagnosis is to provide the clinician with a logical basis for planning TREATMENT and estimating prognosis.

Another purpose of classification is the collection of statistical information on a national and international scale. The primary purpose of the INTERNATIONAL CLASSIFICATION OF DISEASES (ICD), for example, is the enumeration of morbidity and mortality data for public health planning (World Health Organization, 1992). In addition, a good classification will facilitate communication among scientists and provide the basic concepts needed for theory development. Both ICD and DSM have been used extensively to classify persons for scientific research. Classification provides a common frame of reference in communicating scientific findings.

Diagnosis also may serve a variety of administrative purposes. When a patient is suspected of having a substance use disorder, diagnostic procedures are needed to exclude "false positives" (i.e., people who appear to have the disorder but who really do not) and borderline cases. Insurance reimbursement for medical treatment increasingly demands that a formal diagnosis be confirmed according to standard procedures or criteria. The need for uniform reporting of statistical data, as well as the generation of prevalence estimates for epidemiological research, often requires a diagnostic classification of the patient.

CLASSIFICATION SYSTEMS

ALCOHOLISM and drug ADDICTION have been variously defined as medical diseases, mental disorders, social problems, and behavioral conditions. In some cases, they are considered the symptom of an underlying mental disorder (Babor, 1992). Some of these definitions permit the classification of alcoholism and drug dependence within standard nomenclatures such as the *International Classification of Diseases* and the *Diagnostic and Statistical Manual of Mental Disorders*. The recent revisions of both of these diagnostic systems has resulted in a high degree of compatibility between the classification criteria used in the United States and those used internationally. Both systems now diagnose dependence according to the elements first proposed by Edwards and Gross (1976). They also include a residual category (harmful alcohol use [ICD-10]; alcohol abuse [DSM-III-R]) that allows classification of psychological, social, and medical consequences directly related to substance use.

Some diagnostic classification systems are used primarily for epidemiological and clinical research. These include the Feighner Criteria (Feighner et al., 1972) and the Research Diagnostic Criteria (Robins, 1981). Other classifications are intended primarily for clinical care (DSM-III-R; see American Psychiatric Association, 1987) or statistical reporting (ICD-10; see World Health Organization, 1992).

HISTORY TAKING

Obtaining accurate information from patients with alcohol and drug problems is often difficult because of the stigma associated with substance abuse and the fear of legal consequences. At times they want help for the medical complications of substance use (such as injuries, or depression) but are ambivalent about giving up alcohol or drug use entirely. It is often the case that these patients are evasive and attempt to conceal or minimize the extent of their alcohol or drug use. Acquiring accurate information about the presence, severity, duration and effects of alcohol and drug use therefore requires a considerable amount of clinical skill.

The medical model for history taking is the most widely used approach to diagnostic evaluation. This consists of identifying the chief complaint, evaluating the present illness, reviewing past history,

conducting a review of biological systems (e.g., gastrointestinal, cardiovascular), asking about family history of similar disorders, and discussing the patient's psychological and social functioning. A history of the present illness begins with questions on use of alcohol, drugs, and TOBACCO. The questions should cover PRESCRIPTION DRUGS as well as illicit drugs, with additional elaboration of the kind of drugs, the amount used, and the mode of administration (e.g., smoking, injection). Questions about alcohol use should refer specifically to the amount and frequency of use of major beverage types (wine, spirits, BEER). A thorough physical examination is important because each substance has specific pathological effects on certain organs and body systems. For example, alcohol affects the liver, stomach, and cardiovascular system. Drugs often produce abnormalities in "vital signs" such as temperature, pulse, and blood pressure. A mental status examination frequently gives evidence of substance use disorders because of poor personal hygiene, inappropriate affect (sad, euphoric, irritable, ANXIOUS), illogical or delusional thought processes, and memory problems. The physical examination can be supplemented by laboratory tests, which sometimes aid in early diagnosis before severe or irreversible damage has taken place. Laboratory tests are useful in two ways (1) alcohol and drugs can be measured directly in blood, urine, and exhaled air; (2) biochemical and psychological functions known to be affected by substance use can be assessed. Many drugs can be detected in the urine for twelve to forty-eight hours after their consumption. An estimate of BLOOD ALCOHOL CONCENTRATION (BAC) can be made directly by blood test or indirectly by means of a breath or saliva test. Elevated gamma glutamyl transpeptidase (GGTP), a liver enzyme, is a sensitive indicator of chronic, heavy alcohol intake.

In addition to the physical examination and laboratory tests, a variety of diagnostic interview procedures have been developed to provide objective, empirically based, reliable diagnoses of substance-use disorders in various clinical populations. One type, exemplified by the DIAGNOSTIC INTERVIEW SCHEDULE (DIS; see Robins et al., 1981) and the Composite International Diagnostic Interview (CIDI; see Robins et al., 1988), is highly structured and requires a minimum of clinical judgment by the interviewer. These interviews provide information not only about substance-use disorders, but also about

physical conditions and psychiatric disorders that are commonly associated with substance abuse. A second type of diagnostic interview is exemplified by the STRUCTURED CLINICAL INTERVIEW for DSM-III-R (SCID), which is designed for use by mental health professionals (Spitzer et al., 1992). The SCID assesses thirty-three of the more commonly occurring psychiatric disorders described in DSM-III-R. Among these are depression, schizophrenia, and the substance-use disorders. A similar clinical interview, which has been designed for international use, is the Schedules for Clinical Assessment in Neuropsychiatry (SCAN; see Wing et al., 1990). The SCID and SCAN interviews allow the experienced clinician to tailor questions to fit the patient's understanding, to ask additional questions that clarify ambiguities, to challenge inconsistencies, and to make clinical judgments about the seriousness of symptoms. They are both modeled on the standard medical history practiced by many mental-health professionals. Questions about the chief complaint, past episodes of psychiatric disturbance, treatment history, and current functioning all contribute to a thorough and orderly psychiatric history that is extremely useful for diagnosing substance-use disorders.

DIAGNOSIS OF ABUSE AND HARMFUL USE

A major diagnostic category that has received increasing attention in research and clinical practice is substance *abuse* in contrast to *dependence*. This category permits the classification of maladaptive patterns of alcohol or drug use that do not meet criteria for dependence. The diagnosis of *abuse* is designed primarily for persons who have recently begun to experience ALCOHOL or drug problems, and for chronic users whose substance-related consequences develop in the absence of marked dependence symptoms. Examples of situations in which this category would be appropriate include (1) a pregnant woman who keeps drinking alcohol even though her physician has told her that it could be responsible for FETAL damage; (2) a college student whose weekend binges result in missed classes, poor grades, and alcohol-related traffic ACCIDENTS; (3) a middle-aged beer drinker regularly consuming a six-pack each day who develops high blood pressure and fatty liver in the absence of alcohol-dependence symptoms, and (4) an occasional MARIJUANA smoker who has an accidental injury while intoxicated.

In the latest version of the *Diagnostic and Statistical Manual* (DSM IV; see American Psychiatric Association, 1994), *substance abuse* is defined as a maladaptive pattern of alcohol or drug use leading to clinically significant impairment or distress, as manifested by one or more of the symptoms listed in Table 1. For comparative purposes, the table also lists the criteria for harmful use in ICD-10 and for alcohol abuse in DSM-III-R. To assure that the diagnosis is based on clinically meaningful symptoms rather than the results of an occasional excess, the duration criterion specifies how long the symptoms must be present to qualify for a diagnosis.

In ICD-10, the term *harmful use* refers to a pattern of using one or more PSYCHOACTIVE substances that causes damage to health. The damage may be (1) physical (physiological)—such as fatty liver, pancreatitis from alcohol, or hepatitis from needle-injected drugs; or (2) mental (psychological)—such as depression related to heavy drinking or drug use. Adverse social consequences often accompany substance use, but are not in themselves sufficient to result in a diagnosis of harmful use. The key issue in the definition of this term is the distinction between *perceptions* of adverse effects (e.g., wife complaining about husband's drinking) and *actual* health consequences (e.g., trauma due to accidents during drug intoxication). Since the purpose of ICD is to classify diseases, injuries, and causes of death, harmful use is defined as a pattern of use already causing damage to health.

Harmful patterns of use are often criticized by others and sometimes legally prohibited by governments. The fact that alcohol or drug intoxication is disapproved by another person or by the user's culture is not in itself evidence of harmful use—unless socially negative consequences have actually occurred at dosage levels that also result in psychological and physical consequences. This is the major difference that distinguishes ICD-10's harmful use from DSM-IV's substance abuse—the latter category includes social consequences in the diagnosis of abuse.

THE DEPENDENCE SYNDROME CONCEPT

The diagnosis of *substance-use disorders* in ICD-10 and DSM-IV is based on the concept of a *dependence syndrome*, which is distinguished from disabilities caused by substance use (Edwards, Arif, & Hodgson, 1981). An important diagnostic issue is

TABLE 1
Diagnostic Criteria for Harmful Use (ICD-10) and Substance Abuse (DSM-III-R, DSM-IV)

	ICD-10 Criteria for Harmful Use	DSM-III-R Criteria for Abuse	DSM-IV Criteria for Abuse
Symptom Criteria	Clear evidence that alcohol or drug use is responsible for causing actual psychological or physical harm to the user	A maladaptive pattern of alcohol or drug use indicated by at least one of the following: (1) continued use despite knowledge of having a persistent or recurrent social, occupational, psychological, or physical problem that is caused or exacerbated by substance use or (2) recurrent use in situations in which substance use is physically hazardous (e.g., driving while intoxicated)	A maladaptive pattern of alcohol or drug use indicated by at least one of the following: (1) failure to fulfill major role obligations at work, school, or home (e.g., neglect of children or household); (2) use in situations in which it is physically hazardous (e.g., driving an automobile); (3) recurrent substance-related legal problems (e.g., arrests for substance-related disorderly conduct); (4) continued substance use despite having recurrent social or interpersonal problems
Duration Criterion	The pattern of use has persisted for at least 1 month or has occurred repeatedly over the previous 12 months	Some symptoms of the disturbance have persisted for at least 1 month or have occurred repeatedly over a longer period of time	One or more symptoms has occurred during the same 12-month period

the extent to which dependence is sufficiently distinct from abuse or harmful use to be considered a separate condition. In DSM-IV, *substance abuse* is a residual category that allows the clinician to classify clinically meaningful aspects of a patient's behavior when that behavior is not clearly associated with a dependence syndrome. In ICD-10, *harmful substance use* implies identifiable substance-induced medical or psychiatric consequences that occur in the absence of a dependence syndrome. In both classification systems, dependence is conceived as an underlying condition that has much greater clinical significance because of its implications for understanding etiology, predicting course, and planning treatment. This will become clear in the following discussion of the assumptions behind the dependence-syndrome concept.

The dependence syndrome is seen as an interrelated cluster of cognitive, behavioral, and physiological symptoms. Table 2 summarizes the criteria used

to diagnose dependence in ICD-10, DSM-III-R, and DSM-IV. A diagnosis of dependence in all systems is made if three or more of the criteria have been experienced at some time in the previous twelve months.

The dependence syndrome may be present for a specific substance (e.g., tobacco, alcohol, or diazepam), for a class of substances (e.g., opioid drugs), or for a wider range of various substances. A diagnosis of *dependence* does not necessarily imply the presence of physical, psychological, or social consequences, although some form of harm is usually present. There are some differences among these classification systems, but the criteria are very similar, making it unlikely that a patient diagnosed in one system would be diagnosed differently in the other.

The syndrome concept implicit in the diagnosis of alcohol and drug dependence in ICD and DSM is a way of describing the nature and severity of addic-

TABLE 2
ICD and DSM Diagnostic Criteria for Dependence (labeled according to diagnostic level—physiological, cognitive, and behavioral—and underlying dependence elements)

Dependence Element	Diagnostic Level	ICD-10 Symptoms	DSM-III-R Symptoms	DSM-IV Symptoms
Salience	Cognitive, behavioral	Progressive neglect of alternative activities in favor of substance use	Important social, occupational, or recreational activities given up	Important social, occupational, or recreational activities given up
	Behavioral	Persistence with substance use despite harmful consequences	Continued use despite social, psychological, or physical problems	Continued use despite psychological or physical problems
	Behavioral, physiological		Great amount of time devoted to substance use, intoxication, or withdrawal; failure to fulfill major role obligations	
Impaired control	Behavioral, cognitive	A strong desire or sense of compulsion to drink or use drugs	Persistent desire or one or more unsuccessful efforts to cut down or control substance use	
	Behavioral	Evidence of impaired capacity to control substance use in terms of its onset, termination, or levels of use	Substance often taken in larger amounts or over longer period than the person intended	Substance often taken in larger amounts or over a longer period than intended
				Any unsuccessful effort or a persistent desire to cut down or control substance use
Tolerance	Biological, behavioral	Increased doses of substance are required to achieve effects originally produced by lower doses	Marked tolerance; need for markedly increased amounts of substance to achieve intoxication or desired effect or markedly diminished effect with continued use of the same amount	Either (a) increased amounts needed to achieve desired effect; or (b) markedly diminished effect with continued use
Withdrawal and withdrawal relief	Behavioral, biological, cognitive	A physiological withdrawal state. Use to relieve or avoid withdrawal symptoms and subjective awareness that this strategy is effective	Characteristic withdrawal symptoms. Substance often taken to relieve or avoid withdrawal symptoms	Either (a) characteristic withdrawal syndrome for substance; or (b) the same substance taken to relieve or avoid symptoms

tion (Babor, 1992). Table 2 describes four dependence syndrome elements in relation to the criteria for DSM-III-R, DSM-IV, and ICD-10. The same elements apply to the diagnosis of dependence on all psychoactive substances, including alcohol, marijuana, opiates, cocaine, sedatives, phencyclidine, other hallucinogens, and tobacco. The elements represent biological, psychological (cognitive), and behavioral processes. This helps to explain the linkages and interrelationships that account for the coherence of signs and symptoms. The co-occurrence of signs and symptoms is the essential feature of a syndrome. If three or more criteria do occur repeatedly during the same period, it is likely that dependence is responsible for the amount, frequency, and pattern of the person's substance use.

Salience. Salience means that drinking or drug use is given a higher priority than other activities in spite of its negative consequences. This is reflected in the emergence of substance use as the preferred activity from a set of available alternative activities. In addition, the individual does not respond well to the normal processes of social control. For example, when drinking to intoxication goes against the tacit social rules governing the time, place, or amount typically expected by the user's family or friends, this may indicate increased salience.

A characteristic of salience is that drinking or drug use persists in spite of its negative consequences. This implies that substance use has become the preferred activity in the person's life. One indication of this is the amount of time or effort devoted to obtaining, using or recovering from substance use. For example, people who spend a great deal of time at parties, bars, or business lunches give evidence of the increased salience of drinking over nondrinking activities.

Chronic drinking and drug intoxication interfere with the person's ability to conform to tacit social rules governing daily activities, such as keeping appointments, caring for children, or performing a job properly, that are typically expected by the person's reference group. Substance use also results in mental and medical consequences. Thus, a key aspect of the dependence syndrome is the persistence of substance use in spite of social, psychological, or physical harm, such as loss of employment, marital problems, depressive symptoms, accidents, and liver disease. This indicates that substance use is given a higher priority than other activities, in spite of its negative consequences.

One explanation for the salience of drug and alcohol-seeking behavior despite negative consequences is the relative reinforcement value of immediate and long-term consequences. For many alcoholics and drug abusers, the immediate positive reinforcing effects of the substance, such as euphoria or stimulation, far outweigh the delayed negative consequences, which may occur either infrequently or inconsistently.

Impaired Control. The main characteristic of impaired control is the lack of success in limiting the amount or frequency of substance use. For example, the alcoholic wants to stop drinking, but repeated attempts have been unsuccessful. Typically, rules and other stratagems are used to avoid alcohol entirely or to limit the frequency of drinking. Resumption of heavy drinking after receiving professional help for a drinking problem is evidence of lack of success. The symptom is considered present if the drinker has repeatedly failed to abstain or has only been able to control drinking with the help of treatment, mutual-help groups, or removal to a controlled environment (e.g., prison).

In addition to an inability to abstain, impaired control is also reflected in the failure to regulate the amount of alcohol or drug consumed on a given occasion. The cocaine addict vows to snort only a small amount but then continues until the entire supply is used up. For the alcoholic, impaired control includes inability to prevent spontaneous onset of drinking bouts as well as failure to stop drinking before intoxication. This behavior should be distinguished from situations in which the drinker's "control" over the onset or amount of drinking is regulated by social or cultural factors, such as occur during college beer parties or fiesta drinking occasions. One way to judge the degree of impaired control is to determine whether the drinker or drug user has made repeated attempts to limit the quantity of substance use by making rules or imposing limits on access to alcohol or drugs. The more these attempts have failed, the more the impaired control is present.

Tolerance. TOLERANCE is a decrease in response to a psychoactive substance that occurs with continued use. For example, increased doses of heroin are required to achieve effects originally produced by lower doses. Tolerance may be physical, behavioral, or psychological. Physical tolerance is a change in cellular functioning. The effects of a dependence-producing substance are reduced, even though the cells normally affected by the substance

are subjected to the same concentration. A clear example is the finding that alcoholics can drink amounts of alcohol (e.g., a quart of vodka) that would be sufficient to incapacitate or kill nontolerant drinkers. Tolerance may also develop at the psychological and behavioral levels, independent of the biological adaptation that takes place. Psychological tolerance occurs when the marijuana smoker or heroin user no longer experiences a "high" after the initial dose of the substance. Behavioral tolerance is a change in the effect of a substance because the person has learned to compensate for the impairment caused by a substance. Some alcoholics, for example, can operate machinery at moderate doses of alcohol without impairment.

Withdrawal Signs and Symptoms. A withdrawal state is a group of symptoms occurring after cessation of substance use. It usually occurs after repeated, and usually prolonged drinking or drug use. Onset and course of the withdrawal symptoms are related to type of substance and dose being used immediately prior to abstinence. Table 3 lists some common withdrawal symptoms associated with different psychoactive substances. Some drugs, such as *Cannabis* (MARIJUANA) and HALLUCINOGENS do not typically produce a withdrawal syndrome after cessation of use.

Alcohol withdrawal symptoms follow the cessation or reduction of prolonged heavy drinking within hours. These include tremors, hyperactive reflexes, rapid heart beat, hypertension, general malaise, nausea, and vomiting. Seizures and convulsions may occur, particularly in people with a preexisting seizure disorder. Patients may have HALLUCINATIONS, illusions, or vivid nightmares. Sleep is usually disturbed. In addition to physical withdrawal symptoms, anxiety and depression are also common. Some chronic drinkers never have a long enough period of abstinence to permit withdrawal to occur.

The use of a substance with the intention of relieving withdrawal symptoms and with an awareness that this strategy is effective are cardinal symptoms of dependence. Morning drinking to relieve nausea or the "shakes" is one of the most common manifestations of physical dependence in alcoholics.

Other Features of Dependence. To be labeled dependence, symptoms must have persisted for at least one month or must have occurred repeatedly (2 or more times) over a longer period of time. The patient does not need to be using the substance continually to

TABLE 3
Withdrawal Symptoms Associated with Different Psychoactive Substances

	Alcohol	Amphetamine	Caffeine	Cocaine	Opioids	Nicotine
Craving					X	X
Tremor	X					
Sweating, fever	X				X	
Nausea or vomiting	X				X	
Malaise, fatigue	X	X		X		
Hyperactivity, restlessness	X	X	X	X		X
Headache	X					
Insomnia	X	X	X		X	
Hallucinations	X					
Convulsions	X					
Delirium	X					
Irritability	X	X		X		X
Anxiety	X		X	X		X
Depression	X			X		
Difficulty concentrating						X
Gastrointestinal disturbance			X			
Increased appetite						X
Diarrhea					X	

have recurrent or persistent problems. Some symptoms (e.g., the desire to cut down) may occur repeatedly whether the person is using the substance or not.

Many patients with a history of dependence experience rapid reinstatement of the syndrome following resumption of substance use after a period of abstinence. Rapid reinstatement is a powerful diagnostic indicator of dependence. It points to the impairment of control over substance use, the rapid development of tolerance, and (frequently) physical withdrawal symptoms.

Patients who receive OPIATES or other drugs for PAIN relief following surgery (or for malignant disease like cancer) sometimes show signs of a withdrawal state when these drugs are ended. The great majority have no desire to continue taking such drugs and therefore do not fulfill the criteria for dependence. The presence of a physical withdrawal syndrome does not necessarily indicate dependence but rather a state of neuroadaptation to the drug that was being administered.

It is commonly assumed that severe dependence is not reversible—an assumption indicated by the rapid reinstatement of dependence symptoms when drinking or drug use is resumed after a period of detoxification.

(SEE ALSO: *Addiction: Concepts and Definitions; Alcoholism: Origin of the Term; Causes of Substance Abuse; Disease Concept of Alcoholism and Drug Abuse; Tolerance and Physical Dependence; Wikler's Pharmacologic Theory of Drug Addiction*)

BIBLIOGRAPHY

AMERICAN PSYCHIATRIC ASSOCIATION. (1994). *Diagnostic and statistical manual of mental disorders-4th edition.* Washington, DC: Author.

AMERICAN PSYCHIATRIC ASSOCIATION. (1987). *Diagnostic and statistical manual of mental disorders-3rd edition-revised.* Washington, D.C.

BABOR, T. F. (1992). Nosological considerations in the diagnosis of substance use disorders. In M. Glantz & R. Pickens (Eds.), *Vulnerability to drug abuse.* Washington DC: American Psychological Association.

EDWARDS, G., ARIF, A., & HODGSON, R. (1981). Nomenclature and classification of drug- and alcohol-related problems: A WHO memorandum. *Bulletin of the World Health Organization, 59*(2), 225–242.

EDWARDS, G., & GROSS, M. M. (1976). Alcohol dependence: Provisional description of a clinical syndrome. *British Medical Journal, 1,* 1058–1061.

FEIGHNER, J., ET AL. (1972). Diagnostic criteria for use in psychiatric research. *Archives of General Psychiatry, 26,* 57–63.

ROBINS, L. (1981). The diagnosis of alcoholism after DSM III. In R. E. Meyer et al., *Evaluation of the alcoholic: Implications for research, theory and treatment.* NIAAA Research Monograph no. 5, DHHS Publication no. (ADM) 81-1033. Washington, DC: U.S. Government Printing Office.

ROBINS, L. N., ET AL. (1988). The composite international diagnostic interview: An epidemiological instrument suitable for use in conjunction with different diagnostic systems and in different cultures. *Archives of General Psychiatry, 45,* 1069–1077.

ROBINS, L. N., ET AL. (1981). National Institute of Mental Health diagnostic interview schedule. *Archives of General Psychiatry, 38,* 381–389.

SPITZER, R. L., ET AL. (1992). The structured clinical interview for DSM-III-R (SCID), I. History, rationale and description. *Archives of General Psychiatry, 49*(8), 624–629.

WING, J. K., ET AL. (1990). SCAN: Schedules for Clinical Assessment in Neuropsychiatry. *Archives of General Psychiatry, 47,* 589–593.

WORLD HEALTH ORGANIZATION. (1992). *The ICD-10 classification of mental and behavioural disorders; Clinical descriptions and diagnostic guidelines.* Geneva: Author.

THOMAS F. BABOR

DIAGNOSTIC AND STATISTICAL MANUAL (DSM)

Several formal diagnostic systems are available for use in clinical assessment and research on substance-use disorders. The most widely accepted diagnostic system in the United States is the 1987 set of criteria formulated by the American Psychiatric Association (APA). This is the revised third edition of the *Diagnostic and Statistical Manual* (DSM-III-R), and it provides a complete description of the diagnostic criteria used to diagnose psychiatric conditions and substance-use disorders.

Until the third edition of DSM (DSM-III), the U.S. system for classifying psychiatric disorders had been virtually identical to the INTERNATIONAL CLASSIFICATION OF DISEASES (ICD); this lists major causes of mortality and related health problems. During the 1970s, researchers affiliated with the Washington

University School of Medicine (Feighner et al., 1972) developed the "research diagnostic" approach to psychiatric diagnosis, which strongly influenced the classification of substance-use disorders adopted in 1980 by the APA in the DSM-III. This approach emphasized clearly formulated and objectively observable signs and symptoms that could be used for research as well as for clinical practice.

In DSM-III, in contrast to previous editions of DSM, ALCOHOLISM and drug dependence were included within the separate category of Substance Use Disorders—rather than as a subcategory of Personality Disorder. Reflecting a trend toward greater semantic precision, the term *dependence* was used in preference to the more generic terms *alcoholism* or *addiction*; and *dependence* was distinguished from *abuse* by the presence of the symptoms of TOLERANCE or WITHDRAWAL. The separate categories of Alcohol Abuse and Drug Abuse were added to permit greater differentiation along a range of severity for each.

As part of the APA's ongoing program of work on classification, the revised DSM-III was published in 1987. DSM-III-R presents a multiaxial system for diagnostic evaluation to ensure that all relevant clinical information is taken into account. Axis I describes syndromes, such as major DEPRESSION, SCHIZOPHRENIA, and substance-use disorders. Axis II comprises childhood and personality disorders that often persist into adult life. Axis III refers to physical disorders or conditions that are potentially relevant to the understanding or management of the patient. Axis IV provides a rating of the severity of psychosocial stressors that have occurred in the year preceding the current evaluation and that may have contributed to the patient's symptoms. Axis V is a global assessment of psychological, social, and occupational functioning, which should be taken into account in treatment planning.

The most important change to the Substance Use Disorders section of DSM-III (Rounsaville, Spitzer, & Williams, 1986) involved the adoption of a new dependence syndrome concept (Edwards, Arif, & Hodgson, 1981). Dependence is defined as an interrelated cluster of psychological symptoms (e.g., a strong desire [CRAVING] to take the substance), physiological signs (e.g., tolerance and withdrawal) and behavioral indicators (e.g., use of substance to relieve withdrawal discomfort). Significantly, the medical and social consequences of both acute intoxication and chronic substance use (e.g., ACCIDENTS, liver damage) are not among the primary diagnostic criteria of dependence in DSM-III-R. These consequences do, however, play a prominent role in the definition of a residual category termed *substance abuse*.

In 1993, the fourth edition of DSM (DSM-IV) was published (American Psychiatric Association, 1994), with additional changes in the diagnosis of substance-related disorders. These changes were designed to assure compatibility between DSM and ICD. Like ICD, DSM defines substance dependence as a maladaptive pattern of substance use leading to clinically significant impairment or distress, as manifested by three or more of the following symptoms occurring in the same twelve-month period: (1) tolerance (i.e., the need for markedly increased amounts of the substance to achieve intoxication or the desired effect); (2)withdrawal; (3) taking the substance in larger amounts or over a longer period than was intended; (4) unsuccessful attempts to cut down or control substance use; (5) a great deal of time spent in activities related to the procurement or use of the substance; (6) important social, occupational, or recreational activities given up or reduced because of substance use; and (7) continued use despite physical or psychological problems known to be caused by the substance.

The diagnosis of dependence can be made with regard to one or more of the following substances: ALCOHOL, TOBACCO, SEDATIVE-HYPNOTICS-ANXIOLYTICS, CANNABIS (MARIJUANA), STIMULANTS, OPIOIDS, COCAINE, HALLUCINOGENS/PCP (PHENCYCLIDINE), POLYSUBSTANCE, and other substances. The clinician is asked to specify whether dependence is associated with the physiological symptoms of tolerance and withdrawal. If so, this information should be used in treatment planning.

(SEE ALSO: *Addiction: Concepts and Definitions; Alcoholism: Origin of the Term; Comorbidity and Vulnerability; Disease Concept of Alcoholism and Drug Abuse*)

BIBLIOGRAPHY

AMERICAN PSYCHIATRIC ASSOCIATION. (1994). *Diagnostic and statistical manual of mental disorders-4th edition.* Washington, DC: Author.

AMERICAN PSYCHIATRIC ASSOCIATION. (1987). *Diagnostic and statistical manual of mental disorders-3rd edition-revised.* Washington, DC: Author.

AMERICAN PSYCHIATRIC ASSOCIATION. (1980). *Diagnostic and Statistical Manual of Mental Disorders-3rd edition.* Washington DC: Author.

EDWARDS, G., ARIF, A., & HODGSON, R. (1981). Nomenclature and classification of drug- and alcohol-related problems: A WHO memorandum. *Bulletin of the World Health Organization, 59,* 225–242.

FEIGHNER, J., ET AL. (1972). Diagnostic criteria for use in psychiatric research. *Archives of General Psychiatry, 26,* 57–63.

ROUNSAVILLE, B. J., SPITZER, R. L., & WILLIAMS, J. B. (1986). Proposed changes in DSM-III substance use disorders: description and rationale. *American Journal of Psychiatry, 143,* 463–468.

THOMAS F. BABOR

DIAGNOSTIC INTERVIEW SCHEDULE (DIS)

Developed in the late 1970s for use in large-scale studies of the prevalence of mental disorders in the U.S. population (Regier et al., 1984), the Diagnostic Interview Schedule (DIS) is a highly structured psychiatric interview that carefully specifies the questions that the interviewer must ask to make a DIAGNOSIS. Because the DIS requires a minimum of clinical judgment, it can be administered by nonprofessional or nonclinician interviewers who have received a week of intensive training. In addition to alcohol and other substance-use disorders, the DIS provides diagnostic information about DEPRESSION, SCHIZOPHRENIA, ANXIETY disorders, eating disorders, ANTISOCIAL PERSONALITY, and a variety of other psychiatric conditions. The DIS has been the subject of a number of validation studies that show nonclinician interviewers diagnose patients as accurately as trained clinicians using criteria from DSM-III (DIAGNOSTIC AND STATISTICAL MANUAL *of Mental Disorders,* third edition). With the American Psychiatric Association's publication of the revised versions of DSM, major changes were made to the DIS as well. The DIS-III-R includes questions that permit classification of substance-use disorders according to both DSM-III and DSM-III-R criteria.

The DIS has been widely used in research on substance-use disorders (Helzer & Canino, 1992) in part because it can be administered by nonclinician interviewers in population surveys. Interviewers read questions aloud to the subject exactly as they are written in the interview booklet. No deviation from the written format is allowed, except to repeat questions that may have been misunderstood. A set of standard probes is used to determine whether a given symptom was caused by the effects of physical illness. The interviewer also asks for the age of onset and the recency of most symptoms.

A series of thirty questions constitutes the ALCOHOL DEPENDENCE/abuse section of the DIS. The section begins with questions about alcohol consumption and intoxication (e.g., "Have you ever gone on binges or benders where you kept drinking for a couple of days or more without sobering up?"). Additional questions are asked to diagnose the symptoms of dependence (e.g., "Did you ever get tolerant to alcohol, that is, you needed to drink a lot more in order to get an effect, or found that you could no longer get high on the amount you used to drink?"). A third type of question pertains to the symptoms of alcohol abuse (e.g., "Have you ever had trouble driving because of drinking—like having an ACCIDENT or being arrested for drunk driving?").

The drug dependence section of the DIS (version III-R) consists of twenty-four questions that conform to the DSM-III-R criteria for drug use disorders. This section begins by asking if the patient has used any of the following types of drugs "to get high or for other mental effects": MARIJUANA, STIMULANTS (e.g., AMPHETAMINES), SEDATIVES (e.g., BARBITURATES), prescribed drugs (e.g., TRANQUILIZERS), COCAINE, HEROIN, other OPIATES, PSYCHEDELICS, PCP, INHALANTS, and other drugs not previously specified. If the person has used any of these substances more than five times, additional questions are asked to evaluate the mode of ingestion for each drug (e.g., by mouth, smoking, snorting, injecting).

The remaining questions ask about DSM-III-R symptoms of dependence and abuse. For example, patients are asked if they have had difficulty abstaining from drugs ("Have you ever tried to cut down on any of these drugs but found you couldn't?"); experienced WITHDRAWAL symptoms ("Has stopping or cutting down on any of these drugs made you sick?"); or experienced other physical complications ("Did you have any health problems like an accidental OVERDOSE, a persistent cough, a seizure [fit], an infection, a cut, sprain, burn, or other injury as a result of taking any of these drugs?"). The DIS can

be scored manually or by computer to obtain specific drug and alcohol diagnoses in DSM-III-R.

(SEE ALSO: *Addiction: Concepts and Definitions; Diagnosis of Drug Abuse; Disease Concept of Alcoholism and Drug Addiction*)

BIBLIOGRAPHY

AMERICAN PSYCHIATRIC ASSOCIATION. (1987). *Diagnostic and statistical manual of mental disorders-3rd edition-revised.* Washington, DC: Author.

HELZER, J. E., & CANINO, G. J. (EDS.). (1992). *Alcoholism in North America, Europe, and Asia.* New York: Oxford University Press.

REGIER, D. A., ET AL. (1984). The NIMH Epidemiologic Catchment Area (ECA) program: Historical context, major objectives and study population characteristics. *Archives of General Psychiatry, 41,* 934–941.

ROBINS, L., ET AL. (1989). *NIMH diagnostic interview schedule, version III, revised (DIS-III-R).* St. Louis: Washington University.

ROBINS, L. N., ET AL. (1982). Validity of the diagnostic interview schedule, version II: DSM-III diagnoses. *Psychological Medicine, 12,* 855–870.

ROBINS, L. N., ET AL. (1981). National Institute of Mental Health diagnostic interview schedule. *Archives of General Psychiatry, 38,* 381–389.

THOMAS F. BABOR

DIAZEPAM   *See* Benzodiazepines.

DIET PILL   *See* Amphetamine; Anorectic.

DIHYDROMORPHINE   Dihydromorphine is a semisynthetic OPIOID ANALGESIC (painkiller), derived from MORPHINE. Structurally, it is very similar to morphine—the only difference being the reduction of the double bond between positions 7 and 8 in morphine to a single bond. Although slightly more potent than morphine in relieving PAIN, it is not widely used clinically. At standard analgesic doses, it has a side-effect profile very similar to that of morphine. These include constipation and respiratory

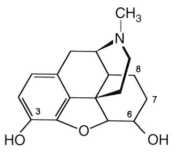

Figure 1
Dihydromorphine

depression. Chronic use will produce TOLERANCE AND PHYSICAL DEPENDENCE.

(SEE ALSO: *Addiction: Concepts and Definitions; Opiates/Opioids; Opioids: Complications and Withdrawal*)

BIBLIOGRAPHY

JAFFE, J. H., & MARTIN, W. R. (1990). Opioid analgesics and antagonists. In A. G. Gilman, et al. (Eds.), *Goodman and Gilman's the pharmacological basis of therapeutics,* 8th ed. New York: Pergamon.

GAVRIL W. PASTERNAK

DILAUDID   *See* Hydromorphone.

DIMETHYLTRYPTAMINE (DMT)   This drug is a member of the HALLUCINOGENIC substances known as indoleamines. These are compounds that are structurally similar to the neurotransmitter SEROTONIN. Although found in certain plants and, according to some evidence, can be formed in the brain, DMT is synthesized for use. Its effects are similar to those produced by LYSERGIC ACID DIETHYLAMIDE (LSD), but unlike LSD, DMT is inactive after oral administration. It must be injected, sniffed, or smoked.

DMT has a rapid onset, usually within one minute, but the effects last for a much shorter period than those produced by LSD—with the user feeling "normal" within thirty to sixty minutes. This is because DMT is very rapidly destroyed by the enzyme monoamine oxidase, which metabolizes structurally related compounds, such as serotonin. The dose

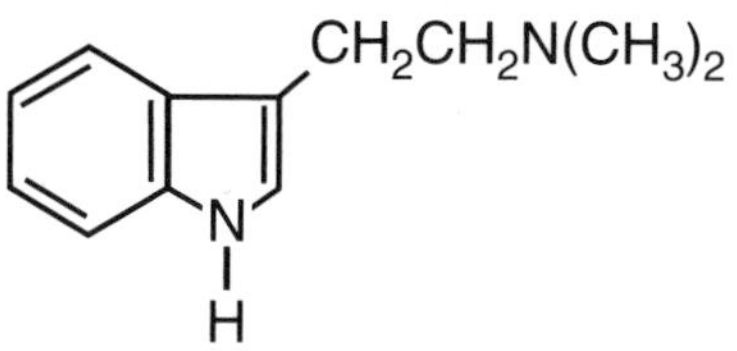

Figure 1
DMT

amount of DMT is critical, since larger doses produce slightly longer, much more intense, and sometimes very uncomfortable "trips" than do lower doses. The sudden and rapid onset of a period of altered perceptions that soon terminates is also disconcerting to some users. DMT was known briefly as the "businessman's LSD"—one could have a PSY-CHEDELIC experience during the lunch hour and be back at work in the afternoon. It has, however, in fact never been a widely available, steadily obtainable, or popular drug on the street.

(SEE ALSO: *DOM; MDMA*)

BIBLIOGRAPHY

HOLLISTER, L. E. (1978). Psychotomimetic drugs in man. *Handbook of Psychopharmacology, 11*, 389–424.

STRASSMAN, R. J., & QUALLS, C. R. (in press). Dose-response study of N,N-Dimethyltryptamine in humans. I. neuroendocrine, autonomic and cardiovascular effects. *Archives of General Psychiatry.*

DANIEL X. FREEDMAN
R. N. PECHNICK

DIS *See* Diagnostic Interview Schedule.

DISEASE CONCEPT OF ALCOHOLISM AND DRUG ABUSE Throughout most of recorded history, excessive use of ALCOHOL was viewed as a willful act leading to intoxication and other sinful behaviors. The Bible warns against drunkenness; Islam bans alcohol use entirely. Since the early nineteenth century, the moral perspective has competed with a conceptualization of excessive use of alcohol as a disease or disorder, not necessarily a moral failing. The disease (or disorder) concept has, in turn, been evolving with considerable controversy since

then, and has itself been challenged by other conceptual models. Because this article is concerned primarily with the disease concept, the other models will be mentioned only briefly.

Among the first to propose that excessive alcohol use might be a disorder, rather than willful or sinful behavior, were the physicians Benjamin Rush, in the United States, and Thomas Trotter, in Great Britain. Both Rush and Trotter believed that some individuals developed a pernicious "habit" of drinking and that it was necessary to undo the habit to restore those individuals to health. Words such as *habit* and *disease* were used to convey interwoven notions. Trotter saw "the habit of drunkenness" as "a disease of the will," while Rush saw drunkenness as a disease in which alcohol was the causal agent, loss of control over drinking behavior the characteristic symptom, and total abstinence the only effective cure. In 1849, a Swedish physician, Magnus Huss, introduced the term *alcoholism* ["alcoholismus"] to designate not only the disorder of excessive use but an entire syndrome, including the multiple somatic consequences of excessive use.

Late-nineteenth-century physicians, although not the first to see habitual use of other drugs (such as OPIATES, TOBACCO, COFFEE) as disorders, are credited with stressing the idea that each was but a subtype of a more *generic disorder of inebriety*. However, they also minimized Trotter's and Rush's notions of learned behavior as a central feature of a generic disorder of inebriety and emphasized instead the idea of a disorder rooted in acquired or inherited biological malfunction or VULNERABILITY. This more biologically based view of inebriety was used in Britain and the United States by advocates of publicly funded treatment facilities—inebriate asylums. Many temperance leaders also supported the establishment of treatment facilities. However, while physicians advocated treatment, temperance leaders, still convinced that alcohol itself was the root of the problem, pushed for its control and, eventually, for its prohibition.

In the United States, the ratification in 1920 of the Eighteenth Amendment, which prohibited the production, sale, and distribution of alcohol, temporarily dampened scientific inquiry into the nature of alcoholism. But concern about the problematic and excessive use of other drugs, such as OPIOIDS, COCAINE, and BARBITURATES, continued to stimulate writings both in the United States and abroad. Was

excessive drug use a disease, a moral failure, or something else—perhaps something in between?

By the mid-twentieth century, the rise of ALCOHOLICS ANONYMOUS (AA), the publications of E. M. Jellinek, and the establishment of the Yale Center for Alcohol Studies revived interest in exploring the nature of ALCOHOLISM. In the early 1960s, the idea reemerged that, for certain "vulnerable" people, alcohol use leads to physical addiction—a true disease.

EARLY MODELS OF THE DISEASE CONCEPT

Central to the disease concept of alcoholism put forward by Jellinek were the roles of TOLERANCE AND PHYSICAL DEPENDENCE, usually considered hallmarks of ADDICTION. *Tolerance* indicates that increased doses of a drug are required to produce effects previously attained at lower doses. *Physical dependence* refers to the occurrence of WITHDRAWAL symptoms following cessation of alcohol or other drug use. Although Jellinek recognized that alcohol problems could occur without alcohol addiction, addiction to alcohol moved to the center of scientific focus.

Despite being couched in the language of science, the reemergence of the disease concept of alcoholism was not a result of new scientific findings. Jellinek believed it was necessary to see alcoholism as a disease in order to increase the availability of services for alcoholics within established medical facilities. He also recognized that efforts to prevent alcoholism would still have to address the complex cultural, demographic, political, and economic issues contributing to the problem. Although he sometimes appeared to take a broad view of the disease concept of alcoholism, he reserved the disease category for those individuals manifesting tolerance, withdrawal symptoms, and either "loss of control" or "inability to abstain" from alcohol. These individuals could not drink in moderation; with continued drinking, their disease was progressive. Others who drank merely in response to psychological stress ("alpha alcoholism") and those who sustained toxic consequences from alcohol but were not physically dependent ("beta alcoholism") did not qualify for his more explicit and restrictive definition of disease. Jellinek's view of alcoholism as a progressive disease is sometimes referred to as the "classic" dis-

ease model to distinguish it from later perspectives of a disorder or syndrome more powerfully influenced by learning and social factors.

Alcohol researcher and theorist Thomas Babor has pointed out that when definitions specify alcohol addiction or dependence as a disease entity, it can be argued more convincingly that "dependence is an organically based entity which produces a characteristic set of signs and symptoms . . . and increases the probability of repetitive drinking behavior."

The American Psychiatric Association included alcoholism in the first edition (1952) of the DIAGNOSTIC AND STATISTICAL MANUAL *of Mental Disorders*. In the second edition (*DSM-II*), published in 1968, the group followed a precedent set by the World Health Organization's INTERNATIONAL CLASSIFICATION OF DISEASES (*ICD-8*) and included three subcategories of alcohol-related disorders: alcohol addiction, episodic excessive drinking, and habitual excessive drinking. Both of these publications included alcoholism among the personality disorders and certain other nonpsychotic disorders, implying that the alcohol use was either secondary to an underlying personality problem or a response to extreme internal distress. This view of excessive drug use as a symptom of some other psychiatric disorder is sometimes referred to as the symptomatic model. According to this concept, drug or alcohol dependence is not really a disorder in and of itself.

Meanwhile, from the late 1950s and throughout the 1960s, the Expert Committee on Addiction-Producing Drugs of the WORLD HEALTH ORGANIZATION (WHO) continued to formulate and refine definitions of addiction and HABITUATION that could facilitate WHO's responsibility (required by international treaties) for control of NARCOTICS, cocaine, and CANNABIS. In the 1950s, the presence of physical dependence was emphasized in the definition of drug dependence, and the WHO Expert Committee was still concerned with differentiating between psychic dependence and physical dependence. At one level, the concept of psychic dependence was compatible with the psychodynamic view that these disorders were a response to psychic distress (such as negative mood states). According to the psychodynamic model, excessive alcohol or drug consumption was merely a response to underlying psychopathology. This model was also consistent with Jellinek's view of one of the "species" of alcoholism, in which individuals drink to relieve emotional pain (alpha alcoholism). In 1969, the committee abandoned the

effort to differentiate *habits* from *addictions* and adopted terminology first proposed by Nathan Eddy and colleagues in 1965, in which the term *drug dependence* designates "those syndromes in which drugs come to control behavior." The committee recognized that dependencies on different classes of drugs (such as alcohol, opiates, cocaine) can differ significantly and that withdrawal symptoms are not always present or necessary aspects of dependence (see table 1).

In 1972, alcoholism was included in a listing of diagnostic criteria for use in psychiatric research published by Feighner and coworkers. The defining criteria for alcoholism included withdrawal symptoms, loss of control, severe medical consequences, and social problems. In the same year the NATIONAL COUNCIL ON ALCOHOLISM also outlined criteria for diagnosing alcoholism, which emphasized tolerance and physical dependence and incorporated certain concepts developed by ALCOHOLICS ANONYMOUS. This definition, and one issued jointly with the American Medical Society on Alcoholism in 1976 (see table 1), represented an attempt to emphasize the seriousness of the disorder, the experience of clinicians and of recovering alcoholics, and the view that alcoholism is a primary or independent disorder, not merely a manifestation of an underlying personality problem. These statements come close to being current definitions of the classic disease model.

PROBLEM DRINKING AS A DISTINCT DIMENSION

The importance of what can now be called the classic "disease model" of alcoholism as a primary focus for health programs was challenged in 1977 by a report of a WHO Expert Committee on alcohol-related disabilities. This report stressed that not everyone who develops a disability related to alcohol use exhibits alcohol dependence or addiction, nor would such an individual necessarily develop a dependence in the future. The report asserted that some alcohol-related disabilities represent a dimension of *problem drinking* distinct from the disease of alcoholism or *alcohol dependence syndrome*. This perspective provided support for policies aimed at reducing overall alcohol consumption, not just at promoting abstinence among vulnerable individuals. The report described the alcohol dependence syndrome itself as a learned phenomenon, not a disease state, which is either present or absent, but "a condition which exists in degrees of severity." It is important to recognize that this syndrome perspective does not take a position on whether alcoholism should be considered a disease.

The concept of dependence as a syndrome is quite similar to that put forward in 1965 by drug-abuse researcher Jerome Jaffe, who viewed addiction as standing at one end of a continuum of involve-

TABLE 1
Some Recent Attempts to Define Alcoholism and/or Drug Dependence

Drug dependence. A state, psychic and sometimes also physical, resulting from the interaction between a living organism and a drug, characterized by behavioural and other responses that always include a compulsion to take the drug on a continuous or periodic basis in order to experience its psychic effects, and sometimes to avoid the discomfort of its absence. Tolerance may or may not be present. A person may be dependent on more than one drug. (*World Health Organization Technical Report Series*, 1969, no. 407, p. 6.) This definition was reaffirmed in the WHO Expert Committee on Drug Dependence Nineteenth Report, *World Health Organization Technical Report Series*, 1973, no. 526, p. 16.

Alcoholism is a chronic, progressive, and potentially fatal disease. It is characterized by tolerance and physical dependency or pathologic organ changes or both, all of

which are the direct or indirect consequences of the alcohol ingested. (National Council on Alcoholism/American Medical Society on Alcoholism, 1976.) (See Flavin & Morse, 1991.)

The 1976 definition was revised and broadened in 1991 to include the concept of *denial:*

Alcoholism is a primary, chronic disease with genetic, psychosocial, and environmental factors influencing its development and manifestations. The disease is often progressive and fatal. It is characterized by continuous or periodic impaired control over drinking, preoccupation with the drug alcohol, use of alcohol despite adverse consequences, and distortions in thinking, most notably denial. (National Council on Alcoholism and Drug Dependence/American Medical Society on Alcoholism, 1976) (See Flavin & Morse, 1991.)

ment in drug use: "In most instances it will not be possible to state with precision at what point [along the continuum] compulsive use should be considered addiction," Jaffe observed. He emphasized that "the term addiction cannot be used interchangeably with physical dependence. It is possible to be physically dependent on drugs without being addicted and . . . to be addicted without being physically dependent." In this view, the behavioral disorder, not physical dependence, is the syndrome. Jaffe defined addiction as *"a behavioral pattern of drug use, characterized by overwhelming involvement with the use of a drug (compulsive use), the securing of its supply, and a high tendency to relapse after withdrawal."* This proposed generic notion of dependence is applicable to STIMULANTS and HALLUCINOGENS (for which physical dependence is not a significant factor), as well as to alcohol, opiates, and SEDATIVE-HYPNOTIC drugs (for which physical dependence is a factor). The *Diagnostic and Statistical Manual of Mental Disorders*, 3rd edition, revised (DSM-III-R), published by the American Psychiatric Association more than twenty years later, in 1987, also used such a generic definition.

FROM PSYCHIC AND PHYSICAL DEPENDENCE TO DEPENDENCE SYNDROME

The changing perspectives on the general concept of drug dependence, given momentum by the 1977 WHO report on alcohol and by other research, were ultimately reflected in changes in the definitions and other positions of the World Health Organization and in its 1980 *International Classification of Diseases*, 9th edition (ICD-9). With its publication, the concept of an alcohol dependence syndrome formally emerged at an international level. The ICD-9 concept of dependence was based on a 1976 proposal by researchers Griffith Edwards and Milton Gross, who defined seven characteristics of the alcohol dependence syndrome and proposed that there are certain implicit assumptions to the syndrome: First, it is a symptom complex involving both biological processes and learning. Second, it should be defined along a continuum of severity, rather than as a discrete category. Third, dependence should be differentiated from alcohol-related disabilities. Both dependence and disabilities exist in degrees, rather than on an all-or-none basis. There is some evidence that people with more severe degrees of alcohol de-

pendence who seek treatment have a different clinical course from those with less severe dependence.

By the late 1970s, the American Psychiatric Association's *Diagnostic and Statistical Manual*, 3rd edition (DSM-III), moved away from more descriptive and psychodynamic orientation toward a nomenclature in which specific diagnostic criteria were laid out for specific syndromes. In the case of alcohol and drug dependence, the original drafts of DSM-III considered inclusion of a dependence syndrome that varied in degree of severity and in which tolerance and physical dependence were important, but not essential, criteria for diagnosis. At the last moment, however, it was decided that tolerance and physical dependence were both necessary and sufficient for a diagnosis of drug dependence; the presence of other criteria listed were by themselves insufficient without tolerance and physical dependence. Nevertheless, by distinguishing drug (or alcohol) *dependence* from drug (or alcohol) *abuse*, DSM-III recognized the two-dimensional conceptualization previously put forth in the WHO report of 1977 and in ICD-9.

In 1980, during the short interval between the publication of DSM-III and the beginning of work on DSM-III-R, a WHO working group met to further refine terminology. One result of the meeting was the publication of a WHO memorandum on nomenclature and classification of drug- and alcohol-related problems that endorsed the concept that drug dependence is a syndrome that exists in degrees and that can be inferred from the way in which drug use takes priority over a drug user's once-held VALUES. The criteria for making this inference included many of those mentioned by Edwards and Gross in their 1976 paper and some that had been developed for DSM-III. The WHO memorandum, while recognizing the importance of tolerance and physical dependence, did not view these phenomena as always essential and required. It endorsed again the two-dimensional perspective—not all drug or alcohol problems are manifestations of dependence; and harmful or hazardous use can occur independently of the decreased flexibility and constricted choice that are the hallmarks of the dependence syndrome. This perspective was underscored by pointing out that the presence of physical dependence per se (as in the case of patients taking drugs for pain) was not in itself sufficient for the diagnosis of dependence. The memorandum also presented a model of dependence emphasizing that the dependence phenomenon is not a property of the individual but resides

in the relationships among the elements in the model—social, psychological, and biological. This view has been called the *biopsychosocial model.*

CRITERIA FOR DIAGNOSIS OF A GENERIC DEPENDENCE DISORDER

The American Psychiatric Association's DSM-III-R, published in 1987, built on both DSM-III and the WHO memorandum. It presented nine criteria for diagnosing a *generic dependence syndrome,* applied to a wide variety of drugs. The user must have experienced at least three criteria in order for the practitioner to consider any degree of dependence to be present. Neither tolerance nor physical dependence was a required criterion. The presence of more than three criteria would indicate a more severe degree of dependence. *Drug abuse* was a residual category used for designating drug-related problems when dependence was not present.

The DSM-III-R conceptualization of dependence was controversial. Because for many years physical dependence and tolerance had been considered evidence of "true disease," many clinicians believed that changing these criteria from the necessary and required status they had had in DSM-III was a mistake that erroneously broadened the category of drug dependence. Much of the focus in the development of DSM-IV, published in 1994, was on how to restore the primacy of these phenomena in the diagnosis of drug and alcohol dependence. DSM-IV defines seven generic criteria for alcohol and other drug dependence. Three are required for a diagnosis of alcohol or other drug dependence. Although tolerance and withdrawal are listed first, they are not required—but the clinician must specify whether either is present.

Despite these concerns, there was little argument about the importance of psychological and sociological factors in the development and perpetuation of the syndrome—that is, there was still consensus about the biopsychosocial model.

At the same time, at the international level, the framers of ICD-10 continued the evolution begun in ICD-9 and adhered closely to the concepts of dependence outlined in the 1977 WHO report and 1981 WHO memorandum. Published in 1992, ICD-10 includes a generic model of drug dependence with similar criteria for alcohol, tobacco, opioids, and other drugs that affect the brain. Like DSM-IV, ICD-10 presents a number of criteria (six) for determining the presence of the alcohol (or drug) *dependence* syndrome; at least three of these must be present for the clinician to judge that the syndrome exists to some degree. The DSM-IV and ICD-10 criteria for drug and alcohol dependence are shown in Table 2.

ICD-10 does not include a diagnostic category of alcohol or drug *abuse* but instead includes a category of *harmful use*—a pattern of use that is causing damage to mental or physical health. Unlike DSM-IV, which defines drug or alcohol (substance) abuse as "a maladaptive pattern of use" causing significant impairment or distress and interpersonal, family, and legal problems (e.g., arrests), ICD-10 does not consider such patterns of use and consequences necessarily to be evidence of harmful use.

ICD-10 and DSM-IV share important characteristics that represent a further evolution in understanding drug and alcohol dependence syndromes. In contrast to some disease-oriented definitions that see alcoholism as uniformly progressive, in ICD-10 and DSM-IV the course of the disorder is not one of uniform progression or predictable cure, but there are a variety of significant states of remission. For example, DSM-IV distinguishes early remission (within the first 12 months) from sustained remission (at least 12 months); within each of these it differentiates full remission from partial remission (i.e., all criteria for dependence have not been met, although at least one has been met intermittently or continuously). DSM-IV also recognizes the circumstances supporting remission and allows for distinctions such as remission while the user is in a controlled environment (where substances are highly restricted) or remission from drug of dependence when the user is maintained on a similar agonist. The categorization of states of remission (abstinence) in ICD-10 is somewhat similar, although the distinction between early and sustained remission is not made.

CHALLENGES TO THE DISEASE CONCEPT

The classic disease model of alcoholism and drug dependence has served as a challenge to some behavioral researchers and social scientists; they have raised a number of questions about biologically based theories of such behaviors. Critics of the disease concept point to studies showing that some former alcoholics could apparently return to normal

TABLE 2
A Comparison of ICD-10 and DSM-IV Criteria for Dependence

ICD-10 Dependence Syndrome	*DSM-IV Substance Dependence*
A cluster of physiological, behavioural, and cognitive phenomena in which the use of a substance or a class of substances takes higher priority for a given individual than other behaviours that once had greater value. A central descriptive characteristic of the syndrome is the desire (often strong, sometimes overpowering) to take psychoactive drugs (medically prescribed or not), alcohol, or tobacco. There may be evidence that return to substance use after a period of abstinence leads to a more rapid reappearance of other features of the syndrome than occurs with nondependent individuals. *Diagnostic guidelines* A definite diagnosis of dependence should usually be made only if three or more of the following have been experienced or exhibited during the previous year: (a) a strong desire or sense of compulsion to take the substance; (b) difficulties in controlling substance-taking behaviour in terms of its onset, termination, or levels of use; (c) a physiological withdrawal state . . . when substance use has ceased or been reduced, as evidenced by: the characteristic withdrawal syndrome for the substance; or use of the same (or a closely related) substance with the intention of relieving or avoiding withdrawal symptoms; (d) evidence of tolerance, such that increased doses of the substance are required in order to achieve effects originally produced by lower doses (clear examples of this are found in alcohol- and opiate-dependent individuals who may take daily doses sufficient to incapacitate or kill nontolerant users); (e) progressive neglect of alternative pleasures or interests because of psychoactive substance use, increased amount of time necessary to obtain or take the substance or to recover from its effects; (f) persisting with substance use despite clear evidence of overtly harmful consequences, such as harm to the liver through excessive drinking, depressive mood states consequent to periods of heavy substance use, or drug-related impairment of cognitive functioning; efforts should be made to determine the extent of the user's awareness of the nature and extent of the harm. Narrowing of the personal repertoire of patterns of psychoactive substance use has also been described as a characteristic feature (e.g. a tendency to drink alcoholic drinks in the same way on weekdays and weekends, regardless of social constraints that determine appropriate drinking behaviour).	A maladaptive pattern of substance use, leading to clinically significant impairment or distress, as manifested by three or more of the following occurring at any time in the same twelve-month period: (1) tolerance, as defined by either of the following: (a) need for markedly increased amounts of the substance to achieve intoxication or desired effect (b) markedly diminished effect with continued use of the same amount of the substance (2) withdrawal, as manifested by either of the following: (a) the characteristic withdrawal syndrome for the substance . . . (b) the same (or closely related) substance is taken to relieve or avoid withdrawal symptoms (3) the substance is often taken in larger amounts or over a longer period than was intended (4) a persistent desire or unsuccessful efforts to cut down or control substance use (5) a great deal of time is spent in activities necessary to obtain the substance (e.g., visiting multiple doctors or driving long distances), use the substance (e.g., chain-smoking), or recover from its effects. (6) important social, occupational, or recreational activities given up or reduced because of substance use (7) continued substance use despite knowledge of having had a persistent or recurrent physical or psychological problem that was likely to have been caused or exacerbated by the substance (e.g., current cocaine use despite recognition of cocaine-induced depression, or continued drinking despite recognition that an ulcer was made worse by alcohol consumption) Specify if: *with physiological dependence:* Evidence of tolerance or withdrawal (i.e., either item [1] or [2] is present): *without physiological dependence:* No evidence of tolerance or withdrawal (i.e., neither item [1] nor [2] is present).

SOURCES

World Health Organization. (1992). *The ICD-10 Classification of Mental and Behavioral Disorders. Clinical Descriptions and Diagnostic Guidelines.* Geneva: Author.

American Psychiatric Association. (1994). *Diagnostic and Statistical Manual of Mental Disorders, 4th ed.* Washington, DC: Author.

drinking. Such findings challenged the concept of alcoholism as a progressive disease. The concept of inevitable "loss of control" over drinking was also challenged by Merry's study (1966) in which alcoholics were given drinks containing either vodka or a placebo (no alcohol) on alternate days and reported having no more desire to drink after consuming the vodka than after the placebo. The results suggested that if "loss of control" did occur in alcoholics, it was not triggered as a biological response to alcohol but rather as a learned response with associated EXPECTANCIES concerning drinking behavior. Researchers Nancy Mello and Jack Mendelson also reported, in 1971, that alcoholics did not manifest "loss of control" in their drinking behavior and did not drink to avoid withdrawal symptoms. The work of Mello and Mendelson and of other researchers led to the conclusion that drinking behavior could be shaped like any other operant in a behavioral paradigm. Other researchers challenged the notion of alcoholism as a distinct entity (with clear differentiations between alcoholics and nonalcoholics), as well as the concepts of inevitable progression to loss of control and of alcoholism as a permanent and irreversible condition precluding the possibility of moderate drinking. (For these and other references, see Meyer, 1992.)

These findings by behavioral researchers in the laboratory had counterparts in large surveys of drinking practices conducted by the RAND Corporation. Evidence in the general population indicated that some alcoholics might be able to drink moderately without relapsing to excessive drinking.

These and other such challenges to the disease concept of alcoholism sharpened the debate and clarified the construct. Efforts to replicate some of these earlier studies sometimes led to conflicting results, calling into question the conclusions they had drawn or leading to refinements. RAND Corporation found at later follow-up that severely dependent alcoholics had to remain abstinent in order to maintain improvement. Several studies appeared to confirm that severely dependent alcoholics might be different from those who were less dependent. Some researchers, such as Hodgson, reported that small doses of alcohol had a "priming" effect (i.e., stimulated a strong urge to drink more), the magnitude of which correlated with the severity of alcohol dependence. Other researchers criticized the methodology used in previous studies. (For references, see Meyer, 1992.)

These findings help to explain why, beginning in the late 1970s, the classic disease concept was being reexamined and redefined as a symptom complex called "dependence" or "dependence syndrome." However, this shift has not satisfied some critics who object to any conceptualization that comes close to viewing compulsive alcohol or other drug use as a disease or disorder. The debate over the disease concept continues to be more heated in the alcohol field than in other areas of addictive disorders, such as compulsive use of opioids. In the early 1990s, however, an analogous and equally heated debate has developed about the conceptualization of tobacco smoking.

While health professionals throughout the world now generally agree that some forms of drug and alcohol use should be seen as disorders (at least for record-keeping and some public policy purposes), dissent from this view persists. The most compelling arguments against the disease concept have come from social and behavioral scientists. This may be partly because behavioral clinicians tend to work with less seriously impaired individuals, while physicians usually deal with people whose dependence has become more severe; and also because the physician's primary-care office may be where early identification of substance-abuse problems and effective behavioral interventions is most likely to take place.

ALTERNATIVE MODELS

Swedish researcher Lars Lindström's summary of current perspectives on the nature of alcoholism is equally applicable to the divergent views about other forms of excessive and/or compulsive drug use. Each of these models attempts to explain why people use alcohol or drugs, why use escalates to excessive and/or harmful levels, why some people continue drug use despite the harmful consequences, how and why they stop using drugs, and why they relapse after a period of abstinence. The perspectives include the *moral model*, which holds that individuals have choice and are accountable for their behavior; the *disease model* (both the classic and its variants); the *symptomatic model*, which views excessive drug or alcohol use as a symptom of underlying psychiatric disorder; the learning model (drug addiction and alcoholism are learned behaviors); the *social model*, which emphasizes the primacy of environmental factors, such as availability, social controls, interpersonal relationships; and the *biopsychosocial model*,

which attempts (in several variants) to synthesize elements of other models, taking into account biology, vulnerability, psychopathology, and cultural, social, economic, and pharmacological factors. The dependence syndrome model is probably best viewed as a variant of the biopsychosocial model.

Lindström points out that these models are now rarely encountered in pure form: each commonly incorporates elements from other perspectives. Furthermore, proponents of a particular model may, in practice, give greater emphasis to the central features of another. For example, ALCOHOLICS ANONYMOUS (AA) generally espouses the disease model. Yet because AA holds people accountable for the consequences of their drug use and emphasizes the central role of spiritual alienation in the perpetuation of alcoholism, AA's approach may also be seen as a variant of the moral model.

Although the term *disease concept* is often used synonymously with *biological* or *medical model,* these terms do not always convey the same ideas, especially with respect to implications for treatment. For example, the medical model of treatment is frequently contrasted with the social or *social recovery model,* now widely used and advocated in California. Medical-model programs are generally characterized not only by a philosophy about the problem but also by hospital-based detoxification, often pharmacologically assisted, and outpatient components in which there are formal treatment plans. Attention is paid to careful record keeping and professional credentials of the treatment staff. Physicians retain medical and legal responsibility for the overall program. In contrast, social-model recovery programs reject the involvement of professional staff and many of the activities of the medical model, such as the data gathering, licensing, and record keeping that link funding to units of service for specific patients. Instead, these programs emphasize the experience and knowledge that staff derive from the recovery process built on TWELVE-STEP mutual-help principles. There are no patients—only participants—and the role of staff is to manage the environment. Yet social models, in emphasizing the critical role that people "in recovery" play in the helping process, are employing a term—*recovery*—that is itself derived from the classic disease concept, which views alcoholism as a permanent disease state for which the only cure is total abstinence and the twelve-step AA program as the best route to such abstinence.

PERSISTENCE OF THE MORAL PERSPECTIVE

Despite the preponderance of medical opinion that some drug and alcohol users have a disorder—a diminished capacity to choose freely whether or not to use a particular substance—the moral models retain some vitality. In 1882, when the disease concept was first gaining momentum, the Reverend J. E. Todd wrote an essay entitled "Drunkenness a Vice, Not a Disease." In the late 1980s, the disease concept critics Fingarette and Peele put forth almost precisely the same thesis. Peele has argued that the disease concept exculpates the individual from responsibility, runs counter to scientific facts, and is perpetuated for the benefit of the treatment industry. However, his thesis has been criticized for using the classic disease model as a "straw man" because it does not take into account the more recent adoption of the biopsychosocial model.

Some sociologists in the United States have noted that the term *alcoholic* is still commonly used as a synonym for *drunkard* rather than as a designation for someone with an illness or disorder. The word *addict* is similarly used in a pejorative way, even when it is used more loosely to refer to a wide range of relatively benign behaviors, such as running or watching television. In the minds of most people, the concept of alcoholism or drug addiction as a disorder or disease can coexist quite comfortably with the concept of drunkenness or drug use as a vice. Since the nature of drug dependence is so closely linked to questions about the nature of free will and human volition—issues that have fascinated philosophers and scientists through the ages—it is likely that the disease concept of addiction will continue to be debated for a long time to come.

(SEE ALSO: *Addiction: Concepts and Definitions; Alcoholism; Causes of Substance Abuse; Tolerance and Physical Dependence; Treatment, History of*)

BIBLIOGRAPHY

AMERICAN PSYCHIATRIC ASSOCIATION. (1952). *Diagnostic and statistical manual of mental disorders;* DSM-II (1968); DSM-III (1978); DSM-III-R (1987); DSM-IV (1994). Washington, DC: Author.

BABOR, T. F., ET AL. (1986). Issues in the definition of diagnosis of alcoholism: Implications for a reformulation.

Progress in Neuropsychopharmacology and Biological Psychiatry, 10, 113–128.

EDDY, N. B., ET AL. (1965). Drug dependence: Its significance and characteristics. *Bulletin of the World Health Organization, 32,* 721–733.

EDWARDS, G. (1992). Problems and dependence: The history of two dimensions. In M. Lader et al. (Eds.), *The nature of alcohol and drug-related problems.* New York: Oxford University Press.

EDWARDS, G., & GROSS, M. M. (1976). Alcohol dependence: Provisional description of a clinical syndrome. *British Medical Journal, 1,* 1058–1061.

EDWARDS, G., ET AL. (1981). Nomenclature and classification of drug- and alcohol-related problems: A WHO memorandum. *Bulletin of the World Health Organization, 50,* 225–242.

EDWARDS, G., ET AL. (EDS.). (1977). *Alcohol-related disabilities.* Geneva: World Health Organization.

FEIGHNER, J. P., ET AL. (1972). Diagnostic criteria for use in psychiatric research. *Archives of General Psychiatry, 26,* 57–63.

FINGARETTE, H. (1988). *Heavy drinking: The myth of alcoholism as a disease.* Berkeley: University of California Press.

FLAVIN, D. K., & MORSE, R. M. (1991). What is alcoholism: Current definitions and diagnostic criteria and their implications for treatment. *Alcohol Health and Research World, 15*(4), 266–271.

GRANT, B. F., & TOWLE, L. H. (1991). A comparison of diagnostic criteria: *DSM-III-R,* proposed *DSM-IV,* and proposed *ICD-10. Alcohol Health and Research World, 15*(4), 284–292.

HODGSON, R., ET AL. (1979). Alcohol dependence and the priming effect. *Behavioral Research Therapy, 17,* 379–387.

JAFFE, J. H. (1992). Current concepts of addiction. In C. P. O'Brien & J. H. Jaffe (Eds.), *Addictive states.* New York: Raven Press.

JAFFE, J. H. (1965). Drug addiction and drug abuse. In L. S. Goodman & A. Gilman (Eds.), *The pharmacological basis of therapeutics,* 3rd ed. New York: Macmillan.

JELLINEK, E. M. (1960). *The disease concept of alcoholism.* New Brunswick, NJ: Hillhouse Press.

KELLER, M., & DORIA, J. (1991). On defining alcoholism. *Alcohol Health and Research World, 15*(4), 253–259.

LINDSTRÖM, L. (1992). *Managing alcoholism.* New York: Oxford University Press.

MELLO, N. K., & MENDELSON, J. H. (1971). Drinking patterns during work: Contingent and noncontingent alcohol acquisition. In Public Health Service HSM 719045, *Recent advances in alcoholism.* Washington, DC: U.S. Government Printing Office.

MERRY, J. (1966). The loss of control myth. *Lancet, 1,* 1257–1258.

MEYER, R. (1992). The concept of disease in alcoholism and drug addiction. In *Drugs and alcohol against life* (Dolentium hominum, no. 19). Vatican City: Vatican Press.

PEELE, S. (1989). *Diseasing of America.* Boston: Houghton Mifflin.

ROOM, R. (1983). Sociology and the disease concept of alcoholism. In R. G. Smart et al. (Eds.), *Research advances in alcohol and drug problems,* vol. 7. New York: Plenum.

SCHUCKIT, M. S., ET AL. (1991). Evolution of the *DSM* diagnostic criteria for alcoholism. *Alcohol Health and Research World, 15*(4), 278–283.

WIKLER, A. (1980). *Opioid dependence: Mechanisms and treatment.* New York: Plenum.

WORLD HEALTH ORGANIZATION. (1992). *The ICD-10 classification of mental and behavioural disorders: Clinical descriptions and diagnostic guidelines.* Geneva: Author.

WORLD HEALTH ORGANIZATION. (1978). *Mental disorders: Glossary and guide to their classification (in accordance with 9th revision of the International Classification of Diseases).* Geneva: Author.

JEROME H. JAFFE
ROGER E. MEYER

DISTILLATION   Distillation is the process of purifying liquid compounds on the basis of different boiling points or the process of separating liquids from compounds that do not vaporize. Since the actual process causes liquids to precipitate in a wet mist or drops that concentrate and drip, the word derives from the Latin *de* (from, down, away) + *stillare* (to drip).

In the simplest form of distillation, saltwater can be purified to yield freshwater by steam distillation, leaving a residue of salt. Distillation is also the process by which alcohol (ethanol, also called ethyl alcohol) as liquors or spirits, are separated from fermenting mashes of grains, fruits, or vegetables. When this process is used to distill alcohol, it is based on the following: Ethyl alcohol (C_2H_6O) has a lower boiling point than does water (78.5°C versus 100°C), so alcohol vapors rise first into the condenser, where cool water circulates around the out-

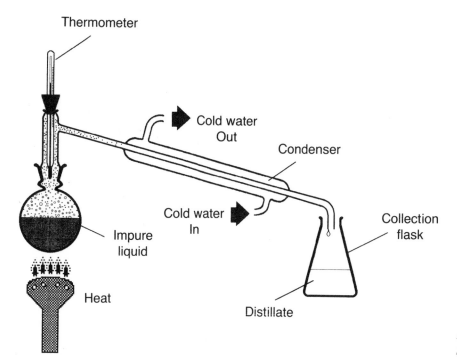

Figure 1
Simple Distillation Apparatus

side of the condenser, causing the alcohol vapors to return to liquid form and drop into the collection flask. The purity of the distillate can be increased by repeating the process several times.

About 800 A.D., the process of distillation was evolved by the Arabian alchemist Jabir (or Geber) ibn Hayyan. He may also have named the distillate *alcohol*, since the word derives from an Arabic root, *al-kuhul*, which refers to powdered antimony (kohl) used as an eye cosmetic in the Mediterranean region; with time and use it came to mean any finely ground substance, then the "essence," and eventually, the essence of wine—its spirit, or alcohol. It came into English from Old Spanish, from the Arabic spoken by the Moors of the Iberian peninsula during their rule there (750–1492 A.D.).

(SEE ALSO: *Beers and Brews; Distilled Spirits; Fermentation*)

BIBLIOGRAPHY

Lucía, S. P. (1963). *Alcohol and civilization.* New York: McGraw-Hill.

SCOTT E. LUKAS

DISTILLED SPIRITS: TYPES OF   Distilled spirits (or, simply, spirits or liquors) are the AL-COHOL-containing fluids (ethanol, also called ethyl alcohol) obtained via DISTILLATION of fermented juices from plants. These juices include wines, distillates of which are termed brandies. The most commonly used plants are sugarcane, potatoes, sugar beets, corn, rye, rice, and barley; various fruits such

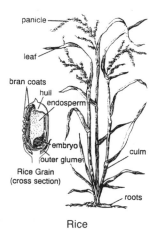

Rice

Corn

Oat

as grapes, peaches, and apples are also used. Flavors may be added to provide distinctive character.

All distilled spirits begin as a colorless liquid, pure ethyl alcohol (as it was called by 1869)—C_2H_6O. This had been called *aqua vitae* (Latin, water of life) by medieval alchemists; today it is often called grain alcohol, and the amount contained in distilled spirits ranges from 30 to 100 percent (60 to 200 proof)—the rest being mainly water.

Examples of distilled spirits include brandy, whiskey, rum, gin, and vodka. Brandy was called *brandewijn* by the Dutch of the 1600s—burned, or distilled, wine. It was originally produced as a means of saving space on trade ships, to increase the value of a cargo. The intent was to add water to the condensate to turn it back into wine, but customers soon preferred the strong brandy to the acidic wines it replaced. Cognac is a special brandy produced in the district around the Charente river towns of Cognac and Jarnac, in France, where wine is usually distilled

twice, then put into oak barrels to age. The spirits draw color and flavor (tannins) from the wood during the required five-year aging process.

Beer and wine were the most popular drinks of the New World colonists. By the mid-1700s, whiskey (from *uisce beathadh* in Irish Gaelic; *uisge beatha* in Scots Gaelic) was introduced into the American colonies by Scottish and Irish settlers to Pennsylvania. Whiskey is distilled off grains—usually corn or rye, but millet, sorghum, and barley are also used. Traditional American whiskeys are bourbons (named after Bourbon county in Kentucky), which are made from a sour mash of rye and corn. Bourbons typically contain 40 to 50 percent ethyl alcohol (called 80 to 100 proof, doubled by the liquor industry). Canadian whiskey is very similar to bourbon and to rye whiskey, while Irish whiskey is dry (has less sugars), with a distinctive austere flavor gained by filtration. All these whiskeys lack the smoky taste of Scotch whiskeys, which get their unique flavor by using malt that

Juniper

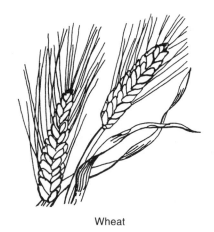

Wheat

had been heated over peat fires. By using less malt and by aging for only a few years in used sherry casks (traditionally), a light flavor is produced; by using more malt and long aging, heavy peaty smoky flavors are produced. Today, some scotches and other whiskeys are blended to achieve uniform taste from batch to batch.

The distillation of fermented sugarcane (*Saccarum officinarum*) results in rum. Of all distilled spirits, rum best retains the natural taste of its base, because (1) the step of turning starch into sugar is unnecessary; (2) it can be distilled at a lower proof; (3) chemical treatment is minimized; and (4) maturing can be done with used casks. The amount of added (sugar-based) caramel gives rum its distinctive flavor and color—which can vary from clear to amber to mahogany. The New England colonists made rum from molasses, which is the thick syrup separated from raw sugar during crystal sugar manufacture. Caribbean colonists grew sugarcane and shipped barrels of molasses to New England. New Englanders shipped back barrels of rum. Both substances were originally ballast for the barrels, which were made from New England's local forests to hold the sugar shipped from the Caribbean to the mother country, England.

Gin is a clear distillate of a grain (or beer) base that is then reprocessed; juniper berries and other herbs are added to give it its traditional taste. Vodka is also clear liquor, often the same as gin without the juniper flavor. Traditional vodkas, made in Russia, Ukraine, Poland, and other Eastern European countries, are made from grain or potatoes at a very high proof; typical ranges are 65 to 95 proof, or about 33 to 43 percent ethyl alcohol. Vodka has no special taste or aroma, although some are slightly flavored with immersed grasses, herbs, flowers, or fruits. The Scandinavian aquavit is clear, like vodka, distilled from either grain or potatoes, and flavored with caraway seed; it is similar to Germany's kümmelvasser (*kümmel* means caraway in German). When any clear liquor is added to fruit syrups, the product is called a cordial or a liqueur. Swiss kirschwasser is, however, a clear high-proof cherry-based brandy (*Kirsche* means cherry in German); and slivovitz is a clear high-proof Slavic plum-based brandy.

The raw grain alcohol distilled in the American South and in Appalachia has been called *white lightning* since the early 1900s; this is also known as moonshine, corn whiskey, or corn liquor—illegally produced in private nonlicensed stills, in very high proofs, to avoid state and federal controls or taxation. The term *firewater* was used along the American frontier after about 1815, to indicate any strong alcoholic beverage; this was often traded, given, or sold to Native Americans, causing cultural disruptions and social problems that continue even today. These include a high rate of ALCOHOLISM and children born with FETAL ALCOHOL SYNDROME.

(SEE ALSO: *Alcohol: History of Drinking; Beer and Brews*)

BIBLIOGRAPHY

BRANDER, M. (1905). *The original scotch.* New York: Clarkson N. Potter.

LUCÍA, S. P. (1963). *Alcohol and civilization.* New York: McGraw-Hill.

SCOTT E. LUKAS

DISTILLED SPIRITS COUNCIL In 1974, the Distilled Spirits Council of the United States, Inc. (DISCUS), was formed by the merger of three organizations—the industrywide Licensed Beverage Industries, Inc. (LBI), the Distilled Spirits Institute (DSI), and the Bourbon Institute. DISCUS, headquartered in Washington, D.C., is supported by the distilled spirits producers, representing 90 percent of the liquor sold in the United States.

DISCUS's primary functions are to maintain legislative relations with state and federal governments (lobbying); to conduct or support economic and statistical research; to promote export and standards of identity for American-made liquors; to maintain a voluntary code of ADVERTISING practices; and to represent the distilling industry on social issues of concern, such as teenage drinking, DRUNK DRIVING, and other forms of ALCOHOL abuse. State government relations activities are conducted by DISCUS regional representatives.

As had its predecessor LBI, DISCUS has supported programs of alcohol abuse PREVENTION and research conducted by independent groups and experts in education, traffic safety, and alcoholism. These projects have included the Grand Rapids, Michigan, study of drunk driving (1961–1965) and

the research led by Harburg and Gomburg (1978–1984) on how drinking may affect the offspring of different types of drinkers.

In addition to supporting the Harvard Medical School course for DIAGNOSIS and TREATMENT of alcoholism, now adopted by eighty medical schools, DISCUS has provided extensive support to national organizations in the alcoholism field since 1970. Its approach is based on the knowledge that alcoholism is an identifiable illness and can respond to intervention and treatment.

In 1978, DISCUS endorsed the "responsible decisions on alcohol" approach developed by the Education Commission of the States, and in 1982 it supported the National Association of State Boards of Education in its nationwide project based on this concept (which includes abstinence). In 1980, DISCUS cooperated with the U.S. Department of Health and Human Services and other sponsors in supporting the Friends of the Family parenting education program.

In 1979, DISCUS became a charter member of the Licensed Beverage Information Council (LBIC), an industrywide consortium (beer, wine, and spirits at the producer, wholesaler, and retailer levels), whose membership includes nine other associations. LBIC has supported varied prevention groups and specialists in conducting medical and public education programs devoted to alcoholism as a treatable illness; FETAL ALCOHOL SYNDROME (FAS); teenage drinking; and drunk driving. The consortium has conducted the nationwide Friends Don't Let Friends Drive Drunk campaign.

DISCUS has discouraged drinking by underage youth; encouraged adults who choose to drink to do so responsibly; and emphasized significant distinctions between normal social drinking and alcohol abuse.

The organization's economic research includes annual compilations of "apparent consumption" data (i.e., distilled spirits entering channels of trade) and an assessment of the liquor industry's contributions to the economy. Total U.S. distilled spirits consumption has declined in recent years, a fact noted by DISCUS as one of the many refutations of the "control of alcohol availability" hypothesis.

The DISCUS Code of Good Practice provides for self-regulation of advertising practices by distillers. An unusually high degree of compliance has been achieved even with nonmembers. The code was applied to radio in 1936, when that was a major medium; it has voluntarily excluded the use of television as an advertising medium by distillers since 1947. Contrary to a widely held impression, spirits advertising on television is not prohibited by law.

Distilled spirits have been the most heavily taxed consumer commodity in the United States. DISCUS and its predecessors have claimed over the years that such taxes are discriminatory and excessive, and they do not reduce chronic alcohol-abuse problems.

As a long-standing policy, DISCUS and its members do not encourage people to start drinking or to drink too much. DISCUS's review of the research literature indicates that there is no scientific evidence that brand advertising either influences or shapes those behaviors. The marketing purpose of product advertising is to build consumer acceptance of specific brands, according to DISCUS.

(SEE ALSO: *Advertising and Alcohol Use; Alcohol: History of Drinking; Legal Regulation of Drugs and Alcohol; Minimum Drinking Age Laws; Prevention; Social Costs of Alcohol and Drug Abuse; Tax Laws and Alcohol*)

BIBLIOGRAPHY

BORKENSTEIN, R. F. (1965). The role of the drinking driver in traffic accidents. Project of the Department of Police Administration, University of Indiana, initiated November 10, 1961.

HARBURG, E., & GOMBERG, E. (1984). Alcohol use/abstinence among men and women in a small town setting. Project of the University of Michigan, initiated November 1, 1978.

PAUL F. GAVAGHAN

DISULFIRAM The registered trademark name for disulfiram is Antabuse—it is the most commonly used medication for the treatment of ALCOHOLISM and the only one approved for use in the United States by the mid-1990s whose only function is for this treatment. Taking it is not intended as a substitute for the counseling alcoholics receive while in treatment; it is meant to be an aid in keeping alcoholics sober, so that they may benefit from coun-

seling. While disulfiram has been in clinical use since the late 1940s, only since the 1980s has its efficacy been studied by appropriate scientific methodology.

Disulfiram is used to deter drinking—by causing an unpleasant reaction if a medicated person drinks ALCOHOL (ethanol). This reaction is called the disulfiram–ethanol reaction (DER); the symptoms include flushing, dizziness, rapid heartbeat, nausea, vomiting, and headache. The DER is of varying severity, and the degree of severity often depends on the dose of disulfiram being taken plus the amount of alcohol that was consumed. A DER can cause hypotension (low blood pressure) and can be so severe that deaths have occurred, although with adjusted dosage regimens this is very rare.

Disulfiram blocks the action of several of the body's enzymes, including aldehyde dehydrogenase (AlDH). The inhibition of AlDH is responsible for the DER; this occurs because ethanol (drinking alcohol) is metabolized in the liver to acetaldehyde. Acetaldehyde, in turn, is converted to acetic acid, which is catabolized to water and carbon dioxide.

Aldehyde dehydrogenase is the enzyme that facilitates catabolizing acetaldehyde to acetate acid. When the action of AlDH is inhibited by disulfiram, acetaldehyde is not converted to acetate but accumulates in the blood. Most of the symptoms of the DER are due to the increased circulating acetaldehyde. Since the inhibition of AlDH by disulfiram is irreversible, a person taking disulfiram cannot stop taking it one day and begin drinking the next—several days (usually four to seven) must go by, because this is the amount of time necessary for the body to produce new enzyme. Only then might one drink alcohol without having a DER.

OTHER MEDICATIONS CAUSING SIMILAR REACTIONS

Other medications cause a mild DER, for example, some antibiotics, such as metronidazole (Flagyl). A medication available in Canada but not in the United States is citrated CALCIUM CARBIMIDE (Temposil), which inhibits AlDH in a mixed reversible-irreversible fashion. When citrated calcium carbimide is discontinued, 80 percent of the AlDH activity is restored within twenty-four hours. Hence, one can drink alcohol as soon as the day after stopping

the use of citrated calcium carbimide without having a reaction.

DOSAGE OF DISULFIRAM

Disulfiram is given orally. The dose is usually 250 milligrams or 500 milligrams daily. Some patients report not experiencing a DER with smaller doses, so larger doses may be required. Clinical experience indicates, however, that doses larger than 500 milligrams are accompanied by a greater risk of serious side effects. A problem that limits the effectiveness of disulfiram is that patients frequently stop taking the medication. To prevent this, disulfiram tablets have been implanted just below the skin of the abdominal wall. However, this has been shown not to be effective (Johnsen et al., 1987) because the absorption of the implanted disulfiram is erratic and poor. This results in very low blood levels of disulfiram and a weak or no DER.

SIDE EFFECTS OF DISULFIRAM

The use of disulfiram may be accompanied by side effects. The most common one is drowsiness; so, initially, it should be taken at bedtime. This is usually sufficient to take care of this problem, but if not, the medication may have to be discontinued, especially for those who drive or work in hazardous environments, for example, on a ladder or with machinery. Idiosyncratic liver toxicity can occur from taking disulfiram. Although this is unusual and is usually reversible if detected soon enough, deaths from liver toxicity have occurred. For this reason, liver function must be monitored closely during the first several months of treatment, and if the tests indicate possible liver damage, disulfiram must be discontinued immediately. Serious psychotic reactions and depressive episodes have occurred in patients taking disulfiram. In a multi-site study of 605 men, admissions for psychiatric problems were uncommon; as many admissions of this type occurred in men taking the placebo or not receiving disulfiram as in those receiving a 250-milligram dose (Branchey et al., 1987). The risk of serious psychoses or of major affective illnesses occurring appears to be worse with higher doses.

(SEE ALSO: *Naltrexone; Relapse Prevention; Treatment: Alcohol; Treatment Types: Aversion Therapy*)

BIBLIOGRAPHY

BRANCHEY, L., ET AL. (1987). Psychiatric complications following disulfiram treatment. *American Journal of Psychiatry, 144,* 1310–1312.

BYAR, D. P., ET AL. (1976). Randomized clinical trials. *New England Journal of Medicine, 295,* 74–80.

FULLER, R. K., ET AL. (1986). Disulfiram treatment of alcoholism: A Veterans Administration Cooperative Study. *Journal of the American Medical Association, 256,* 1449–1455.

FULLER, R. K., LEE, K. K., & GORDIS, E. (1988). Validity of self-report in alcoholism research: Results of a Veterans Administration Cooperative Study. *Alcoholism: Clinical and Experimental Research, 12,* 201–205.

HAYNES, R. B., TAYLOR, D. W., & SACKETT, D. L. (1979). *Compliance in health care.* Baltimore: Johns Hopkins University Press.

JOHNSEN, J., ET AL. (1987). A double-blind placebo controlled study of male alcoholics given a subcutaneous disulfiram implantation. *British Journal of Addiction, 82,* 607–613.

RICHARD K. FULLER
RAYE Z. LITTEN

DIVERSION   *See* Treatment Alternatives to Street Crime.

DMT   *See* Dimethyltryptamine.

DOGS IN DRUG DETECTION   In 1970, the U.S. CUSTOMS SERVICE faced a shrinking inspectional staff, a flood of illegal NARCOTICS, and an increasing load of vehicles and passengers entering the United States. In that same year a manager in the U.S. Customs Service thought that dogs could be used to detect illegal narcotics. The manager's name has been lost in the corporate history of the Customs Service, yet years later not only are dogs used to detect narcotics but also currency, weapons, explosives, fruits, and meats. Dogs could be trained to detect anything that produces an odor. Although the idea of narcotic detector dogs originated in the U.S. Customs Service, Customs' managers had to go to the U.S. Air Force for the technical expertise—not in narcotic detection, because it did not exist, but dog training in general. The air force loaned the Customs Service five instructors to develop the program. Those instructors, using the age-old method of trial and error, developed a training method for narcotic detection that is still used in the 1990s. Through the years, several key aspects of the training program were identified and became the basis for a very successful program—dog selection; development of a conditioned response; and odor integrity.

It became evident that to make the training program successful the instructors had to start with a dog that displayed certain natural traits. Those traits were retrieval motivation and self-confidence. The instructors soon realized that a dog displaying a natural desire to retrieve was the easiest to condition for response to the narcotic odor. They used the retrieval method just as the Russian physiologist Ivan Pavlov (1849–1936) used a bell: In Pavlov's experiments, he had observed that dogs salivate when food is placed in their mouths. He would give the dogs food while providing another stimulus such as a bell ringing. After a few repetitions, the dogs would salivate when they heard the bell ring—even without the food being present. The dogs had learned to associate the bell ringing with food. This response was called a *conditioned response.*

The Customs' instructors used a similar method to create a conditioned response to a drug odor. A dog was subjected to a series of retrieval exercises with a specific drug's odor. After each retrieval, the dog played a game of tug-of-war with its handler and would receive physical praise. The dog soon associated the specific drug odor with the game and the physical praise.

Using the dogs' natural desire to retrieve as a selection criterion limited the number of breeds that could be considered for this type of training. It was obvious that most of the sporting breeds fit this criterion—golden retrievers; Labrador retrievers; German short-hair retrievers; and mixed breeds of these types. They have had the retrieval drive bred into them over the centuries. In addition, these breeds predominated in the dog shelters and humane societies used by U.S. Customs as the primary source for its dog procurement, which has not only benefited the Customs' program but the local dog shelters too. Local shelters must by law destroy stray dogs after a certain time period if no one selects or adopts the dog. The Customs' instructors select dogs scheduled to be put to sleep.

The Customs' training method is based on the natural behavior of these retriever breeds of dogs. By using a dog's natural behavior, the instructors can adjust certain aspects of their training program to deal with the individual personality of each dog. Although each dog that entered the training program possessed the same basic qualifications, each then displayed them in varying degrees of intensity because of personality differences.

During the training process, the other aspect of the program that ensured success has been maintaining the integrity of the narcotic odor. During the development of the training program there were several incidents when the detector dog would respond to nondrug odors. In those incidents a common factor was identified: The nondrug odor was present in the training program. To the dog, the materials that were used in the scent-association process (a process by which the dog identifies the narcotic odor with the tug-of-war game) combined with the drug odor represented a completely different odor picture (a combination of odors that the dog associates with a positive reward). This situation became apparent when the drug was separated from the other materials; the dog would not respond to it alone or would display a considerable amount of confusion when confronted with it. This problem was eliminated by ensuring that all materials used in the training process smelled like the specific drug in question.

In summary, the key factors in narcotic detector-dog training is (1) dog selection; (2) development of a conditioned response; and (3) the integrity of the narcotic odor. This information is a very small segment of the overall training methodology. Further information about this type of dog training can be obtained through the U.S. Customs Service's Office of Canine Enforcement Programs, Washington, D.C.

(SEE ALSO: *Drug Interdiction*)

BIBLIOGRAPHY

U.S. CUSTOMS SERVICE (1978). *Detector dog training manual.* Washington, DC: U.S. Government Printing Office.

CARL A. NEWCOMBE

DOLOPHINE   *See Methadone.*

DOM   This drug's street name is STP. During the hippie drug culture of the VIETNAM war period, its name referred to "serenity, tranquility, and peace." This was also a taunt and a spoof, since the initials were the same as a widely available oil additive that made an automobile engine run smoothly. The drug DOM is a member of a family of HALLUCINOGENIC substances based on molecular additions to phenethylamine. This is a group of compounds that have structural similarities to the catecholamine-type NEUROTRANSMITTERS, such as NOREPINEPHRINE, epinephrine, and DOPAMINE. While our bodies make these catecholamines from dietary amino acids, they do not make the chemical substitutions that produce a PSYCHEDELIC compound. MESCALINE is the best and longest known of this family of HALLUCINOGENS.

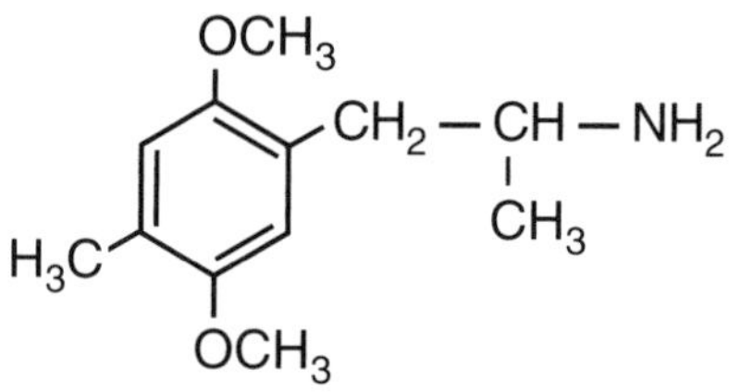

Figure 1
DOM

DOM is a synthesized compound that produces effects similar to mescaline and LYSERGIC ACID DIETHYLAMIDE (LSD), but the effects of DOM can last for fourteen to twenty hours, much longer than those of LSD. In addition, the effects of DOM have a very slow onset. Some of the initial street users of DOM had previous experience with LSD, a drug with a much more rapid onset. When the typical LSD-type effects were not found soon after taking DOM, some users took more drug, which led to a very intense and long-lasting psychedelic experience. Ironically, DOM was originally manufactured in the hope of producing a shorter, less-intense trip than LSD, which, it was thought, might be more useful and manageable in producing a period of insight and self-reflection in psychotherapy. This hope was never achieved.

(SEE ALSO: *Designer Drugs; Dimethyltryptamine*)

BIBLIOGRAPHY

SHULGIN, A., & SHULGIN, A. (1991). *PIHKAL: A chemical love story*. Berkeley, CA: Transform Press.

DANIEL X. FREEDMAN
R. N. PECHNICK

DOPAMINE

Dopamine (DA) is a decarboxylated form of dopa (an amino acid) found especially in the basal ganglia. Chemically known as 3,4 dihydroxyphenylethylamine, DA arises from dihydroxyphenylacetic acid (dopa), by the action of the enzyme dopa decarboxylase. Dopamine-containing NEURONS (nerve cells) are widespread in the brain and the body. Small interneurons are found in the autonomic ganglia, retina, hypothalamus, and medulla. Long axon neurons are found in two extensive midbrain circuits: (1) the nigrostriatal pathway links the substantia nigra neurons to the basal ganglia neurons and regulates locomotor events; (2) the mesocortical and mesolimbic circuits arise in the ventrotegmental area and project to the neocortex and limbic cortices, where they regulate emotional events, including several forms of drug addiction, reinforcement, or reward. DA is also found in minute amounts in other catecholamine neurons as a precursor.

(SEE ALSO: *Brain Structures and Drugs; Causes of Substance Abuse: Drug Effects and Biological Responses; Neurotransmission; Research, Animal Model; Reward Pathways and Drugs*)

BIBLIOGRAPHY

COOPER, J. R., BLOOM, F. E., & ROTH, R. H. (1991). *The biochemical basis of neuropharmacology*, 6th ed. New York: Oxford University Press.

FLOYD BLOOM

DOPE/DOPE FIEND   *See* Slang and Jargon.

DORIDEN   *See* Glutethimide.

DOSE-RESPONSE RELATIONSHIP

The relationship between the dose (amount) of a drug and the response observed can often be extremely complex, depending on a variety of factors including the absorption, metabolism, and elimination of the drug; the site of action of the drug in the body; and the presence of other drugs or disease. In general, however, at relatively low doses, the response to a drug generally increases in direct proportion to increases in the dose. At higher doses of the drug, the amount of change in response to an increase in the dose gradually decreases until a dose is reached that produces no further increase in the observed response (i.e., a plateau). The relationship between the concentration of the drug and the observed effect can therefore be graphically represented as a hyperbolic curve (see Figure 1).

Often, however, the response (ordinate) is plotted against the logarithm of the drug concentration (abscissa) to transform the dose-response relationship into a sigmoidal curve. This transformation makes it easier to compare different dose-response curves—since the scale of the drug concentration axis is expanded at low concentrations where the effect is rapidly changing, while compressing the scale at higher doses where the effect is changing more slowly (see Figure 2).

Finally, there are two basic types of dose-response relationships. A graded dose-response curve plots the degree of a given response against the concentration of the drug as described above. The second type of dose-response curve is the quantal dose-effect curve. In this case, a given quantal effect is chosen (e.g., a certain degree of cough suppression), and the concentration of the drug is plotted against

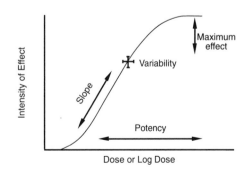

Figure 1
Representative Dose-Effect Curve, with Its Four Characteristics

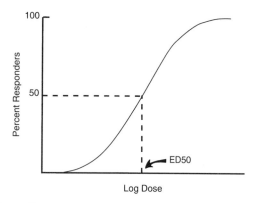

Figure 2

Representative Dose-Percent Curve, Showing a Median Effective Dose (ED50)

the percentage of a specific population in which the drug produces the effect. The median effective dose (ED50 or the dose at which 50% of the individuals exhibit the specified quantal effect) and the median lethal dose (LD50 or the dose at which death is produced in 50% of the experimental animals in preclinical studies) can be estimated from quantal dose-effect curves. With this type of curve, the relative effectiveness of various drugs for producing a desired or undesired effect, as well as the relative safety between various drugs, can be determined. The ratio of the LD50 to the ED50 for a given effect indicates the therapeutic index of a drug for that effect and suggests how selective the drug is in producing its desired effects. In clinical studies, the concentration of the drug required to produce toxic effects can be compared to the concentration required for a specific therapeutic effect in the population to estimate the clinical therapeutic index.

(SEE ALSO: *Drug Metabolism*)

BIBLIOGRAPHY

GILMAN, A. G., ET AL. (EDS.). (1990). *Goodman and Gilman's the pharmacological basis of therapeutics*, 8th ed. New York: Pergamon.

NICK E. GOEDERS

DOVER'S POWDER *Dover's Powder*, developed and described by the British physician Thomas Dover in 1732, was one of the more popular and enduring of the opium-based medications that were widely used in the United States and Europe prior to the twentieth century. The medication combined OPIUM with what we know today as ipecac (ipecacuanha), a substance that induces vomiting. The result was a pain-reducing potion that might induce a sense of euphoria but could not be ingested in large quantities because of its emetic properties. Taken as a nonprescription medicine by the general public for over 200 years, it was also prescribed by physicians for home and hospital use. Versions of the preparation are still listed in pharmaceutical formularies in which "Dover's Powder" commonly denotes any opium-based mixture that includes ipecacuanha. The wide use of Dover's Powder declined in the early 1900s largely because of the addiction that resulted from the prolonged use of OPIATES, because of the introduction of other nonaddicting ANALGESICS (painkillers), mainly aspirin, and because of laws regulating sales of opium products.

Thomas Dover (1662–1742) studied medicine at Oxford University in the 1680s. He claimed to have served an apprenticeship with Dr. Thomas Sydenham, the illustrious seventeenth-century practitioner and teacher, who originated the formula for LAUDANUM, another early and popular opium-based medicine. Dover practiced medicine for over fifty years, although during his lifetime he was more famous for his exploits as an adventurer and privateer. His involvement in the early slave trade and in the plundering of Spanish settlements off the coast of South America brought him fortune and fame. On one of his voyages he found the shipwrecked Alexander Selkirk, who, on being returned to London, created a sensation and was to become the inspiration for Daniel Defoe's *Robinson Crusoe*. Dover retired from his merchant sailing career a wealthy man, but poor investments led him to resume his medical career first in Gloucestershire and later in London.

In 1732, probably to attract patients to his new practice in London, Dover published *An Ancient Physician's Legacy to His Country*, one of the earliest medical treatises written for the general public. The book listed forty-two ailments with successful treatments used by Dover, and included the testimonial letters of many "cured" patients. The book enjoyed popular success and was reprinted eight times, the last in 1771, nearly thirty years after his death. One remedy described in the book, the use of mercury, earned him the nickname during his lifetime of the Quicksilver Doctor, but the formula for Dover's Pow-

der, which appears unchanged in all eight editions, has proven to be his most enduring legacy. Appearing on page 18 of the original edition as a treatment for gout, the directions read:

> Take Opium one ounce, Salt-Petre and Tartar vitriolated each four ounces, Ipecacuana one ounce, Liquorish one ounce. Put the Salt Petre and Tartar into a red hot mortar, stirring them with a spoon until they have done flaming. Then powder them very fine; after that slice your opium, grind them into powder, and then mix the other powders with these. Dose from forty to sixty or seventy grains in a glass of white wine Posset going to bed, covering up warm and drinking a quart or three pints of the Posset—Drink while sweating.

Dover's familiarity with opium most probably resulted from his association with Thomas Sydenham and thereby his acquaintance with the benefits of laudanum (an alcoholic tincture of opium). Dover's ingenious use of opium with ipecacuanha seems to have been original. His unique formula, Pulvis Ipecacuanha Compositus, with its specificity of ingredients, produced a relatively reliable and consistent potion in an era when there was no regulation of medications and little standardization in their preparation. Medications could be purchased at apothecary shops with or without doctors' prescriptions or at back-street stores that sold drugs along with food, clothing, and other necessities of life. The major issue at the time in the use of opiate-based medications was not that they contained what we now know to be a NARCOTIC, but whether the consistency of the formula or the misuse by the patient caused overdoses of what could be poisonous ingredients. Dover's Powder provided a stable product that, because of the ipecacuanha, could not be taken in excess at any one time. The powder came to be trusted by the general public and widely prescribed by physicians. It was considered such a safe remedy that it was even prescribed for children.

Although Dover originated his powder as treatment for gout, it was used throughout the eighteenth and nineteenth centuries along with many other opium-based patent and official preparations by large numbers of people for a wide variety of disorders. Opium, used as a healing plant for over 6,000 years, was an ingredient in countless formulas that were openly available and credited with curing the most common disorders of the time. Mixed in a tincture, it was found in laudanum; in a camphorated formula it became PAREGORIC; and it was also included in preparations for lozenges, plasters, enemas, liniments, and other general medications.

Opium-based medicines were used for many disorders, including insomnia, diarrhea, bronchitis, tuberculosis, chronic headache, insanity, menstrual disorders, pain, malaria, syphilis, and smallpox. Often both physicians and patients mistook its narcotic properties, which relieved pain and created a sense of well-being, as curative rather than palliative, and little was understood of the darker side of opiate medications—the destructive nature of addiction—until well into the nineteenth century. By this time, it was common for middle- and upper-class people, especially women and those with chronic diseases, to be addicted to opiates that were frequently seen in innocuous health elixirs or in remedies that had been originally prescribed by physicians. Widespread prescribing by physicians and easy availability of the opiate medications made addiction a frequent result of medical therapeutics.

By the middle of the nineteenth century, the issue of opium addiction began to appear with more frequency in the medical literature, and in both the United States and England there were pressures to regulate both the pharmacy trade and the use of narcotic medications, especially the patent medicines containing opiates. Even then, it was not until the end of the century—as a result of better education of physicians and pharmacists, advances in diagnosis and therapeutics, and a growing understanding of the nature of addiction—that opium-based medications were supplanted by other curative treatments and by nonaddictive salicylate analgesics such as aspirin. Opium-based medicines used today, such as MORPHINE and CODEINE, are government-regulated and can be purchased legally only by prescription.

Dover's Powder in its original form is now an obsolete medication. It should be recognized for its place in the history of pharmacology as a relatively reputable medicine used from 1732 until the 1930s, an era in which opium-based medications were one of the few remedies that brought relief to suffering patients. Many of these medications came to be misused by both patients and physicians who had little understanding of addiction and few options for PAIN relief. Thomas Dover, seen as an adventurer and opportunist by many during his lifetime, developed a preparation that allowed patients to use a narcotic while limiting its ingestion. More precise knowledge of the healing as well as the addictive properties of narcotics allows modern physicians and pharmacol-

ogists to deal specifically with the dosage of narcotic medications. Nevertheless, Dover's Powder, an ingenious and effective solution to a thorny problem, became a household name long after its originator's medical career had ended and his medical treatise had been published.

(SEE ALSO: *Addiction: Concepts and Definitions; Britain, Drug Use in; Disease Concept of Alcoholism and Drug Abuse; Opioids and Opioid Control*)

BIBLIOGRAPHY

BERRIDGE, V., & EDWARDS, G. (1981). *Opium and the people: Opiate use in nineteenth century England.* London: Allen Lane.

BOYES, J. H. (1931). Dover's Powder and Robinson Crusoe. *New England Journal of Medicine, 204,* 440–443.

COURTWRIGHT, D. T. (1982). *Dark paradise: Opiate addiction in America before 1940.* Cambridge, MA: Harvard University Press.

DUKE, M. (1985). Thomas Dover—physician, pirate and powder, as seen through the looking glass of the 20th-century physician. *Connecticut Medicine, 49,* 179–182.

OSLER, W. (1896). Thomas Dover, M. B. (of Dover's Powder), physician and buccaneer. *Bulletin of the Johns Hopkins Hospital, 7,* 1–6.

VERNER STILLNER

DRAMSHOP LIABILITY LAWS

Dramshops are taverns, saloons, bars, and drinking establishments. They have been the subject of laws since before the time of Hammurabi's code in ancient Babylon (ruled 1792–1750 B.C.). Two of the four dramshop laws in Hammurabi's code were enforced by the death penalty; they prohibited priestesses from becoming tavern keepers, and they also sought to have tavern keepers arrest outlawed patrons. A third law specified punishments for bartenders who gave a short measure to their customers, a crime punished by having the bartender thrown into the water. The fourth law dealt with selling beer or wine on credit and stipulated that the ALCOHOL should cost less if the full account was paid up at the time it was being sold.

Today's dramshop liability laws, enacted in more than 50 percent of the United States, are extensions of Hammurabi's code as it applied to tavern keepers who were required to arrest outlawed patrons—but there are some significant and important differences. The modern dramshop liability laws are intended to reduce the consequences of excessive drinking and underage drinking, and they actually make the tavern keepers liable for damage done by an underage or "obviously intoxicated" patron to whom they have served alcoholic beverages. That is, in addition to the underage or "obviously intoxicated" patron being liable, the tavern keeper can be sued as well.

Quite clearly, lawmakers intend that these penalties will curb any tendency to make a profit on the unethical sale of alcoholic beverages whenever possible. In this sense, the dramshop liability laws strengthen other alcohol-related laws that impose fines and other punishments when alcohol is sold to "visibly intoxicated" customers, to known alcoholics or habitual drunkards, and to minors.

Unfortunately, dramshop liability laws may not be as effective as lawmakers and the public might hope. Although it may be relatively easy to document when a drinker is under age, it is more difficult to prove that someone was obviously intoxicated when alcohol was sold.

Difficulties of this type have led to some constraints on the extent of bartenders' and tavern keepers' liability for damages. For example, in Illinois, the limit on damages is set at 20,000 dollars and the effective statute of limitations has been reduced: a lawsuit must be filed within one year of the incident (versus two years under an older Illinois dramshop liability law). In some jurisdictions, judges can disallow the liability of tavern keepers who have attempted to follow the legal standards and who clearly made reasonable efforts to restrain their intoxicated patrons.

Even if dramshop liability laws are difficult to apply in practice, and without regard to their actual effectiveness, it is likely that their long tradition will be maintained as part of an overall societal effort to reduce and control the damages that can be caused by excessive alcohol consumption. A similar concept of seller's liability could be extended to legal and illicit drug sales, making dispensers, drug-sharing peers, or drug dealers liable for what happens when users cause damage. Even if a seller's liability would not deter drug dealers who seek enormous profits by trading in illicit drugs, this type of liability might make friends hesitate before sharing drugs with their previously uninitiated peers, with a resulting beneficial reduction in the spread of drug use throughout

peer groups. It might also make physicians and pharmacists pause before prescribing or endorsing habit-forming medications and OVER-THE-COUNTER (OTC) remedies.

(SEE ALSO: *Alcohol: History of Drinking; Driving, Alcohol, and Drugs; Driving Under the Influence; Drug Interactions and Alcohol; Drunk Driving; Legal Regulation of Drugs and Alcohol; Mothers Against Drunk Driving; Students Against Driving Drunk*)

BIBLIOGRAPHY

GAINES, L. S., & HARFORD, T. C. (1979). *Social drinking contexts*. Rockville, MD: National Institute on Alcohol Abuse and Alcoholism.

GERSTEIN, D. R., & OLSON, S. (1985). *Alcohol in America: Taking action to prevent abuse*. Washington, DC: National Academy Press.

GRANT, M. (1985). *Alcohol policies*. Copenhagen: World Health Organization Regional Publications.

MOSER, J. (1985). *Alcohol policies in national health and development planning*. Geneva: World Health Organization.

PITTMAN, D. J., & SNYDER, C. R. (1973). *Society, culture, and drinking patterns*. Carbondale, IL: Southern Illinois University Press.

RUBINGTON, E. (1973). *Alcohol problems and social control*. Columbus, OH: C. E. Merrill.

WALSH, D. (1982). *Alcohol-related medicosocial problems and their prevention*. Copenhagen: World Health Organization Regional Publications.

CHRISTOPHER B. ANTHONY

DRIVING, ALCOHOL, AND DRUGS

Injuries, especially from motor vehicle accidents, are the leading cause of death for people under age 44. The most frequent factor associated with traffic, drowning, fire, and fall fatalities is the presence of ALCOHOL (Special Report, 1987). In 1968 alone, more than 50 percent of fatal traffic ACCIDENTS and 33 percent of serious-injury traffic accidents had been alcohol-related (U.S. Department of Transportation, 1968).

Although the association between alcohol consumption and traffic accidents had been recognized by the beginning of the twentieth century, the magnitude of the problem did not capture general attention until the 1970s. Public tolerance of DRIVING UNDER THE INFLUENCE of alcohol decreased sharply—a shift in attitude that, combined with increased legal countermeasures, resulted in a significant decline in alcohol-related fatalities from a high of 57 percent in 1982 to 45 percent in 1992.

Lund and Wolfe (1991) compared the frequency of driving under the influence of alcohol in two U.S. nationwide roadside surveys, carried out in 1973 and 1986 in thirty-two localities. Drivers were stopped at random and asked to provide breath tests. The BLOOD ALCOHOL CONCENTRATION (BAC) levels were compared for the two surveys as a function of time, day, region, gender, age, ethnicity, employment, and educational status. Across nearly all population subgroups, alcohol-impaired driving decreased by 33 percent or more from 1973 to 1986. Nevertheless, in the mid 1990s, alcohol remains the single most significant factor in traffic injuries and deaths.

Epidemiological studies have compared BAC levels in accident-involved drivers with BAC levels from randomly selected drivers passing the accident site at similar times. These studies have demonstrated that the chance of an accident increases with any departure from zero BAC and increases exponentially with increasing BAC levels. By the time blood levels of alcohol exceed 0.2 grams per deciliter (200 mg per 100 ml), the chance of an accident increases more than 100 times (i.e., 10,000%).

All types of human activity are impaired eventually by increasing alcohol levels. However, examination of alcohol-related accident data from governmental investigations and from police accident reports indicates that information-processing errors are responsible for the majority of alcohol-related traffic accidents. Information-processing deficits include impairment of attention, visual search, and perception. The second largest category of errors involves judgment, such as speed selection. Failure to control the car because of decreasing motor skills remains a distant third category, despite the popular assumption that identifies driving impairment with the appearance of intoxication and motor incapacitation.

The results seen in survey studies are supported by extensive experimental studies in which driver behavior is examined under controlled conditions. These laboratory studies examine one or two behaviors relevant to driving at a time. More complex studies of driving-related behavior use driver simulators

and closed-course driving situations that preserve the safety of the driver.

Moskowitz and Robinson (1988) reviewed 177 experimental studies of alcohol effects on driving behaviors that met a certain level of scientific merit. The behavior that they found most affected by alcohol was divided attention with impairment seen even at alcohol levels below 0.02 percent (20 mg/100 ml). Divided-attention tasks involve simultaneously monitoring and responding to more than one source of information, which is characteristic of many complex interaction systems such as driving and flying. While operating the vehicle, drivers under the influence of alcohol fail to detect significant potential threats in the environment.

Similarly, studies on information processing and perception were affected at low BAC levels. Tracking, which is analogous to car control functions such as maintaining heading and lane position, was shown to be impaired at low BACs when performed simultaneously with other functions in divided-attention situations, but less impaired when the tracking task was performed by itself. Similarly, complex reaction-time tasks involving several competing stimuli and responses were impaired at low BACs, whereas simple reaction time requiring little information processing was more resistant to alcohol effects.

Studies of psychomotor skills, performance in driving simulators, and closed-course driving studies showed considerable variation in the alcohol levels at which impairment appeared. These variations are likely explained by the differences in information-processing requirements among these varied tasks. The review concluded that no minimum threshold point for alcohol impairment of complex human–machine tasks exists. Any reliable measure of alcohol in the human system produces some impairment.

Other areas that have been suggested as leading to alcohol-related accidents, such as poor judgment and violent or aggressive behavior, have been infrequently examined by researchers—mainly because of the difficulty of developing laboratory techniques to measure them.

The low BAC levels at which laboratory studies have shown significant impairment and epidemiological studies have shown increased accident frequency are below the levels at which the majority of the population would exhibit intoxication symptoms such as slurred speech and unsteady gait.

OTHER DRUGS

The massive involvement of alcohol in traffic and other injuries is well documented. What conclusions can we draw about the role of drugs other than alcohol in traffic safety? While laboratory studies on the effects of many drugs and alcohol are similar in demonstrating impairment of performance skills, there have been difficulties in executing epidemiological studies on drugs in driving. For example, few nonaccident control drivers volunteer to provide blood samples to compare their drug levels with drug levels in blood samples obtained from accident victims.

While several studies have been done in hospitals with drug blood levels from trauma patients involved in driving accidents compared with samples from volunteers who were in the hospital for other reasons, questions arise regarding the representativeness of the control group.

Another problem in relating drug use to accidents has been the difficulty of evaluating the behavioral significance of the drug blood levels. Unlike alcohol, where levels in venous blood samples or breath samples are essentially equivalent to those from blood in the brain, the site of drug action, nearly all other drugs have a complex relationship between blood plasma level and degree of resulting behavioral impairment. Many drugs remain present in plasma for weeks beyond any period in which behavioral effects are seen. In other cases, drug levels in plasma drop extremely rapidly and become difficult to detect while behavioral impairment remains. Thus most epidemiological studies of drugs and driving report the presence or absence of the drug rather than the level.

One technique to circumvent control-group problems has been to assign responsibility or nonresponsibility to accident-involved drivers and then correlate the presence of drugs with the frequency of accident responsibility. Within the constraints of these epidemiological studies, researchers have often concluded that tranquilizers, antihistamines, and antidepressants are overrepresented in accident-involved drivers.

Terhune and colleagues (1992) examined the presence of drugs in blood specimens from 1,882 fatally injured drivers. Drugs, both illicit and prescription, were found in 18 percent of the fatalities. MARIJUANA was found in 6.7 percent, COCAINE in 5.3

percent, tranquilizers in 2.9 percent, and AMPHET-AMINES in 1.9 percent of these fatally injured drivers.

While crash responsibility was assigned and correlated with drug use, the small number of individuals in each separate drug classification made obtaining statistical significance difficult to evaluate, despite the fact that several of the drug categories were associated with crash responsibility. Moreover, crash-responsibility rates increased significantly as the number of drugs in the driver increased. That is, many of the drug users were multiple consumers. In addition, alcohol was found in 52 percent of the fatalities, with more than 90 percent of the drivers with BACs over 0.08 percent considered responsible for the crash.

In contrast to the sparcity of information available about accident causation from epidemiological sources, extensive experimental research has been performed to evaluate the effects of drugs on skill performance. Regulatory agencies in various countries have frequently required evaluation of the skill performance side effects of prescription drugs. Also, the effects of illicit and abused drugs on skill performance have been extensively studied in the laboratory.

Thus the evaluation of the effects of drugs on driving and other human–machine situations has depended primarily on experimental studies where changes in behavior can be observed as a function of differences in administered doses and time after administration. However, no other drug has been evaluated in as extensive a range of behavioral studies as has alcohol. Nevertheless, many drugs have been studied with respect to some important variables required for driving.

The emphasis in these drug studies has tended to be on the evaluation of vision, attention, vigilance, and psychomotor skills. Driving-simulator studies have also been done on occasion. The psychomotor skill most often examined has been some form of tracking.

Reviewing this literature in a short space presents considerable difficulties since there are differences between classes of drugs as well as individual drugs within the same drug classification. For example, many minor tranquilizers, especially BENZODIAZE-PINES, have been shown to impair attention and tracking in a wide variety of studies. However, recently introduced tranquilizers, such as buspirone, exhibited little evidence of impairment effects.

Conclusions about drugs are likely to change because of the pressures exerted by drug regulatory agencies to develop drugs that do not impair skills performance. For example, hypnotics often exhibit residual skills impairment the day following use. New drugs have been introduced whose duration of effects are shortened so there will be less residual impairment after awakening.

Another class of psychoactive drugs, the ANTIDE-PRESSANTS, has been long known to impair performance on a variety of skills, especially amitriptyline. Again, recently introduced types of antidepressants do not produce the same degree of impairment.

While narcotic ANALGESICS derived from OPIUM (OPIATES) have been shown experimentally to lead to decreased alertness, there have been reports that chronic use produces considerable tolerance to some of these side effects, which may explain why epidemiological studies have not found differences in accident rates between NARCOTIC users and control groups. Moreover, those maintained on METHADONE, a synthesized narcotic, and who have been stabilized on it have shown little evidence of impairment in a wide variety of experimental and epidemiological studies.

Another drug category that shows evidence of impairing skills performance in laboratory studies is the antihistamines, many of which have produced impairment of performance accompanied by complaints of drowsiness and lack of alertness. Again, recent pharmacological improvements have produced antihistamine drugs, such as terfenadine (Seldane), which maintain antihistamine actions but have difficulty crossing the blood-brain barrier and produce little impairment.

The most frequently used illicit drug in the United States is marijuana, and epidemiological studies have indicated it is frequently consumed by drivers. Of all illicit drugs, marijuana has had the largest number of experimental studies performed to examine its effects. Many studies indicate that marijuana impairs coordination, tracking, perception, and vigilance, as well as performance in driving simulators and on-the-road studies.

While there has been concern over increased driver use of STIMULANTS, such as AMPHETAMINES and COCAINE, there is little experimental evidence demonstrating driving impairment by these drugs. On the contrary, most studies of these stimulants, as well as of CAFFEINE, indicate an improvement in

skills performance. However, with chronic (long-term) use of stimulants increased dosage is taken as TOLERANCE develops. Thus the dose levels examined in the laboratory may not reflect those found among drivers. Secondly, after the stimulation phase, a subsequent depressed phase occurs (the "crash") with increased drowsiness and lack of alertness. The stimulant drugs have not been well studied in relation to driving and should be. Further study is needed—both for acute (one-time) use and chronic use.

(SEE ALSO: *Dramshop Liability Laws; Drunk Driving; Minimum Drinking Age Laws; Mothers Against Drunk Driving; Psychomotor Effects of Alcohol and Drugs; Social Costs of Alcohol and Drug Abuse; Students Against Driving Drunk*)

BIBLIOGRAPHY

LUND, A., & WOLFE, A. (1991). Changes in the incidence of alcohol impaired driving in the United States, 1973–1986. *Journal of Studies on Alcohol, 52*(4), 293–301.

MOSKOWITZ, H., & ROBINSON, C. D. (1988). *Effects of low doses of alcohol on driving-related skills: A review of the evidence.* Report no. DOT HS-807-280. U.S. Department of Transportation, Washington, DC.

TERHUNE, K. W., ET AL. (1992). *The incidence and role of drugs in fatally injured drivers.* Report no. DOT HS-808-065. U.S. Department of Transportation, Washington, DC.

U.S. DEPARTMENT OF HEALTH AND HUMAN SERVICES. (1987). *Sixth special report to the U.S. Congress on alcohol and health.* Rockville, MD: Author.

U.S. DEPARTMENT OF TRANSPORTATION. (1968). *Alcohol and highway safety:* A report to the Congress. Washington, DC: U.S. Government Printing Office.

HERBERT MOSKOWITZ

DRIVING UNDER THE INFLUENCE (DUI)
This is a term that refers to the operation of a motor vehicle after consuming alcohol and being affected by it in some way. It may be used as a legal term denoting a lesser offense than *driving while intoxicated* (DWI). Specific blood-alcohol concentration (BAC) limits are associated with a DUI offense. These vary among states and countries but are often between 0.05 percent and 0.10 percent (50 milligrams per deciliter [mg/dl] and 100 mg/dl). There is a strong correlation between a BAC greater than 0.05 percent and risk of serious injury or death while operating a motor vehicle. Fifty percent of single-vehicle fatalities and 25 to 35 percent of all serious motor vehicle ACCIDENTS involve drivers under the influence of alcohol.

(SEE ALSO: *Breathalyzer; Dramshop Liability Laws; Driving, Alcohol, and Drugs; Drug Interactions and Alcohol; Drunk Driving; Mothers Against Drunk Driving; Students Against Driving Drunk*)

BIBLIOGRAPHY

AMERICAN MEDICAL ASSOCIATION COUNCIL ON SCIENTIFIC AFFAIRS. (1986). Alcohol and the Driver. *Journal of the American Medical Association, 255*(4), 522–527.

MYROSLAVA ROMACH
KAREN PARKER

DROPOUTS AND SUBSTANCE USE
The Monitoring the Future project (HIGH SCHOOL SENIOR SURVEY) and other studies of school-age youths have helped us to understand the substance-use patterns of ADOLESCENTS who remain in and graduate from high school. In contrast, not nearly as much is known about the substance use of those who become high school dropouts. Nonetheless, by putting together evidence from a variety of sources, including the NATIONAL HOUSEHOLD SURVEY ON DRUG ABUSE and the Epidemiologic Catchment Area surveys sponsored by the U.S. government, it is possible to say that high school dropouts are much more likely to have started using TOBACCO, ALCOHOL, and other drugs, as compared with their peers who remained in school. There also is some evidence that dropping out of high school is associated with an increased risk of adult-onset alcohol-dependence syndromes, even among persons whose dropping out could not have been caused by the consequences of starting to drink during the adolescent years. Whether this conclusion also holds for adult-onset DEPENDENCE on other drugs such as COCAINE or MARIJUANA is not yet clear but is under study.

In trying to understand how it might happen that dropouts are more likely to be substance users, the possibility should be considered that substance use has caused some people to drop out of school, as well as the possibility that some schools suspend or expel students for smoking tobacco, drinking alcohol, or

using other drugs. By themselves, these circumstances could be enough to explain why high school dropouts are more likely to have taken illicit drugs or started underage smoking or drinking.

In addition, when students drop out before graduating from high school, they often begin spending more time with older youths and adults, some of whom serve as role models for substance use and may give the dropouts cigarettes or offer them opportunities to try alcohol or other drugs for the first time. As a result, not only is there the possibility that substance use may lead to dropping out, but it is also possible that dropping out may lead to substance use.

More complicated possibilities must also be taken into account. A developmental perspective makes it possible to imagine that the greater frequency of substance use among high school dropouts might have its origins in the earlier years of childhood, so that it is not a simple matter of substance use leading to dropping out, or dropping out leading to substance use. For example, it has been found that youngsters who frequently broke rules, got into fights, and had trouble adapting to elementary school were more likely to become heavy drug users ten or more years later. CONDUCT problems at school are other signs that help to predict who will drop out before completing high school. This provides some evidence that sometimes there can be an underlying common cause in earlier stages of growth and development and that it is essential to consider these earlier stages in observing a link between later substance use and dropping out of school.

Working along these lines, Brook and her research team have looked into school grades and poor school achievement with the idea that students who did not do well in elementary school might be at greater risk for later drug use, in the same way that not doing well in school was a sign of greater risk for dropping out. Her research group also wished to know whether later improvements in academic performance might modify the risk profiles of low achievers in primary school.

In studying a large sample of school-age youth from primary school through high school, this team of researchers has been finding a moderately strong linkage between early poor school achievement and later drug use, but it also discovered that the linkage was substantially weaker when low achievers in primary school did much better in later school years. This type of developmental evidence is important and needs to be replicated by others before strong

conclusions can be drawn. It does suggest that it might be possible to use achievement-strengthening programs not only to reduce school dropout rates but also to reduce rates and levels of teenage drug use.

Several research groups are carrying out rigorous field experiments to see whether intervention programs directed at entire classrooms of first- and second-graders might change their risks for later drug use, conduct problems, and dropout rates. These very early interventions are not drug-education classes. The first- and second-grade teachers are working with the students to help promote their learning and school achievement in new ways and to help them behave themselves and adapt better to the rules of the elementary school classroom.

Other research groups are trying to reduce dropout rates and substance use by targeting the more vulnerable or higher-risk elementary school and middle school students, and giving them and their FAMILIES special programs to promote learning and a sense of mastery over schoolwork; sometimes this is being done in connection with social-influences intervention programs. In contrast with interventions directed at all students in the first- and second-grade classrooms, these involve "pull-out" programs for the specially targeted higher-risk students.

Research in the 1990s will provide more definitive evidence on the underlying mechanisms that account for observed associations between dropping out of high school and substance use, as well as for the very new suspected associations between dropping out of school and the risk for adult-onset alcohol-dependence syndromes (Crum et al., 1993). If the intervention programs are found to reduce dropout rates and also levels of alcohol, tobacco, and other drug involvement, this will provide some powerful evidence on the causal significance of the early developmental antecedents and will help to explain why dropouts are more likely to be drug users.

In the meantime, the broad range of unfortunate effects of dropping out of school makes it important to sustain and increase the vigor of stay-in-school programs as well as outreach programs for youths who are chronically absent from school or who actually have dropped out before graduation. These programs may help the individual youths, their families, and society in many ways; they may not only confer benefits in relation to schooling and better preparation for adult life, but also reduce the amount of substance use in the teenage years, prevent the occurrence of alcohol and drug problems in

adulthood, and possibly prevent other psychiatric disorders such as major DEPRESSION.

(SEE ALSO: *Appendix, Volume 4; Attention Deficit Disorder; Coping and Drug Use; Education and Prevention; Epidemiology of Drug Abuse; Vulnerability As Cause of Substance Abuse*)

BIBLIOGRAPHY

BROOKS, J. S., ET AL. (1988). Personality, family and ecological influences on adolescent drug use. A developmental analysis. *J Chem Dep Treatment, 1,* 123–162.

CRUM, R. M., HELZER, J. E., & ANTHONY, J. C. (1993). Level of education and alcohol abuse and dependence in adulthood: A further inquiry. *American Journal of Public Health, 83*(6), 830–837.

DISHION, T. J., REID, J. B., & PATTERSON, G. R. (1988). Empirical guidelines for a family intervention for adolescent drug use. *J Chem Dep Treatment, 1*(2), 189–224.

DRYFOOS, J. D. (1991). *Adolescents at risk.* New York: Oxford University Press.

GALLO, J. J., ROYALL, D. R. & ANTHONY, J. C. (1993). Risk factors for the onset of depression in middle age and later life. *Soc Psych and Psych Epi, 28*(3), 101–108.

HAWKINS, J. D., VON CLEVE E., & CATALANO, R. F., JR. (1991). Reducing early childhood aggression. Results of a primary prevention program. *J Amer Child Adol Psych, 30*(2), 208–217.

JOHNSTON, L. D., O'MALLEY, P. M., & BACHMAN, J. G. (1992). *National survey results on drug use from the Monitoring the Future Study, 1975–1992* (vol. 1). Rockville, MD: National Institute on Drug Abuse.

KELLAM, S. G., ET AL. (1994). The course and malleability of aggressive behavior from early first grade into middle school: Results of a developmental epidemiologically-based preventive trial. *J Child Psyc Psychiatry, 35*(2), 259–281.

WERTHAMER-LARSSON, L. A. (1994). Methodological issues in school-based services research. Special section: Mental health services research on children, adolescents and their families. *J Clinical and Child Psych, 23*(2), 121–132.

ERIC O. JOHNSON

DRUG As a therapeutic agent, a drug is any substance, other than food, used in the prevention, diagnosis, alleviation, treatment, or cure of disease.

It is also a general term for any substance, stimulating or depressing, that can be habituating. According to the U.S. Food, Drug, and Cosmetic Act, a drug is (1) a substance recognized in an official pharmacopoeia or formulary; (2) a substance intended for use in the diagnosis, cure, mitigation, treatment, or prevention of disease; (3) a substance other than food intended to affect the structure or function of the body; (4) a substance intended for use as a component of a medicine but not a device or a component, part, or accessory of a device.

Pharmacologists consider a drug to be any molecule that, when introduced into the body, affects living processes through interactions at the molecular level. Hormones can be considered to be drugs, whether they are administered from outside the body or their release is stimulated endogenously. Although drug molecules vary in size, the molecular weight of most drugs falls within the range of 100–1,000, since to be a drug it must be absorbed and distributed to a target organ. Efficient absorption and distribution may be more difficult when drugs have a molecular weight greater than 1,000. The drug's molecular shape is also important, since most drugs interact with specific RECEPTORS to produce their biological effects. The shape of the receptor determines which drug molecules are capable of binding. The shape of the drug molecule must be complementary to that of the receptor to produce an optimal fit and, therefore, a physiological response.

Within this general definition, most POISONS would be considered to be drugs. Although water and oxygen technically fit this general definition and are used therapeutically and discussed in pharmacology textbooks, they are rarely considered to be drugs. Efforts have been made to develop a more restricted definition, but because so many molecules and substances can affect living tissue, it is difficult to draw a sharp line.

(SEE ALSO: *Inhalants; Plants, Drugs from; Vitamins*)

BIBLIOGRAPHY

GILMAN, A. G., ET AL. (EDS.) (1990). *Goodman and Gilman's the pharmacological basis of therapeutics,* 8th ed. New York: Pergamon.

Stedman's medical dictionary. (1990). 25th ed. Baltimore: Williams & Wilkins.

NICK E. GOEDERS

DRUG ABUSE *See* Addiction: Concepts and Definitions.

DRUG ABUSE REPORTING PROGRAM (DARP)

The Drug Abuse Reporting Program began in 1969 as a comprehensive data system that included intake and during-treatment information on individuals entering drug treatment programs funded by the United States government. Over time, it was the basis for carrying out the first national evaluation study of community-based treatment programs. It was conducted at Texas Christian University over a period of twenty years and included four distinct phases of research: (1) describing major treatment modalities and the characteristics of drug abusers entering them in the early 1970s; (2) describing during-treatment performance measures and how they related to differences in treatments and clients; (3) describing posttreatment outcomes and how they related to differences in treatments and clients; and (4) describing important elements of long-term addiction careers.

The DARP data system contained records on almost 44,000 admissions to 52 federally supported treatment agencies from 1969 to 1973—the years during which the current community treatment delivery system first emerged in the United States. The study population consisted of clients from major TREATMENT modalities—METHADONE MAINTENANCE, THERAPEUTIC COMMUNITY, outpatient drug-free, and DETOXIFICATION—as well as a comparison intake-only group.

THE EFFECTIVENESS OF DRUG-ABUSE TREATMENT

Initial research in this 20-year project focused on ways of measuring characteristics of treatments, clients, and behavioral outcomes (see Sells & Associates, 1975). It was found that drug use and criminal activities decreased significantly during treatment, including outpatient as well as residential programs. More important, the effects continued after treatment was ended. A sample of 6,402 clients located across the United States were selected for follow-up an average of 3 years after leaving DARP treatment (and 83% were located). Methadone maintenance, therapeutic communities, and outpatient drug-free programs were associated with more favorable outcomes among opioid addicts than outpatient detoxification and intake-only comparison groups; however, only clients who remained in treatment 3 months or longer showed significant improvements after treatment. Numerous studies of these data helped establish that "treatment works" and that the longer clients stay in treatment, the better they function after treatment (see Simpson & Sells, 1982).

LONG-TERM OPIOID ADDICTION CAREERS

To study long-term addiction careers, a sample of 697 daily OPIOID (primarily HEROIN) users were followed up again, at 12 years after entering treatment (and 80% were located). It was found that about 25 percent of the sample was still addicted to daily opioid use in year 12. Length of addiction (defined as the time between first and last daily opioid use) ranged from 1 to 34 years. Of the total sample, 50 percent was addicted 9.5 years or longer, yet 59 percent never had a period of continuous daily use that exceeded 2 years. Only 27 percent reported continuous addiction periods that lasted more than 3 years.

Three-fourths of the addicts studied had experienced at least one "relapse" to daily opioid use after they had temporarily quit. Among those who had ever temporarily quit daily opioid use at least once, 85 percent had done so while in a drug abuse treatment, 78 percent had quit while in a jail or prison, 69 percent had temporarily quit "on their own" (without treatment), and 41 percent had quit while in a hospital for medical treatment. The most frequent reasons cited for quitting addiction the last time involved psychological and emotional problems. Ex-addicts reported they had "become tired of the hustle" (rated as being important by 83% of the sample) and needed a change after "hitting bottom" (considered important by 80%). Other reasons cited as being important were "personal or special" events such as a marriage or the death of a friend (64%), fear of being sent to jail (56%), and the need to meet family responsibilities (54%). (See Simpson & Sells, 1990, for further details.)

SUMMARY

The DARP findings have been widely used to support continued public funding of drug-abuse treatments and to influence federal drug policy in

the United States. Other similar national treatment evaluation studies have been planned and undertaken at the beginning of each decade since the 1970s. Current research efforts focus on increasing understanding of the particular elements of treatment that are most effective and how they can be improved.

(SEE ALSO: *Drug Abuse Treatment Outcome Study; Narcotic Addict Rehabilitation Act*)

BIBLIOGRAPHY

SELLS, S. B., ET AL. (1975). The DARP research program and data system. *American Journal of Drug and Alcohol Abuse, 2*, 1–136.

SIMPSON, D. D., & SELLS, S. B. (1990). Opioid addiction and treatment: A 12-year follow-up. Malabar, FL: Krieger.

SIMPSON, D. D., & SELLS, S. B. (1982). Effectiveness of treatment for drug abuse: An overview of the DARP research program. *Advances in Alcohol and Substance Abuse, 2*(1), 7–29.

DWAYNE SIMPSON

DRUG ABUSE TREATMENT OUTCOME STUDY (DATOS)

This study is designed to provide comprehensive information on continuing and new questions about the effectiveness of the drug abuse treatment that is available in a variety of publicly funded and private programs. The study data will update and augment the information available from earlier national studies, such as the DRUG ABUSE REPORTING PROGRAM (DARP) study, which began in the late 1960s, and the TREATMENT OUTCOME PROSPECTIVE STUDY (TOPS) of clients entering treatment in the 1970s. The study is sponsored by the NATIONAL INSTITUTE ON DRUG ABUSE (NIDA) and conducted by the Research Triangle Institute (RTI).

The major objective of DATOS is to examine the effectiveness of drug abuse treatment by conducting a multisite, prospective clinical and epidemiological longitudinal study of drug abuse treatment. Effectiveness is being examined by using data from client interviews conducted from 1991 to 1993 at entry to treatment each three months, during treatment, and one year after leaving treatment. Interviews of clients at admission are being supplemented with comprehensive clinical assessments of psychological, social, and physical impairments in addition to drug and alcohol dependence. Treatment outcomes will be compared for clients who are entering treatment with varied patterns of drug abuse and levels of psychosocial impairment and who are undergoing varied types and lengths of treatment.

A secondary objective is to investigate the drug-abuse treatment itself. A detailed examination is being conducted of the treatments and services available and provided to each client, how these treatments and services are delivered to the client, and how the client responds to treatment in terms of cognitive and behavioral changes.

The study population includes 1,540 clients from 29 outpatient methadone programs; 2,774 clients from long-term residential or THERAPEUTIC COMMUNITY programs; 2,574 clients from 36 outpatient drug-free programs; and 3,122 clients from short-term, inpatient, or chemical dependency programs. In addition, we are comparing the clients' treatment and outcomes for DATOS with those for TOPS a decade earlier and for DARP two decades earlier in order to determine how drug abuse treatment programs have changed and what the changes imply for the provision of effective treatment approaches and services.

The DATOS research builds upon and expands the knowledge generated from previous research on the effectiveness of treatment. Several events, however, have necessitated a continuing nationally based multisite study of drug abuse treatment and treatment effectiveness. Major changes have occurred in the nation's drug-abusing population and treatment system. The OMNIBUS Recommendation Act of 1981 shifted the administration of treatment programs from the federal government to the states. The AIDS epidemic has intensified interest in drug-abuse treatment as a strategy to reduce exposure to the HUMAN IMMUNODEFICIENCY VIRUS (HIV), which causes AIDS. COCAINE use rather than OPIOID use was the major drug problem of the late 1980s and early 1990s. Efforts to contain health-care costs may dramatically transform both the public and private treatment systems. It has therefore become necessary to update information so as to reexamine what we have learned about treatment effectiveness, and so as to augment the types of data that are available for exploring new issues about the nature, effectiveness, and costs of treatment approaches.

The initial comparison of data from the DATOS and from the TOPS shows the following about the clients in DATOS: They are older, a greater percentage of them are women, they have more years of education, more of them are married, fewer are fully employed, and a lower percentage of them report that they have considered or attempted suicide. A higher proportion of the clients on METHADONE are entering treatment for the first time, and the proportion of criminal-justice referrals is higher in long-term residential and outpatient drug-free programs than they are in short-term inpatient programs.

In all types of programs, cocaine abuse predominates in the early 1990s, compared to heroin abuse in the past, but the cocaine use is usually combined with extensive use of ALCOHOL. Multiple abuse of psychotherapeutic agents has decreased, so less than 10 percent of clients report that they regularly use these agents as opposed to treatment services. Outpatient programs have less early dropouts, but this may reflect better screening and longer, more extensive intake processes. The influence of cost-containment measures and managed care became evident with shortened durations of treatment for short-term inpatients. Short-term inpatient programs also admit more public sector patients than in the 1980s. Early analyses from the DATOS indicate that rates of drug use toward the start of outpatient treatment are two to three times higher than those that were found in TOPS.

The early results of DATOS also show that clients entering drug treatment are a diverse group who have multiple problems. Two-thirds of the clients are men, and approximately half have previously been in treatment. Those who have health insurance that covers treatment are, by far, in the minority. Although the clients entering short-term inpatient chemical-dependency programs appear to have a higher rate of private insurance coverage (40 percent) than any other classification of clients, their rate is considerably lower than that observed among the same type of clients in the 1980s. Depending on the type of treatment, 25 to 50 percent of clients reported predatory criminal activity in the previous year, and less than 20 percent were fully employed. The clients have a variety of health problems, and many report significant psychiatric impairment. Few have received mental health services, however, and about one in every three clients report that they use emergency rooms as their primary health-care provider. Taken together, these data indicate that most clients have deficits in many areas of their lives and have multiple needs in addition to those directly related to their drug abuse (e.g., medical services and vocational needs).

Patterns of drug use vary markedly by type of treatment in DATOS clients. Although cocaine is the most frequently cited drug of abuse, most clients abuse multiple drugs and exhibit complex patterns of drug use. Frequent alcohol use is also common among many of the clients, as is weekly use of marijuana. Multiple abuse of psychotherapeutic agents is reported by less than 10 percent of clients.

Drug-treatment programs are focusing on providing drug-counseling services to meet the multiple problems of clients, but fewer specialized services, such as medical or psychological services, are being provided to meet clients' other needs. Only a third to a half of clients who report a need for medical services are receiving them, and the situation is much worse in regard to psychological, family, legal, employment, and financial services. Less than 10 percent of clients who report a need actually receive the service while they are in methadone and outpatient drug-free programs. The percentage of clients who receive specialized services other than drug counseling (e.g., medical or psychological attention) has declined dramatically since the mid-1980s. The impact of cost-containment measures and managed care is evident in the shorter stays of clients, particularly those enrolled in short-term inpatient programs.

Limited information on treatment outcomes has provided a mixed indication of the outcomes of treatment. The combination of more severe impairment and less extensive services suggests the potential for poorer outcomes. On the positive side, clients in treatment are being retained in treatment. However, compared to earlier findings, findings from the early 1990s indicate that a higher percentage of clients are actively using drugs during the first months of treatment.

Two other studies have been included under the DATOS program of research. The Drug Abuse Treatment Outcome Study–Adolescent (DATOS–Adolescent) research study is designed to examine the effectiveness of drug abuse treatment for adolescents through a multisite prospective longitudinal study of youth entering treatment programs that focus on adolescents. Effectiveness will be examined by using

interview data from youth under 18 supplemented by interviews with parents or guardians conducted at entry to treatment, during treatment, and one year after leaving treatment. Treatment outcome will also be assessed by using such measures as changes in the use of the primary problem drug; the use of other drugs; antisocial, delinquent, or criminal behavior; school attendance and achievement; vocational training and employment; family and social functioning; and treatment retention. A secondary objective of DATOS–Adolescent is to investigate the drug-abuse treatment that adolescents receive. A sample of thirty long-term residential, outpatient drug-free, and short-term inpatient programs will be used to accomplish these objectives. The proposed sample design will include three thousand clients.

The Early Retrospective Study of Cocaine Treatment Outcomes is an accelerated retrospective study of clients with a primary diagnosis of cocaine dependence who had been admitted to DATOS–Adult programs prior to or in the early stages of DATOS. The research will provide data about outcomes for cocaine abusers during the first year of treatment, describe the treatment received by these clients through a study of treatment process, and establish a client data base for future follow-up studies. A sample of 2,000 records for cocaine-abusing clients discharged from residential, hospital-based, and outpatient nonmethadone programs will be abstracted to obtain baseline data. A sample of 1,200 face-to-face, follow-up interviews will be completed, after 12 months of treatment, with the discharged clients whose records were reviewed during the record-abstraction phase of the project.

Data analyses of the project will be targeted at describing the posttreatment outcomes for cocaine abusers by detailing their treatment experiences and investigating their posttreatment experiences. This description will include the type, intensity, and duration of services received and an examination of the interrelationships between client and treatment characteristics and posttreatment outcomes. Along with cocaine use, other outcomes that will be considered include the use of drugs other than cocaine, economic functioning, illegal activities, and psychological status. Analytic methods will include univariate and descriptive statistics as well as multivariate methods. The collection of follow-up data for this retrospective study will be conducted simultaneously with the collection of data for the twelve-month follow-up of DATOS clients. A close coordination with the DATOS–Adult is designed to permit comparison across studies.

(SEE ALSO: *Drug Abuse Reporting Program; Methadone Maintenance Programs; Opioid Dependence; Treatment, History of; Treatment Outcome Prospective Study; Treatment Types*)

ROBERT HUBBARD

DRUG ABUSE WARNING NETWORK (DAWN)

This is a nationwide system of hospitals and medical examiners that report on emergency-room cases and deaths that are related to drug abuse. The DAWN system includes two components—an emergency room component and a medical examiner component. The emergency room component tracks drug emergencies in a sample of 24-hour, nonfederal, short-term, general hospitals—representative of the nation as a whole—plus a sample of hospitals in twenty-one major metropolitan areas. The medical examiner component consists of a sample of medical examiners' reports for twenty-seven metropolitan areas.

The data collected by the DAWN system represent one of the most widely used national indicators of drug abuse—frequently used by researchers and policymakers to determine the nature and extent of medical consequences of drug use nationally and in the participating metropolitan areas. Although the data are widely used to monitor patterns and trends of drug abuse, DAWN should be used with caution. The DAWN data represent information only on individuals who enter an emergency room because of their drug use, or those whose deaths are determined to be drug-related. Therefore, the data reflect only the most serious cases of drug abuse. Consequences of drug use that are less severe are not represented in the data. In addition, analysis of the data requires familiarity with the types of cases reportable to DAWN. For example, the emergency-room system contains data not only on OVERDOSE cases, but on individuals seeking DETOXIFICATION in an emergency room or those suffering from chronic effects of drug use—situations that do not necessarily require emergency treatment. Therefore, DAWN cases may reflect other phenomena than medical consequences (e.g., changes in nonhospital medical treatment availability).

HISTORY

DAWN was created in 1972 by the U.S. Department of Justice, DRUG ENFORCEMENT ADMINISTRATION (DEA), as a surveillance system for new drugs of abuse. Since 1980, the DAWN system has been sponsored by the NATIONAL INSTITUTE ON DRUG ABUSE (NIDA) and, more recently, the Office of Applied Studies of the SUBSTANCE ABUSE AND MENTAL HEALTH SERVICES ADMINISTRATION (SAMHSA), which currently oversees it.

The Emergency-Room System. At its outset, the DAWN emergency-room system consisted of a sample of hospitals in twenty metropolitan areas and

TABLE 1
DAWN: Weighted Emergency Room Estimates (Drugs mentioned most frequently)

Rank	Drug Name	Number of Mentions	Percent of Total Episodes	Rank	Drug Name	Number of Mentions	Percent of Total Episodes
1	Alcohol-in-combination	123,758	30.93	39	Erythromycin	2,931	0.73
2	Cocaine	102,727	25.68	40	Theophylline	2,878	0.72
3	Heroin/Morphine	36,576	9.14	41	Thioridazine	2,715	0.68
4	Acetaminophen	30,885	7.72	42	Chlordiazepoxide	2,699	0.67
5	Aspirin	21,982	5.49	43	Methadone	2,680	0.67
6	Marijuana/Hashish	16,492	4.12	44	Amphetamine	2,331	0.58
7	Alprazolam	16,465	4.12	45	Butalbital Combinations	2,285	0.57
8	Ibuprofen	15,628	3.91	46	Caffeine	2,257	0.56
9	Diazepam	14,852	3.71	47	Benztropine	2,248	0.56
10	Amitriptyline	8,785	2.20	48	Hydroxyzine	2,247	0.56
11	D-Propoxyphene	7,919	1.98	49	Pseudoephedrine	2,123	0.53
12	Acetaminophen w Codeine	7,236	1.81	50	Meperidine HCl	2,012	0.50
13	Lorazepam	7,009	1.75	51	Desipramine	1,978	0.49
14	Fluoxetine	6,954	1.74	52	Codeine	1,842	0.46
15	Diphenhydramine	6,836	1.71	53	Propanolol HCl	1,826	0.46
16	Clonazepam	6,563	1.64	54	Amoxicillin	1,772	0.44
17	OTC Sleep Aids	6,434	1.61	55	Chlorpromazine	1,709	0.43
18	Hydrocodone	5,089	1.27	56	Amitriptyline Combs.	1,488	0.37
19	Methamphetamine/Speed	4,980	1.24	57	Clorazepate	1,473	0.37
20	Unspec. Benzodiazepine	4,937	1.23	58	Thiothixene	1,470	0.37
21	Lithium Carbonate	4,568	1.14	59	Methocarbamol	1,249	0.31
22	Trazodone	4,316	1.08	60	Insulin	1,208	0.30
23	Carisoprodol	4,291	1.07	61	Fluphenazine HCl	1,163	0.29
24	Oxycodone	4,001	1.00	62	Trifluoperazine	1,141	0.29
25	LSD	3,912	0.98	63	Trimethoprim/Sulfamethox	1,127	0.28
26	Doxepin	3,791	0.95	64	OTC Diet Aids	1,065	0.27
27	PCP/PCP Combinations	3,492	0.87	65	Butabarbital Combination	1,059	0.26
28	Naproxen	3,473	0.87	66	Perphenazine	1,053	0.26
29	Imipramine	3,440	0.86	67	Penicillin G Potassium	1,020	0.25
30	Carbamazepine	3,435	0.86	68	Ampicillin	979	0.24
31	Triazolam	3,411	0.85	69	Ephedrine	955	0.24
32	Haloperidol	3,218	0.80	70	Cephalexin	905	0.23
33	Hydantoin	3,194	0.80	71	Metoprolol	884	0.22
34	Temazepam	3,189	0.80	72	Doxycycline	851	0.21
35	Cyclobenzaprine	3,138	0.78	73	Promethazine	833	0.21
36	Nortriptyline	3,066	0.77	74	Tetracycline HCl	828	0.21
37	Phenobarbital	3,062	0.77	75	Brompheniramine Maleate	812	0.20
38	Flurazepam	3,047	0.76	76	Cimetidine	791	0.20

SOURCE: SAMHSA, Drug Abuse Warning Network.

a panel of hospitals representative of the nation as a whole. In subsequent years, however, the sample became nonrepresentative, as changes in hospital participation skewed both the national sample and the metropolitan samples. Thus, data collected by the DAWN system before 1989 measure drug-related cases only for those hospitals included in the sample; actual estimates of the number of drug deaths and hospital emergencies in the metropolitan areas or across the nation are not available for those years (1972–1988). The primary utility of the data prior to 1989, then, is for examining trends in drug emergencies and deaths in the participating hospitals over time, rather than deriving estimates of actual numbers of cases in the United States.

Beginning in 1989, NIDA instituted a nationally representative sample as well as representative samples from twenty-one metropolitan areas. This sample design, which currently includes hospitals from twenty-one metropolitan areas and the national panel of hospitals, corrects the limitations of the pre-1989 sample. The new design allows for inferences about the actual number of drug-abuse episodes in the nation and in the oversampled metropolitan areas.

The Medical-Examiner System.   Initially, the DAWN medical-examiner system consisted of a sample of medical examiners in thirteen metropolitan areas. The system currently consists of samples of medical examiners in twenty-seven metropolitan areas. Unlike the emergency-room system, the medical-examiner system does not have a national sample, and so estimates of nationwide drug-related deaths are not possible with this system. Therefore, the primary utility of the medical-examiner data is for analysis of trends in the various metropolitan areas from year to year.

DATA COLLECTION

Data for both the hospital emergency rooms and the medical examiners are collected weekly. In each participating facility, a reporter, usually a staff member of the hospital or medical examiner's office, is assigned to conduct data collection. The reporter reviews each record for the time period and records demographic and drug-related information for each case appropriate for inclusion in the DAWN system. A report for each case is then submitted to SAMHSA for data entry.

The following criteria are used in determining whether a case is reportable to DAWN. For each record, all of the following criteria must be met:

Emergency-Room Cases:

1. The patient was treated in the hospital's emergency department.
2. The patient's presenting problem(s) had been induced by or related to drug abuse, regardless of whether the drug ingestion occurred minutes or hours before the visit.
3. The case involved the nonmedical use of a legal drug or any use of an illegal drug.
4. The patient's reason for taking the substance(s) included one of the following: (1) DEPENDENCE, (2) SUICIDE attempt or gesture, or (3) psychic effects.

Medical-Examiner Cases:

1. The death was induced by or related to drug abuse, regardless of whether the drug ingestion occurred minutes or months before the death and regardless of manner of death.
2. The case involved the nonmedical use of a legal drug or any use of an illegal drug.
3. The decedent's reason for taking the substance(s), if known, included one of the following: (1) dependence, (2) suicide, (3) psychic effects, or, if the death was a homicide, (4) the homicide was induced by or directly related to use of a drug/substance(s).

Up to six substances can be reported for any drug-abuse death or emergency-room case. Any licit or illicit drug is reported to DAWN, but alcohol is reportable only when it is used in combination with another licit or illicit drug (U.S. Department of Health and Human Services, *Drug Abuse Warning Network: Instruction Manual for Hospital Emergency Departments*, July 1991; *Drug Abuse Warning Network: Instruction Manual for Medical Examiners*, July 1991; Washington, DC: U.S. Government Printing Office).

TRENDS IN DRUG-RELATED EMERGENCIES AND DEATHS, 1990–1991

Drug-related Emergencies. In 1991, there were an estimated 400,079 drug-abuse episodes in emergency rooms in the United States: 48 percent of

emergency-room episodes were for males and 51 percent for females; 56 percent were white, 27 percent black, and 8 percent Hispanic.

Suicide was the motive for drug use for the highest proportion of emergency-room visits, representing 44 percent of all visits. Dependence was the motive for use for 29 percent of emergency room visits; obtaining psychic effects was the motive for use for 15 percent of the visits.

When analyzed by reason for contact in the emergency room, the data show that overdoses (ODs) were responsible for the majority of episodes (57%), while the remainder of episodes were due to another cause (chronic effects of use, unexpected reactions from use, seeking detoxification, ACCIDENT/INJURY, or WITHDRAWAL).

The largest number of episodes (123,758, or 31% of all episodes) were due to use of ALCOHOL in combination with other drugs; 102,727 (26%) were due to COCAINE use, and 36,576 (9%) were due to use of HEROIN/MORPHINE.

The number of drug-abuse emergencies rose 8 percent from 1990 to 1991, from 371,208 to 400,079. The number of episodes for males rose faster than for females, rising 11 percent over the period. The number of emergency-room episodes for blacks rose faster than for any other racial or ethnic group, rising 23 percent over the period. The number for whites rose 4 percent, and the number for Hispanics rose 13 percent. Emergencies for those over age 25 rose significantly over the time period, while there was no significant change in episodes for those 25 and under.

The number of emergency-room episodes rose significantly for those using drugs for psychic effects or for dependence but did not change significantly for those attempting suicide. The number admitted for overdoses did not change significantly, but those admitted suffering chronic effects of use or unexpected reactions, or because of an accident or injury, did rise significantly (see Figure 1).

Emergency-room episodes involving cocaine rose 28 percent from 1990 to 1991, while episodes involving alcohol-in-combination rose 7 percent (see Figure 2). The number of heroin/morphine episodes did not change significantly over the period. However, the 1991 to 1992 data indicate that the number of heroin/morphine episodes increased significantly—34 percent.

Drug-Related and Drug-Induced Deaths. In 1991, a total of 6,601 drug abuse deaths were re-

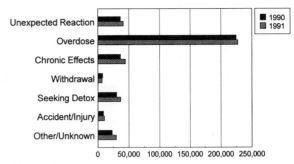

Figure 1

Emergency Room Drug-Related Episodes, U.S., by Reason for Emergency Room Visit, 1990–1991

SOURCE: National Institute on Drug Abuse: Annual Emergency Room Data, 1991. Statistical Series I, Number 11-A. Rockville, MD. (Estimates based on a sample of hospitals.)

ported to DAWN by participating hospitals. Accidents were the cause of the majority of deaths (55%) while suicide represented 23 percent of deaths. The manner of death for the remaining 22 percent was unknown.

In 1991, 4,242 (64%) of deaths were due to an overdose (see Figure 3); most of these involved use of more than one drug; 2,298 deaths (35%) were drug-related, meaning that drug use was a contributory factor in the death, but the death was not due to an overdose.

Heroin/morphine was involved in 2,333 deaths (35% of reported deaths). Cocaine was involved in

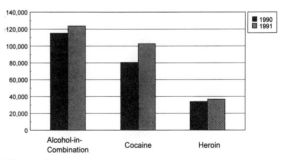

Figure 2

Emergency Room Drug-Related Episodes, U.S., by Drug Involved, 1990–1991, Top Three Reported Drugs

SOURCE: National Institute on Drug Abuse, Annual Emergency Room Data, 1991. Series I, Number 11-A. Rockville, MD. (Estimates based on a sample of hospitals.)

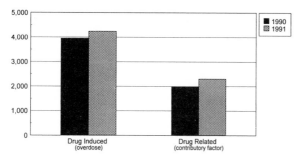

Figure 3
Drug-Abuse Deaths Reported to DAWN, U.S., by Cause of Death, 1990–1991

SOURCE: National Institute on Drug Abuse: Annual Medical Examiner Data, 1991. Statistical Series I, Number 11-B. Rockville, MD.

3.020 deaths (46%), and alcohol, when used in combination with at least one other drug, was involved in 2,436 deaths (37%).

From 1990 to 1991, the number of reported drug-abuse deaths rose 10 percent, from 5,984 to 6,601. The number of deaths for males rose 11 percent and for females 8 percent. The number of deaths for whites rose 14 percent, for blacks 19 percent, but for Hispanics dropped 15 percent. In the same time period, the number of deaths for those aged 6 to 17 rose higher than any other age group, although the numbers were small (rising from 58 to 70 deaths reported). Deaths among those 18 to 25 rose 13 percent, among those 26 to 34, 6 percent, and among those 35 and older, 12 percent.

From 1990 to 1991, the number of drug-related deaths (deaths in which drug abuse was a contributory factor but for which overdose was not the cause of death) rose faster than the number of drug-induced deaths (deaths from an overdose). Drug-related deaths rose to 16 percent, while drug-induced deaths rose 7 percent.

Cocaine deaths rose 20 percent from 1990 to 1991; alcohol-in-combination deaths rose 4 percent; and heroin/morphine deaths rose 16 percent (see Figure 4).

(SEE ALSO: *Abuse Liability of Drugs; Complications: An Overview; Drug Interactions and Alcohol; Epidemiology of Drug Abuse; National Household Survey on Drug Abuse*)

BIBLIOGRAPHY

COLLIVER, J. D. (1991). Characteristics and implications of the new DAWN emergency room sample. In *Epidemiologic trends in drug abuse, Community Epidemiology Workgroup Proceedings, July*. Rockville, MD: National Institute on Drug Abuse.

NATIONAL INSTITUTE ON DRUG ABUSE. (1991). *Statistical series: Annual emergency room data, 1991*. Data from the Drug Abuse Warning Network. Series I, Number 11-A. Rockville, MD: Author.

NATIONAL INSTITUTE ON DRUG ABUSE. (1991). *Statistical Series: Annual medical examiner data, 1991*. Data from the Drug Abuse Warning Network. Series I, Number 11-B. Rockville, MD: Author.

CLARE MUNDELL

DRUG ADDICTION *See* Addiction: Concepts and Definitions; Disease Concept of Alcoholism and Drug Abuse.

DRUG CARTEL *See* Colombia As Drug Source.

DRUG CONTROLS *See* Controls: Scheduled Drugs/Drug Schedules, U.S.

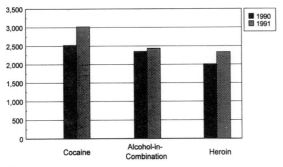

Figure 4
Drug-Abuse Deaths Reported to DAWN, U.S., by Drug Involved in Death, 1990–1991, Top Three Reported Drugs

SOURCE: National Institute on Drug Abuse: Annual Medical Examiner Data, 1991. Statistical Series I, Number 11-B. Rockville, MD.

DRUG COUNSELOR   *See* Professional Credentialing.

DRUG ENFORCEMENT ADMINISTRATION   *See* U.S. Government.

DRUG-FREE SCHOOLS   *See* Parents Movement; Prevention Movement.

DRUG INTERACTION AND THE BRAIN

When two or more drugs are taken at the same time, complex interactions may occur. Drugs can interact to change biological functions within the body through PHARMACOKINETIC or PHARMACODYNAMIC mechanisms or through their combined toxic effects. Changes in the pharmacokinetic properties of a drug can include changes in absorption, distribution, metabolism, and excretion of the drug, and each of these can affect blood and plasma concentrations and, ultimately, brain levels of the drug. Although a change in the speed at which a drug reaches the bloodstream is rarely clinically relevant, a change in the amount of drug absorbed can be important, because this can lead to changes in the plasma levels of the drug, which, in turn, can influence the amount of drug that reaches the brain.

The distribution of a drug throughout the body can be affected by changes in the binding of the drug to proteins in the bloodstream or by displacing the drug from tissue binding sites, both of which can affect the plasma concentration of the drug and potentially affect the amount of drug that reaches the brain. Drug metabolism can be either stimulated or inhibited, resulting in decreased or increased plasma concentrations of the drug, respectively. The stimulation (induction) of drug-metabolizing enzymes in the liver can be produced by drugs such as the BARBITURATES, but a week or more is often required before maximal effects on drug metabolism are observed. As drug metabolism increases, the amount of drug available to enter the brain decreases.

The inhibition of drug metabolism often occurs much more rapidly than the stimulation, usually as soon as a sufficient concentration of the metabolic inhibitor is achieved, which results in increased plasma and brain concentrations of the drug. The renal (kidney) excretion of drugs that are weak acids or weak bases can be influenced by drugs that alter urinary pH to change the reabsorption of the drug from urine into the kidney. The active secretion of the drug into the urine can also be affected. Both processes can ultimately affect the plasma and subsequent brain concentrations of the drug. Pharmacodynamic mechanisms can either enhance or reduce the response of a given drug. For example, if two drugs are agonists for the same receptor site—DIAZEPAM (Valium) and CHLORDIAZEPOXIDE (Librium) for BENZODIAZEPINE receptor-binding sites—then an additive biological response is likely to occur unless a maximum response is already present. If, however, an AGONIST competes with an ANTAGONIST for the same binding site (e.g., see MORPHINE and NALOXONE in Opioids, discussed below), then a decreased biological response is likely.

Enhanced or diminished biological responses can be observed even if the drugs do not interact with the same receptor-binding sites. In this case, the net effect is the sum of the pharmacological properties of the drugs. For example, if two drugs share a similar biological response (e.g., central nervous system depression) even though they produce their effects at different sites, then the concurrent ingestion of both drugs can result in an enhanced depression of the central nervous system (see the Alcohol [ethanol] and Valium discussion below). Finally, the concurrent ingestion of two or more drugs, each with toxic effects on the same organ system, can increase the chance for extensive organ damage.

DEPRESSANTS

Alcohol (Ethanol) and Valium.   Reactions that are additive (combined) or synergistic (cooperative effects greater than the sum of the independent effects of the drugs taken alone) are common side effects that result from the consumption of two or more drugs with similar pharmacological properties. For example, although ALCOHOL (ethanol) is considered by many to be a stimulant drug because, typically, it releases an individual's latent behavioral inhibitions (i.e., it produces disinhibition), alcohol actually produces a powerful depression of the central nervous system similar to that seen with general anesthetics. The subsequent impairment of muscular coordination and judgment associated with alco-

hol intoxication can be enhanced by the concurrent administration of other central nervous system depressants. Often, Valium or Librium (benzodiazepines that are considered relatively safe drugs) may be purposely ingested along with ethanol in an attempt to "feel drunk" faster or more easily. Since ethanol actually increases the absorption of benzodiazepines, and also enhances the depression of the central nervous system, the potential toxic side effects of the two drugs are augmented. Ethanol is often a common contributor to benzodiazepine-induced coma as well as to benzodiazepine-related deaths, demonstrating that interactions of these drugs with alcohol can be especially serious. Furthermore, the combination of alcohol with the SEDATIVE-HYPNOTIC BARBITURATES (e.g., pentobarbital, secobarbital) can also produce a severe depression of the central nervous system, with decreased respiration (shallow breathing to none). In fact, the intentional ingestion of ethanol and secobarbital (or Valium) is a relatively common means of SUICIDE ("snuffing").

Alcohol (Ethanol) and Opioids. Alcohol can also enhance the respiratory depression, sedation, and hypotensive effects of MORPHINE and related OPIOID drugs. Therefore, the concurrent ingestion of the legal and socially acceptable drug ethanol with other sedatives, hypnotics, anticonvulsants, ANTIDEPRESSANTS, antianxiety drugs, or with an ANALGESIC agent such as morphine can result in serious and potentially fatal drug interactions through a potentiation of the depressant effects of these drugs on the central nervous system. Since the 1960s, a significant number of musicians, actors, and other high-profile personalities have either accidentally or intentionally overdosed from the combination of alcohol and other central nervous system depressants. A few notable examples include Marilyn Monroe, Jimi Hendrix, Janis Joplin, Jim Morrison, Keith Moon, and John Bonham, to name a few.

STIMULANTS

Stimulants and Toxic Effects. Synergistic toxic effects are also often observed with other classes of drugs. For example, the concurrent ingestion of central nervous system stimulants (e.g., AMPHETAMINE, COCAINE, CAFFEINE) can also produce additive side effects, especially with respect to toxic reactions in-

volving the heart and cardiovascular system. These toxic reactions are often manifested as irregular heartbeats, stroke, heart attacks, and even death. Sometimes, however, drugs with apparently different mechanisms of action can result in dangerous and unexpected synergistic side effects with fatal consequences. For example, some amphetamine and cocaine users often attempt to self-medicate their feelings of "overamp," or the excessive STIMULANT high resulting from prolonged central nervous system stimulation, through the concurrent administration of central nervous system depressants such as alcohol, barbiturates, or heroin (i.e., a "speedball"). The rationale behind this potentially dangerous practice is that a few beers or a Quaalude or perhaps a shot of heroin will help the individual "mellow out" for a while before inducing a stimulant high again. High doses of cocaine or amphetamine can, however, result in respiratory depression from actions on the medullary respiratory center. Therefore, the concurrent ingestion of a central nervous system stimulant (e.g., cocaine) with a depressant (e.g., heroin) can result in increased toxicity or death from the enhanced respiratory depression produced by the combination of the two drugs. The most well-known casualty from this type of pharmacological practice was the comedian John Belushi.

CLINICAL USES

The principles of drug interactions can be used clinically for the treatment of acute INTOXICATION and for WITHDRAWAL—by transforming, reducing, or blocking the pharmacological properties and/or the toxic effects of drugs used and abused for nonmedical purposes. Although these interactions often involve a competition with the abused drug for similar central nervous system RECEPTOR sites, other mechanisms are also clinically relevant.

Disulfiram and Alcohol (Ethanol). One such nonreceptor-mediated interaction involves DISULFIRAM (Antabuse) and ethanol (alcohol). Since an ethanol-receptor site has not yet been conclusively identified, specific receptor agonists and antagonists are not yet available for the treatment of ethanol intoxication, withdrawal, and abstinence (as they are for opioids). Disulfiram is sometimes used in the treatment of chronic ALCOHOLISM, although the drug does not cure alcoholism; rather, it interacts with ethanol in such a way that it helps to strengthen an

individual's desire to stop drinking. Although disulfiram by itself is relatively nontoxic, it significantly alters the intermediate metabolism of ethanol, resulting in a five- to tenfold increase in plasma acetaldehyde concentrations. This acetaldehyde syndrome results in vasodilatation, headache, difficulty breathing, nausea, vomiting, sweating, faintness, weakness, and vertigo. All of these reactions are obviously unpleasant, especially at the same time, thus well worth avoiding. The acetaldehyde syndrome therefore helps to persuade alcoholics to remain abstinent, since they realize that they cannot drink ethanol for at least three or four days after taking disulfiram. The consumption of even small or moderate amounts of ethanol following disulfiram pretreatment can result in extremely unpleasant drug interactions through the acetaldehyde syndrome, which makes the drink hardly worthwhile.

Opioids. Drug interactions involving opioids (morphine-like drugs) and opioid receptors are classic examples of how knowledge of the molecular mechanisms of the actions of a class of drugs can assist in the treatment of acute intoxication, withdrawal, and/or abstinence. Naloxone, the opioid-receptor antagonist, can be used as a diagnostic aid in emergency rooms. In the case of a comatose patient with unknown medical history, the intravenous administration of naloxone can provide information on whether or not the coma is the result of an opioid overdose. The antagonist competes with the agonist (usually heroin or morphine) for the opioid-receptor sites, displacing the agonist from the binding sites to reverse the symptoms of an overdose effectively and rapidly. Continued naloxone therapy and supportive treatment are often still necessary.

If, however, naloxone is administered to an individual dependent on opioids but not in a coma, a severe withdrawal syndrome develops within a few minutes and peaks after about thirty minutes. Depending on the individual, such precipitated withdrawal can be more severe than that following the abrupt withdrawal of the opioid-receptor agonist (e.g., heroin). In the former instance, the binding of the agonist to opioid receptors is suddenly inhibited by the presence of the antagonist (e.g., naloxone); even relatively large doses of the agonist (e.g., heroin) cannot effectively overcome the binding of the antagonist. Quite the contrary, respiratory depression can develop if higher doses of the agonist are administered. Therefore, opioid-receptor antagonists are not recommended for the pharmacological treatment of opioid withdrawal. Rather, longer acting, less potent, opioid receptor-agonists, such as METHADONE, are more commonly prescribed.

Methadone. The symptoms associated with methadone withdrawal are milder, although more protracted, than those observed with morphine or heroin. Therefore, methadone therapy can be gradually discontinued in some heroin-dependent people. If the patient refuses to withdraw from methadone, the person can be maintained on methadone relatively indefinitely. TOLERANCE develops to some of the pharmacological effects of methadone, including any reinforcing or rewarding effects (e.g., the euphoria or "high"). Therefore, the patient cannot attain the same magnitude of euphoria with continued methadone therapy, although the symptoms associated with opioid withdrawal will be prevented or attenuated. Furthermore, cross-tolerance also develops to other opioid drugs, so the patient will not feel the same high if heroin is again used on the street.

This type of maintenance program makes those heroin dependent more likely to accept other psychiatric or rehabilitative therapy. It also reduces the possibility that methadone patients will continue to seek heroin or morphine outside the clinic. Therefore, the principles of drug interactions involving opioid receptors in the central nervous system have helped to stabilize TREATMENT strategies for opioid withdrawal and abstinence.

(SEE ALSO: *Accidents and Injuries; Complications: Neurological; Drug Abuse Warning Network*)

BIBLIOGRAPHY

CLONINGER, C. R., DINWIDDIE, S. H., & REICH, T. (1989). Epidemiology and genetics of alcoholism. *Annual Review of Psychiatry 8*, 331–346.

CREGLER, L. L. (1989). Adverse consequences of cocaine abuse. *Journal of the National Medical Association 81*, 27–38.

KORSTEN, M. A., & LIEBER, C. S. (1985). Medical complications of alcoholism. In J. H. Mendelson & N. K. Mello (Eds.), *The diagnosis and treatment of alcoholism*, pp. 21–64. New York: McGraw-Hill.

REDMOND, D. E., JR., & KRYSTAL, J. H. (1984). Multiple mechanisms of withdrawal from opioid drugs. *Annual Review of Neuroscience 7*, 443–478.

NICK E. GOEDERS

DRUG INTERACTIONS AND ALCOHOL

The term *alcohol–drug interaction* refers to the possibility that alcohol may alter the intensity of the pharmacological effect of a drug, so that the overall actions of the combination of alcohol plus drug are additive, potentiated, or antagonistic. Such interactions can be divided into two broad categories—PHARMACOKINETIC and PHARMACODYNAMIC. Pharmacokinetics are concerned with the extent and rate of absorption of the drugs, their distribution within the body, binding to tissues, biotransformation (metabolism), and excretion. Pharmacokinetic interactions refer to the ability of alcohol to alter the plasma and tissue concentration of the drug and/or the drug metabolites, such that the effective concentration of the drug at its target site of action is significantly decreased or increased. Pharmacodynamics are concerned with the biochemical and physiological effects of drugs and their mechanisms of action. Pharmacodynamic interactions refer to the combined actions of alcohol and the drug at the target site of action, for example, binding to enzyme, receptor, carrier, or macromolecules. Pharmacodynamic interactions may occur with or without a pharmacokinetic component. For many drugs acting on the central nervous system, which exhibit cross-tolerance (a similar tolerance level) with alcohol, pharmacodynamic interactions with alcohol are especially notable and important.

Most drugs are metabolized in the liver by an enzyme system usually designated as the cytochrome P450 mixed-function oxidase system, and the liver is the principal site of many alcohol–drug pharmacokinetic interactions. Two major factors—blood flow to the liver and the activity of drug-metabolizing enzymes—strongly influence the overall metabolism of drugs. Biotransformation of drugs that are actively metabolized by liver enzymes mainly depends on the rate of delivery of the drug to the liver. These may be flow-limited drugs, where the liver can transform as much drug as it receives, or capacity-limited drugs, which have a low liver-extraction ratio—their clearance (removal from the blood) primarily depends on the rate of their metabolism by the liver.

ALCOHOL–DRUG INTERACTIONS

Alcohol–drug interactions are complex because acute and chronic alcohol treatment can have both pharmacokinetic and pharmacodynamic interactions with drugs. Alcohol–drug interactions are most important with drugs that have a steep DOSE-RESPONSE CURVE and a small therapeutic ratio—so that small quantitative changes at the target site of action lead to significant changes in drug action. In alcoholics, changes in susceptibility to drugs are due to changes in their rates of metabolism (pharmacokinetics) and the adaptive and synergistic effects on their organs, such as the central nervous system (pharmacodynamics). The clinical interactions of alcohol and drugs often appear paradoxical: Sensitivity to many drugs, especially sedatives and tranquilizers, is strikingly increased when alcohol is present at the same time; however, alcoholics, when abstinent, are tolerant to many drugs. These acute and chronic actions of alcohol have been attributed, respectively, to additive and adaptive responses in the central nervous system (pharmacodynamic interactions).

It is now recognized that alcohol can also interact with the cytochrome P450 drug-metabolizing system, binding to P450, being oxidized to acetaldehyde by P450, increasing the content of P450, and inducing (causing an increase in the activity of) a unique isozyme of P450. Inhibition of drug oxidation when alcohol is present at the active site of P450 is due to displacement of the drug by alcohol and competition for metabolism; this increases the half-life and circulating concentration of drugs. Induction of P450 by chronic-alcohol treatment can result in the increased metabolism of drugs, as long as alcohol is not present to compete for oxidation. These pharmacokinetic interactions may contribute to either increased sensitivity or the tolerance observed with alcohol–drug interactions.

Alcohol can affect drug pharmacokinetics by altering drug absorption from the alimentary tract. For example, diazepam absorption is enhanced by the effects of alcohol on gastric emptying. Alcohol placed in the stomach at concentrations of 1 percent to 10 percent increases the absorption of pentobarbital, PHENOBARBITAL, and theophylline, whereas drugs such as DISULFIRAM and CAFFEINE decrease alcohol absorption by decreasing gastric emptying. Cimetidine (Tagamet)—a drug used to treat stomach ulcers—increases blood alcohol concentrations by inhibiting stomach alcohol dehydrogenase (an enzyme that oxidizes alcohol to acetaldehyde) and first-pass metabolism of alcohol. Binding of a drug to plasma proteins will change the effective therapeutic level of the drug, because when the drug is linked to the proteins, it is not available to act on the tissue.

Alcohol itself and alcohol-induced liver disease cause a decreased synthesis and release of plasma proteins, such as albumin. The resulting hypoproteinemia can result in decreased plasma-protein binding of drugs such as quinidine, dapsone, triamterene, and fluorescein. Alcohol may also directly displace drugs from plasma proteins.

The effects of alcohol on hepatic blood flow are controversial, although most recent reports suggest an increase; this could be significant with respect to metabolism of flow-limited drugs. Alcohol, at higher concentrations, can act as an organic solvent and "fluidize" cellular membranes, which may increase uptake or diffusion of drugs into the cell.

METABOLISM

Many alcohol–drug interactions occur at the level of actual metabolism. Ethanol (ethyl alcohol—common in wines and liquors) will compete with other alcohols, such as METHANOL (methyl alcohol—called wood alcohol) or ethylene glycol (antifreeze), for oxidation via alcohol dehydrogenase. In fact, treatment against poisoning by methanol or ethylene glycol involves the administration of ethanol—as the competitive inhibitor—or the addition of inhibitors of alcohol dehydrogenase such as pyrazole or 4-methylpyrazole.

As discussed above, the presence of alcohol will inhibit the oxidation of drugs by cytochrome P450. Alcohol has been shown to inhibit oxidation of representative drugs such as aniline, pentobarbital, benzphentamine, benzpyrene, aminopyrine, ethylmorphine, METHADONE, meprobamate, phenytoin, propranolol, caffeine, tolbutamide, warfarin, phenothiazine, BENZODIAZEPINE, CHLORDIAZEPOXIDE, amitriptyline, chlormethiazole, chlorpromazine, isoniazid, imipramine, dextropropoxyphene, triazolam, industrial solvents, and acetaminophen. As this partial list indicates, oxidation of many classes of drugs can be inhibited in the presence of alcohol; these include HYPNOTICS, OPIOIDS, psychotropic drugs, anticonvulsants, vasodilators, antidiabetics, anticoagulants, ANALGESICS, and antibacterials. Chronic consumption of alcohol induces the P450 drug-metabolizing system, which could increase oxidation of drugs in sober or abstinent alcoholics. Among the drugs that may be more rapidly metabolized in abstinent alcoholics are ethoxycoumarin, ethylmorphine, aminopyrine, antipyrine, pentobarbital, meprobamate, methadone, theophylline, tolbutamide,

propranolol, rifamycin, warfarin, acetaminophen, phenytoin, deoxycline, and ethanol itself. An important consequence of this ability of chronic ethanol intake to increase drug-clearance rates is that the effective therapeutic level of a drug will be different in an abstaining alcoholic than it is in a nondrinker. This metabolic drug tolerance can persist for several days to weeks after alcohol WITHDRAWAL.

PHARMACODYNAMIC IMPLICATIONS

These alcohol–drug pharmacokinetic interactions can have major pharmacodynamic implications. Some examples include the following. The concurrent administration of alcohol plus amitriptyline to healthy volunteers resulted in an increase in the plasma-free concentration of amitriptyline, since the alcohol inhibited drug clearance. At peak concentrations of amitriptyline, mean postural sway was increased and short-term memory and alertness were decreased by the presence of alcohol. Other pharmacodynamic interactions between alcohol and amitriptyline include decreased driving skills (and other psychomotor skills), greater than additive loss of righting reflex, unexpected blackouts, and even death. Laisi et al. (1979) showed that plasma levels of the tranquilizer DIAZEPAM (Valium—an antianxiety drug) were increased in the presence of beer and wine, so the combination of alcohol plus diazepam produced impaired tracking skills, increased nystagmus (nodding off), and impaired oculomotor (eye) coordination, as compared to diazepam alone. The authors concluded that impairment of psychomotor skills was due primarily to pharmacodynamic interactions between alcohol and diazepam at the receptor level. Therapeutic doses of the tranquilizers diazepam or chlordiazepoxide (Librium) plus alcohol have been consistently shown to produce impairment of many mental and psychomotor skills; EEG (electroencephalogram) abnormalities could still be detected sixteen hours after administration of fluorazepam in the presence of alcohol to volunteers. Alcohol also decreases the rates of elimination of several benzodiazepines in humans. Phenothiazines and alcohol compete for metabolism by P450, resulting in the decreased clearance of chlorpromazine, for example, and enhanced sedative effects, impaired coordination, and a severe potentially fatal respiratory depression. Alcohol inhibits the metabolism of BARBITURATES, prolonging the time and increasing the concentration of these drugs in the

bloodstream, so that central nervous system interactions are intensified. In humans, alcohol doubles the half-life of pentobarbital; this is associated with a 10 to 50 percent lower concentration of barbiturate sufficient to cause death by respiratory depression, as compared to the lethal dose in the absence of alcohol. Striking pharmacokinetic and pharmacodynamic interactions occur between alcohol and the hypnotic drug CHLORAL HYDRATE—the so-called Mickey Finn or knockout drops. Alcohol inhibition of MORPHINE metabolism increases morphine accumulation, potentiates central nervous system actions, and increases the probability of death.

OTHER CONSEQUENCES

Pharmacokinetic interactions between alcohol and drugs also have important toxicological and carcinogenic consequences. The metabolism of certain drugs produces reactive metabolites; these are much more toxic than the parent compound. The induction of P450, especially the P4502E1 isozyme by alcohol, results in the increased activation of drugs and SOLVENTS to toxic reactive intermediates—such as carbon tetrachloride, acetaminophen, benzene, halothane, enflurane, COCAINE, and isoniazid. In a similar manner, procarcinogens—such as aflatoxins, nitrosamines, and aniline dyes—are activated to carcinogenic metabolites after alcohol induction of P4502E1. Since P4202E1 is localized largely in the perivenous zone of the liver cell, the increased activation of these toxins (and alcohol itself) after induction by alcohol may explain the preferential perivenous toxicity of several hepatotoxins, carcinogens, and alcohol itself.

(SEE ALSO: *Complications; Drug Interaction and the Brain; Drug Metabolism; Psychomotor Effects of Alcohol and Drugs*)

BIBLIOGRAPHY

DEITRICH, R. A., & PETERSEN, D. R. (1981). Interaction of ethanol with other drugs In B. Tabakoff, P. B. Sutker, and C. L. Randall (Eds.), *Medical and social aspects of alcohol abuse.* New York: Plenum Press.

LAISI, U., ET AL. (1979). Pharmacokinetic and pharmacodynamic interactions of diazepam with different alcoholic beverages. *European Journal of Clinical Pharmacology, 16,* 263–270.

LIEBER, C. S. (1990). Interaction of ethanol with drugs, hepatotoxic agents, carcinogens and vitamins. *Alcohol and Alcoholism, 25,* 157–176.

LIEBER, C. S. (1982). Medical disorders of alcoholism: Pathogenesis and treatment. Philadelphia: W. B. Saunders.

RUBIN, E., & LIEBER, C. S. (1971). Alcoholism, alcohol and drugs. *Science, 172,* 1097–1102.

ARTHUR I. CEDERBAUM

DRUG INTERDICTION The interdiction of illicit drugs into the United States is the effort to seize them, together with the transport and/or persons that carry them on their way from the producing country to the importing country; many of the SEIZURES occur just as the drugs are brought across the border. The principal drugs subject to U.S. interdiction are COCAINE and MARIJUANA, both of which are imported primarily from LATIN AMERICA. The United States, uniquely among modern nations, has made interdiction a significant part of its effort to control the supply of drugs, at least for cocaine and marijuana, since about 1975. In addition to other federal agencies, it has also involved the military in this effort. Interdictors have seized large quantities of drugs, but there have been numerous questions about the effectiveness of the program as a method of reducing the use of drugs, particularly cocaine.

GOALS

Interdiction has two general goals. The primary one is to reduce the consumption of specific drugs within the nation by making it more expensive and risky for smugglers to conduct their business. Drug seizures raise costs by increasing the amount that has to be shipped in order to ensure that a given quantity will reach the market. In addition to seizing drugs, an effective interdiction program will, among other things, raise the probability that a courier is arrested; then, in order to induce individuals to become couriers, smugglers will have to pay more to those who undertake the task. The higher fees, of course, raise smugglers' costs of doing business and thus the price they must charge their customers, the importers. Finally, the increased costs lead to a higher retail price and hence a lower consumption of the drug.

At one time it was thought that interdiction could impose a physical limit on the quantity of drugs available in this country. With a fixed supply available in the producing nations, each kilogram seized on its way to the United States would be one less kilogram available for consumption here. However, with ongoing experience, it is now generally accepted that production is expandable and that increased seizures can be compensated for with increases in production, although farmers may have to receive higher prices to provide greater production.

A second, more modest, general goal is to increase the difficulty of smuggling itself and to provide suitable punishment. Smugglers, or at least the principals in smuggling organizations, are among the most highly rewarded participants in the drug trades. There is support for programs that conspicuously make their lives less easy and that subject them to the risks of punishment.

Three illegal drugs have traditionally dominated imports: cocaine, HEROIN, and marijuana. Heroin is subject to only modest interdiction efforts, because it is usually smuggled in conventional commercial cargo, or it is carried on (or within) the person of the smugglers who travel by commercial traffic; seizures are made only in the course of routine inspection of cargo and traffic. Some 10 tons of heroin are smuggled into the United States each year, and seizures of more than 10 kilograms are rare. Cocaine and marijuana have been the primary targets of interdiction, although an effective program of interdiction against Colombian maritime smuggling has led to a sharp rise in the share of the U.S. marijuana market served by domestic producers.

TECHNIQUES

The techniques of interdiction inevitably mirror those of smugglers. Drugs enter the United States by air, land, and sea, by private vessel and commercial carrier. Interdiction must, if it is to have any substantial effect on the drug trade, act against all the modes of smuggling; otherwise smugglers will rely on the mode that is not subject to interdiction.

Interdiction has three separate elements: monitoring; detection and sorting; pursuit and apprehension. For example, U.S. COAST GUARD ships supported by an extensive radar system patrol the Caribbean, which constitutes the major thoroughfare for smuggling from Latin America. The Coast Guard patrol vessels attempt to see, either directly or

through radar, all ships moving along certain routes. This constitutes the monitoring activity. The interdictors must then sort, from all that traffic, the relatively small number that are carrying illegal drugs. Finally, they must pursue the smugglers that have been detected, arrest the personnel, and seize the drugs and the ship itself. The interdiction system is as weak as its weakest component; for example, a system that has good pursuit capacities but is unable to sort smuggler from innocent effectively will waste much of that pursuit capacity in chasing nonsmugglers. Similarly, good detection will lead to few captures without effective monitoring capabilities.

In the late 1980s the U.S. government invested heavily in improving the monitoring capacity of the system, particularly through the creation of a "picket line" of tethered balloons carrying radars (aerostats) along the nation's southern border. The U.S. CUSTOMS SERVICE acquired a large fleet of planes to provide pursuit capacity in support of the aerostats; many of the planes were those seized from smugglers.

It is widely believed that the aerostats have been ineffective, though no meaningful evaluation has yet been conducted. The aerostats have a history of technical problems that render them inoperable for a significant fraction of the time. Even after three or more years of operation, some aerostats have yet to claim successful detection of a single smuggling flight, and it seems that smugglers have simply shifted to other routes and other modes that are not subject to this interdiction asset; they may also have successfully exploited the downtime of the aerostats. This experience points to a chronic systemic problem of interdiction—namely, that it is reactive and often involves investments in fixed assets that smugglers can render almost valueless with appropriate adaptations.

The Coast Guard and Customs Service share primary responsibility for marine and air interdiction. The Coast Guard patrols more distant routes, with Customs having a greater role in the U.S. coastal zone. Both agencies also conduct interdiction against private planes, with Customs having primary responsibility over the Mexican land border, a major trafficking area. The Border Patrol, a unit of the Immigration and Naturalization Service, has primary responsibility for the interdiction of drugs carried in cars or on persons crossing the land border.

Both Customs and the Border Patrol make many seizures and arrests in the course of routine inspec-

tion. For example, Customs may find a shipment of cocaine concealed inside a cargo container being unloaded in the Miami port; the Border Patrol, in the course of pursing illegal immigrants, might find a "mule" (a person) carrying a backpack full of cocaine or heroin. Drugs are shipped in an amazing array of forms; for example, suspended in frozen fruit pulp being imported from Ecuador or in hollowed lumber from Brazil.

MILITARY INVOLVEMENT

For a variety of reasons, there was pressure throughout the 1980s to increase the extent of military involvement in drug interdiction. The drug problem was viewed as a national crisis with an important international element. The military was seen as having unique capabilities, in terms both of equipment and of training, to protect the borders. Indeed, in 1988 the House of Representatives passed a bill requiring that the Department of Defense seal the borders against drug smuggling, within sixty days of the passage of the legislation; however, the Senate voted overwhelmingly against this bill, and it has not been introduced since.

The military has been ambivalent about entering the drug interdiction arena in any significant degree, seeing it as potentially corrupting and an inappropriate diversion from its primary mission. With the collapse of its principal strategic enemy, the Soviet Union, the U.S. military has become more willing to play a major interdiction role, though confined to detection and monitoring rather than pursuit and apprehension. Current law prohibits arrests by military personnel.

The U.S. Navy provides a number of ships for interdiction patrols in both the Caribbean and the Pacific, combining training with a useful mission. The military runs the integrated radar and communication system that links the Customs Service, the Coast Guard, the Border Patrol, and other agencies. There have been no reports of significant problems of corruption associated with the military role in drug interdiction, but relations between the military and the civilian law-enforcement agencies with primary jurisdiction have sometimes been strained, the result largely of differences in organizational cultures.

The element of the military most enthusiastic about participating in drug interdiction has been the National Guard. National Guard units routinely assist the Customs Service in searching container ships; in some border states in the early 1990s, this activity constituted a significant share of all National Guard time provided by the weekend volunteers who operate those units.

EVALUATION TECHNIQUES

Evaluation of the effectiveness of interdiction has been a vexed issue ever since the activity became prominent, in the late 1970s. Very large quantities of drugs, particularly of cocaine, have been seized, but the activity has been cited as evidence both of success and of failure. Is more cocaine being seized because interdictors are getting better at their job or because more cocaine is being shipped?

At a minimum, it would be desirable to express seizures as a fraction of total shipments (consumption plus seizures), but, unfortunately, estimates of consumption lack any systematic basis; indeed, the U.S. government has published only one such estimate in the last decade (Office of National Drug Control Policy, 1991). But even expressed as a fraction of shipments, seizures are clearly an inadequate measure of the effectiveness of interdiction, since the program imposes two other costs on smugglers— namely, seizure of assets (e.g., boats, planes, real estate, and financial holdings) and the arrest and imprisonment of smuggling agents (e.g., crew members on ships, pilots, couriers for financial transactions).

Reuter et al. (1988) suggest that the most appropriate measure is the price increase in the smuggling sector of drug distribution. Effective interdiction should raise smugglers' costs; the increase will be reflected in the difference between the price at which smugglers purchase drugs in the producer country (export price) and that at which they sell it in the importing country (import price). However, the process cannot serve as an operational criterion for any individual component of interdiction, since prices are set in a national market serviced by all modes and routes of smuggling. Anderberg (1992) concluded that the available data supported only inappropriate and/or inadequate measures of effectiveness, while any more cogent measure requires data that are not available and are not likely to be readily obtained.

One negative consequence of interdiction identified by Reuter et al. (1988) has received little attention. By seizing drugs on their way from the source country, interdiction may actually increase export demand for those drugs. As noted earlier, more strin-

gent interdiction has two effects; it raises prices and thus reduces final demand in the United States, but it also increases the amount that must be shipped to meet a given consumption (because of a higher replacement rate). It turns out that under reasonable assumptions about the cost structure of the cocaine trade, the second effect is greater than the first.

THE EFFECTIVENESS OF INTERDICTION

Interdiction clearly has had some important consequences for the drug trade in the United States. In contrast to the 1970s, little marijuana is now imported from COLOMBIA, though that nation remains a low-cost producer. Interdiction, particularly against marine traffic from Colombia, has imposed such high costs that now both Mexican and U.S. producers have come to dominate the U.S. market, despite the fact that the "farmgate" price in Mexico is 100 dollars per pound, compared with less than 10 dollars in Colombia. Interdiction against Mexican-produced drugs is more difficult and thus the import price of Mexican marijuana is less than that of Colombian. Marijuana has become an expensive drug in the United States, in part because of interdiction; marijuana consumption has declined sharply since the late 1970s.

For cocaine there is much less evidence of success. Interdictors have certainly forced changes in modes of smuggling. In the early 1980s much of the cocaine was brought up by private plane directly from Colombia, but now most of it seems to enter either by transshipment through MEXICO or by commercial cargo. However, though interdictors now seize a large share of all shipments, they have not managed to prevent a massive decline in the landed price of the drug, from 60,000 dollars per kilogram in 1981 to 20,000 dollars per kilogram in 1990.

The reasons for this limited success are not hard to find. Smugglers defray the risks of getting caught across very large quantities, so that the risks per unit smuggled are low. A pilot who charges 250,000 dollars for the risks (imprisonment, suffering injury or death in the course of landing) involved in bringing across a shipment of 250 kilograms is asking for only 1 dollar per gram, less than 1 percent of the retail price. Even if interdictors make smuggling much more risky, so that the pilot doubles the demand to 500,000 dollars, the higher fee still adds only another 1 percent to the retail price.

Moreover, it is difficult to make smuggling very risky when the nation is determined also to maintain the free flow of commerce and traffic. Hundreds of millions of people enter the country each year; cargo imports also amount to hundreds of millions of tons. Only a few hundred tons of cocaine need to be concealed in that mountain of goods and only a few thousand of those who enter need be in the smuggling business to ensure an adequate and modestly priced supply of cocaine.

Interdiction has accounted for between 20 and 30 percent of federal government expenditures on drug control, amounting to 2.9 billion dollars out of 12.7 billion dollars in fiscal year 1992, for example. If price is taken as the appropriate criterion for effectiveness, then the program can justify its current levels quite comfortably. It costs about 15,000 dollars per kilogram for a smuggler to deliver cocaine from Colombia to the United States. While this figure is barely one-seventh of the retail price of cocaine, a useful comparison is that the cost of sending the same kilogram by Federal Express would be less than 50 dollars. Interdiction has, however, been a chronic disappointment because it has promised so much more than it has delivered. Moreover, there is little reason to believe that a large expansion of the effort would increase smugglers' costs further.

(SEE ALSO: *Border Management*; *Dogs in Drug Detection*; *Foreign Policy and Drugs*; *International Drug Supply Systems*; *Operation Intercept*; *U.S. Government: The Organization of U.S. Drug Policy*)

BIBLIOGRAPHY

ANDERBERG, M. (1992). *Measures of effectiveness in drug interdiction.*

OFFICE OF NATIONAL DRUG CONTROL POLICY. (1991). *What America's users spend on illicit drugs.* Washington, DC: U.S. Government Printing Office.

REUTER, P., CRAWFORD, G., & CAVE, J. (1988). *Sealing the borders: The effects of increased military participation in drug interdiction.*

PETER REUTER

DRUG LAWS: FINANCIAL ANALYSIS IN ENFORCEMENT

The application of financial investigative techniques to sophisticated forms of CRIME began decades ago in campaigns to bring un-

derworld bosses to justice. They were charged not with the underlying offenses of bootlegging or extortion, but for reaping financial windfalls from activities that either were not federal offenses at the time or that prosecutors just could not prove. Beginning with the federal tax case against Al Capone in 1931, Treasury investigators had to find ways around both the lack of federal laws proscribing racketeering activity and the difficulties in catching underworld bosses for their offenses. The approach was creative but simple: Internal Revenue agents gathered evidence to prove that the racketeers spent more income than they reported on their tax returns. The differential between what was reported and what the government alleged they earned would establish that their target received substantial amounts of unreported income. In an underworld without pay stubs and annual wage statements, how did the government know what the racketeers earned? To tax investigators, it was simple: Show how much the person spent—or at least the portion of income spent that could be substantiated.

As Prohibition gave way to different forms of industrial racketeering, syndicated gambling, and drug trafficking, federal agents grew more frustrated over their poor showing against criminals who were developing increased sophistication. Investigators turned more and more to financial analysis as an alternative. They reasoned that what worked against Al Capone and his cohorts would probably work against other high-profile racketeers too insulated by their underlings to be implicated in syndicate transactions.

Proving that individuals—whether they were Mafia bosses or Colombian drug importers—received more income than they could substantiate was easier said than done. Typically, there were no records that acted as a smoking gun by pointing directly to one large unreported sum of yearly income. Rather, evidence of unreported income was gathered from a variety of sources and was traced to documented purchases that left a paper trail of deposit slips, bank statements, advices, credit card receipts, and mortgages. As investigators soon came to find out, moreover, financial analyses frequently turned up large amounts of money in the possession of people who recently had approved plans for a lucrative drug deal or some other illegal transaction.

For investigators struggling to tie drug traffickers to crimes they only planned, finding the proceeds of those transactions was welcome evidence. For one thing, it could tie their target to the drug or other transactions that other evidence showed they had planned or approved. Drug traffickers and other racketeers who never touch drugs do touch, or otherwise control, the money that was exchanged for the drugs. Hundreds of criminals have been sent to prison on the basis of financial analyses tying large sums in question to the defendants and alleged criminal transactions.

As organized crime began to wane in national prominence in the 1970s, its place was quickly taken by an amalgam of homegrown and foreign-based drug traffickers. Often just as smart and insulated as Mafia bosses, drug traffickers were surprised to find themselves equally vulnerable to cases built on financial evidence. Passage of a number of federal drug reform laws (in 1970, 1978, 1984, 1986, and 1988) added the remedy of asset forfeiture to the government's arsenal of weapons. In order to show that their targets acquired assets with tainted funds that rendered them forfeitable, investigators resorted to the same financial investigative techniques that had helped build criminal tax cases against the same kinds of underworld leaders.

In the mid-1990s, virtually all federal enforcement agencies provide some type of basic training in financial investigation, and several—such as the Drug Enforcement Administration, Federal Bureau of Investigation, and Internal Revenue Service—have highly specialized programs at their academies. DEA and FBI, expert investigators and financial analysts support major drug-trafficking cases by providing evidence of unexplained income to prove the drug charges, and to tie the money to drug activity for the purpose of forfeiture.

(SEE ALSO: *International Drug Supply Systems; Money Laundering*)

CLIFFORD L. KARCHMER

DRUG LAWS: PROSECUTION OF Drug arrests in the United States involve a wide variety of controlled substances, including MARIJUANA, COCAINE, HEROIN, PHENCYCLIDINE (PCP), and others, and a number of different charges, including possession, dealing (selling), and conspiracy to sell. After arrest, the prosecutor (who in some states is called a

district attorney) exercises the discretion to choose among this broad range of legal options in deciding whether to bring a charge and for what activity.

Drug offenses can violate either federal or state laws. Since the majority of arrests are made by local law-enforcement officials, most defendants are charged in state courts. The cases received by federal prosecutors, called U.S. attorneys, from such federal enforcement agencies as the Federal Bureau of Investigation (FBI) or the DRUG ENFORCEMENT ADMINISTRATION (DEA), frequently involve more complex matters.

In determining what charges should be filed against the offender, the prosecutor looks to many factors: the criminal history of the defendant, the seriousness of the drug involved, and the quality of the evidence. Most states give the district attorney the discretion to charge an enhanced-penalty crime for a repeat offender.

The vast majority of the cases lead to guilty pleas, through some form of bargaining (plea bargaining) between the prosecutor and the defense attorney. In these agreements, which must be approved by the court, the defendant pleads guilty, often in return for a fine, court-ordered counseling, or a lessened prison term. Repeat offenders face tougher agreements.

In deciding what plea to accept, prosecutors consider many of the same factors they did when they brought the original charges. A critical factor is the quality of the evidence. Many drug cases are very easy to prove, because the defendant purchased or sold the drugs directly to a police officer or because a search warrant leads to the discovery of drugs in an area controlled by the defendant. District attorneys face much more difficult challenges in convicting suspects involved in complicated conspiracy charges such as those associated with the shipment or distribution of drugs. In many drug prosecutions, motions to suppress evidence are filed by defense attorneys to determine whether the search that turned up the drugs was conducted in a legal manner. Rulings by the U.S. Supreme Court provide wider latitude to officers who have secured a search warrant.

Another important factor involves the level of cooperation provided by the defendant. The district attorney often accepts a more lenient agreement for defendants who assist law-enforcement officers and/ or testify in court concerning who sold them the drugs they possessed or resold. These plea agreements allow the police to target other offenders and also relieve the pressure on the courts. Plea bargaining does, however, raise serious questions in the public's mind about the dangers of leniency; it raises other questions, among defendants and their attorneys, about equity and fairness. Additionally, narcotic officers and prosecutors often disagree about the outcome or the handling of a case. These differences are often mediated by task forces in which prosecutors with specialized drug experience are assigned to work with a select group of narcotics officers.

Generally less than 10 percent of drug cases go to trial. In a trial the police officer is a witness in the case brought by the prosecutor. By questioning the officer, the prosecutor, as lawyer for the state, illicits evidence designed to show that the defendant possessed or sold drugs.

Ultimately the judge determines the actual sentence. But commonly in a plea agreement or after trial, the prosecutor makes a recommendation concerning sentencing. The recommendation raises several issues, the most significant of which are uncertainty about what really deters a drug offender and the overcrowded conditions of most courts, jails, and prisons. Across the country, and even within large counties, great differences occur in sentencing and in sanction recommendations.

(SEE ALSO: *Exclusionary Rule; Mandatory Sentencing; Rockefeller Drug Law*)

STEPHEN GOLDSMITH

DRUG METABOLISM

Most drugs are taken by mouth and, in order to be absorbed through the stomach and intestine, they need to be lipid-soluble. This solubility permits them to easily cross the membrane barrier. After absorption, organs with plentiful blood-flow such as the brain, liver, lungs, and kidneys are first exposed to the drug. Only highly lipid-soluble drugs can enter the brain by crossing the blood-brain barrier.

Drug concentration at the target organ is an important index for therapy and generally has an optimal range. The drug level can be raised by increasing dose, or by more frequent administration, but too high a level could cause toxicity. The drug level at the target organ can also be lowered by elim-

DRUG (lipid-soluble)

↓   Absorption into the blood stream

↓   Metabolism usually by liver

METABOLITE(S) (water-soluble)

↓   Elimination by kidneys, bile

↓

Figure 1
Drug Metabolism

ination through the urine or by metabolic steps that convert the drug to more water-soluble forms. Water-soluble metabolites are eliminated quickly in the urine. Most drugs given orally are lipid-soluble enough to be reabsorbed in the kidneys and are eliminated only slowly in small amounts in the unchanged form in urine (see Figure 1). Therefore drug metabolism is an important factor that controls drug levels in the body, because without the metabolic step the drug usually remains in the body or accu-

mulates if it continues to be taken. Drug metabolism is a biochemical process and involves enzymes; drugs are metabolized sequentially or by parallel pathways to various products called metabolites. Many enzymes have been identified and some are very specific for drugs or substrates, whereas others have broad or less stringent structure requirements (see Table 1).

Many factors can modify drug metabolism. Genetic factors or inherited deficiency of an enzyme could cause accumulation of certain drugs. Increased levels and increased toxicity may be caused by inhibition of drug metabolism by other concurrently administered drugs. Decreased plasma levels of drugs after repeated administration have been observed and this is attributed to increased enzyme activity by a process called induction; autoinduction causes the increased metabolism of the inducing drug and cross-induction refers to the accelerated metabolism of other drugs.

DRUG-METABOLIZING ENZYMES

Drug-metabolizing enzymes change the chemical nature of drugs by inserting oxygen, hydrogen, water, or small molecules such as amino acids and sugar

TABLE 1
Drug-Metabolizing Enzymes

Reaction Class	Representative Drug Substrates
OXIDATIONS	
(monooxygenases, dehydrogenases)	
Hydroxylation	phenobarbital, imipramine
Dealkylation	codeine, diazepam, caffeine
Deamination	amphetamine
Dehydrogenation	alcohol, acetaldehyde
HYDROLYSIS	
(esterases)	
Ester hydrolysis	cocaine, heroin
REDUCTIONS	
(reductases)	
Nitro reduction	nitrazepam
Carbonyl reduction	naloxone, hydrocodone
CONJUGATIONS	
(transferases)	
Glucuronidation, sulfation	acetaminophen, morphine
Glycine, glutathione conj.	salicylic acid, ethacrynic acid

molecules. The resulting metabolites may thus contain hydroxyl (the univalent group or ion OH), or hydrogenated or hydrolysis products, or be conjugated with sugar or other functional groups. By far the most commonly occurring metabolic step is hydroxylation (the addition of oxygen) by the enzyme oxygenase—and this will be discussed in detail.

OXIDATION BY CYTOCHROME P450 MONOOXYGENASE

Oxygen is vital for living organisms, and enzymatic reactions involving this molecule for drug metabolism are numerous and well characterized. Lipid-solubility is an important factor for absorption across the stomach and intestinal wall, and the insertion of an oxygen atom to lipid-soluble compounds results in hydroxylated groups (-OH) that are more water-soluble than the parent compound. The pioneering work on the oxygenation reaction involved the metabolism of BARBITURATES, a class of centrally acting drugs very popular in the 1950s. A long-acting barbiturate, PHENOBARBITAL, very slowly hydroxylates compared to other barbiturates, such as hexobarbital, pentobarbital, and secobarbital. The oxygenation enzymes involved were named cytochrome P450 after the wavelength of light they absorbed in a spectrophotometer (Peak at 450 nanometers [nm]). Subcellular fractionation by centrifugation yielded "microsome" pellets which contained the cytochrome P450 activity. Cytochrome P450 is most abundant in the liver and, before the full nature of cytochrome P450 was known, the microsomal oxygenase was often called mixed function oxidase. Cytochrome P450 consists of a superfamily of enzymes, with wide and sometimes overlapping substrate specificities.

Although phenobarbital is no longer widely used for therapeutic purposes, because of better alternatives with fewer side effects, it is an excellent inducer of certain forms of cytochrome P450 (e.g., the CYP2B family).

Other important drugs of abuse that are metabolized by cytochrome P450 include BENZODIAZEPINES (tranquilizers such as DIAZEPAM [Valium], CHLORDIAZEPOXIDE, alprazolam, triazolam) and OPIOIDS (CODEINE, oxycodone, dextromethorphan). The first group of drugs is hydroxylated and the second group is metabolized by loss of a carbon moiety (dealkylation). The dealkylation reactions are also mediated by cytochrome P450.

Many cytochrome P450 enzymes have been isolated and characterized. With molecular biology techniques, the genetic code DNA has been identified for many cytochrome P450 enzymes. Among these, two forms of cytochrome P450 are known to be deficient in certain individuals. In the mid 1970s, a deficiency of the specific cytochrome P450 called CYP2D6 was independently reported for sparteine (a labor-inducing or antiarrhythmic drug) and for debrisoquine (an antihypertensive agent). Since then, more than thirty clinically useful drugs have been shown to be metabolized by this enzyme. The presence of this cytochrome P450 in a population is polymorphic, that is, some people lack this enzyme. A simple urine test using dextromethorphan, a cough suppressant, is commonly used to identify the enzyme deficiency in a patient. Another cytochrome P450 deficiency involves metabolism of mephenytoin (CYP2C type) but not many drugs are metabolized by this enzyme. The frequency of both deficiencies were first established in Caucasians, and CYP2D6 deficiency was reported to be 7 percent while CYP2C deficiency was 3 percent. Because of the presence of deficient subjects, the population data do not show a bell-shaped normal distribution curve but rather a bimodal distribution indicating polymorphism.

ALCOHOL METABOLISM

ALCOHOL (ethanol) metabolism predominantly involves a type of oxidation called dehydrogenation (loss of hydrogen) and the subcellular fraction called the mitochondria is the major site. Alcohol is metabolized by successive dehydrogenation steps, first producing acetaldehyde and secondly acetic acid. The major organ for alcohol metabolism is the liver. In heavy drinkers, however, alcohol induces another enzyme, cytochrome P450, and the proportion of the metabolism by this route compared to dehydrogenation becomes significant. Because the amount of alcohol ingested must be relatively large to have pharmacological effects, the amount of alcohol exceeds the amount of enzyme, resulting in saturation. Acetaldehyde, in general, is toxic because it is reactive and forms a covalent bond with proteins. When the enzyme that metabolizes acetaldehyde to acetic acid is inhibited by an external agent acetalaldehyde levels increase and produce a toxic syndrome. Inhibitions of this enzyme, such as DISULFIRAM (Anta-

buse), have been used in the treatment of excessive drinking.

TRANSFERASES FOR CONJUGATION/ SYNTHETIC REACTIONS

Products formed by oxidation (e.g., by cytochrome P450) are often metabolized further with small molecules such as glucuronic acid (glucose metabolite) or sulphate. The enzymes involved are called transferases. Other conjugation reactions are carried out by transferases linking glutathione with reactive metabolic products, acetyl-CoA with an amino group on aromatic rings, and glycine (amino acid) with salicylate.

Glucuronic-acid conjugations are catalyzed by various forms of glucuronyl transferases, which appear to have broad substrate specificity. Glucuronide conjugates are very water-soluble and likely to be eliminated via the kidneys very quickly. The plasma levels of glucuronide conjugates of oxazepam (a benzodiazepine antianxiety agent) are, however, several-fold higher than the parent drug. This can be explained by the relatively rapid process of conjugation reaction in the liver compared to the renal (kidney) clearance of its conjugate. Because glucuronidation involves a glucose metabolite, which is abundant, the transferase would not reach saturation easily, although sulfo-transferase utilizes the sulphate which is of limited supply via foods and can be saturated. For example, ACETAMINOPHEN (Tylenol) forms both glucuronide and sulfate conjugates and the sulfation process can be easily saturated after a few tablets.

Glutathione conjugation is very important as a detoxification pathway. Unstable or reactive metabolites formed from other metabolic reactions may cause toxicity by reacting with so-called house-keeping enzymes in the body. Glutathione, because of its abundance, can react with these metabolites instead and acts as a scavenger; an epoxide whose formation is catalyzed by cytochrome P450 is detoxified, except in an overdose case, by glutathione transferase. Some epoxide intermediary metabolites have been shown to be ultimate carcinogens, and detoxification by gluthione would be beneficial.

Glycine is the smallest amino acid and the conjugation with salicylic acid (formed rapidly from aspirin) is the major metabolic pathway for salicylates. Salicylate poisoning, especially in children, was very common before the introduction of the child-proof cap for drug containers in the 1960s. The difficulty of treating the salicylate poisoning was due to saturable glycine conjugation; the higher the dose, the slower was the rate of elimination.

Acetylation is also important for the detoxification of carcinogens containing aromatic amines. One form of N-acetyltransferase is polymorphic (people have different forms of the enzyme). The frequency of slow acetylator types shows a large variation ranging from 5 to 10 percent in Oriental and Inuit (Eskimo) subjects to as high as 50 percent in Caucasians and Africans. Drugs affected by this genetic polymorphism are isoniazid (antituberculosis), procainamide (antiarrhythmic), sulfamethazole (antibiotic), and other amine-containing compounds.

CLINICAL CONSEQUENCES

Drug metabolites are often pharmacologically less active than the parent drug. Yet some biotransformation products are active—for example CODEINE is relatively inactive but is metabolized to the active drug MORPHINE. Because the liver is the major site of drug metabolism, acute or chronic liver diseases would alter drug metabolism, resulting in prolonged drug half-lives and effects.

(SEE ALSO: *Complications: Liver (Alcohol)*; *Drug Interaction and the Brain*; *Drug Interactions and Alcohol*)

BIBLIOGRAPHY

JAKOBY, W. B. (ED.). (1980). *Enzymatic basis of detoxication*. New York: Academic Press.

KALOW, W. (ED.). (1992). *Pharmacogenetics of drug metabolism*. New York: Pergamon.

KATSUNG, B. G. (ED.). (1992). *Basic and clinical pharmacology*, 5th edition. Norwalk, CT: Appleton & Lange.

TED INABA

DRUG POLICY FOUNDATION (DPF)

The Drug Policy Foundation (4455 Connecticut Avenue, NW, Washington, DC) is a not-for-profit organization established to stimulate debate about drug policy in the United States and to oppose the current "war on drugs" approach. It favors shifting from policies emphasizing law enforcement and drug prohibition to ones that either legalize drug use entirely or at least medicalize the distribution of certain

drugs. Founded in 1986 by Arnold Trebach, a lawyer who is also a professor at American University in Washington, DC, the foundation reports that its membership had grown to more than 13,000 by 1994. Listed as members of its advisory board are well-known individuals long associated with drug-law liberalization, such as Thomas Szasz, Andrew Weil, Ethan Nadelmann, and N.O.R.M.L. president Richard C. Cowan, as well as other prominent academicians, such as Carl Sagan of Cornell University.

The Drug Policy Foundation sponsors annual conferences, at which speakers discuss their perspectives on current drug policy, and has enlisted in its cause many individuals prominent in public life. Kurt Schmoke, the mayor of Baltimore, Maryland, was keynote speaker at the 1993 conference and joined with the foundation in sponsoring an international conference attended by mayors of cities around the world.

The Drug Policy Foundation publishes a newsletter, *The Drug Policy Letter*, and the Drug Policy Foundation Press publishes books and papers that support viewpoints of the foundation. The DPF also receives support from several other foundations not ordinarily associated with advocacy of drug legalization, such as the John D. and Catherine T. Mac-Arthur Foundation, as well as from a few that are, such as the Grateful Dead's Rex Foundation. A major contributor since 1992 has been the Open Society Foundation, a charitable organization that receives its funding from the financier George Soros.

(SEE ALSO: *Prevention Movement; Policy Alternatives*)

JEROME H. JAFFE

DRUG RESPONSE *See* Causes of Substance Abuse: Drug Effects and Biological Responses.

DRUG TESTING AND ANALYSIS As interest increases in employment-related drug testing, the technologies and the interpretive skills of analysts continue to evolve. Although recent literature indicates that significant refinements and modifications to drug-testing technology have been made, the complexity of drug effects is so great that many problems exist in interpretation of the test results. The most frequent problems that confront the toxicology

laboratory relate to developing technology that can determine how much and when the drug was taken, how long after use the tests are capable of showing positive results, the causes and rates of false positive and false negatives, and how tests can be "beaten" by employees. These problems will be discussed and the various laboratory procedures that are used to combat these problems will be examined.

DRUG PROPERTIES

Absorption, Distribution, and Elimination Phases. Detection of a drug depends largely on its absorption, distribution, and elimination properties. There are various routes of drug administration; oral (e.g., drinking ALCOHOL or swallowing pills), intravenous (e.g., HEROIN injected into a vein), and inhalation (e.g., smoking MARIJUANA; snorting COCAINE; sniffing GLUE). Drugs taken orally are usually the slowest to be absorbed (i.e., the speed at which the drug reaches the brain and other body organs), whereas intravenous and inhalation routes result in the fastest absorption. Once the absorbed drug enters the blood stream it is rapidly distributed to the various tissues in the body. The amount of drug stored depends on the nature of the drug, the quantity, duration of ingestion, the tissue holding the drug, and the frequency of use.

Some drugs are fat-soluble and are deposited in fat tissues. For example, δ^9-TETRAHYDROCANNABINOL (THC), the active ingredient in marijuana, is highly fat-soluble, resulting in rapid reductions in blood levels as the drug is being distributed to the various tissues. Blood levels of δ^9-THC peak and start to decline in half the time it takes to smoke a marijuana "joint." Concentrations in blood are known to fall by almost 90 percent in the first hour. Depending on the amount of drug stored in the fat tissues, detection may be possible in the urine for many days after last use. There are cases where marijuana metabolites have been detected for as long as sixty days after last use, since small amounts from fat go back into blood and then appear in the urine. Ethanol or ethyl ALCOHOL (the beverage alcohol) is not fat-soluble but is distributed in the total body water. Since blood is mostly made up of water, the presence of alcohol in blood is easier to detect than fat-soluble drugs like δ^9-THC.

The "absorption" and "distribution" phases are followed by an "elimination" phase. The liver is the major detoxification center in the body, where the

drugs are metabolized as blood circulates through this organ. The metabolites are then excreted into the urine through the kidneys. At the same time, drugs deposited in fat tissues are also slowly released into the bloodstream and metabolized.

Drugs vary by their elimination half-life, which is the time required for the blood levels to decline by 50 percent (see Table 1). The half-life of a drug is heavily influenced by a variety of factors including the individual's age, sex, physical condition as well as clinical status. A compromised liver and concurrent presence of another disease or drug have the potential of enhancing the toxic effects of the drug by slowing down the elimination process. Under differing clinical conditions, however, this process may be speeded up. Therefore, great variation can be found in the half-lives of the same drug.

Approximately six half-lives are required to eliminate 99 percent of any drug. Because cocaine's half-life is relatively short, averaging one hour, only six hours are needed for elimination of 99 percent of the drug. Cocaine's metabolites have a longer half-life, however, and can be detected for a considerably longer period through urine drug assays. Compared to cocaine, PHENOBARBITAL has a much longer half-life, of 80–120 hours, so at least 480 hours (or 20 days) are required to eliminate 99 percent of the drug. Since there is much variation in the half-life of different drugs and the absolute amount of drug present can be very small, it is crucial that the appropriate body fluid for analysis is selected for testing.

Elimination of ethanol (alcohol) follows a very different pattern. Its levels decline almost linearly over time—the average elimination rate being between 15 milligrams (mg)/100 milliliters (ml) to 20mg/100ml per hour, although ranges between 10mg/100ml to 30mg/100ml per hour have also

TABLE 1
Drug Half-Lives and Approximate Urine Detection Periods

Drug	Half-life (t $\frac{1}{2}$)	Detection Period*
Methamphetamine	12–34 hours	2–3 days
Amphetamine (metabolite of methamphetamine)	7–34 hours	
Heroin	60–90 minutes	in minutes
Morphine (metabolite of heroin)	1.3–6.7 hours	1–3 days
Phencyclidine (PCP)	7–16 hours	2–3 days
Cocaine	0.5–1.5 hours	few hours
Benzoylecgonine (metabolite of cocaine)	5–7 hours	3–5 days
δ^9-tetrahydrocannabinol (THC)	14–38 hours	90% fall in 1 hour (blood)
δ^9-tetrahydrocannaboic acid (marijuana metabolite in urine)		(depending on use few days to many weeks)
Alcohol (ethanol)	Blood levels fall by an average of 15–18mg/100ml/hour	1.5 > 12 hours (depending on the peak blood level): urine typically positive for an additional 1–2 hours

*The detection period is very much dose dependent. The larger the dose, the longer the period the drug/metabolite can be detected in the urine.

been observed. In the alcoholic patient, the elimination rate is generally higher. In forensic calculations, a rate of 15mg/100ml/hour is usually used.

SELECTION OF DRUGS TO BE TESTED

A number of criteria can be applied to the drug(s) or category of drugs that should be tested or monitored. Drug availability, clinical effects, and robustness of the analytical method(s) used for analysis are probably the most important.

Availability. Prescription patterns and the availability of illicit drugs vary from place to place. Abuse of the BENZODIAZEPINE nitrazepam is common in Europe but almost unknown in North America, since it is not sold here. The PSYCHOACTIVE chemical CATHINONE (cathine), the active ingredient in the leaves of the KHAT plant that are chewed in northeast Africa, is not a problem in North America. CODEINE, an OPIOID, is available in CANADA as an OVER-THE-COUNTER preparation, but it is sold only by prescription in the United States.

A wide availability of "legal" STIMULANTS poses an interesting problem since they are a common finding in ACCIDENT victims. A study carried out by the U.S. National Transportation Safety Board from October 1987 to September 1988 showed that over-the-counter stimulants—such as ephedrine, pseudoephedrine, and phenylpropanolamine—were common findings among drivers killed in heavy-truck accidents. Among the eight states that participated in this safety study almost all AMPHETAMINE use was in the California region. Similar findings are also reported from emergency rooms over the past five years as well as from admissions in a trauma unit for motor-vehicle accidents. All this suggests that drug use varies not only from place to place but also region to region within a given country.

Thus, the selection of a drug to be tested for and monitored, appropriate for one country and place, may not necessarily be appropriate for another.

Clinical Effects. Drugs that manifest abuse potential and impair behavior such that job performance can be affected are prime candidates for testing or monitoring in the workplace. Alcohol and cocaine are examples of this.

Analytical Methods. A false positive finding can have serious impact on the livelihood of the person being tested. Therefore, special attention needs to be paid to the testing methods. Ideally the analytical method should be specific for the drug being tested (i.e., no false positive), easy and inexpensive to perform. Confirmation methods should also be readily available. Technical and scientific expertise to perform the tests are also essential.

Interpretation of the analytical results needs to be carefully considered, since even a normal diet can sometimes result in a positive drug identification. For example, poppy seed ingestion can result in a true positive *analytical* result (OPIATES, like heroin, are derived from the poppy plant PAPAVER SOMNIFERUM) but is a false positive for drug use. Some ethnic diets may also lead to these confounding problems, as when food containing poppy seed is eaten during Ramadan.

What should be analyzed? Ideally the analysis should look for the parent drug rather than its metabolite, although this may not always be possible, as some drugs are very rapidly metabolized (e.g., heroin metabolism to MORPHINE). Sensitivity of the analytical procedure should be dictated by the drug's psychoactive pharmacological properties. If the drug is shown to be devoid of abuse potential, then its detection beyond the time of pharmacological activity, although important in the clinical management of the patient, does not necessarily serve a useful purpose for workplace drug-screening programs.

The guidelines developed by the NATIONAL INSTITUTE ON DRUG ABUSE in April 1988 address five "illegal" drugs: marijuana, PHENCYCLIDINE (PCP), AMPHETAMINE, cocaine, and heroin. Rapid-screening methods that allowed for "mass screening" were available at that time, as were the confirmation methods for these five drugs. Mood-altering substances such as benzodiazepines, BARBITURATES, and some stimulants and certain antihistamines are at present excluded from these regulations in the United States. This is probably due to the wide availability of these drugs as medications within the general population and the technological requirements for screening and monitoring these drugs.

TYPES OF TESTING

Blood, Urine, and Hair Specimens. Blood and urine are the most commonly used biological fluids in the analysis for drugs other than alcohol. Blood, obtained by an invasive procedure, is available only in small quantities and drug concentration levels in blood are typically low. Urine is the pre-

ferred sample of choice as it is available in larger volumes, contains the metabolite, and requires less invasive procedures in its collection. Both sampling procedures, however, are limited in their ability as they only determine the absolute amount of drug present in the fluid being examined. This quantity is dependent upon the amount of the drug used, when it was last used, as well as the half-life of the drug.

Recently, hair samples have been suggested and are used to detect drug use. A number of technical problems must be overcome before hair can be used as a definitive proof of drug use. Hair treatment and environmental adsorption are but two of the many concerns and problems that have been cited. An advisory committee of the Society of Forensic Toxicology has recently reported, "The committee concluded that, because of these deficiencies, results of HAIR ANALYSIS alone do not constitute sufficient evidence of drug use for application in the workplace."

Various body fluids, such as sweat, saliva, blood, urine, and breath, have been used for alcohol analysis. Breath, though not a body fluid, is commonly used by law-enforcement authorities. Although a number of variables can affect breath/blood ratio, 2100:1 alveolar breath/blood conversion ratio has been used and accepted for the BREATHALYZERS. Breath-testing equipment calibrated with a blood: breath conversion factor of 2100 consistently underestimate actual BLOOD ALCOHOL CONCENTRATIONS (BAC). Accuracy of breath analysis result is subject to various instruments and biological factors. Potential errors in breath analysis can also be caused by the presence of residual alcohol in the mouth. Immediately after drinking there is enough alcohol vapor in the mouth to give artificially high concentrations on breath analysis. Generally, this effect disappears twenty minutes after drinking but high values for as long as forty-five minutes have been reported.

As of the early 1990s, all existing technologies are limited in terms of determining how much or when the drug was consumed.

Blood and saliva concentrations reflect the current blood alcohol concentration, but generally a blood sample is used in hospitals to assess the patient in the casualty wards. In programs that require the monitoring of alcohol use, urine is probably the sample of choice. Urine alcohol concentration, which represents the average blood alcohol concentration between voiding, has the potential of being "positive" while the blood may be "negative."

MEASURING IMPAIRMENT

Except for alcohol, the degree to which a person is influenced or impaired by a drug at the time of the test *cannot* be determined from test results alone. Correlations between positive blood levels and degree of impairment are usually stronger than correlations between urine levels and degree of impairment; however, neither blood nor urine tests are sufficiently accurate to indicate impairment even at high levels of concentration. Human studies using marijuana and cocaine have shown that a "perceived high" is reached *after* the drug concentration has peaked in the blood. Generally, blood can only show positive results for a short time after drug consumption, whereas urine can be positive for a few days to weeks after last use. For example, metabolites of δ^9-THC (the active ingredient in marijuana) that are lipid-soluble can be detected in the urine from a few days to many weeks, depending on the drug habit of the user. Excretion of the drug in urine and its concentrations are also affected by several factors, such as dilution and pH (acidity) of the urine. I have seen many cases where a strong, positive urine sample for CANNABINOIDS was found in the morning, a borderline positive in the afternoon, followed by a strong positive the next morning; I have also seen similar cases with respect to phenobarbital.

A positive urine test cannot reveal the form in which the drug was originally taken—or when and how much was taken. For example, CRACK-cocaine, impure cocaine powder, or cocaine paste (which can be smoked, inhaled, injected, or chewed) all give the same result in the urine test. The consumption of poppy seeds has been reported to give positive results for opiate use, because some seeds contain traces of opiates and some have been known to be contaminated with OPIUM derivatives. Similarly, consumption of herbal COCA tea has resulted in positive results for cocaine use. These diverse incidences only begin to illustrate the difficulties involved in measuring impairment using urine results.

The problem of interpreting urine-test results is one of the major bases of concern for restricting their use in the employment setting. Even the effectiveness of preemployment drug-screening tests, due to the difficulties in interpretation, is being questioned. Based on a study of 2,229 preemployment drug screening tests and follow-up, one group of researchers came to the following conclusion, "our findings raise the possibility that a preemployment drug

screening may be decreasingly effective in predicting adverse outcomes associated with marijuana use after the first year of employment." They make a similar comment about cocaine.

There is no threshold for alcohol effects on performance or motor-vehicle-accident risk. Although the effects of alcohol on impairment and crash risk appear more dramatically above 80mg/100ml, a review of the literature would suggest that impairment may be observed at levels as low at 15mg/100ml. It is not possible to specify a blood alcohol concentration level above which all drivers are dangerous and below which they are safe or at "normal" risk.

"Legal" BAC levels differ in different countries. Some even have more than one legal limit over which the driver of a vehicle is considered as "impaired." Some European countries have 50mg/100ml others have 80mg/100ml as their legal limits. In the United States, the legal limits vary from 80mg/100ml to 100mg/100ml in different states, but employees regulated by the U.S. Department of Transportation have a BAC legal limit of 40mg/100ml. In Canada, there are also two limits: 50mg/100ml and 80mg/100ml. BAC levels between 50mg/100ml and 80mg/100ml call for suspension of driving privileges but above 80mg/100ml are subject to criminal charges.

URINE-TESTING METHODS

Urine is the most commonly used fluid for drug screening. The methods most commonly used in toxicology laboratories are: *immunoassay, chromatographic and chromatography coupled with mass spectrometry.* These methods vary considerably with respect to their sensitivity and reliability. Thin-layer chromatography is least expensive, gas chromatography coupled with mass spectrometry (GC/MS) which is considered as nearly perfect or "gold standard" is the most expensive. Table 2 summarizes the various methods.

Immunoassays (EIA, EMIT, FPIA). IMMUNOASSAY methods are used for preliminary screening (i.e., initial screening). Since these methods are based on an antibody-antigen reaction, small amounts of the drug or metabolite(s) can be detected. Antibodies specific to a particular drug are produced by injecting laboratory animals with the drug. These antibodies are then tagged with markers such as an enzyme (enzyme immunoassay, EIA), a radio isotope (radioimmunoassay, RIA), or a fluores-

TABLE 2
Common Drug-Testing Methods

Immunoassays:
Enzyme Immunoassay (EIA)
Enzyme Multiplied Immunoassay Technique (EMIT)
Fluorescence Polarization Immunoassay (FPIA)
Radio Immunoassay (RIA)

Chromatographic methods:
Thin Layer Chromatography (TLC)
Liquid Chromatography (HPLC)
Gas Chromatography (GC)

Chromatography/Mass Spectrometry:
Gas Chromatography/Mass Spectrometry (GC/MS)
Liquid Chromatography/Mass Spectrometry (HPLC/MS)

cence (fluorescence polarization immunoassay, FPIA) label. Reagents containing these labeled antibodies can then be introduced into urine samples, and if the specific drug is present against which the antibody was made—a reaction will occur. RIA is the oldest immunoassay method used to detect drugs. The major drawback of this method is that it requires a separation step and generates radioactive waste. RIA also requires special equipment to measure radioactivity.

Typically, immunoassays are designed for a *class* of drugs. Thus, their specificity (the ability to detect the presence of a *specific* drug) is not very good, since substances that have similar chemical structures will "cross react" and give a false positive reaction. For example, the immunoassay method for cannabinoids was developed to detect the carboxylic acid metabolite of δ^9-THC. Yet, there is a suggestion in literature, that some nonsteroidal anti-inflammatory drugs, such as ibuprofen (a nonprescription drug in the U.S. and Canada) and naproxyn give random or sporadic false positive results for cannabinoids. Cough-syrup codeine will also give a positive reaction for the morphine (a metabolic product of heroin use) immunoassay and many antihistamines that are over-the-counter drugs may yield positive reactions for amphetamines. While some reagent manufacturers claim to have overcome many of these cross-reactivity problems, confirmation by a nonimmunoassay method is very important.

Urine drug-assay kits, designed to detect drugs, have been available in North America for the past few years. More recently, single and multiple test im-

munoassay kits designed for home and on-site testing have also been introduced. These kits generally carry a cautionary disclaimer that positive test results must be confirmed by the reference GC/MS method. When used in the nonlaboratory environment, they are prone to procedural inaccuracies, poor quality control, abuse, and misinterpretations. Therefore, these kits should be used with great caution. The risk of labeling a person with a false positive is high without the accompanying confirmatory analysis. Table 3 summarizes the advantages and disadvantages of immunoassay testing.

Chromatographic Methods. Separation of a mixture is the main outcome of the chromatographic method. For illustrative purposes, if one were to put a drop of ink on a blotting paper and hold the tip of the paper in water, one would observe the water rise in the paper. After a period of time and under the right conditions, the single ink spot would separate into many different compounds (spots) of different colors (blue ink is a mixture of many dyes). This process, where a mixture of substances is separated in a stationary medium (filter paper), is called chromatography. The types of chromatographic processes used in the analysis of drugs include thin-layer, gas, and liquid chromatography as well as a combination of gas or liquid chromatography with mass spectrometry.

Of the several chromatographic methods, thin-layer chromatography (TLC) is the one most similar to the ink separation example mentioned above. This method requires extensive sample preparation and technical expertise on the part of the analyst, but it is inexpensive and very powerful if used properly. With the exception of *Cannabis*, which requires separate sample preparation, a large number of drugs (e.g., cocaine, amphetamine, codeine, and morphine) can be screened at the same time. By combining different TLC systems a high degree of specificity can be obtained, although the training of the analyst is crucial because of the subjectivity involved in interpreting the results. To identify positive TLC "spots," the technologist looks for the drugs and or its metabolite pattern, often after spraying with

---

TABLE 3
Advantages and Disadvantages of Immunoassays

Advantages

- Screening tests can be done quickly, because automation and batch processing are possible.

- Technologists doing routine clinical chemistry testing can be easily trained.

- Detection limits are low and can be tailored to meet the program screening requirements. For example, low detection thresholds can be raised to eliminate positive results from passive inhalation of marijuana smoke.

- Immunoasays are relatively inexpensive, although the single test kits can be very expensive when quality assurance and quality control samples are included.

- Immunoassays do not require a specialized laboratory. Most clinical laboratories have automated instruments to do the procedures.

Disadvantages

- Although the tests are useful for detecting classes of drugs, specificity for individual drugs is weak.

- Since the antibody is generated from laboratory animals, there can be a lot-to-lot or batch-to-batch variation in the antibody reagents.

- Results must be confirmed by another nonimmunoassay method.

- A radioactive isotope is used in RIA that requires compliance with special licensing procedures, use of gamma counters to measure radioactivity, and disposal of the radioactive waste.

- Only a single drug can be tested for at one time.

reagents that react to form different colors with different drugs. The trained technologist can comfortably identify more than forty drugs.

Similar to TLC, gas chromatography (GC) requires extensive sample preparation. In GC, the sample to be analyzed is introduced via a syringe into a narrow bore (capillary) column which sits in an oven. The column, which typically contains a liquid adsorbed onto an inert surface, is flushed with a carrier gas such as helium or nitrogen. (GC is also sometimes referred to as gas-liquid chromatography [GLC].) In a properly set up GC system, a mixture of substances introduced into the carrier gas is volatilized, and the individual components of the mixture migrate through the column at different speeds. Detection takes place at the end of the heated column

and is generally a destructive process. Very often the substance to be analyzed is "derivatized" to make it volatile or to change its chromatographic characteristics.

In contrast to GC, high pressure liquid chromatography (HPLC) uses a liquid under high pressure to flush the column rather than a gas. Typically, the column operates at room or slightly above room temperature. This method is generally used for substances that are difficult to volatilize (e.g., STEROIDS) or are heat labile (e.g., BENZODIAZEPINES).

Gas chromatography/mass spectrometry (GC/MS) is a combination of two sophisticated technologies. GC physically separates (chromatographs or purifies) the compound, and MS fragments it so that a fingerprint of the chemical (drug) can be obtained. Al-

TABLE 4
Summary of Chromatographic Methods

Advantages

All the chromatographic methods are specific and sensitive and can screen a large number of drugs at the same time.

• TLC	Negligible capital outlay is needed.
• GC	The procedure can be automated.
• HPLC	Of the chromatographic procedures, this has the easiest sample preparation requirements. The procedure can be automated.
• GC/MS	This is the "gold standard" test. Computerized identification of fingerprint patterns makes identification easy. The procedure can be automated. This is currently the preferred method for defence in the legal system.

Disadvantages

All chromatographic methods are labour intensive and require highly trained staff. Although all the chromatographic methods are specific, confirmation is still desirable.

• TLC	Interpretation is subjective, hence training and experience in interpretation capabilities of the technologist are crucial.
• HPLC or GC	Equipment costs are high, ranging between $25,000 and $60,000, depending on the type of detector and automation selected (1994 $).
• GC/MS	Equipment costs are the highest, ranging between $120,000 and $2,000,000 depending on the degree of sophistication required (1994 $). Due to the complexity of the instrument, highly trained operators and technologists are required.

though sample preparation is extensive, when the methods are used together the combination is regarded as the "gold standard" by most authorities. This combination is sensitive i.e., can detect low levels, is specific, and can identify all types of drugs in any body fluid. Furthermore, assay sensitivity can be enhanced by treating the test substance with reagents. When coupled with MS, HPLC/MS is the method of choice for substances that are difficult to volatilize (e.g. steroids).

Given the higher costs associated with GC/MS, urine samples are usually tested in batches for broad classes of drugs by immunoassays, and positive screens are later subjected to confirmation by this more expensive technique.

Table 4 gives a summary of the advantages and disadvantages of each method of chromatographic drug testing and Table 5 compares all the methods of testing. The initial minimal immunoassay and GC/MS (cut-off) levels for five drugs or classes of drugs, as suggested by the U.S. NATIONAL INSTITUTE ON DRUG ABUSE, are listed in Table 6.

Procedures for Alcohol Testing. Since the introduction of the micro method for alcohol analysis in blood by Widmark in 1922, many new methods and modifications have been introduced. The distillation/oxidation methods are generally *nonspecific* for alcohol (ethanol), whereas biochemical methods (spectrophotometric) using alcohol dehydrogenase (ADH) obtained from yeast and the gas chromatographic method that are currently used are specific for ethanol. The radiative attenuation energy technique and those using the alcohol oxidase method are nonspecific and will detect not only ethanol but also other alcohols. The recently introduced alcohol dipstick, based on the ADH enzyme system, is not only specific of ethanol but is also sensitive and does not require instrumentation. It can be used for the detection of ethanol in all body fluids and can provide semi-quantitative results in ranges of pharmacological-toxicological interest. Alcohol dipsticks are being used in many alcohol treatment programs as well as in a number of laboratories as a screening device.

Breath can be analyzed by using a variety of instruments. Most of the instruments used today detect ethanol by using thermal conductivity, colorimetry, fuel cell, infrared or gas chromatography. Typically in most countries, local statutes define the instrument and method that can be used for evidentiary

TABLE 5
Comparison of All Testing Methods

	EMIT FPIA	RIA	TLC	GC HPLC	GC/MS
Ease of sample preparation	✔	✔		✔	
Less highly trained technologists required	✔	✔			
Limited equipment requirement	✔	✔	✔		
Low detection limits	✔	✔	✔	✔	✔
Adjustable lower threshold	✔	✔			
Highly specific and sensitive			✔	✔	✔
Computerized identification possible					✔
Screen for several drugs at a time			✔	✔	✔
Procedure can be automated	✔	✔		✔	✔
Special atomic energy license required		✔			
Confirmation of results required	✔	✔	✔	✔	
Interpretation is subjective			✔		

TABLE 6
Cut-Off Levels for Initial and Confirmatory Tests*

Test	Initial Test	Confirmatory Test
THC metabolite**	100 ng/ml	15 ng/ml
Cocaine metabolites***	300 ng/ml	150 ng/ml
Opiate metabolites****	300 ng/ml	
morphine		300 ng/ml
codeine		300 ng/ml
Phencyclidine (PCP)	25 ng/ml	25 ng/ml
Amphetamines	1000 ng/ml	
amphetamine		500 ng/ml
methamphetamine		500 ng/ml
Alcohol	10 mg/100ml	10 mg/100ml

*National Institute of Drug Abuse (NIDA) Guidelines, April 1988
**THC metabolite is 11-nor-delta-9 THC carboxylic acid
***Cocaine metabolite is benzoylecognine
****25 nanograms per milliliter (ng/ml) if immunoassay is specific for free morphine

purposes. A variety of breathalyzer instruments ranging in cost from $100 to $1,000 are available to do the test. These instruments are compact and portable. Canadian law-enforcement authorities use the breathalyzer "Alert," which can give a "pass" or "fail" result as a roadside alcohol-screening device. The "failed" person is generally subjected to a "Borkenstein" breathalyzer to measure the BAC before any charges are made. Many devices are available to preserve the breath sample for later analysis if a breathalyzer is not available immediately. In forensic laboratories, gas chromatography (North America) or biochemical procedures (many European countries) are used to analyze biological samples.

Blood samples that cannot be analyzed soon after collection have sodium fluoride (NaF) added as a preservative. Alcohol dehydrogenase (ADH), the enzyme responsible for the oxidation of alcohol, is also present in red blood cells and will slowly metabolize the alcohol, causing its concentration to drop if the preservative is not added. Large amounts of alcohol can be produced *in vitro* in the urine samples of diabetic people if samples are not processed immediately or properly preserved.

INTERPRETATIONS OF TEST RESULTS

False Negatives. A positive or negative result is highly dependent on the sensitivity of the drug-detection method. A false negative occurs when the drug is present but is not found because the detection limit of the method used is too high or the absolute quantity of the drug in the specimen is too low.

Large amounts of fluids consumed prior to obtaining a sample for analysis can affect detection of drugs in urine samples. Under conditions of dilution, although the absolute amount of drug or metabolite excreted may be the same over a period of time, the final concentration per milliliter will be reduced and may give a false negative result. Acidity levels in the urine may also affect the excretion of the drug into the urine. In some cases, elimination is enhanced, whereas in other cases, the drug is reabsorbed.

Several measures can be used to decrease the likelihood of obtaining a false negative result. First, sensitivity of the method can be enhanced by analyzing for the drug's metabolites. Heroin use, for example, is determined by the presence of its metabolite, morphine. Increasing the specimen volume used for analysis or treating it with chemicals can also make laboratory methods more sensitive. Studies have shown that a 5 milligram dose of Valium is usually detected for three to four days; however, when these improved methods are utilized, sensitivity can be increased, such that, the same dose can be detected for up to twenty days. One important drawback of such high sensitivities is that estimates of when the drug was taken are far less accurate.

False Positives. A false positive occurs when results show that the drug is present, when in fact it is not. False-positive tests are obtained if an interfering drug or substance is present in the biological fluid and it cross-reacts with the reagents. As dis-

cussed in the previous section on immunoassay, an initially positive test based on immunoassay technique should always be confirmed with a nonimmunoassay method. A confirmed positive finding only implies that the urine sample contains the detected drug and nothing more.

At times, false positives are attributable to ingested substances, such as allergy medications. Some authors have suggested that employees subject to drug screening refrain from using popular over-the-counter medications, such as Alka-Seltzer Plus and Sudafed, because they have caused false-positive results. Some natural substances such as herbal teas and poppy seeds can also give positive responses to screens. These may be analytically true positives but they need to be distinguished from those due to illegal drug use. In some instances, false positives have been caused by mistakes or sabotage of the chain of custody for urine samples.

COMMON ADULTERATION METHODS

The method of switching "clean" urine for "dirty" urine is the most common way to fool the drug-screening system. A number of entrepreneurs have attempted to bypass urine-specimen inspection by substituting clean urine. For example, a company in Florida sells lyophilized (freeze-dried) clean urine samples through newspaper and magazine advertisements. Hiding condoms containing "clean" urine on the body or inside the vagina is another common trick.

Some have substituted apple juice or tea in samples for analysis. Patients are known to add everything from bleach and liquid soap to eyedrops and many other household products, hoping that their drug use will be masked. Others may hide a masking substance under their fingernails and release it into the urine specimen. Another method is to poke a small hole into the container with a pin so that the sample leaks out by the time it reaches the laboratory.

Since addition of table salt (NaCl) or bleach to the urine is a common practice, many laboratories routinely test for Na and Cl in the urine. Liquid soap and crystalline drain cleaners that are strong alkaline products containing sodium hydroxide (NaOH) are also used to adulterate the urine sample. These contaminators can be detected by checking for high levels of pH in the urine sample. In vivo alkalizing or acidifying of the urine pH can also change the

excretion pattern of some drugs, including amphetamines, barbiturates, and phencyclidine (PCP).

Water-loading (drinking large amounts of water prior to voiding) poses an interesting challenge to the testing laboratories. Specific gravity has been used to detect dilution; however, the measurement range is limited so it is not yet useful. Creatinine levels on random urine samples appear to be a promising method in detection of water-loading.

Drug users are very resourceful and their ingenuity should not be underestimated. To reduce the opportunities for specimen contamination, some workplaces require that employees provide a urine sample under direct supervision. Another technique used to detect any sample adulteration is to take the temperature of the sample. In a study, we took the temperature of urine samples within one minute of voiding; it falls between 36.5°C and 34°C, reflecting the inner body core temperature. It is very difficult to achieve this narrow temperature range by hiding a condom filled with urine under the armpit or adding water from a tap or toilet bowl to the urine sample. For security, it is important that the temperature of the specimen be measured immediately after the sample is taken, since it can drop rapidly.

LABORATORY PROCEDURAL AND SECURITY STANDARDS

It is important that the laboratory drug testing facility has qualified individuals who follow a specific set of laboratory procedures and meet recommended security standards.

SUMMARY AND CONCLUSION

In this paper, major issues related to drug testing are discussed. For example, drug-testing techniques measure drug presence but are not sophisticated enough to measure impairment from drug use. It is also very difficult to determine the route of drug administration, quantity, frequency, or when the drug was taken.

Selection of the drug to be tested should depend on the local availability of the drug, its ABUSE POTENTIAL, and clinical effects—as well as the available analytical technology and expertise in the testing and interpretation of the laboratory results. The most sophisticated drug-testing approach, gas chromatography in combination with mass spectrometry, is considered as a gold standard and thus

utilized in confirmatory testing. Typically GC/MS is preceded by a rapid immunoassay method to eliminate the majority of negative samples.

Despite the existence of sophisticated drug-testing methods, incorrect test results can still occur. These can be due to the presence of interfering substances or adulteration of the urine sample. Patients have been known to adulterate urine samples to avoid drug detection. A number of techniques can be employed to reduce the likelihood of obtaining erroneous results, as well as detect adulterated urine samples.

A "positive" drug finding can have a serious impact on the livelihood of an individual, therefore the performance of these tests should adhere to the strictest laboratory standards of performance. Only qualified and experienced individuals with proper laboratory equipment should perform these analyses. Standards of laboratory performance must meet local legal and forensic requirements. Access to the patient samples as well as laboratory records must be restricted to prevent tampering with samples and results. To maintain confidentiality, the results must be communicated only to the physician reviewing the case/patient. Chain of custody and all documents pertaining to the urine sample must be maintained so that they can be examined in case of a legal challenge. The laboratory must have a complete record on quality control. Finally, specific initial and confirmatory testing requirements should be met.

(SEE ALSO: *Drug Metabolism; Drug Use Forecasting Program; Drunk Driving; Industry and Workplace, Drug Use in; Military, Drug and Alcohol Abuse in the U.S.; Pharmacokinetics; Vietnam: Drug Use in*)

BHUSHAN M. KAPUR

DRUG TRAFFIC CONTROL. *See* Drug Interdiction.

DRUG TRAFFICKING *See* International Drug Supply Systems.

DRUG TYPES There are many ways to classify drugs, depending on the purposes for the classification. For example, a classification can be based on the chemical properties of drugs and may actually

disregard the effects the drugs have on the body, or it may be based on legal principles, such as legal versus illegal, or prescription versus over-the-counter (prescription not needed). For purposes of discussion and teaching, the various drugs that are used and abused by humans for nonmedical purposes are usually grouped into several major categories, each based on their pharmacological actions and their subjective effects. Although the mechanisms of action may vary among drugs within a single category, the general subjective effects of the drugs are similar.

The major categories include: (1) ethanol (ALCOHOL); (2) NICOTINE and tobacco; (3) central nervous system depressants (BARBITURATES, BENZODIAZEPINES); (4) central nervous system stimulants (AMPHETAMINES, COCAINE); (5) cannabinoids; (6) OPIOIDS (MORPHINE, HEROIN, METHADONE); (7) psychedelics (LSD, MESCALINE); (8) INHALANTS (glue, nitrous oxide); (9) arylcyclohexylamines (PCP). Some categorizers might put cocaine and the amphetamines into separate categories and group alcohol and the central nervous system depressants together. Some might have a separate category for CAFFEINE; others, one for "DESIGNER DRUGS" (such as MDMA), and refer to them as *entactogens*. There might also be a miscellaneous category, where drugs such as BETEL NUT, KAVA-KAVA, or NUTMEG would be included.

Drugs from the various categories are described below in terms of their pharmacology, abuse, DEPENDENCE, and WITHDRAWAL, as well as their toxicity. The legal and readily available drugs (alcohol and tobacco) are described first because the worldwide use and abuse of these drugs is far more widespread than *all* the other categories of abused drugs combined. The ill health associated with the ongoing use of alcohol and tobacco has become a far-reaching problem—not only because of the vast numbers of people who suffer and die each year from the toxic effects of these drugs but also because of the financial drain in terms of employee absenteeism as well as the staggering increases in annual health-care costs. Prescription drugs are covered next, since more prescriptions are written for diazepam (Valium) and the related benzodiazepines each year than for any other drug.

The illegal drugs are then discussed. Although the illicit use of heroin, cocaine, and other drugs remains a major social, legal, financial and health problem in the United States today, the proportion of the population physically dependent on these drugs is actually relatively low when compared to the

legal drugs listed above. Finally, it is important to take into consideration the fact that individuals often do not restrict their drug use only to drugs within a single category. Alcoholics typically smoke cigarettes and often use benzodiazepines as well. Heroin users also smoke and may consume alcohol and other sedatives, as well as CANNABIS and stimulants in some instances. Multiple-drug use is, therefore, a relatively common occurrence for those using legal and/or illegal drugs.

ETHANOL

Although alcohol (ethyl alcohol, called ethanol) has been in use since prehistory and worldwide throughout recorded history, it is now generally accepted that its therapeutic value is extremely limited and that chronic ALCOHOLISM is a major social and medical problem. Perhaps 65 percent of all adults in the United States use alcohol occasionally. Hundreds of thousands of individuals suffer and die each year, however, from complications associated with chronic alcoholism—and tens of thousands of innocent individuals are injured or killed each year in alcohol-related traffic ACCIDENTS. Therefore, alcoholism is a far-reaching problem, affecting the lives of individuals who consume ethanol as well as those who do not. Although alcohol is considered by many people to be a stimulant drug because it typically releases an individual's latent behavioral inhibitions (i.e., through disinhibition), alcohol actually produces a powerful primary and continuous depression of the central nervous system similar to that seen with general anesthetics. In general, the effects of alcohol on the central nervous system are proportional to the blood concentrations of the drug. Initially, MEMORY and the ability to concentrate decrease, and mood swings become more evident. As the intoxication increases, so does the impairment of nervous function until a condition of general anesthesia is reached ("passing out" or "sleeping it off"). There is little margin of safety, however, between an anesthetic dose of ethanol and severe respiratory depression (unconsciousness or coma).

In chronic (long-term) alcoholism, brain damage, memory loss, sleep disturbances, psychoses, and increased seizure susceptibility often occur. Chronic alcoholism is also one of the major causes of cardiomyopathy (heart disease) in the United States due to ethanol-induced, irreversible damage to the heart muscle. Ethanol also stimulates the secretion of gastric acid in the stomach and can contribute to the production of ulcers of the stomach and gastrointestinal system. One of the primary metabolic products of ethanol is acetaldehyde. In chronic alcoholism, acetaldehyde can accumulate in the liver, resulting in hepatitis and cirrhosis of the liver. Finally, the long-term use of alcohol can result in a state of PHYSICAL DEPENDENCE. With relatively low levels of dependence, withdrawal from alcohol may be associated with rather minor problems such as SLEEP disturbances, ANXIETY, weakness, and mild tremors. In more severe dependence, the alcohol withdrawal syndrome includes more pronounced tremors, seizures, and DELIRIUM, as well as a number of other physiological and psychological effects. In some cases, this withdrawal can be life-threatening.

Since alcohol has CROSS-TOLERANCE with other central nervous system (CNS) depressants (i.e., ethanol shares many of the same biological effects as other CNS depressants), benzodiazepines or barbiturates can be substituted for ethanol to successfully decrease the severity of the alcohol withdrawal syndrome. Longer-acting benzodiazepines and related drugs can be used as an ethanol substitute, and the dose of the benzodiazepine can then be gradually reduced over time to attenuate or prevent the occurrence of convulsions and other potentially life-threatening toxic reactions generally associated with alcohol withdrawal.

As outlined above, the chronic use of ethanol can result in a wide range of toxic effects on a variety of organ systems; however, the mechanisms through which ethanol produces its varied effects are not clearly understood. The anesthetic or central nervous system depressant effects may result, in part, from general changes in the function of ion channels that occur when ethanol dissolves in lipid (fat) membranes. Other research suggests that alcohol may interact with specific receptors—binding sites associated with the inhibitory NEUROTRANSMITTER GAMMA-AMINOBUTYRIC ACID (GABA), in a manner somewhat analogous to other central nervous system depressants (e.g., benzodiazepines or barbiturates). Since an ethanol RECEPTOR site has not yet been conclusively identified, specific receptor AGONISTS and ANTAGONISTS are not yet available for the treatment of ethanol intoxication, WITHDRAWAL, and abstinence. The drug DISULFIRAM (Antabuse) is sometimes used in the treatment of chronic ALCOHOLISM, although it

does not cure alcoholism. Rather, disulfiram interacts with ethanol to alter the intermediate metabolism of ethanol, resulting in a five- to tenfold increase in plasma acetaldehyde concentrations. Those who drink while on disulfiram experience the acetaldehyde syndrome—vasodilatation, headache, difficulty breathing, nausea, vomiting, sweating, faintness, weakness, and vertigo. Taking the drug helps persuade alcoholics to remain abstinent, since they realize that they cannot drink ethanol for at least three or four days without provoking ill-effects.

NICOTINE AND TOBACCO

TOBACCO was first introduced to Europe by the crews that accompanied Columbus to the New World, and by the middle of the nineteenth century, tobacco use had become widespread. By the 1990s, almost 30 percent of the adults in the United States are still regular tobacco smokers. This relatively high use of tobacco continues despite the growing warnings that are based on a wealth of scientific evidence linking cigarette smoking to numerous life-threatening health disorders, including lung CANCER and heart disease. The constituents of tobacco smoke that most likely contribute to these health problems include carbon monoxide, NICOTINE, and "tar." However, nicotine also appears to be the primary component of tobacco smoke that promotes smoking. In regular cigarette smokers, nicotine facilitates memory, reduces aggression, and decreases weight gain. Each of these effects could, by itself, provide a rationale for continued tobacco use since most individuals find increased alertness and memory, decreased irritability, and decreased weight gain to be somewhat pleasant or desirable; however, these effects may actually be secondary to the primary reinforcing effects of nicotine itself. Nicotine is self-administered by laboratory animals, and in laboratory settings, smokers report that the intravenous injection of nicotine produces a pleasant feeling on its own. It is of interest to note, however, that nicotine is aversive to nonsmokers, often resulting in dizziness, nausea, and vomiting. TOLERANCE rapidly develops to these unpleasant effects in tobacco smokers, however.

Although nicotine obviously binds to nicotinic receptors associated with the neurotransmitter ACETYLCHOLINE, there is evidence that the reinforcing or rewarding properties of nicotine may result from an activation of ascending limbic neurons, which release the neurotransmitter DOPAMINE (i.e., the mesocorticolimbic dopaminergic system). Interestingly, this same system has been implicated in the reinforcing properties of a variety of drugs, including stimulants and opiates. As stated above, tobacco smoking has been associated with a wide variety of serious health effects, including cancer and heart disease; however, the chances of developing them decrease once smoking is terminated. Although some of the smoking-induced damage is irreversible, the incidence rates for cancer and heart disease gradually become more similar between smokers and nonsmokers the longer the smoker refrains from smoking. However, withdrawal from tobacco smoking results in a withdrawal syndrome that varies in intensity from individual to individual and often leads to a relapse of smoking. This syndrome consists of cravings for tobacco, irritability, weight gain, difficulty concentrating, drowsiness, and sleep disturbances. The recent introduction of nicotine-containing chewing gum and transdermal patches have significantly helped to facilitate abstinence from smoking in a number of individuals by delivering nicotine through a relatively less toxic route of administration.

CENTRAL NERVOUS SYSTEM DEPRESSANTS

In general, the incidence and prevalence of the nonmedical use of central nervous system depressants (approximately 6 to 8% of young adults) exceeds that of the opioids. These drugs include the barbiturates, benzodiazepines, and related drugs. The shorter-acting barbiturates, such as pentobarbital ("yellow jackets") or SECOBARBITAL ("red devils"), are usually preferred to the longer-acting drugs such as phenobarbital. Nonbarbiturates such as MEPROBAMATE, GLUTETHIMIDE, methyprylon, METHAQUALONE (Quaalude) and some of the shorter-acting benzodiazepines are also abused. Presumably, the quicker the onset of action for a particular central nervous system depressant, the better the "high."

There is no general rule that can be used to predict the pattern of use of a central nervous system depressant for a given individual. Often there is a fine line between appropriate therapy for insomnia or ANXIETY and drug dependence. Some individuals exhibit cyclic patterns of use with gross intoxication for a few days interspersed with periods of absti-

nence. Other barbiturate or benzodiazepine users maintain a chronic low level of intoxication without observable signs of impairment because of the development of tolerance to many of the actions of these drugs. When higher doses are used, however, the intoxication may resemble alcohol intoxication, with slurred speech, difficulty thinking, memory impairment, sluggish behavior, and emotional instability. Withdrawal from chronic barbiturate or benzodiazepine use can also be manifested to varying degrees. In the mildest form, the individual may only experience mild anxiety or insomnia. With greater degrees of physical dependence, tremors and weakness may also be included. In severe withdrawal, delirium and tonic-clonic (epileptic) seizures may also be present. This severe withdrawal syndrome can be life-threatening. The degree of severity of the withdrawal syndrome appears to be related to the pharmacokinetics of the drug used. Shorter-acting benzodiazepines and barbiturates produce much more severe cases of withdrawal than do the longer-acting drugs. Therefore, in the case of severe withdrawal symptoms associated with the chronic use of a short-acting drug, a longer-acting drug should be substituted. The dose of this drug can be gradually decreased so that the individual experiences a much milder and less threatening withdrawal.

Receptor-binding sites for benzodiazepines and barbiturates are part of a macromolecular complex associated with chloride ion channels and the inhibitory neurotransmitter GABA. The interaction of these drugs with their distinct binding sites results in a facilitation of GABAergic neurotransmission, producing an inhibitory effect on neuronal impulse flow in the central nervous system.

CENTRAL NERVOUS SYSTEM STIMULANTS

Central nervous system stimulants include caffeine, cocaine, and amphetamine, although the use and abuse of amphetamines and cocaine represent a much greater health risk, with deviation from social norms.

Caffeine. Perhaps 80 percent of the world's population ingests caffeine in the form of TEA, COFFEE, COLA-flavored drinks, and CHOCOLATE. In the central nervous system, caffeine decreases drowsiness and fatigue and produces a more rapid and clearer flow of thought. With higher doses, however,

nervousness, restlessness, insomnia, and tremors may result. Cardiac and gastrointestinal disturbances may also be observed. Tolerance typically develops to the anxiety and dysphoria experienced by some individuals. Some degree of PHYSICAL DEPENDENCE has, however, been associated with the chronic consumption of caffeine. The most characteristic symptom of caffeine withdrawal is a long throbbing headache, although fatigue, lethargy, and some degree of anxiety are also common. In general, the long-term consequences of chronic caffeine consumption are relatively minor.

COCAINE AND AMPHETAMINE

The problems associated with chronic cocaine and amphetamine use and withdrawal are much more serious than those associated with caffeine. By the mid-1980s, more than 20 million people had used cocaine in the United States. With the recent introduction of cocaine in the free alkaloid base ("FREEBASE" or "CRACK") form, there has been a significant increase in cocaine-related medical, economic, social, and legal problems. In the free-base form, cocaine can be smoked, resulting in blood levels and brain concentrations of the drug that compare to those observed when the drug is injected intravenously. In non-user subjects in a laboratory setting, the administration of cocaine or amphetamine produces an elevation of mood, an increase in energy and alertness, and a decrease in fatigue and boredom. In some individuals, however, anxiety, irritability, and insomnia may be observed.

In nonlaboratory settings, heavy users of cocaine often take the drug in bouts or binges, only stopping when their supply runs out or they collapse from exhaustion. Immediately following the intravenous administration or inhalation of cocaine, the individual experiences an intense pleasurable sensation known as a "rush" or "flash," followed by euphoria. Cocaine rapidly penetrates into the brain to produce these effects, but then is rapidly redistributed to other tissues. In many cases, the intense pleasure followed by the rapid decline in the cocaine-induced elevation of mood is sufficient for the individual to begin to immediately seek out and use more of the drug to prolong these pleasurable effects. Following the intranasal administration of cocaine, the pleasure is less intense and the decline in brain concentrations of the drug progresses much more slowly, so

that the craving for more of the drug is less pronounced. Cocaine and amphetamine appear to produce their reinforcing or pleasurable effects through interactions with the neurotransmitter dopamine, especially in limbic and cortical regions of the brain (i.e., within the mesocorticolimbic dopaminergic system). Both cocaine and amphetamine block the reabsorption of dopamine into the NEURONS, where it was released, thereby prolonging the action of dopamine in the synapse—the space between nerve cells. Amphetamine can also cause the direct release of dopamine from nerve cells and can inhibit the metabolism of the neurotransmitter. It is important to note, however, that every drug that augments the action of dopamine does not produce pleasurable or rewarding subjective effects.

The toxicity associated with cocaine or amphetamine use can be quite severe; it is often unrelated to the duration of use or to preexisting medical conditions in the individual. This potential for serious toxic side effects is amplified by the fact that tolerance usually develops to the subjective feelings of the cocaine-induced rush and euphoria, but not to some of the other central nervous system effects of the drug (especially seizure susceptibility). Some of the more minor toxic reactions include dizziness, confusion, nausea, headache, sweating, and mild tremors. These symptoms are experienced by virtually all cocaine and amphetamine users to some degree, as a result of stimulation of the sympathetic nervous system. More serious reactions are also frequently observed. These serious toxic effects can include irregular heartbeats, convulsions and seizures, heart attacks, liver failure, kidney failure, heart failure, respiratory depression, stroke, coma, and death. The effects on the heart and cardiovascular system can sometimes be treated with alpha and beta noradrenergic-receptor antagonists or calcium channel blockers, although even prompt medical attention is not always successful. The convulsions can sometimes be controlled with diazepam (Valium), and ventilation (oxygen) may be required for the respiratory depression. In addition to the effects described above for cocaine, amphetamine has been reported to produce direct and irreversible neuronal damage to dopaminergic neurons. A similar effect for cocaine has not yet been identified.

Psychiatric abnormalities resulting from chronic central nervous system stimulant abuse can include anxiety, DEPRESSION, HALLUCINATIONS, and, in some cases, a paranoid psychosis that is virtually indistin-guishable from a paranoid SCHIZOPHRENIC psychosis. A withdrawal syndrome is also observed following the abrupt cessation of chronic cocaine or amphetamine use. This syndrome begins with exhaustion during the "crash" phase and is followed by prolonged periods of anxiety, depression, anhedonia (loss of pleasure), hyperphagia (gluttony), and high craving for the drug. This craving may persist for several weeks, depending on the individual. The administration of dopaminergic agonists or tricyclic ANTIDEPRESSANTS may have some utility in decreasing the severity of the withdrawal symptoms.

CANNABINOIDS

Marijuana is probably still the most commonly used illicit drug in the United States, with about 55 percent of young adults reporting some experience with the drug during their lifetimes. The active ingredient in MARIJUANA is delta^9-TETRAHYDROCAN-NABINOL (Δ^9-THC), which exerts its most prominent effects on the central nervous system and the cardiovascular system. A marijuana cigarette that contains approximately 2 percent of the active ingredient can produce an increase in feelings of well-being, euphoria, and relaxation when smoked; however, short-term memory can be impaired as is the ability to carry out goal-directed behavior. The ability to drive or operate machinery is similarly impaired, often much longer than the persistence of subjective effects. With higher doses, paranoia, hallucinations, and anxiety or panic may be manifested.

Chronic marijuana users sometimes exhibit what is called the AMOTIVATIONAL SYNDROME—which consists of apathy, impairment of judgment, and a loss of interest in personal appearance and the pursuit of conventional goals. However, it is not clear whether this syndrome results from the use of marijuana alone or from other factors. Although this is seldom severe, Δ^9-THC also produces a dose-related increase in heart rate. Tolerance does develop to the effects of marijuana, and in some countries, regular users of HASHISH (a concentrated resin containing increased amounts of Δ^9-THC) consume quantities of the drug that would be toxic to most marijuana users in the United States. The withdrawal associated with the cessation of marijuana smoking is relatively mild—consisting of irritability, restlessness, nervousness, insomnia, weight loss, chills, and increased body temperature.

OPIOIDS

The use of opioids in the United States is much less prevalent than reported for the other drugs discussed above. For example, as of the 1990s, less than 0.5 percent of young adults have reported trying heroin at some time during their lives. There are three basic patterns of opioid use and dependence in the United States. The first group constitutes the smallest percentage of opioid users—those who initially began using morphinelike drugs medically, for the relief of PAIN. The second group began using illegal drugs through experimentation and then progressed to chronic use and dependence. The third group represents physically addicted individuals who eventually switched to oral METHADONE, obtained through organized treatment centers. Interestingly, the incidence of opioid addiction is greater among physicians, nurses, and related health-care professionals (who have access to these drugs) than in any group with a comparable educational background. In many instances (but not all), those addicted either to heroin (usually purchased illegally on the street) or to methadone (usually from treatment centers) are able to hold jobs and raise a family. Opioids reduce pain, aggression, and sexual drives, so the use of these drugs is unlikely to induce crime. Those who cannot afford opioids, those who like the "drug life," and those who are unable or unwilling to hold a job, resort to crime to support their drug habits.

Opioid drugs produce their pharmacological effects by binding to opiate RECEPTORS. The euphoria associated with the use of opioids results from interactions of these drugs with the mu-opiate receptor, possibly resulting in the stimulation of mesocorticolimbic dopaminergic neuronal activity. The rapid intravenous injection of morphine (or heroin, which is converted to morphine once it enters the brain) results in a warm flushing of the skin and sensations in the lower abdomen that are often described as being similar in intensity and quality to sexual orgasm. This initial rush ("kick" or "thrill") lasts for about 45 seconds and is followed by a high—described as a state of dreamy indifference. Depending on the individual and the social circumstances, good health and productive work are not incompatible with the regular use of opioids. Tolerance can develop to the ANALGESIC, respiratory depressant, sedative, and reinforcing properties of opioids, but the degree and extent of tolerance depends largely on the pattern of use. Desired analgesia can often be maintained through the intermittent use of morphine. Tolerance develops more rapidly with more continuous opioid administration.

The abrupt discontinuation of opioid use can lead to a withdrawal syndrome that varies in degree and severity depending on the individual as well as the particular opioid used. Watery eyes (lacrimation), a runny nose (rhinorrhea), yawning and sweating occur within twelve hours from the last dose of the opioid. As the syndrome progresses, dilated pupils, anorexia, gooseflesh ("cold turkey"), restlessness, irritability, and tremor can develop. As the syndrome intensifies, weakness and depression are pronounced, and nausea, vomiting, diarrhea, and intestinal spasms are common. Muscle cramps and spasms, including involuntary kicking movements ("kicking the habit"), are also characteristic of opioid withdrawal; however, seizures do not occur and the withdrawal syndrome is rarely life-threatening. Without treatment, the morphine-induced withdrawal syndrome usually runs its course within seven to ten days. Opiate-receptor antagonists (e.g., NALOXONE) are contraindicated in opioid withdrawal, since these drugs can precipitate a more severe withdrawal on their own. Rather, longer-acting, less potent, opiate-receptor agonists such as methadone are more commonly prescribed. The symptoms associated with methadone withdrawal are milder, although more protracted, than those observed with morphine or heroin. Therefore, methadone therapy can be gradually discontinued in some heroin-dependent individuals. If the patient is unwilling or unable to withdraw from methadone, the individual can be maintained on methadone indefinitely.

PSYCHEDELICS

Psychedelics include drugs related to the indole-alkylamines, such as lysergic acid diethylamide (LSD), PSILOCYBIN, psilocin, DIMETHYLTRYPTAMINE (DMT) and diethyltryptamine (DET), to the phenylethylamines, such as mescaline, or to the phenylisopropylamines, such as 2,5-dimethoxy-4-methylamphetamine (DOM or "STP") as well as 3,4-methylenedioxyamphetamine (MDA) and 3,4-methylene-dioxymethamphetamine (MDMA or "ecstasy"). The feature that distinguishes these psychedelic agents from other classes of drugs is their capacity to reliably induce states of altered perception, thought, and feeling. There is a heightened awareness of sensory input accompanied by an enhanced sense of clarity,

but a diminished control over what is experienced. The effects of LSD and related psychedelic drugs appear to be mediated through a subclass of receptors associated with the inhibitory neurotransmitter serotonin (i.e., serotonin 5HT$_2$ receptors). Immediately after the administration of LSD, somatic symptoms such as dizziness, weakness and nausea are present, although euphoric effects usually predominate. Within two to three hours, visual perceptions become distorted; colors are heard and sounds may be seen. Vivid visual hallucinations are also often present. Many times this loss of control is disconcerting to the individual, resulting in the need for structure—in the form of experienced companions during the "trip." The entire syndrome begins to clear after about twelve hours.

Little evidence exists for long-term changes in personality, beliefs, values, or behavior produced by the drug. Tolerance rapidly develops to the behavioral effects of LSD after three or four daily doses of the drug. In general, however, the psychedelic drugs do not give rise to patterns of continued use over extended periods. The use of these drugs is generally restricted to the occasional trip. Withdrawal phenomena are not observed after the abrupt discontinuation of LSD-like drugs, and deaths directly related to the pharmacological effects of LSD are unreported in humans—however, fatal ACCIDENTS and SUICIDES have occurred during periods of LSD intoxication.

INHALANTS

The intoxicating and euphorigenic properties of nitrous oxide and ethyl ether were well known even before their potential as anesthetics was recognized. Physicians, nurses and other health-care professionals have been known to inhale anesthetic gases even though they have access to a wide variety of other drugs. Adolescents with restricted access to alcohol often resort to "glue sniffing" or the inhalation of vapors from substances with marked toxicity, such as gasoline, paint thinners, or other industrial solvents. The alkyl nitrites (butyl, isobutyl, and amyl) have been used as aphrodisiacs, since the inhalation of these agents is suggested to intensify and prolong orgasm. At least 12 percent of young adults have reported some experience with inhalants—however, fatal toxic reactions (usually due to cardiac arrhythmias) are often associated with the inhalation of many of these drugs. Inhalation from a plastic bag

can result in hypoxia (too little oxygen) as well as an extremely high concentration of vapor. Fluorinated hydrocarbons can produce cardiac arrhythmias and ischemia (localized anemia). Chlorinated solvents depress heart muscle (myocardial) contractility. Ketones can produce pulmonary (lung) hypertension. Neurological impairment can also occur with a variety of solvents.

ARYLCYCLOHEXYLAMINES

Arylcyclohexylamines include phencyclidine (PCP or "angel dust") and related drugs that possess central nervous system stimulant, central nervous system depressant, hallucinogenic, and analgesic properties. These drugs (also known as dissociative anesthetics) are well absorbed following all routes of administration. With even small doses, intoxication is produced, with associated staggering gait, slurred speech, and numbness in the extremities. PCP users may also exhibit sweating, catatonia, and a blank stare as well as hostile and bizarre behavior. Amnesia during the intoxication may also occur. In higher doses, anesthesia, stupor, convulsions, and coma may appear. The typical high from a single dose can last four to six hours and is followed by a prolonged period of "coming down."

PCP and related compounds bind with high affinity to a number of distinct sites in the central nervous system, although it is not certain which site(s) is responsible for the primary pharmacological effects of these drugs. PCP binds to the sigma site, which also has a high affinity for some selected opioids, although the function of the sigma site is unknown. PCP blocks the cation channel (e.g., Ca^{2+}) that is regulated by N-methyl-D-aspartate (NMDA), one type of receptor for excitatory amino acid neurotransmitters such as glutamate or aspartate. PCP also blocks the reabsorption of the neurotransmitter dopamine into the neurons, where it was released, resulting in a prolonged action of the neurotransmitter, especially within the mesocorticolimbic dopaminergic neuronal system.

There appears to be some degree of tolerance to the effects of PCP, and some chronic users of PCP complain of cravings and difficulties with recent memory, thinking, and speech after discontinuing the use of the drug. Personality changes can range from social withdrawal and isolation to severe anxiety, nervousness, and depression. Although the frequency is uncertain, deaths due to direct toxicity,

violent behavior, and accidents have been reported following the use of PCP. PCP can also produce acute behavioral toxicity—consisting of intoxication, aggression, and confusion, as well as coma, convulsions, and psychoses. A PCP-induced psychosis can persist for several weeks following a single dose of the drug.

(SEE ALSO: *Addiction: Concepts and Definitions; Complications; Epidemiology of Drug Abuse; National Household Survey on Drug Abuse; Treatment*)

BIBLIOGRAPHY

ANILINE, O., & PITTS, F. N., JR. (1982). Phencyclidine (PCP): A review and perspectives. *CRC Critical Review of Toxicology, 10*, 145–177.

BENOWITZ, N. L. (1988). Pharmacologic aspects of cigarette smoking and nicotine addiction. *New England Journal of Medicine, 319*, 1318–1330.

BLOOM, F. E. (1989). Neurobiology of alcohol action and alcoholism. *Annual Review of Psychiatry, 8*, 347–360.

CLONINGER, C. R., DUNWIDDIE, S. H., & REICH, T. (1989). Epidemiology and genetics of alcoholism. *Annual Review of Psychiatry, 8*, 331–346.

DEWEY, W. L. (1986). Cannabinoid pharmacology. *Pharmacology Review, 38*, 151–178.

FREEDMAN, D. X. (1969). The psychopharmacology of hallucinogenic agents. *Annual Review of Medicine, 20*, 409–418.

GAWIN, F. H., & ELLINWOOD, E. H., JR. (1988). Cocaine and other stimulants: Actions, abuse, and treatment. *New England Journal of Medicine, 318*, 1173–1182.

GRIFFITHS, R. R., & WOODSON, P. P. (1988). Caffeine physical dependence: A review of human and laboratory animal studies. *Psychopharmacology* (Berlin) *94*, 437–451.

HOLLISTER, L. E. (1986). Health aspects of cannabis. *Pharmacology Review, 38*, 1–20.

JAFFE, J. H. (1990). Drug addiction and drug abuse. In A. G. Gilman et al. (Eds.), *Goodman and Gilman's the pharmacological basis of therapeutics*, 8th ed. New York: Pergamon.

KRANZLER, H. R., & ORROK, B. (1989). The pharmacotherapy of alcoholism. *Annual Review of Psychiatry, 8*, 397–417.

NUTT, D., ADINOFF, B., & LINNOILA, M. (1989). Benzodiazepines in the treatment of alcoholism. *Recent Developments in Alcohol, 7*, 283–313.

ROBINSON, G. M., SELLERS, E. M., & JANECEK, E. (1981). Barbiturate and hypnosedative withdrawal by a multiple oral phenobarbital loading dose technique. *Clinical and Pharmacological Therapeutics, 30*, 71–76.

WOODS, J. H., KATZ, J. L., & WINGER, G. (1987). Abuse liability of benzodiazepines. *Pharmacology Review, 39*, 251–419.

NICK E. GOEDERS

DRUG USE FORECASTING PROGRAM (DUF)

In early 1987, the National Institute of Justice (NIJ), the primary research arm of the U.S. Department of Justice, established the DUF program. It was intended to provide local estimates of recent drug use in arrested persons in the largest cities in the United States. The DUF program provides a unique opportunity to measure drug use in a highly deviant part of the population, those likely to be missed in surveys of school and household populations.

The roots of the DUF program extend back to the pretrial drug-testing program established for the District of Columbia by Dr. Robert DuPont in 1970. Since that time, all persons arrested for a criminal offense have been subject to urine testing. The test results were used by the court to make decisions regarding referral to drug-treatment programs for the period during which the person was released to the community, pending trial.

It seemed reasonable that if illegal drugs became available in a community, they would first be used by persons already involved in other illegal behaviors. If cities were to periodically test arrestees for drugs, the information might be useful for predicting drug epidemics. The information from the D.C. drug-testing program provided unique evidence that urine-test results from arrestees could serve as a leading indicator of community drug trends. The tests results for arrestees in the District of Columbia increased markedly in the mid 1970s (HEROIN) and early 1980s (COCAINE), foreshadowing the subsequent heroin and cocaine epidemics during those periods.

The DUF program was also based on the growing body of research showing that CRIME and drug use were closely linked. A study of addicts in Baltimore, Maryland, reported a sixfold increase in crime rates when addicts used heroin frequently; it helped to focus the research community on drug use in the criminal population.

Another important reason for the DUF program was the growing realization by researchers that self-report surveys might underestimate drug use. Many believed that as drug use became more stigmatized in the 1980s with the advent of the crack-cocaine epidemic, that respondees might have become more reluctant to disclose recent drug use. The DUF program was unique in being the first large-scale drug-monitoring system to use objective urinalysis tests to measure drug use.

In late 1986, James K. Stewart, the director of NIJ supported Dr. Eric Wish in becoming a visiting fellow, to supervise the establishment of the DUF program. The launching of the new drug-monitoring system resulted from Stewart's vision and his commitment to the project.

The main goals of the DUF program are to (1) provide local jurisdictions with quarterly estimates of recent drug use in samples of booked arrestees; (2) enable local researchers to have access to highly deviant offenders; (3) provide the federal government with information for describing and assessing drug trends in arrestees.

The DUF methodology was designed to fit in with the hectic operations of large city booking facilities. Each quarter, interviewers spend about 14 to 20 evenings in the participating booking facility: A sample of approximately 200 to 250 adult males is sought, and in some, about 100 females are sampled. A few sites study juvenile arrestees. In 1993, there were 23 sites participating in the DUF program.

The hectic booking environment has precluded the selection of respondents by a systematic sampling plan. After making sure to eliminate any known selection biases, interviewers approach all available arrestees. If several persons are available,

TABLE 1
DUF Estimates of Recent Drug Use in Adults and Juveniles, 1993 (Percent Positive*)

DUF Site	Adult Males	Adult Females	Juvenile Males
Atlanta	72	74	—
Birmingham	68	55	24
Chicago	81	—	—
Cleveland	64	77	36
Dallas	62	61	—
Denver	64	66	54
Detroit	63	76	—
Fort Lauderdale	61	60	—
Houston	59	53	—
Indianapolis	60	58	19
Los Angeles	66	77	34
Manhattan	78	83	—
Miami	70	—	—
New Orleans	62	47	—
Omaha	54	—	—
Philadelphia	76	79	—
Phoenix	62	62	36
Portland, Or.	63	74	18
Saint Louis	68	69	20
San Antonio	55	42	32
San Diego	78	78	43
San Jose	54	51	30
Washington, D.C.	60	71	51

*Tested positive for any of the following: opiates, cocaine, methadone, marijuana, amphetamines, PCP, benzodiazepines, barbiturates, methaqualone, propoxyphene.
SOURCE: U.S. Department of Justice, *Drug Use Forecasting 1993 annual report*, Washington, DC: National Institute of Justice.

then the interviewers select first those charged with serious nondrug offenses. In the quarterly samples, the DUF program has routinely limited to about 25 percent persons with drug charges, so that the samples would represent a diverse set of offenses. Because of this limit, and the fact that persons charged with the sale or possession of drugs are most likely to test positive at arrest, the DUF statistics are actually conservative estimates of recent drug use among all arrestees.

Participation in the study is voluntary, and no names or identifying information are retained. Response rates have typically been high. Approximately 95 percent of approached arrestees agree to be interviewed, and about 80 percent of the interviews provide the urine specimen. The urine specimens are sent to a standard laboratory for analysis for ten drugs: OPIATES, COCAINE, PCP, METHADONE, BENZO-DIAZEPINES, MARIJUANA, propoxyphene, BARBITU-RATES, METHAQUALONE, and AMPHETAMINES. For all drugs except marijuana and PCP, both of which can be detected for weeks in heavy users, the URINE TEST is sensitive to use in the prior 24 to 72 hours.

Since the DUF program began in 1987, the average percentage of arrestees testing positive across all DUF sites has been stable, between 50 percent and 70 percent. In 1993, the percentage of arrestees who tested positive for an illicit drug, usually cocaine, ranged from 54 percent of adult males processed in the Omaha, Nebraska, facility to 81 percent of those processed in Chicago, Illinois (see Table 1). These rates are even more dramatic when one remembers that the overwhelming majority of the DUF samples contain arrestees who were charged with offenses *other* than the sale or possession of drugs. High rates of recent use were also found among female arrestees, ranging from 42 percent in the San Antonio, Texas, sample to 79 percent in Philadelphia, Pennsylvania. In 1993, the 12 DUF sites that sampled male juveniles found positive tests ranging from 18 percent in Portland, Oregon, to 51 percent in Washington, D.C. In 8 of the 12 sites, 30 percent or more of the juveniles tested positive for a drug.

The DUF findings have been used by local jurisdictions to promote the funding of drug-intervention programs and to focus policymakers on the drug problem among convicted criminals. The DUF findings have helped refine the estimates of cocaine consumption in the United States. Researchers have also taken the opportunity of access to this highly deviant population to extend the DUF interview; it now covers such topics as HIV-risk behaviors, route of use of heroin, and the purchase and use of weapons.

In 1994, NIJ launched the computerized administration of the DUF interview with the AutoDUF program. And in 1994, the federal CENTER FOR SUBSTANCE ABUSE TREATMENT (CSAP) funded many states to modify the DUF methodology in studying the need for drug treatment among arrestees.

(SEE ALSO: *Drug Abuse Warning Network; Drug Testing and Analysis; Epidemiology of Drug Abuse; National Household Survey on Drug Abuse; Prisons and Jails*)

BIBLIOGRAPHY

DuPont, R. L., & Wish, E. D. (1992). Operation tripwire revisited. In E. D. Wish (Ed.), *The annals of the American Academy of Political and Social Science*, vol. 521 (May).

U.S. DEPARTMENT OF JUSTICE. (1994). *Drug Use Forecasting 1993 Annual Report*. Washington, DC: National Institute of Justice.

WISH, E. D. (1990/91). U.S. drug policy in the 1990's: Insights from new data from arrestees. *International Journal of the Addictions*, 25 (3A), 377–409.

ERIC D. WISH

DRUNK DRIVING   Drunk driving results in one of the most costly social consequences of ALCOHOL abuse. The toll on human life and health exacted by drunk drivers can, on its own, make alcohol abuse one of the most serious U.S. social problems. The extent and consequences of drunk driving demonstrate the challenges of harmonizing a drinking culture with a modern industrial society.

The combination of drinking and driving has been recognized as a serious problem since the invention of the automobile in the 1880s. In 1904, the *Quarterly Journal on Inebriety* editorialized that "the precaution of railroad companies to have only total abstainers guide their engines will soon extend to the owners of these new motor wagons . . . with the increased popularity of these wagons, accidents of this kind will multiply rapidly." By 1910, drunk driving had already been codified as a misdemeanor offense. Moreover, the dangerous mixture of alcohol and

driving was a key point in the Prohibitionists' argument in favor of the Eighteenth Amendment.

During the 1950s and 1960s, with postwar prosperity and a developing highway network, both alcohol abuse and traffic safety became serious national widespread issues. The Highway Safety Act of 1966 was crucial to mobilizing attention and resources in an attack against drunk driving. In effect, it established a federal (not just a state and local) jurisdiction by creating the National Highway Safety Bureau, the precursor of the National Highway Safety Administration (NHTSA), and it authorized the U.S. Department of Education's 1968 Report, *Alcohol and Highway Safety*. This report found that "the use of alcohol by drivers and pedestrians leads to some 25,000 deaths and a total of at least 800,000 crashes in the United States each year." The report warned that "this major source of human morbidity will continue to plague our mechanically powered society until its ramifications and many present questions have been exhaustively explored and the precise possibilities for truly effective countermeasures determined."

NHTSA became the main sponsor of research and action projects aimed at reducing drunk driving. In 1970, NHTSA launched the Alcohol Safety Action Project (ASAP), the first major U.S. initiative against drunk driving. The ASAPs, established in thirty-five communities, sought to achieve a significant reduction in drunk driving through a mixture of intensive countermeasures—including law enforcement, rehabilitation, and education. These programs were rigorously monitored. Unfortunately, despite huge increases in arrests and tens of thousands of referrals to drunk-driver schools and rehabilitation programs, a significant decrease in drunk driving could not be confirmed, and the ASAP was terminated in 1977.

The attack on drunk driving did not subside. In the late 1970s, there emerged a remarkable grassroots antidrunk-driving movement comprised of victims, their families, and many other concerned citizens. MOTHERS AGAINST DRUNK DRIVING (MADD, STUDENTS AGAINST DRIVING DRUNK (SADD), and REMOVE INTOXICATED DRIVERS (RID) opened local chapters throughout the United States and vigorously campaigned for new and tougher drunk-driving countermeasures. The crusade launched by these groups attracted a great deal of media attention, vaulting drunk driving to the top of the nation's social problems agenda. In 1982, President Ronald W. Reagan appointed a Presidential Commission on Drunk Driving. Congress linked state highway funds to the states' passage of specified antidrunk-driving measures, including a minimum drinking age of twenty-one. Ultimately, every state raised its drinking age accordingly. The states passed a deluge of legislation, providing for more and better law enforcement and more severe criminal penalties of a greater range—including mandatory jail terms, automatic license forfeiture, public education, drinking-driver schools, and rehabilitation.

MAGNITUDE OF THE PROBLEM

More than 2 million people are arrested each year for drunk driving. The actual number of offenses, while unknown and unknowable, must be far greater, since only a fraction of all violators are apprehended. A few researchers have mounted roadside surveys in which drivers are stopped and asked to voluntarily provide a breath sample from which the amount of alcohol in the blood can be calculated. While this is the best strategy for determining the actual amount of drunk driving, there are many problems with this methodology (which roads? what times? how many refusals?). A 1985 Minnesota roadside survey found that of 838 drivers on the road between 8:00 P.M. and 3:00 A.M. (prime time for drunk driving), 82.3 percent tested negative for any alcohol, 6 percent tested at BLOOD ALCOHOL CONCENTRATION (BAC) 0.05–0.09 percent (included as a lesser Driving While Intoxicated [DWI] offense in some states), and 2.4 percent tested above the drunk-driving threshold of BAC 0.1 percent.

Drunk drivers do not pose a uniform risk to themselves, their passengers, other motorists, and pedestrians. The most dangerous of the drunk drivers are the vehicular equivalent of the "fighting drunk"; they drive far in excess of the speed limit, weave in and out of traffic, and cross into lanes of traffic going in the opposite direction. At the low end of the continuum are drunk drivers who make an impaired effort to drive safely; although operating with diminished skill and judgment, they pose less of a risk than the agressive drunk drivers.

The most impressive experimental study of the causal role of alcohol in traffic crashes was carried out during the 1960s by Professor Robert Borkenstein (inventor of the BREATHALYZER) and his University of Indiana colleagues. The researchers ob-

tained breath samples from 6,000 accident-involved drivers and, as controls, from 7,500 nonaccident-involved drivers. They found that 6.3 percent of the accident-involved drivers, but fewer than 1 percent of the control drivers, had BACs equal to or greater than 0.1 percent (the prevailing definition of drunk driving in the 1980s). Moreover, each higher BAC level included a disproportionate number from the accident-involved group. Thus Borkenstein and his colleagues concluded that "BACs above .04% are definitely associated with an increased accident rate. The probability of accident-involvement increases rapidly at BACs above .15%. When drivers with BACs over 0.08% have accidents, they disproportionately involve only the driver's vehicle, and are more costly in terms of personal injury and property damage."

While most drunk-driving episodes do not result in a crash or injuries, the aggregate personal and property damage perpetrated by drunk drivers is staggering. A good deal of methodological controversy exists about the percentage of the approximately 45,000 annual traffic fatalities in the United States that can be attributed to drunk drivers. NHTSA's Fatal Accident Reporting System, which has been operating since the mid-1970s, presents important information about alcohol and traffic fatalities but does not attempt to estimate what proportion of all traffic deaths were *caused by* drunken driving. James Fell and Terry Klein, using statistical modeling techniques, have estimated that approximately 30 percent of all traffic fatalities can be attributed to drunk driving; other analyses have put this estimate at 50 percent.

Drunk drivers themselves, often in single-car collisions, comprise a large proportion of those who are killed, giving fatal drunk-driving episodes as much resemblance to suicide as to homicide. Nevertheless, each year thousands of innocent pedestrians and motorists are killed by drunk drivers, and tens of thousands are badly injured. There is also a huge amount of property damage.

THE OFFENDER

It is difficult to find reliable data on which to base a profile of the drunk driver. Using arrest data, we find that the vast majority, 90 percent, of drunk drivers are male and white. Despite the common belief that teenage drivers are most likely to be drunk, it is the mid-twenties age group that deserves this noto-

riety. Since it takes heavy drinking (from four to six drinks in two hours, depending on the drinker's weight) to reach the prohibited level, it is unlikely that light drinkers very often commit drunk-driving offenses. Thus, people who drive drunk are likely to be heavy drinkers and alcohol abusers. Nevertheless, light and moderate drinkers may on occasion drive drunk, perhaps due to a binge. Since light and moderate drinkers greatly outnumber heavy and abusive drinkers, they may in fact comprise a substantial proportion of arrested drunk drivers.

The consensus of studies based on screening tests of drunk drivers is that about 50 percent arrested for this offense are alcohol abusers, about 35 percent are social drinkers, and the remainder fall in between. While the categories *alcohol abuser* and *social drinker* are amorphous, a disproportionately high percentage of those arrested for drunk driving are actually heavy drinkers.

The large majority of drunk drivers arrested in any given year have not been arrested before. A well-executed Minnesota study found that only 7 percent of drivers involved in fatal accidents had been convicted of drunk driving in the preceding three years; of drunk drivers involved in fatal accidents, 13 percent had a DWI conviction in the previous 3 years, and 25 percent had a license revocation during the preceding 8 years. The low official rate of recidivism probably means that the chance of a drunk driver's being caught is extremely small.

THE CRIME OF DRUNK DRIVING

The first drunk-driving laws made it an offense to drive while intoxicated or to drive under the influence (DUI) of alcohol. Starting in the 1950s, states began to pass *per se* laws, which made it an offense to operate a motor vehicle with a BAC that exceeds certain levels. When a suspect is arrested for drunk driving, he or she is asked to take a deep breath and blow into a machine (the Breathalyzer is one model) that measures the amount of alcohol in the breath and converts it into a measure of the amount of alcohol in the blood. Pursuant to *implied consent* laws, suspects who refuse to provide a deep-lung breath sample are penalized by loss of driver's license and sometimes by other sanctions as well. The evolution of breath-testing equipment, including hand-held devices (like the Intoxilyzer), has greatly eased the identification and conviction of drunk drivers. Despite folklore to the contrary, it is extremely rare that

suspects who "fail" the breath test obtain acquittals. Indeed, the conviction rate for drunk driving is well over 90 percent.

In most states, a first drunk-driving offense is a misdemeanor, and a second offense within a specified time period (up to ten years in some states) is a felony. In a few states, a first offense is treated as a traffic violation, a second offense a misdemeanor, and a third offense a felony. Punishments vary from state to state; however, the usual range of punishments includes forfeiture of a driver's license for up to 1 year, fines of 500 to 1,000 dollars, and incarceration up to 30 days. In the late 1980s, spurred by the antidrunk-driving citizens' groups and federal financial incentives, several states passed laws mandating at least forty-eight hours of incarceration for a first DWI offense and a longer time for a second or subsequent offense. Another penalty that has been gaining popularity is the automatic and immediate forfeiture of the driver's license at the police station when the suspect fails or refuses to take the breath test (administrative per se law). In the early 1990s, and once again in response to federal pressure, states began lowering the prohibited BAC from 0.1 percent to 0.08 percent.

At least since the early 1970s, the criminal justice system's processing of drunk drivers has been linked to alcohol-treatment programs. In many jurisdictions, all drunk-driving offenders are routinely screened for ALCOHOLISM and alcohol abuse. Alcoholics and abusers may be diverted from prosecution to TREATMENT. More likely, however, the judge will require the offender to participate in treatment as a condition of probation or in order to obtain a provisional or regular driver's license. In some jurisdictions, the criminal-justice system is the largest source of clients flowing into alcohol-treatment programs. Thus, enforcement of the drunk-driving laws is one of the major ways that alcohol abusers are brought into the alcohol-treatment matrix.

In addition to the standard alcohol treatments, the attack on drunk driving has produced one unique kind of treatment—the drinking-driver school, which several million people have passed through since the mid-1970s. States and localities that have such schools often require all drunk drivers to attend. New York's school consists of five two-hour sessions and two three-hour sessions. The classes provide information about such matters as the deterioration of driving skills at different BAC levels, the inability to counteract intoxication with coffee or cold showers, and criminal penalties for drunk driving. People taking these classes are also required to fill out the Michigan Alcohol Screening Test (MAST) to determine whether they are alcohol abusers.

ENFORCEMENT

Enforcement of drunk-driving laws is the responsibility of local police, county sheriffs, and the state police or highway patrol. The Fourth Amendment to the U.S. Constitution prevents police from stopping cars at random and requiring drivers to take breath tests. Police must have probable cause to believe that drunk driving or some other offense (including traffic offenses) has been, is, or is about to be committed. Once a driver has been legitimately stopped, the police officer can order the driver to submit to a field sobriety test, which may consist of walking heel-to-toe, counting backwards, or performing other tasks that reveal intoxication. If the driver's performance on the test gives the officer probable cause to believe that the driver is intoxicated, the officer will arrest the driver. At the station, drivers will be told that they are required by the implied-consent law to submit to a breath test; refusal to cooperate will lead to license revocation.

In the 1980s, as states and localities searched for more effective antidrunk-driving strategies, some police departments mounted roadblocks at which every car (or every nth car) was briefly stopped and the driver briefly observed and sometimes questioned. If the officer detected alcohol, the driver was pulled over and required to submit to a breath test. Since these drunk-driving roadblocks were not based upon probable cause, they were challenged. In 1990, however, the U.S. Supreme Court upheld drunk-driving roadblocks under its administrative search doctrine (Michigan Dept. of State Police v. Sitz, 110 S. Ct. 2481). The Court ruled that as long as the roadblocks are situated in a fixed location, overseen by high-level officials, and operated nondiscriminatorily, they do not violate the Fourth Amendment.

DETERRING THE DRINKING DRIVER

Deterrence based on the threat of arrest, conviction, and punishment remains the chief strategy in the attack on drunk driving. During the 1980s, state and local governments have established dozens of strike forces and passed hundreds of laws aiming to raise the costs to the offender of driving while intox-

icated. In a series of empirical evaluations of police crackdowns and elevated maximum punishments in the United States and abroad, the sociologist H. Laurence Ross found that this type of law-enforcement escalation usually produces a reduction in drunk driving (as measured by single-vehicle fatalities), but not a long-term reduction. "No such policies have been scientifically demonstrated to work over time under conditions achieved in any jurisdiction . . . the option of merely increasing penalties for drinking and driving has been strongly discredited by experience to date."

While Ross has done far more empirical research than anybody else on deterring the drunk driver, his conclusion is not uncontroversial. One criticism is that he uses single-vehicle automobile fatalities to measure the amount of drunk driving; however, this kind of accident might not be strongly associated with the full range of drunk driving, but only with a narrow group, the most drunken and reckless of drunk driving. Possibly, while law-enforcement escalations cannot affect the kind of drunk drivers who kill themselves in single-vehicle crashes, they might be effective in the far more numerous nonfatal drunk-driving episodes that are engaged in by less pathological alcohol abusers and sociopathological persons.

The number of traffic fatalities has fallen from the late 1980s into the 1990s; drunk-driving fatalities seem to have fallen more than non-alcohol-related accidents. There may be reasons for this other than deterrence, including general reductions in alcohol consumption and abuse and more responsible public attitudes toward sober driving—however, a marginal deterrence effect cannot be ruled out.

OTHER ANTIDRUNK-DRIVING STRATEGIES

In addition to deterrence, states and localities have implemented many other antidrunk-driving strategies. Since all these strategies are being used simultaneously, it is impossible to attribute any reductions to one strategy over another.

Some courts have made punitive damages available in drunk-driving cases. This allows the victim of a drunk driver to recover any amount of money a jury deems appropriate for punishment. Some states permit insurance coverage of punitive damages, thereby negating whatever deterrent effect such damages might produce, but not negating a windfall for the victim.

In some states, legislatures and courts have expanded civil (tort) liability for causing drunk-driving injuries to include commercial hosts and package sellers of alcohol. While these DRAMSHOP LAWS vary from state to state, they essentially make purveyors of alcohol to underage or intoxicated persons liable for the injuries caused by such persons to themselves or others. A few state courts have even made social hosts liable for the alcohol-related traffic injuries caused by their guests.

An essential strategy for incapacitating drunk drivers is taking away their licenses to drive. Several studies have shown that drunk drivers who lose their driver's licenses are less likely to have a recurrence than drunk drivers who are fined, sent to jail, or assigned to mandatory treatment programs (actually, all these sanctions can be imposed together). Nevertheless, when licenses are suspended or revoked, a good deal of licenseless driving takes place—which is not surprising in a society where people depend on automobiles for their economic and social lives. Several states also have laws that authorize vehicle impoundment or forfeiture, but these sanctions are rarely used, perhaps because of the sacred status of the automobile as expensive private property.

Opportunity blocking refers to anticrime strategies that change the environment to reduce the opportunities of committing particular offenses. The best opportunity-blocking strategy for drunk driving involves fixing the defendant's vehicle so that it cannot be started until he blows alcohol-free breath into a tube affixed to the vehicle. Such equipment is now available, and several jurisdictions have implemented experimental programs. Other opportunity-blocking strategies include the twenty-one-year-old drinking age and a spate of new laws and regulations on bars, taverns, and package-goods stores.

The antidrunk-driving movement has spawned a large number of educational strategies. These include public-service announcements on radio and television and educational materials for primary and secondary schools. The effects of such programs are very difficult to evaluate. Rehabilitation strategies for drunk drivers are closely linked to the matrix of community alcoholism and alcohol-abuse services. Drunk drivers are regularly screened for alcohol problems, and those who are identified as abusers

are typically channeled into treatment through a probationary sentence.

(SEE ALSO: *Accidents and Injuries; Addiction: Concepts and Definitions; Distilled Spirits Council; Driving, Alcohol, and Drugs; Minimum Drinking Age Laws; Prevention Movement; Psychomotor Effects of Alcohol and Drugs; Social Costs of Alcohol and Drug Abuse*)

BIBLIOGRAPHY

GUSFIELD, J. (1981). *The culture of public problems: Drinking—driving and the symbolic order.* Chicago: University of Chicago Press.

JACOBS, J. B. (1989). *Drunk driving: An American dilemma.* Chicago: University of Chicago Press.

LABELL, A. (1992). *John Barleycorn must pay: Compensating the victims of drinking drivers.* Urbana, IL: University of Illinois Press.

MOORE, H., & GERSTEIN, D. (EDS.) (1981). *Alcohol and public policy: Beyond the shadow of prohibition.* Washington, DC: National Academy Press.

ROSS, H. L. (1992). *Confronting drunk driving.* New Haven: Yale University Press.

ROSS, H. L. (1982). *Deterring the drinking driver: Legal policy and social control.* Lexington, MA: Lexington Books.

JAMES B. JACOBS

DSM *See* Diagnostic and Statistical Manual.

DTs *See* Delirium Tremens.

DUAL DIAGNOSIS *See* Comorbidity and Vulnerability; Complications: Mental Disorders.

DUI/DWI *See* Driving, Alcohol, and Drugs; Driving Under the Influence; Drunk Driving.